LONDON MATHEMATICAL SOCIETY LECTURE NOTE SERIES

Managing Editor: Professor Endre Süli, Mathematical Institute, University of Oxford,
Woodstock Road, Oxford OX2 6GG, United Kingdom

The titles below are available from booksellers, or from Cambridge University Press at
www.cambridge.org/mathematics

372 Moonshine: The first quarter century and beyond, J. LEPOWSKY, J. MCKAY & M.P. TUITE (eds)
373 Smoothness, regularity and complete intersection, J. MAJADAS & A. G. RODICIO
374 Geometric analysis of hyperbolic differential equations: An introduction, S. ALINHAC
375 Triangulated categories, T. HOLM, P. JØRGENSEN & R. ROUQUIER (eds)
376 Permutation patterns, S. LINTON, N. RUŠKUC & V. VATTER (eds)
377 An introduction to Galois cohomology and its applications, G. BERHUY
378 Probability and mathematical genetics, N. H. BINGHAM & C. M. GOLDIE (eds)
379 Finite and algorithmic model theory, J. ESPARZA, C. MICHAUX & C. STEINHORN (eds)
380 Real and complex singularities, M. MANOEL, M.C. ROMERO FUSTER & C.T.C WALL (eds)
381 Symmetries and integrability of difference equations, D. LEVI, P. OLVER, Z. THOMOVA & P. WINTERNITZ (eds)
382 Forcing with random variables and proof complexity, J. KRAJÍČEK
383 Motivic integration and its interactions with model theory and non-Archimedean geometry I, R. CLUCKERS, J. NICAISE & J. SEBAG (eds)
384 Motivic integration and its interactions with model theory and non-Archimedean geometry II, R. CLUCKERS, J. NICAISE & J. SEBAG (eds)
385 Entropy of hidden Markov processes and connections to dynamical systems, B. MARCUS, K. PETERSEN & T. WEISSMAN (eds)
386 Independence-friendly logic, A.L. MANN, G. SANDU & M. SEVENSTER
387 Groups St Andrews 2009 in Bath I, C.M. CAMPBELL et al (eds)
388 Groups St Andrews 2009 in Bath II, C.M. CAMPBELL et al (eds)
389 Random fields on the sphere, D. MARINUCCI & G. PECCATI
390 Localization in periodic potentials, D.E. PELINOVSKY
391 Fusion systems in algebra and topology, M. ASCHBACHER, R. KESSAR & B. OLIVER
392 Surveys in combinatorics 2011, R. CHAPMAN (ed)
393 Non-abelian fundamental groups and Iwasawa theory, J. COATES et al (eds)
394 Variational problems in differential geometry, R. BIELAWSKI, K. HOUSTON & M. SPEIGHT (eds)
395 How groups grow, A. MANN
396 Arithmetic differential operators over the p-adic integers, C.C. RALPH & S.R. SIMANCA
397 Hyperbolic geometry and applications in quantum chaos and cosmology, J. BOLTE & F. STEINER (eds)
398 Mathematical models in contact mechanics, M. SOFONEA & A. MATEI
399 Circuit double cover of graphs, C.-Q. ZHANG
400 Dense sphere packings: a blueprint for formal proofs, T. HALES
401 A double Hall algebra approach to affine quantum Schur–Weyl theory, B. DENG, J. DU & Q. FU
402 Mathematical aspects of fluid mechanics, J.C. ROBINSON, J.L. RODRIGO & W. SADOWSKI (eds)
403 Foundations of computational mathematics, Budapest 2011, F. CUCKER, T. KRICK, A. PINKUS & A. SZANTO (eds)
404 Operator methods for boundary value problems, S. HASSI, H.S.V. DE SNOO & F.H. SZAFRANIEC (eds)
405 Torsors, étale homotopy and applications to rational points, A.N. SKOROBOGATOV (ed)
406 Appalachian set theory, J. CUMMINGS & E. SCHIMMERLING (eds)
407 The maximal subgroups of the low-dimensional finite classical groups, J.N. BRAY, D.F. HOLT & C.M. RONEY-DOUGAL
408 Complexity science: the Warwick master's course, R. BALL, V. KOLOKOLTSOV & R.S. MACKAY (eds)
409 Surveys in combinatorics 2013, S.R. BLACKBURN, S. GERKE & M. WILDON (eds)
410 Representation theory and harmonic analysis of wreath products of finite groups, T. CECCHERINI-SILBERSTEIN, F. SCARABOTTI & F. TOLLI
411 Moduli spaces, L. BRAMBILA-PAZ, O. GARCÍA-PRADA, P. NEWSTEAD & R.P. THOMAS (eds)
412 Automorphisms and equivalence relations in topological dynamics, D.B. ELLIS & R. ELLIS
413 Optimal transportation, Y. OLLIVIER, H. PAJOT & C. VILLANI (eds)
414 Automorphic forms and Galois representations I, F. DIAMOND, P.L. KASSAEI & M. KIM (eds)
415 Automorphic forms and Galois representations II, F. DIAMOND, P.L. KASSAEI & M. KIM (eds)
416 Reversibility in dynamics and group theory, A.G. O'FARRELL & I. SHORT
417 Recent advances in algebraic geometry, C.D. HACON, M. MUSTAŢĂ & M. POPA (eds)
418 The Bloch–Kato conjecture for the Riemann zeta function, J. COATES, A. RAGHURAM, A. SAIKIA & R. SUJATHA (eds)
419 The Cauchy problem for non-Lipschitz semi-linear parabolic partial differential equations, J.C. MEYER & D.J. NEEDHAM
420 Arithmetic and geometry, L. DIEULEFAIT et al (eds)
421 O-minimality and Diophantine geometry, G.O. JONES & A.J. WILKIE (eds)
422 Groups St Andrews 2013, C.M. CAMPBELL et al (eds)
423 Inequalities for graph eigenvalues, Z. STANIĆ
424 Surveys in combinatorics 2015, A. CZUMAJ et al (eds)

425	Geometry, topology and dynamics in negative curvature, C.S. ARAVINDA, F.T. FARRELL & J.-F. LAFONT (eds)
426	Lectures on the theory of water waves, T. BRIDGES, M. GROVES & D. NICHOLLS (eds)
427	Recent advances in Hodge theory, M. KERR & G. PEARLSTEIN (eds)
428	Geometry in a Fréchet context, C.T.J. DODSON, G. GALANIS & E. VASSILIOU
429	Sheaves and functions modulo p, L. TAELMAN
430	Recent progress in the theory of the Euler and Navier–Stokes equations, J.C. ROBINSON, J.L. RODRIGO, W. SADOWSKI & A. VIDAL-LÓPEZ (eds)
431	Harmonic and subharmonic function theory on the real hyperbolic ball, M. STOLL
432	Topics in graph automorphisms and reconstruction (2nd Edition), J. LAURI & R. SCAPELLATO
433	Regular and irregular holonomic D-modules, M. KASHIWARA & P. SCHAPIRA
434	Analytic semigroups and semilinear initial boundary value problems (2nd Edition), K. TAIRA
435	Graded rings and graded Grothendieck groups, R. HAZRAT
436	Groups, graphs and random walks, T. CECCHERINI-SILBERSTEIN, M. SALVATORI & E. SAVA-HUSS (eds)
437	Dynamics and analytic number theory, D. BADZIAHIN, A. GORODNIK & N. PEYERIMHOFF (eds)
438	Random walks and heat kernels on graphs, M.T. BARLOW
439	Evolution equations, K. AMMARI & S. GERBI (eds)
440	Surveys in combinatorics 2017, A. CLAESSON et al (eds)
441	Polynomials and the mod 2 Steenrod algebra I, G. WALKER & R.M.W. WOOD
442	Polynomials and the mod 2 Steenrod algebra II, G. WALKER & R.M.W. WOOD
443	Asymptotic analysis in general relativity, T. DAUDÉ, D. HÄFNER & J.-P. NICOLAS (eds)
444	Geometric and cohomological group theory, P.H. KROPHOLLER, I.J. LEARY, C. MARTÍNEZ-PÉREZ & B.E.A. NUCINKIS (eds)
445	Introduction to hidden semi-Markov models, J. VAN DER HOEK & R.J. ELLIOTT
446	Advances in two-dimensional homotopy and combinatorial group theory, W. METZLER & S. ROSEBROCK (eds)
447	New directions in locally compact groups, P.-E. CAPRACE & N. MONOD (eds)
448	Synthetic differential topology, M.C. BUNGE, F. GAGO & A.M. SAN LUIS
449	Permutation groups and cartesian decompositions, C.E. PRAEGER & C. SCHNEIDER
450	Partial differential equations arising from physics and geometry, M. BEN AYED et al (eds)
451	Topological methods in group theory, N. BROADDUS, M. DAVIS, J.-F. LAFONT & I. ORTIZ (eds)
452	Partial differential equations in fluid mechanics, C.L. FEFFERMAN, J.C. ROBINSON & J.L. RODRIGO (eds)
453	Stochastic stability of differential equations in abstract spaces, K. LIU
454	Beyond hyperbolicity, M. HAGEN, R. WEBB & H. WILTON (eds)
455	Groups St Andrews 2017 in Birmingham, C.M. CAMPBELL et al (eds)
456	Surveys in combinatorics 2019, A. LO, R. MYCROFT, G. PERARNAU & A. TREGLOWN (eds)
457	Shimura varieties, T. HAINES & M. HARRIS (eds)
458	Integrable systems and algebraic geometry I, R. DONAGI & T. SHASKA (eds)
459	Integrable systems and algebraic geometry II, R. DONAGI & T. SHASKA (eds)
460	Wigner-type theorems for Hilbert Grassmannians, M. PANKOV
461	Analysis and geometry on graphs and manifolds, M. KELLER, D. LENZ & R.K. WOJCIECHOWSKI
462	Zeta and L-functions of varieties and motives, B. KAHN
463	Differential geometry in the large, O. DEARRICOTT et al (eds)
464	Lectures on orthogonal polynomials and special functions, H.S. COHL & M.E.H. ISMAIL (eds)
465	Constrained Willmore surfaces, Á.C. QUINTINO
466	Invariance of modules under automorphisms of their envelopes and covers, A.K. SRIVASTAVA, A. TUGANBAEV & P.A. GUIL ASENSIO
467	The genesis of the Langlands program, J. MUELLER & F. SHAHIDI
468	(Co)end calculus, F. LOREGIAN
469	Computational cryptography, J.W. BOS & M. STAM (eds)
470	Surveys in combinatorics 2021, K.K. DABROWSKI et al (eds)
471	Matrix analysis and entrywise positivity preservers, A. KHARE
472	Facets of algebraic geometry I, P. ALUFFI et al (eds)
473	Facets of algebraic geometry II, P. ALUFFI et al (eds)
474	Equivariant topology and derived algebra, S. BALCHIN, D. BARNES, M. KĘDZIOREK & M. SZYMIK (eds)
475	Effective results and methods for Diophantine equations over finitely generated domains, J.-H. EVERTSE & K. GYŐRY
476	An indefinite excursion in operator theory, A. GHEONDEA
477	Elliptic regularity theory by approximation methods, E.A. PIMENTEL
478	Recent developments in algebraic geometry, H. ABBAN, G. BROWN, A. KASPRZYK & S. MORI (eds)
479	Bounded cohomology and simplicial volume, C. CAMPAGNOLO, F. FOURNIER-FACIO, N. HEUER & M. MORASCHINI (eds)
480	Stacks Project Expository Collection (SPEC), P. BELMANS, W. HO & A.J. DE JONG (eds)
481	Surveys in combinatorics 2022, A. NIXON & S. PRENDIVILLE (eds)

London Mathematical Society Lecture Note Series: 476

An Indefinite Excursion in Operator Theory
Geometric and Spectral Treks in Kreĭn Spaces

AURELIAN GHEONDEA
Bilkent University, Ankara
and IMAR, Bucharest

CAMBRIDGE
UNIVERSITY PRESS

University Printing House, Cambridge CB2 8BS, United Kingdom

One Liberty Plaza, 20th Floor, New York, NY 10006, USA

477 Williamstown Road, Port Melbourne, VIC 3207, Australia

314–321, 3rd Floor, Plot 3, Splendor Forum, Jasola District Centre, New Delhi – 110025, India

103 Penang Road, #05–06/07, Visioncrest Commercial, Singapore 238467

Cambridge University Press is part of the University of Cambridge.

It furthers the University's mission by disseminating knowledge in the pursuit of education, learning, and research at the highest international levels of excellence.

www.cambridge.org
Information on this title: www.cambridge.org/9781108969031
DOI: 10.1017/9781108979061

© Aurelian Gheondea 2022

This publication is in copyright. Subject to statutory exception and to the provisions of relevant collective licensing agreements, no reproduction of any part may take place without the written permission of Cambridge University Press.

First published 2022

Printed in the United Kingdom by TJ Books Limited, Padstow Cornwall

A catalogue record for this publication is available from the British Library.

ISBN 978-1-108-96903-1 Paperback

Cambridge University Press has no responsibility for the persistence or accuracy of URLs for external or third-party internet websites referred to in this publication and does not guarantee that any content on such websites is, or will remain, accurate or appropriate.

To Tiberiu Constantinescu (1955–2005) and Peter Jonas (1941–2007)

who shared with me the joy of trekking through

the indefinite realm of operator theory

as well as

To my beautiful and generous country, Roumania

Contents

Preface   **xi**

1   Inner Product Spaces   **1**
    1.1   Basic Definitions and Properties 1
    1.2   The Weak Topology 10
    1.3   Normed Topologies 13
    1.4   Kreĭn Spaces 16
    1.5   Pre-Kreĭn Spaces 22
    1.6   Notes 27

2   Angular Operators   **28**
    2.1   Semidefinite Subspaces and Angular Operators 28
    2.2   Extensions of Semidefinite Subspaces 35
    2.3   Intermezzo: Maximal Accretive and Selfadjoint Extensions 45
    2.4   A Universality Property of Kreĭn Spaces 55
    2.5   Generalised Angular Operators 57
    2.6   Notes 60

3   Subspaces of Kreĭn Spaces   **62**
    3.1   Regular Subspaces 62
    3.2   Pseudo-Regular Subspaces 66
    3.3   Strong Duality of Subspaces 71
    3.4   Fredholm Subspaces 75
    3.5   Index Formulae for Linear Relations 81
    3.6   Notes 84

4   Linear Operators on Kreĭn Spaces   **85**
    4.1   The Adjoint Operator 85
    4.2   Some Classes of Bounded Operators 90
    4.3   Continuity of Isometric Operators 96

4.4	Dissipative Operators	101
4.5	Cayley Transformations	107
4.6	Notes	112

5 Selfadjoint Projections and Unitary Operators — 113

5.1	Selfadjoint Projections	113
5.2	Monotone Nets of Selfadjoint Projections	118
5.3	Unitary Operators	123
5.4	Dense Operator Ranges	130
5.5	Notes	133

6 Techniques of Induced Kreĭn Spaces — 134

6.1	Kreĭn Spaces Induced by Selfadjoint Operators	134
6.2	Nevanlinna Type Representations	143
6.3	Linearisation of Selfadjoint Operator Pencils	153
6.4	Carathéodory Type Representations	156
6.5	Elementary Rotations	169
6.6	Isometric and Unitary Dilations	178
6.7	Notes	181

7 Plus/Minus-Operators — 182

7.1	Spaces with Two Inner Products	182
7.2	Minus-Operators	184
7.3	Uniform Minus-Operators	189
7.4	Extensions of Uniform Minus-Operators	194
7.5	Notes	198

8 Geometry of Contractive Operators — 199

8.1	Contractions in Kreĭn Spaces	199
8.2	Boundedness of Contractions in Kreĭn Spaces	204
8.3	The Adjoint of a Contraction	205
8.4	The Scattering Transform	209
8.5	Linear Fractional Transformations	213
8.6	Notes	218

9 Invariant Maximal Semidefinite Subspaces — 219
- 9.1 Questions and Discussions 219
- 9.2 Spectral Methods 223
- 9.3 Fixed Point Approach 232
- 9.4 Fundamental Reducibility 237
- 9.5 Strong Stability 245
- 9.6 Notes 248

10 Hankel Operators and Interpolation Problems — 250
- 10.1 A Generalised Nehari Problem 250
- 10.2 More or Less Classical Hankel Operators 256
- 10.3 Intertwining Dilations 261
- 10.4 Generalised Interpolation 263
- 10.5 The Bitangential Nevanlinna–Pick Problem 268
- 10.6 Notes 271

11 Spectral Theory for Selfadjoint Operators — 273
- 11.1 Eigenvalues and Root Manifolds 273
- 11.2 Jordan Canonical Forms 279
- 11.3 Definitisable Selfadjoint Operators 285
- 11.4 Herglotz's Theorems 290
- 11.5 The Resolvent Function Representation 298
- 11.6 Stieltjes Inversion Formulae 303
- 11.7 The Spectral Function 308
- 11.8 Definitisable Positive Operators 318
- 11.9 Notes 322

12 Quasi-Contractions — 323
- 12.1 Geometric Properties of Quasi-Contractions 323
- 12.2 Double Quasi-Contractions 335
- 12.3 Polar Decompositions of Contractions 339
- 12.4 A Spectral Characterisation of Double Quasi-Contractions . . . 342
- 12.5 Notes 346

13 More on Definitisable Operators — **347**
- 13.1 Critical Points 347
- 13.2 Functional Calculus 353
- 13.3 Regularity of Critical Points 363
- 13.4 The Inverse Spectral Problem 369
- 13.5 Invariant Maximal Semidefinite Subspaces 380
- 13.6 Definitisable Unitary Operators 386
- 13.7 Notes 393

Appendix — **394**
- A.1 General Topology 394
- A.2 Measure and Integration 402
- A.3 Topological Vector Spaces 413
- A.4 Banach and Hilbert Spaces 417
- A.5 Functions of One Complex Variable 429
- A.6 Banach Algebras 448
- A.7 Banach Algebras with Involution 454
- A.8 Linear Operators on Banach Spaces 457
- A.9 Linear Operators on Hilbert Spaces 462

References — **476**

Symbol Index — **484**

Index — **487**

Preface

The idea of writing a monograph on operator theory on indefinite inner product spaces occurred to me about twenty years ago, after teaching a graduate course on the spectral theory of definitisable operators at the Faculty of Mathematics and Informatics of the University of Bucharest. With a lot of enthusiasm I wrote the core of the chapters on spectral theory and invited Heinz Langer to join me in this enterprise. At the beginning, Heinz showed his interest and encouraged me by performing a careful reading of those notes and providing very pertinent observations, but in the end he declined my invitation. That left me alone in this enterprise and offered me more freedom in choosing the subjects to be included, but this path proved to be dangerous in the end, as I explain below.

After performing research on the geometry of Kreĭn spaces and the spectral theory of selfadjoint definitisable operators in the 1980s as part of my PhD programme, I became more focused on studying dilation theory of operators on Kreĭn spaces and started to work together with my friend and collaborator Tiberiu Constantinescu. Dilation theory is a heterogeneous domain full of concepts and results that may look connected in some way but not explicitly, and because of that it became clear, at least to us, that it needs some unification. Considering dilation theory from a rather general perspective, we soon realised that one way to unify most of dilation theory is to consider Hermitian kernels and their linearisations, or Kolmogorov decompositions, which turn out to be yet another face of reproducing kernel Kreĭn spaces. Moreover, using some older ideas from mathematical physics, we also understood that some invariance under group, or $*$-semigroup, actions would open rather large avenues and connections with problems of a wider interest.

So, I made a very general and ambitious plan for a book that should contain and explain the extraordinary connections that I have observed between the geometry of Kreĭn spaces, the spectral theory of their linear operators, the dilation theory of Hermitian kernels and reproducing kernel Kreĭn spaces. The material grew over the years, but challenging problems showed up because important parts of this programme had not yet been investigated. Then, at some moment I had to admit that this plan was too ambitious and I should split it: on the one hand, I should dedicate my energy and time to exploring the new territory at the level of research articles and, on the other hand, I should reduce the plan of the book to a more realistic one. Therefore, I eliminated most of the dilation theory, harmonic analysis on Kreĭn spaces, and Hermitian kernels from the plan and I ended up with what is now this monograph: an excursion in the realm of operator theory on Kreĭn spaces from the point of view of the interplay of geometry and of spectral theory.

* * *

This monograph is a gentle and modern introduction to spaces with indefinite inner products and their operator theory and it is supposed to continue and complement the two existing monographs on this subject, that of J. Bognár (1974) and that of T. Ya. Azizov and I. S. Iokvidov (1986), and also gather some important results that were not included in those books or were obtained after 1980. Operator theory on indefinite inner product spaces has developed rapidly during the last thirty years and, unfortunately, there are many directions of research that have been explored and many interesting results that have been obtained that I had to leave out. This monograph is a selection of those topics that I consider to be representative for this theory and, of course, this selection is subjective.

Operator theory on Kreĭn spaces is, of course, a part of operator theory and a natural question is what novelty can it bring into a domain that is already so diverse and sophisticated. A possible answer to this question comes from the geometry of Kreĭn spaces, when compared to that of Hilbert spaces, and here lie both the power and the weakness of this theory. The simplest, intuitive, view of a Kreĭn space is as an infinite dimensional and complex analogue of Minkowski space, the space-time that lies at the foundation of relativistic physics. In a classical Minkowski space one encounters both positive and negative "lengths" of vectors, and also null "lengths" of nontrivial vectors. In the case of Minkowski space, the latter vectors, called neutral, make the so-called light cone which separates space-time into two distinct regions, with rather different physical interpretations. So, in a Kreĭn space, one has all these "anomalies" and even more, due to the intricate geometry of indefiniteness combined with the topological complications that show up in infinite dimensional spaces.

Continuing with this analogy, let us recall that the class of displacements that leave invariant Minkowski space is made by the Lorentz group $O(3, 1)$, which is not compact. Reasoning by similarity, we expect that the group of unitary operators in a Kreĭn space will play a central rôle. This indeed happens but, even more than that, in a genuine Kreĭn space the natural generalisation of a unitary operator leads to unbounded operators. This simple fact, that appears right from the beginning, of having to deal with unbounded isometric operators when considering indefinite inner product spaces, gives us just a pale idea of the novelty and the difficulties of operator theory in Kreĭn spaces.

From the point of view of applications, spectral theoretical aspects prevail in operator theory but, from my experience, the best way to explain the difficulties and, why not, the beauty, of the spectral theory of operators in Kreĭn spaces is to connect it with the more intricate geometry that comes together with indefiniteness. What the theory of linear operators on indefinite inner product spaces brings into play is a certain inner symmetry that, when emphasised, may or may not solve many difficulties, but at least it brings more geometry to the problems that we deal with.

The idea of merging the geometry of indefinite inner product spaces with the spectral theory of linear operators on these spaces is substantiated in Chapter 9 which presents a panoply of situations under which operators, or families of operators, have invariant maximal semidefinite subspaces. This problem is central to the theory of operators in indefinite inner product spaces and turns out to catch the core of this theory. Although it is not my aim to extensively explore applications of operator theory on indefinite inner product spaces to other domains of mathematics, we dedicate Chapter 10 to applications of invariant maximal semidefinite subspaces to interpolation for meromorphic functions, to Nehari type problems, and to Hankel type operators. In this way, the importance and the technical difficulties encountered when dealing with problems on invariant maximal semidefinite spaces provide a consistent justification for further exploration of the spectral theory of selfadjoint operators and unitary operators, on the one hand and, on the other hand, exploration of contractions and their generalisations, quasi-contractions, from both spectral and geometric points of view. It comes then as no surprise that, in order to obtain satisfactory results on the spectral theory of selfadjoint or unitary operators on Kreĭn spaces, imposing more technical assumptions is unavoidable. In the final chapters we employ the condition of definitisability in the sense of Heinz Langer which has proved to be remarkably successful. At first glance, the concept of definitisability may look too narrow and hence we felt obliged to justify its power by showing how useful it is in producing rather strong results on quasi-contractions. In this way we go back, from spectral theory to geometry, thus closing the circle.

There is one more characteristic of the approach that was used in this book which, at first glance, might not be visible. The applications that are included in this volume refer mostly to problems related to complex functions: the realisation theorems from Chapter 6 which concern kernels of holomorphic functions of Nevanlinna, Carathéodory, and Schur type as well as the problems of interpolation of meromorphic functions that occupy an important part of Chapter 10. But this is only one face of this characteristic. The other face can be seen when taking a closer look at the techniques that allow us to perform spectral theory and that require sophisticated results of holomorphic functions, like the Herglotz representations theorem, for example. So, in this monograph, the theory of complex functions plays a central rôle not only in applications but also in the essential tools that allow us to approach spectral theory in its large diversity.

In dealing with Kreĭn spaces there are, historically, two different points of view. One point of view, that was used, for example, by J. Bognár [18] and that we follow in this book, is to consider an inner product space $(\mathcal{X}; [\cdot, \cdot])$ onto which a certain Hilbert space topology that makes the indefinite inner product jointly continuous can be defined, as in Theorem 1.4.1. In this respect, the associated concepts such as fundamental symmetry, fundamental decomposition, and fundamental norm, are not fixed and they can be changed, within well specified classes, according to the requirements of the problems we are interested in. I consider that this approach offers a lot of flexibility and points out the prevailing properties of the indefinite inner product and the underlying geometry. It also points out the fact that the topology on a Kreĭn space is an extrinsic object and not intrinsic, as is the case with Hilbert spaces.

The other point of view, which was used by T. Ya. Azizov and I. S. Iokhvidov [10], is to start with a Hilbert space $(\mathcal{H}; \langle \cdot, \cdot \rangle)$ onto which a symmetry J, that is, a bounded linear operator on $\mathcal{H}$ such that $J = J^* = J^{-1}$, equivalently, J is a unitary selfadjoint operator on $\mathcal{H}$, is given and then introduce the indefinite inner product $[x, y] = \langle Jx, y \rangle$. This approach is used in many articles and has some merits but it can also lead to confusion and to some difficulties in understanding the geometry of a Kreĭn space. Moreover, as the results in Chapter 6 show, for most applications we do not have an underlying Hilbert structure available beforehand and we have to build one from the indefinite structure by an inducing construction. In this respect, we have to draw attention to the fact that the notation and the terminology of these two approaches are rather different.

In this monograph we intend to offer a presentation that, on the one hand, can be read and understood by a wider audience and, nevertheless, has some unity and harmony and, on the other hand, includes some of the most comprehensive results and the most relevant examples that point out the power and the applicability of operator theory on indefinite inner product spaces. Thus, we gradually explore the geometry of indefinite inner product spaces and their linear operators and introduce the spectral theoretical aspects only after sufficient experience and examples have been given.

We tried to increase the level of completeness of this book by avoiding sending the reader in search of books or articles that may be difficult to find. On the one hand, by imagining this monograph as an excursion, on a few occasions we take the freedom to step aside and we dedicate some pages to a careful presentation of the technical results that are needed, for example, the Herglotz theorems on representations of holomorphic functions mapping the upper half plane into itself, the Stieltjes inversion formulae, and the fixed point theorems in locally convex spaces. On the other hand, we want to make this book useful not only for the research mathematician, but also for the interested physicist or engineer, and especially for graduate students. For this reason, we have added nine appendices that review the basics of general topology, measure theory, topological vector spaces, functional analysis, complex functions, operator theory in Banach spaces with an emphasis on operator theory in Hilbert spaces, for both bounded and unbounded operators, and Fredholm theory. Of course, there are other prerequisites that the reader is supposed to be aware of, for example, basic linear algebra, a good command of differential and integral calculus at the level of advanced calculus, and some basic abstract algebra. Anyhow, before embarking on this excursion, readers are kindly advised to take a quick look at the appendices and check that all these concepts and results are present in their equipment.

On one occasion in Chapter 2, after performing a rather elaborate exploration of lifting of contractions in connection with R. S. Phillips's theorem of extensions of pairs of semidefinite subspaces to maximal ones, the excursion takes a side path in order to apply these results to extensions of dual pairs of accretive operators to maximal dual pairs and to extensions of positive operators to positive selfadjoint operators in Hilbert space. In this way we make available an older manuscript [28] that was not published. Also, there is a deeper connection here: the theory of positive selfadjoint extensions of positive operators was initiated by Mark G. Kreĭn in [92] and the original proof of Phillips's theorem [120] was inspired exactly by that article.

Operator theory on indefinite inner product spaces was originally motivated by and obtained remarkable success when applied to the spectral theory of ordinary and partial differential equations with certain symmetry properties. In this excursion we do not visit this kind of application because it would require a rather long preparation that would devi-

Preface   xv

ate considerably from the main course and would make this book too large. Although we think that a monograph dedicated to the spectral theory of ordinary and partial differential equations that employs techniques of operator theory on indefinite inner product spaces is necessary, this should be the topic of another book and, most likely, will be the project of another author.

* * *

I have to express my gratitude to many mathematicians who over the years influenced my research in operator theory on indefinite inner product spaces which eventually led me to write this monograph. As a student, I learned the basics of operator theory from Ion Colojoară, the basics of functional analysis from Ciprian Foiaş, then more sophisticated operator theory from Constantin Apostol and Dan-Virgil Voiculescu, and the basics of operator algebras from Şerban Strătilă. As a junior researcher in Bucharest, I was introduced to operator theory on indefinite inner product spaces by Grigore Arsene who at that time was in the group of mathematicians who worked in dilation theory. Then, I met Heinz Langer and Peter Jonas, from whom I learned a lot about the spectral theory of operators in spaces with indefinite inner product. Although my proposed collaboration with Heinz Langer on this monograph did not happen in the end, he made a careful reading of the chapters on spectral theory and provided valuable observations. An important part of the research in dilation theory in indefinite inner product spaces, which is present only to a very small extent in this monograph, was done in collaboration with my friend Tiberiu Constantinescu. Finally, by working in the domain of operator theory on indefinite inner product spaces I met more and more mathematicians and I became a part of the larger family of operator theorists that from time to time gathered in professional meetings, especially during the series of conferences on operator theory held in Timişoara, Roumania.

During the last year, I distributed the draft of this monograph to some of the specialists in operator theory on indefinite inner product spaces and I got very important feedback from them. Jim Rovnyak read very carefully the first six chapters and made valuable corrections and observations. Aad Dijksma provided a long list of corrections and remarks covering almost the whole book while Henk de Snoo drew my attention to some results that are not contained in this book but that deserve to be mentioned in the notes. Michael Kältenback provided another long list of corrections and observations on all chapters which helped me improve the presentation considerably. My former student Serdar Ay provided a list of corrections of the appendix. I want to express my deep gratitude to all of them.

I want to thank my parents, Elena and Ştefan, who took good care of me in difficult times and supported my decisions throughout their lives, although I was never able to explain to them what I am really doing. Finally, I want to thank my family, my wife Cristina and my children Alexandra and Sabin, who always supported me and, although we encountered difficult situations, they accepted that I needed to spend time and effort that kept me away from them, while I was wandering through the indefinite realm of operator theory.

Aurelian Gheondea   Bucharest, August 2021

Chapter 1

Inner Product Spaces

The first section of this introductory chapter contains more definitions and terminology than results. Although many of the definitions used in connection with inner product spaces are encountered in Hilbert space theory, there are also many new definitions that are specific to indefinite inner products. Many of these concepts come from the more general setting of duality theory but, in the context we use, they have special significance.

We then introduce the weak topology and discuss the existence of topologies making the inner product separately or jointly continuous. Among the inner product spaces admitting a normed topology that makes the inner product jointly continuous, an important rôle is played by Kreĭn spaces. In this chapter we consider only a general characterisation and leave to the next two chapters a more detailed description of their geometry. More tractable indefinite inner product spaces are the Pontryagin spaces, which we describe from both geometric and topological points of view.

One of the major differences between Kreĭn spaces and Hilbert spaces comes from the strong topology. In a Hilbert space, the strong topology is intrinsic to the inner product while, in the case of a genuine Kreĭn space, that is, one that is not of Pontryagin type, the strong topology has to be introduced independently. As a middle of the road we find the Pontryagin spaces which, although sharing similarities with Kreĭn spaces, have an intrinsic strong topology and hence these spaces are closer to Hilbert spaces.

Given an indefinite inner product space, a problem that often appears in applications is whether it can be "embedded" into a Kreĭn space. We give some characterisations of the affirmative case and some examples of the negative case.

1.1 Basic Definitions and Properties

Let $\mathcal{X}$ be a complex vector space. An *inner product* on $\mathcal{X}$ is, by definition, a mapping $[\cdot,\cdot]\colon \mathcal{X} \times \mathcal{X} \to \mathbb{C}$, where $\mathbb{C}$ denotes the field of complex numbers, with the following properties

$$[\alpha_1 x_1 + \alpha_2 x_2, y] = \alpha_1 [x_1, y] + \alpha_2 [x_2, y], \quad x_1, x_2, y \in \mathcal{X},\ \alpha_1, \alpha_2 \in \mathbb{C}, \qquad (1.1)$$

and

$$[x, y] = \overline{[y, x]}, \quad x, y \in \mathcal{X}, \qquad (1.2)$$

that is, it is *linear* with respect to the first variable and *conjugate symmetric*. It follows that the inner product is *conjugate linear* with respect to the second variable, that is,

$$[x, \beta_1 y_1 + \beta_1 y_2] = \overline{\beta}_1 [x, y_1] + \overline{\beta}_2 [x, y_2], \quad x, y_1, y_2 \in \mathcal{X}, \ \beta_1, \beta_2 \in \mathbb{C}. \tag{1.3}$$

If the complex vector space $\mathcal{X}$ has a fixed inner product $[\cdot, \cdot]$ then we call it an *inner product space* and use the notation $(\mathcal{X}, [\cdot, \cdot])$. Sometimes, in order to avoid confusion, we use the notation $(\mathcal{X}; [\cdot, \cdot]_\mathcal{X})$.

Let us first observe that $[x, x] \in \mathbb{R}$ for all $x \in \mathcal{X}$ and that an inner product $[\cdot, \cdot]$ always satisfies the *polarisation formula*, or *polarisation identity*

$$[x, y] = \frac{1}{4} \sum_{k=0}^{3} \mathrm{i}^k [x + \mathrm{i}^k y, x + \mathrm{i}^k y], \quad x, y \in \mathcal{X}, \tag{1.4}$$

where i denotes the imaginary complex number $\mathrm{i}^2 = -1$. This shows that an inner product $[\cdot, \cdot]$ on $\mathcal{X}$ is fully determined by the associated *quadratic form* $\mathcal{X} \ni x \mapsto [x, x] \in \mathbb{R}$.

A vector $x \in \mathcal{X}$ is *positive, neutral,* or *negative*, by definition, if the corresponding inner square $[x, x]$, which is always a real number, is positive $[x, x] > 0$, null $[x, x] = 0$, or negative $[x, x] < 0$, respectively. An inner product space $(\mathcal{X}, [\cdot, \cdot])$ is called *indefinite* if there exist both positive and negative vectors in $\mathcal{X}$, it is called *semidefinite* if it is not indefinite, and it is called *definite* if $\mathcal{X}$ contains no nontrivial neutral vectors. The following result justifies the last definition.

Lemma 1.1.1. *If the inner product space $(\mathcal{X}, [\cdot, \cdot])$ is indefinite then it contains nontrivial neutral vectors.*

Proof. Let $x, y \in \mathcal{X}$ be such that x is positive and y is negative. Since $[x, x] \cdot [y, y] < 0$, there always exists a real solution λ of the equation

$$[x, x] + 2\lambda \operatorname{Re}[x, y] + \lambda^2 [y, y] = 0,$$

hence the vector $z = x + \lambda y$ is neutral. If $z = 0$ then $0 < [x, x] = |\lambda|^2 [y, y] < 0$, a contradiction, hence $z \neq 0$. ∎

So, an inner product space can be, exclusively, either positive semidefinite, or negative semidefinite, or indefinite. On the other hand, a definite inner product space can be, exclusively, either positive definite or negative definite.

For semidefinite inner product spaces we recall the celebrated Schwarz Inequality.

Lemma 1.1.2 (Schwarz Inequality). *If the inner product space $(\mathcal{X}, [\cdot, \cdot])$ is semidefinite then*

$$|[x, y]|^2 \leq [x, x] \cdot [y, y], \quad x, y \in \mathcal{X}.$$

Proof. To make a choice, assume that $(\mathcal{X}, [\cdot, \cdot])$ is positive semidefinite. Then, for arbitrary $x, y \in \mathcal{X}$ and all real λ, the vector $z = \lambda x + y$ is nonnegative, hence

$$[z, z] = \lambda^2 [x, x] + 2\lambda \operatorname{Re}[x, y] + [y, y] \geq 0.$$

1.1 Basic Definitions and Properties

This implies that the discriminant of the second order real polynomial, in the variable λ, from above is nonnegative and hence

$$(\operatorname{Re}[x,y])^2 \leqslant [x,x] \cdot [y,y].$$

Choosing $\theta \in \mathbb{R}$ such that $[x, e^{i\theta}y] = \operatorname{Re}[x,y]$, we get the desired inequality.

In the negative semidefinite case, either observe that the proof is similar or consider the positive semidefinite inner product space $(\mathcal{X}; -[\cdot,\cdot])$. ∎

Lemma 1.1.3. *If the inner product space $(\mathcal{X}, [\cdot,\cdot])$ has at least one positive (negative) vector then each vector of $\mathcal{X}$ can be written as a sum of two positive (negative) vectors in $\mathcal{X}$.*

Proof. Let $x_0 \in \mathcal{X}$ be positive and $x \in \mathcal{X}$ arbitrary. Then, for $\lambda > 0$ and sufficiently large, we have

$$[x + \lambda x_0, x + \lambda x_0] = [x,x] + 2\lambda \operatorname{Re}[x,x_0] + \lambda^2[x_0,x_0] > 0,$$

hence $x = x_1 + x_2$, where the vectors $x_1 = x + \lambda x_0$ and $x_2 = -\lambda x_0$ are positive. The statement corresponding to negative vectors has a similar proof. ∎

Throughout this monograph, a subset $\mathcal{A}$ of $\mathcal{X}$ is called a *linear manifold* if it is nonempty and stable under linear operations $\alpha x + \beta y$, for any $\alpha, \beta \in \mathbb{C}$ and any $x, y \in \mathcal{A}$. In order to make things clearer, we reserve the word *subspace* for a linear manifold that is closed, when a certain linear topology on $\mathcal{X}$ is specified.

A subset $\mathcal{A} \subseteq \mathcal{X}$ is called *positive* if $[x,x] \geqslant 0$ for all $x \in \mathcal{A}$ and it is called *strictly positive* if $[x,x] > 0$ for all $x \in \mathcal{A} \setminus \{0\}$. In this respect, a linear manifold $\mathcal{A} \subseteq \mathcal{X}$ is strictly positive if and only if the inner product space $(\mathcal{A}; [\cdot,\cdot]_\mathcal{A})$ is positive definite, where $[x,y]_\mathcal{A} = [x,y]$ for all $x, y \in \mathcal{A}$. The definitions of *negative* and *strictly negative* sets are now clear. $\mathcal{A}$ is *neutral* if $[x,x] = 0$ for all $x \in \mathcal{A}$.

Since there is no general agreement on the terminology, some remarks are in order. What we call here a positive linear manifold other authors call a *nonnegative* linear manifold and what we call here a strictly positive linear manifold other authors call a *positive* linear manifold, for example. One reason for the terminology that we have adopted in this monograph is that, actually, in agreement with the previous definitions, a positive linear manifold is just an abbreviation for positive semidefinite linear manifold. Another reason for this convention comes from the fact that later we will establish some connections with linear operators for which positive traditionally means positive semidefinite.

Two vectors $x, y \in \mathcal{X}$ are called *orthogonal* if $[x,y] = 0$ and, in this case, we write $x \perp y$. Sometimes, in order to avoid confusion, we may use the more involved notation $[\perp]$, in order to make clear to which inner product the orthogonality is referring. Two subsets $\mathcal{A}, \mathcal{B} \in \mathcal{X}$ are *orthogonal* if $x \perp y$ for all $x \in \mathcal{A}$ and $y \in \mathcal{B}$. The *orthogonal companion* of a subset $\mathcal{A} \subseteq \mathcal{X}$ is, by definition, the linear manifold $\mathcal{A}^\perp \subseteq \mathcal{X}$ defined by

$$\mathcal{A}^\perp = \{x \in \mathcal{X} \mid x \perp y, \, y \in \mathcal{A}\}. \tag{1.5}$$

If $\mathcal{A}$ and $\mathcal{B}$ are nonempty subsets of the vector space $\mathcal{X}$ then we define $\mathcal{A}+\mathcal{B} = \{a+b \mid a \in \mathcal{A},\ b \in \mathcal{B}\}$. If both $\mathcal{A}$ and $\mathcal{B}$ are linear manifolds then $\mathcal{A} + \mathcal{B}$ is a linear manifold as well and coincides with the linear manifold spanned by the vectors in $\mathcal{A} \cup \mathcal{B}$. If $\mathcal{A}$ and $\mathcal{B}$ are linear manifolds and $\mathcal{A} \cap \mathcal{B} = \{0\}$ then we use the notation $\mathcal{A}\dotplus\mathcal{B}$ for $\mathcal{A} + \mathcal{B}$ and call it a *direct sum*.

If $\mathcal{A} \subseteq \mathcal{B}$ then $\mathcal{B}^\perp \subseteq \mathcal{A}^\perp$. If both subsets $\mathcal{A}$ and $\mathcal{B}$ contain the vector 0 then

$$(\mathcal{A} + \mathcal{B})^\perp = \mathcal{A}^\perp \cap \mathcal{B}^\perp. \tag{1.6}$$

Denoting $\mathcal{A}^{\perp\perp} = (\mathcal{A}^\perp)^\perp$ it follows readily that $\mathcal{A} \subseteq \mathcal{A}^{\perp\perp}$ and $\mathcal{A}^\perp = \mathcal{A}^{\perp\perp\perp}$.

Let $\mathcal{L}$ be a linear manifold in $\mathcal{X}$. The linear manifold $\mathcal{L}^0 = \mathcal{L} \cap \mathcal{L}^\perp$ is called the *isotropic part* of $\mathcal{L}$, and the vectors $x \in \mathcal{L}^0$ are called the *isotropic vectors* of $\mathcal{L}$. We have $\mathcal{X}^0 = \mathcal{X}^\perp$. The linear manifold $\mathcal{L} \subseteq \mathcal{X}$ is called *degenerate* if $\mathcal{L}^0 \neq \{0\}$, it is called *nondegenerate* in the opposite case, and it is called *maximal nondegenerate* if whenever $\mathcal{M}$ is a nondegenerate linear manifold in $\mathcal{X}$ such that $\mathcal{L} \subseteq \mathcal{M}$ it follows that $\mathcal{L} = \mathcal{M}$.

Example 1.1.4. Let $(\omega_n)_{n \geqslant 1}$ be a sequence of real numbers and denote by $\mathcal{X}$ the set of all sequences $(x_n)_{n \geqslant 1}$ of complex numbers such that

$$\sum_{n=1}^{\infty} |\omega_n|\, |x_n|^2 < \infty.$$

Then the formula

$$[x, y] = \sum_{n=1}^{\infty} \omega_n x_n \overline{y}_n, \quad x = (x_n)_{n \geqslant 1},\ y = (y_n)_{n \geqslant 1} \in \mathcal{X},$$

defines an inner product space $(\mathcal{X}, [\cdot, \cdot])$. The inner product space $(\mathcal{X}, [\cdot, \cdot])$ is indefinite if and only if the sequence $(\omega_n)_{n \geqslant 1}$ contains both positive and negative numbers. It is degenerate if and only if the sequence $(\omega_n)_{n \geqslant 1}$ has some null elements. ∎

Remark 1.1.5. Let $(\mathcal{X}, [\cdot, \cdot])$ be an inner product space. Then, a new inner product space can be defined by considering the quotient space $\widehat{\mathcal{X}} = \mathcal{X}/\mathcal{X}^0$ on which a natural inner product is defined

$$[\hat{x}, \hat{y}] = [x, y], \quad x \in \mathcal{X},\ y \in \mathcal{X}.$$

This definition is correct and $(\widehat{\mathcal{X}}, [\cdot, \cdot])$ is nondegenerate. ∎

Here and in the following we use the symbol $\dotplus$ any time we have a direct sum of two linear manifolds, that is, if $\mathcal{L}_1$ and $\mathcal{L}_2$ are two linear manifolds such that $\mathcal{L}_1 \cap \mathcal{L}_2 = \{0\}$ then $\mathcal{L}_1 \dotplus \mathcal{L}_2 := \{x_1 + x_2 \mid x_1 \in \mathcal{L}_1,\ x_2 \in \mathcal{L}_2\}$.

Proposition 1.1.6. *Let $(\mathcal{X}; [\cdot, \cdot])$ be an inner product space.* (a) *Every nondegenerate linear manifold of $\mathcal{X}$ is contained in a maximal nondegenerate linear manifold in $\mathcal{X}$.*

(b) *A linear manifold $\mathcal{L}$ of the inner product space $(\mathcal{X}, [\cdot, \cdot])$ is a direct summand of $\mathcal{X}^0$, that is, $\mathcal{X} = \mathcal{L} \dotplus \mathcal{X}^0$, if and only if it is maximal nondegenerate.*

1.1 BASIC DEFINITIONS AND PROPERTIES

Proof. (a) This assertion is a consequence of Zorn's Lemma. Briefly, assuming that $\mathcal{N}$ is a nondegenerate linear manifold in $\mathcal{X}$, let $\mathfrak{X}_\mathcal{N}$ denote the set of all nondegenerate linear manifolds $\mathcal{L}$ in $\mathcal{X}$ such that $\mathcal{N} \subseteq \mathcal{L}$, and ordered by inclusion. Since $\mathcal{N} \in \mathfrak{X}_\mathcal{N}$ is such a linear manifold $\mathfrak{X}_\mathcal{N}$ is nonvoid. It is easy to see that any chain in $\mathfrak{X}_\mathcal{N}$ has an upper bound, in the sense that, whenever $\{\mathcal{L}_\alpha\}_{\alpha \in A}$ is a family of elements $\mathcal{L}_\alpha \in \mathfrak{X}_\mathcal{N}$ indexed by a totally ordered set $(A; \leqslant)$ and such that $\alpha \leqslant \beta$ implies $\mathcal{L}_\alpha \subseteq \mathcal{L}_\beta$, then $\mathcal{L} = \bigcup_{\alpha \in A} \mathcal{L}_\alpha$ is an element in $\mathfrak{X}_\mathcal{N}$ such that $\mathcal{L} \supseteq \mathcal{L}_\alpha$ for all $\alpha \in A$. Thus, by Zorn's Lemma, $\mathfrak{X}_\mathcal{N}$ has a maximal element $\mathcal{M}$ which turns out to be a maximal nondegenerate linear manifold in $\mathcal{X}$ such that $\mathcal{N} \subseteq \mathcal{M}$.

(b) If $\mathcal{X} = \mathcal{L} \dotplus \mathcal{X}^0$ holds then $\mathcal{L}$ is nondegenerate and if $\mathcal{M}$ is another linear manifold strictly containing $\mathcal{L}$ then $\mathcal{X}^0 \cap \mathcal{M} \neq \{0\}$, hence $\mathcal{M}$ is degenerate.

Conversely, if $\mathcal{L}$ is maximal nondegenerate then $\mathcal{X}^0 \cap \mathcal{L} = \{0\}$. Let $\mathcal{N}$ be a direct summand of $\mathcal{X}^0 \dotplus \mathcal{L}$, that is, $\mathcal{X} = \mathcal{X}^0 \dotplus \mathcal{L} \dotplus \mathcal{N}$. Then $\mathcal{L} \dotplus \mathcal{N}$ is also nondegenerate and, on account of the maximality of $\mathcal{L}$, we get $\mathcal{N} = \{0\}$. ∎

If $\mathcal{L}$ is a linear manifold of the inner product space $(\mathcal{X}; [\cdot, \cdot])$, we denote by $\mathcal{L}_0 = \{x \in \mathcal{L} \mid [x, x] = 0\}$ the *neutral part* of $\mathcal{L}$.

Proposition 1.1.7. *Let $\mathcal{L}$ be a linear manifold in the inner product space $(\mathcal{X}; [\cdot, \cdot])$. Then,*

(a) $\mathcal{L}^0 \subseteq \mathcal{L}_0$.
(b) *The following assertions are equivalent.*

   (i) $\mathcal{L}_0$ *is a linear manifold.*
   (ii) $\mathcal{L}$ *is semidefinite, that is, either positive or negative.*
   (iii) $\mathcal{L}_0 = \mathcal{L}^0$.

(c) $\mathcal{L}_0 = \{0\}$ *if and only if $\mathcal{L}$ is definite, that is, either strictly positive or strictly negative.*

Proof. (a) This is clear, by definition.

(b) (iii)⇒(ii). Assume that $\mathcal{L}$ is not semidefinite, that is, it is indefinite. By Proposition 1.1.6, there exists a maximal nondegenerate linear manifold $\mathcal{M} \subseteq \mathcal{L}$ and hence $\mathcal{L} = \mathcal{L}^0 + \mathcal{M}$, with $\mathcal{L}^0 \cap \mathcal{M} = \{0\}$. Since $\mathcal{L}$ is indefinite, the nondegenerate linear manifold $\mathcal{M}$ is indefinite as well and then, by Lemma 1.1.1, there exists $x \in \mathcal{M} \subseteq \mathcal{L}$ a nontrivial neutral vector, hence $x \in \mathcal{L}_0 \setminus \mathcal{L}^0$.

(ii)⇒(i). This is a consequence of the Schwarz Inequality, see Lemma 1.1.2.

(i)⇒(iii). If $\mathcal{L}_0 = \mathcal{L}$ then $\mathcal{L}^0 = \mathcal{L}$ and hence $\mathcal{L}_0 = \mathcal{L}^0$ so we can assume, without restricting the generality, that $\mathcal{L}_0 \neq \mathcal{L}$. By (a), we only need to prove $\mathcal{L}_0 \subseteq \mathcal{L}^0$. Let us assume, by contradiction, that there exists $y \in \mathcal{L}_0 \setminus \mathcal{L}^0$, hence $[y, y] = 0$ and there exists $x \in \mathcal{L}$ such that $[x, y] \neq 0$. By changing x with $\mathrm{e}^{\mathrm{i}\theta} x$ for a suitable real number θ we can assume that $[x, y] \in \mathbb{R}$. Then consider the vectors

$$x_\lambda = (1 - \lambda)x + \lambda y, \quad \lambda \in \mathbb{C}. \tag{1.7}$$

Depending on the sign of the real number $[x, x]$ we distinguish three possible cases.

If $[x, x] = 0$ then, since $\mathcal{L}_0$ is assumed to be a linear manifold and both $x, y \in \mathcal{L}_0$, it follows that $x_\lambda \in \mathcal{L}_0$ for all $\lambda \in \mathbb{C}$, hence, taking into account that $[x, x] = [y, y] = 0$ and $[x, y] \in \mathbb{R}$, we have

$$0 = [x_\lambda, x_\lambda] = 2\operatorname{Re}(1-\lambda)\overline{\lambda}[x, y], \quad \lambda \in \mathbb{C},$$

which implies $[x, y] = 0$, a contradiction.

If $[x, x] > 0$ then, we consider the real polynomial

$$p(t) = [x_t, x_t] = (1-t)^2[x, x] + 2(1-t)t[x, y], \quad t \in \mathbb{R}, \tag{1.8}$$

which has a root for $t_1 = 1$ and a second root for $t_2 = [x, x]/([x, x] - 2[x, y])$. Without any loss of generality we can always assume that $[x, x] \neq 2[x, y]$ and, since $[x, y] \neq 0$ it follows that the two roots t_1 and t_2 are distinct. This implies that x_{t_1} and x_{t_2} are neutral vectors hence the whole linear manifold generated by them is neutral, by hypothesis, in particular $x = x_0$ is neutral, a contradiction.

If $[x, x] < 0$ then we proceed as in the previous case and get the same contradiction. In conclusion, we have proven that $\mathcal{L}_0 = \mathcal{L}^0$.

(c) If $\mathcal{L}_0 = \{0\}$ then it is a linear manifold and, by assertion (b), it follows that $\mathcal{L}$ is a nondegenerate semidefinite linear manifold, hence definite. The converse implication is obvious. ∎

In the following we fix an inner product space $(\mathcal{X}, [\cdot, \cdot])$. For any linear manifold $\mathcal{L}$ in $\mathcal{X}$ we set $\dim(\mathcal{L})$, the *algebraic dimension* of $\mathcal{L}$, that is, $\dim(\mathcal{L})$ is either 0, if $\mathcal{L} = \{0\}$, or a natural number n, if n is the maximal number of linearly independent vectors in $\mathcal{L}$, or the symbol ∞, if $\mathcal{L}$ has linearly independent subsets of vectors of any finite cardinality.

Lemma 1.1.8. *For any linear manifolds $\mathcal{L}$ and $\mathcal{M}$ of $\mathcal{X}$ we have*

$$\dim(\mathcal{L} \cap \mathcal{M}^\perp) + \dim(\mathcal{M}) \geqslant \dim(\mathcal{L}).$$

Proof. If either $\dim(\mathcal{M}) = \infty$ or $\dim(\mathcal{L}) \leqslant \dim(\mathcal{M})$ holds, then the inequality is true. Let $\dim(\mathcal{M}) = m < \infty$ and assume that for $l > m$ there exists a linearly independent system of vectors $\{e_j\}_{j=1}^l$ in $\mathcal{L}$. Let $\{f_k\}_{k=1}^m$ be a basis for $\mathcal{M}$. Then there exist at least $l - m$ linearly independent vectors of the form $x = \sum_{j=1}^{l} \alpha_j e_j$ such that

$$\sum_{j=1}^{l} \alpha_j [e_j, f_k] = 0, \quad k = 1, \ldots, m,$$

equivalently, $x \in \mathcal{L} \cap \mathcal{M}^\perp$, and hence $\dim(\mathcal{L} \cap \mathcal{M}^\perp) \geqslant l - m$. ∎

Corollary 1.1.9. *If $\mathcal{L} \cap \mathcal{M}^\perp = \{0\}$ then $\dim(\mathcal{L}) \leqslant \dim(\mathcal{M})$.*

1.1 BASIC DEFINITIONS AND PROPERTIES

By definition, the *algebraic ranks of negativity, isotropy* and *positivity* of $\mathcal{X}$ are, respectively,

$$\kappa_-(\mathcal{X}) = \sup\{\dim(\mathcal{L}) \mid \mathcal{L} \text{ is a strictly negative linear manifold in } \mathcal{X}\},$$
$$\kappa_0(\mathcal{X}) = \dim(\mathcal{X}^0), \tag{1.9}$$
$$\kappa_+(\mathcal{X}) = \sup\{\dim(\mathcal{L}) \mid \mathcal{L} \text{ is a strictly positive linear manifold in } \mathcal{X}\}.$$

Clearly, these ranks are either natural numbers or the symbol ∞. Due to the polarisation formula (1.4), these ranks are also called the *number of negative (null, positive) squares* of the associated quadratic form $\mathcal{X} \ni x \mapsto [x,x]$. In addition, the *rank of indefiniteness* of an inner product space $(\mathcal{X}; [\cdot,\cdot])$ is, by definition, $\kappa(\mathcal{X}) = \min\{\kappa_-(\mathcal{X}), \kappa_+(\mathcal{X})\}$. The triple $(\kappa_-(\mathcal{X}), \kappa_0(\mathcal{X}), \kappa_+(\mathcal{X}))$ is also called the *inertia* of the inner product space $(\mathcal{X}; [\cdot,\cdot])$.

Example 1.1.10. Let $(\mathcal{H}, \langle\cdot,\cdot\rangle)$ be a Hilbert space and let $A \in \mathcal{B}(\mathcal{H})$ be selfadjoint. Recall that, for a Hilbert space $\mathcal{H}$, we denote by $\mathcal{B}(\mathcal{H})$ the collection of all linear bounded operators $T\colon \mathcal{H} \to \mathcal{H}$. Set $\mathcal{X}_A = \mathcal{H}$, define an inner product $[\cdot,\cdot]_A$ by

$$[x,y]_A = \langle Ax, y \rangle, \quad x,y \in \mathcal{X}_A,$$

and consider the inner product space $(\mathcal{X}_A, [\cdot,\cdot]_A)$. Then $\kappa_-(\mathcal{X}_A)$ $(\kappa_+(\mathcal{X}_A))$ is equal to the algebraic dimension of the spectral subspace of A corresponding to the negative (positive) semi-axis and $\kappa_0(\mathcal{X}_A) = \dim(\operatorname{Ker}(A))$. If $\kappa_-(\mathcal{X}_A) < \infty$ ($\kappa_+(\mathcal{X}_A) < \infty$) then this number is equal to the number of negative (positive) eigenvalues of A, counted with their multiplicities. ∎

Let $(\mathcal{X}; [\cdot,\cdot]_\mathcal{X})$ and $(\mathcal{Y}; [\cdot,\cdot]_\mathcal{Y})$ be two inner product spaces. A linear mapping $T\colon \mathcal{X} \to \mathcal{Y}$ is called *isometric* if

$$[Tx_1, Tx_2]_\mathcal{Y} = [x_1, x_2]_\mathcal{X}, \quad x_1, x_2 \in \mathcal{X}.$$

If, in addition, the linear isometric operator T is a bijection, then we call it an *isomorphism* of inner product spaces.

Remark 1.1.11. Let $T\colon \mathcal{X} \to \mathcal{Y}$ be linear and isometric. Then for an arbitrary strictly positive linear manifold $\mathcal{L} \subseteq \mathcal{X}$ the mapping $T|\mathcal{L}$ is injective. This shows that $\kappa_+(\mathcal{X}) \leqslant \kappa_+(\mathcal{Y})$. Similarly we have $\kappa_-(\mathcal{X}) \leqslant \kappa_-(\mathcal{Y})$. In particular, if an isometric linear bijection $T\colon \mathcal{X} \to \mathcal{Y}$ exists then $\kappa_\pm(\mathcal{X}) = \kappa_\pm(\mathcal{Y})$ and, since $T(\mathcal{X}^0) = \mathcal{Y}^0$ we also have $\kappa_0(\mathcal{X}) = \kappa_0(\mathcal{Y})$. Consequently, isomorphisms of inner product spaces preserve the inertia. ∎

Proposition 1.1.12. *Given an inner product space $(\mathcal{X}; [\cdot,\cdot]_\mathcal{X})$, $\kappa_-(\mathcal{X})$ $(\kappa_+(\mathcal{X}))$ is equal to the supremum of the number of negative (positive) eigenvalues, counted with their multiplicities, of the matrices $([x_i, x_j])_{i,j=1}^n$, where $n \geqslant 1$ and $\{x_i\}_{i=1}^n \subset \mathcal{X}$.*

Proof. Let $\mathcal{L}$ be a strictly negative (strictly positive) linear manifold of $\mathcal{X}$ of dimension $m < \infty$. By the Gram–Schmidt orthogonalisation procedure we obtain a system of vectors $\{e_i\}_{i=1}^m$ of $\mathcal{L}$ such that $[e_i, e_j] = -\delta_{ij}$ ($[e_i, e_j] = \delta_{ij}$), $i, j = 1, \ldots, m$. Hence the matrix $([e_i, e_j])_{i,j=1}^m$ has exactly m negative (positive) eigenvalues, counted with their multiplicities.

Conversely, let $\{x_i\}_{i=1}^n$ be a finite system of vectors of $\mathcal{X}$. Without restricting the generality, we can assume that this system of vectors is linearly independent. If $\mathcal{L}$ denotes the linear span of $\{x_i\}_{i=1}^n$ then $\dim(\mathcal{L}) = n$. Consider the matrix $A = ([x_i, x_j])_{i,j=1}^n$ as a selfadjoint operator on the Hilbert space $\mathbb{C}^n$. With notation as in Example 1.1.10, we have

$$[x, y] = [\alpha, \beta]_A, \quad x, y \in \mathcal{L}, \tag{1.10}$$

where $x = \sum_{i=1}^n \alpha_i x_i$, $y = \sum_{j=1}^n \beta_j x_j$, $\alpha = (\alpha_1, \ldots, \alpha_n) \in \mathbb{C}^n$, and $\beta = (\beta_1, \ldots, \beta_n) \in \mathbb{C}^n$. The mapping

$$\mathcal{L} \ni x = \sum_{i=1}^n \alpha_i x_i \mapsto \alpha = (\alpha_1, \ldots, \alpha_n) \in \mathcal{X}_A,$$

is a linear bijection of $\mathcal{L}$ onto $\mathcal{X}_A$ and (1.10) shows that it preserves the inner products. Hence $\kappa_-(\mathcal{L})$ ($\kappa_+(\mathcal{L})$) coincides with the number of negative (positive) eigenvalues of the matrix A, counted with their multiplicities. ∎

An inner product space $(\mathcal{X}; [\cdot, \cdot])$ is called *decomposable* if there exist $\mathcal{L}$ a strictly positive subspace of $\mathcal{X}$ and $\mathcal{M}$ a strictly negative subspace of $\mathcal{X}$ such that $\mathcal{L} \perp \mathcal{M}$ and

$$\mathcal{X} = \mathcal{L} + \mathcal{X}^0 + \mathcal{M}. \tag{1.11}$$

Clearly, in this case we have $\mathcal{X} = \mathcal{L} \dotplus \mathcal{X}^0 \dotplus \mathcal{M}$. In this case, a decomposition as in (1.11) is called a *fundamental decomposition*.

A positive linear manifold $\mathcal{L} \in \mathcal{X}$ is called *maximal positive* if there exists no positive linear manifold $\mathcal{L}' \in \mathcal{X}$, different from $\mathcal{L}$, such that $\mathcal{L} \subset \mathcal{L}'$. Similarly one can define *maximal strictly positive, maximal negative*, and *maximal strictly negative* linear manifolds.

Remark 1.1.13. It is easy to see that, in a decomposition as in (1.11), the sum is direct, $\mathcal{L}^\perp = \mathcal{X}^0 + \mathcal{M}$, and $\mathcal{M}^\perp = \mathcal{X}^0 + \mathcal{L}$. Also, the subspace $\mathcal{L}$ is maximal strictly positive and the subspace $\mathcal{M}$ is maximal strictly negative. ∎

Proposition 1.1.14. *In any inner product space, all maximal positive linear manifolds have the same dimension. The same is true for all maximal strictly positive linear manifolds, all maximal negative linear manifolds, and all maximal strictly negative linear manifolds, respectively.*

Proof. Let $\mathcal{L}$ be a maximal strictly positive linear manifold of the inner product space $\mathcal{X}$. Then $\mathcal{L}^\perp$ is a negative linear manifold and hence, if $\mathcal{M}$ is another maximal strictly positive linear manifold, then $\mathcal{M} \cap \mathcal{L}^\perp = \{0\}$ and then, by Lemma 1.1.8, it follows that $\dim(\mathcal{L}) \leqslant \dim(\mathcal{M})$. By symmetry the converse inequality must hold as well, hence

1.1 Basic Definitions and Properties

$\dim(\mathcal{L}) = \dim(\mathcal{M})$. For maximal strictly negative linear manifolds the reasoning is similar.

Let now $\mathcal{L}$ and $\mathcal{M}$ be two maximal positive linear manifolds of $\mathcal{X}$. Then $\mathcal{M}^\perp$ is a negative linear manifold and hence $\mathcal{L} \cap \mathcal{M}^\perp$ is neutral, hence $\mathcal{M} + \mathcal{L} \cap \mathcal{M}^\perp$ is a positive linear manifold and, since $\mathcal{M}$ is maximal positive, it follows that $\mathcal{L} \cap \mathcal{M}^\perp \subseteq \mathcal{M}$. In view of Lemma 1.1.8 it follows that $\mathcal{L}$ and $\mathcal{M}$ are simultaneously infinite dimensional.

It remains to investigate the case when both $\mathcal{L}$ and $\mathcal{M}$ are finite dimensional. In this case, in view of maximality we have $\mathcal{X}^\circ \subseteq \mathcal{L}, \mathcal{M}$. Hence $\mathcal{X}^\circ$ is finite dimensional as well and, by factoring out $\mathcal{X}^\circ$, without loss of generality we can assume that $\mathcal{X}^\circ = \{0\}$. In addition, in view of Proposition 1.1.12, it follows that $\kappa_+(\mathcal{X})$ is finite. Then $\mathcal{X} = \mathcal{X}_- \dotplus \mathcal{X}_+$, where $\mathcal{X}_+$ is a finite dimensional maximal strictly positive linear manifold and $\mathcal{X}_-$ is a maximal strictly negative linear manifold. The proof will be complete if we show that $\dim(\mathcal{L}) = \kappa_+(\mathcal{X}) = \dim(\mathcal{X}_+)$.

To see this, in view of the decomposition $\mathcal{X} = \mathcal{X}_- \dotplus \mathcal{X}_+$, each vector $x \in \mathcal{L}$ is uniquely decomposed as $x = x_+ + x_-$, with $x_+ \in \mathcal{X}_+$ and $x_- \in \mathcal{X}_-$. Since $\mathcal{L}$ is positive, we have $0 \leqslant [x,x] = [x_+, x_+] + [x_-, x_-]$, hence $[x_+, x_+] \geqslant -[x_-, x_-]$ and, consequently, the mapping $\mathcal{L} \ni x \mapsto x_+ \in \mathcal{X}_+$ is injective and, in view of the maximality of $\mathcal{L}$, it is surjective as well. This shows that $\dim(\mathcal{L}) = \dim(\mathcal{X}_+)$. ∎

A linear manifold $\mathcal{L}$ of an inner product space $(\mathcal{X}; [\cdot, \cdot])$ is called *orthocomplemented* if there exists a linear manifold $\mathcal{M}$ in $\mathcal{X}$ such $\mathcal{L} \perp \mathcal{M}$, $\mathcal{M} \cap \mathcal{L} = \{0\}$ and $\mathcal{X} = \mathcal{L} + \mathcal{M}$. The linear manifold $\mathcal{M}$ with this property is called an *orthocomplement* of $\mathcal{L}$.

Remark 1.1.15. We fix an inner product space $(\mathcal{X}; [\cdot, \cdot])$ and a linear manifold $\mathcal{L}$ in $\mathcal{X}$.

(a) $\mathcal{L}$ is orthocomplemented in $\mathcal{X}$ if and only if there exists a linear operator $P \colon \mathcal{X} \to \mathcal{X}$ such that it is *idempotent*, that is, $P^2 = P$, *Hermitian*, equivalently, that is, $[Px, y] = [x, Py]$ for all $x, y \in \mathcal{X}$, and $\mathrm{Ran}(P) = \mathcal{L}$. The operator P is uniquely determined by $\mathcal{L}$ and $\mathcal{M}$ and it is called the *orthogonal projection* on $\mathcal{L}$ along $\mathcal{M}$. More precisely, if $\mathcal{M}$ denotes an orthocomplement of $\mathcal{L}$, then for any $x \in \mathcal{X}$ there exist unique $x_1 \in \mathcal{L}$ and $x_2 \in \mathcal{M}$ such that $x = x_1 + x_2$ and we let $Px = x_1$. In this case, $\mathrm{Ker}(P) = \mathcal{M}$ and $Q = I - P$ is the orthogonal projection on $\mathcal{M}$.

(b) If $\mathcal{L}$ is orthocomplemented in $\mathcal{X}$ and $\mathcal{M}$ is its orthocomplement then $\mathcal{X}^0 = \mathcal{L}^0 \dotplus \mathcal{M}^0$. In particular, this shows that, in general, we do not have uniqueness for the orthocomplement and hence we do not have uniqueness for the orthogonal projection.

(c) Assume that $\mathcal{X}$ is nondegenerate. If $\mathcal{L}$ is orthocomplemented then, from (b) it follows that $\mathcal{L}$ and any of its orthocomplements are nondegenerate. Also, in this case $\mathcal{L}^\perp$ is the unique orthocomplement of $\mathcal{L}$ and the orthogonal projection on $\mathcal{L}$ is unique as well.

(d) Assume that $\mathcal{L}$ is nondegenerate and that $\kappa_\pm(\mathcal{L}) < \infty$. Then $\mathcal{L}$ is orthocomplemented. Indeed, since $\kappa = \dim(\mathcal{L}) = \kappa_-(\mathcal{L}) + \kappa_+(\mathcal{L}) < \infty$, by the Gram–Schmidt orthonormalisation procedure there exists an *orthonormal basis* $\{e_j\}_{j=1}^\kappa$ of $\mathcal{L}$, that is, $[e_j, e_k] = \pm \delta_{j,k}$ for all $j, k \in \{1, \ldots, \kappa\}$. Let $P \colon \mathcal{X} \to \mathcal{X}$ be defined by $Px = \sum_{j=1}^\kappa [x, e_j] e_j$, for all $x \in \mathcal{X}$. Then P is an orthogonal projection with $\mathrm{Ran}(P) = \mathcal{L}$ and then we apply the remark at item (a). ∎

Proposition 1.1.16. *Let $\mathcal{L}$ be a strictly positive (strictly negative) linear manifold in the inner product space $(\mathcal{X}; [\cdot, \cdot])$. The following assertions are equivalent.*

(i) *$\mathcal{X}$ has a fundamental decomposition $\mathcal{X} = \mathcal{L} + \mathcal{X}^0 + \mathcal{M}$.*
(ii) *$\mathcal{L}$ is maximal strictly positive and orthocomplemented.*

Proof. (i)$\Rightarrow$(ii). To make a choice, assume that $\mathcal{L}$ is strictly positive. If $\mathcal{X}$ has a fundamental decomposition $\mathcal{X} = \mathcal{L} + \mathcal{X}^0 + \mathcal{M}$ then $\mathcal{X}^0 + \mathcal{M}$ is a negative linear manifold and an orthocomplement of $\mathcal{L}$. Then $\mathcal{L}$ is a maximal strictly positive linear manifold.

(ii)$\Rightarrow$(i). If $\mathcal{L}$ is an orthocomplemented maximal strictly positive linear manifold of $\mathcal{X}$ then $\mathcal{X} = \mathcal{L} + \mathcal{L}^\perp$ and $\mathcal{L}^\perp$ is a negative linear manifold. By Proposition 1.1.6 it follows that $\mathcal{L}^\perp = (\mathcal{L}^\perp)^0 + \mathcal{M}$, for some strictly negative linear manifold $\mathcal{M}$. It is easy to see that $(\mathcal{L}^\perp)^0 = \mathcal{X}^0$ and that $\mathcal{X} = \mathcal{L} + \mathcal{X}^0 + \mathcal{M}$, hence $\mathcal{X}$ has a fundamental decomposition of the required type. ∎

Proposition 1.1.17. *Let $(\mathcal{X}; [\cdot, \cdot])$ be an inner product space with finite rank of indefiniteness $\kappa(\mathcal{X}) = \min\{\kappa_+(\mathcal{X}), \kappa_-(\mathcal{X})\} < \infty$. Then it is decomposable.*

Proof. To make a choice, let us assume that $\kappa_-(\mathcal{X}) < \infty$ and let $\mathcal{L}$ be a maximal strictly negative linear manifold of $\mathcal{X}$. Then, by Proposition 1.1.14 we have $\dim(\mathcal{L}) = \kappa_-(\mathcal{X})$ and, by Remark 1.1.15.(d), $\mathcal{L}$ is orthocomplemented, hence Proposition 1.1.16 shows that $\mathcal{X}$ is decomposable. ∎

1.2 The Weak Topology

Let $(\mathcal{X}, [\cdot, \cdot])$ be an inner product space. The locally convex topology on $\mathcal{X}$ defined by the family of seminorms $\{p_y\}_{y \in \mathcal{X}}$

$$p_y(x) = |[x, y]|, \quad x \in \mathcal{X},$$

is called *the weak topology*. Since

$$\mathcal{X}^0 = \bigcap_{y \in \mathcal{X}} \operatorname{Ker}(p_y), \qquad (2.1)$$

it follows that the weak topology is separated if and only if the inner product space $(\mathcal{X}, [\cdot, \cdot])$ is nondegenerate. Also, the inner product $[\cdot, \cdot]$ is always *separately continuous*, that is, the linear functionals $\mathcal{X} \ni x \mapsto [x, y]$ are continuous for all $y \in \mathcal{X}$, with respect to the weak topology.

Theorem 1.2.1. *A linear functional φ on the inner product space $\mathcal{X}$ is weakly continuous if and only if there exists a vector $y_0 \in \mathcal{X}$ such that*

$$\varphi(x) = [x, y_0], \quad x \in \mathcal{X}.$$

1.2 The Weak Topology

Proof. The weak continuity of φ implies the existence of $y_1, \ldots, y_n \in \mathcal{X}$ such that

$$|\varphi(x)| \leqslant \sum_{k=1}^{n} |[x, y_k]|, \quad x \in \mathcal{X}.$$

Consider the complex vector space $\mathcal{X}^n$ (the direct sum of n copies of $\mathcal{X}$), the linear manifold $\mathcal{D} = \{(x_k)_{k=1}^n \mid x_i = x_j, 1 \leqslant i, j \leqslant n\}$ in $\mathcal{X}^n$ and the seminorm p on $\mathcal{X}^n$

$$p(x_1, \ldots, x_n) = \sum_{k=1}^{n} |[x_k, y_k]|, \quad (x_1, \ldots, x_n) \in \mathcal{X}^n.$$

We can define the linear functional φ_0 on $\mathcal{D}$ by

$$\varphi_0(x, \ldots, x) = \varphi(x), \quad x \in \mathcal{X}.$$

It follows, by the complex version of the Hahn–Banach Theorem, that there exists a linear functional $\tilde{\varphi}$ on $\mathcal{X}^n$ such that

$$\tilde{\varphi}(x, \ldots, x) = \varphi(x), \quad x \in \mathcal{X},$$

and

$$|\tilde{\varphi}(x_1, \ldots, x_n)| \leqslant p(x_1, \ldots, x_n), \quad (x_1, \ldots, x_n) \in \mathcal{X}^n.$$

Let φ_k, $1 \leqslant k \leqslant n$, be the linear functionals on $\mathcal{X}$ defined by

$$\varphi_k(x) = \tilde{\varphi}(0, \ldots, 0, x, 0, \ldots, 0), \quad x \in \mathcal{X},$$

where, on the right-hand side of this equality, x is on the kth position. Then

$$\varphi(x) = \sum_{k=1}^{n} \varphi_k(x), \quad x \in \mathcal{X},$$

$$|\varphi_k(x)| \leqslant |[x, y_k]|, \quad x \in \mathcal{X}, \ 1 \leqslant k \leqslant n.$$

If $y_k \in \mathcal{X}^0$ then $\varphi_k = 0$ and, in this case, we set $x_k = 0$; if $y_k \notin \mathcal{X}^0$ then there exists $x_k \in \mathcal{X}$ such that $[x_k, y_k] = 1$. Hence

$$|\varphi_k(x - [x, y_k]x_k)| \leqslant |[x - [x, y_k]x_k, y_k]| = 0, \quad x \in \mathcal{X},$$

that is,

$$\varphi_k(x) = \varphi_k(x_k)[x, y_k], \quad x \in \mathcal{X}.$$

The vector $y_0 = \sum_{k=1}^{n} \overline{\varphi_k(x_k)} y_k$ satisfies the required property. ∎

The weak topology provides a characterisation of those linear manifolds $\mathcal{L}$ in $\mathcal{X}$ such that $\mathcal{L} = \mathcal{L}^{\perp\perp}$. Here and in the following, we denote by $\text{Clos}_\text{w} \mathcal{L}$ the weak closure of $\mathcal{L}$.

Lemma 1.2.2. *Let $\mathcal{L}$ be a linear manifold of $\mathcal{X}$ and denote by $\mathrm{Clos_w}\,\mathcal{L}$ its weak closure. Then $\mathcal{L}^\perp$ is weakly closed and $\mathcal{L}^\perp = (\mathrm{Clos_w}\,\mathcal{L})^\perp$.*

Proof. If $x \notin \mathcal{L}^\perp$ then there exists $y \in \mathcal{L}$ such that $[x, y] \neq 0$. Since the inner product is weakly continuous in the first variable there exists a neighbourhood V of x_0, with respect to the weak topology, such that $[x, y] \neq 0$ for all $x \in V \cap \mathcal{L}^\perp$. Hence $\mathcal{L}^\perp$ is weakly closed.

Since $\mathcal{L} \subseteq \mathrm{Clos_w}\,\mathcal{L}$ we obtain $\mathcal{L}^\perp \supseteq (\mathrm{Clos_w}\,\mathcal{L})^\perp$. Conversely, if $x \notin (\mathrm{Clos_w}\,\mathcal{L})^\perp$ there exists $y_0 \in \mathrm{Clos_w}\,\mathcal{L}$ such that $[x, y_0] \neq 0$. Then $[x, y] \neq 0$ for all y in a neighbourhood U of y_0. Since $U \cap \mathcal{L} \neq \emptyset$ it follows that $x \notin \mathcal{L}^\perp$. ∎

Proposition 1.2.3. *A linear manifold $\mathcal{L}$ of the inner product space $\mathcal{X}$ is weakly closed if and only if $\mathcal{L} = \mathcal{L}^{\perp\perp}$.*

Proof. From Lemma 1.2.2 we obtain $\mathrm{Clos_w}\,\mathcal{L} \subseteq \mathcal{L}^{\perp\perp}$. Conversely, let $x_0 \notin \mathrm{Clos_w}\,\mathcal{L}$. Then there exists $\varepsilon > 0$ and $\{y_1, \ldots, y_n\} \subset \mathcal{X}$ such that

$$\{x \mid |[x - x_0, y_k]| < \varepsilon,\ 1 \leqslant k \leqslant n\} \cap \mathcal{L} = \emptyset.$$

Let us consider the seminorms on $\mathcal{X}$

$$p(x) = \max_{k=1}^{n} |[x, y_k]|, \quad x \in \mathcal{X},$$

$$q(x) = \inf_{z \in \mathcal{L}} p(x - z), \quad x \in \mathcal{X}.$$

Then $q(x_0) \geqslant \varepsilon$. By the complex version of the Hahn–Banach Theorem we obtain a linear functional φ on $\mathcal{X}$ such that $\varphi(x_0) = \varepsilon$ and

$$|\varphi(x)| \leqslant q(x), \quad x \in \mathcal{X}. \tag{2.2}$$

By definition, q is weakly continuous hence φ is weakly continuous. Thus, by Theorem 1.2.1, there exists $y_0 \in \mathcal{X}$ such that

$$\varphi(x) = [x, y_0], \quad x \in \mathcal{X}.$$

From (2.2) and the definition of q, it follows that $y_0 \in \mathcal{L}^\perp$ while $[x_0, y_0] = \varepsilon > 0$, hence $x_0 \notin \mathcal{L}^{\perp\perp}$. ∎

It is clear from the definition that the inner product $[\cdot, \cdot]$ is always separately continuous with respect to the weak topology. We consider now the *joint weak continuity* of the inner product $[\cdot, \cdot]$, that is, the continuity with respect to the product of the weak topologies on $\mathcal{X} \times \mathcal{X}$.

Proposition 1.2.4. *The inner product $[\cdot, \cdot]$ is jointly continuous with respect to the weak topology if and only if both of $\kappa_+(\mathcal{X})$ and $\kappa_-(\mathcal{X})$ are finite.*

1.3 NORMED TOPOLOGIES

Proof. From Remark 1.1.5, Proposition 1.1.6, and (2.1), it follows that, without restricting the generality, we can assume that the inner product space $(\mathcal{X}, [\cdot, \cdot])$ is nondegenerate.

If both of $\kappa_+(\mathcal{X})$ and $\kappa_-(\mathcal{X})$ are finite then, by the Gram–Schmidt orthogonalisation procedure we obtain two linear manifolds $\mathcal{X}^+$ and $\mathcal{X}^-$ in $\mathcal{X}$, such that $\mathcal{X}^+$ is positive, $\dim(\mathcal{X}^+) = \kappa_+(\mathcal{X})$, $\mathcal{X}^-$ is negative, $\dim(\mathcal{X}^-) = \kappa_-(\mathcal{X})$, and

$$\mathcal{X} = \mathcal{X}^+ \dotplus \mathcal{X}^-.$$

The space $\mathcal{X}$ is finite dimensional and the weak topology is a separated linear topology on $\mathcal{X}$. Representing the inner product $[\cdot, \cdot]$ with respect to the orthogonal basis it follows that it is jointly continuous.

Conversely, let us assume that $[\cdot, \cdot]$ is jointly continuous with respect to the weak topology. Then there exists $\varepsilon > 0$ and $z_1, \ldots, z_k \in \mathcal{X}$ such that $|[x, y]| < 1$, whenever $|[x, z_k]| < \varepsilon$ and $|[y, z_k]| < \varepsilon$, for all $k \in \{1, \ldots, n\}$. Let $\mathcal{L}$ be the span of all $z_1, \ldots, z_n$. If $x \in \mathcal{L}^\perp$ and $y \in \mathcal{X}$ then, for all $\lambda \in \mathbb{C}$ we have

$$|[\lambda x, z_j]| = 0 < \varepsilon, \quad 1 \leqslant j \leqslant n,$$

and for a certain $\alpha > 0$ we have

$$|[\alpha y, z_j]| < \varepsilon, \quad 1 \leqslant j \leqslant n.$$

Then

$$|[\lambda x, \alpha y]| < 1, \quad \lambda \in \mathbb{C},$$

and hence $[x, y] = 0$. It follows that $\mathcal{L}^\perp \subset \mathcal{X}^\perp = \{0\}$ which implies $\mathcal{X} = \mathcal{L}^{\perp\perp}$. By Proposition 1.2.3 this means that $\mathcal{L}$ is weakly dense in $\mathcal{X}$. Since $\mathcal{L}$ is finite dimensional and $\mathcal{X}$ is nondegenerate, $\mathcal{L}$ is weakly closed, hence $\mathcal{L} = \mathcal{X}$, in particular both of $\kappa_-(\mathcal{X})$ and $\kappa_+(\mathcal{X})$ are finite. ∎

Corollary 1.2.5. *Let $(\mathcal{X}; [\cdot, \cdot])$ be a nondegenerate inner product space. Then the inner product $[\cdot, \cdot]$ is jointly continuous with respect to the weak topology if and only if $\mathcal{X}$ is finite dimensional.*

1.3 Normed Topologies

We pass now to considerations on normed topologies on an inner product space $(\mathcal{X}; [\cdot, \cdot])$ making the inner product continuous.

Lemma 1.3.1. *Assume that on $\mathcal{X}$ there is defined a norm $\|\cdot\|$ such that $(\mathcal{X}, \|\cdot\|)$ is a Banach space. Then, with respect to this norm, the inner product $[\cdot, \cdot]$ is separately continuous if and only if it is jointly continuous.*

Proof. If $[\cdot,\cdot]$ is separately continuous with respect to the norm $\|\cdot\|$ then for each fixed $y \in \mathcal{X}$ there exists $\alpha_y \geqslant 0$ such that $|[x,y]| \leqslant \alpha_y \|x\|$, $x \in \mathcal{X}$. From the Uniform Boundedness Principle in Banach spaces, it follows that there exists $\alpha \geqslant 0$ such that $|[x,y]| \leqslant \alpha \|x\| \|y\|$, $x,y \in \mathcal{X}$, hence $[\cdot,\cdot]$ is jointly continuous. The converse implication is always true. ∎

Theorem 1.3.2. *Let $\|\cdot\|_1$ and $\|\cdot\|_2$ be two complete norms on the nondegenerate inner product space $(\mathcal{X}; [\cdot,\cdot])$, such that $[\cdot,\cdot]$ is continuous with respect to each norm. Then $\|\cdot\|_1 \sim \|\cdot\|_2$, that is, the two norms are equivalent, in the sense that $\|x\|_1 \leqslant \alpha \|x\|_2 \leqslant \beta \|x\|_1$ for some $\alpha, \beta > 0$ and all $x \in \mathcal{X}$.*

Proof. Define a new norm on $\mathcal{X}$ by
$$\|x\| = \sup\{\|x\|_1, \|x\|_2\}, \quad x \in \mathcal{X},$$
and observe that this norm is also complete. Indeed, let $(x_n)_{n \in \mathbb{N}}$ be a Cauchy sequence in $\mathcal{X}$ with respect to $\|\cdot\|$. Since $\|x\| \geqslant \|x\|_j$, $x \in \mathcal{X}, j = 1, 2$, it follows that this sequence is Cauchy with respect to each norm $\|\cdot\|_j$, $j = 1, 2$. Let y_j be its limit with respect to $\|\cdot\|_j$, $j = 1, 2$, respectively. Since each of the norms $\|\cdot\|_j$, $j = 1, 2$ makes $[\cdot,\cdot]$ continuous (separately is equivalent with jointly in this case, by Lemma 1.3.1) we have
$$\lim_{n \to \infty} [x_n, z] = [y_j, z], \quad z \in \mathcal{X}, j = 1, 2,$$
hence, since $\mathcal{X}$ is nondegenerate, $y_1 = y_2$. Then this vector is also the limit of $(x_n)_{n \in \mathbb{N}}$ with respect to $\|\cdot\|$.

Consider the inclusion mapping of the Banach space $(\mathcal{X}, \|\cdot\|)$ onto the Banach space $(\mathcal{X}, \|\cdot\|_j)$, $j = 1, 2$. From the Closed Graph Priniciple in Banach spaces, we obtain that $\|\cdot\|$ and $\|\cdot\|_j$ are equivalent, hence $\|\cdot\|_1$ and $\|\cdot\|_2$ are equivalent as well. ∎

Remark 1.3.3. Assume that a positive inner product $\langle\cdot,\cdot\rangle$ is given on $\mathcal{X}$ such that the corresponding norm $\|\cdot\|$ is complete (that is, $(\mathcal{X}, \langle\cdot,\cdot\rangle)$ is a Hilbert space) and the inner product $[\cdot,\cdot]$ is $\|\cdot\|$-continuous. Then, by the Riesz Representation Theorem there exists a unique selfadjoint operator $G \in \mathcal{B}(\mathcal{X})$, in the sense of Hilbert spaces, such that
$$[x,y] = \langle Gx, y\rangle, \quad x, y \in \mathcal{X}.$$
The operator G is called the *Gram operator* of the inner product $[\cdot,\cdot]$ with respect to the positive inner product $\langle\cdot,\cdot\rangle$. Moreover, $\mathcal{X}^0 = \mathrm{Ker}(G)$ holds. Denoting by E the spectral measure of G, let $\mathcal{X}_- = E(-\infty, 0)\mathcal{X}$ and $\mathcal{X}_+ = E(0, \infty)\mathcal{X}$. $(\mathcal{X}_+, [\cdot,\cdot])$ is strictly positive, $(\mathcal{X}_-, [\cdot,\cdot])$ is strictly negative, $\mathcal{X}_- \perp \mathcal{X}_+$ and the following decomposition holds
$$\mathcal{X} = \mathcal{X}_- \dotplus \mathcal{X}^0 \dotplus \mathcal{X}_+. \tag{3.1}$$
In particular, we have shown that $\mathcal{X}$ is decomposable, $\kappa_-(\mathcal{X}) = \mathrm{rank}\, E(-\infty, 0)$ and $\kappa_+(\mathcal{X}) = \mathrm{rank}\, E(0, \infty)$. This is a special case of Example 1.1.10. ∎

1.3 NORMED TOPOLOGIES 15

The next example shows that, in Remark 1.3.3, the assumption on the completeness of $(\mathcal{X}, \langle \cdot, \cdot \rangle)$ is essential.

Example 1.3.4. Consider $\ell^2(\mathbb{Z})$ the Hilbert space of complex valued sequences $x = (x_n)_{n \in \mathbb{Z}}$ such that $\sum_{n \in \mathbb{Z}} |x_n|^2 < \infty$, and denote by $\langle \cdot, \cdot \rangle$ the canonical positive definite inner product $\langle x, y \rangle = \sum_{n \in \mathbb{Z}} x_n \overline{y}_n$, for $x = (x_n)_{n \in \mathbb{Z}}$ and $y = (y_n)_{n \in \mathbb{Z}}$. Let $\mathcal{X}$ be the space of complex sequences $x = (x_n)_{n \in \mathbb{Z}}$ in $\ell^2(\mathbb{Z})$, such that, for some $n(x) \in \mathbb{Z}$ we have $x_n = 0$ for all $n \leqslant n(x)$. On $\mathcal{X}$ we consider the indefinite inner product $[\cdot, \cdot]$ defined by

$$[x, y] = \sum_{n \in \mathbb{Z}} x_{-n} \overline{y}_n, \quad x = (x_n)_{n \in \mathbb{Z}} \in \mathcal{X}, \ y = (y_n)_{n \in \mathbb{Z}} \in \mathcal{X}. \tag{3.2}$$

Then we have

$$|[x, y]|^2 \leqslant \langle x, x \rangle \langle y, y \rangle, \quad x, y \in \mathcal{X}.$$

It is easy to see that $(\mathcal{X}, [\cdot, \cdot])$ is nondegenerate, and let us assume that it is decomposable, that is, there exist two mutually orthogonal subspaces $\mathcal{X}_\pm$ such that $\mathcal{X}_+$ is positive, $\mathcal{X}_-$ is negative and $\mathcal{X} = \mathcal{X}_+ \dotplus \mathcal{X}_-$. Consider the linear operator $T \colon \mathcal{X} \to \mathcal{X}$ defined as follows: if $x = (x_n)_{n \in \mathbb{Z}}$ is an arbitrary vector in $\mathcal{X}$, then $Tx = (x'_n)_{n \in \mathbb{Z}}$, where $x'_n = x_n$ for all $n < 0$ and $x'_n = 0$ for all $n \geqslant 0$.

We first note that, if $\mathcal{A}$ is a definite linear submanifold of $\mathcal{X}$, then $T|\mathcal{A}$ is injective. To see this, let $x \in \mathcal{X}$ be such that $Tx = 0$. By (3.2) it follows that $[x, x] = 0$, hence, $x = 0$ since, due to the Schwarz Inequality, there exists no neutral vectors except the null vector. Since the space of complex vectors with finite support has a countable algebraic basis, this implies that both $\mathcal{X}_\pm$ have countable algebraic bases, and hence $\mathcal{X} = \mathcal{X}_- \dotplus \mathcal{X}_+$ has a countable algebraic basis.

Let now $\mathcal{Y}$ be the linear manifold of vectors $y = (y_n)_{n \in \mathbb{Z}}$ in $\ell^2(\mathbb{Z})$ such that for some $n(y) \in \mathbb{Z}$ we have $y_n = 0$ for all $n \geqslant n(y)$. It is easy to see that $\mathcal{X}$ and $\mathcal{Y}$ are isomorphic, by the symmetrisation map on sequences: $\mathcal{X} \ni x \mapsto y \in \mathcal{Y}$, where $y_n = x_{-n}$ for $n \in \mathbb{Z}$. Thus, under our assumption that $\mathcal{X}$ is decomposable, it follows that both $\mathcal{X}$ and $\mathcal{Y}$ have countable algebraic bases, hence $\ell^2(\mathbb{Z}) = \mathcal{X} \vee \mathcal{Y}$ has a countable algebraic basis. But this is a contradiction since, due to the second Baire Category Theorem, there exists no Hilbert space admitting a countable algebraic basis. ∎

Finally we give an example of an inner product space $(\mathcal{X}, [\cdot, \cdot])$ which admits no norm topology with respect to which the inner product $[\cdot, \cdot]$ is continuous.

Example 1.3.5. Let $(\mathcal{X}, [\cdot, \cdot])$ be the vector space of complex sequences $x = (x_n)_{n \in \mathbb{Z}}$ such that $x_n = 0$ for $n \leqslant n_x$ and the inner product $[\cdot, \cdot]$ is defined as in (3.2). Then $(\mathcal{X}, [\cdot, \cdot])$ is nondegenerate. Assume that $\|\cdot\|$ is a norm on $\mathcal{X}$ with respect to which the inner product $[\cdot, \cdot]$ is separately continuous. Then there exists the seminorm α on $\mathcal{X}$

$$\mathcal{X} \ni y \mapsto \alpha(y) = \sup_{\|x\|=1} |[x, y]| \in \mathbb{R},$$

such that

$$|[x, y]| \leqslant \alpha(y) \|x\|, \quad x, y \in \mathcal{X}.$$

With $e_{-n} = (\delta_{j,-n})_{j \in \mathbb{Z}}$, $n \in \mathbb{N}$ we obtain

$$|[x, e_{-n}]| = |x_n| \leqslant \alpha(e_{-n})\|x\|, \quad n \in \mathbb{N}, \ x \in \mathcal{X}. \tag{3.3}$$

Now choose $x = (x_j)_{j \in \mathbb{Z}}$ as follows:

$$x_j = \begin{cases} 0, & j < 0, \\ j\alpha(e_{-j}), & j \leqslant 0. \end{cases}$$

From (3.3) we obtain

$$n\alpha(e_{-n}) \leqslant \alpha(e_{-n})\|x\|, \quad n \in \mathbb{N}.$$

Since $\alpha(e_n) \neq 0$, this is a contradiction. ∎

1.4 Kreĭn Spaces

We introduce the class of Kreĭn spaces by means of the following result.

Theorem 1.4.1. *Let $(\mathcal{H}, [\cdot, \cdot])$ be an indefinite inner product space. Then the following statements are equivalent.*

(a) There exist two linear manifolds $\mathcal{H}^+$ and $\mathcal{H}^-$ such that $\mathcal{H}^+ \perp \mathcal{H}^-$, $\mathcal{H} = \mathcal{H}^+ + \mathcal{H}^-$, and both $(\mathcal{H}^+, [\cdot, \cdot])$ and $(\mathcal{H}^-, -[\cdot, \cdot])$ are Hilbert spaces.

(b) There exists a linear operator J on $\mathcal{H}$ such that $J^2 = I$ and the equality

$$\langle x, y \rangle_J = [Jx, y], \quad x, y \in \mathcal{H}, \tag{4.1}$$

defines a positive definite inner product in $\mathcal{H}$ such that $(\mathcal{H}, \langle \cdot, \cdot \rangle_J)$ is a Hilbert space.

(c) There exists a positive definite inner product $\langle \cdot, \cdot \rangle$ on $\mathcal{H}$ such that $(\mathcal{H}, \langle \cdot, \cdot \rangle)$ is a Hilbert space and the associated norm $\| \cdot \|$ satisfies

$$\|x\| = \sup_{\|y\| \leqslant 1} |[x, y]|, \quad x \in \mathcal{H}. \tag{4.2}$$

Proof. (a)⇒(b). Let $\mathcal{H} = \mathcal{H}^+ + \mathcal{H}^-$ be a decomposition of $\mathcal{H}$ with the properties required in (a). We define the linear operator J:

$$J(x_+ + x_-) = x_+ - x_-, \quad x_+ \in \mathcal{H}^+, \ x_- \in \mathcal{H}^-.$$

Then

$$[Jx, y] = [x_+ - x_-, y_+ + y_-] = [x_+, y_+] - [x_-, y_-], \quad x = x_+ + x_-, \ y = y_+ + y_-,$$

and hence the inner product $\langle \cdot, \cdot \rangle_J$ is positive definite and coincides with the inner product of the direct sum Hilbert space $\mathcal{H}^+ \oplus \mathcal{H}^-$. The equality $J^2 = I$ follows immediately from the definition of J.

1.4 Kreĭn Spaces

(b)⇒(a). Let J be a linear operator with the properties required at (b). Then, $J = J^*$ with respect to the Hilbert space $(\mathcal{H}; \langle \cdot, \cdot \rangle_J)$ and consequently, letting $J^+ = \frac{1}{2}(I + J)$ and $J^- = \frac{1}{2}(I - J)$ we have $J = J^+ - J^-$, $I = J^+ + J^-$, $J^+ J^- = 0$, and J^+ and J^- are the orthogonal projections onto the Hilbert spaces $\mathcal{H}^+ = J^+ \mathcal{H}$ and $\mathcal{H}^- = J^- \mathcal{H}$. The assertion (a) follows.

(b)⇒(c). If the operator J has the properties required at assertion (b) then, with respect to the Hilbert space $(\mathcal{H}, \langle \cdot, \cdot \rangle_J)$, where $\langle \cdot, \cdot \rangle_J$ is defined at (4.1), we have $J = J^* = J^{-1}$, that is, the operator J is a *symmetry*. In particular, J is unitary and, for any $x \in \mathcal{H}$,

$$\|x\| = \|Jx\| = \sup_{\|y\| \leq 1} |\langle Jx, y\rangle_J| = \sup_{\|y\| \leq 1} |[J^2 x, y]| = \sup_{\|y\| \leq 1} |[x, y]|,$$

where $\|\cdot\|$ denotes the norm associated to the inner product $\langle \cdot, \cdot \rangle_J$.

(c)⇒(b). Let J be the Gram operator (see Remark 1.3.3) of the inner product $[\cdot, \cdot]$ with respect to the inner product $\langle \cdot, \cdot \rangle$. Then

$$\|x\| = \sup_{\|y\| \leq 1} |[x, y]| = \sup_{\|y\| \leq 1} |\langle Jx, y\rangle| = \|Jx\|, \quad x \in \mathcal{H},$$

which means that J is an isometry in the Hilbert space $(\mathcal{H}, \langle \cdot, \cdot \rangle)$. Since $J = J^*$ also holds, it follows that $J^2 = I$. ∎

A *Kreĭn space* is, by definition, an inner product space $(\mathcal{H}, [\cdot, \cdot])$ satisfying one (hence all) of the statements in Theorem 1.4.1.

If $(\mathcal{H}, [\cdot, \cdot])$ is a Kreĭn space then a decomposition $\mathcal{H} = \mathcal{H}^+ + \mathcal{H}^-$, with the properties in (a), is called a *fundamental decomposition*; a linear operator J on $\mathcal{H}$ satisfying the requirements in (b) is called a *fundamental symmetry*; a norm $\|\cdot\|$ with the properties as in (c) is called a *fundamental norm*.

Remark 1.4.2. From the proof of Theorem 1.4.1 it follows that the class of fundamental symmetries, the class of fundamental decompositions, and the class of fundamental norms of a given Kreĭn space, are in bijective correspondences, respectively. ∎

Corollary 1.4.3. *Any two fundamental norms of a Kreĭn space are equivalent.*

Proof. Since a Kreĭn space is always nondegenerate this statement is a consequence of Theorem 1.3.2. ∎

The *strong topology* of a Kreĭn space $\mathcal{K}$ is, by definition, the topology induced by an arbitrary fundamental norm of $\mathcal{K}$. By Corollary 1.4.3, this definition is correct.

In a Kreĭn space the Riesz Theorem of representation of bounded linear functionals works as in the Hilbert space.

Proposition 1.4.4. *For any bounded linear functional $\varphi \colon \mathcal{K} \to \mathbb{C}$, where $\mathcal{K}$ is a Kreĭn space, there exists a unique vector $y \in \mathcal{K}$ such that $\varphi(x) = [x, y]$ for all $x \in \mathcal{K}$.*

Proof. Let J be a fundamental symmetry of $\mathcal{K}$. Then, since the strong topology of $\mathcal{K}$ is the same as that of the Hilbert space $(\mathcal{K}; \langle \cdot, \cdot \rangle_J)$, φ is a bounded linear functional on this Hilbert space and then, by the Riesz Theorem of representation, there exists $z \in \mathcal{K}$ such that $\varphi(x) = \langle x, z \rangle_J$ for all $x \in \mathcal{K}$. Hence, letting $y = Jz$ we have $\varphi(x) = \langle x, z \rangle_J = [x, Jz] = [x, y]$ for all $x \in \mathcal{K}$. The uniqueness of y follows due to the nondegeneracy of the inner product $[\cdot, \cdot]$. ∎

Remark 1.4.5. Given a Kreĭn space $(\mathcal{K}; [\cdot, \cdot])$, it follows that the weak topology of $\mathcal{K}$ is equivalently defined by any inner product $\langle \cdot, \cdot \rangle_J$ defined as in (4.1), where J is an arbitrary fundamental symmetry, and it coincides with the weak topology defined by the indefinite inner product (see Section 1.2). ∎

The fundamental symmetries of Kreĭn spaces are special types of Gram operators, hence it is natural to have a characterisation of Kreĭn spaces from this point of view. The answer to this question is that this is related to the bounded invertibility of the Gram operator and it will be proven in Proposition 2.1.9 since some more tools are needed for its proof.

Remark 1.4.6. A rather general example of a Kreĭn space is obtained as follows: if $(\mathcal{H}, \langle \cdot, \cdot \rangle)$ is a Hilbert space and J is a linear operator on $\mathcal{H}$ such that $J = J^* = J^{-1}$ (that is, J is a *symmetry*) then the inner product

$$[x, y] = \langle Jx, y \rangle, \quad x, y \in \mathcal{H},$$

turns $\mathcal{H}$ into the Kreĭn space $(\mathcal{H}; [\cdot, \cdot])$ on which J is a fundamental symmetry.

In addition, any symmetry J has its spectrum $\sigma(J) = \{-1, 1\}$, unless J is $\pm I$, and the spectral subspaces $\mathcal{H}^{\pm}$ corresponding to the eigenvalues ± 1 make a decomposition $\mathcal{H} = \mathcal{H}^+ \oplus \mathcal{H}^-$ which is exactly the fundamental decomposition corresponding to the fundamental symmetry J in the Kreĭn space $(\mathcal{H}; [\cdot, \cdot])$. On the other hand, letting $J^{\pm}$ denote the orthogonal projection of $\mathcal{H}$ onto $\mathcal{H}^{\pm}$, we have $J = J^+ - J^- = 2J^+ - I = I - 2J^-$ and, equivalently, $J^{\pm} = (I \pm J)/2$.

However, we stress that, although this example can be seen in many applications, there are others that require the construction of a Kreĭn space from an indefinite inner product space on which some norm topology can be found and then a certain procedure of factorisation and completion should be performed, see Section 1.5. ∎

The underlying notion of *isomorphism of Kreĭn spaces* has to be clarified. Two Kreĭn spaces $\mathcal{K}_1$ and $\mathcal{K}_2$ are *isomorphic* if there exists a bijective linear operator $U \colon \mathcal{K}_1 \to \mathcal{K}_2$, such that it is *isometric*, that is, $[Ux, Uy]_{\mathcal{K}_2} = [x, y]_{\mathcal{K}_1}$ for all $x, y \in \mathcal{K}_1$. This agrees with the general definition of an isomorphism of inner product spaces given before. Here two remarks are in order. Firstly, we stress that, unlike the case of Hilbert spaces, the strong continuity of a linear operator $T \colon \mathcal{K}_1 \to \mathcal{K}_2$ does not follow from the isometry property, see Example 4.3.5. This anomaly comes from the fact that the strong topology of a Kreĭn space is not intrinsic, that is, it is not defined by the inner product. Secondly, any isomorphism of Kreĭn spaces is automatically strongly continuous.

1.4 Kreĭn Spaces

Proposition 1.4.7. *If $\mathcal{K}_1$ and $\mathcal{K}_2$ are Kreĭn spaces and $T\colon \mathcal{K}_1 \to \mathcal{K}_2$ is an isomorphism of Kreĭn spaces then T is bounded.*

Proof. Since T is isometric and bijective it is easy to see that we have

$$[Tx, z] = [x, T^{-1}z], \quad x \in \mathcal{K}_1,\ z \in \mathcal{K}_2. \tag{4.3}$$

Then, in order to finish the proof, by the Closed Graph Principle it is sufficient to prove that T is a closed operator. To this end, let $(x_n)_n$ be a sequence in $\mathcal{K}_1$ such that $x_n \xrightarrow[n\to\infty]{} x$ and $Tx_n \xrightarrow[n\to\infty]{} y$, for some $x \in \mathcal{K}_1$ and $y \in \mathcal{K}_2$, with respect to the strong topology on $\mathcal{K}_1$ and on $\mathcal{K}_2$, respectively. Then, in view of (4.3), for any $z \in \mathcal{K}_2$ we have

$$[y, z] = \lim_{n\to\infty}[Tx_n, z] = \lim_{n\to\infty}[x_n, T^{-1}z] = [x, T^{-1}z] = [Tx, z],$$

hence $y = Tx$. This shows that T is closed, hence bounded. ∎

A more general statement on continuity of isometric operators is proven in Proposition 4.3.4. Isomorphisms of Kreĭn spaces are also called *bounded unitary* operators (or, simply, *unitary* operators, if their boundedness is implicit) of Kreĭn spaces. The concept of a unitary operator in Kreĭn spaces is more general and will be treated more carefully in the next chapters.

Remark 1.4.8. The previous discussion on isomorphisms of Kreĭn spaces makes natural another general procedure of construction of a Kreĭn space. Let $(\mathcal{K}; [\cdot, \cdot]_\mathcal{K})$ be a Kreĭn space and let $(\mathcal{H}; [\cdot, \cdot]_\mathcal{H})$ be an inner product space such that there exists an isometric linear bijection $T\colon \mathcal{K} \to \mathcal{H}$. In the following we show that the Kreĭn space structure of $(\mathcal{K}; [\cdot, \cdot]_\mathcal{K})$ can be transported in a natural way onto $(\mathcal{H}; [\cdot, \cdot]_\mathcal{H})$ such that T becomes an isomorphism of Kreĭn spaces. If $\|\cdot\|_\mathcal{K}$ denotes an arbitrary fundamental norm on $\mathcal{K}$ with associated positive definite inner product denoted by $\langle\cdot, \cdot\rangle_\mathcal{K}$, then one can define the positive definite inner product $\langle\cdot, \cdot\rangle_\mathcal{H}$ by letting

$$\langle h_1, h_2\rangle_\mathcal{H} = \langle T^{-1}h_1, T^{-1}h_2\rangle_\mathcal{K}, \quad h_1, h_2 \in \mathcal{H},$$

equivalently, one can define the quadratic norm $\|\cdot\|_\mathcal{H}$ by letting

$$\|h\|_\mathcal{H} = \|T^{-1}h\|_\mathcal{K}, \quad h \in \mathcal{H}.$$

Then, it is easy to show that the norm $\|\cdot\|_\mathcal{H}$ turns $\mathcal{H}$ into a Hilbert space and that

$$\|h\|_\mathcal{H} = \sup_{\|k\|_\mathcal{H} \leqslant 1} |[h, k]_\mathcal{H}|, \quad h \in \mathcal{H},$$

hence, in view of Theorem 1.4.1, $(\mathcal{H}; [\cdot, \cdot]_\mathcal{H})$ is a Kreĭn space. Moreover, the isometric linear bijection $T\colon \mathcal{K} \to \mathcal{H}$ is an isomorphism of Kreĭn spaces. ∎

Remark 1.4.9. Let $(\mathcal{K}, [\cdot, \cdot])$ be a Kreĭn space, consider a fundamental symmetry J on $\mathcal{K}$, and let $\langle \cdot, \cdot \rangle_J$ denote the associated positive inner product as in (4.1).

(a) Two elements $x, y \in \mathcal{K}$ are called *J-orthogonal* if $\langle x, y \rangle_J = 0$ and we denote this by $x \perp_J y$. For a subset $\mathcal{A} \subseteq \mathcal{K}$ let us denote its J-orthogonal companion by $\mathcal{A}^{\perp_J} = \{x \in \mathcal{K} \mid x \perp_J y, \, y \in \mathcal{A}\}$. Then

$$\mathcal{A}^\perp = J \mathcal{A}^{\perp_J}. \tag{4.4}$$

(b) Consider a linear manifold $\mathcal{L} \subseteq \mathcal{K}$. Since the weak topology on the Kreĭn space $\mathcal{K}$ coincides with the weak topology on the Hilbert space $(\mathcal{K}, \langle \cdot, \cdot \rangle_J)$, it follows that the closure of $\mathcal{L}$ with respect to the weak topology coincides with the closure of $\mathcal{L}$ with respect to the strong topology. Thus, by Proposition 1.2.3 it follows that $\mathrm{Clos}(\mathcal{L}) = \mathcal{L}^{\perp\perp}$, where we denoted by $\mathrm{Clos}(\mathcal{L})$ the closure of $\mathcal{L}$ with respect to the strong topology.

(c) Let $\mathcal{L}_1$ and $\mathcal{L}_2$ be two subspaces of $\mathcal{K}$. Then

$$(\mathcal{L}_1 + \mathcal{L}_2)^\perp = \mathcal{L}_1^\perp \cap \mathcal{L}_2^\perp, \tag{4.5}$$

and

$$(\mathcal{L}_1 \cap \mathcal{L}_2)^\perp = \mathrm{Clos}(\mathcal{L}_1^\perp + \mathcal{L}_2^\perp). \tag{4.6}$$

It follows that

$$\mathrm{Clos}(\mathcal{L} + \mathcal{L}^\perp) = \mathcal{L}^{0\perp}, \tag{4.7}$$

where $\mathcal{L}^0 = \mathcal{L} \cap \mathcal{L}^\perp$ denotes the isotropic subspace of $\mathcal{L}$. In particular, $\mathcal{L}$ is nondegenerate if and only if $\mathcal{L} + \mathcal{L}^\perp$ is dense in $\mathcal{K}$. ∎

Throughout this book, if $\mathcal{M}$ and $\mathcal{N}$ are subspaces of the Kreĭn space $\mathcal{K}$, we use the symbol $\mathcal{M}[+]\mathcal{N}$ to denote the subspace generated by $\mathcal{M}$ and $\mathcal{N}$ whenever $\mathcal{M} \perp \mathcal{N}$, $\mathcal{M} \cap \mathcal{N} = \{0\}$, and $\mathcal{M} + \mathcal{N}$ is closed. In particular, a fundamental decomposition of $\mathcal{K}$ will be written $\mathcal{K} = \mathcal{K}^+[+]\mathcal{K}^-$.

A Kreĭn space $\mathcal{K}$ is called a *Pontryagin space* if it has a finite rank of indefiniteness $\kappa(\mathcal{K}) = \min\{\kappa_+(\mathcal{K}), \kappa_-(\mathcal{K})\} < \infty$. Equivalently, there is either only a finite number of linearly independent positive vectors in $\mathcal{K}$ or only a finite number of linearly independent negative vectors in $\mathcal{K}$.

A Pontryagin space is also encountered in the literature under the name of Π_κ *space*, where $\kappa < \infty$ denotes the rank of indefiniteness and hence, with a certain ambiguity, some times it is the negative rank and other times the positive rank, depending on the adopted point of view and the peculiarities of the problems to be investigated. Clearly, a nondegenerate inner product space of finite dimension is a Pontryagin space. Very often, in applications, Pontryagin spaces appear as completions of inner product spaces $\mathcal{X}$ in which either $\kappa_+(\mathcal{X})$ or $\kappa_-(\mathcal{X})$ is finite (see Corollary 1.5.4). This makes interesting the following characterisation of dense linear manifolds.

Lemma 1.4.10. *Let $\mathcal{D}$ be a dense linear manifold of the Pontryagin space $\mathcal{K}$. If $\kappa_+(\mathcal{K}) < \infty$ ($\kappa_-(\mathcal{K}) < \infty$) then there exists a fundamental decomposition $\mathcal{K} = \mathcal{K}^+[+]\mathcal{K}^-$ such that $\mathcal{K}^+ \subseteq \mathcal{D}$ (respectively, $\mathcal{K}^- \subseteq \mathcal{D}$).*

1.4 Kreĭn Spaces

Proof. Assume $\kappa = \kappa_+(\mathcal{K}) < \infty$. If $\kappa = 0$ then the statement is trivial so we assume $\kappa > 0$. Then for an arbitrary fundamental decomposition $\mathcal{K} = \mathcal{K}^+[+]\mathcal{K}^-$ we have $\dim(\mathcal{K}^+) = \kappa$ and let $\{e_j\}_{j=1}^{\kappa}$ be an orthonormal basis of $\mathcal{K}^+$. Denote by $\|\cdot\|$ the fundamental norm associated to this fundamental decomposition. Since $\mathcal{D}$ is dense in $\mathcal{K}$, for any $\varepsilon > 0$ there exists a system of vectors $\{f_j\}_{j=1}^{\kappa}$ such that $\|f_j - e_j\| < \varepsilon$ for all $j \in \{1, 2, \ldots, \kappa\}$. If x and y are two vectors such that

$$x = \sum_{j=1}^{\kappa} \alpha_j e_j, \quad y = \sum_{j=1}^{\kappa} \alpha_j f_j, \quad \sum_{j=1}^{\kappa} |\alpha_j|^2 = 1,$$

then $[x, x] = \|x\|^2 = 1$ and

$$\|x - y\| \leqslant \sum_{j=1}^{\kappa} |\alpha_j| \|e_j - f_j\| \leqslant \Big(\sum_{j=1}^{\kappa} |\alpha_j|^2\Big)^{1/2} \Big(\sum_{j=1}^{\kappa} \|e_j - f_j\|^2\Big)^{1/2} < \varepsilon\sqrt{\kappa},$$

and hence

$$\|y\| \leqslant \|y - x + x\| \leqslant \|x\| + \|x - y\| < 1 + \varepsilon\sqrt{\kappa}.$$

The previous inequalities imply

$$\begin{aligned} |[y, y] - 1| = |[y, y] - [x, x]| &= |[y, x - y] + [y - x, x]| \\ &\leqslant |[y, y - x]| + |[y - x, x]| \\ &\leqslant \|y\| \|y - x\| + \|y - x\| \|x\| < (1 + \varepsilon\sqrt{\kappa})\varepsilon\sqrt{\kappa}. \end{aligned}$$

Taking $\varepsilon < \frac{1}{3\kappa}$, from above it follows that $[y, y] > 0$. If a nonzero vector z is considered

$$z = \sum_{j=1}^{\kappa} \beta_j f_j, \quad \sum_{j=1}^{\kappa} |\beta_j|^2 = \beta^2 \neq 0,$$

we apply the above conclusion to $y = \frac{1}{\beta}z$ and get $[z, z] > 0$. We have thus proven that the subspace $\mathcal{L}^+ \subseteq \mathcal{D}$ generated by the system of vectors $\{f_j\}_{j=1}^{\kappa}$ is strictly positive and has dimension κ. Then, see Remark 1.1.15 (d), $\mathcal{L}^+$ is maximal strictly positive, and denoting $\mathcal{L}^- = \mathcal{L}^{+\perp}$ it follows that $\mathcal{K} = \mathcal{L}^+[+]\mathcal{L}^-$ is a fundamental decomposition, by Proposition 1.1.14.

The proof in the case $\kappa_-(\mathcal{K}) < \infty$ is similar or can be reduced to the above situation by considering the inner product space $(\mathcal{K}, -[\cdot, \cdot])$. ∎

The next result shows that in a Pontryagin space, unlike the case of a genuine Kreĭn space, the strong topology is intrinsic, that is, it can be characterised exclusively in terms of the inner product.

Theorem 1.4.11. *Let $\mathcal{K}$ be a Pontryagin space and let $(x_n)_{n \in \mathbb{N}}$ be a sequence of vectors in $\mathcal{K}$.*

(a) *The sequence $(x_n)_{n\in\mathbb{N}}$ converges with respect to the strong topology to a vector $x \in \mathcal{K}$ if and only if $\lim_{n\to\infty} [x_n, x_n] = [x, x]$ and $\lim_{n\to\infty} [x_n, y] = [x, y]$ for all y in some total subset $\mathcal{T}$ of $\mathcal{K}$.*

(b) *The sequence $(x_n)_{n\in\mathbb{N}}$ is Cauchy with respect to the strong topology if and only if $\lim_{n\to\infty} [x_n, x_n]$ exists and $([x_n, y])_{n\in\mathbb{N}}$ is a Cauchy numerical sequence for all y in some total subset $\mathcal{T}$ of $\mathcal{K}$.*

Proof. (a). Let us assume $\kappa = \kappa_+(\mathcal{K})$. If $\mathcal{D}$ denotes the linear manifold generated by the total subset $\mathcal{T}$ then

$$\lim_{n\to\infty} [x_n, y] = [x, y], \quad y \in \mathcal{D}. \tag{4.8}$$

By Lemma 1.4.10 there exists a fundamental decomposition $\mathcal{K} = \mathcal{K}^+[+]\mathcal{K}^-$ such that $\mathcal{K}^+ \subseteq \mathcal{D}$. Let $\{e_j\}_{j=1}^{\kappa}$ be an orthonormal system of $\mathcal{K}^+$ and denote by $\|\cdot\|$ the fundamental norm corresponding to this fundamental decomposition. Let us represent

$$x = x_+ + x_- = \xi_1 e_1 + \cdots + \xi_\kappa e_\kappa + x_-,$$

$$x_n = x_n^+ + x_n^- = \sum_{j=1}^{\kappa} \xi_{nj} e_j + x_n^-.$$

From (4.8) it follows that

$$\lim_{n\to\infty} \xi_{nj} = \xi_j, \quad j \in \{1, \ldots, \kappa\}$$

and hence $\lim_{n\to\infty} \|x_n^+ - x_+\| = 0$. Since $\lim_{n\to\infty} [x_n, x_n] = [x, x]$ this implies $\lim_{n\to\infty} \|x_n^-\| = \|x_-\|$ and then, taking into account that $\mathcal{D} \cap \mathcal{K}^-$ is dense in $\mathcal{K}^-$ and of (4.8), we obtain that for any $y \in \mathcal{K}^-$ we have

$$\lim_{n\to\infty} \|x_n^- - x_-\| = \lim_{n\to\infty} -[x_n^- - x_-, x_n^- - x_-]$$
$$= \lim_{n\to\infty} \|x_n^-\|^2 + \lim_{n\to\infty} 2\operatorname{Re}[x_n^-, x_-] + \|x_-\|^2 = 0.$$

We thus obtained $\lim_{n\to\infty} \|x_n - x\| = 0$. The converse implication is trivial.

(b) The argument is similar to that used in the proof of (a) and we leave it to the reader. ∎

1.5 Pre-Kreĭn Spaces

We are now interested in a question that will show its importance throughout this book, namely, whether an inner product space can be "completed" to a Kreĭn space. The term "completed" should be explained since, unlike the completion of a positive semidefinite inner product space to a Hilbert space, in the case of indefinite inner product spaces the completion is performed with respect to some exterior topology.

1.5 Pre-Kreĭn Spaces

Let $(\mathcal{X}, [\cdot, \cdot])$ be an inner product space. A pair $(\mathcal{K}, \Pi)$ is called a *Kreĭn space induced by* $(\mathcal{X}, [\cdot, \cdot])$, if $(\mathcal{K}, [\cdot, \cdot]_{\mathcal{K}})$ is a Kreĭn space and $\Pi \colon \mathcal{X} \to \mathcal{K}$ is a linear mapping with dense range, such that $[\Pi x, \Pi y]_{\mathcal{K}} = [x, y]$ for all $x, y \in \mathcal{X}$.

Before considering the problem of existence of Kreĭn spaces induced by an inner product space $(\mathcal{X}, [\cdot, \cdot])$, we firstly explain the meaning of this concept.

Remark 1.5.1. The first observation is that the isotropic part plays no role here. Indeed, if $(\mathcal{K}, [\cdot, \cdot]_{\mathcal{K}})$ is a Kreĭn space induced by $(\mathcal{X}, [\cdot, \cdot])$, since a Kreĭn space is, by definition, nondegenerate, we always have $\mathcal{X}^0 = \operatorname{Ker}(\Pi)$. Therefore, by factoring out $\mathcal{X}^0$ as in Remark 1.1.5, we get a Kreĭn space $(\mathcal{K}, \widehat{\Pi})$ induced by the nondegenerate inner product space $(\widehat{\mathcal{X}}, [\cdot, \cdot])$; here $\widehat{\Pi} \colon \widehat{\mathcal{X}} \to \mathcal{K}$ is the quotient mapping, that is, $\widehat{\Pi}(x + \mathcal{X}^0) = \Pi x$, for arbitrary $x \in \mathcal{X}$. This induced Kreĭn space has the additional property that $\widehat{\Pi}$ is injective and hence, we can identify $\widehat{\mathcal{X}}$ with a dense linear manifold in $\mathcal{K}$, such that $[x, y] = [x, y]_{\mathcal{K}}$, for all $x, y \in \widehat{\mathcal{X}}$. Thus, a nondegenerate inner product space $(\mathcal{X}, [\cdot, \cdot])$ that induces a Kreĭn spaces can be called a *pre-Kreĭn space*, by analogy with a pre-Hilbert space. ∎

Proposition 1.5.2. *Let $(\mathcal{X}, [\cdot, \cdot])$ be a nondegenerate inner product space. The following two assertions are equivalent.*

(a) *$(\mathcal{X}, [\cdot, \cdot])$ is decomposable, that is, there exists a decomposition*

$$\mathcal{X} = \mathcal{X}_+ + \mathcal{X}_- \tag{5.1}$$

such that the linear manifolds $\mathcal{X}_+$ and $\mathcal{X}_-$ are strictly positive and, respectively, strictly negative and $\mathcal{X}_+ \perp \mathcal{X}_-$.

(b) *There exists a linear operator J on $\mathcal{X}$ such that $J^2 = I$ and the equality*

$$\langle x, y \rangle_J = [Jx, y], \quad x, y \in \mathcal{X} \tag{5.2}$$

defines a positive definite inner product on $\mathcal{X}$.

Any of the previous assertions implies the following.

(c) *There exists a positive definite inner product $\langle \cdot, \cdot \rangle$ on $\mathcal{X}$ with the property*

$$\|x\| = \sup_{\|y\| \leqslant 1} |[x, y]|, \quad x \in \mathcal{X}, \tag{5.3}$$

where $\|\cdot\|$ denotes the norm associated to $\langle \cdot, \cdot \rangle$.

Moreover, if any of these three statements is satisfied, then $(\mathcal{X}, [\cdot, \cdot])$ induces a Kreĭn space.

Proof. (a)⇒(b). If the decomposition (5.1) is given, define the operator J by

$$J(x_+ + x_-) = x_+ - x_-, \quad x_+ \in \mathcal{X}_+,\ x_- \in \mathcal{X}_-.$$

Then $J^2 = I$ and $\langle \cdot, \cdot \rangle$ from (5.2) is a positive definite inner product.

(b)⇒(a). If the operator J is given then we take

$$\mathcal{X}_+ = (I + J)\mathcal{X}, \quad \mathcal{X}_- = (I - J)\mathcal{X}.$$

The verifications are trivial.

(a)⇒(c). Given the decomposition (5.1), define the norm

$$\|x\| = \sqrt{[x_+, x_+] - [x_-, x_-]}, \quad x = x_+ + x_-, \ x_+ \in \mathcal{X}_+, x_- \in \mathcal{X}_-.$$

Then (5.3) holds. The completion $\tilde{\mathcal{X}}$ of $\mathcal{X}$ under this norm is a Hilbert space, the inner product $[\cdot, \cdot]$ is continuous with respect to this norm and can be uniquely extended to $\tilde{\mathcal{X}}$. Since (5.3) holds also for $\tilde{\mathcal{X}}$ it follows that $(\tilde{\mathcal{X}}, [\cdot, \cdot])$ is a Kreĭn space. We let $\Pi\colon \mathcal{X} \to \mathcal{K}$ be the embedding of $\mathcal{X}$ into $\mathcal{K}$ and it is easy to see that $(\mathcal{K}; \Pi)$ is a Kreĭn space induced by $(\mathcal{X}; [\cdot, \cdot])$. ∎

The next theorem is a comprehensive characterisation of those inner product spaces that admit induced Kreĭn spaces.

Theorem 1.5.3. *For any inner product space $(\mathcal{X}, [\cdot, \cdot])$, the following assertions are equivalent.*

(1) *There exists a positive semidefinite inner product $\langle \cdot, \cdot \rangle$ on $\mathcal{X}$ such that $-\langle x, x \rangle \leqslant [x, x] \leqslant \langle x, x \rangle$ for all $x \in \mathcal{X}$.*

(1)′ *There exists a positive semidefinite inner product $\langle \cdot, \cdot \rangle$ on $\mathcal{X}$ such that $[x, x] \leqslant \langle x, x \rangle$ for all $x \in \mathcal{X}$.*

(2) *There exists a positive semidefinite inner product $\langle \cdot, \cdot \rangle$ on $\mathcal{X}$ such that*

$$|[x, y]|^2 \leqslant \langle x, x \rangle \langle y, y \rangle, \quad x, y \in \mathcal{X}. \tag{5.4}$$

(3) *There exist two positive semidefinite inner products $\langle \cdot, \cdot \rangle_\pm$ on $\mathcal{X}$ such that $[x, y] = \langle x, y \rangle_+ - \langle x, y \rangle_-$ for all $x, y \in \mathcal{X}$.*

(4) *There exist two positive semidefinite inner products $\langle \cdot, \cdot \rangle_\pm$ on $\mathcal{X}$ such that $[x, y] = \langle x, y \rangle_+ - \langle x, y \rangle_-$ for all $x, y \in \mathcal{X}$, and, in addition, if $\langle \cdot, \cdot \rangle$ is a positive semidefinite inner product on $\mathcal{X}$ such that $\langle x, x \rangle \leqslant \langle x, x \rangle_\pm$ for all $x \in \mathcal{X}$, then $\langle x, x \rangle = 0$ for all $x \in \mathcal{X}$.*

(5) *There exists a Kreĭn space induced by $(\mathcal{X}, [\cdot, \cdot])$.*

Proof. (1)⇒(1)′. Obvious.

(1)⇒ (2). Let $\langle \cdot, \cdot \rangle$ be a positive semidefinite inner product on $\mathcal{X}$ such that $-\langle x, x \rangle \leqslant [x, x] \leqslant \langle x, x \rangle$ for all $x \in \mathcal{X}$, that is,

$$|[x, x]| \leqslant \langle x, x \rangle, \quad x \in \mathcal{X}.$$

Let $x, y \in \mathcal{X}$ be fixed. Since $[\cdot, \cdot]$ is symmetric we have

$$4\operatorname{Re}[x, y] = [x + y, x + y] - [x - y, x - y],$$

and hence

$$4|\operatorname{Re}[x, y]| \leqslant \langle x + y, x + y \rangle + \langle x - y, x - y \rangle = 2\langle x, x \rangle + 2\langle y, y \rangle.$$

1.5 Pre-Kreĭn Spaces

Let $\lambda \in \mathbb{C}$ be chosen such that $|\lambda| = 1$ and $\operatorname{Re}[x, \lambda y] = [x, \lambda y]$. Then

$$|[x, \lambda y]| \leq \frac{1}{2}\langle x, x \rangle + \frac{1}{2}\langle y, y \rangle. \tag{5.5}$$

To prove (5.4), we distinguish two possible cases. First, assume that either $\langle x, x \rangle = 0$ or $\langle y, y \rangle = 0$. To make a choice, let $\langle x, x \rangle = 0$. We consider the inequality (5.5) with y replaced by ty for $t > 0$. Then

$$|[x, y]| \leq \frac{t}{2}\langle y, y \rangle.$$

Letting $t \to 0$ we get $[x, y] = 0$. In the second situation, assume that both $\langle x, x \rangle$ and $\langle y, y \rangle$ are nontrivial. In (5.5) we replace x by $\langle x, x \rangle^{-1/2} x$ and y by $\langle y, y \rangle^{-1/2} y$ and get (5.4).

(2)⇒(1). Clear, letting $x = y$.

(2)⇒(4). Let $(\mathcal{H}, \langle \cdot, \cdot \rangle)$ be the quotient completion of the positive inner product space $(\mathcal{X}, \langle \cdot, \cdot \rangle)$ to a Hilbert space. More precisely, we factor $\mathcal{X}$ through $\mathcal{N} = \{x \in \mathcal{X} \mid \langle x, x \rangle = 0\}$ and take the completion to a Hilbert space. The inequality (5.4) guarantees that $\mathcal{N} \subseteq \mathcal{X}^0$ and hence the inner product $[\cdot, \cdot]$ factors by $\mathcal{N}$ and is uniquely extended by continuity to $\mathcal{H}$, such that (5.4) holds. Let $G \in \mathcal{B}(\mathcal{H})$ be the Gram operator of $[\cdot, \cdot]$ with respect to $\langle \cdot, \cdot \rangle$, that is

$$[x, y] = \langle Gx, y \rangle, \quad x, y \in \mathcal{H}. \tag{5.6}$$

Note that G is contractive and selfadjoint in the Hilbert space $\mathcal{H}$. Let $G = G_+ - G_-$ be the Jordan decomposition of G. Then $G_\pm$ are contractive positive operators in the Hilbert space $(\mathcal{H}, \langle \cdot, \cdot \rangle)$. We consider the positive semidefinite inner products $\langle \cdot, \cdot \rangle_\pm$ on $\mathcal{H}$ defined by

$$\langle x, y \rangle_\pm = \langle G_\pm x, y \rangle, \quad x, y \in \mathcal{H}.$$

We consider the restrictions of $\langle \cdot, \cdot \rangle_\pm$ to $\mathcal{X}/\mathcal{N}$ and then the extensions to $\mathcal{X}$ by letting them be null on $\mathcal{N}$. Thus, we obtain two positive semidefinite inner products $\langle \cdot, \cdot \rangle_\pm$ on $\mathcal{X}$ and, from $G = G_+ - G_-$, it is obvious that $[x, y] = \langle x, y \rangle_+ - \langle x, y \rangle_-$ for all $x, y \in \mathcal{X}$.

Taking into account that $G_+ G_- = 0$ we can prove easily that, if $\langle \cdot, \cdot \rangle$ is a positive semidefinite inner product on $\mathcal{X}$ such that $\langle x, x \rangle \leq \langle x, x \rangle_\pm$ for all $x \in \mathcal{X}$, then $\langle x, x \rangle = 0$ for all $x \in \mathcal{X}$.

(4)⇒(3). Obvious.

(3)⇒(1). If there exist two positive semidefinite inner products $\langle \cdot, \cdot \rangle_\pm$ on $\mathcal{X}$ such that $[x, y] = \langle x, y \rangle_+ - \langle x, y \rangle_-$ for all $x, y \in \mathcal{X}$, then letting $\langle x, y \rangle = \langle x, y \rangle_+ + \langle x, y \rangle_-$ we get a positive semidefinite inner product $\langle \cdot, \cdot \rangle$ on $\mathcal{X}$ such that $-\langle x, x \rangle \leq [x, x] \leq \langle x, x \rangle$ for all $x \in \mathcal{X}$.

(1)′ ⇒(3). Indeed, letting $\langle x, y \rangle_1 = \langle x, y \rangle - [x, y]$, for all $x, y \in \mathcal{X}$, it follows that $\langle \cdot, \cdot \rangle_1$ is a positive semidefinite inner product, hence $[\cdot, \cdot]$ can be written as the difference of two positive semidefinite inner products, and hence assertion (3) holds.

(2)⇒(5). As in the proof of (2)⇒(4), let $(\mathcal{H}, \langle \cdot, \cdot \rangle)$ be the quotient-completion of $(\mathcal{X}, \langle \cdot, \cdot \rangle)$ to a Hilbert space, so the representation (5.6), as well as the Jordan decomposition $G = G_+ - G_-$, hold. We obtain a Kreĭn space $(\mathcal{K}, [\cdot, \cdot]_\mathcal{K})$ in the following way: we decompose $\mathcal{H} = \mathcal{H}^+ \oplus \operatorname{Ker}(G) \oplus \mathcal{H}^-$, where $\mathcal{H}^\pm$ are the spectral subspaces of G corresponding to the positive/negative semiaxis, and then take the Hilbert space completions $\mathcal{K}^\pm$ of the pre-Hilbert spaces $(\mathcal{H}^\pm, [G_\pm \cdot, \cdot])$. Since, by (5.6), $\operatorname{Ker}(A)$ coincides with the isotropic part of $(\mathcal{H}, [\cdot, \cdot])$, the inner product $[\cdot, \cdot]$ factors to $\mathcal{H}/\operatorname{Ker}(G)$ and, by (5.4), it is uniquely extended by continuity to an inner product $[\cdot, \cdot]_\mathcal{K}$. Now it is easy to see that $\mathcal{K} = \mathcal{K}^+[+]\mathcal{K}^-$ is a fundamental decomposition of $(\mathcal{K}, [\cdot, \cdot])$, hence it is a Kreĭn space. The operator $\Pi \colon \mathcal{X} \to \mathcal{K}$ is defined as the composition of the quotients and the embeddings in this construction, and it is easy to see that it has dense range and that $[\Pi x, \Pi y]_\mathcal{K} = [x, y]$ for all $x, y \in \mathcal{X}$.

(5)⇒(1). This implication is clear from Remark 1.5.1: the existence of a positive semidefinite inner product $\langle \cdot, \cdot \rangle$ on $\mathcal{X}$ such that $-\langle x, x \rangle \leqslant [x, x] \leqslant \langle x, x \rangle$, for all $x \in \mathcal{X}$, is obvious whenever $(\mathcal{X}/\mathcal{X}^0, [\cdot, \cdot])$ is embedded in a Kreĭn space in such a way that the indefinite inner products coincide on $\mathcal{X}$. ∎

Remark 1.3.3 presents a situation when an indefinite inner product space is decomposable and hence it induces a Kreĭn space. Example 1.3.4 shows that there exists a gap between the situations depicted in Proposition 1.5.2 and Theorem 1.5.3, more precisely, there exists a nondegenerate inner product space $(\mathcal{X}, [\cdot, \cdot])$ that is not decomposable but which induces a Kreĭn space.

Another question of interest is when an inner product space induces a Pontryagin space. In the case of a Pontryagin space induced by an indefinite inner product space, the uniqueness up to unitary equivalence holds too. To make things precise, two Kreĭn spaces $(\mathcal{K}_i, \Pi_i)$, $i = 1, 2$, induced by the same inner product space $(\mathcal{X}, [\cdot, \cdot])$, are called *unitary equivalent* if there exists a unitary operator $U \in \mathcal{B}(\mathcal{K}_1, \mathcal{K}_2)$ such that $U\Pi_1 = \Pi_2$.

Corollary 1.5.4. *If the inner product space $(\mathcal{X}, [\cdot, \cdot])$ has finite rank of indefiniteness $\kappa(\mathcal{X}) := \min\{\kappa_+(\mathcal{X}), \kappa_-(\mathcal{X})\} < \infty$, then it induces a Kreĭn space $(\mathcal{K}; \Pi)$, with either $\kappa_+(\mathcal{K}) = \kappa_+(\mathcal{X}) < \infty$ or $\kappa_-(\mathcal{K}) = \kappa_-(\mathcal{X}) < \infty$, hence a Pontryagin space, that is unique up to unitary equivalence.*

Proof. As shown in Remark 1.5.1, without loss of generality we can assume that $\mathcal{X}$ is nondegenerate. If $\kappa_-(\mathcal{X}) < \infty$, let $\mathcal{L}$ be a maximal negative subspace of $\mathcal{X}$. It is easy to see that $\dim(\mathcal{L}) = \kappa_-(\mathcal{X})$ and $\mathcal{L}^\perp$ is a strictly positive subspace. Since $\mathcal{L}$ is finite dimensional and $\mathcal{X}$ is nondegenerate, we have $\mathcal{X} = \mathcal{L}\dot{+}\mathcal{L}^\perp$, hence $\mathcal{X}$ is decomposable. The proof of Proposition 1.5.2 also shows that the Kreĭn space $\mathcal{K}$ induced by $\mathcal{X}$ is a Pontryagin space with $\kappa_-(\mathcal{K}) = \kappa_-(\mathcal{X})$.

Let $\mathcal{H}$ be another Kreĭn space which contains $\mathcal{X}$ densely and isometric. Clearly, $\kappa^-(\mathcal{H}) \geqslant \kappa$. Assuming $\kappa^-(\mathcal{H}) > \kappa$, from Lemma 1.4.10 it follows that $\mathcal{X}$ contains a negative subspace of dimension $> \kappa$, a contradiction. We have thus proven that $\kappa^-(\mathcal{H}) = \kappa$. Again by Lemma 1.4.10, it follows that there exists a fundamental decomposition $\mathcal{H} = \mathcal{H}^-[+]\mathcal{H}^+$ such that $\mathcal{H}^- \subseteq \mathcal{X}$ and, since $\dim(\mathcal{K}^-) = \dim(\mathcal{H}^-) = \kappa$, there exists a (Hilbert space) unitary operator $V_-\colon \mathcal{K}^- \to \mathcal{H}^-$. On the other hand, $\mathcal{H}^+ \cap \mathcal{X}$ is dense in $\mathcal{H}^+$ and hence the Hilbert space dimensions of $\mathcal{K}^+$ and of $\dim(\mathcal{H}^+)$ are the same. Then there exists a Hilbert space unitary operator $V_+\colon \mathcal{K}^+ \to \mathcal{H}^+$ and the direct sum of V_- and V_+ is a bounded unitary operator of Kreĭn spaces from $\mathcal{K}$ onto $\mathcal{H}$. ∎

An important question is asking for equivalent characterisations of uniqueness of induced Kreĭn spaces. This question requires more sophisticated tools than those we have at this moment and it will be treated in Section 6.1.

1.6 Notes

I learned most of the material presented in this chapter from, and in this order, J. Bognár [18], M. G. Kreĭn [93], I. S. Iokhvidov, M. G. Kreĭn, and H. Langer [79], T. Ando [4], M. Tomita [142], and F. Hansen [73]. The idea of an induced Kreĭn space used in Section 1.5 follows our joint work with T. Constantinescu [32] and part of the proof of Theorem 1.5.3 follows an argument of L. Schwartz [134]. For the most comprehensive presentation of the foundations of the theory of indefinite inner product spaces and their rich history we recommend J. Bognár's monograph [18]: the definition of an orthocomplemented linear manifold which we use here is different from that used in [18].

In this respect, the pioneering work of R. Nevanlinna [112, 113, 114, 115] and of I. S. Iokhvidov and M. G. Kreĭn [77, 78], should be emphasised. The name of Kreĭn space was given by J. Bognár and it was rapidly assumed by the community, taking into account that the most active research groups were in the Soviet Union. The name of Pontryagin space for a Π_k space is justified by the fact that it was L. S. Pontryagin in [121] who first considered these spaces and obtained a landmark result in this theory, even though it was S. L. Sobolev in 1940 (the results were kept secret for about twenty years and published only in 1960 in [136]) who first considered a space of type Π_1 and, apparently, inspired and motivated Pontryagin's research.

Chapter 2

Angular Operators

Kreĭn spaces turned out to be the relevant class of indefinite inner product spaces for which a reasonable operator theory can be developed. One of the features that distinguishes these spaces from Hilbert spaces is the geometry of their subspaces, which is firstly clarified. We start with semidefinite subspaces and their representations by means of angular operators. This makes possible an approach to the extension theorem of alternating pairs of subspaces by means of the lifting of contractions in Hilbert spaces and provides an opportunity to consider, as an application, maximal accretive and positive selfadjoint extensions of operators in Hilbert spaces. Also, this makes it possible to prove a universality property of Kreĭn spaces among the class of inner product spaces admitting a Hilbert space norm that makes the inner product jointly continuous. Finally, we generalise the notion of angular operator within a certain class of subspaces of Kreĭn spaces.

2.1 Semidefinite Subspaces and Angular Operators

Let $(\mathcal{K}, [\cdot, \cdot])$ be a Kreĭn space. Throughout this section we fix a fundamental symmetry J and denote by $\|\cdot\|$ the corresponding fundamental norm. Also, let $J = J^+ - J^-$ be the Jordan decomposition of J and let $\mathcal{K} = \mathcal{K}^+[+]\mathcal{K}^-$ be the associated fundamental decomposition. A linear manifold $\mathcal{S}$ of $\mathcal{K}$ is called a *subspace* if it is closed. Let us observe that, since the strong topology of a Kreĭn space is a Hilbert space topology, it follows that a linear manifold is strongly closed if and only if it is weakly closed.

Lemma 2.1.1. *Let $\mathcal{M}$ be a positive subspace of $\mathcal{K}$. Then, for any $x \in \mathcal{M}$ we have*

$$\|J^+ x\| \leqslant \|x\| \leqslant \sqrt{2}\, \|J^+ x\|. \tag{1.1}$$

As a consequence, $J^+\mathcal{M}$ is a subspace of $\mathcal{K}^+$, the operator $J^+|\mathcal{M}\colon \mathcal{M} \to J^+\mathcal{M}$ is boundedly invertible, the mapping K

$$K\colon J^+\mathcal{M} \ni J^+ x \mapsto J^- x \in \mathcal{K}^-, \quad x \in \mathcal{M}, \tag{1.2}$$

is a linear contraction on $J^+\mathcal{M}$, and $\mathcal{M}$ coincides with the graph of the operator K, that is,

$$\mathcal{M} = G(K) = \{x_+ + Kx_+ \mid x_+ \in J^+\mathcal{M}\}. \tag{1.3}$$

2.1 Semidefinite Subspaces and Angular Operators

Proof. Let x be a vector in $\mathcal{M}$. Then
$$\|x\|^2 = \|J^+x\|^2 + \|J^-x\|^2 \geq \|J^+x\|^2,$$
and
$$0 \leq [x,x] = \|J^+x\|^2 - \|J^-x\|^2,$$
and hence
$$\|x\|^2 \leq \|J^+x\|^2 + \|J^-x\|^2 \leq 2\|J^+x\|^2.$$
Thus we have proven (1.1). The other assertions follow immediately from (1.1). ∎

If $\mathcal{M}$ is a positive subspace of $\mathcal{K}$ then the operator K defined in (1.2) is called the *angular operator* of $\mathcal{M}$ and the subspace $J^+\mathcal{M}$ is called the *shadow* of $\mathcal{M}$ onto $\mathcal{K}^+$.

In a similar way one associates angular operators to negative subspaces, more precisely, for a negative subspace $\mathcal{N}$ one can define the angular operator L of $\mathcal{N}$

$$L\colon J^-\mathcal{N} \ni J^-x \mapsto J^+x \in \mathcal{K}^+, \quad x \in \mathcal{N}, \tag{1.4}$$

which is a linear contraction with domain the subspace $J^-\mathcal{N}$ of $\mathcal{K}^-$ and such that $\mathcal{N}$ coincides with the graph of L in the sense

$$\mathcal{N} = G(L) = \{Lx_- + x_- \mid x_- \in J^-\mathcal{N}\}. \tag{1.5}$$

Recall that a subspace $\mathcal{L}$ of the Kreĭn space $\mathcal{K}$ is called strictly positive if $[x,x] > 0$ for all $x \in \mathcal{L}$. Clearly, a strictly positive subspace is nondegenerate. A subspace $\mathcal{L}$ of $\mathcal{K}$ is called *uniformly positive* if there exists $\delta > 0$ such that $[x,x] \geq \delta\|x\|^2$, $x \in \mathcal{L}$. Correspondingly, a subspace $\mathcal{M}$ is called *uniformly negative* if $[x,x] \leq -\delta\|x\|^2$, $x \in \mathcal{M}$, for some $\delta > 0$.

Note that, since any two fundamental norms of $\mathcal{K}$ are equivalent (see Theorem 1.3.2) it follows that the definition of uniformly definite subspaces does not depend on the particularly fixed fundamental norm.

Lemma 2.1.2. *Let $\mathcal{M}$ be a positive subspace of $\mathcal{K}$ and denote by K its angular operator with respect to the fundamental decomposition $\mathcal{K}^+[+]\mathcal{K}^-$. Then $\mathcal{M}$ is uniformly positive (strictly positive) if and only if K is a uniform contraction, that is, $\|K\| < 1$ (respectively, strict contraction, that is, $\|Kx\| < \|x\|$, $x \in J^+\mathcal{M} \setminus \{0\}$).*

Proof. Let $\delta > 0$ be such that $[x,x] \geq \delta\|x\|^2$, $x \in \mathcal{M}$. Then

$$\|J^-x\|^2 \leq \frac{1-\delta}{1+\delta}\|J^+x\|^2, \quad x \in \mathcal{M}.$$

Therefore, by the definition of the angular operator K, we have

$$\|K\| \leq \sqrt{\frac{1-\delta}{1+\delta}} < 1.$$

Conversely, if $K < 1$ then we take $\delta = \sqrt{\frac{1-\|K\|}{1+\|K\|}}$. The other assertions follow in a similar way. ∎

A positive (strictly positive, uniformly positive) subspace $\mathcal{M}$ of $\mathcal{K}$ is called *maximal positive (maximal strictly positive, maximal uniformly positive)* if there exists no positive (strictly positive, uniformly positive) subspace $\widetilde{\mathcal{M}}$ properly containing $\mathcal{M}$. Since one of the main problems in operator theory on indefinite inner product spaces is the problem of extensions of semidefinite subspaces, it is really helpful to translate these in terms of angular operators.

Lemma 2.1.3. *Let $\mathcal{M}$ be a positive (strictly positive, uniformly positive) subspace of $\mathcal{K}$.*

(a) *Let K be the angular operator of $\mathcal{M}$ defined as in (1.2) and consider the representation $\mathcal{M}$ as in (1.3). Then $\mathcal{M}$ is maximal positive (maximal strictly positive, maximal uniformly positive) if and only if $J^+\mathcal{M} = \mathcal{K}^+$.*

(b) *There exists a maximal positive (maximal strictly positive, maximal uniformly positive) subspace $\widetilde{\mathcal{M}}$ of $\mathcal{K}$ such that $\widetilde{\mathcal{M}} \supseteq \mathcal{M}$.*

Proof. Let $\mathcal{M}'$ be another positive (strictly positive, uniformly positive) subspace of $\mathcal{K}$, with angular operator $K'\colon J^+\mathcal{M}'(\subseteq \mathcal{K}^+) \to \mathcal{K}^-$ and such that $\mathcal{M}' = \{x + K'x \mid x \in J^+\mathcal{M}'\}$. Then $\mathcal{M} \subseteq \mathcal{M}'$ if and only if $J^+\mathcal{M} \subseteq J^+\mathcal{M}'$ and $K'x = Kx$ for all $x \in J^+\mathcal{M}$, that is, $K \subseteq K'$. Consequently, if $J^+\mathcal{M} = \mathcal{K}^+$ then $\mathcal{M}$ is maximal positive (maximal strictly positive, maximal uniformly positive).

On the other hand, assuming that $J^+\mathcal{M} \neq \mathcal{K}^+$, let $\widetilde{K} \in \mathcal{B}(\mathcal{K}^+, \mathcal{K}^-)$ be the extension of K such that $\widetilde{K}$ is null on the subspace $\mathcal{K}^+ \ominus J^+\mathcal{M}$. Then $\widetilde{K}$ is again a contraction hence the space $\widetilde{\mathcal{M}} = \{x + \widetilde{K}x \mid x \in \mathcal{K}^+\}$ is positive and $\widetilde{\mathcal{M}} \supseteq \mathcal{M}$ holds. If $\mathcal{M}$ is strictly positive (uniformly positive) then by Lemma 2.1.2, the operator K is a strict contraction, (uniform contraction), $\widetilde{K}$ is also a strict contraction (uniform contraction, that is, $\|\widetilde{K}\| = \|K\| < 1$) and hence $\widetilde{\mathcal{M}}$ is a maximal strictly positive (maximal uniformly positive) subspace. ∎

For reasons of symmetry and later cross referencing we translate the previous lemma to negative subspaces.

Lemma 2.1.4. *Let $\mathcal{N}$ be a negative (strictly negative, uniformly negative) subspace of $\mathcal{K}$.*

(a) *Let L be the angular operator of $\mathcal{N}$ defined as in (1.4) and the representation as in (1.5). Then $\mathcal{N}$ is maximal negative (maximal strictly negative, maximal uniformly negative) if and only if $J^-\mathcal{N} = \mathcal{K}^-$.*

(b) *There exists a maximal negative (maximal strictly negative, maximal uniformly negative) subspace $\widetilde{\mathcal{N}}$ of $\mathcal{K}$ such that $\widetilde{\mathcal{N}} \supseteq \mathcal{N}$.*

We now consider pairs of semidefinite subspaces that have orthogonal components and translate this property in terms of their angular operators. Here and in the following, we denote by T^* the adjoint operator with respect to inner products of Hilbert spaces.

2.1 Semidefinite Subspaces and Angular Operators

Lemma 2.1.5. *Let $\mathcal{M}$ be a positive subspace of $\mathcal{K}$ with angular operator $K\colon J^+\mathcal{M}(\subseteq \mathcal{K}^+) \to \mathcal{K}^-$ and let $\mathcal{N}$ be a negative subspace of $\mathcal{K}$ with angular operator $L\colon J^-\mathcal{N}(\subseteq \mathcal{K}^-) \to \mathcal{K}^+$. Then $\mathcal{M} \perp \mathcal{N}$ if and only if*

$$K^*|J^-\mathcal{N} = P^{\mathcal{K}^+}_{J^+\mathcal{M}} L, \tag{1.6}$$

where $P^{\mathcal{K}^+}_{J^+\mathcal{M}}$ denotes the orthogonal projection of $\mathcal{K}^+$ onto its subspace $J^+\mathcal{M}$, in the Hilbert space $\mathcal{K}^+$. In particular, if $\mathcal{M}$ is maximal positive and $\mathcal{N}$ is maximal negative, then $\mathcal{M} \perp \mathcal{N}$ if and only if $K^* = L$.

Proof. Taking into account (1.2), (1.3), (1.4) and (1.5), it follows that $\mathcal{M} \perp \mathcal{N}$ if and only if
$$[x + Kx, Ly + y] = 0, \quad \text{for all } x \in J^+\mathcal{M}, \text{ and all } y \in J^-\mathcal{N},$$
equivalently
$$\langle x, Ly \rangle = \langle Kx, y \rangle, \quad \text{for all } x \in J^+\mathcal{M}, \text{ and all } y \in J^-\mathcal{N},$$
which is exactly (1.6).

In the case that $\mathcal{M}$ is maximal positive and $\mathcal{N}$ is maximal negative, equivalently, $J^+\mathcal{M} = \mathcal{K}^+$ and $J^-\mathcal{N} = \mathcal{K}^-$, (1.6) becomes simply $K^* = L$. ∎

We record some immediate consequences of these results on extensions of semidefinite subspaces.

Corollary 2.1.6. *If $\mathcal{M}$ is a positive subspace and $\mathcal{N}$ is a negative subspace of the Kreĭn space $\mathcal{K}$ such that $\mathcal{K} = \mathcal{M} + \mathcal{N}$, then $\mathcal{M}$ is a maximal positive subspace and $\mathcal{N}$ is a maximal negative subspace. If, in addition, $\mathcal{M} \perp \mathcal{N}$ then $\mathcal{M}$ is uniformly positive, $\mathcal{N}$ is uniformly negative and $\mathcal{N} = \mathcal{M}^\perp$.*

Proof. According to Lemma 2.1.3 we can consider a maximal positive subspace $\widetilde{\mathcal{M}} \supseteq \mathcal{M}$ and a maximal negative subspace $\widetilde{\mathcal{N}} \supseteq \mathcal{N}$. It suffices to prove $\widetilde{\mathcal{M}} \cap \widetilde{\mathcal{N}} = \{0\}$. Let x be a vector in $\widetilde{\mathcal{M}} \cap \widetilde{\mathcal{N}}$. Then x is neutral and applying the Schwarz Inequality (see Lemma 1.1.2) we obtain that $x \in \widetilde{\mathcal{M}}^0$ and $x \in \widetilde{\mathcal{N}}^0$. Since $\mathcal{K} = \widetilde{\mathcal{M}} + \widetilde{\mathcal{N}}$ we get $x \in \mathcal{K}^0 = \{0\}$.

If, in addition, $\mathcal{M} \perp \mathcal{N}$ then both of $\mathcal{M}$ and $\mathcal{N}$ are nondegenerate. Define the linear operator G in $\mathcal{K}$ by
$$G(x+y) = x - y, \quad x \in \mathcal{M}, y \in \mathcal{N}.$$
We leave to the reader to show that G is a closed operator and, since $\mathcal{M} + \mathcal{N} = \mathcal{K}$ is closed, by the Closed Graph Principle we obtain that G is bounded. Moreover, $G^2 = I$ and hence G is boundedly invertible. Since the inner product $(x, y)_G = [Gx, y]$, $x, y \in \mathcal{K}$, is positive definite, it follows that G is a fundamental symmetry of $\mathcal{K}$ and hence $\mathcal{M}$ is uniformly positive, $\mathcal{N}$ is uniformly negative and $\mathcal{N} = \mathcal{M}^\perp$. ∎

Corollary 2.1.7. *The mapping $\mathcal{M} \mapsto \mathcal{M}^\perp$ is a bijective correspondence between the class of maximal positive (maximal strictly positive, maximal uniformly positive) subspaces of $\mathcal{K}$ and the class of maximal negative (maximal strictly negative, maximal uniformly negative) subspaces of $\mathcal{K}$.*

Proof. We use the angular operator correspondence as in Lemmas 2.1.1 through 2.1.5, the characterisation of maximality, and the fact that a maximal positive subspace $\mathcal{M}$ of $\mathcal{K}$ has the angular operator $K \in \mathcal{B}(\mathcal{K}^+, \mathcal{K}^-)$ if and only if the subspace $\mathcal{M}^\perp$ has the angular operator K^*. ∎

The angular operator representation has some important consequences for a better understanding of fundamental decompositions.

Lemma 2.1.8. *A subspace $\mathcal{L}$ is maximal uniformly positive if and only if the decomposition $\mathcal{K} = \mathcal{L}[+]\mathcal{L}^\perp$ is a fundamental decomposition of $\mathcal{K}$.*

Proof. Assume that $\mathcal{L}$ is maximal uniformly positive. Since $\mathcal{L}$ is nondegenerate it follows that $\mathcal{L} + \mathcal{L}^\perp$ is dense in $\mathcal{K}$, see Remark 1.3.3. Let z be an arbitrary vector in $\mathcal{K}$. Applying the Riesz Representation Theorem in the Hilbert spaces $(\mathcal{L}, [\cdot,\cdot])$ and $(\mathcal{L}^\perp, [\cdot,\cdot])$ we get
$$[x_+, z] = [x_+, z_+], \quad x_+ \in \mathcal{L},$$
and
$$[x_-, z] = [x_-, z_-], \quad x_- \in \mathcal{L}^\perp,$$
for some vectors $z_+ \in \mathcal{L}$ and $z_- \in \mathcal{L}^\perp$. Thus, for any $x = x_+ + x_-, x_+ \in \mathcal{L}, x_- \in \mathcal{L}^\perp$ we have
$$[z - (z_+ + z_-), x] = [z, x_+] - [z_+, x_+] - [z_-, x_-] = 0$$
and hence $z = z_+ + z_-$. We thus have proven $\mathcal{K} = \mathcal{L}[+]\mathcal{L}^\perp$ and, invoking Theorem 1.4.1, this is a fundamental decomposition.

The converse implication is contained in Theorem 1.4.1. ∎

We are now in a position to prove a characterisation of Kreĭn spaces in terms of Gram operators.

Proposition 2.1.9. *Let $(\mathcal{X}, [\cdot,\cdot])$ be an inner product space on which there exists a positive definite inner product $\langle\cdot,\cdot\rangle$ whose underlying norm $\|\cdot\|$ is complete, such that the inner product $[\cdot,\cdot]$ is continuous (jointly is the same as separately, by Lemma 1.3.1) with respect to the norm $\|\cdot\|$, and let G be the Gram operator of $[\cdot,\cdot]$ with respect to $\langle\cdot,\cdot\rangle$, see Remark 1.3.3. Then $(\mathcal{X}, [\cdot,\cdot])$ is a Kreĭn space such that its strong topology coincides with that defined by the norm $\|\cdot\|$ if and only if the operator G is boundedly invertible.*

Proof. Since Kreĭn spaces are nondegenerate it follows that, without loss of generality we can assume that G is injective. Letting $G = G_+ - G_-$ be the Jordan decomposition of G and $\mathcal{X}_\pm = \operatorname{Clos} \operatorname{Ran}(G_\pm)$ we have $\mathcal{X} = \mathcal{X}_+ + \mathcal{X}_-$, $\mathcal{X}_+ \perp \mathcal{X}_-$, $\mathcal{X}_+$ is strictly positive and $\mathcal{X}_-$ is strictly negative. In addition,
$$[x, y] = \pm \langle G_\pm x, y \rangle, \quad x, y \in \mathcal{X}_\pm. \tag{1.7}$$

2.1 Semidefinite Subspaces and Angular Operators

Since $|G| = G_+ + G_-$ and $|G|^{1/2} = G_+^{1/2} + G_-^{1/2}$, from (1.7) we have

$$[x,x] = \pm \|G_\pm^{1/2} x\|, \quad x \in \mathcal{X}_\pm. \tag{1.8}$$

Assume that $(\mathcal{X}, [\cdot,\cdot])$ is a Kreĭn space whose strong topology coincides with that defined by the norm $\|\cdot\|$. Then, by Corollary 2.1.6, $\mathcal{X}_+$ is maximal uniformly positive and $\mathcal{X}_-$ is maximal uniformly negative. From (1.8) it follows that the norm $\|G_\pm \cdot \|$ is equivalent with the norm $\|\cdot\|$ on $\mathcal{X}_\pm$, hence $G_\pm$ has closed range which implies that G has closed range and, since it is injective, it follows that it is boundedly invertible.

Conversely, if G is boundedly invertible, then $(-\varepsilon, \varepsilon) \subset \rho(G)$, the resolvent set of G, for some $\varepsilon > 0$ and hence both $G_\pm$ have closed range and, consequently, the normed spaces $(\mathcal{X}_\pm; \|G_\pm^{1/2} \cdot \|)$ are complete, hence the normed space $(\mathcal{X}; \||G|^{1/2} \cdot \|)$ is complete as well. Considering the polar decomposition $G = J|G|$, it follows that J is a selfadjoint isometry, hence a symmetry, in particular $J^2 = I$. Since

$$[x,y] = \langle Gx, y \rangle = \langle J|G|x, y \rangle, \quad x, y \in \mathcal{X},$$

and the positive definite inner product $\langle |G| \cdot, \cdot \rangle$ has its associated norm $\||G|^{1/2} \cdot \|$ complete, by Theorem 1.4.1 (b), it follows that $(\mathcal{X}; [\cdot, \cdot])$ is a Kreĭn space, with the strong topology that coincides with that defined by the norm $\|\cdot\|$. ∎

As a consequence of the discussion on angular operators, we have the following picture of a Kreĭn space $\mathcal{K}$. The neutral part $\mathcal{K}_0 = \{x \in \mathcal{K} \mid [x,x] = 0\}$ splits $\mathcal{K}$ into two connected regions $\mathcal{K}_{0,\pm} = \{x \in \mathcal{K} \mid \pm[x,x] > 0\}$, which are distinguished with respect to an arbitrary fundamental decomposition $\mathcal{K} = \mathcal{K}^+[+]\mathcal{K}^-$, in the sense that $\mathcal{K}^\pm \subset \mathcal{K}_{0,\pm}$. Moreover, all strictly negative subspaces are necessarily contained inside $\mathcal{K}_{0,-}$ and all strictly positive subspaces are inside $\mathcal{K}_{0,+}$. On the other hand, positive/negative subspaces $\mathcal{L}$ may have some intersection with $\mathcal{K}_{0,\pm}$ while their isotropic parts $\mathcal{L}^0 \subseteq \mathcal{K}_0$.

We can apply the angular operator method in order to define the *geometric ranks* of positivity and negativity of a Kreĭn space $\mathcal{K}$. We first note that in a fundamental decomposition $\mathcal{K} = \mathcal{K}^+[+]\mathcal{K}^-$ the subspace $\mathcal{K}^+$ is uniformly positive and the subspace $\mathcal{K}^-$ is uniformly negative. Further, if $\mathcal{K} = \mathcal{K}_1^+[+]\mathcal{K}_1^-$ is another fundamental decomposition then $J^+|\mathcal{K}_1^+\colon \mathcal{K}_1^+ \to \mathcal{K}^+$ and $J^-|\mathcal{K}_1^-\colon \mathcal{K}_1^- \to \mathcal{K}^-$ are boundedly invertible and hence, the Hilbert space dimensions $\dim^{\mathrm{h}}(\mathcal{K}^+)$ and $\dim^{\mathrm{h}}(\mathcal{K}^-)$ are independent of the particularly chosen fundamental decomposition. Here and in the following we denote by $\dim^{\mathrm{h}}$ the *Hilbert space dimension*, which is a cardinal number, in order to distinguish it from the algebraic dimension. The cardinal numbers $\kappa^+(\mathcal{K}) = \dim^{\mathrm{h}}(\mathcal{K}^+)$ and $\kappa^-(\mathcal{K}) = \dim^{\mathrm{h}}(\mathcal{K}^-)$ are called the *geometric ranks of positivity* and, respectively, of *negativity* of the Kreĭn space $\mathcal{K}$.

The above considered ranks can be defined for subspaces, too. Let $\mathcal{M}$ be a subspace of the Kreĭn space $\mathcal{K}$. A decomposition

$$\mathcal{M} = \mathcal{M}^+[+]\mathcal{M}^0[+]\mathcal{M}^- \tag{1.9}$$

is called a *fundamental decomposition* of $\mathcal{M}$ if $\mathcal{M}^+$ is a strictly positive subspace, $\mathcal{M}^-$ is a strictly negative subspace, $\mathcal{M}^+ \perp \mathcal{M}^-$, and the algebraic sum of any two of the three subspaces $\mathcal{M}^+$, $\mathcal{M}^0$, and $\mathcal{M}^-$ is closed. By Remark 1.3.3, any subspace of a Kreĭn space admits a fundamental decomposition.

Lemma 2.1.10. *If $\mathcal{M}$ is an arbitrary subspace of the Kreĭn space $\mathcal{K}$ and (1.9) is an arbitrary fundamental decomposition of $\mathcal{M}$ then the cardinal numbers $\dim^h(\mathcal{M}^+)$ and $\dim^h(\mathcal{M}^-)$ do not depend on the fundamental decomposition.*

Proof. Let $\mathcal{M} = \mathcal{M}_1^+[+]\mathcal{M}^0[+]\mathcal{M}_1^-$ be another fundamental decomposition of $\mathcal{M}$. If P denotes the projection of $\mathcal{M}$ onto $\mathcal{M}^+$ along $\mathcal{M}^0[+]\mathcal{M}^-$ then P is a bounded linear operator in $\mathcal{M}$. Consider the operator $P|\mathcal{M}_1^+ \colon \mathcal{M}_1^+ \to \mathcal{M}^+$ and note that it is one-to-one. Indeed, $x \in \mathcal{M}_1^+$ and $Px = 0$ yields $x \in (\mathcal{M}^0[+]\mathcal{M}^-) \cap \mathcal{M}_1^+ = \{0\}$, the second equality being true since $\mathcal{M}_1^+$ is strictly positive and $\mathcal{M}^0[+]\mathcal{M}^-$ is negative. Therefore $\dim^h(\mathcal{M}_1^+) \leqslant \dim^h(\mathcal{M}^+)$ and, by symmetry, we obtain $\dim^h(\mathcal{M}_1^+) = \dim^h(\mathcal{M}^+)$. In a similar way we obtain $\dim^h(\mathcal{M}_1^-) = \dim^h(\mathcal{M}^-)$. ∎

Lemma 2.1.10 makes consistent the following definition. Given a subspace $\mathcal{M}$ of the Kreĭn space $\mathcal{K}$, the cardinal numbers

$$\kappa^+(\mathcal{M}) = \dim^h(\mathcal{M}^+), \quad \kappa^0(\mathcal{M}) = \dim^h(\mathcal{M}^0), \quad \kappa^-(\mathcal{M}) = \dim^h(\mathcal{M}^-),$$

are called, respectively, the *geometric ranks of positivity, isotropy*, and *negativity* of $\mathcal{M}$.

Proposition 2.1.11. *Let $\mathcal{M}$ be a negative (positive) subspace of $\mathcal{K}$ and $\widetilde{\mathcal{M}}$ a maximal negative (maximal positive) subspace such that $\mathcal{M} \subseteq \widetilde{\mathcal{M}}$. Then $\operatorname{codim}_{\widetilde{\mathcal{M}}}^h \mathcal{M} = \kappa^-(\mathcal{M}^\perp)$ (respectively, $\operatorname{codim}_{\widetilde{\mathcal{M}}}^h \mathcal{M} = \kappa^+(\mathcal{M}^\perp)$).*

Proof. Fix the fundamental decomposition $\mathcal{K} = \mathcal{K}^+[+]\mathcal{K}^-$, let J be the corresponding fundamental symmetry and $\langle \cdot, \cdot \rangle$ the associate positive definite inner product. Let $T \colon J^-\mathcal{M} \to \mathcal{K}^+$ be the angular operator of the negative subspace $\mathcal{M}$, and hence $\mathcal{M} = \{x + Tx \mid x \in J^-\mathcal{M}\}$. It is easy to see that

$$\mathcal{M}^\perp = \{y + T^*y \mid y \in \mathcal{K}^+\}[+](\mathcal{K}^- \ominus J^-\mathcal{M}). \tag{1.10}$$

If $\widetilde{\mathcal{M}}$ is a maximal negative extension of $\mathcal{M}$ and $\widetilde{T} \colon \mathcal{K}^- \to \mathcal{K}^+$ is its angular operator, then with respect to the decomposition $\mathcal{K}^- = J^-\mathcal{M} \oplus (\mathcal{K}^- \ominus J^-\mathcal{M})$ we have the operator block row matrix representation

$$\widetilde{T} = [T \quad X],$$

where $X \in \mathcal{B}(\mathcal{K}^- \ominus J^-\mathcal{M}, \mathcal{K}^+)$. Then we consider the subspace

$$\mathcal{N} = \{Xx + x \mid x \in \mathcal{K} \ominus J^-\mathcal{M}\},$$

and get that $\widetilde{\mathcal{M}} = \mathcal{M} \dotplus \mathcal{N}$. Therefore, $\operatorname{codim}_{\widetilde{\mathcal{M}}} \mathcal{M} = \dim^h(\mathcal{K}^- \ominus J^-\mathcal{M})$. From (1.10) we get $\kappa^-(\mathcal{M}^\perp) = \dim^h(\mathcal{K}^- \ominus J^-\mathcal{M})$ hence $\operatorname{codim}_{\widetilde{\mathcal{M}}}^h \mathcal{M} = \kappa^-(\mathcal{M}^\perp)$. ∎

2.2 Extensions of Semidefinite Subspaces

A pair of subspaces $(\mathcal{M}, \mathcal{N})$ of the Kreĭn space $\mathcal{K}$ is called an *alternating pair* if $\mathcal{M}$ is positive, $\mathcal{N}$ is negative, and $\mathcal{M} \perp \mathcal{N}$. If, in addition, $\mathcal{M}$ is strictly positive (uniformly positive) and $\mathcal{N}$ is strictly negative (uniformly negative) then we call $(\mathcal{M}, \mathcal{N})$ a *strictly* (respectively, *uniformly*) *alternating pair*. The alternating pair is called *maximal* if $\mathcal{M}$ is maximal positive and $\mathcal{N}$ is maximal negative. From Corollary 2.1.7 it follows that the alternating pair $(\mathcal{M}, \mathcal{N})$ is maximal if and only if $\mathcal{M} = \mathcal{N}^\perp$.

It is natural to ask whether any alternating pair of subspaces in a Kreĭn space does admit an extension to a maximal alternating pair. One possible approach to this problems is to make use of Zorn's Lemma on the existence of maximal elements, with the remark that, one still has to prove that the maximal element produces maximal semidefinite subspaces. We will not pursue this way. Our approach will be more constructive: we will describe precisely the maximal semidefinite subspaces in terms of angular operators and, moreover, we will parametrise all maximal extensions. This approach makes intensive use of angular operators, that is, of Hilbert space contractions and hence, the problem in question will be equivalently formulated in the language of a lifting problem for contractions on Hilbert spaces.

In the preceding section, the existence of maximal extensions for semidefinite subspaces was considered in Lemma 2.1.3, Lemma 2.1.4, and Corollary 2.1.6, in terms of angular operators. We reconsider this from the point of view of the parametrisation of all maximal extensions.

Let $\mathcal{H}_1$ and $\mathcal{H}_2$ be Hilbert spaces and let $T \in \mathcal{B}(\mathcal{H}_1, \mathcal{H}_2)$ be a linear contraction. Consider the *defect operators* $D_T \in \mathcal{L}(\mathcal{H}_1)$ and $D_{T^*} \in \mathcal{L}(\mathcal{H}_2)$ defined by

$$D_T = (I - T^*T)^{1/2}, \quad D_{T^*} = (I - TT^*)^{1/2}, \tag{2.1}$$

and the *defect spaces* $\mathcal{D}_T \subseteq \mathcal{H}_1$ and $\mathcal{D}_{T^*} \subseteq \mathcal{H}_2$ defined by

$$\mathcal{D}_T = \operatorname{Clos}\operatorname{Ran}(D_T), \quad \mathcal{D}_{T^*} = \operatorname{Clos}\operatorname{Ran}(D_{T^*}). \tag{2.2}$$

Lemma 2.2.1. *The following identity holds:*

$$TD_T = D_{T^*}T, \tag{2.3}$$

and, as a consequence of this, the operator

$$R(T) = \begin{bmatrix} T & D_{T^*} \\ D_T & -T^*|\mathcal{D}_{T^*} \end{bmatrix} : \begin{matrix} \mathcal{H}_1 \\ \oplus \\ \mathcal{D}_{T^*} \end{matrix} \longrightarrow \begin{matrix} \mathcal{H}_2 \\ \oplus \\ \mathcal{D}_T \end{matrix} \tag{2.4}$$

is unitary.

Proof. A straightforward induction argument shows that for all $n \in \mathbb{N}$ we have $T(I - T^*T)^n = (I - TT^*)^n T$ and hence, for any complex polynomial p that vanishes at 0, we have

$$Tp(I - T^*T) = p(I - TT^*)T.$$

By the Weierstrass Approximation Theorem, the continuous function $f(t) = \sqrt{t}$ can be approximated by polynomials that vanish at 0, uniformly on the interval $[0, 1]$, and from here and the functional calculus for selfadjoint operators in Hilbert space we obtain (2.3). Then the unitarity of the operator in (2.4) follows from (2.3). ∎

Lemma 2.2.2. *The linear contraction $T \in \mathcal{B}(\mathcal{H}_1, \mathcal{H}_2)$ is a strict (uniform) contraction, that is, $\|Tx\| < \|x\|$ for all $x \in \mathcal{H}_1 \setminus \{0\}$ ($\|T\| < 1$), if and only if any of the operators $I_1 - T^*T$, D_T, $I_2 - TT^*$, or D_{T^*}, is injective (respectively, boundedly invertible). Moreover, if one of these operators is injective (boundedly invertible) then so are all the others.*

Proof. To see this, let us first consider the positive operator $H \in \mathcal{B}(\mathcal{H}_1 \oplus \mathcal{H}_2)$ defined as a 2×2 operator block matrix

$$H = \begin{bmatrix} I_1 & T^* \\ T & I_2 \end{bmatrix}.$$

Note that the following factorisations hold:

$$H = \begin{bmatrix} I_1 & 0 \\ T & I_2 \end{bmatrix} \begin{bmatrix} I_1 & 0 \\ 0 & I_2 - TT^* \end{bmatrix} \begin{bmatrix} I_1 & T^* \\ 0 & I_2 \end{bmatrix}, \quad (2.5)$$

and

$$H = \begin{bmatrix} I_1 & T^* \\ 0 & I_2 \end{bmatrix} \begin{bmatrix} I_1 - T^*T & 0 \\ 0 & I_2 \end{bmatrix} \begin{bmatrix} I_1 & 0 \\ T & I_2 \end{bmatrix}, \quad (2.6)$$

which are special cases of the *Frobenius-Schur factorisation formulae*. Any of these factorisations shows that H is a positive operator. It is easy to see that the 2×2 operator block matrices on the extremal positions on the right-hand sides of (2.5) and (2.6) are boundedly invertible on $\mathcal{H}_1 \oplus \mathcal{H}_2$, more precisely

$$\begin{bmatrix} I_1 & 0 \\ T & I_2 \end{bmatrix}^{-1} = \begin{bmatrix} I_1 & 0 \\ -T & I_2 \end{bmatrix}, \quad \begin{bmatrix} I_1 & T^* \\ 0 & I_2 \end{bmatrix}^{-1} = \begin{bmatrix} I_1 & -T^* \\ 0 & I_2 \end{bmatrix}.$$

Therefore, H is injective (boundedly invertible) if and only if either (equivalently, both of) $I_1 - T^*T$ or (and) $I_2 - TT^*$ are injective (boundedly invertible). Taking into account the definitions of the defect operators D_T and D_{T^*}, see (2.1), it is clear that $I_1 - T^*T$ is injective (boundedly invertible) if and only if D_T is the same, and similarly, $I_2 - TT^*$ is invertible if and only if D_{T^*} is the same.

The fact that T is a strict (uniform) contraction if and only if $I_1 - T^*T$ is injective (respectively, boundedly invertible) is a simple interpretation of the definitions. ∎

In the following we fix a Kreĭn space $\mathcal{K}$, a fundamental decomposition $\mathcal{K} = \mathcal{K}^+[+]\mathcal{K}^-$ and let J be the corresponding fundamental symmetry. Let $\mathcal{M}$ be a positive subspace of $\mathcal{K}$. According to the angular operator representation, see Lemma 2.1.3, in order to

2.2 EXTENSIONS OF SEMIDEFINITE SUBSPACES

describe all the maximal positive subspaces extending $\mathcal{M}$ we have to describe all the contractive extensions to the whole subspace $\mathcal{K}^+$ of a given contraction defined on a subspace of $\mathcal{K}^+$. Thus, we are led to consider the following problem:

Contractive Row Extensions. *Let $\mathcal{H}_1$, $\mathcal{H}_1'$ and $\mathcal{H}_2$ be Hilbert spaces and let $T\colon \mathcal{H}_1 \to \mathcal{H}_2$ be a (strict, uniform) contraction, $\|T\| \leqslant 1$. Give a description of all (strictly, uniformly) contractive extensions $T_{\mathrm{r}}\colon \mathcal{H}_1 \oplus \mathcal{H}_1' \to \mathcal{H}_2$ of T.*

In the following we will often use two facts on factorisations and majorisations of operators in Hilbert spaces, that we list below.

Remarks 2.2.3. Let $\mathcal{H}_1$, $\mathcal{H}_2$, $\mathcal{H}_3$ be three Hilbert spaces, $A \in \mathcal{B}(\mathcal{H}_2, \mathcal{H}_1)$, and $B \in \mathcal{B}(\mathcal{H}_3, \mathcal{H}_1)$.

(a) $BB^* \leqslant AA^*$ if and only if $B = AC$ where $C \in \mathcal{B}(\mathcal{H}_3, \mathcal{H}_2)$ is a contraction, that is, $\|C\| \leqslant 1$. If this is the case, then C can be chosen such that $\operatorname{Clos}\operatorname{Ran}(C^*) = \operatorname{Clos}\operatorname{Ran}(A^*)$ and $\operatorname{Clos}\operatorname{Ran}(C) = \operatorname{Clos}\operatorname{Ran}(B^*)$, and it is unique with these properties.

Indeed, if $B = AC$ where $C \in \mathcal{B}(\mathcal{H}_2, \mathcal{H}_3)$ and $\|C\| \leqslant 1$, equivalently, $CC^* \leqslant I_{\mathcal{H}_2}$, then $BB^* = ACC^*A^* \leqslant AA^*$. Conversely, if $BB^* \leqslant AA^*$ then let the linear operator $T\colon \operatorname{Ran}(A^*) \to \operatorname{Ran}(B^*)$ be defined by $TA^*h = B^*h$, and observe that this definition is correct since, whenever $A^*h = 0$ we have

$$\|B^*h\|^2 = \langle B^*h, B^*h \rangle = \langle BB^*h, h \rangle \leqslant \langle AA^*h, h \rangle = \|A^*h\|^2 = 0,$$

hence $B^*h = 0$. The previous inequality shows that $\|Tg\| \leqslant \|g\|$ for all $g \in \operatorname{Ran}(B^*)$, hence T is a contraction that can be uniquely extended to a contraction, denoted by T as well, from $\operatorname{Clos}\operatorname{Ran}(A^*) \to \operatorname{Clos}\operatorname{Ran}(B^*)$. Then, we extend T on $\mathcal{H}_3$ by letting $T|\operatorname{Ker}(A) = 0$ and view $T\colon \mathcal{H}_3 \to \mathcal{H}_2$. Then $\|T\| \leqslant 1$ and let $C = T^*$.

(b) $BB^* = AA^*$ if and only if $B = AV$ where $V \in \mathcal{B}(\mathcal{H}_3, \mathcal{H}_2)$ is a partial isometry such that V^*VV^* is the orthogonal projection on $\operatorname{Clos}\operatorname{Ran} B^*$ and VV^*V is the orthogonal projection on $\operatorname{Clos}\operatorname{Ran}(A^*)$. If this is the case, then V is unique with these properties.

Indeed, if $AA^* = BB^*$ we use (a) twice and get that $B = AC$ and $A = BD$ for unique contractions $C \in \mathcal{B}(\mathcal{H}_2, \mathcal{H}_3)$ and $D \in \mathcal{B}(\mathcal{H}_3, \mathcal{H}_2)$ with additional requirements that $\operatorname{Clos}\operatorname{Ran} C^* = \operatorname{Clos}\operatorname{Ran}(A^*)$, $\operatorname{Clos}\operatorname{Ran}(C) = \operatorname{Clos}\operatorname{Ran}(B^*)$, $\operatorname{Clos}\operatorname{Ran}(D) = \operatorname{Clos}\operatorname{Ran}(B^*)$, and $\operatorname{Clos}\operatorname{Ran} D^* = \operatorname{Clos}\operatorname{Ran} A^*$. From $AA^* = BB^*$ it follows that C and D are partial isometries, $C = D^*$, and let $V = C$. The converse implication is clear. ∎

Lemma 2.2.4. *The operator block row matrix*

$$T_{\mathrm{r}} = [T \quad D_{T^*}\Gamma] \tag{2.7}$$

establishes a bijective correspondence between the set of all solutions T_{r} of the Contractive Row Extension Problem and all contractions $\Gamma\colon \mathcal{H}_1' \to \mathcal{D}_{T^*}$. Moreover, in the above *situation, the operators U and U_**

$$\begin{cases} U\colon \mathcal{D}_{T_{\mathrm{r}}} \to \mathcal{D}_T \oplus \mathcal{D}_\Gamma, \\ U D_{T_{\mathrm{r}}} = \begin{bmatrix} D_T & -T^*\Gamma \\ 0 & D_\Gamma \end{bmatrix}, \end{cases} \qquad (2.8)$$

and

$$\begin{cases} U_*\colon \mathcal{D}_{T_{\mathrm{r}}^*} \to \mathcal{D}_{\Gamma^*}, \\ U_* D_{T_{\mathrm{r}}^*} = D_{\Gamma^*} D_{T^*}, \end{cases} \qquad (2.9)$$

are unitary operators. In particular, assuming T is a strict (uniform) contraction, then the row extension T_{r} is a strict (uniform) contraction if and only if the operator Γ is a strict (respectively, uniform) contraction.

Proof. An operator $T_{\mathrm{r}}\colon \mathcal{H}_1 \oplus \mathcal{H}_1' \to \mathcal{H}_2$ extending T has the operator block row matrix representation

$$T_{\mathrm{r}} = [T \ \ X]\colon \mathcal{H}_1 \oplus \mathcal{H}_1' \to \mathcal{H}_2.$$

Then T_{r} is contractive if and only if

$$I_2 - T_{\mathrm{r}} T_{\mathrm{r}}^* = I_2 - TT^* - XX^* \geqslant 0$$

or, equivalently,

$$XX^* \leqslant I_2 - TT^* = D_{T^*}^2. \qquad (2.10)$$

The latter is equivalent with the representation of $X = D_{T^*}\Gamma$ for a uniquely determined contraction $\Gamma\colon \mathcal{H}_1' \to \mathcal{D}_{T^*}$. Indeed, we first define $Y\colon \mathrm{Ran}(D_{T^*}) \to \mathcal{H}_1'$ by $Y D_{T^*} h = X^* h$ for arbitrary $h \in \mathcal{H}_2$ and observe that, by (2.10), this definition is correct and that Y is a Hilbert space contraction $Y\colon \mathcal{D}_{T^*} \to \mathcal{H}_1'$, and finally let $\Gamma = Y^*$. Thus, we obtain the description of all solutions of the Contractive Row Extension Problem as in (2.7).

Further, let T_{r} be as in (2.7). Then

$$I_2 - T_{\mathrm{r}} T_{\mathrm{r}}^* = D_{T^*}(I - \Gamma\Gamma^*) D_{T^*} = D_{T^*} D_{\Gamma^*}^2 D_{T^*}.$$

This shows that the definition of the operator U_* in (2.9) is correct and that it is isometric. Since the operator $D_{\Gamma^*} D_{T^*}$ has dense range in the defect space $\mathcal{D}_{\Gamma^*}$ it follows that U_* uniquely extends to a unitary operator $U_*\colon \mathcal{D}_{T_{\mathrm{r}}^*} \to \mathcal{D}_{\Gamma^*}$.

Similarly, we note that the defect of T_{r} has the following factorisation

$$\begin{bmatrix} I_1 & 0 \\ 0 & I \end{bmatrix} - T_{\mathrm{r}}^* T_{\mathrm{r}} = \begin{bmatrix} I_1 - T^*T & -T^* D_{T^*}\Gamma \\ -\Gamma^* D_{T^*} T & I - \Gamma^* D_{T^*}^2 \Gamma \end{bmatrix}$$
$$= \begin{bmatrix} D_T & 0 \\ -\Gamma^* T & D_\Gamma \end{bmatrix} \begin{bmatrix} D_T & -T^*\Gamma \\ 0 & D_\Gamma \end{bmatrix}.$$

Therefore, the operator U is correctly defined as in (2.8) and it is isometric. Since the 2×2 operator block matrix on the right side of (2.8) has dense range in $\mathcal{D}_T \oplus \mathcal{D}_\Gamma$, it follows that U uniquely extends to a unitary operator $U\colon \mathcal{D}_{T_{\mathrm{r}}} \to \mathcal{D}_T \oplus \mathcal{D}_\Gamma$.

2.2 Extensions of Semidefinite Subspaces

For the last assertion in the statement we only have to take into account Lemma 2.2.2 and either of the unitary operators U or U_*. ∎

We are now in a position to get the description of maximal extensions of a semidefinite subspace of a Kreĭn space, as a consequence of Lemma 2.2.4 and Lemma 2.1.1.

Lemma 2.2.5. *Let $\mathcal{M}$ and $\widetilde{\mathcal{M}}$ be positive (strictly positive, uniformly positive) subspaces of $\mathcal{K}$ and let K and $\widetilde{K}$ be their angular operators, respectively, with respect to a fixed fundamental decomposition $\mathcal{K} = \mathcal{K}^+[+]\mathcal{K}^-$ with corresponding fundamental symmetry $J = J^+ - J^-$. Then, $\mathcal{M} \subseteq \widetilde{\mathcal{M}}$ if and only if $J^+\mathcal{M} \subseteq J^+\widetilde{\mathcal{M}}$ and there exists a uniquely determined linear contraction (strict contraction, uniform contraction) $\Gamma \in \mathcal{B}(J^+\widetilde{\mathcal{M}} \ominus J^+\mathcal{M}, \mathcal{D}_{T^*})$ such that*

$$\widetilde{K} = [K \quad D_{T^*}\Gamma]\colon J^+\mathcal{M} \oplus (J^+\widetilde{\mathcal{M}} \ominus J^+\mathcal{M}) \to \mathcal{K}^-. \tag{2.11}$$

For reasons of symmetry, we record now a dual problem of extensions of contractions on Hilbert spaces.

Contractive Column Extensions. *Let $\mathcal{H}_1$, $\mathcal{H}_2$, and $\mathcal{H}'_2$ be Hilbert spaces and let $T \in \mathcal{B}(\mathcal{H}_1, \mathcal{H}_2)$ be a (strict, uniform) contraction. Give a description of all (strict, uniform) contractions $T_c\colon \mathcal{H}_1 \to \mathcal{H}_2 \oplus \mathcal{H}'_2$ such that $P_{\mathcal{H}_2} T_c = T$.*

By passing to the adjoints in Lemma 2.2.4 we obtain the following result.

Lemma 2.2.6. *The formula*

$$T_c = \begin{bmatrix} T \\ \Gamma D_T \end{bmatrix} \tag{2.12}$$

establishes a bijective correspondence between the set of all solutions T_c of the Contractive Column Extension Problem and the set of all contractions $\Gamma\colon \mathcal{D}_T \to \mathcal{H}'_2$. In this situation, the operators V and V_* defined by

$$\begin{cases} V\colon \mathcal{D}_{T_c} \to \mathcal{D}_\Gamma, \\ V D_{T_c} = D_\Gamma D_T, \end{cases} \tag{2.13}$$

and

$$\begin{cases} V_*\colon \mathcal{D}_{T_c^*} \to \mathcal{D}_{T^*} \oplus \mathcal{D}_{\Gamma^*}, \\ V_* D_{T_c^*} = \begin{bmatrix} D_{T^*} & -T\Gamma^* \\ 0 & D_{\Gamma^*} \end{bmatrix}, \end{cases} \tag{2.14}$$

are unitary. In particular, assuming T is a strict (uniform) contraction, then the row extension T_c is a strict (uniform) contraction if and only if the operator Γ is a strict (respectively, uniform) contraction.

Let now $(\mathcal{M}, \mathcal{N})$ be an alternating pair of subspaces in the Kreĭn space $\mathcal{K}$. Let K and L denote the angular operators of $\mathcal{M}$ and $\mathcal{N}$, respectively, see (1.2) and (1.4). From $\mathcal{M} \perp \mathcal{N}$ it follows that

$$[Kx, y] = [x, Ly], \quad x \in J^+\mathcal{M},\ y \in J^-\mathcal{N}. \tag{2.15}$$

This relation enables us to define a linear contraction $T\colon J^+\mathcal{M} \to J^-\mathcal{N}$ by

$$T = P^{\mathcal{K}^-}_{J^-\mathcal{N}} K = L^*|J^+\mathcal{M},$$

where $P^{\mathcal{K}^-}_{J^-\mathcal{N}}$ denotes the orthogonal projection of the Hilbert space $\mathcal{K}^-$ onto its subspace $J^-\mathcal{N}$. We are looking for the maximal alternating pairs $(\widetilde{\mathcal{M}}, \widetilde{\mathcal{N}})$ extending the alternating pair $(\mathcal{M}, \mathcal{N})$. As noted before, see Lemma 2.1.5, if $\widetilde{K}$ and $\widetilde{L}$ are the angular operators of $\widetilde{\mathcal{M}}$ and $\widetilde{\mathcal{N}}$, respectively, we must have $\widetilde{L}^* = \widetilde{K}$ and hence, only the determination of $\widetilde{K}$ matters. Further, writing down that $\mathcal{M} \subseteq \widetilde{\mathcal{M}}$ we obtain $\widetilde{K}|J^+\mathcal{M} = K$ and from $\widetilde{\mathcal{M}} \perp \mathcal{N}$ we get $P_{J^-\mathcal{N}}\widetilde{K} = P_{J^-\mathcal{N}}L$. We have thus showed that the problem of maximal extensions of alternating pairs of subspaces in a Kreĭn space is equivalent with the following problem.

Lifting of Contractions. Let $\mathcal{H}_1$, $\mathcal{H}'_1$, $\mathcal{H}_2$, and $\mathcal{H}'_2$ be Hilbert spaces and let $T_{\mathrm{r}}\colon \mathcal{H}_1 \oplus \mathcal{H}'_1 \to \mathcal{H}_2$ and $T_{\mathrm{c}}\colon \mathcal{H}_1 \to \mathcal{H}_2 \oplus \mathcal{H}'_2$ be contractions such that

$$T_{\mathrm{r}}|\mathcal{H}_1 = P_{\mathcal{H}_2}T_{\mathrm{c}} = T\colon \mathcal{H}_1 \to \mathcal{H}_2. \tag{2.16}$$

Give a description of all contractions $\widetilde{T}\colon \mathcal{H}_1 \oplus \mathcal{H}'_1 \to \mathcal{H}_2 \oplus \mathcal{H}'_2$ such that $\widetilde{T}|\mathcal{H}_1 = T_{\mathrm{c}}$ and $P_{\mathcal{H}_2}\widetilde{T} = T_{\mathrm{r}}$.

Theorem 2.2.7. *According to Lemmas 2.2.4 and 2.2.6, let the contractions T_{r} and T_{c} be represented by*

$$T_{\mathrm{r}} = [T \quad D_{T^*}\Gamma_1], \quad T_{\mathrm{c}} = \begin{bmatrix} T \\ \Gamma_2 D_T \end{bmatrix}, \tag{2.17}$$

for uniquely determined contractions $\Gamma_1\colon \mathcal{H}'_1 \to \mathcal{D}_{T^*}$ and $\Gamma_2\colon \mathcal{D}_T \to \mathcal{H}'_2$. Then, the formula

$$\widetilde{T} = \begin{bmatrix} T & D_{T^*}\Gamma_1 \\ \Gamma_2 D_T & -\Gamma_2 T^* \Gamma_1 + D_{\Gamma_2^*}\Gamma D_{\Gamma_1} \end{bmatrix} \tag{2.18}$$

establishes a bijective correspondence between the set of all solutions $\widetilde{T}$ of the Lifting of Contractions Problem and the set of all contractions $\Gamma\colon \mathcal{D}_{\Gamma_1} \to \mathcal{D}_{\Gamma_2^*}$. Moreover, in this situation, the operators U and U_* defined by

$$\begin{cases} U\colon \mathcal{D}_{\widetilde{T}} \to \mathcal{D}_{\Gamma_2} \oplus \mathcal{D}_\Gamma, \\ UD_{\widetilde{T}} = \begin{bmatrix} D_{\Gamma_2}D_T & -(D_{\Gamma_2}T^*\Gamma_1 + \Gamma_2^*\Gamma D_{\Gamma_1}) \\ 0 & D_\Gamma D_{\Gamma_1} \end{bmatrix}, \end{cases} \tag{2.19}$$

and

$$\begin{cases} U_*\colon \mathcal{D}_{\widetilde{T}^*} \to \mathcal{D}_{\Gamma_1^*} \oplus \mathcal{D}_{\Gamma^*}, \\ U_*D_{\widetilde{T}^*} = \begin{bmatrix} D_{\Gamma_1^*}D_{T^*} & -(D_{\Gamma_1^*}T\Gamma_2^* + \Gamma_1\Gamma^* D_{\Gamma_2^*}) \\ 0 & D_{\Gamma^*}D_{\Gamma_2^*} \end{bmatrix}, \end{cases} \tag{2.20}$$

are unitary.

2.2 Extensions of Semidefinite Subspaces

Proof. Let $\widetilde{T}$ be a solution of the Lifting of Contractions Problem. From (2.16) and (2.17) we have that $\widetilde{T}$ is represented by the 2×2 operator block matrix

$$\widetilde{T} = \begin{bmatrix} T & D_{T^*}\Gamma_1 \\ \Gamma_2 D_T & X \end{bmatrix}, \tag{2.21}$$

where $X \colon \mathcal{H}'_1 \to \mathcal{H}'_2$ has to be determined. To this end, we regard $\widetilde{T}$ as a row contractive extension of the contraction T_c and applying Lemma 2.2.4 it follows that

$$\widetilde{T} = [T_c \quad D_{T_c^*}\Delta], \tag{2.22}$$

where $\Delta \colon \mathcal{H}'_1 \to \mathcal{D}_{T_c^*}$ is a uniquely determined contraction. Moreover, by the second part of Lemma 2.2.4 there exists the unitary operator $V \colon \mathcal{D}_{T_c^*} \to \mathcal{D}_{T^*} \oplus \mathcal{D}_{\Gamma_2^*}$ defined by

$$V D_{T_c^*} = \begin{bmatrix} D_{T^*} & -T\Gamma_2^* \\ 0 & D_{\Gamma_2^*} \end{bmatrix}. \tag{2.23}$$

Denote by $\Lambda = V\Delta \colon \mathcal{H}'_1 \to \mathcal{D}_{T^*} \oplus \mathcal{D}_{\Gamma_2^*}$ and note that Λ is a contraction. We represent

$$\Lambda = \begin{bmatrix} \Lambda_1 \\ \Lambda_2 \end{bmatrix}, \tag{2.24}$$

where $\Lambda_1 \colon \mathcal{H}'_1 \to \mathcal{D}_{T^*}$ and $\Lambda_2 \colon \mathcal{H}'_1 \to \mathcal{D}_{\Gamma_2^*}$. Then, by (2.23) we have

$$D_{T_c^*}\Delta = D_{T_c^*}V^*\Lambda = \begin{bmatrix} D_{T^*}\Lambda_1 \\ -\Gamma_2 T^* \Lambda_1 + D_{\Gamma_2^*}\Lambda_2 \end{bmatrix}. \tag{2.25}$$

Taking into account that $P_{\mathcal{H}_2}\widetilde{T}|\mathcal{H}'_1 = T_{\mathrm{r}}|\mathcal{H}'_1 = D_{T^*}\Gamma_1$ and the uniqueness part in Lemma 2.2.4, it follows that $\Lambda_1 = \Gamma_1$. We now regard Λ as a column extension of $\Lambda_1 = \Gamma_1$ and, by Lemma 2.2.6, it follows that $\Lambda_2 = \Gamma D_{\Gamma_1}$ for a uniquely determined contraction $\Gamma \colon \mathcal{D}_{\Gamma_1} \to \mathcal{D}_{\Gamma_2^*}$. Plugging this representation of Λ_2 in (2.25), from (2.22) and (2.21) it follows that the operator $X = -\Gamma_2 T^* \Gamma_1 + D_{\Gamma_2^*}\Gamma D_{\Gamma_1}$. This proves that $\widetilde{T}$ has the representation (2.18).

Conversely, if $\Gamma \colon \mathcal{D}_{\Gamma_1} \to \mathcal{D}_{\Gamma_2^*}$ is a contraction and $\widetilde{T}$ is as in (2.18) then we can reverse the arrows in the above reasoning to conclude that $\widetilde{T}$ is a solution of the Lifting of Contraction Problem.

Let now $\widetilde{T}$ be as in (2.18). Then, taking account of Lemma 2.2.1, a rather long but straightforward calculation proves the following factorisation

$$I - \widetilde{T}^*\widetilde{T} = \begin{bmatrix} D_T D_{\Gamma_2} & 0 \\ (-D_{\Gamma_2} T\Gamma_1^* + D_{\Gamma_1}\Gamma^*\Gamma_1) & D_{\Gamma_1}D_{\Gamma} \end{bmatrix}$$
$$\times \begin{bmatrix} D_{\Gamma_2}D_T & -(D_{\Gamma_2}T^*\Gamma_1 + \Gamma_2^*\Gamma D_{\Gamma_1}) \\ 0 & D_{\Gamma}D_{\Gamma_1} \end{bmatrix}.$$

This shows that the operator U is correctly defined as in (2.19) and it is isometric. Since the operator block 2×2 matrix in (2.19) has dense range in $\mathcal{D}_{\Gamma_2} \oplus \mathcal{D}_\Gamma$ it follows that the operator U uniquely extends to a unitary operator $U \colon \mathcal{D}_{\widetilde{T}} \to \mathcal{D}_{\Gamma_2} \oplus \mathcal{D}_\Gamma$.

In a similar way one proves that the operator U_* is unitary. ∎

Corollary 2.2.8. *With notation as in Theorem 2.2.7, let the operator T be a strict contraction (uniform contraction) and let $\widetilde{T}$ be defined as in (2.18). Then $\widetilde{T}$ is a strict contraction (uniform contraction) if and only if all parameters Γ_1, Γ_2 and Γ are strict contractions (uniform contractions).*

In order to apply the previous theorem to the extension problem of alternating pairs of subspaces in Kreĭn spaces, some notation is necessary. Let $(\mathcal{M}, \mathcal{N})$ be an alternating pair of subspaces of the Kreĭn space $\mathcal{K}$ and, with respect to the angular operator representation, consider the angular operator $K \colon J^+\mathcal{M} \to \mathcal{K}^-$ of $\mathcal{M}$ and the angular operator $L \colon J^-\mathcal{N} \to \mathcal{K}^+$ of $\mathcal{N}$. Note that, since $\mathcal{M} \perp \mathcal{N}$, by (1.6) we have

$$L^* | J^+\mathcal{M} = P^{\mathcal{K}^-} K = T, \qquad (2.26)$$

where $T \colon J^\mathcal{M} \to J^-\mathcal{N}$ is a uniquely determined contraction.

Since L^* has the property (2.26), we use Lemma 2.2.4 and get that, with respect to the decomposition $\mathcal{K}^- = J^-\mathcal{N} \oplus (\mathcal{K}^- \ominus J^-\mathcal{N})$, we have the representation

$$L^* = [T \quad D_{T^*} \Gamma_1], \qquad (2.27)$$

where $\Gamma_1 \colon \mathcal{K}^+ \ominus J^+\mathcal{M} \to \mathcal{D}_{T^*}$ is a contraction uniquely determined by L.

Similarly, since K is a contraction extending T, we use Lemma 2.2.6 and get that, with respect to the decomposition $\mathcal{K}^+ = J^+\mathcal{M} \oplus (\mathcal{K}^+ \ominus J^+\mathcal{M})$, we have the representation

$$K = \begin{bmatrix} T \\ \Gamma_2 D_T \end{bmatrix}, \qquad (2.28)$$

where $\Gamma_2 \colon \mathcal{D}_T \to (\mathcal{K}^- \ominus J^-\mathcal{N})$ is a contraction uniquely determined by K.

Theorem 2.2.9. (a) *Let $(\mathcal{M}, \mathcal{N})$ be an alternating pair of subspaces of the Kreĭn space $\mathcal{K}$ and consider the notation as in (2.26), (2.28), and (2.27). Then, the formula*

$$\widetilde{K} = \begin{bmatrix} T & D_{T^*}\Gamma_1 \\ \Gamma_2 D_T & -\Gamma_2 T^* \Gamma_1 + D_{\Gamma_2^*} \Gamma D_{\Gamma_1} \end{bmatrix} \qquad (2.29)$$

establishes a one-to-one correspondence between the set of all maximal alternating pairs $(\widetilde{\mathcal{M}}, \widetilde{\mathcal{N}})$ extending $(\mathcal{M}, \mathcal{N})$, where

$$\widetilde{\mathcal{M}} = \{x + \widetilde{K}x \mid x \in \mathcal{K}^+\}, \quad \widetilde{\mathcal{N}} = \{\widetilde{K}^*y + y \mid y \in \mathcal{K}^-\}, \qquad (2.30)$$

and the set of all contractions $\Gamma \colon \mathcal{D}_{\Gamma_1} \to \mathcal{D}_{\Gamma_2^*}$.

2.2 Extensions of Semidefinite Subspaces

(b) *With respect to notation as in item (a), if $(\mathcal{M},\mathcal{N})$ is a strictly (uniformly) alternating pair of subspaces of $\mathcal{K}$, the formulae (2.29) and (2.30) establish a one-to-one correspondence between the set of maximal strictly (uniformly) alternating pairs $(\widetilde{\mathcal{M}},\widetilde{\mathcal{N}})$ and the set of all strictly (uniformly) contractions $\Gamma \colon \mathcal{D}_{\Gamma_1} \to \colon \mathcal{D}_{\Gamma_2^*}$.*

(c) *For any (strictly, uniformly) alternating pair of subspaces $(\mathcal{M},\mathcal{N})$ of the Kreĭn space $\mathcal{K}$ there exists a maximal (strictly, uniformly) alternating pair of subspaces $(\widetilde{\mathcal{M}},\widetilde{\mathcal{N}})$ of $\mathcal{K}$ such that $\widetilde{\mathcal{M}} \supseteq \mathcal{M}$ and $\widetilde{\mathcal{N}} \supseteq \mathcal{N}$.*

Proof. (a) With notation as in (2.26), (2.28), and (2.27), it follows from Lemma 2.1.3, Lemma 2.1.4, and Lemma 2.1.5, that Theorem 2.2.7 provides the solution to the parametrisation of all maximal alternating pairs $(\widetilde{\mathcal{M}},\widetilde{\mathcal{N}})$ extending the alternating pair $(\mathcal{M},\mathcal{N})$.

(b) If $(\mathcal{M},\mathcal{N})$ is a strictly alternating pair of subspaces of $\mathcal{K}$, then by Lemma 2.2.4 and Lemma 2.2.6, the operator T is a strict contraction and similarly Γ_1 and Γ_2 are strict contractions as well. Then, by Corollary 2.2.8 the set of all maximal strictly alternating pairs $(\widetilde{\mathcal{M}},\widetilde{\mathcal{N}})$ extending the alternating pair $(\mathcal{M},\mathcal{N})$ is parametrised by taking the parameter Γ a strict contraction. A similar reasoning applies in the case when $(\mathcal{M},\mathcal{N})$ is a maximal uniformly alternating pair of subspaces of $\mathcal{K}$.

(c) This assertion follows from the previous assertions (a) and (b) by the observation that it is always possible to take $\Gamma = 0$ in which case we obtain a maximal (strictly, uniformly) alternating pair $(\widetilde{\mathcal{M}},\widetilde{\mathcal{N}})$ extending $(\mathcal{M},\mathcal{N})$. ∎

As a first application of Theorem 2.2.9 we present a criterion for a subspace to be maximal uniformly definite.

Corollary 2.2.10. *Let $\mathcal{M}$ and $\mathcal{N}$ be subspaces of the Kreĭn space $\mathcal{K}$ such that $\mathcal{M}$ is uniformly positive, $\mathcal{N}$ is negative, $\mathcal{M} \perp \mathcal{N}$, and $\mathcal{M} + \mathcal{N}$ is dense in $\mathcal{K}$. Then $\mathcal{M}$ is a maximal uniformly positive subspace and $\mathcal{N} = \mathcal{M}^\perp$.*

Proof. Let us first remark that the subspace $\mathcal{N}$ is strictly negative. Indeed, let $x \in \mathcal{N}$ be such that $x \perp \mathcal{N}$. Then $x \perp \mathcal{M} + \mathcal{N} = \mathcal{K}$ and hence $x = 0$. Theorem 2.2.9 implies the existence of a maximal alternating pair $(\widetilde{\mathcal{M}},\widetilde{\mathcal{N}})$ extending the alternating pair $(\mathcal{M},\mathcal{N})$. Since $\mathcal{M}$ is uniformly positive, the same Theorem 2.2.9 (or its particular case Lemma 2.1.3) implies the existence of a fundamental decomposition $\mathcal{K} = \mathcal{K}^+[+]\mathcal{K}^-$ such that $\mathcal{K}^+ \supseteq \mathcal{M}$. Let $K \in \mathcal{B}(\mathcal{K}^+,\mathcal{K}^-)$ be the angular operator of $\widetilde{\mathcal{M}}$ with respect to this fundamental decomposition. If $x \in \mathcal{K}^+ \ominus \mathcal{M}$ then $x + Kx \in \widetilde{\mathcal{M}} \cap \mathcal{M}^\perp$ and hence $x + Kx \in \mathcal{M} + \widetilde{\mathcal{N}}$. Since $\mathcal{M} + \widetilde{\mathcal{N}} \supseteq \mathcal{M} + \mathcal{N}$ is dense in $\mathcal{K}$, it follows that $x = 0$. We have proven that $\mathcal{M} = \mathcal{K}^+$ is a maximal uniformly positive subspace. This implies that $\mathcal{N} \subseteq \mathcal{K}^-$ is uniformly negative, in particular, $\mathcal{M}[+]\mathcal{N} = \mathcal{K}$ and hence, $\mathcal{N} = \mathcal{M}^\perp = \mathcal{K}^-$. ∎

In connection with the previous results we present an example of an alternating pair $(\mathcal{M},\mathcal{N})$ such that $\mathcal{M} + \mathcal{N}$ is dense in $\mathcal{K}$ and neither $\mathcal{M}$ nor $\mathcal{N}$ is maximal.

Example 2.2.11. Let us consider the set $X = \{\pm\frac{n-1}{n} \mid n \geq 2\}$ and the measure σ on X defined by
$$\sigma(\pm\frac{n-1}{n}) = \frac{1}{n(n-1)}, \quad n \geq 2.$$
Let $\mathcal{H}$ denote the Hilbert space $L^2(X;\sigma)$ and consider the linear operator $T \in \mathcal{B}(\mathcal{H})$ defined by $(Tx)(t) = tx(t)$, $t \in X$. Clearly, T is a strict contraction and selfadjoint. Considering the vector $x_0 = (1-t)^{1/4}$, $t \in X$, we claim that

$$x_0, \; Tx_0, \; \text{and} \; (I - T^2)\mathcal{H} \; \text{are linearly independent.} \tag{2.31}$$

Indeed, let $\alpha_1, \alpha_2 \in \mathbb{C}$ and $z \in L^2(X;\sigma)$ be such that
$$\alpha_1(1-t)^{1/4} + \alpha_2 t(1-t)^{1/4} = (1-t^2)z(t), \quad t \in X,$$
that is,
$$(\alpha_1 + \alpha_2 t)(1+t)^{-1}(1-t)^{-3/4} \in L^2(X;\sigma).$$
Making this explicit, by means of the definition of the measure σ, we get that the series
$$\sum_{n \geq 2} \left(\frac{(\alpha_1 n + \alpha_2(n-1))^2 n^{1/2}}{(2n-1)^2(n-1)} + \frac{(\alpha_1 n - \alpha_2(n-1))^2 n^{1/2}}{(2n-1)^{3/2}(n-1)} \right)$$
must converge. But this holds only for $\alpha_1 = \alpha_2 = 0$, and hence the claim is proved.

Let $\mathcal{H}_0$ denote the subspace of $\mathcal{H}$ orthogonal to x_0. We consider now the space $\mathcal{K} = \mathcal{H} \oplus \mathcal{H}$ endowed with the indefinite inner product $[\cdot, \cdot]$ corresponding to the symmetry
$$J = \begin{bmatrix} I & 0 \\ 0 & -I \end{bmatrix}, \quad \text{with respect to } \mathcal{K} = \mathcal{H} \oplus \mathcal{H},$$
that is,
$$[x_1 \oplus y_1, x_2 \oplus y_2] = \langle x_1, x_2 \rangle - \langle y_1, y_2 \rangle, \quad x_1, x_2, y_1, y_2 \in \mathcal{H}.$$
Then $(\mathcal{K}, [\cdot, \cdot])$ is a Kreĭn space. If $K \in \mathcal{B}(\mathcal{H}_0, \mathcal{H})$ is defined by $K = T|\mathcal{H}_0$ and we introduce the subspaces $\mathcal{M}$ and $\mathcal{N}$ of $\mathcal{K}$ by
$$\mathcal{M} = \{x \oplus Kx \mid x \in \mathcal{H}_0 \oplus 0\}, \quad \mathcal{N} = \{Kx \oplus x \mid x \in 0 \oplus \mathcal{H}_0\},$$
then $(\mathcal{M}, \mathcal{N})$ is an alternating pair which is not maximal (cf. Corollary 2.1.7). However, we prove subsequently that $\mathcal{M} + \mathcal{N}$ is dense in $\mathcal{K}$. To this end, let $z = z_1 \oplus z_2$ be orthogonal to $\mathcal{M} + \mathcal{N}$. This means
$$\langle z_1, x_1 \rangle - \langle z_2, Tx_1 \rangle = \langle z_1, Tx_2 \rangle - \langle z_2, x_2 \rangle = 0, \quad x_1, x_2 \in \mathcal{H}_0.$$
Further, this yields $z_1 - Tz_2 = \alpha_1 x_0$ and $Tz_1 - z_2 = \alpha_2 x_0$ for some $\alpha_1, \alpha_2 \in \mathbb{C}$. These imply
$$z_1 - T^2 z_1 + \alpha_2 Tx_0 - \alpha_1 x_0 = 0.$$
On the grounds of (2.31) it follows that $\alpha_1 = \alpha_2 = 0$ and then $z_1 = T^2 z_1$ and $z_2 = T^2 z_2$. But then necessarily $z_1 = z_2 = 0$, since T^2 (as well as T) is a strict contraction. We thus have proven that $\mathcal{M} + \mathcal{N}$ is dense in $\mathcal{K}$. ∎

Recall that a subspace $\mathcal{N}$ is called neutral if $[x,x]=0$ for all vectors $x\in\mathcal{N}$. Clearly, the subspace $\mathcal{N}$ is neutral if and only if $\mathcal{N}\subseteq\mathcal{N}^\perp$, or, what is the same, $\mathcal{N}=\mathcal{N}^0$. By definition, a subspace $\mathcal{N}$ is *maximal neutral* if it is neutral and has no neutral extension different from $\mathcal{N}$. The subspace $\mathcal{N}$ is called *hypermaximal neutral* if $\mathcal{N}^\perp=\mathcal{N}$.

Remark 2.2.12. Let $\mathcal{N}$ be a neutral subspace. Clearly $\mathcal{N}$ is also a positive subspace, in particular we can consider its angular operator $V\colon \mathrm{Dom}(V)=J^+\mathcal{N}\to J^-\mathcal{N}\subseteq\mathcal{K}^-$. Since $\mathcal{N}$ is neutral it follows that V is isometric. Conversely, to any isometric angular operator V there corresponds the neutral subspace $\mathcal{N}=\{x+Vx\mid x\in\mathrm{Dom}(V)\}$. In this angular operator representation, neutral extensions of the subspace $\mathcal{N}$ correspond to isometric extensions of the operator V.

Using for example Lemma 2.2.4, it follows that $\mathcal{N}$ is maximal neutral if and only if its angular operator V is maximal isometric, that is, either $J^+\mathcal{N}=\mathcal{K}^+$ or $J^-\mathcal{N}=\mathcal{K}^-$. Similarly, $\mathcal{N}$ is hypermaximal neutral if and only if its angular operator $V\colon\mathcal{K}^+\to\mathcal{K}^-$ is unitary. From here it follows that for the existence of hypermaximal neutral subspaces it is necessary and sufficient that $\kappa^-(\mathcal{K})=\kappa^+(\mathcal{K})$.

Concerning maximal neutral extensions of a given neutral subspace $\mathcal{N}$ it is now clear that they always exist and are described by the maximal isometric extensions of its angular operator V. Lemma 2.2.4 gives us all the necessary formulae. Moreover, the neutral subspace $\mathcal{N}$ admits hypermaximal neutral extensions if and only if its *defect indices* $\dim^{\mathrm{h}}(\mathcal{K}^+\ominus J^+\mathcal{N})$ and $\dim^{\mathrm{h}}(\mathcal{K}^-\ominus J^-\mathcal{N})$ coincide and, in this case, all its hypermaximal neutral extensions can be parametrised by the set of unitary operators $\mathcal{K}^+\ominus J^+\mathcal{N}\to\mathcal{K}^-\ominus J^-\mathcal{N}$. ∎

2.3 Intermezzo: Maximal Accretive and Selfadjoint Extensions

Let $(\mathcal{H};\langle\cdot,\cdot\rangle)$ be a Hilbert space. A closed densely defined operator S in $\mathcal{H}$ is called *accretive* if

$$\mathrm{Re}\langle Sx,x\rangle=\frac{\langle Sx,x\rangle+\langle x,Sx\rangle}{2}\geq 0,\quad x\in\mathrm{Dom}(S). \tag{3.1}$$

The accretive operator S is called *maximal accretive* if it admits no proper accretive extension.

A closely related notion is that of dissipative operator: a closed and densely defined operator A in $\mathcal{H}$ is called *dissipative* if

$$\mathrm{Im}\langle Ax,x\rangle=\frac{\langle Ax,x\rangle-\langle x,Ax\rangle}{2\mathrm{i}}\geq 0,\quad x\in\mathrm{Dom}(A). \tag{3.2}$$

Clearly, T is accretive if and only if $-\mathrm{i}T$ is dissipative.

Remark 2.3.1. The assumption, in the definition of an accretive operator S, that it is closed is not essential. Indeed, it is easy to show that the closure of the graph of a densely defined and closable operator S with the property (3.1) is the graph of an operator with the same property. In addition, a maximal accretive operator is always closed.

We first give an application of the extensions of alternating pairs of subspaces in Kreĭn spaces to extensions of alternating pairs of accretive operators.

Let S_1 and S_2 be accretive operators in $\mathcal{H}$. If

$$\langle S_1 x, y \rangle = \langle x, S_2 y \rangle, \quad x \in \mathrm{Dom}(S_1), \ y \in \mathrm{Dom}(S_2), \tag{3.3}$$

or, equivalently, $S_1 \subseteq S_2^*$ and $S_2 \subseteq S_1^*$, we call (S_1, S_2) an *alternating pair of accretive operators*.

Theorem 2.3.2. *Let (S_1, S_2) be an alternating pair of accretive operators in $\mathcal{H}$. Then there exists a maximal alternating pair of accretive operators extending (S_1, S_2), that is, two maximal accretive operators $\widetilde{S}_1 \supset S_1$, $\widetilde{S}_2 \supset S_2$ such that*

$$\langle \widetilde{S}_1 x, y \rangle = \langle x, \widetilde{S}_2 y \rangle, \quad x \in \mathrm{Dom}(\widetilde{S}_1), \ y \in \mathrm{Dom}(\widetilde{S}_2). \tag{3.4}$$

Proof. We consider the Hilbert space $\mathcal{K} = \mathcal{H} \oplus \mathcal{H}$ onto which we define the inner product $[\cdot, \cdot]$ by

$$[x_1 \oplus x_2, y_1 \oplus y_2] = \langle x_1, y_2 \rangle + \langle x_2, y_1 \rangle, \quad x_1, x_2, y_1, y_2 \in \mathcal{H}. \tag{3.5}$$

This corresponds to the Gram operator J

$$J = \begin{bmatrix} 0 & I \\ I & 0 \end{bmatrix}, \tag{3.6}$$

and hence $(\mathcal{K}, [\cdot, \cdot])$ is a Kreĭn space. We consider the subspaces $\mathcal{M} = \mathcal{G}(S_1) = \{x \oplus S_1 x \mid x \in \mathrm{Dom}(S_1)\}$ and $\mathcal{N} = \mathcal{G}(-S_2) = \{y \oplus -S_2 y \mid y \in \mathrm{Dom}(S_2)\}$, the graphs of the operators S_1 and $-S_2$, respectively. Since S_1 and S_2 are accretive it follows that $\mathcal{M}$ is a positive subspace and $\mathcal{N}$ is a negative subspace in $\mathcal{K}$. From the assumption (3.3) we get that $\mathcal{M} \perp \mathcal{N}$ and hence $(\mathcal{M}, \mathcal{N})$ is an alternating pair of subspaces in the Kreĭn space $\mathcal{K}$. By Theorem 2.2.9 there exists $(\widetilde{\mathcal{M}}, \widetilde{\mathcal{N}})$ a maximal alternating pair of subspaces extending $(\mathcal{M}, \mathcal{N})$.

We prove that $\widetilde{\mathcal{M}}$ is the graph of a maximal accretive extension of S_1. Indeed, if $0 \oplus y \in \widetilde{\mathcal{M}}$ then, by (3.5), it follows that the vector $0 \oplus y$ is neutral in $\mathcal{K}$. Since $\widetilde{\mathcal{M}}$ is a positive subspace, by means of the Schwarz Inequality, it follows that $0 \oplus y \perp \widetilde{\mathcal{M}}$ and hence $\langle y, x \rangle = 0$ for all $x \in \mathrm{Dom}(S_2)$. This implies $y = 0$, since $\mathrm{Dom}(S_2)$ is dense in $\mathcal{H}$. We have thus proven that $\widetilde{\mathcal{M}}$ is the graph of some closed operator $\widetilde{S}_1 \supseteq S_1$. Since $\widetilde{\mathcal{M}}$ is positive we get that $\widetilde{S}_1$ is an accretive operator. S_1 is maximal accretive since $\widetilde{\mathcal{M}}$ is a maximal positive subspace.

In a similar way we prove that $\widetilde{\mathcal{N}}$ is the graph of some maximal accretive operator $\widetilde{S}_2 \supseteq S_2$. Writing down explicitly $\widetilde{\mathcal{M}} \perp \widetilde{\mathcal{N}}$ we obtain (3.4). ∎

Remark 2.3.3. Let us consider an accretive operator S on the Hilbert space $\mathcal{H}$ and let $\mathcal{K}$ be the Kreĭn space defined in the proof of Theorem 2.3.2 with inner product as in (3.5).

2.3 INTERMEZZO: MAXIMAL ACCRETIVE AND SELFADJOINT EXTENSIONS

It is easy to see that the fundamental decomposition $\mathcal{K} = \mathcal{K}^+[+]\mathcal{K}^-$ corresponding to the fundamental decomposition J as in (3.6) is given by

$$\mathcal{K}^+ = \{x \oplus x \mid x \in \mathcal{H}\}, \quad \mathcal{K}^- = \{x \oplus -x \mid x \in \mathcal{H}\}.$$

Let $\mathcal{M} = \{x \oplus Sx \mid x \in \text{Dom}(S)\}$ be the graph of S considered as a positive subspace in the Kreĭn space $\mathcal{K}$. Calculating the angular operator of $\mathcal{M}$ we get

$$T(S) = (I - S)(I + S)^{-1}, \quad \text{Dom}(T) = \text{Ran}(I + S). \tag{3.7}$$

Since $\mathcal{M}$ is positive $T(S)$ is a Hilbert space contraction and $\text{Dom}(T) = \text{Ran}(I + S)$ is closed. We recognise immediately that $T(S)$ is a Cayley type transformation of S. The formalism developed in Lemma 2.2.4, Lemma 2.2.6, and Theorem 2.2.7 provides an explicit description of all maximal accretive extensions required in Theorem 2.3.2. ∎

Let now S be a densely defined positive operator in the Hilbert space $\mathcal{H}$. Without restricting the generality we also assume that S is closed (otherwise, we replace S with its closure which shares the same properties). It is easy to see that this is equivalent with saying that the pair (S, S) is an alternating pair of accretive operators, in particular $S \subseteq S^*$. The problem we are interested in is to describe all positive selfadjoint extensions of S. This problem is equivalently formulated as requiring a description of all maximal accretive pairs of type $(\widetilde{S}, \widetilde{S})$ of the given pair (S, S). We first notice that Theorem 2.2.9 does not give any answer in this case but, fortunately, the angular operator approach, which we used in its counterpart Theorem 2.2.7, will do the job.

Lemma 2.3.4. *The Cayley transformation (3.7) establishes a bijective correspondence between the class of all densely defined, closed, and positive operators S on the Hilbert space $\mathcal{H}$ and the class of all symmetric (that is, $\langle Tx, y \rangle = \langle x, Ty \rangle$ for all $x, y \in \text{Dom}(T)$) contractions $T\colon \text{Dom}(T) \subseteq \mathcal{H} \to \mathcal{H}$ with $\text{Dom}(T)$ closed and $\text{Ran}(I + T)$ dense in $\mathcal{H}$.*

Proof. Let S be a densely defined, closed, positive operator on $\mathcal{H}$. As noticed before, S is accretive and hence the operator $T = T(S)$, defined as in (3.7), is well defined, contractive, and $\text{Dom}(T) = \text{Ran}(I + S)$ is closed. In addition, taking into account that $S \subseteq S^*$, for all $h, g \in \text{Dom}(S)$ we have

$$\begin{aligned}\langle T(I+S)h, (I+S)g \rangle &= \langle (I-S)h, (I+S)g \rangle \\ &= \langle h, g \rangle - \langle Sh, g \rangle + \langle h, Sg \rangle - \langle Sh, Sg \rangle \\ &= \langle h, g \rangle - \langle h, Sg \rangle + \langle Sh, g \rangle - \langle Sh, Sg \rangle \\ &= \langle (I+S)h, (I-S)g \rangle = \langle (I+S)h, T(I+S)g \rangle,\end{aligned}$$

and hence the operator T is symmetric. In addition, $\text{Ran}(I + T) = \text{Dom}(S)$ is dense in $\mathcal{H}$.

Conversely, suppose T is a symmetric contraction with $\text{Dom}(T)$ closed and such that $\text{Ran}(I + T)$ is dense in $\mathcal{H}$. Then, letting $T^*\colon \mathcal{H} \to \text{Dom}(T)$ be defined by $\langle Th, g \rangle =$

$\langle h, T^*g \rangle$ for all $h \in \text{Dom}(T)$ and all $g \in \mathcal{H}$, it follows that $I + T^*$ is injective and hence, since T is contractive, $I + T$ is also injective. Then it makes sense to define

$$S = (I - T)(I + T)^{-1}, \quad \text{Dom}(S) = \text{Ran}(I + T). \tag{3.8}$$

For arbitrary $h \in \text{Dom}(S)$ we find $x \in \text{Dom}(T)$ such that $h = (I + T)x$ and then

$$\langle Sh, h \rangle = \langle (I - T)x, (I + T)x \rangle = \langle x, x \rangle - \langle Tx, Tx \rangle \geq 0.$$

Thus S is positive. Clearly it is also closed.

We leave to the reader to verify that the transformations (3.7) and (3.8) are inverse one to each other. ∎

Lemma 2.3.5. *Let S be a densely defined, closed, and positive operator on $\mathcal{H}$, T its Cayley transformation as in (3.7) and consider the sets*

$$\mathfrak{S} = \{\widetilde{S} \mid \widetilde{S} \text{ positive selfadjoint on } \mathcal{H}, \widetilde{S} \supseteq S\}, \tag{3.9}$$

$$\mathfrak{T} = \{\widetilde{T} \in \mathcal{B}(\mathcal{H}) \mid \widetilde{T} \text{ selfadjoint contraction, } \widetilde{T} \supseteq T\}. \tag{3.10}$$

Then the Cayley transformation as in (3.7) is bijective as a map from $: \mathfrak{S}$ *to* $\mathfrak{T}$.

Proof. As noted before, (S, S) is an alternating pair of accretive operators and $\widetilde{S}$ is a positive selfadjoint extension of S if and only if $(\widetilde{S}, \widetilde{S})$ is a maximal alternating pair of accretive operators. Then we use Theorem 2.3.2 and the angular operator interpretation of the operators in the set $\mathfrak{T}$ to conclude the proof. ∎

Remark 2.3.6. Note that Lemma 2.3.5 provides also a proof of the classical fact that a densely defined positive operator A in a Hilbert space $\mathcal{H}$ is positive selfadjoint if and only if $(I + A)\mathcal{H} = \mathcal{H}$. ∎

These results show that, in order to describe all positive selfadjoint extensions of the given positive operator S, we have to describe all contractive and selfadjoint extensions of a given symmetric contraction, its Cayley transformation T as in (3.7). Recall that $\mathcal{B}(\mathcal{H})^+$ denotes the cone of positive bounded operators on the Hilbert space $\mathcal{H}$.

Lemma 2.3.7. *Let $T: \text{Dom}(T) \to \mathcal{H}$ be a symmetric contraction with $\text{Dom}(T)$ closed in $\mathcal{H}$. Then the set $\mathfrak{T}$ defined as in (3.10) is nonvoid and there exist $\widetilde{T}_{-1}, \widetilde{T}_1 \in \mathfrak{T}$, $\widetilde{T}_{-1} \leq \widetilde{T}_1$, such that $\mathfrak{T} = \{B \in \mathcal{B}(\mathcal{H})^+ \mid \widetilde{T}_{-1} \leq B \leq \widetilde{T}_1\}$.*

Proof. Splitting the space $\mathcal{H} = \text{Dom}(T) \oplus (\mathcal{H} \ominus \text{Dom}(T))$ and using Lemma 2.2.6, the symmetric contraction T is uniquely represented as

$$T = \begin{bmatrix} A \\ \Gamma_1^* D_A \end{bmatrix}, \tag{3.11}$$

2.3 Intermezzo: Maximal Accretive and Selfadjoint Extensions

with $A \colon \operatorname{Dom}(T) \to \operatorname{Dom}(T)$ a selfadjoint contraction and $\Gamma_1 \colon \mathcal{H} \ominus \operatorname{Dom}(T) \to \mathcal{D}_A$ a contraction. We search for elements $\widetilde{T} \in \mathfrak{T}$. Since $\widetilde{T}$ is symmetric and $\widetilde{T}|\operatorname{Dom}(T) = T$, it must have the following operator matrix representation

$$\widetilde{T} = \begin{bmatrix} A & D_A \Gamma_1 \\ \Gamma_1^* D_A & X \end{bmatrix},$$

with $X \in \mathcal{B}(\mathcal{H} \ominus \operatorname{Dom}(T))$ selfadjoint. Since $\widetilde{T}$ is a contraction, we use Theorem 2.2.7 and get

$$\widetilde{T}(\Gamma) = \begin{bmatrix} A & D_A \Gamma_1 \\ \Gamma_1^* D_A & -\Gamma_1^* A \Gamma_1 + D_{\Gamma_1} \Gamma D_{\Gamma_1} \end{bmatrix}, \qquad (3.12)$$

where $\Gamma \colon \mathcal{D}_{\Gamma_1} \to \mathcal{D}_{\Gamma_1}$ is a uniquely determined selfadjoint contraction. The existence of at least one choice for Γ (e.g. $\Gamma = 0$) proves that the set $\mathfrak{T}$ is nonvoid.

If $\mathcal{D}_{\Gamma_1} = 0$, that is, Γ_1 is isometric, we put $\widetilde{T}_{-1} = \widetilde{T}_1 = \widetilde{T}(0)$ and in this case the set $\mathfrak{T}$ has a single element. If $\mathcal{D}_{\Gamma_1} \neq 0$ then define

$$\widetilde{T}_{-1} = \widetilde{T}(-I) = \begin{bmatrix} A & D_A \Gamma_1 \\ \Gamma_1^* D_A & -\Gamma_1^* A \Gamma_1 - I + \Gamma_1^* \Gamma_1 \end{bmatrix}, \qquad (3.13)$$

$$\widetilde{T}_1 = \widetilde{T}(I) = \begin{bmatrix} A & D_A \Gamma_1 \\ \Gamma_1^* D_A & -\Gamma_1^* A \Gamma_1 + I - \Gamma_1^* \Gamma_1 \end{bmatrix}, \qquad (3.14)$$

and it is easy to show now that $\widetilde{T}_{-1} \leqslant \widetilde{T}(\Gamma) \leqslant \widetilde{T}_1$ for all selfadjoint contractions $\Gamma \in \mathcal{B}(\mathcal{D}_{\Gamma_1})$.

Conversely, let $B \in \mathcal{B}(\mathcal{H})$ be selfadjoint and such that $\widetilde{T}_{-1} \leqslant B \leqslant \widetilde{T}_1$. Since both $\widetilde{T}_{-1}$ and $\widetilde{T}_1$ are contractions it follows that B itself is contractive, too. Therefore, by Theorem 2.2.7 it is of the form

$$B = \begin{bmatrix} C & D_C \Gamma_2 \\ \Gamma_2^* D_C & -\Gamma_2^* C \Gamma_2 + D_{\Gamma_2} \Delta D_{\Gamma_1} \end{bmatrix},$$

with respect to the decomposition

$$\mathcal{H} = \operatorname{Dom}(T) \oplus (\mathcal{H} \ominus \operatorname{Dom}(T)),$$

where $C \in \mathcal{B}(\operatorname{Dom}(T))$ is a selfadjoint contraction, $\Gamma_2 \in \mathcal{B}(\mathcal{H} \ominus \operatorname{Dom}(T), \mathcal{D}_C)$ is a contraction and $\Delta \in \mathcal{B}(\mathcal{D}_{\Gamma_2})$ is a selfadjoint contraction, all of them uniquely determined by B. For an arbitrary vector $x \in \operatorname{Dom}(T)$ we have

$$\langle Ax, x \rangle = \langle \widetilde{T}_{-1} x, x \rangle \leqslant \langle Bx, x \rangle \leqslant \langle \widetilde{T}_1 x, x \rangle = \langle Ax, x \rangle,$$

and hence $C = A$. Then we get

$$B - \widetilde{T}_{-1} = \begin{bmatrix} 0 & D_A(\Gamma_2 - \Gamma_1) \\ (\Gamma_2 - \Gamma_1)^* D_A & -\Gamma_2^* A \Gamma_2 + D_{\Gamma_2} \Delta D_{\Gamma_1} + \Gamma_1^* A \Gamma_1 + D_{\Gamma_1}^2 \end{bmatrix} \geqslant 0,$$

$$\tilde{T}_1 - B = \begin{bmatrix} 0 & D_A(\Gamma_1 - \Gamma_2) \\ (\Gamma_1 - \Gamma_2)^* D_A & -\Gamma_1^* A \Gamma_1 + D_{\Gamma_1}^2 + \Gamma_2^* A \Gamma_2 - D_{\Gamma_2} \Delta D_{\Gamma_1} \end{bmatrix} \geq 0,$$

whence taking account of the elementary fact that an operator matrix of the form

$$\begin{bmatrix} 0 & M \\ M^* & N \end{bmatrix}$$

is positive if and only if $M = 0$ and $N \geq 0$, we get $\Gamma_2 = \Gamma_1$. This proves that B is of the form (3.12) and hence $B \in \mathfrak{T}$. ∎

The proof of Lemma 2.3.7 gives actually more than is stated: it provides a parametrisation of the set $\mathfrak{T}$, made explicit by (3.12), in terms of the parameters Γ running through the set

$$\mathfrak{C} = \{\Gamma \in \mathcal{B}(\mathcal{D}_{\Gamma_1}) \mid -I \leq \Gamma \leq I\}. \tag{3.15}$$

But this parametrisation has even more properties.

Lemma 2.3.8. *The mapping $\mathfrak{C} \ni \Gamma \mapsto \tilde{T}(\Gamma) \in \mathfrak{T}$ defined by (3.12) is a continuous bijection such that $\Gamma' \leq \Gamma''$ if and only if $\tilde{T}(\Gamma') \leq \tilde{T}(\Gamma'')$.*

Proof. We already found during the proof of Lemma 2.3.7 that the mapping $\mathfrak{C} \ni \Gamma \mapsto \tilde{T}(\Gamma) \in \mathfrak{T}$ defined as in (3.12) is bijective. Taking into account of (3.12), for $\Gamma', \Gamma'' \in \mathfrak{C}$ we have

$$\tilde{T}(\Gamma') - \tilde{T}(\Gamma'') = \begin{bmatrix} 0 & 0 \\ 0 & D_{\Gamma_1}(\Gamma' - \Gamma'')D_{\Gamma_1} \end{bmatrix}.$$

This implies immediately the continuity of the mapping $\tilde{T}(\cdot)$ as well as the order preserving property. ∎

In order to describe the set of all positive selfadjoint extensions of the positive operator S we also need to investigate some order and continuity properties of the inverse of the Cayley transformation (3.7). We need first to recall some definitions. Let S_1 and S_2 be two selfadjoint positive operators in the Hilbert space $\mathcal{H}$. Then $S_1 \leq S_2$ is considered in the sense of quadratic forms, that is, $\operatorname{Dom}(S_2^{1/2}) \subseteq \operatorname{Dom}(S_1^{1/2})$ and $\|S_1^{1/2}h\| \leq \|S_2^{1/2}h\|$ for all $h \in \operatorname{Dom}(S_2^{1/2})$. According to Theorem VI.2.21 in [84], $S_1 \leq S_2$ if and only if $(I + S_1)^{-1} \geq (I + S_2)^{-1}$.

Also, suppose $(R_n)_{n \in \mathbb{N}}$ and R are selfadjoint operators on the same Hilbert space $\mathcal{H}$. The sequence $(R_n)_{n \in \mathbb{N}}$ *converges in the strong resolvent sense* to R if for every, equivalently, for some, $\lambda \in \mathbb{C} \setminus \mathbb{R}$, see Theorem IV.2.25 in [84],

$$(\lambda I - R_n)^{-1}\xi \xrightarrow[n \to \infty]{} (\lambda I - R)^{-1}\xi, \quad \xi \in \mathcal{H}.$$

Lemma 2.3.9. *The injective mapping $\mathfrak{T} \ni \tilde{T} \mapsto \tilde{S}(\tilde{T}) = (I - \tilde{T})(I + \tilde{T})^{-1} \in \mathfrak{S}$, the inverse of the mapping in (3.7), is nonincreasing, that is, for $S_1, S_2 \in \mathfrak{S}$ we have $S_1 \leq S_2$ if and only if $T(S_1) \geq T(S_2)$, and it is continuous when $\mathfrak{T}$ is endowed with the operator norm topology and $\mathfrak{S}$ is endowed with the strong resolvent convergence.*

2.3 Intermezzo: Maximal Accretive and Selfadjoint Extensions

Proof. Suppose that $S_1, S_2 \in \mathfrak{S}$ and $S_1 \leqslant S_2$. We have to prove that this is equivalent with

$$\langle (I - S_1)(I + S_1)^{-1}x, x \rangle \geqslant \langle (I - S_2)(I + S_2)^{-1}x, x \rangle, \quad x \in \mathcal{H}. \tag{3.16}$$

For arbitrary $x \in \mathcal{H}$, there exist uniquely determined vectors $h \in \mathrm{Dom}(S_1)$ and $g \in \mathrm{Dom}(S_2)$, such that

$$(I + S_1)h = x = (I + S_2)g, \tag{3.17}$$

and hence (3.16) is equivalent with

$$\langle (I - S_1)h, (I + S_1)h \rangle \geqslant \langle (I - S_2)g, (I + S_2)g \rangle,$$

or, equivalently,

$$\|h\|^2 - \|S_1 h\|^2 \geqslant \|g\|^2 - \|S_2 g\|^2. \tag{3.18}$$

Making use of (3.17) it follows that $\|(I + S_1)h\| = \|(I + S_2)g\|$ and

$$\|S_2 g\|^2 - \|S_1 h\|^2 = 2(\langle S_1 h, h \rangle - \langle S_2 g, g \rangle) + (\|h\|^2 - \|g\|^2).$$

Therefore, (3.18) holds if and only if

$$\langle (I + S_1)h, h \rangle \geqslant \langle (I + S_2)g, g \rangle.$$

But, on the grounds of (3.17), this is equivalent with

$$\langle x, (I + S_1)^{-1}x \rangle \geqslant \langle x, (I + S_2)^{-1}x \rangle,$$

that is, $(I+S_1)^{-1} \geqslant (I+S_2)^{-1}$. According to Theorem VI.2.21 in [84], this is equivalent with $S_1 \leqslant S_2$.

Let $(\widetilde{T}_n)_{n \in \mathbb{N}} \in \mathfrak{T}$ be a sequence of operators from $\mathfrak{T}$ such that $\widetilde{T}_n \to \widetilde{T} \in \mathfrak{T}$ uniformly as $n \to \infty$. Denote $\widetilde{S}_n = (I - \widetilde{T}_n)(I + \widetilde{T}_n)^{-1}$ and $\widetilde{S} = (I - \widetilde{T})(I + \widetilde{T})^{-1}$. Since $(I + \widetilde{S})^{-1} = (I + \widetilde{T})/2$ and $(I + \widetilde{S}_n)^{-1} = (I + \widetilde{T}_n)/2$ for all $n \in \mathbb{N}$, it follows that

$$\lim_{n \to \infty}(I + \widetilde{S}_n)^{-1} = \lim_{n \to \infty}(I + \widetilde{S})^{-1}, \tag{3.19}$$

whence, taking into account Theorem IV.2.25 in [84], we obtain that $\widetilde{S}_n$ converges in the strong resolvent sense to $\widetilde{S}$. ∎

Summing up the results obtained until now we can describe the set $\mathfrak{S}$ of all positive selfadjoint extensions of the positive operator S.

Theorem 2.3.10. *Let S be a densely defined, closed and positive operator on the Hilbert space $\mathcal{H}$ and consider the sets $\mathfrak{S}$, $\mathfrak{T}$, and $\mathfrak{C}$, defined as in (3.9), (3.10), and (3.15), as well as the mapping $\mathfrak{C} \ni \Gamma \mapsto \widetilde{T}(\Gamma) \in \mathfrak{T}$ as in (3.12) and the mapping $\mathfrak{T} \ni \widetilde{T} \mapsto \widetilde{S} = (I - \widetilde{T})(I + \widetilde{T})^{-1} \in \mathfrak{S}$. Then the compound mapping*

$$\mathfrak{C} \ni \Gamma \mapsto \widetilde{S}(\Gamma) = (I - \widetilde{T}(\Gamma))(I + \widetilde{T}(\Gamma))^{-1} \in \mathfrak{S} \tag{3.20}$$

is bijective, nonincreasing and continuous, when $\mathfrak{C}$ is endowed with the operator norm topology and $\mathfrak{S}$ is endowed with the strong resolvent convergence. In particular the set $\mathfrak{S}$ is always nonvoid, that is, there always exist positive selfadjoint extensions of S.

Moreover, there exist two distinguished positive selfadjoint extensions $\widetilde{S}_K = \widetilde{S}(I) = (I - \widetilde{T}_1)(I + \widetilde{T}_1)^{-1}$ and $\widetilde{S}_F = \widetilde{S}(-I) = (I - \widetilde{T}_{-1})(I + \widetilde{T}_{-1})^{-1}$, with notation as in (3.13) and (3.14), such that, a positive selfadjoint operator R in $\mathcal{H}$ extends S if and only if $\widetilde{S}_K \leqslant R \leqslant \widetilde{S}_F$.

With notation as in Theorem 2.3.10, if $\widetilde{S}(\Gamma)$ is an arbitrary positive selfadjoint extension of S then clearly $\widetilde{S}(\Gamma) \subseteq S^*$ so in order to determine the operator $\widetilde{S}(\Gamma)$ only its domain must be calculated. Thanks to the parametrisation in (3.20) it is easy to see that

$$\mathrm{Dom}(\widetilde{S}(\Gamma)) = \mathrm{Dom}(S) + \left[\frac{D_A \Gamma_1}{I - \Gamma_1^* A \Gamma_1 + D_{\Gamma_1} \Gamma D_{\Gamma_1}} \right] \mathrm{Ker}(I + S^*). \quad (3.21)$$

In the following we will characterise in a different way the distinguished positive selfadjoint extension $\widetilde{S}(-I)$ and, as a consequence, we will find a different description of the domains $\mathrm{Dom}(\widetilde{S}(\Gamma))$.

We now consider the *Friedrichs extension* F of S defined by

$$\mathrm{Dom}(F) = \{x \in \mathrm{Dom}(S^*) \mid \exists (x_n)_{n \in \mathbb{N}} \subset \mathrm{Dom}(S), \lim_{n \to \infty} x_n = x \text{ and} \quad (3.22)$$

$$\lim_{m,n \to \infty} \langle x_n - x_m, S(x_n - x_m) \rangle = 0\},$$

and $Fx = S^*x$ for all $x \in \mathrm{Dom}(F)$.

For the beginning we obtain a different characterisation of the operator F, more precisely, of its domain, in terms of the symmetric contraction $T = (I - S)(I + S)^{-1}$, $\mathrm{Dom}(T) = \mathrm{Ran}(I + S)$, and the selfadjoint contraction $A = P_{\mathrm{Dom}(T)} T \in \mathcal{B}(\mathrm{Dom}(T))$.

Lemma 2.3.11. *Suppose $x \in \mathrm{Dom}(S^*)$. Then $x \in \mathrm{Dom}(F)$ if and only if there exists a sequence $(y_n)_{n \in \mathbb{N}}$ of vectors from $\mathrm{Dom}(T)$ such that*

$$\lim_{n \to \infty} (I + T)y_n = x, \text{ and } \lim_{n,m \to \infty} (I + A)^{1/2}(y_n - y_m) = 0.$$

Proof. Since $T = (I - S)(I + S)^{-1}$ and taking into account (3.22) we obtain that a vector $x \in \mathrm{Dom}(S^*)$ is in $\mathrm{Dom}(F)$ if and only if there exists a sequence $(y_n)_{n \in \mathbb{N}}$ of vectors from $\mathrm{Ran}(I + S)$ such that

$$\lim_{n \to \infty} (I + T)y_n = x, \quad (3.23)$$

and

$$\lim_{n,m \to \infty} \langle (I + T)(y_n - y_m), (I - T)(y_n - y_m) \rangle = 0. \quad (3.24)$$

Taking into account (3.23) we get that (3.24) holds if and only if

$$\lim_{n,m \to \infty} \langle (I + T)(y_n - y_m), (I + T)(y_n - y_m) \rangle = 0. \quad (3.25)$$

2.3 Intermezzo: Maximal Accretive and Selfadjoint Extensions 53

On the other hand

$$\langle (I+T)(y_n - y_m), (I-T)(y_n - y_m)\rangle + \langle (I+T)(y_n - y_m), (I+T)(y_n - y_m)\rangle$$
$$= 2\langle (I+T)(y_n - y_m), y_n - y_m\rangle$$
$$= 2\langle (I+T)(y_n - y_m), P_{\mathrm{Dom}(T)}(y_n - y_m)\rangle$$
$$= 2\|(I+A)^{1/2}(y_n - y_m)\|^2,$$

and hence, by adding side by side (3.24) and (3.25), we get

$$\lim_{n,m\to\infty} \|(I+A)^{1/2}(y_n - y_m)\| = 0. \tag{3.26}$$

Conversely, if there exists a sequence $(y_n)_{n\in\mathbb{N}}$ of vectors from $\mathrm{Dom}(T)$ such that (3.23) and (3.26) hold then, reversing the procedure from above, we obtain (3.23) and (3.24) and hence $x \in \mathrm{Dom}(F)$. ∎

Theorem 2.3.12. *The Friedrichs extension F coincides with the distinguished positive selfadjoint extension $\widetilde{S}(-I)$.*

Proof. We first prove that

$$\mathrm{Ran}(I+A) \subseteq \mathrm{Ran}(I+T^*) \subseteq \mathrm{Ran}((I+A)^{1/2}). \tag{3.27}$$

Indeed, we have

$$I+T = \begin{bmatrix} I+A \\ \Gamma_1^* D_A \end{bmatrix}, \quad I+T^* = [I+A \quad D_A\Gamma_1].$$

Therefore

$$(I+T^*)(I+T) = (I+A)^2 + D_A\Gamma_1\Gamma_1^* D_A \leqslant (I+A)^2 + D_A^2 \tag{3.28}$$
$$= (I+A)^2 + (I-A)^2 = 2(I+A) \tag{3.29}$$
$$= 2(I+A)^{1/2}(I+A)^{1/2}. \tag{3.30}$$

This proves the first inclusion. As for the latter inclusion, note that (3.28) yields the factorisation $I+T^* = (I+A)^{1/2}X$ with some bounded operator X and hence $\mathrm{Ran}(I+T^*) \subseteq \mathrm{Ran}((I+A)^{1/2})$.

We first prove that $\mathrm{Dom}(\widetilde{S}(-I)) \subseteq \mathrm{Dom}(F)$. To see this, let $x \in \mathrm{Dom}(\widetilde{S}(-I)) = \mathrm{Ran}(I+\widetilde{T}(-I))$ be arbitrary. There exists $y \in \mathcal{H}$ such that $x = (I+\widetilde{T}(-I))y$. Since $I+T$ is injective it follows that $\mathrm{Ran}(I+T^*)$ is dense in $\mathrm{Dom}(T)$. By (3.27) this implies that $\mathrm{Ran}((I+A)^{1/2})$ is dense in $\mathrm{Dom}(T)$. Again by (3.27) there exists a sequence $(y_n)_{n\in\mathbb{N}}$ of vectors from $\mathrm{Dom}(T)$ such that

$$\lim_{n\to\infty} (I+A)^{1/2}y_n = (I+A)^{-1/2}(I+T^*)y, \tag{3.31}$$

and hence
$$\lim_{n\to\infty} (I+A)y_n = (I+T^*)y. \tag{3.32}$$

Applying the bounded operator $(I-A)^{1/2}$ to (3.31) and denoting $P = P_{\mathrm{Dom}(T)}$ we get

$$\begin{aligned}
\lim_{n\to\infty} D_A y_n &= \lim_{n\to\infty} (I-A)^{1/2}(I+A)^{1/2} y_n \\
&= (I-A)^{1/2}(I+A)^{1/2}((I+A)Py + (I+A)^{1/2}(I-A)^{1/2}\Gamma_1(I-A)y) \\
&\quad \times (I-A)^{1/2}(I+A)^{1/2} Py + (I-A)\Gamma_1(I-P)y \\
&= D_A Py + (I-A)\Gamma_1(I-P)y.
\end{aligned}$$

Therefore,
$$\lim_{n\to\infty} \Gamma_1^* D_A y_n = \Gamma_1^* D_A Py + \Gamma_1^*(I-A)\Gamma_1(I-P)y. \tag{3.33}$$

We write (3.31) and (3.33) in one formula

$$\lim_{n\to\infty} y_n = \lim_{n\to\infty} \begin{bmatrix} (I+A)y_n \\ \Gamma_1^* D_A y_n \end{bmatrix} = \begin{bmatrix} (I+A)Py + D_A\Gamma_1(I-P)y \\ \Gamma_1^* D_A Py + \Gamma_1^*(I-A)\Gamma_1(I-P)y \end{bmatrix} \tag{3.34}$$
$$= (I+\widetilde{T}(-I))y = x.$$

Taking into account Lemma 2.3.11, from (3.31) and (3.34) we get $x \in \mathrm{Dom}(F)$.

Conversely, if $x \in \mathrm{Dom}(F)$, let the sequence $(y_n)_{n\in\mathbb{N}}$ of vectors from $\mathrm{Dom}(T)$ be as in Lemma 2.3.11 and denote $z = \lim_{n\to\infty} (I+A)^{1/2} y_n$. Then

$$\lim_{n\to\infty} (I+A)y_n = (I+A)^{1/2}z, \quad \lim_{n\to\infty} \Gamma_1^* D_A y_n = \Gamma_1^*(I-A)^{1/2}z.$$

Therefore,
$$\lim_{n\to\infty} (I+T)y_n = \begin{bmatrix} (I+A)^{1/2}z \\ \Gamma_1^*(I-A)^{1/2}z \end{bmatrix} = x,$$

and then, taking into account (3.12), we obtain $x = (I+\widetilde{T}(-I))z \in \mathrm{Dom}(\widetilde{S}(-I))$. ∎

Finally, we get a parametric description of the domains of all selfadjoint extensions in terms of the domain of the Friedrichs extension.

Corollary 2.3.13. *For every* $\Gamma \in \mathfrak{C}$ *we have*

$$\mathrm{Dom}(\widetilde{S}(\Gamma)) = \mathrm{Dom}(F) + D_{\Gamma_1}(I+\Gamma)D_{\Gamma_1}\,\mathrm{Ker}(I+S^*).$$

Proof. If $\Gamma \in \mathfrak{C}$ then by (3.12) we get

$$\begin{aligned}
I + \widetilde{T}(\Gamma) &= \begin{bmatrix} I+A & D_A\Gamma_1 \\ \Gamma_1^* D_A & \Gamma_1^*(I-A)\Gamma_1 + D_{\Gamma_1}(I+\Gamma)D_{\Gamma_1} \end{bmatrix} \\
&= I + \widetilde{T}(-I) + \begin{bmatrix} 0 & 0 \\ 0 & D_{\Gamma_1}(I+\Gamma)D_{\Gamma_1} \end{bmatrix},
\end{aligned}$$

and hence by Theorem 2.3.12 we get

$$\begin{aligned}\operatorname{Dom}(\widetilde{S}(\Gamma)) &= \operatorname{Ran}(I + \widetilde{T}(\Gamma))\\ &= \operatorname{Ran}(I + \widetilde{T}(-I)) + D_{\Gamma_1}(I+\Gamma)D_{\Gamma_1}(\mathcal{G} \ominus \operatorname{Dom}(T))\\ &= \operatorname{Dom}(F) + D_{\Gamma_1}(I+\Gamma)D_{\Gamma_1}\operatorname{Ker}(I+S^*).\end{aligned}$$ ∎

2.4 A Universality Property of Kreĭn Spaces

The class of Kreĭn spaces seems to be narrow within the class of indefinite inner product spaces. However, we will see in this section that, if some natural requirements are imposed on an indefinite inner product space, such as the presence of a Hilbert space topology making the indefinite inner product jointly continuous, then it can be realised as a subspace of a Kreĭn space. In other words, Kreĭn spaces and their subspaces have a universality property, from this point of view.

An indefinite inner product space $(\mathcal{H}, [\cdot, \cdot])$ is called a *G-space* if there exists a positive definite inner product $\langle \cdot, \cdot \rangle$ with respect to which $(\mathcal{H}, \langle \cdot, \cdot \rangle)$ is a Hilbert space and the inner product $[\cdot, \cdot]$ is jointly continuous. This is equivalent with saying that $(\mathcal{H}, \langle \cdot, \cdot \rangle)$ is a Hilbert space, $G \in \mathcal{B}(\mathcal{H})$ is a bounded selfadjoint operator (the Gram operator, see Remark 1.3.3), and

$$[x, y] = \langle Gx, y\rangle \quad \text{for all } x, y \in \mathcal{H}. \tag{4.1}$$

As a consequence of Theorem 1.3.2, the Hilbert space topology of a nondegenerate G-space is unique with the property that it makes the indefinite inner product $[\cdot, \cdot]$ jointly continuous. Thus, without any ambiguity, we call this topology the *strong topology* and all the continuity properties will be referred to it.

We now show that a G-space $\mathcal{H}$ can always be identified with a subspace of a Kreĭn space $\mathcal{K}$ and that, among all the possible Kreĭn spaces that contain $\mathcal{H}$ as a subspace, there exists one which is minimal and uniquely determined, modulo isometric isomorphisms, with this property.

Theorem 2.4.1. *Let $(\mathcal{H}, [\cdot, \cdot])$ be a G-space. Then there exists a Kreĭn space $(\mathcal{K}, [\cdot, \cdot])$ which contains $(\mathcal{H}, [\cdot, \cdot])$ as a subspace and is minimal in the following sense: if $(\widetilde{\mathcal{K}}, [\cdot, \cdot])$ is another Kreĭn space containing $(\mathcal{H}, [\cdot, \cdot])$ as a subspace, then there exists a bounded operator $U\colon \mathcal{K} \to \widetilde{\mathcal{K}}$ such that $Ux = x$ for all $x \in \mathcal{H}$ and it is isometric, that is, $[Ux, Uy] = [x, y]$ for all $x, y \in \mathcal{K}$.*

In addition, this minimal Kreĭn space $(\mathcal{K}, [\cdot, \cdot])$ is unique in the following sense: if $(\widetilde{\mathcal{K}}, [\cdot, \cdot])$ is another Kreĭn space that contains $(\mathcal{H}, [\cdot, \cdot])$ as a subspace and with the minimality property as before, then there exists an isometric isomorphism $W\colon \mathcal{K} \to \widetilde{\mathcal{K}}$ such that $Wx = x$ for all $x \in \mathcal{H}$.

Proof. Let $\langle \cdot, \cdot \rangle$ be a positive definite inner product with respect to which $(\mathcal{H}, \langle \cdot, \cdot \rangle)$ is a Hilbert space and the inner product $[\cdot, \cdot]$ is jointly continuous. By the Riesz Representation

Theorem, there exists a bounded selfadjoint operator $G\colon \mathcal{H} \to \mathcal{H}$ (the Gram operator, see Remark 1.3.3) such that
$$[x, y] = \langle Gx, y\rangle, \quad x, y \in \mathcal{H}.$$
Modulo the multiplication of the positive inner product $\langle \cdot, \cdot \rangle$ with a positive constant we can assume that
$$|[x, y]| \leqslant \langle x, x\rangle^{1/2}\langle y, y\rangle^{1/2}, \quad x, y \in \mathcal{H},$$
or equivalently, that G is a contraction, $|\langle Gx, x\rangle| \leqslant \langle x, x\rangle$.

Consider now the defect operator $D_G = (I - G^2)^{1/2}$ and the defect space $\mathcal{D}_G = \operatorname{Clos}\operatorname{Ran}(D_G)$, see (2.1), and let $\mathcal{K} = \mathcal{H} \oplus \mathcal{D}_G$ as Hilbert spaces. On the Hilbert space $\mathcal{K}$ we define the operator
$$J = \begin{bmatrix} G & D_G \\ D_G & -G|\mathcal{D}_G \end{bmatrix}. \tag{4.2}$$

It is a straightforward calculation to prove that J is a symmetry, that is, $J^* = J = J^{-1}$ (the selfadjointness of J is clear, while the unitarity follows for example from Lemma 2.2.1). Thus, defining the indefinite inner product $[\cdot, \cdot]_\mathcal{K}$ by
$$[x, y]_\mathcal{K} = \langle Jx, y\rangle, \quad x, y \in \mathcal{K},$$
this turns $\mathcal{K}$ into a Kreĭn space and J is a fundamental symmetry on it. $\mathcal{H}$ is naturally identified with the subspace $\mathcal{H} \oplus 0$ of $\mathcal{K}$ as Hilbert spaces. If $P_\mathcal{H}$ denotes the orthogonal projection of the Hilbert space $\mathcal{K}$ onto its subspace $\mathcal{H}$ then, taking into account of (4.2), for all $x, y \in \mathcal{H}$ we have
$$[x, y]_\mathcal{K} = \langle Jx, y\rangle = \langle P_\mathcal{H} J P_\mathcal{H} x, y\rangle = \langle Gx, y\rangle = [x, y],$$
and hence this identification holds also with respect to the indefinite inner products.

Let $(\widetilde{\mathcal{K}}, [\cdot, \cdot]_{\widetilde{\mathcal{K}}})$ be another Kreĭn space which contains $(\mathcal{H}, [\cdot, \cdot])$ as a subspace and let $\widetilde{J}$ be a fundamental symmetry of $\widetilde{\mathcal{K}}$. This implicitly fixes the positive inner product $\langle \cdot, \cdot \rangle_{\widetilde{J}}$ on $\widetilde{\mathcal{K}}$. As above, G denotes the Gram operator of $[\cdot, \cdot]$ with respect to the positive inner product induced by $\langle \cdot, \cdot \rangle_{\widetilde{J}}$, that is,
$$[x, y] = \langle Gx, y\rangle_{\widetilde{J}}, \quad x, y \in \mathcal{H}.$$

This means that, with respect to the orthogonal decompositions of the Hilbert spaces $\widetilde{\mathcal{K}} = \mathcal{H} \oplus \mathcal{L}$ we have $G = P_\mathcal{H} \widetilde{J}|\mathcal{H}$, or, in other words, $\widetilde{J}$ is a lifting of G to a symmetry on $\widetilde{\mathcal{K}}$. By Theorem 2.2.7 it follows that
$$\widetilde{J} = \begin{bmatrix} G & D_G \Gamma \\ \Gamma^* D_G & -\Gamma^* G \Gamma + D_\Gamma \Delta D_\Gamma \end{bmatrix}, \tag{4.3}$$

where $\Gamma\colon \mathcal{L} \to \mathcal{D}_G$ is coisometric, that is, its adjoint $\Gamma^*\colon \mathcal{D}_G \to \mathcal{L}$ is isometric and $\Delta \in \mathcal{B}(\mathcal{D}_\Gamma)$ is a symmetry, that is, unitary and selfadjoint. In particular, $\Gamma^* = \Gamma^{-1}\colon \mathcal{D}_G \to \mathcal{L}$

is a bounded isometry and $D_\Gamma = P_{\operatorname{Ran}(\Gamma^*)}$ is the orthogonal projection onto the subspace $\operatorname{Ran}(\Gamma^*)$. Therefore, representing the symmetry $\widetilde{J}$ with respect to the decomposition

$$\widetilde{\mathcal{K}} = \mathcal{H} \oplus \operatorname{Ran}(\Gamma^*) \oplus \operatorname{Ker}(\Gamma),$$

we have

$$\widetilde{J} = \begin{bmatrix} G & D_G \Gamma & 0 \\ \Gamma^* D_G & -\Gamma^* G \Gamma & 0 \\ 0 & 0 & \Delta \end{bmatrix}. \tag{4.4}$$

With respect to the decompositions $\mathcal{K} = \mathcal{H} \oplus \mathcal{D}_G$ and $\widetilde{\mathcal{K}} = \mathcal{H} \oplus \mathcal{L}$, define the bounded operator $U \colon \mathcal{K} \to \widetilde{\mathcal{K}}$ by

$$U = \begin{bmatrix} I & 0 \\ 0 & \Gamma \end{bmatrix}. \tag{4.5}$$

Then, from (4.2), (4.4), and (4.5) we get

$$U^* \widetilde{J} U = J.$$

Let x, y be arbitrary in $\mathcal{K}$. Then

$$[Ux, Uy]_{\widetilde{\mathcal{K}}} = \langle U^* \widetilde{J} Ux, y \rangle_{\widetilde{J}} = \langle U^* \widetilde{J} Ux, y \rangle_J = \langle Jx, y \rangle_J = [x, y]_{\mathcal{K}},$$

that is, U is isometric. By (4.5) we have $Ux = x$ for all $x \in \mathcal{H}$. This proves the minimality property of the Kreĭn space $\mathcal{K}$.

In order to prove the uniqueness of the Kreĭn space $\mathcal{K}$ as above, let us consider another Kreĭn space $\widetilde{\mathcal{K}}$ with the same minimality property. We then have the representation (4.4). The minimality property of the Kreĭn space $\widetilde{\mathcal{K}}$ implies that the operator Γ as in (4.3) is a coisometric operator and injective, and hence the operator U defined in (4.5) is an isometric isomorphism of Kreĭn spaces. ∎

2.5 Generalised Angular Operators

In the following we generalise the angular operator representation, as developed in Section 2.1 for semidefinite subspaces of a Kreĭn space, to the case of semidefinite subspaces of certain G-spaces. Recall that, according to Theorem 2.4.1, G-spaces are always embeddable as subspaces of Kreĭn spaces and hence, the notion of angular operator that we investigate here is actually a generalisation to the case when the ambient indefinite space is a subspace of a Kreĭn space.

Let $\mathcal{H}$ be a G-space and, with respect to some positive definite inner product $\langle \cdot, \cdot \rangle$ turning $\mathcal{H}$ into a Hilbert space and making the indefinite inner product $[\cdot, \cdot]$ jointly continuous, let $G \in \mathcal{B}(\mathcal{H})$ be the Gram operator, that is, (4.1) holds. Consider $G = G_+ - G_-$ the Jordan decomposition of G, let $\mathcal{H}_+$ denote the spectral subspace of G corresponding to the positive semiaxis $[0, +\infty)$ and $\mathcal{H}_-$ be the spectral subspace of G corresponding to the negative semiaxis $(-\infty, 0)$. Clearly, we have the decomposition

$$\mathcal{H} = \mathcal{H}_+ \oplus \mathcal{H}_- \tag{5.1}$$

and, if $x = x_+ + x_-$ and $y = y_+ + y_-$ are the corresponding representations of arbitrary vectors $x, y \in \mathcal{H}$, then

$$[x, y] = \langle G_+ x_+, x_+ \rangle - \langle G_- x_-, y_- \rangle.$$

In the following, we will use the notions of positivity, negativity, neutrality, etc. with respect to the indefinite inner product $(\mathcal{H}, [\cdot, \cdot])$, and fix the decomposition (5.1).

Let $\mathcal{M}$ be a positive subspace of $\mathcal{H}$, that is, a closed linear manifold such that $[x, x] \geqslant 0$ for all $x \in \mathcal{M}$. With respect to the decomposition (5.1) this means

$$\langle G_+ x_+, x_+ \rangle \geqslant \langle G_- x_-, x_- \rangle, \quad x = x_+ + x_- \in \mathcal{M}. \tag{5.2}$$

As in the case of Kreĭn spaces this enables us to introduce an angular operator. Let $P_\pm$ denote the projection of $\mathcal{H}_\pm$ with respect to the decomposition (5.1). Clearly, $P_\pm$ are orthogonal projections in the Hilbert space $\mathcal{H}$, in particular, their norms are $\leqslant 1$. Let us define an operator $K_\mathcal{M} \colon P_+\mathcal{M} \to \mathcal{H}_-$ by

$$K_\mathcal{M} \colon P_+ x \mapsto P_- x, \quad x \in \mathcal{M}. \tag{5.3}$$

By (5.2) and taking into account that G_- is injective on $\mathcal{H}_-$, this definition is correct and

$$\mathcal{M} = \{x + K_\mathcal{M} x \mid x \in P_+\mathcal{M}\}. \tag{5.4}$$

Since $\mathcal{M}$ is closed, this implies that the operator $K_\mathcal{M}$ is closed. The operator $K_\mathcal{M}$ is called the *generalised angular operator* of the positive subspace $\mathcal{M}$. Note that, in this general setting, there is no reason to conclude that $P_+\mathcal{M}$ is a (closed) subspace and/or that the operator $K_\mathcal{M}$ is bounded. This is remedied if an extra condition is imposed.

Remark 2.5.1. The condition that, in the Jordan decomposition of the Gram operator G, the operator G_- has closed range is equivalent with the condition that the spectrum of G has a gap $(-\varepsilon, 0)$. In particular, this shows that this condition is independent of which admissible positive definite inner product $\langle \cdot, \cdot \rangle$ we consider on the G-space $\mathcal{H}$, since, by changing it with another Gram operator, say B, we have $B = C^*GC$ for some bounded and boundedly invertible $C \colon \mathcal{H} \to \mathcal{H}$ (the inner products on the incoming Hilbert space $\mathcal{H}$ and the outgoing Hilbert space $\mathcal{H}$ are different) and this transformation preserves the topology of the spectrum.

Viewing this from the perspective of Proposition 2.1.9, this means that if $\mathcal{M}$ is a subspace of a Kreĭn space $\mathcal{K}$ and $\mathcal{M} = \mathcal{M}_+[+]\mathcal{M}^0[+]\mathcal{M}_-$ is a fundamental decomposition of $\mathcal{M}$ as in (1.9), if $\mathcal{M}_-$ is uniformly negative then the same holds for any other fundamental decomposition of $\mathcal{M}$. ∎

2.5 GENERALISED ANGULAR OPERATORS

Theorem 2.5.2. *Let $(\mathcal{H}; [\cdot,\cdot])$ be a G-space and, letting G denote the Gram operator of $[\cdot,\cdot]$ with respect to a fixed Hilbert inner product $\langle\cdot,\cdot\rangle$ with Jordan decomposition $G = G_+ - G_-$, assume that the operator G_- has closed range.*

(1) If $\mathcal{M}$ is a positive subspace of $\mathcal{H}$, with notation as in (5.3) and (5.4), then $P_+\mathcal{M}$ is closed, $K_\mathcal{M}$ is bounded and the following inequality holds:

$$K_\mathcal{M}^* G_- K_\mathcal{M} \leqslant P_{P_+\mathcal{M}} G_+ | P_+\mathcal{M}. \tag{5.5}$$

(2) Let $\mathcal{M}$ and $\mathcal{N}$ be positive subspaces of $\mathcal{H}$. Then $\mathcal{M} \subseteq \mathcal{N}$ if and only if $K_\mathcal{M} \subseteq K_\mathcal{N}$, that is, $P_+\mathcal{M} \subseteq P_+\mathcal{N}$ and $K_\mathcal{M} x = K_\mathcal{N} x$ for all $x \in P_+\mathcal{M}$.

(3) For any positive subspace $\mathcal{M}$ of $\mathcal{H}$ there exists a maximal positive subspace $\widetilde{\mathcal{M}}$ such that $\mathcal{M} \subseteq \widetilde{\mathcal{M}}$.

(4) A positive subspace $\mathcal{M}$ of $\mathcal{H}$ is maximal if and only if $P_+\mathcal{M} = \mathcal{H}_+$.

Proof. (1) Let $\mathcal{M}$ be a positive subspace and consider $K_\mathcal{M}$ its generalised angular operator as in (5.3). We first prove that the linear manifold $P_+\mathcal{M}$ is closed.

To this end, let $(x_n)_{n\geqslant 1}$ be a sequence of vectors from $\mathcal{M}$ such that $P_+ x_n \to y$ as $n \to \infty$, for some vector $y \in \mathcal{H}_+$. Then,

$$\|G_+^{1/2} P_+ x_n\|^2 = \langle G_+ P_+ x_n, P_+ x_n \rangle \geqslant \langle G_- P_- x_n, P_- x_n \rangle = \|G_-^{1/2} P_- x_n\|^2.$$

This implies that $(G_-^{1/2} P_- x_n)_{n\geqslant 1}$ is a Cauchy sequence and hence there exists $z \in \mathcal{H}_-$ such that $G_-^{1/2} P_- x_n \to z$ $(n \to \infty)$. Since G_- has closed range, it is invertible on $\mathcal{H}_-$ and the same is its square root $G_-^{1/2}$. Therefore, $P_- x_n \to G_-^{-1/2} z$ $(n \to \infty)$ and hence

$$x_n = P_+ x_n + P_- x_n \to y + G_-^{-1/2} z = x, \quad (n \to \infty).$$

Since the projection P_+ is bounded, this implies that $P_+ x_n \to P_+ x = y$ $(n \to \infty)$ and hence $y \in P_+\mathcal{M}$. Thus $P_+\mathcal{M}$ is closed.

Since the generalised angular operator $K_\mathcal{M}$ is closed and its domain $P_+\mathcal{M}$ is closed, by the Closed Graph Theorem we get that the operator $K_\mathcal{M}$ is bounded. Further, the inequality (5.2) can be written as

$$\langle G_+ x_+, x_+ \rangle \geqslant \langle G_- K_\mathcal{M} x_+, K_\mathcal{M} x_+ \rangle, \quad x_+ \in P_+\mathcal{M}, \tag{5.6}$$

and, since the operator $K_\mathcal{M}$ is bounded, it is equivalent with the inequality (5.5). This equivalence proves also the converse implication.

(2) If $\mathcal{M} \subseteq \mathcal{N}$ are positive subspaces then $P_+\mathcal{M} \subseteq P_+\mathcal{N}$ and the inclusion $K_\mathcal{M} \subseteq K_\mathcal{N}$ comes directly from the definition of the generalised angular operator. Conversely, if $K_\mathcal{M} \subseteq K_\mathcal{N}$ then

$$\mathcal{M} = \{x + K_\mathcal{M} x \mid x \in P_+\mathcal{M}\} \subseteq \{x + K_\mathcal{N} x \mid x \in P_+\mathcal{N}\} = \mathcal{N}.$$

(3) and (4). Let $\mathcal{M}$ be a positive subspace and $K_{\mathcal{M}}$ its angular operator. As in (1), $P_+\mathcal{M}$ is closed, $K_{\mathcal{M}}$ is bounded and the inequality (5.5) holds. We first remark that the inequality (5.5) is equivalent with

$$K_{\mathcal{M}} = G_-^{-1/2} C G_+^{1/2} | P_+\mathcal{M}, \qquad (5.7)$$

for a uniquely determined contraction $C\colon \operatorname{Clos}\operatorname{Ran}(G_+^{1/2} P_+ | \mathcal{M}) \to \mathcal{H}_-$. We then remark that

$$\|C G_+^{1/2} P_+ h\| \leqslant \|G_+^{1/2} P_+ h\|, \quad h \in \mathcal{M}. \qquad (5.8)$$

The inequality (5.8) enables us to define the operator $\widetilde{C}\colon \mathcal{H}_+ \to \mathcal{H}_-$ as follows:

$$\begin{cases} \widetilde{C} G_+^{1/2} P_+ h = C G_+^{1/2} P_+ h, & h \in \mathcal{M}, \\ \widetilde{C} x = 0, & x \in \mathcal{H}_+ \ominus (G_+^{1/2} P_+\mathcal{M}). \end{cases}$$

Again by (5.8), the operator $\widetilde{C}$ is contractive. Define now the operator $\widetilde{K}\colon \mathcal{H}_+ \to \mathcal{H}_-$ by

$$\widetilde{K} = G_-^{-1/2} \widetilde{C} G_+^{1/2}.$$

Then the subspace $\widetilde{\mathcal{M}} = \{x + \widetilde{K} x \mid x \in \mathcal{H}_+\}$ is maximal positive. From the definition of the operator $\widetilde{K}$ it follows that $K_{\mathcal{M}} \subseteq \widetilde{K}$ and hence $\mathcal{M} \subseteq \widetilde{\mathcal{M}}$. ∎

Corollary 2.5.3. *If the operator G_- has closed range then the set*

$$\mathcal{X} = \{K_{\mathcal{M}} \mid \mathcal{M} \text{ maximal positive subspace}\}$$

is convex and compact with respect to the weak operator topology on $\mathcal{B}(\mathcal{H}_+, \mathcal{H}_-)$.

Proof. It follows from the proof of Theorem 2.5.2 (see (5.7)) that

$$\mathcal{X} = \{G_-^{-1/2} C G_+^{1/2} \mid C\colon \mathcal{H}_+ \ominus \operatorname{Ker}(G) \to \mathcal{H}_-,\ \|C\| \leqslant 1\}.$$

This shows that $\mathcal{X}$ is convex and, in view of the Alaoglu Theorem, it is also compact with respect to the weak operator topology on $\mathcal{B}(\mathcal{H}_+, \mathcal{H}_-)$. ∎

Clearly, similar assertions to those in Theorem 2.5.2 and Corollary 2.5.3 can be stated if one replaces the closed range condition of G_- with the closed range condition of G_+ and, instead of positive subspaces, one considers negative subspaces. The details are left to the reader.

2.6 Notes

The idea of using angular operators for semidefinite subspaces in Kreĭn spaces appears, in a slightly different formulation, in the seminal paper of R. S. Phillips [120], and explicitly in Yu. P. Ginzburg [66]. The existence part of Theorem 2.2.9 was proven also in [120] by an adaptation of an idea of M. G. Kreĭn from [92]. The material in Section 2.2 that leads to an explicit description of maximal extensions of alternating pairs of subspaces follows the approach of Gr. Arsene and the author in [8]. Example 2.2.11 was communicated to us by H. Langer.

The material in Section 2.3 refers to the M. G. Kreĭn theory of positive selfadjoint extensions of positive operators in [92] and Phillips' theory of maximal dissipative extensions in [118, 119, 120], but our approach follows the explicit description in [8]. In the present formulation, most of the proofs from this section follow an unpublished manuscript of T. Constantinescu and the author [28].

The concept of G-spaces was considered by T. Ya. Azizov and I. S. Iokhvidov [10], but their terminology is slightly different.

Chapter 3

Subspaces of Kreĭn Spaces

The geometry of subspaces of Kreĭn spaces is much more involved than that of Hilbert spaces. We first discuss the geometry of regular subspaces, which give the right idea for Kreĭn subspaces, and of their generalisations, pseudo-regular subspaces. At the other extreme there are the neutral subspaces on which the inner product completely vanishes. The strong duality of subspaces is the notion that unifies both these extreme types of subspaces.

In a genuine Kreĭn space, that is, the inner product is indefinite, subspaces that contain only finite dimensional uniformly definite subspaces give rise to a geometric Fredholm theory and the possibility to define an index. As a consequence, an index formula relating the ranks of negativity/positivity of a subspace and its orthogonal companion can be obtained. We apply this formula to linear relations in Kreĭn spaces and their scattering transform.

3.1 Regular Subspaces

A special role in the geometry of Kreĭn spaces is played by those subspaces $\mathcal{L}$ of a given Kreĭn space $\mathcal{K}$ which become Kreĭn spaces under the induced inner product and strong topology. It is clear that such a subspace must be nondegenerate but, it turns out that, for the case of genuine Kreĭn spaces, nondegeneracy is not sufficient. In this section we present several characterisations of these subspaces and develop their geometry.

Recall that a subspace of a Kreĭn space is a closed linear manifold. For two subspaces $\mathcal{L}$ and $\mathcal{M}$ we use the notation $\mathcal{L}[+]\mathcal{M}$ only when $\mathcal{L} \perp \mathcal{M}$, $\mathcal{L} \cap \mathcal{M} = \{0\}$, and the algebraic sum $\mathcal{L} + \mathcal{M}$ is closed. Also, the notation $\mathcal{L} \oplus \mathcal{M}$, the Hilbert space orthogonal sum, is used only with respect to a specified Hilbertian inner product.

Theorem 3.1.1. *Let $\mathcal{L}$ be a subspace of the Kreĭn space $\mathcal{K}$. The following statements are equivalent.*

(i) $\mathcal{L} = \mathcal{L}_+[+]\mathcal{L}_-$, *where $\mathcal{L}_+$ is a uniformly positive subspace and $\mathcal{L}_-$ is a uniformly negative subspace.*

(ii) $\mathcal{K} = \mathcal{L}[+]\mathcal{L}^\perp$.

(iii) *There exists a fundamental norm $\|\cdot\|$ on $\mathcal{K}$ such that*

$$\|x\| = \sup_{\|y\|\leqslant 1, y\in\mathcal{L}} |[x,y]|, \quad x \in \mathcal{L}.$$

3.1 REGULAR SUBSPACES

(iv) *There exists a fundamental symmetry J on $\mathcal{K}$ such that $J\mathcal{L} \subseteq \mathcal{L}$ (equivalently, $J\mathcal{L} = \mathcal{L}$).*

Proof. (i)$\Rightarrow$(iv). According to Theorem 2.2.9 there exist a maximal uniformly positive subspace $\mathcal{K}_+ \supseteq \mathcal{L}_+$ and a maximal uniformly negative subspace $\mathcal{K}_- \supseteq \mathcal{L}_-$ such that $\mathcal{K}_+ \perp \mathcal{K}_-$. By Lemma 2.1.8 we conclude that $\mathcal{K} = \mathcal{K}_+[+]\mathcal{K}_+$ is a fundamental decomposition and letting J denote the corresponding fundamental symmetry, $J\mathcal{L} = \mathcal{L}$ follows.

(iv)$\Rightarrow$(iii). Let J be a fundamental symmetry of $\mathcal{K}$ such that $J\mathcal{L} \subseteq \mathcal{L}$. Then $J\mathcal{L} = \mathcal{L}$ holds (since $J^2 = I$). If $\|\cdot\|$ denotes the fundamental norm determined by J we also have

$$\|Jy\| = \|y\|, \quad y \in \mathcal{K}.$$

For any $x \in \mathcal{L}$ we have

$$\|x\| = \sup_{\|y\|\leqslant 1,\, y\in\mathcal{L}} |\langle x, y\rangle_J| = \sup_{\|y\|\leqslant 1,\, y\in\mathcal{L}} |[x, Jy]| = \sup_{\|y\|\leqslant 1,\, y\in\mathcal{L}} |[x, y]|.$$

(iii)$\Rightarrow$(i). The statement (iii) means that the inner product space $(\mathcal{L}, [\cdot,\cdot])$ is a Kreĭn space. Therefore, its strong topology coincides with the induced strong topology of $\mathcal{K}$ (cf. Theorem 1.4.1). Thus, it is sufficient to consider a fundamental decomposition $\mathcal{L} = \mathcal{L}^+[+]\mathcal{L}^-$ of the Kreĭn space $(\mathcal{L}, [\cdot,\cdot])$.

(iv)$\Rightarrow$(ii). Let J be a fundamental symmetry of $\mathcal{K}$ such that $J\mathcal{L} = \mathcal{L}$. By (4.4) it follows that $\mathcal{L}^\perp = \mathcal{L}^{\perp J}$ and hence $\mathcal{L} + \mathcal{L}^\perp = \mathcal{L} \oplus \mathcal{L}^{\perp J} = \mathcal{K}$.

(ii)$\Rightarrow$(i). Assuming $\mathcal{K} = \mathcal{L}[+]\mathcal{L}^\perp$ it follows (see (4.7) in Remark 1.4.9.(c)) that $\mathcal{L}$ and $\mathcal{L}^\perp$ are nondegenerate subspaces. Considering two fundamental decompositions, in the sense of (1.9), $\mathcal{L} = \mathcal{L}_+[+]\mathcal{L}_-$ and $\mathcal{L}^\perp = \mathcal{L}'_+[+]\mathcal{L}'_-$, it follows that

$$\mathcal{K} = (\mathcal{L}_+[+]\mathcal{L}'_+)[+](\mathcal{L}_-[+]\mathcal{L}'_-).$$

Clearly, $\mathcal{L}_+[+]\mathcal{L}'_+$ is a positive subspace and $\mathcal{L}_-[+]\mathcal{L}'_-$ is a negative subspace. From Corollary 2.1.6 it follows that $\mathcal{L}_+[+]\mathcal{L}'_+$ is a maximal uniformly positive subspace and $\mathcal{L}_-[+]\mathcal{L}'_-$ is a maximal uniformly negative subspace and hence, $\mathcal{L}_+$ and $\mathcal{L}_-$ are uniformly definite subspaces. ∎

A subspace $\mathcal{L}$ of the Kreĭn space $\mathcal{K}$ which satisfies one (hence all) of the conditions (i)–(iv) in Theorem 3.1.1 is called *regular*. Regular subspaces are also called *Kreĭn subspaces* (compare Theorem 3.1.1 with Theorem 1.4.1).

Remark 3.1.2. We formulate some immediate consequences of Theorem 3.1.1.

(a) A positive (negative) subspace is regular if and only if it is uniformly positive (uniformly negative).

(b) A subspace is regular if and only if its orthogonal companion is regular.

(c) Let $\mathcal{L}$ be a regular subspace of $\mathcal{K}$ and let $\mathcal{L} = \mathcal{L}_+[+]\mathcal{L}_-$ be a fundamental decomposition of $\mathcal{L}$ in the sense of (1.9). Then $\mathcal{L}_+$ is uniformly positive and $\mathcal{L}_-$ is uniformly negative. ∎

Proposition 3.1.3. *Let $\mathcal{L}$ and $\mathcal{M}$ be two orthogonal subspaces.*
(a) *If either $\mathcal{L}$ or $\mathcal{M}$ is regular then the linear manifold $\mathcal{L} + \mathcal{M}$ is closed.*
(b) *Both of $\mathcal{L}$ and $\mathcal{M}$ are regular if and only if $\mathcal{L}[+]\mathcal{M}$ is regular.*

Proof. (a) If $\mathcal{L}$ is regular let J be a fundamental symmetry such that $J\mathcal{L} = \mathcal{L}$, hence $\mathcal{L}^\perp = \mathcal{L}^{\perp_J}$. Since $\mathcal{M} \subseteq \mathcal{L}^{\perp_J}$ it follows that $\mathcal{L} + \mathcal{M} = \mathcal{L} \oplus \mathcal{M}$.

(b) If both $\mathcal{L}$ and $\mathcal{M}$ are regular, let J be a fundamental symmetry of $\mathcal{K}$ such that $J\mathcal{L} = \mathcal{L}$ and hence $J\mathcal{L}^\perp = \mathcal{L}^\perp$. Since $\mathcal{M} \subseteq \mathcal{L}^\perp$ holds, modifying the fundamental symmetry $J|\mathcal{L}^\perp$ of the Kreĭn space $\mathcal{L}^\perp$, we can assume $J\mathcal{M} = \mathcal{M}$. Then $J(\mathcal{L} + \mathcal{M}) = \mathcal{L} + \mathcal{M}$ and hence $\mathcal{L} + \mathcal{M} = \mathcal{L}[+]\mathcal{M}$ is a regular subspace.

Conversely, if the subspace $\mathcal{L}[+]\mathcal{M}$ is regular then, without restricting the generality, we can assume that $\mathcal{L}[+]\mathcal{M} = \mathcal{K}$. Then $\mathcal{M} = \mathcal{L}^\perp$ follows and hence $\mathcal{L}$ and $\mathcal{M}$ are regular. ∎

Corollary 3.1.4. *If $(\mathcal{L}_i)_{i=1}^n$ is a finite family of mutually orthogonal regular subspaces then the subspace $\mathcal{L}_1[+]\mathcal{L}_2[+]\cdots[+]\mathcal{L}_n$ is regular.*

Two subspaces $\mathcal{L}$ and $\mathcal{M}$ of (possibly different) Kreĭn spaces are called *isometric isomorphic* if there exists a linear operator $V\colon \mathcal{L} \to \mathcal{M}$ which is bounded and boundedly invertible such that $[Vx, Vy] = [x, y]$, $x, y \in \mathcal{L}$.

Proposition 3.1.5. *Let $\mathcal{R}$ be a regular subspace. If the subspace $\mathcal{L}$ is isometric isomorphic with $\mathcal{R}$ then $\mathcal{L}$ is regular as well.*

Proof. Let $V\colon \mathcal{R} \to \mathcal{L}$ be an isometric isomorphism and let $\mathcal{R} = \mathcal{R}^+[+]\mathcal{R}^-$ be a fundamental decomposition, hence $\mathcal{R}^+$ is uniformly positive and $\mathcal{R}^-$ is uniformly negative. Let $\|\cdot\|$ denote a fundamental norm and $\alpha > 0$ be such that

$$[x, x] \geqslant \alpha \|x\|^2, \quad x \in \mathcal{R}^+.$$

Then

$$[Vx, Vx] = [x, x] \geqslant \alpha \|x\|^2 \geqslant \frac{\alpha}{\|V\|^2} \|Vx\|^2, \quad x \in \mathcal{R}^+,$$

that is, $V\mathcal{R}^+$ is uniformly positive. Similarly one proves that $V\mathcal{R}^-$ is uniformly negative. Then, $\mathcal{L} = V\mathcal{R}^+ + V\mathcal{R}^-$, $V\mathcal{R}^+ \perp V\mathcal{R}^-$ and hence, by Theorem 3.1.1, we obtain that $\mathcal{L}$ is regular. ∎

Corollary 3.1.6. *Two regular subspaces $\mathcal{L}$ and $\mathcal{M}$ are isometric isomorphic if and only if $\kappa^\pm(\mathcal{L}) = \kappa^\pm(\mathcal{M})$.*

Proof. If the regular subspaces $\mathcal{L}$ and $\mathcal{M}$ are isometric isomorphic then $\kappa^\pm(\mathcal{L}) = \kappa^\pm(\mathcal{M})$ follows from the proof of Proposition 3.1.5. Conversely, if $\kappa^\pm(\mathcal{L}) = \kappa^\pm(\mathcal{M})$ holds then we consider two fundamental decompositions $\mathcal{L} = \mathcal{L}^+[+]\mathcal{L}^-$ and $\mathcal{M} = \mathcal{M}^+[+]\mathcal{M}^-$ and there exist unitary operators (in the Hilbert space sense) $V_\pm \colon \mathcal{L}^\pm \to \mathcal{M}^\pm$. Define $V\colon \mathcal{L} \to \mathcal{M}$ as the direct sum of V_+ and V_- and then V is an isometric isomorphism. ∎

3.1 REGULAR SUBSPACES

Since any subspace of a Kreĭn space gives rise to Gram operators, see Remark 1.3.3, it is natural to ask whether regularity can be characterised in terms of these Gram operators. The proof of the next proposition is a straightforward consequence of Remark 1.3.3 and Theorem 3.1.1, and we omit it.

Proposition 3.1.7. *Let $\mathcal{M}$ be a subspace of a Kreĭn space $\mathcal{K}$. The following assertions are equivalent.*

(i) $\mathcal{M}$ *is a regular subspace of* $\mathcal{K}$.

(ii) *For any fundamental symmetry J, the Gram operator G of the inner product space $(\mathcal{M};[\cdot,\cdot])$ with respect to the Hilbertian inner product $\langle\cdot,\cdot\rangle_J$ is boundedly invertible.*

(ii)' *For some fundamental symmetry J, the Gram operator G of the inner product space $(\mathcal{M};[\cdot,\cdot])$ with respect to the Hilbertian inner product $\langle\cdot,\cdot\rangle_J$ is boundedly invertible.*

In Pontryagin spaces some problems concerning the geometry of regular subspaces simplify considerably, as the following proposition shows.

Proposition 3.1.8. *Let $\mathcal{K}$ be a Kreĭn space. The following assertions are equivalent.*

(i) $\mathcal{K}$ *is a Pontryagin space.*
(ii) *Every definite subspace of $\mathcal{K}$ is uniformly definite.*
(iii) *Every nondegenerate subspace of $\mathcal{K}$ is regular.*

Proof. (i)⇒(ii). Let $\mathcal{L}$ be a positive subspace of $\mathcal{K}$ and let $\mathcal{K} = \mathcal{K}^-[+]\mathcal{K}^+$ be a fundamental decomposition with $\|\cdot\|$ the corresponding fundamental norm. If K denotes the angular operator of $\mathcal{L}$ then K is a strict contraction, that is, $\|Kx\| < \|x\|$, $x \in J^+\mathcal{L}$. Since either $\mathcal{K}^+$ or $\mathcal{K}^-$ is finite dimensional it follows that K has finite rank and hence $\|K\| < 1$, that is, $\mathcal{L}$ is uniformly positive (cf. Lemma 2.1.2).

(ii)⇒(i). Assume that $\mathcal{K}$ is not a Pontryagin space. Without restricting the generality we can also assume that $\mathcal{K}$ is separable. Let $\mathcal{K} = \mathcal{K}^+[+]\mathcal{K}^-$ be a fundamental decomposition and, since both $\mathcal{K}^+$ and $\mathcal{K}^-$ are separable and infinite dimensional Hilbert spaces, there exist $(e_j)_{j\geq 1}$ and $(f_j)_{j\geq 1}$ orthonormal bases of $\mathcal{K}^+$ and of $\mathcal{K}^-$, respectively. Let $(\alpha_j)_{j\geq 1}$ be a numerical sequence with the following properties:

$$|\alpha_j| < 1, \quad j \in \mathbb{N}; \quad \lim_{j\to\infty} \alpha_j = \alpha, \quad |\alpha| = 1.$$

Define a linear operator $K \in \mathcal{B}(\mathcal{K}^+,\mathcal{K}^-)$ by

$$Ke_j = \alpha_j f_j, \quad j \in \mathbb{N}.$$

Then K is a strict contraction but $\|K\| = 1$. By Lemma 2.1.2 the subspace $\mathcal{L} = \{x + Kx \mid x \in \mathcal{K}^+\}$ is positive but not uniformly positive.

(ii)⇒(iii). A nondegenerate subspace $\mathcal{L}$ can always be represented as $\mathcal{L} = \mathcal{L}^+[+]\mathcal{L}^-$ with $\mathcal{L}^\pm$ positive/negative (cf. Remark 1.3.3). Then $\mathcal{L}^\pm$ is uniformly positive/negative and hence $\mathcal{L}$ is regular, by definition. ∎

We conclude this section with examples which emphasise some of the difficulties that the indefiniteness brings, even in the case of a Pontryagin space. The first example shows that an indefinite and infinite dimensional Pontryagin space always contains a nondegenerate linear manifold whose closure is degenerate.

Example 3.1.9. Let $\mathcal{H}$ be a separable Pontryagin space with $0 \neq \kappa^-(\mathcal{H}) < \infty$. If $\mathcal{H} = \mathcal{H}^+[+]\mathcal{H}^-$ is a fundamental decomposition let $(e_j)_{j=1}^\infty$ be an orthogonal basis of $\mathcal{H}^+$ and let $f \in \mathcal{H}^-$ with $[f, f] = -1$. Consider the linear manifold

$$\mathcal{L} = \{x \in \mathcal{H} \mid x = \sum_{j=1}^\infty \alpha_j e_j + \beta f, \ \mathrm{supp}(\alpha_j)_{j=1}^\infty < \infty, \ \alpha_1 = \sum_{j=2}^\infty \alpha_j = \beta\}, \quad (1.1)$$

and observe that $\mathcal{L}$ is strictly positive and hence its closure is a positive subspace of $\mathcal{H}$. The sequence of vectors in $\mathcal{L}$ defined by

$$x_n = e_1 + \frac{1}{n-1}\sum_{j=2}^n e_j + f, \quad n \in \mathbb{N},$$

converges to $e_1 + f$, which is a nonzero neutral vector. Applying the Schwarz Inequality it follows that $e_1 + f \in \mathrm{Clos}(\mathcal{L})^0$ and hence $\mathrm{Clos}(\mathcal{L})$ is a degenerate subspace. ∎

The next example shows that an infinite, mutually orthogonal family of regular subspaces can span a degenerate subspace.

Example 3.1.10. Let $\mathcal{H}$ and $\mathcal{L}$ be as in Example 3.1.9. $\mathcal{L}$ is separable and by the Gram–Schmidt procedure there exists a complete orthonormal system $(f_j)_{j=1}^\infty$ of vectors in $\mathcal{L}$. Denote $\mathcal{L}_j = \mathbb{C}f_j$. Then the one dimensional subspaces $\mathcal{L}_j$ are uniformly positive and $\mathcal{L}_j \perp \mathcal{L}_k$, $j \neq k$. The closed span of $(\mathcal{L}_j)_{j=1}^\infty$ is $\mathrm{Clos}(\mathcal{L})$ which is a degenerate subspace. ∎

3.2 Pseudo-Regular Subspaces

A subspace $\mathcal{M}$ of the Kreĭn space $\mathcal{K}$ is called *pseudo-regular* if the linear manifold $\mathcal{M} + \mathcal{M}^\perp$ is closed. It follows, for example by Theorem 3.1.1, that any regular subspace is pseudo-regular.

Theorem 3.2.1. *Let $\mathcal{M}$ be a subspace of $\mathcal{K}$. The following conditions are equivalent.*
  (i) *$\mathcal{M}$ is pseudo-regular.*
  (ii) *$\mathcal{M} + \mathcal{M}^\perp = \mathcal{M}^{0\perp}$.*
  (iii) *There exists a regular subspace $\mathcal{R}$ such that $\mathcal{M} = \mathcal{R}[+]\mathcal{M}^0$.*
  (iv) *If $\mathcal{L}$ is a nondegenerate subspace of $\mathcal{M}$ such that $\mathcal{M} = \mathcal{L}[+]\mathcal{M}^0$, then $\mathcal{L}$ is regular.*
  (v) *There exists a regular subspace $\mathcal{S} \supseteq \mathcal{M}$ such that $\mathcal{S} \cap \mathcal{M}^\perp = \mathcal{M}^0$.*
  Moreover, the subspaces $\mathcal{R}$ from (iii) and $\mathcal{S}$ from (v) are uniquely determined by $\mathcal{M}$ up to isometric isomorphisms.

3.2 PSEUDO-REGULAR SUBSPACES

Proof. (iii)⇒(ii). This equivalence follows from the relation $\mathcal{M}^{0\perp} = \mathrm{Clos}(\mathcal{M} + \mathcal{M}^\perp)$, see Remark 1.4.9(b).

(iii)⇒(iv). Let $\mathcal{L}$ and $\mathcal{R}$ be two subspaces such that $\mathcal{M} = \mathcal{L}[+]\mathcal{M}^0 = \mathcal{R}[+]\mathcal{M}^0$, with $\mathcal{R}$ a regular subspace. Then $\mathcal{L}$ is nondegenerate and hence the equality

$$x + x_0 = y + y_0, \quad x \in \mathcal{R}, \ y \in \mathcal{L}, \ x_0, y_0 \in \mathcal{M}^0,$$

defines a linear operator $V: \mathcal{R} \ni x \mapsto y \in \mathcal{L}$ and, after a moment of thought, we see that V is bijective and isometric. We prove subsequently that V is bounded.

Let us consider a sequence $(x_n)_{n \in \mathbb{N}}$ of vectors in $\mathcal{R}$ such that

$$\lim_{n \to \infty} x_n = x, \quad \lim_{n \to \infty} V x_n = y.$$

Then

$$\lim_{n \to \infty} [V x_n, z] = [y, z], \quad z \in \mathcal{L}.$$

On the other hand, since $V^{-1}: \mathcal{L} \to \mathcal{R}$ is isometric as well, we have

$$\lim_{n \to \infty} [V x_n, z] = \lim_{n \to \infty} [V^{-1} V x_n, V^{-1} z]$$
$$= \lim_{n \to \infty} [x_n, V^{-1} z] = [V x, V V^{-1} z] = [V x, z], \quad z \in \mathcal{L},$$

and hence

$$[V x - y, z] = 0, \quad z \in \mathcal{L}.$$

Since $\mathcal{L}$ is nondegenerate we get $V x = y$. We have thus proven that V is closed and then by the Closed Graph Principle it follows that V is bounded. From Proposition 3.1.5 it follows that $\mathcal{L}$ is regular.

(iv)⇒(iii). Let J be a fundamental symmetry of $\mathcal{K}$. Letting $\mathcal{R} = \mathcal{M} \ominus \mathcal{M}^0$ it follows that $\mathcal{M} = \mathcal{R}[+]\mathcal{M}^0$.

(ii)⇒(iii). We first show that we can additionally assume $\mathcal{M}^\perp = \mathcal{M}^0$. Indeed, if this is not the case, replace $\mathcal{M}$ by $\mathcal{M} + \mathcal{M}^\perp$ and notice that $(\mathcal{M} + \mathcal{M}^\perp)^\perp = (\mathcal{M} + \mathcal{M}^\perp)^0 = \mathcal{M}^0$. Further, if $\mathcal{N}$ is a regular subspace such that $\mathcal{N}[+]\mathcal{M}^0 = \mathcal{M} + \mathcal{M}^\perp$ and J is a fundamental symmetry, the above proved implication (iii)⇒(iv) shows that, without restricting the generality, we can assume $\mathcal{N} = (\mathcal{M} + \mathcal{M}^\perp) \ominus \mathcal{M}^0$. Then $\mathcal{N} = (\mathcal{M} \ominus \mathcal{M}^0)[+](\mathcal{M}^\perp \ominus \mathcal{M}^0)$ and by Proposition 3.1.3 it follows that $\mathcal{N} \ominus \mathcal{M}^0$ is regular.

So, let us suppose that $\mathcal{M}^\perp = \mathcal{M}^0$ holds and consider a fundamental symmetry J of $\mathcal{K}$ and $\mathcal{R} = \mathcal{M} \ominus \mathcal{M}^0$. We claim that $J\mathcal{R} \subseteq \mathcal{M}$. Indeed, $J\mathcal{R} = J(\mathcal{R}^{\perp_J})^{\perp_J} = (\mathcal{R}^{\perp_J})^\perp$ and hence $J\mathcal{R} \subseteq \mathcal{M}$ if and only if $\mathcal{M}^\perp \subseteq \mathcal{R}^{\perp_J}$, the last inclusion being true since $\mathcal{M}^\perp = \mathcal{M}^0 \perp_J \mathcal{R}$. Now, for arbitrary $x \in \mathcal{R}$ we have

$$\|x\| = \|Jx\| = \sup\{|\langle Jx, y \rangle_J| \mid \|y\| \leq 1, \ y \in \mathcal{M}\}$$
$$= \sup\{|[x, y]| \mid \|y\| \leq 1, \ y \in \mathcal{M}\}$$
$$\leq \sup\{|[x, y + y_0]| \mid \|y\|, \|y_0\| \leq 1, \ y \in \mathcal{R}, \ y_0 \in \mathcal{M}_0\}$$
$$= \sup\{|[x, y]| \mid \|y\| \leq 1, \ y \in \mathcal{R}\} \leq \|x\|,$$

where $\|\cdot\|$ denotes the fundamental norm associated to J. By Theorem 3.1.1 this shows that $\mathcal{R}$ is regular.

(iii)$\Rightarrow$(v). Represent $\mathcal{M} = \mathcal{M}_+[+]\mathcal{M}^0[+]\mathcal{M}_-$, where $\mathcal{M}_+$ is uniformly positive and $\mathcal{M}_-$ is uniformly negative. Let $\mathcal{K} = \mathcal{K}^+[+]\mathcal{K}^-$ be a fundamental decomposition such that $\mathcal{M}_\pm \subseteq \mathcal{K}^\pm$ (see Theorem 2.2.9). If J denotes the corresponding fundamental symmetry and $J = J^+ - J^-$ is its Jordan decomposition then $\mathcal{M}_\pm \perp_J J^\pm \mathcal{M}^0$ and the subspace $\mathcal{S} = (\mathcal{M}_+ \oplus J^+ \mathcal{M}^0)[+](\mathcal{M}_- \oplus J^- \mathcal{M}^0)$ is regular, contains $\mathcal{M}$, and $\mathcal{S} \cap \mathcal{M}^\perp = \mathcal{M}^0$.

In order to prove the uniqueness of the subspace $\mathcal{S}$ let $\mathcal{S}_1$ be another regular subspace such that $\mathcal{S}_1 \cap \mathcal{M}^\perp = \mathcal{M}^0$. Then

$$\mathcal{M}^{0\perp} = \mathrm{Clos}(\mathcal{M} + \mathcal{S}_1) = \mathcal{M}[+]\mathcal{S}_1^\perp,$$

and hence

$$\mathcal{M}^0 \cap \mathcal{S}_1 = (\mathcal{M}[+]\mathcal{S}_1) \cap \mathcal{S}_1 = \mathcal{M}.$$

Considering the regular subspace $\mathcal{P} = \mathcal{S}_1 \cap (\mathcal{M}_+[+]\mathcal{M}_-)^\perp$ we get

$$\mathcal{M}^0 \cap \mathcal{P} = \mathcal{M}^{0\perp} \cap \mathcal{S}_1 \cap (\mathcal{M}_+[+]\mathcal{M}_-)^\perp = \mathcal{M}^0.$$

Let G be a fundamental symmetry of the Kreĭn space $\mathcal{P}$, $\mathcal{P} = \mathcal{P}_+[+]\mathcal{P}_-$ be the associated fundamental decomposition and $G = G_+ - G_-$ be the Jordan decomposition of G. Then $G_+|\mathcal{M}^0\colon \mathcal{M}^0 \to G_+\mathcal{M}^0 = \mathcal{P}_+$ and $G_+|\mathcal{M}^0\colon \mathcal{M}^0 \to G_+\mathcal{M}^0 = \mathcal{P}_+$ are unitary operators with respect to the inner product $\langle \cdot, \cdot \rangle_G$. Define a linear operator $U\colon \mathcal{S} \to \mathcal{S}_1$ by

$$U\colon x + J_+ x_0 + J_- y_0 \mapsto x + G_+ x_0 + G_- y_0, \quad x \in \mathcal{M}_+[+]\mathcal{M}_-, \; x_0, y_0 \in \mathcal{M}^0.$$

The operator U is the desired isometric isomorphism from $\mathcal{S}$ to $\mathcal{S}_1$.

(v)$\Rightarrow$(i). Let $\mathcal{S}$ be a regular subspace of $\mathcal{K}$ such that $\mathcal{S} \supseteq \mathcal{M}$ and $\mathcal{S} \cap \mathcal{M}^\perp = \mathcal{M}^0$. Then $\mathcal{S} \cap \mathcal{M}^{0\perp} = \mathcal{M}$ which is equivalent to $\mathcal{M}^\perp = \mathcal{S}^\perp[+]\mathcal{M}^0$ and hence $\mathcal{M} + \mathcal{M}^\perp = \mathcal{M} + (\mathcal{M}^0[+]\mathcal{S}^\perp) = \mathcal{M}[+]\mathcal{S}^\perp$. ∎

Remark 3.2.2. Let $\mathcal{M}$ be a subspace of the Kreĭn space $\mathcal{K}$ and let $\|\cdot\|$ be a fundamental norm. On the inner product space $(\mathcal{M}/\mathcal{M}^0, [\cdot, \cdot])$ (see Remark 1.1.5) we consider the quotient norm, also denoted by $\|\cdot\|$. Then $\mathcal{M}$ is a pseudo-regular subspace if and only if

$$\|x + \mathcal{M}^0\| = \sup_{\substack{\|y + \mathcal{M}^0\| \leq 1 \\ y \in \mathcal{M}}} |[x + \mathcal{M}^0, y + \mathcal{M}^0]|, \quad x \in \mathcal{M},$$

that is, $(\mathcal{M}/\mathcal{M}^0, [\cdot, \cdot])$ is a Kreĭn space with the induced strong topology. ∎

Since any subspace of a Kreĭn space gives rise to Gram operators, see Remark 1.3.3, it is natural to ask whether pseudo-regularity can be characterised in terms of these Gram operators.

3.2 Pseudo-Regular Subspaces

Proposition 3.2.3. *Let $\mathcal{M}$ be a subspace of a Kreĭn space $\mathcal{K}$. The following assertions are equivalent.*

(i) *$\mathcal{M}$ is a pseudo-regular subspace of $\mathcal{K}$.*
(ii) *For any fundamental symmetry J, the Gram operator G of the inner product space $(\mathcal{M};[\cdot,\cdot])$, with respect to the Hilbertian inner product $\langle\cdot,\cdot\rangle_J$, has closed range.*
(ii)′ *For some fundamental symmetry J, the Gram operator G of the inner product space $(\mathcal{M};[\cdot,\cdot])$, with respect to the Hilbertian inner product $\langle\cdot,\cdot\rangle_J$, has closed range.*
(iii) *For any fundamental symmetry J, if the number 0 is in the spectrum of the Gram operator G of the inner product space $(\mathcal{M};[\cdot,\cdot])$ with respect to the Hilbertian inner product $\langle\cdot,\cdot\rangle_J$, then it is an isolated point.*
(iii)′ *For some fundamental symmetry J, if the number 0 is in the spectrum of the Gram operator G of the inner product space $(\mathcal{M};[\cdot,\cdot])$ with respect to the Hilbertian inner product $\langle\cdot,\cdot\rangle_J$, then it is an isolated point.*

Proof. In view of Remark 3.2.2, this proposition is a direct consequence of Theorem 3.2.1 and Proposition 3.1.7. ∎

Corollary 3.2.4. *Let $\mathcal{M}$ and $\mathcal{N}$ be pseudo-regular subspaces such that $\mathcal{M} \perp \mathcal{N}$. Then $\mathrm{Clos}(\mathcal{M}+\mathcal{N})$ is also a pseudo-regular subspace.*

Proof. We represent $\mathcal{M} = \mathcal{R}[+]\mathcal{M}^0$ and $\mathcal{N} = \mathcal{P}[+]\mathcal{N}^0$, with $\mathcal{P}$ and $\mathcal{R}$ regular. Then $\mathcal{M} \perp \mathcal{N}$ yields $\mathrm{Clos}(\mathcal{M}^0 + \mathcal{N}^0) = \mathrm{Clos}(\mathcal{M}+\mathcal{N})^0$ and $\mathcal{R} \perp \mathcal{P}$, hence

$$\mathrm{Clos}(\mathcal{M}+\mathcal{N}) = \mathcal{R}[+]\mathcal{P}[+]\big(\mathrm{Clos}(\mathcal{M}^0+\mathcal{N}^0)\big) = \mathcal{R}[+]\mathcal{P}[+]\mathrm{Clos}\big((\mathcal{M}+\mathcal{N})^0\big),$$

where $\mathcal{R}[+]\mathcal{P}$ is a regular subspace, by Proposition 3.1.3. ∎

Corollary 3.2.5. *Let $\mathcal{M}$ and $\mathcal{N}$ be subspaces such that the subspace $\mathcal{M}[+]\mathcal{N}$ is pseudo-regular. Then both of $\mathcal{M}$ and $\mathcal{N}$ are pseudo-regular.*

Proof. Let $\mathcal{P}$ and $\mathcal{R}$ be nondegenerate subspaces such that $\mathcal{M} = \mathcal{P}[+]\mathcal{M}^0$ and $\mathcal{N} = \mathcal{R}[+]\mathcal{N}^0$. Then

$$\mathcal{M}[+]\mathcal{N} = (\mathcal{P}[+]\mathcal{R})[+](\mathcal{M}^0[+]\mathcal{N}^0) = (\mathcal{P}[+]\mathcal{R})[+](\mathcal{M}[+]\mathcal{N})^0,$$

and hence $\mathcal{P}[+]\mathcal{R}$ is regular. Then we take into account Proposition 3.1.3 and conclude that both of $\mathcal{P}$ and $\mathcal{R}$ are regular. ∎

Corollary 3.2.6. *If $\mathcal{M}$ is a pseudo-regular subspace and $\mathcal{N}$ is a subspace of $\mathcal{M}$ then*

$$\mathcal{M} \cap (\mathcal{M} \cap \mathcal{N}^\perp)^\perp = \mathrm{Clos}(\mathcal{N} + \mathcal{M}^0).$$

Proof. There exists a regular subspace $\mathcal{S} \supseteq \mathcal{M}$ such that $\mathcal{S} \cap \mathcal{M}^\perp = \mathcal{M}$ and hence, without restricting the generality, we can assume $\mathcal{M}^\perp = \mathcal{M}^0$. Then $\mathcal{M} \cap (\mathcal{M} \cap \mathcal{N}^\perp)^\perp = \mathcal{M} \cap \mathrm{Clos}(\mathcal{M}^\perp + \mathcal{N}) = \mathcal{M} \cap \mathrm{Clos}(\mathcal{M}^0 + \mathcal{N}) = \mathrm{Clos}(\mathcal{M}^0 + \mathcal{N})$. ∎

Corollary 3.2.7. *If $\mathcal{M}$ and $\mathcal{N}$ are two pseudo-regular subspaces such that $\mathcal{N} \subseteq \mathcal{M}$ then the subspace $\mathcal{M} \cap \mathcal{N}^\perp$ is also pseudo-regular.*

Proof. As in the proof of the previous corollary, we assume $\mathcal{M}^\perp = \mathcal{M}^0$. Then, by Corollary 3.2.4, the subspace $\text{Clos}(\mathcal{N} + \mathcal{M}^0)$ is also pseudo-regular and hence its orthogonal companion $\mathcal{N}^\perp \cap \mathcal{M}^{0\perp} = \mathcal{N}^\perp \cap \mathcal{M}$ is the same. ∎

Proposition 3.2.8. *Let $\mathcal{M}$ be a pseudo-regular subspace of the Kreĭn space $\mathcal{K}$, and let $\mathcal{L} \subseteq \mathcal{M}$ be an $\mathcal{M}$-maximal negative (positive) subspace. Then $\kappa^-(\mathcal{L}^\perp) = \kappa^-(\mathcal{M}^\perp)$ (respectively, $\kappa^+(\mathcal{L}^\perp) = \kappa^+(\mathcal{M}^\perp)$).*

Proof. If $\mathcal{M}$ is pseudo-regular then $\mathcal{M}^\perp$ is pseudo-regular and hence, by Theorem 3.2.1, we have
$$\mathcal{M}^\perp = \mathcal{M}'_-[+]\mathcal{M}^0[+]\mathcal{M}'_+,$$
where $\mathcal{M}'_\pm$ are uniformly positive/negative subspaces.

Let $\mathcal{L} \subseteq \mathcal{M}$ be an $\mathcal{M}$-maximal negative subspace. Then $\mathcal{M} \cap \mathcal{L}^\perp$ is a positive subspace. By Proposition 2.1.11, in order to prove $\kappa^-(\mathcal{L}^\perp) = \kappa^-(\mathcal{M}^\perp)$ it is sufficient to prove that $\mathcal{L} \dotplus \mathcal{N}$ is $\mathcal{K}$-maximal negative, for some negative subspace $\mathcal{N}$ with $\dim \mathcal{N} = \kappa^-(\mathcal{M}^\perp)$. Thus, our proof would be finished if we prove that $\mathcal{L}[+]\mathcal{M}'_-$ is a $\mathcal{K}$-maximal negative subspace.

To see this, note first that $\mathcal{L} \perp \mathcal{M}'_-$. Since $\mathcal{M}'_-$ is uniformly negative, it follows that $\mathcal{L} + \mathcal{M}'_-$ is closed and negative. Therefore, by Corollary 2.1.7, it remains to prove that the subspace $(\mathcal{L} + \mathcal{M}'_-)^\perp = \mathcal{L}^\perp \cap \mathcal{M}'^\perp_-$ is positive. To this end, let us first note that, since $\mathcal{L}$ is $\mathcal{M}$-maximal negative, we have $\mathcal{M}^0 \subseteq \mathcal{L}$ and hence $\mathcal{L}^\perp \subseteq \mathcal{M}^{0\perp} = \mathcal{M} + \mathcal{M}^\perp$. Thus, taking into account that $\mathcal{M}^\perp \subseteq \mathcal{L}^\perp$, we have
$$\mathcal{L}^\perp = \mathcal{L}^\perp \cap (\mathcal{M} + \mathcal{M}^\perp) = (\mathcal{L}^\perp \cap \mathcal{M}) + \mathcal{M}^\perp. \tag{2.1}$$

On the other hand, it is easy to see that
$$\mathcal{M}'^\perp_- \cap (\mathcal{M} + \mathcal{M}^\perp) = \mathcal{M}[+]\mathcal{M}'_+. \tag{2.2}$$

Since $\mathcal{M}^0 \subseteq \mathcal{L} + \mathcal{M}'_-$ it follows that $(\mathcal{L} + \mathcal{M}'_-)^\perp \subseteq \mathcal{M}^{0\perp} = \mathcal{M} + \mathcal{M}^\perp$. Therefore, from (2.1) and (2.2), we get
$$(\mathcal{L} + \mathcal{M}'_-)^\perp = \mathcal{L}^\perp \cap \mathcal{M}'^\perp_- \cap (\mathcal{M} + \mathcal{M}^\perp) = (\mathcal{L}^\perp \cap \mathcal{M})[+]\mathcal{M}'_-.$$

Since the subspace $\mathcal{L}^\perp \cap \mathcal{M}$ is positive and the subspace $\mathcal{M}'_+$ is uniformly positive, it follows that the subspace $(\mathcal{L} + \mathcal{M}'_-)^\perp$ is positive. ∎

Pseudo-regular subspaces can be used to add one more equivalent characterisation of Pontryagin spaces, to those stated in Proposition 3.1.8.

Proposition 3.2.9. *A Kreĭn space $\mathcal{K}$ is a Pontryagin space if and only if every subspace of $\mathcal{K}$ is pseudo-regular.*

3.3 Strong Duality of Subspaces

Proof. We use Proposition 3.1.8. If $\mathcal{L}$ is a definite subspace of $\mathcal{K}$ then it is nondegenerate and hence regular. By Remark 3.1.2, $\mathcal{L}$ is uniformly definite. The converse implication is clear. ∎

Finally we show that a positive linear manifold which is complete under the norm $x \mapsto [x,x]^{1/2}$ may not be uniformly positive.

Example 3.2.10. Let $\mathcal{H}$ and $\mathcal{L}$ be as in Example 3.1.9. By Proposition 1.1.6 there exists a linear manifold $\mathcal{M} \supseteq \mathcal{L}$ which is maximal nondegenerate in $\text{Clos}(\mathcal{L})$ and hence

$$\text{Clos}(\mathcal{L}) = \mathcal{M} \dotplus \text{Clos}(\mathcal{L}^0).$$

Since $\text{Clos}(\mathcal{L})$ is pseudo-regular there exists a uniformly positive subspace $\mathcal{R}$ such that

$$\text{Clos}(\mathcal{L}) = \mathcal{R}[+]\text{Clos}(\mathcal{L}^0).$$

Define a linear operator $V\colon \mathcal{R} \to \mathcal{M}$ as follows: for any $x \in \mathcal{R}$ represent $x = x_1 + x_0$ with $x_1 \in \mathcal{M}$, $x_0 \in \text{Clos}(\mathcal{L}^0)$ and put $Vx = x_1$. Then V is correctly defined, preserves the inner product $[\cdot,\cdot]$ and is bijective. Thus $\mathcal{M}$ is complete under the norm $x \mapsto [x,x]^{1/2}$ but it is not uniformly positive since its closure $\text{Clos}(\mathcal{M}) = \text{Clos}(\mathcal{L})$ is degenerate. ∎

3.3 Strong Duality of Subspaces

Let $\mathcal{L}_1$ and $\mathcal{L}_2$ be subspaces of the Kreĭn space $\mathcal{K}$. We fix on $\mathcal{K}$ a fundamental norm $\|\cdot\|$ and denote by $(\cdot,\cdot)$ the corresponding positive definite inner product. Letting $(\cdot,\cdot)_{\mathcal{L}_i}$ denote the restriction of the inner product $(\cdot,\cdot)$ to $\mathcal{L}_i$, consider the Hilbert space $(\mathcal{L}_i,(\cdot,\cdot)_{\mathcal{L}_i})$, $i=1,2$. For any $x \in \mathcal{L}_1$ the mapping $\mathcal{L}_2 \ni y \mapsto [y,x] \in \mathbb{C}$ is a bounded functional on $\mathcal{L}_2$ and hence the Riesz–Fréchet Representation Theorem implies the existence of a bounded linear operator $T_{12} \in \mathcal{B}(\mathcal{L}_1,\mathcal{L}_2)$ such that

$$[x,y] = (T_{12}x,y)_{\mathcal{L}_2}, \quad x \in \mathcal{L}_1,\ y \in \mathcal{L}_2. \tag{3.1}$$

Denote

$$p_1(x) = \|T_{12}x\| = \sup_{\substack{\|y\| \leqslant 1 \\ y \in \mathcal{L}_2}} |[x,y]|, \quad x \in \mathcal{L}_1. \tag{3.2}$$

Then p_1 is a seminorm on $\mathcal{L}_1$. Define the positive inner product $\langle\cdot,\cdot\rangle_{\mathcal{L}_1}$ on $\mathcal{L}_1$ by

$$\langle x,y \rangle_{\mathcal{L}_1} = (T_{12}^* T_{12} x, y)_{\mathcal{L}_1}, \quad x,y \in \mathcal{L}_1. \tag{3.3}$$

Then p_1 is the seminorm associated to $\langle\cdot,\cdot\rangle_{\mathcal{L}_1}$, that is,

$$p_1(x) = (\langle x,x \rangle_{\mathcal{L}_1})^{1/2}, \quad x \in \mathcal{L}_1. \tag{3.4}$$

Moreover,

$$\text{Ker}(p_1) = \text{Ker}(T_{12}) = \mathcal{L}_1 \cap \mathcal{L}_2^{\perp}. \tag{3.5}$$

Similarly there exists a bounded linear operator $T_{21} = T_{12}^* \in \mathcal{B}(\mathcal{L}_2, \mathcal{L}_1)$ such that

$$[x, y] = (T_{21}x, y)_{\mathcal{L}_1}, \quad x \in \mathcal{L}_2, \ y \in \mathcal{L}_1. \tag{3.6}$$

The positive inner product on $\mathcal{L}_2$

$$\langle x, y \rangle_{\mathcal{L}_2} = (T_{21}^* T_{21} x, y)_{\mathcal{L}_2}, \quad x, y \in \mathcal{L}_2, \tag{3.7}$$

gives rise to the seminorm

$$p_2(x) = \|T_{21}x\| = \sup_{\substack{\|y\| \leqslant 1 \\ y \in \mathcal{L}_1}} |[x, y]|, \quad x \in \mathcal{L}_2, \tag{3.8}$$

which satisfies

$$\operatorname{Ker}(p_2) = \operatorname{Ker}(T_{21}) = \mathcal{L}_2 \cap \mathcal{L}_1^\perp. \tag{3.9}$$

From (3.2) and (3.8) it follows that T_{12} and T_{21} are contractions or, equivalently,

$$p_i(x) \leqslant \|x\|, \quad x \in \mathcal{L}_i, \ i = 1, 2, \tag{3.10}$$

in particular the topology induced by p_i on $\mathcal{L}_i$ is weaker than the strong topology.

Proposition 3.3.1. *The following statements are equivalent.*

(i) *There exists an $\alpha > 0$ such that*

$$\alpha \|x\| \leqslant p_i(x), \quad x \in \mathcal{L}_i, \ i = 1, 2. \tag{3.11}$$

(ii) T_{12} *(or equivalently, $T_{21} = T_{21}^*$) is boundedly invertible.*
(iii) $\mathcal{L}_1 \cap \mathcal{L}_2^\perp = 0$ *and* $\mathcal{L}_1 + \mathcal{L}_2^\perp = \mathcal{K}$.
(iv) $\mathcal{L}_2 \cap \mathcal{L}_1^\perp = 0$ *and* $\mathcal{L}_2 + \mathcal{L}_1^\perp = \mathcal{K}$.

Proof. (i)⇒(ii). This implication is trivial, since (3.11) for $i = 1$ means that T_{12} is injective and has closed range, while the same inequality for $i = 2$ yields T_{12}^* injective, and hence T_{12} has dense range.

(ii)⇒(iii). If T_{12} is boundedly invertible then $\mathcal{L}_1 \cap \mathcal{L}_2^\perp = 0$ follows from (3.5). Let $z \in \mathcal{K}$ be arbitrary. Considering the linear form $\mathcal{L}_2 \ni y \mapsto [y, z]$, $y \in \mathcal{L}_2$, from the Riesz–Fréchet Representation Theorem in the Hilbert space $(\mathcal{L}_2, (\cdot, \cdot))$ we get a vector $t \in \mathcal{L}_2$ such that $[y, z] = (y, t)$, $y \in \mathcal{L}_2$. For $z_1 = T_1^{-1} t \in \mathcal{L}_1$, by (3.1) it follows that $[y, z] = [y, z_1]$, $y \in \mathcal{L}_2$ and hence $z_2 = z - z_1 \in \mathcal{L}_2^\perp$. We have proven that $\mathcal{L}_1 + \mathcal{L}_2^\perp = \mathcal{K}$.

(iii)⇒(iv). If $\mathcal{L}_1 \cap \mathcal{L}_2^\perp = 0$ holds then $\mathcal{L}_2 + \mathcal{L}_1^\perp$ is dense in $\mathcal{K}$. From $\mathcal{L}_1 + \mathcal{L}_2^\perp = \mathcal{K}$ we first obtain $\mathcal{L}_2 \cap \mathcal{L}_1^\perp = 0$ and then, by Proposition 3.4.2, it follows that $\mathcal{L}_2 + \mathcal{L}_1^\perp$ is closed, hence $\mathcal{L}_2 + \mathcal{L}_1^\perp = \mathcal{K}$.

(iv)⇒(i). Let P_2 be the projection of $\mathcal{K}$ onto $\mathcal{L}_2$ along $\mathcal{L}_1^\perp$. By assumption, P_2 is bounded and hence, for any $x \in \mathcal{L}_1$ we have

$$\|x\| = \sup_{\|y\| \leqslant 1} |[x, y]| \leqslant \sup_{\substack{y_2 \in \mathcal{L}_2 \\ \|y_2\| \leqslant \|P_2\|}} |[x, y_2]| \leqslant \|P_2\| \, p_1(x).$$

3.3 Strong Duality of Subspaces

We remark that $\mathcal{L}_1 = 0$ if and only if $\mathcal{L}_2 = 0$ and in this case the proof is trivial. Assuming that both of $\mathcal{L}_1$ and $\mathcal{L}_2$ are nonnull, we notice that by proving (iii)⇒(iv) we already proved (iv)⇒(iii) (interchange the roles of $\mathcal{L}_1$ and $\mathcal{L}_2$) and then (3.11) follows with

$$\alpha = \min\{\|P_1\|^{-1}, \|P_2\|^{-1}\},$$

where P_1 denotes the projection of $\mathcal{K}$ onto $\mathcal{L}_1$ along $\mathcal{L}_2$. ∎

By definition, the subspaces $\mathcal{L}_1$ and $\mathcal{L}_2$ are in *strong duality*, or, equivalently, $(\mathcal{L}_1, \mathcal{L}_2)$ is a *strongly dual pair*, if one (and hence all) of the statements of Proposition 3.3.1 is satisfied. The operators T_{12} and T_{21} are called the *duality operators* associated to the pair $(\mathcal{L}_1, \mathcal{L}_2)$.

Remark 3.3.2. (a) The strong duality of subspaces of a Kreĭn space does not depend on the fixed fundamental norm, it depends only on the strong topology. Moreover, this notion of strong duality is the analogue of the duality of abstract Banach spaces.

(b) If the subspaces $\mathcal{L}_1$ and $\mathcal{L}_2$ are in strong duality then they are isomorphic as Hilbert spaces, in particular they have the same dimension.

(c) $\mathcal{L}_1$ and $\mathcal{L}_2$ are in strong duality if and only if $\mathcal{L}_1^\perp$ and $\mathcal{L}_2^\perp$ are in strong duality.

(d) If $\mathcal{L}_1$ and $\mathcal{L}_2$ are in strong duality then they are also in *weak duality*, that is, by definition, that $\mathcal{L}_1 \cap \mathcal{L}_2^\perp = \mathcal{L}_2 \cap \mathcal{L}_1^\perp = \{0\}$ (this definition makes sense in general inner product spaces). If $\mathcal{L}_1$ and $\mathcal{L}_2$ have either finite dimension or they have finite codimension, then their strong duality is equivalent with their weak duality.

(e) If $\mathcal{L}$ is a subspace of $\mathcal{K}$ then the duality operator associated to the pair $(\mathcal{L}, \mathcal{L})$ is the Gram operator (see Remark 1.3.3) of the inner product space $(\mathcal{L}, [\cdot, \cdot])$ with respect to the positive definite inner product $(\cdot, \cdot)$. In particular, the subspace $\mathcal{L}$ is regular if and only if $\mathcal{L}$ is in strong duality with itself.

(f) Let $\mathcal{L}$ be a subspace and S be a fundamental symmetry of $\mathcal{K}$. Then $\mathcal{L}$ is in strong duality with $S\mathcal{L}$. However, this is not a generic situation: if $\mathcal{M}$ is a maximal uniformly positive subspace and $\mathcal{N}$ is a maximal positive subspace of $\mathcal{K}$ then $\mathcal{M} \cap \mathcal{N}^\perp = \{0\}$ and we also have $\mathcal{M} + \mathcal{N}^\perp = \mathcal{K}$, hence $\mathcal{M}$ is in strong duality with $\mathcal{N}$. On the other hand, there exists a fundamental symmetry S of $\mathcal{K}$ such that $\mathcal{N} = S\mathcal{M}$ if and only if $\mathcal{N}$ is uniformly positive. ∎

In connection with Remark 3.3.2 (f) we consider the special case of neutral subspaces.

Theorem 3.3.3. *Let $\mathcal{L}_1$ and $\mathcal{L}_2$ be neutral subspaces of the Kreĭn space $\mathcal{K}$. The following statements are equivalent.*

(a) *$\mathcal{L}_1$ and $\mathcal{L}_2$ are in strong duality.*

(b) *$\mathcal{L}_1 + \mathcal{L}_2$ (the algebraic sum) is a regular subspace.*

(c) *There exists a fundamental symmetry S of $\mathcal{K}$ such that $\mathcal{L}_2 = S\mathcal{L}_1$ (equivalently, $\mathcal{L}_2 = S\mathcal{L}_1$).*

Proof. (a)⇒(b). By definition $\mathcal{L}_1 \cap \mathcal{L}_2^\perp = \{0\} = \mathcal{L}_2 \cap \mathcal{L}_1^\perp$ and

$$\mathcal{L}_1 + \mathcal{L}_2^\perp = \mathcal{K} = \mathcal{L}_2 + \mathcal{L}_1^\perp. \tag{3.12}$$

Since $\mathcal{L}_1 \subseteq \mathcal{L}_1^\perp$, from the first equality in (3.12) we obtain

$$\mathcal{L}_1^\perp = \mathcal{L}_1 + \mathcal{L}_1^\perp \cap \mathcal{L}_2^\perp, \qquad (3.13)$$

and inserting this in the second equality of (3.12) we obtain

$$\mathcal{L}_1 + \mathcal{L}_2 + \mathcal{L}_1^\perp \cap \mathcal{L}_2^\perp = \mathcal{K}, \qquad (3.14)$$

which shows that $\mathcal{L}_1 + \mathcal{L}_2$ is regular (recall that $\mathcal{L}_1^\perp \cap \mathcal{L}_2^\perp = (\mathcal{L}_1 + \mathcal{L}_2)^\perp$).

(b)$\Rightarrow$(a). If $\mathcal{L}_1 + \mathcal{L}_2$ is regular, the decomposition (3.14) holds and then we obtain (3.13), hence $\mathcal{L}_2 \cap \mathcal{L}_1^\perp = 0$ and, using once more (3.14), we get $\mathcal{L}_1 + \mathcal{L}_2^\perp = \mathcal{K}$.

(b)$\Rightarrow$(c). Let the subspace $\mathcal{L} = \mathcal{L}_1 + \mathcal{L}_2$ be regular. From the above proved implication (b)$\Rightarrow$(a), it follows that $\mathcal{L}_1$ and $\mathcal{L}_2$ are in strong duality and hence the operators T_{12} and T_{21} are boundedly invertible. We define a linear operator G on $\mathcal{L}$ by

$$G = \begin{bmatrix} 0 & T_{12}^{-1} \\ T_{21}^{-1} & 0 \end{bmatrix} \quad \text{with respect to } \mathcal{L} = \mathcal{L}_1 + \mathcal{L}_2. \qquad (3.15)$$

On $\mathcal{L}$ we consider the positive definite inner product $\langle \cdot, \cdot \rangle$ defined by

$$\langle x_1 + x_2, y_1 + y_2 \rangle = \langle x_1, y_1 \rangle_{\mathcal{L}_1} + \langle x_2, y_2 \rangle_{\mathcal{L}_2}, \quad x_i, y_i \in \mathcal{L}_i, \quad i = 1, 2, \qquad (3.16)$$

where the positive inner products $\langle \cdot, \cdot \rangle_i$, $i = 1, 2$ are defined in (3.3) and (3.7). The corresponding norm is s

$$s(x_1 + x_2) = (p_1(x_1)^2 + p_2(x_2)^2)^{1/2}, \quad x_i \in \mathcal{L}_i, \quad i = 1, 2,$$

and is equivalent with $\| \cdot \|$. $\mathcal{L}_1$ and $\mathcal{L}_2$ are orthogonal with respect to the inner product $\langle \cdot, \cdot \rangle$ and the operator G is bounded and boundedly invertible with respect to s.

Let $x_i, y_i \in \mathcal{L}_i$ be arbitrary. Then

$$\begin{aligned} \langle G(x_1 + x_2), y_1 + y_2 \rangle &= \langle T_{21}^{-1} x_1 + T_{12}^{-1} x_2, y_1 + y_2 \rangle \\ &= \langle T_{12}^{-1} x_1, y_1 \rangle + \langle T_{21}^{-1} x_2, y_2 \rangle \\ &= (T_{12}^* T_{12} T_{12}^{-1} x_2, y_1) + (T_{21}^* T_{21} T_{21}^{-1} x_1, y_2) \\ &= (T_{12}^* x_2, y_1) + (T_{21}^* x_1, y_2) \\ &= [x_2, y_1] + [x_1, y_2] = [x_1 + x_2, y_1 + y_2], \end{aligned}$$

that is,

$$\langle Gx, y \rangle = [x, y], \quad x, y \in \mathcal{L}, \qquad (3.17)$$

and hence G is the Gram operator of the inner product space $(\mathcal{L}, [\cdot, \cdot])$ with respect to $\langle \cdot, \cdot \rangle$. In particular G is selfadjoint with respect to the Hilbert space $(\mathcal{L}, (\cdot, \cdot))$.

With respect to the Hilbert spaces $(\mathcal{L}_i, \langle \cdot, \cdot \rangle)$ we consider the left polar decompositions

$$T_{12}^{-1} = U_1 |T_{12}^{-1}|, \quad T_{21}^{-1} = U_2 |T_{21}^{-1}|.$$

3.4 FREDHOLM SUBSPACES

Then it is easy to see that

$$G = \begin{bmatrix} 0 & U_1 \\ U_2 & 0 \end{bmatrix} \begin{bmatrix} |T_{21}^{-1}| & 0 \\ 0 & |T_{12}^{-1}| \end{bmatrix}$$

is the polar decomposition of G with respect to the Hilbert space $(\mathcal{L}, \langle \cdot, \cdot \rangle)$. Denote

$$S = \begin{bmatrix} 0 & U_1 \\ U_2 & 0 \end{bmatrix}, \quad |G| = \begin{bmatrix} |T_{21}^{-1}| & 0 \\ 0 & |T_{12}^{-1}| \end{bmatrix}.$$

Then (3.17) can be written as

$$\langle S|G|x, y \rangle = [x, y], \quad x, y \in \mathcal{L},$$

and S is a fundamental symmetry of the Kreĭn space $(\mathcal{L}, [\cdot, \cdot])$. By construction, $\mathcal{L}_2 = S\mathcal{L}_1$ and then S can be extended to a fundamental symmetry on $\mathcal{K}$ with the same property. ∎

3.4 Fredholm Subspaces

In this section we investigate a class of subspaces for which a certain notion of index can be associated and thus leads to a certain geometric Fredholm theory. We need first two preliminary results.

Lemma 3.4.1. *Let $\mathcal{M}$ and $\mathcal{N}$ be two subspaces of the Kreĭn space $\mathcal{K}$. If $\mathcal{M} + \mathcal{N}$ is closed then*

$$\mathcal{M}^\perp + \mathcal{N}^\perp = (\mathcal{M} \cap \mathcal{N})^\perp.$$

Proof. Clearly, we always have the inclusion

$$\mathcal{M}^\perp + \mathcal{N}^\perp \subseteq (\mathcal{M} \cap \mathcal{N})^\perp.$$

Conversely, let us consider a vector $x \in (\mathcal{M} \cap \mathcal{N})^\perp$. Let J be a fundamental symmetry of $\mathcal{K}$, $\|\cdot\|$ the corresponding fundamental norm, and consider the following decomposition

$$\mathcal{M} + \mathcal{N} = \mathcal{M} \cap (\mathcal{M} \cap \mathcal{N})^{\perp_J} \dotplus \mathcal{M} \cap \mathcal{N} \dotplus \mathcal{N} \cap (\mathcal{M} \cap \mathcal{N})^{\perp_J}.$$

Since $\mathcal{M} + \mathcal{N}$ is closed, we have two bounded linear functionals on the space $\mathcal{M} + \mathcal{N}$ defined as follows:

$$f \colon \mathcal{M} + \mathcal{N} \ni u + v + w \mapsto [u, x],$$

$$g \colon \mathcal{M} + \mathcal{N} \ni u + v + w \mapsto [w, x].$$

Then $f|\mathcal{N} = 0$, $g|\mathcal{M} = 0$, and

$$f(y) + g(y) = [y, x], \quad y \in \mathcal{M} + \mathcal{N}.$$

By means of the complex variant of the Hahn–Banach Theorem we obtain two linear functionals $\tilde f$ and $\tilde g$ such that $\tilde f|(\mathcal M + \mathcal N) = f$, $\|\tilde f\| = \|f\|$ and $\tilde g|(\mathcal M + \mathcal N) = g$, $\|\tilde g\| = \|g\|$. In particular, $\tilde f$ and $\tilde g$ are bounded and hence, by Theorem 1.2.1, it follows that there exist $x_1, x_2 \in \mathcal K$ such that

$$\tilde f(y) = [y, x_1], \quad \tilde g(y) = [y, x_2], \quad y \in \mathcal K.$$

Then $x_1 \in \mathcal N^\perp$, $x_2 \in \mathcal M^\perp$ and

$$[y, x_1 + x_2] = \tilde f(y) + \tilde g(y) = f(y) + g(y) = [y, x], \quad y \in \mathcal M + \mathcal N,$$

hence $x - x_1 - x_2 \in (\mathcal M + \mathcal N)^\perp = \mathcal M^\perp \cap \mathcal N^\perp$ and then

$$x \in \mathcal M^\perp + \mathcal N^\perp + \mathcal M^\perp \cap \mathcal N^\perp = \mathcal M^\perp + \mathcal N^\perp. \blacksquare$$

Proposition 3.4.2. *Let $\mathcal M$ and $\mathcal N$ be two subspaces of the Kreĭn space $\mathcal K$. Then $\mathcal M + \mathcal N$ is closed if and only if $\mathcal M^\perp + \mathcal N^\perp$ is closed.*

Proof. We use Lemma 3.4.1 and the fact that $(\mathcal M \cap \mathcal N)^\perp$ is always closed. For the converse implication interchange the roles of $\mathcal M$ and $\mathcal N$, respectively, the roles of $\mathcal M^\perp$ and $\mathcal N^\perp$. $\blacksquare$

A subspace $\mathcal S$ of the Kreĭn space $\mathcal K$ is called *positively Fredholm* (respectively, *positively semi-Fredholm*) if, (either) for every fundamental decomposition $\mathcal K = \mathcal K^+[+]\mathcal K^-$ the subspace $\mathcal S \cap \mathcal K^+$ is finite dimensional, and (or) for every fundamental decomposition $\mathcal K = \mathcal K^+[+]\mathcal K^-$, the subspace $\mathcal S^\perp \cap \mathcal K^-$ is finite dimensional. Clearly, any positively Fredholm subspace is positively semi-Fredholm.

In a similar fashion we define *negatively semi-Fredholm* subspaces and *negatively Fredholm* subspaces.

Using the fact that every uniformly positive subspace can be extended to a maximal uniformly positive subspace (cf. Section 2.2) it follows that the subspace $\mathcal S$ is positively semi-Fredholm (respectively, positively Fredholm) if and only if (either) every uniformly positive subspace of $\mathcal S$ is finite dimensional, and (or) every uniformly negative subspace of $\mathcal S^\perp$ is finite dimensional. These definitions raise the interesting question of whether an intrinsic characterisation of those subspaces containing only finite dimensional uniformly definite subspaces can be found.

Let J be a fundamental symmetry of $\mathcal K$ and consider G the Gram operator of the inner product space $(\mathcal S; [\cdot, \cdot])$ with respect to the positive definite inner product $\langle \cdot, \cdot \rangle_J$ restricted to $\mathcal S$, that is $G = P_{\mathcal S} J | P_{\mathcal S}$, where $P_{\mathcal S}$ is the projection of $\mathcal K$ onto $\mathcal S$ along $J\mathcal S^\perp$. Then G is a selfadjoint contraction onto the Hilbert space $(\mathcal S, \langle \cdot, \cdot \rangle_J)$, and let $G = G^+ - G^-$ be the Jordan decomposition of G.

Lemma 3.4.3. *With notation as before, every uniformly positive (uniformly negative) subspace of $\mathcal S$ is finite dimensional if and only if the operator G^+ (respectively, G^-) is compact.*

3.4 Fredholm Subspaces

Proof. Assume that every uniformly positive subspace of $\mathcal{S}$ is finite dimensional. Let E denote the spectral measure of the Gram operator G. For every $\alpha > 0$ the subspace $E[\alpha, 1]G\mathcal{S}$ is uniformly positive, hence rank $E[\alpha, 1]G < \infty$. Since G^+ is the uniform limit of operators $E[\alpha, 1]G$ as $\alpha \downarrow 0$, it follows that G^+ is a compact operator.

Conversely, assume that the operator G^+ is compact and let $\mathcal{L}$ be a uniformly positive subspace of $\mathcal{S}$. This means that for some $\alpha >$ we have $P_\mathcal{L} G|\mathcal{L} \geqslant \alpha I_\mathcal{M}$, where $P_\mathcal{L}$ denotes the projection onto $\mathcal{L}$ along $J\mathcal{L}^\perp$ and $I_\mathcal{M}$ is the identity operator on $\mathcal{M}$. Therefore

$$P_\mathcal{L} G^+|\mathcal{L} \geqslant P_\mathcal{B}(G^+ - G_-)|\mathcal{L} = P_\mathcal{L} G|\mathcal{L} \geqslant \alpha I_\mathcal{L}.$$

Since G^+ is a compact operator it follows that $I_\mathcal{L}$ is compact and hence $\mathcal{L}$ is finite dimensional.

The statement corresponding to uniformly negative subspaces follows in a similar way. ∎

Lemma 3.4.4. *Let* $\mathcal{K} = \mathcal{K}^+[+]\mathcal{K}^-$ *be a fundamental decomposition, J the corresponding fundamental symmetry of $\mathcal{K}$, and $\mathcal{S}$ a subspace of $\mathcal{K}$. The following assertions are mutually equivalent.*

(i) *The linear manifold* $\mathcal{S} + \mathcal{K}^+$ *is closed.*

(ii) *The linear manifold* $J^-\mathcal{S}$ *is closed.*

(i)' *The linear manifold* $\mathcal{S}^\perp + \mathcal{K}^-$ *is closed.*

(ii)' *The linear manifold* $J^+\mathcal{S}^\perp$ *is closed.*

Proof. The equivalences (i)⇔(i)' and (ii)⇔(ii)' follow from Proposition 3.4.2, so we only have to prove the equivalence of (i) with (ii). But this follows by the simple observation that $\mathcal{S} + \mathcal{K}^+ = J^-\mathcal{S} + \mathcal{K}^+$. ∎

Lemma 3.4.5. *If every uniformly positive subspace of $\mathcal{S}$ is finite dimensional then for every fundamental decomposition $\mathcal{K} = \mathcal{K}^+[+]\mathcal{K}^-$ the conditions* (i), (ii), (i)', *and* (ii)' *in Lemma 3.4.4 are satisfied.*

Proof. Let $\mathcal{K} = \mathcal{K}^+[+]\mathcal{K}^-$ be a fundamental decomposition and J the corresponding fundamental symmetry. Denote by $G = P_\mathcal{S} J|\mathcal{S}$ the Gram operator corresponding to the inner products $[\cdot,\cdot]$ and $\langle\cdot,\cdot\rangle_J$ on $\mathcal{S}$. Then note that $\mathcal{S} \cap \mathcal{K}^+ = \operatorname{Ker}(G - I) = \operatorname{Ker}(G^+ - I)$, where $G = G^+ - G^-$ is the Jordan decomposition of G. Since $\mathcal{S} \cap \mathcal{K}^+$ is finite dimensional, without restricting the generality we can assume that $\mathcal{S} \cap \mathcal{K}^+ = 0$. In this case it follows that the operator $G - I$ is injective. By Lemma 3.4.3 it follows that G^+ is compact and hence the operator $G - I$ has closed range. Since the latter operator is injective, it follows that $G - I$ is boundedly invertible on $\mathcal{S}$.

The fundamental symmetry J is a lifting of G. Taking into account that $2J^- = J - I$ it follows that

$$J^-\mathcal{S} = \left\{ \frac{1}{2}\begin{bmatrix}(G-I)x \\ Hx\end{bmatrix} \bigg| \, x \in \mathcal{S}\right\}.$$

Since $G - I$ is boundedly invertible it follows that
$$J^- S = \{y + H(G - I)^{-1} y \mid y \in S\},$$
and hence the linear manifold $J^- S$ is closed. ∎

From now on and throughout this section, the dimensions will be understood in the algebraic sense, that is, $\dim(\mathcal{L})$ is either a positive integer or the symbol ∞.

Theorem 3.4.6. *If S is a positively semi-Fredholm subspace then the number*
$$\mathrm{ind}_+(S) = \dim(S \cap \mathcal{K}^+) - \dim(S^\perp \cap \mathcal{K}^-)$$
does not depend on the fundamental decomposition $\mathcal{K} = \mathcal{K}^+[+]\mathcal{K}^-$.

Proof. Let S be a positively semi-Fredholm subspace and fix a fundamental decomposition $\mathcal{K} = \mathcal{K}^+[+]\mathcal{K}^-$ and the corresponding fundamental symmetry J. Let P_+ be the orthogonal projection of $\mathcal{K}^+$ onto $S \cap \mathcal{K}^+$ and P_- be the orthogonal projection of $\mathcal{K}^-$ onto $S^\perp \cap \mathcal{K}^-$. As in the proof of Lemma 3.4.5 it follows that
$$S = (S \cap \mathcal{K}^+)[+]\{Kx + x \mid x \in (I_- - P_-)\mathcal{K}^-\},$$
$$S^\perp = (S^\perp \cap \mathcal{K}^-)[+]\{z + K^*z \mid z \in (I_+ - P_+)\mathcal{K}^+\},$$
where $K \in \mathcal{B}((I_- - P_-)\mathcal{K}^-, (I_+ - P_+)\mathcal{K}^+)$.

Let now $\mathcal{K} = \mathcal{L}[+]\mathcal{L}^\perp$ be another fundamental decomposition of $\mathcal{K}$. If T is the angular operator of the maximal uniformly positive subspace $\mathcal{L}$ with respect to the fundamental decomposition $\mathcal{K} = \mathcal{K}^+[+]\mathcal{K}^-$, then
$$S \cap \mathcal{L} = \{y + x \mid TKx - x = Ty, \ x \in (I_- - P_-)\mathcal{K}^-, \ y \in P_+\mathcal{K}^+\}$$
and
$$S^\perp \cap \mathcal{L}^\perp = \{z + t \mid T^*K^*z - z = T^*t, \ z \in (I_+ - P_+)\mathcal{K}^+, \ t \in P_-\mathcal{K}^-\}.$$

Considering the operator block matrix representation of the operator
$$T \colon (I_+ - P_+)\mathcal{K}^+ \oplus P_+\mathcal{K}^+ \to (I_- - P_-)\mathcal{K}^- \oplus P_-\mathcal{K}^-,$$
$$T = \begin{bmatrix} T_{11} & T_{12} \\ T_{21} & T_{22} \end{bmatrix},$$
it follows that
$$S \cap \mathcal{L} = \{x + y \mid -x - T_{11}Kx = T_{12}y, \ T_{21}Kx = -T_{22}y,$$
$$x \in (I_- - P_-)\mathcal{K}^-, \ y \in P_+\mathcal{K}^+\}.$$

3.4 Fredholm Subspaces

Now remark that if $\|T_{11}\| < \|K\|^{-1}$ the operator $I - T_{11}K$ is boundedly invertible on $(I_- - P_-)\mathcal{K}^-$ and hence

$$\dim(\mathcal{S} \cap \mathcal{L}) = \dim(\mathrm{Ker}(T_{22} + T_{21}K(I - T_{11}K)^{-1}T_{12})). \tag{4.1}$$

Similarly, if $\|T_{11}\| < \|K\|^{-1}$ then

$$\dim(\mathcal{S}^\perp \cap \mathcal{L}^\perp) = \dim(\mathrm{Ker}(T_{22}^* + T_{12}^*K^*(I - T_{11}^*K^*)^{-1}T_{21}^*)). \tag{4.2}$$

Taking into account that $K^*(I - T_{11}^*K^*)^{-1} = (I - K^*T_{11}^*)^{-1}K^*$ and that the operator $T_{22} + T_{21}K(I - T_{11}K)^{-1}T_{12}$ has finite rank (since either $(I_+ - P_+)\mathcal{K}^+$ or $(I_- - P_-)\mathcal{K}^-$ is finite dimensional) from (4.1) and (4.2) it follows that

$$\begin{aligned}\dim(\mathcal{S} \cap \mathcal{L}) - \dim(\mathcal{S}^\perp \cap \mathcal{L}^\perp) &= \mathrm{ind}(T_{22} + T_{21}K(I - T_{11}K)^{-1}T_{12}) \\ &= \dim(I_+ - P_+)\mathcal{K}^+ - \dim(I_- - P_-)\mathcal{K}^- \\ &= \dim(\mathcal{S} \cap \mathcal{K}^+) - \dim(\mathcal{S}^\perp \cap \mathcal{K}^-).\end{aligned}$$

We thus have proved that the function $\mathrm{ind}_+(S)$ is locally constant, with respect to a small perturbation of the fundamental decomposition and the topology given by the norm of the angular operator. Since the open unit ball of $\mathcal{B}(\mathcal{K}^+, \mathcal{K}^-)$ is connected, it follows that the function $\mathrm{ind}_+(S)$ is constant, that is, its definition does not depend on the fundamental symmetry. ∎

Remark 3.4.7. Theorem 3.4.6 implies that if for every fundamental decomposition $\mathcal{K} = \mathcal{K}^+[+]\mathcal{K}^-$ at least one of the subspaces $\mathcal{S} \cap \mathcal{K}^+$ and $\mathcal{S}^\perp \cap \mathcal{K}^-$ is finite dimensional then the subspace $\mathcal{S}$ is a positively semi-Fredholm operator; compare with the definition at the beginning of this section. ∎

Similarly, the subspace $\mathcal{S}$ is called *negatively Fredholm* (*negatively semi-Fredholm*) if (either) for every fundamental decomposition $\mathcal{K} = \mathcal{K}^+[+]\mathcal{K}^-$ the subspace $\mathcal{S} \cap \mathcal{K}^-$ is finite dimensional and (or) for every fundamental decomposition $\mathcal{K} = \mathcal{K}^+[+]\mathcal{K}^-$ the subspace $\mathcal{S}^\perp \cap \mathcal{K}^+$ is finite dimensional. Clearly, a subspace $\mathcal{S}$ is negatively semi-Fredholm (negatively Fredholm) if and only if $\mathcal{S}^\perp$ is positively semi-Fredholm (respectively, positively semi-Fredholm). Therefore, a consequence of Theorem 3.4.6 is that for an arbitrary negatively semi-Fredholm subspace $\mathcal{S}$ the number

$$\mathrm{ind}_-(\mathcal{S}) = \dim(\mathcal{S} \cap \mathcal{K}^-) - \dim(\mathcal{S}^\perp \cap \mathcal{K}^+)$$

does not depend on the fundamental decomposition $\mathcal{K} = \mathcal{K}^+[+]\mathcal{K}^-$.

We now apply Theorem 3.4.6 in order to obtain an index formula relating the algebraic positive/negative ranks of a subspace and its orthogonal companion. Recall that by $\kappa_\pm(\mathcal{L})$ we denote the algebraic rank of positivity/negativity of $\mathcal{L}$ which is either a nonnegative integer or the symbol ∞, see (1.9) in Section 1.1.

Theorem 3.4.8. *If S is a subspace of the Kreĭn space $\mathcal{K}$ then for any fundamental decomposition $\mathcal{K} = \mathcal{K}^+[+]\mathcal{K}^-$ we have that*

$$\kappa_\mp(S) + \dim(S^\perp \cap \mathcal{K}^\pm) = \kappa_\pm(S^\perp) + \dim(S \cap \mathcal{K}^\mp). \tag{4.3}$$

Proof. We first note that, by interchanging $+$ and $-$, in order to prove (4.3) it is sufficient to prove that

$$\kappa_-(S) + \dim(S^\perp \cap \mathcal{K}^+) = \kappa_+(S^\perp) + \dim(S \cap \mathcal{K}^-). \tag{4.4}$$

Without restricting the generality we can assume that $\kappa_-(S)$ is finite. Indeed, if both of $\kappa_-(S)$ and $\kappa_+(S^\perp)$ are infinite then the formula (4.4) is trivially true. On the other hand, if (4.4) holds for every S with $\kappa_-(S) < \infty$, then by interchanging S with $S^\perp$ and $+$ with $-$ we also have the formula proven in the case that $\kappa_+(S^\perp) < \infty$ holds.

Further, if $\kappa_-(S)$ is finite then S is a negatively semi-Fredholm subspace. Moreover, the formula (4.4) holds if both of $\kappa_-(S)$ and $\kappa_+(S^\perp)$ are finite. To see this, if both of the numbers $\kappa_-(S)$ and $\kappa_+(S^\perp)$ are finite then $S = \mathcal{P}[+]S^-$, where S^- is a finite dimensional negative subspace and $\mathcal{P}$ is a positive subspace, and $S^\perp = S_1^+[+]\mathcal{N}$, where S_1^+ is a finite dimensional positive subspace and $\mathcal{N}$ is a negative subspace. Then by Theorem 2.2.9 there exists a fundamental decomposition $\mathcal{K} = \mathcal{G}^+[+]\mathcal{G}^-$ such that $S^- \subseteq \mathcal{G}^-$ and $S_1^+ \subseteq \mathcal{G}^+$ and hence

$$\kappa_-(S) = \dim(S \cap \mathcal{G}^-), \quad \kappa_+(S^\perp) = \dim(S^\perp \cap \mathcal{G}^+). \tag{4.5}$$

From Theorem 3.4.6 and (4.5) the formula (4.4) follows.

We now prove that formula (4.4) holds if $\kappa_-(S) = 0$ and $\kappa_+(S^\perp) = \infty$. Indeed, if $\kappa_-(S) = 0$ then S is a positive subspace and $S \cap \mathcal{K}^- = \{0\}$. Let $K \in \mathcal{B}(\mathcal{D}_+, \mathcal{K}^-)$ be the angular operator of S with respect to $\mathcal{K} = \mathcal{K}^+[+]\mathcal{K}^-$, where $\mathcal{D}_+ \subseteq \mathcal{K}^+$ is a closed subspace. Then

$$S^\perp = \{K^*x + x \mid x \in \mathcal{K}^-\}[+](\mathcal{K}^+ \ominus \mathcal{D}_+), \tag{4.6}$$

and $S^\perp \cap \mathcal{K}^+ = \mathcal{K}^+ \ominus \mathcal{D}_+$. If $\kappa_+(S^\perp) = \infty$ then (4.6) also implies that $\dim(\mathcal{K}^+ \ominus \mathcal{D}_+) = \infty$ and hence the formula (4.4) holds.

Finally, we prove that the formula (4.4) holds if $\kappa_-(S)$ is finite and $\kappa_+(S^\perp) = \infty$. Indeed, if $\kappa_-(S) < \infty$ then $\dim(S \cap \mathcal{K}^-) < \infty$ and (4.4) is equivalent with

$$\kappa_-(S) - \kappa_+(S^\perp) = \dim(S \cap \mathcal{K}^-) - \dim(S^\perp \cap \mathcal{K}^+). \tag{4.7}$$

S is a negatively semi-Fredholm subspace and hence, on account of Theorem 3.4.6, the right side in (4.7) is independent of the fundamental decomposition of $\mathcal{K}$. Thus, we can choose the fundamental decomposition $\mathcal{K} = \mathcal{K}^+[+]\mathcal{K}^-$ to our convenience.

Let $S = \mathcal{P}[+]S^-$ be a decomposition of S, where S^- is a finite dimensional negative subspace and $\mathcal{P}$ is a positive subspace. Then $\mathcal{K}_1 = (S^-)^\perp$ is a regular subspace, so we can view $\mathcal{K}_1$ as a Kreĭn space containing $\mathcal{P}$. Let $\mathcal{K}_1 = \mathcal{K}_1^+[+]\mathcal{K}_1^-$ be a fundamental decomposition of $\mathcal{K}_1$. Then, since $\mathcal{P}$ is positive, $\kappa_-(\mathcal{P}) = 0$, $\dim(\mathcal{P} \cap \mathcal{K}_1^-) = 0$, and applying the previous steps to this situation it follows

$$\kappa_+(\mathcal{P}^\perp \cap \mathcal{K}_1) = \dim(\mathcal{P}^\perp \cap \mathcal{K}_1^+). \tag{4.8}$$

Here we may also take $\mathcal{P}^\perp$ to be the orthogonal complement of $\mathcal{P}$ in the ambient Kreĭn space $\mathcal{K}$. In $\mathcal{K}$ we choose the fundamental decomposition

$$\mathcal{K} = \mathcal{K}^+[+]\mathcal{K}^-, \quad \mathcal{K}^+ = \mathcal{K}_1^+, \quad \mathcal{K}^- = \mathcal{K}_1^-[+]\mathcal{S}^-.$$

Then, since $\mathcal{S}^- = \mathcal{S} \cap \mathcal{K}^-$, we have that

$$\kappa_-(\mathcal{S}) = \dim(\mathcal{S} \cap \mathcal{K}^-). \tag{4.9}$$

Since $\mathcal{S}^\perp = \mathcal{P}^\perp \cap \mathcal{K}_1$, we have that $\mathcal{S}^\perp \cap \mathcal{K}^+ = \mathcal{P}^\perp \cap \mathcal{K}_1^+$, and, in view of (4.8), this implies that

$$\kappa_+(\mathcal{S}^\perp) = \dim(\mathcal{S}^\perp \cap \mathcal{K}^+). \tag{4.10}$$

The formulae (4.9) and (4.10) readily imply (4.7), and hence (4.4) is proven. ∎

3.5 Index Formulae for Linear Relations

We indicate in this section some applications to linear relations of the index formula in Theorem 3.4.8.

A *(closed) linear relation* T from a Kreĭn space $\mathcal{K}_1$ to a Kreĭn space $\mathcal{K}_2$ is a (closed) linear manifold of $\mathcal{K}_1[+]\mathcal{K}_2$ whose elements we denote by (f, g), $f \in \mathcal{K}_1, g \in \mathcal{K}_2$. If T is such a relation then T^{-1} and $T^\sharp$ are defined by

$$T^{-1} = \{(g, f) \in \mathcal{K}_2[+]\mathcal{K}_1 \mid (f, g) \in T\},$$

$$T^\sharp = \{(k, h) \in \mathcal{K}_2[+]\mathcal{K}_1 \mid [h, f]_{\mathcal{K}_1} - [k, g]_{\mathcal{K}_2} = 0, \forall (f, g) \in T\}$$

are linear relations in $\mathcal{K}_2[+]\mathcal{K}_1$, called the *inverse* and *adjoint* of T, respectively.

The graph of an operator T is an example of a linear relation and, conversely, a linear relation is the graph of an operator if and only if $(0, g) \in T$ implies $g = 0$. We frequently identify the operator T with its graph and we use the same letter for both. For T a densely defined operator the adjoint $T^\sharp$ will have a different definition in Chapter 4; for the moment we only mention that $T^\sharp$ defined above is the graph of the adjoint operator of T, so the two definitions will be consistent.

We view linear relations as "multivalued operators" and define the domain $\mathrm{Dom}(T)$, the range $\mathrm{Ran}(T)$ and the kernel $\mathrm{Ker}(T)$ in the usual way:

$$\mathrm{Dom}(T) = \{f \in \mathcal{K}_1 \mid (f, g) \in T \text{ for some } g \in \mathcal{K}_2\}, \tag{5.1}$$

$$\mathrm{Ran}(T) = \{g \in \mathcal{K}_2 \mid (f, g) \in T \text{ for some } f \in \mathcal{K}_1\}, \tag{5.2}$$

and
$$\text{Ker}(T) = \{f \in \mathcal{K}_1 \mid (f, 0) \in T\}. \tag{5.3}$$
New in this connection is the multivalued part $T(0) = \{g \mid (0,g) \in T\}$ which can also be written as $T(0) = \text{Ker}(T^{-1})$. For a linear relation T from $\mathcal{K}_1$ to $\mathcal{K}_2$, the *positive/negative signature of defect* $\nu_\pm(T)$ is, by definition, the algebraic positive/negative rank of the quadratic form
$$T \ni (f, g) \mapsto [f, f]_1 - [g, g]_2. \tag{5.4}$$
The terminology is justified by the observation that, in the case $T \in \mathcal{B}(\mathcal{K}_1, \mathcal{K}_2)$, then $\nu_\pm(T) = \kappa_\pm(I - T^\# T)$, where $I - T^\# T$ is the so-called *defect operator* of T and by $\kappa_\pm(I - T^\# T)$ we understand the algebraic ranks of positivity/negativity of the inner product $[(I - T^\# T)\cdot, \cdot]$.

With this definition, note that $\nu_\pm(T) = \nu_\mp(T^{-1})$.

As a first application of Theorem 3.4.8 we give an index formula for linear relations in Hilbert spaces. Note that for Hilbert spaces we use the traditional $*$ to denote the adjoint operation on linear relations.

Corollary 3.5.1. *Let S be a linear relation from the Hilbert space $\mathcal{H}_1$ to the Hilbert space $\mathcal{H}_2$. Then*
$$\nu_-(S) + \dim S^*(0) = \nu_-(S^*) + \dim S(0). \tag{5.5}$$
Moreover, if either $\nu_-(S) < \infty$ or $\nu_-(S^*) < \infty$ then
$$\nu_-(S) - \nu_-(S^*) = \dim S(0) - \dim S^*(0). \tag{5.6}$$

Proof. Denote the inner product on $\mathcal{H}_i$ by $\langle \cdot, \cdot \rangle_i$, $i = 1, 2$, and consider the Kreĭn space
$$\mathcal{K} = \mathcal{H}_1[+] - \mathcal{H}_2, \tag{5.7}$$
that is, $\mathcal{K} = \mathcal{H}_1 \oplus \mathcal{H}_2$ endowed with the indefinite inner product $[\cdot, \cdot]$ defined by
$$[h_1 + h_2, k_1 + k_2] = \langle h_1, k_1 \rangle_1 - \langle h_2, k_2 \rangle_{\mathcal{L}_2}, \quad h_1, k_1 \in \mathcal{H}_1, \; h_2, k_2 \in \mathcal{H}_2.$$
Identify the linear relation S with the subspace $\mathcal{S}$ in $\mathcal{K}$ by
$$\mathcal{S} = \{f + g \mid (f, g) \in S\}.$$
Then
$$\nu_-(S) = \kappa_-(\mathcal{S}), \quad \nu_-(S^*) = \kappa_+(\mathcal{S}). \tag{5.8}$$
The decomposition in (5.7) is a fundamental decomposition of $\mathcal{K}$,
$$\mathcal{K} = \mathcal{K}^+[+]\mathcal{K}^-, \quad \mathcal{K}^+ = \mathcal{H}_1 \oplus \{0\}, \; \mathcal{K}^- = \{0\} \oplus \mathcal{H}_2.$$
It follows that
$$\dim S(0) = \dim(\mathcal{S} \cap \mathcal{K}^-), \quad \dim S^*(0) = \dim(\mathcal{S}^\perp \cap \mathcal{K}^+). \tag{5.9}$$

3.5 Index Formulae for Linear Relations

Inserting (5.8) and (5.9) in (4.4) we obtain the formula (5.5).

The inequalities
$$\dim S(0) \leqslant \nu_-(S), \quad \dim S^*(0) \leqslant \nu_-(S^*)$$
always hold. They imply that, if either $\nu_-(S) < \infty$ or $\nu_-(S^*) < \infty$, then both sides in (5.6) make sense and the formula (5.6) is equivalent with the formula (5.5). ∎

Let $\mathcal{K}_1$ and $\mathcal{K}_2$ be Kreĭn spaces with fundamental decompositions
$$\mathcal{K}_1 = \mathcal{K}_1^+[+]\mathcal{K}_1^-, \quad \mathcal{K}_2 = \mathcal{K}_2^+[+]\mathcal{K}_2^-, \tag{5.10}$$
and let $J_1 = J_1^+ - J_1^-$ and $J_2 = J_2^+ - J_2^-$ be the corresponding fundamental symmetries. Define the Hilbert spaces $\mathcal{H}_1$ and $\mathcal{H}_2$ by
$$\mathcal{H}_1 = \mathcal{K}_1^+ \oplus \mathcal{K}_2^-, \quad \mathcal{H}_2 = \mathcal{K}_1^- \oplus \mathcal{K}_2^+.$$
The mapping $\Pi \colon \mathcal{K}_1[+]\mathcal{K}_2 \to \mathcal{H}_1 \oplus \mathcal{H}_2$ defined by
$$\Pi((f,g)) = \{J_1^+ f + J_2^- g, J_1^- f + J_2^+ g\} \tag{5.11}$$
is called the *scattering* or *Potapov–Ginzburg transform*. If T is a closed linear relation from $\mathcal{K}_1$ to $\mathcal{K}_2$, then $\Pi(T)$ is a closed linear relation from $\mathcal{H}_1$ to $\mathcal{H}_2$ and we have that $\nu_\pm(T) = \nu_\pm(\Pi(T))$ and
$$\Pi(T)^{-1} = \Pi(T^{-1}), \quad \Pi(T)^* = \Pi(T^\sharp), \tag{5.12}$$
where the operator Π on the right-hand side acts from $\mathcal{K}_2[+]\mathcal{K}_1$ to $\mathcal{H}_2 \oplus \mathcal{H}_1$ and the adjoint $\Pi(T)^*$ stands for the adjoint of $\Pi(T)$ in the Hilbert space $\mathcal{H}_1 \oplus \mathcal{H}_2$.

Theorem 3.5.2. *Let T be a closed linear relation from the Kreĭn space $\mathcal{K}_1$ to the Kreĭn space $\mathcal{K}_2$. Then*
$$\nu_-(T) + \dim(\mathcal{H}_1 \ominus \mathrm{Dom}(\Pi(T))) = \nu_-(T^\sharp) + \dim(\mathcal{H}_2 \ominus \mathrm{Dom}(\Pi(T^\sharp))), \tag{5.13}$$
$$\nu_+(T) + \dim(\mathcal{H}_2 \ominus \mathrm{Ran}(\Pi(T))) = \nu_+(T^\sharp) + \dim(\mathcal{H}_1 \ominus \mathrm{Ran}(\Pi(T^\sharp))). \tag{5.14}$$

Proof. If, in formula (5.13), we replace T by $(T^\sharp)^{-1}$ we obtain (5.14). So we only have to prove (5.13).

Consider the Kreĭn space $\mathcal{K} = \mathcal{K}_1[+] - \mathcal{K}_2$, that is, the direct sum Kreĭn space of $\mathcal{K}_1$ with the Kreĭn space $\mathcal{K}_2$ on which we have changed the sign of the inner product. Then
$$\mathcal{K} = \mathcal{K}^+[+]\mathcal{K}^- \quad \mathcal{K}^+ = \mathcal{K}_1^+ \oplus \mathcal{K}_2^-, \quad \mathcal{K}^- = \mathcal{K}_1^- \oplus \mathcal{K}_2^+, \tag{5.15}$$
is a fundamental decomposition of $\mathcal{K}$. Identify the linear relation T with the subspace $\mathcal{S}$ of $\mathcal{L}$ by
$$\mathcal{S} = \{f + g \mid (f,g) \in T\}. \tag{5.16}$$

Then
$$\nu_-(T) = \kappa_-(\mathcal{S}). \tag{5.17}$$
Since $(T^\sharp)^{-1} = \mathcal{S}^\perp$ we have that
$$\nu_-(T^\sharp) = \kappa_+(\mathcal{S}^\perp). \tag{5.18}$$
We claim that
$$\mathcal{H}_1 \ominus \operatorname{Dom}(\Pi(T)) = \mathcal{S}^\perp \cap \mathcal{K}^-, \quad \mathcal{H}_2 \ominus \operatorname{Dom}(\Pi(T^\sharp)) = \mathcal{S} \cap \mathcal{K}^+. \tag{5.19}$$

The equalities (5.16) through (5.19) and Theorem 3.4.8 imply the equality (5.13).

We only prove the second equality in (5.19); the first equality can be proven similarly. Assume that $u \in \mathcal{K}_1^-$, $v \in \mathcal{K}_2^+$. Then, for all $(f, g) \in \mathcal{K}_2 \oplus \mathcal{K}_1$ we have
$$[v, J_2^+ f]_{\mathcal{K}_2^+} + [u, J_1^- g]_{|\mathcal{K}_1^-|} = [v, f]_{\mathcal{K}_2} - [u, g]_{\mathcal{K}_1}, \tag{5.20}$$
where the second inner product on the left-hand side is the positive definite inner product on $\mathcal{K}_1^-$. Hence the left-hand side is equal to zero for all $(f, g) \in T^\sharp$ if and only if $(u, v) \in T^{\sharp\sharp} = T$, that is, $u + v \in \mathcal{S} \cap (\mathcal{K}_1^- + \mathcal{K}_2^+) = \mathcal{S} \cap \mathcal{K}^-$. It follows that
$$\mathcal{H}_2 \ominus \operatorname{Dom}(\Pi(T^\sharp)) = \{u + v \in \mathcal{K}_1^- + \mathcal{K}_2^+ \mid \forall (f, g) \in T^\sharp, [v, J_2^+ f]_{\mathcal{K}_2^+} + [u, J_1^- g]_{\mathcal{K}_1^-} = 0\}$$
$$= \mathcal{S} \cap \mathcal{K}^-.$$

The claim is proven and the theorem as well. ∎

3.6 Notes

Regular subspaces were introduced by Yu. P. Ginzburg [66] and are one of the main ingredients in operator theory of Kreĭn spaces, see J. Bognár [18] and T. Ya. Azizov and I. S. Iokhvidov [10] for their rich history. Pseudo-regular subspaces were introduced by J. A. Ball and J. W. Helton [12]. Our presentation follows our article [58]. The notion of weak duality and its history can be found in J. Bognár [18] and, inspired by this and motivated by certain problems, we have considered strong duality in [60] as unifying regularity with the most unpleasant situation, that of a neutral subspace. For neutral subspaces the notion of dual pairs (these were called *skew linked* by some authors) was considered by I. S. Iokhvidov and M. G. Kreĭn in [77] and used in many research articles.

The notions of positively/negatively Fredholm subspaces have appeared implicitly, in certain particular cases, in N. J. Young [145] and T. Ya. Azizov [9], cf. [10], in investigations related to the action of certain isometric or unitary operators on subspaces of Kreĭn space, cf. N. J. Young [144]. Our presentation follows our joint article with A. Dijksma [43].

The scattering transform for the general case of linear relations was introduced by Yu. L. Shmulyan [135], by generalising the investigations of the finite dimensional case of V. P. Potapov [122] and the infinite dimensional operator case of Yu. P. Ginzburg [65].

Chapter 4

Linear Operators on Kreĭn Spaces

In a similar fashion as in the case of Hilbert spaces, to any densely defined operator acting between Kreĭn spaces, an adjoint operator can be associated. This enables us to introduce classes of operators such as dissipative, symmetric, selfadjoint, unitary, partial isometries, and all the others. In this chapter we look at some general properties of these operators.

We first consider some general properties of bounded operators such as orthogonal projections, unitary operators, and partial isometries and explain what can be obtained similarly to the case of Hilbert spaces and what is problematic. We explain that operators of type $T^\sharp T$ have rather different behaviour, when compared to the case of Hilbert spaces, and usually yield difficulties and anomalies.

One of the main difficulties in operator theory on Kreĭn spaces or on Pontryagin spaces is that isometric operators are not necessarily bounded. We discuss in detail the question of continuity of isometric operators by proving criteria of continuity and by providing examples and counter examples.

A very general and yet very useful class of operators in Kreĭn spaces is that of dissipative operators, and some of their spectral properties are considered. We show that the Cayley transformation can be used to translate results between the classes of contractive, isometric and unitary operators and those of dissipative, symmetric and, selfadjoint operators, respectively.

4.1 The Adjoint Operator

Let $(\mathcal{K}_1, [\cdot, \cdot])$ and $(\mathcal{K}_2, [\cdot, \cdot])$ be Kreĭn spaces. In the following we often omit the lower index of the inner products, hence $[\cdot, \cdot]$ denotes the inner product in $\mathcal{K}_1$ as well as in $\mathcal{K}_2$, and it will be clear from the context which inner product we have in mind. We consider a linear operator T with domain $\mathrm{Dom}(T)$ dense in $\mathcal{K}_1$ and range $\mathrm{Ran}(T)$ in $\mathcal{K}_2$. The *(Kreĭn space) adjoint operator* $T^\sharp$ of T is defined as follows: $\mathrm{Dom}(T^\sharp)$ is the set of all $y \in \mathcal{K}_2$ for which there exists a $z \in \mathcal{K}_1$ such that $[Tx, y] = [x, z]$, for all $x \in \mathrm{Dom}(T)$, and for such a $y \in \mathcal{K}_2$ we let $T^\sharp y = z$. Equivalently,

$$\mathrm{Dom}(T^\sharp) = \{y \in \mathcal{K}_2 \mid [Tx, y] = [x, z] \text{ for some } z \in \mathcal{K}_1 \text{ and all } x \in \mathrm{Dom}(T)\}, \quad (1.1)$$

and then, since $\mathrm{Dom}(T)$ is dense, the element $z \in \mathcal{K}_1$ as in (1.1) is uniquely determined by y and we let $T^\sharp y = z$.

Note that this definition is a special case of the definition of the adjoint of a linear relation given in Section 3.5 of Chapter 3. It is easy to see (as in the Hilbert space situation) that this definition is correct, in the sense that $T^\sharp$ is an operator, provided that $\mathrm{Dom}(T)$ is dense in $\mathcal{K}_1$. The elements y of $\mathrm{Dom}(T^\sharp)$ can also be characterised by the property that the mapping $\mathrm{Dom}(T) \ni x \mapsto [Tx, y] \in \mathbb{C}$ is continuous.

Remark 4.1.1. Since a Hilbert space $\mathcal{H}$ is a Kreĭn space with $\kappa^-(\mathcal{H}) = 0$, the definition of $T^\sharp$ can be considered as an extension of the definition of the adjoint T^* of a Hilbert space operator. On the other hand, as each Kreĭn space turns into a Hilbert space if we choose a fundamental symmetry J and introduce the Hilbert inner product $\langle \cdot, \cdot \rangle_J$, the definition of $T^\sharp$ can be reduced to the definition of the Hilbert space adjoint. Indeed, we fix two fundamental symmetries J_1 and J_2 in $\mathcal{K}_1$ and $\mathcal{K}_2$, respectively. Then T can be considered as a densely defined linear operator from the Hilbert space $(\mathcal{K}_1, \langle \cdot, \cdot \rangle_{J_1})$ to the Hilbert space $(\mathcal{K}_2, \langle \cdot, \cdot \rangle_{J_2})$. Denote its Hilbert space adjoint by T^*. Then we have, for $x \in \mathrm{Dom}(T)$, $y \in \mathrm{Dom}(T^\sharp)$,

$$[Tx, y] = \langle Tx, J_2 y \rangle_{J_2} = [x, T^\sharp y] = \langle x, J_1 T^\sharp y \rangle_{J_1},$$

and hence $J_2 y \in \mathrm{Dom}(T^*)$ and $T^* J_2 y = J_1 T^\sharp y$. Similarly, for $x \in \mathrm{Dom}(T)$ and $y' \in \mathrm{Dom}(T^*)$,

$$\langle Tx, y' \rangle_{J_2} = [Tx, J_2 y'] = \langle x, T^* y' \rangle_{J_1} = [x, J_1 T^* y'],$$

and hence $J_2 y' \in \mathrm{Dom}(T^\sharp)$ and $T^\sharp J_2 y' = J_1 T^* y'$. Thus, we have proven

$$\mathrm{Dom}(T^\sharp) = J_2 \mathrm{Dom}(T^*) \text{ and } T^\sharp = J_1 T^* J_2. \tag{1.2}$$

Formula (1.2) shows that many of the properties of Hilbert space adjoints can be translated to the Kreĭn space setting, but there are many other properties that do not translate easily or they do not hold at all, in general. ∎

We list now some properties of the adjoint operators in Kreĭn spaces whose proofs can be easily obtained either directly from the corresponding results in the Hilbert space theory, by observing that in these statements the positive definiteness plays no role, or by using (1.2).

Remark 4.1.2. (1) If the corresponding adjoint operators and the algebraic operations make sense then: $(\lambda T)^\sharp = \overline{\lambda} T^\sharp$ for all $\lambda \in \mathbb{C}$; $(T_1 + T_2)^\sharp \supseteq T_1^\sharp + T_2^\sharp$; $(T_1 T_2)^\sharp \supseteq T_2^\sharp T_1^\sharp$.
(2) If T_1 is densely defined and $T_1 \subseteq T_2$ then $T_1^\sharp \supseteq T_2^\sharp$.
(3) If T is densely defined, injective and T^{-1} is also densely defined (that is, T has dense range) then $T^\sharp$ is also injective and $(T^{-1})^\sharp = (T^\sharp)^{-1}$.
(4) $T^\sharp$ is a closed operator. Moreover, the densely defined operator T is closable if and only if $T^\sharp$ is densely defined; in this case, $T^{\sharp\sharp}$ coincides with the closure of T.
(5) The densely defined operator T is bounded if and only if $T^\sharp$ is everywhere defined; in this case $T^\sharp$ is also a bounded operator.

4.1 The Adjoint Operator

(6) If T is densely defined then $\operatorname{Ker}(T^\sharp) = \operatorname{Ran}(T)^\perp$ and $\operatorname{Clos}\operatorname{Ran}(T^\sharp) \subseteq \operatorname{Ker}(T)^\perp$ (equality holds provided that T is closed).

(7) For a closed and densely defined operator T, $\operatorname{Ran}(T)$ is closed if and only if $\operatorname{Ran}(T^\sharp)$ is closed.

(8) Recall that by $\sigma_{\mathrm{p}}(T)$, $\sigma_{\mathrm{c}}(T)$, $\sigma_{\mathrm{r}}(T)$, and $\rho(T)$ we denote, respectively, the point spectrum, the continuous spectrum, the residual spectrum, and the resolvent set of a linear operator $T\colon \operatorname{Dom}(T)(\subseteq \mathcal{H}) \to \mathcal{H}$, for some Hilbert space $\mathcal{H}$. If T is a closed and densely defined operator in the Kreĭn space $\mathcal{K}$ (that is, both of $\operatorname{Dom}(T)$ and $\operatorname{Ran}(T)$ are contained in $\mathcal{K}$) and λ is a complex number, then

(i) $\lambda \in \sigma_{\mathrm{p}}(T)$ implies $\overline{\lambda} \in \sigma_{\mathrm{p}}(T^\sharp) \cup \sigma_{\mathrm{r}}(T^\sharp)$,
(ii) $\lambda \in \sigma_{\mathrm{r}}(T)$ implies $\overline{\lambda} \in \sigma_{\mathrm{p}}(T^\sharp)$,
(iii) $\lambda \in \sigma_{\mathrm{c}}(T)$ implies $\overline{\lambda} \in \sigma_{\mathrm{c}}(T^\sharp)$,
(iv) $\lambda \in \rho(T)$ implies $\overline{\lambda} \in \rho(T^\sharp)$. ∎

We consider now the indices of defect $\nu_\pm$ defined as in Section 3.5 for linear relations, and hence, for any operator T, that is, we let $\nu_\pm(T)$ denote the algebraic rank of positivity/negativity of the inner product $\operatorname{Dom}(T) \times \operatorname{Dom}(T) \ni (f,g) \mapsto [f,g] - [Tf,Tg]$.

Theorem 4.1.3. *Let T be a closed densely defined operator from the Kreĭn space $\mathcal{K}_1$ to the Kreĭn space $\mathcal{K}_2$. If $\mathcal{K}_i = \mathcal{K}_i^-[+]\mathcal{K}_i^+$ are fundamental decompositions of $\mathcal{K}_i$ with corresponding fundamental symmetries $J_i = J_i^+ - J_i^-$, $i=1,2$, then:*

$$\nu_-(T) + \dim \operatorname{Ker}(J_1^- T^\sharp|\operatorname{Dom}(T^\sharp) \cap \mathcal{K}_2^-) = \nu_-(T^\sharp) + \dim \operatorname{Ker}(J_2^- T|\operatorname{Dom}(T) \cap \mathcal{K}_1^-). \tag{1.3}$$

Proof. Let the Kreĭn space $\mathcal{K}$ be defined as in the formula (5.15) at the beginning of the proof of Theorem 3.5.2 and consider the subspace

$$\mathcal{S} = \{f + Tf \mid f \in \operatorname{Dom}(T)\} \subseteq \mathcal{K}.$$

Then we have

$$\kappa_-(\mathcal{S}) = \nu_-(T), \quad \kappa_+(\mathcal{S}) = \nu_-(T^\sharp), \tag{1.4}$$

and

$$\mathcal{S} \cap \mathcal{K}^- = \{f + Tf \mid f \in \operatorname{Ker}(J_2^- T|\operatorname{Dom}(T) \cap \mathcal{K}_1^-)\},$$

and hence

$$\dim(\mathcal{S} \cap \mathcal{K}^-) = \dim \operatorname{Ker}(J_2^- T|\operatorname{Dom}(T) \cap \mathcal{K}_1^-). \tag{1.5}$$

Similarly we obtain

$$\dim(\mathcal{S}^\perp \cap \mathcal{K}^+) = \dim \operatorname{Ker}(J_1^- T^\sharp|\operatorname{Dom}(T^\sharp) \cap \mathcal{K}_2^-). \tag{1.6}$$

Finally, from (1.4), (1.5), (1.6), and Theorem 3.4.8 we obtain the formula (1.3). ∎

As in the case of Hilbert spaces, the adjoint operation allows us to introduce some classes of linear operators in Kreĭn spaces. A linear operator A (not necessarily densely defined) in $(\mathcal{K},[\cdot,\cdot])$ is called *symmetric* or *Hermitian* if $[Ax,y] = [x,Ay]$, $x,y \in \operatorname{Dom}(T)$.

Remark 4.1.4. By the polarisation formula (1.4) in Chapter 1, A is symmetric if and only if the quadratic form $\mathrm{Dom}(A) \ni x \mapsto [Ax, x]$ is real. If, in addition, A is densely defined then it is symmetric if and only if $A \subseteq A^\sharp$. It follows that a densely defined symmetric operator is always closable, its closure being also a symmetric operator. ∎

A densely defined operator A in $\mathcal{K}$ is called *selfadjoint* if $A = A^\sharp$ holds. Any selfadjoint operator is symmetric and closed. A closable operator is called *essentially selfadjoint* if its closure is a selfadjoint operator. The following statement is a consequence of Remark 4.1.2, item (8).

Proposition 4.1.5. *Let A be a selfadjoint operator in $\mathcal{K}$. If λ denotes a complex number, then:*

  (i) $\lambda \in \sigma_\mathrm{p}(A)$ *implies* $\overline{\lambda} \in \sigma_\mathrm{p}(A) \cup \sigma_\mathrm{r}(A)$,
 (ii) $\lambda \in \sigma_\mathrm{r}(A)$ *implies* $\overline{\lambda} \in \sigma_\mathrm{p}(A)$,
(iii) $\lambda \in \sigma_\mathrm{c}(A)$ *implies* $\overline{\lambda} \in \sigma_\mathrm{c}(A)$,
(iv) $\lambda \in \rho(A)$ *implies* $\overline{\lambda} \in \rho(A)$.

In particular, the spectrum $\sigma(A)$ is symmetric with respect to the real axis and the residual spectrum $\sigma_\mathrm{r}(A)$ does not intersect the real line.

The next example shows that apart from the symmetry with respect to the real axis, the spectrum of a selfadjoint operator in a Kreĭn space can be fairly arbitrary.

Example 4.1.6. Let $\mathcal{H}$ be a Hilbert space and let $B \in \mathcal{B}(\mathcal{H})$ be arbitrary. Consider the Kreĭn space $\mathcal{K} = \mathcal{H} \oplus \mathcal{H}$ determined by the symmetry

$$J = \begin{bmatrix} 0 & I \\ I & 0 \end{bmatrix}.$$

Then the operator $A = \begin{bmatrix} B & 0 \\ 0 & B^* \end{bmatrix}$ is bounded and selfadjoint in $\mathcal{K}$ and $\sigma(A) = \sigma(B) \cup \overline{\sigma(B)}$. ∎

A linear operator $A \colon \mathrm{Dom}(A)(\subseteq \mathcal{K}) \to \mathcal{K}$, for some Kreĭn space $\mathcal{K}$, is called *positive* if $[Ax, x] \geqslant 0$ for all $x \in \mathrm{Dom}(A)$. Bounded positive operators in Kreĭn spaces have a better spectral theory.

Proposition 4.1.7. *Let A be a bounded positive operator in the Kreĭn space $\mathcal{K}$. Then $\sigma(A) \subseteq \mathbb{R}$.*

If, in addition, the bounded positive operator has a bounded inverse, then there exists a fundamental decomposition $\mathcal{K} = \mathcal{K}^+[+]\mathcal{K}^-$ such that $A\mathcal{K}^\pm \subseteq \mathcal{K}^\pm$.

Proof. Let J be a fundamental symmetry of $\mathcal{K}$. If $A \in \mathcal{B}(\mathcal{K})$ is positive then $D = JA$ is positive in the Hilbert space $(\mathcal{K}, (\cdot, \cdot)_J)$ and hence there exists the positive operator $D^{1/2}$. We recall now that for two bounded operators $X, Y \in \mathcal{B}(\mathcal{K})$ we have

$$\sigma(XY) \setminus \{0\} = \sigma(YX) \setminus \{0\}.$$

4.1 THE ADJOINT OPERATOR

Therefore, since $A = JD = JD^{1/2}D^{1/2}$ we have

$$\sigma(A) \setminus \{0\} = \sigma(D^{1/2}JD^{1/2}) \setminus \{0\} \subset \mathbb{R},$$

and hence $\sigma(A) \subseteq \mathbb{R}$.

If, in addition, A has a bounded inverse then its spectrum splits $\sigma(A) = \sigma_- \cap \sigma_+$, where $\sigma_+ \subset (0, +\infty)$ and $\sigma_- \subset (-\infty)$, are two spectral subsets of A. Letting $P_\pm$ denote the corresponding spectral projections, we observe that they are selfadjoint projections such that $I = P_+ + P_-$, $P_+P_- = 0$, P_+ and $-P_-$ are positive and then, letting $\mathcal{K}^\pm = \mathrm{Ran}(P_\pm)$ it follows that $\mathcal{K} = \mathcal{K}^+[+]\mathcal{K}^-$ is a fundamental decomposition of $\mathcal{K}$ such that $A\mathcal{K}^\pm \subseteq \mathcal{K}^\pm$. ∎

A bounded positive operator that has a bounded inverse is called *uniformly positive*. This concept will play an important rôle later on, for example see Section 7.3. An operator $A \in \mathcal{B}(\mathcal{K})$ with the property that $A\mathcal{K}^\pm \subseteq \mathcal{K}^\pm$ for some fundamental decomposition of the Kreĭn space $\mathcal{K}$ is called *fundamentally decomposable*. So, the previous proposition says that any uniformly positive operator A is *fundamentally reducible*, that is, both A and $A^\sharp$ are fundamentally decomposable.

A linear operator V with domain in the Kreĭn space $\mathcal{K}_1$ and the range in the Kreĭn space $\mathcal{K}_2$ is called *isometric* if

$$[Vx, Vy] = [x, y], \quad x, y \in \mathrm{Dom}(V).$$

Remark 4.1.8. By the polarisation formula it follows that the operator V is isometric if and only if $[Vx, Vx] = [x, x]$, $x \in \mathrm{Dom}(V)$. Isometric operators in Kreĭn spaces need be neither injective nor bounded; see the next section for boundedness criteria of isometric operators and examples of unbounded ones. However, if V is an isometric operator and $\mathrm{Dom}(V)$ is nondegenerate (in particular, if $\mathrm{Dom}(V)$ is dense) then V is injective. On the other hand, an injective and densely defined operator V is isometric if and only if $V^{-1} \subseteq V^\sharp$. ∎

A densely defined and injective linear operator U from $\mathcal{K}_1$ into $\mathcal{K}_2$ is called *unitary* if $U^{-1} = U^\sharp$ holds.

Remark 4.1.9. Any unitary operator is closed and has dense range but in general it is not bounded. For example, if $\mathcal{K} = \mathcal{H} \oplus \mathcal{H}$ for some infinite dimensional Hilbert space $\mathcal{H}$, let $J \colon \mathcal{K} \to \mathcal{K}$ be defined by $J(x \oplus y) = x \oplus -y$, for all $x, y \in \mathcal{H}$, and let $[f, g] = \langle Jf, g \rangle$ for all $f, g \in \mathcal{K}$. Then $(\mathcal{K}; [\cdot, \cdot])$ is a Kreĭn space. Then there exists a maximal strictly positive subspace $\mathcal{M} \subset \mathcal{K}$ that is not uniformly positive, for example see Lemma 2.2.2 and Corollary 2.1.7. Then $\mathcal{M} + \mathcal{M}^\perp$ is dense in $\mathcal{K}$ but not equal to $\mathcal{K}$. Letting $\mathrm{Dom}(S) = \mathcal{M} + \mathcal{M}^\perp$ and $S(f + g) = f - g$ for arbitrary $f \in \mathcal{M}$ and $g \in \mathcal{M}^\perp$ it follows that S is unitary and unbounded.

In Theorem 5.3.6 and its dual Theorem 5.3.9 we present a description of all unitary operators between Kreĭn spaces and explicit criteria to distinguish between the bounded and unbounded ones. ∎

The following lemma is very useful in applications.

Proposition 4.1.10. *Let U be an isometric operator with dense domain in $\mathcal{K}_1$ and with dense range in $\mathcal{K}_2$. Then U is closable and its closure has the same properties.*

Proof. Since U is densely defined it follows that its domain is nondegenerate and taking into account that it is isometric it follows that U is injective. Hence U^{-1} is a densely defined operator, with dense range, and isometric as well. Then, for any $h \in \mathrm{Ran}(U) = \mathrm{Dom}(U^{-1})$ and $k \in \mathrm{Dom}(U)$ we have

$$[U^{-1}h, k]_{\mathcal{K}_1} = [UU^{-1}h, Uk]_{\mathcal{K}_1} = [h, Uk]_{\mathcal{K}_1}, \tag{1.7}$$

which shows that $U \subseteq (U^{-1})^\sharp$. This implies that U is closable and then its closure $\widetilde{U}$ is a closed and isometric operator, with dense domain and dense range. In particular, $\widetilde{U}$ is invertible and its inverse $\widetilde{U}^{-1}$ is closed and isometric, with dense domain and dense range. ∎

In connection with Proposition 4.1.10, a natural question is whether a closed, densely defined isometry, and with dense range is necessarily a unitary operator. The answer is, in general, no but, since more sophisticated tools are needed, it will be provided later in Example 5.3.3.

The counter-part of Proposition 4.1.5 for the case of unitary operators is the following.

Proposition 4.1.11. *Let U be a unitary operator in $\mathcal{K}$. If $\lambda \neq 0$ denotes a complex number and $\lambda^* = \frac{1}{\bar{\lambda}}$ then:*

(i) $\lambda \in \sigma_\mathrm{p}(U)$ *implies* $\lambda^* \in \sigma_\mathrm{p}(U) \cup \sigma_\mathrm{r}(U)$,
(ii) $\lambda \in \sigma_\mathrm{r}(U)$ *implies* $\lambda^* \in \sigma_\mathrm{p}(U)$,
(iii) $\lambda \in \sigma_\mathrm{c}(U)$ *implies* $\lambda^* \in \sigma_\mathrm{c}(U)$,
(iv) $\lambda \in \rho(U)$ *implies* $\lambda^* \in \rho(U)$.

In particular, the spectrum $\sigma(U)$ is symmetric with respect to the unit circle and the residual spectrum $\sigma_\mathrm{r}(U)$ does not intersect the unit circle.

As in the case of selfadjoint operators, except for the symmetry with respect to the unit circle, the spectrum of a unitary operator in a Kreĭn space can be fairly arbitrary. This will be seen in all possible situations as a consequence of the Cayley transformation that is discussed in Section 4.5.

4.2 Some Classes of Bounded Operators

In this section we record some properties of bounded linear operators between Kreĭn spaces. Firstly, the Riesz Representation Theorem of bounded sesquilinear forms works as for Hilbert spaces.

4.2 Some Classes of Bounded Operators

Proposition 4.2.1. *Let $\mathcal{K}_1$ and $\mathcal{K}_2$ be two Kreĭn spaces and $\psi\colon \mathcal{K}_1\times\mathcal{K}_2 \to \mathbb{C}$ a sesquilinear form, that is, linear in the first variable and conjugate linear in the second variable, bounded in the sense that, for some fundamental norms $\|\cdot\|_j$ on $\mathcal{K}_j$, $j=1,2$ we have $|\psi(x,y)| \leqslant C\|x\|_1\|y\|_2$ for some constant $C \geqslant 0$ and all $x \in \mathcal{K}_1$ and all $y \in \mathcal{K}_2$. Then there exists uniquely a bounded linear operator $T \in \mathcal{B}(\mathcal{K}_2,\mathcal{K}_1)$ such that $\psi(x,y) = [x, Ty]_{\mathcal{K}_1}$, for all $x \in \mathcal{K}_1$ and all $y \in \mathcal{K}_2$.*

Proof. Indeed, fix J_j a fundamental symmetry of $\mathcal{K}_j$ and consider the Hilbert spaces $(\mathcal{K}_1,\langle\cdot,\cdot\rangle_{J_j})$, $j=1,2$. Then ψ is a bounded sesquilinear form with respect to these Hilbert spaces and hence, by the Riesz Theorem there exists $S\colon \mathcal{K}_2 \to \mathcal{K}_1$ a unique bounded linear operator such that $\psi(x,y) = \langle x, Sy\rangle_{J_1}$ for all $x \in \mathcal{K}_1$ and all $y \in \mathcal{K}_2$. Then, letting $T = J_1 S \in \mathcal{B}(\mathcal{K}_2,\mathcal{K}_1)$ we have $\psi(x,y) = \langle x, Sy\rangle_{J_1} = [x, J_1 Sy]_{\mathcal{K}_1}$ for all $x \in \mathcal{K}_1$ and all $y \in \mathcal{K}_2$. ■

If $T \in \mathcal{B}(\mathcal{K}_1,\mathcal{K}_2)$ then, with definition as in (1.1), $T^\sharp \in \mathcal{B}(\mathcal{K}_2,\mathcal{K}_1)$. This can be seen as a consequence of Proposition 4.2.1. Correspondingly, we have the usual classes of bounded *selfadjoint operators*, *orthogonal projections*, and all the other classes that can be defined by means of the involution $\sharp$. There are many common features between the involution $*$ in Hilbert spaces and the involution $\sharp$ in Kreĭn spaces, but there are significant differences. One of the major differences is the behaviour of operators of type $T^\sharp T$. This type of factorisation will be studied from the geometric point of view in Chapter 6. Here we only point out what difficulty indefiniteness brings.

Proposition 4.2.2. *Let $\mathcal{K}_1$ and $\mathcal{K}_2$ be Kreĭn spaces.*

(a) *For any operator norm $\|\cdot\|$ associated to fundamental norms on $\mathcal{K}_1$ and $\mathcal{K}_2$ we have $\|T^\sharp\| = \|T\|$ for all $T \in \mathcal{B}(\mathcal{K}_1,\mathcal{K}_2)$. In particular, $\|T^\sharp T\| \leqslant \|T\|^2$.*

(b) *Given $T \in \mathcal{B}(\mathcal{K}_1,\mathcal{K}_2)$ we have $T^\sharp T = 0$ if and only if $\operatorname{Ran}(T)$ is a neutral linear manifold of $\mathcal{K}_2$.*

(c) *If $\mathcal{K}_1 \neq \{0\}$ and $\mathcal{K}_2$ is indefinite, then there exists $T \in \mathcal{B}(\mathcal{K}_1,\mathcal{K}_2)$, $T \neq 0$, such that $T^\sharp T = 0$. In particular, with notation as in item (a), we have $0 = \|T^\sharp T\| < \|T\|^2$.*

Proof. (a) Let J_j be fundamental symmetries on $\mathcal{K}_j$ and consider the fundamental norms associated to them. Then $T^\sharp = J_1 T^* J_2$, where $*$ denotes the involution with respect to the Hilbert spaces $(\mathcal{K}_j; \langle\cdot,\cdot\rangle_{J_j})$, $j=1,2$. Denote by $\|\cdot\|_j$ the fundamental norm defined by J_j, $j=1,2$. Since J_j is a Hilbert space unitary operator in $(\mathcal{K}_j;\langle\cdot,\cdot\rangle_{J_j})$, $j=1,2$, we have

$$\|T^\sharp\| = \|J_1 T^* J_2\| = \sup_{\|y\|_2\leqslant 1,\ y\in\mathcal{K}_2} \|J_1 T^* J_2 y\|_1$$
$$= \sup_{\|y\|_2\leqslant 1,\ y\in\mathcal{K}_2} \|T^* J_2 y\|_1 = \sup_{\|z\|_2\leqslant 1,\ z\in\mathcal{K}_2} \|T^* z\| = \|T^*\| = \|T\|.$$

Then, $\|T^\sharp T\| \leqslant \|T^\sharp\|\|T\| = \|T\|^2$.

(b) We consider the linear manifold $\operatorname{Ran}(T)$ in $\mathcal{K}_2$ and observe that

$$[T^\sharp T x, y] = [Tx, Ty], \quad x, y \in \mathcal{K}_1,$$

hence, the inner product $[\cdot,\cdot]$ of $\mathcal{K}_2$ vanishes on $\mathrm{Ran}(T)$ if and only if $[T^\sharp Tx, y] = 0$ for all $x, y \in \mathcal{K}_1$. Since $\mathcal{K}_2$ is nondegenerate, this is equivalent with $T^\sharp Tx = 0$ for all $x \in \mathcal{K}_1$, that is, $T^\sharp T = 0$.

(c) If $\mathcal{K}_1 \neq \{0\}$, there exists $x_0 \in \mathcal{K}_1 \setminus \{0\}$ and let $\varphi \colon \mathcal{K}_1 \to \mathbb{C}$ be the bounded linear functional $\varphi = [\cdot, x_0]$. Since $\mathcal{K}_1$ is nondegenerate it follows that $\varphi \neq 0$. On the other hand, since $\mathcal{K}_2$ is indefinite, there exists $y_0 \in \mathcal{K}_2 \setminus \{0\}$ a neutral vector, see Lemma 1.1.1. Define the operator $T \colon \mathcal{K}_1 \to \mathcal{K}_2$ by

$$Tx = \varphi(x) y_0, \quad x \in \mathcal{K}_1. \tag{2.1}$$

Then $T \neq 0$, T is bounded, and $\mathrm{Ran}(T)$ is a neutral subspace of $\mathcal{K}_2$ hence, by item (b), we have $T^\sharp T = 0$. ∎

Given a Kreĭn space $\mathcal{K}$, an operator $P \in \mathcal{B}(\mathcal{K})$ is called an *orthogonal projection* if $P = P^\sharp = P^2$. We stress that in the Kreĭn space setting, we distinguish between orthogonal projections, which are always bounded linear operators, and selfadjoint projections, which may be unbounded.

Proposition 4.2.3. *Let $\mathcal{K}$ be a Kreĭn space and $P \in \mathcal{B}(\mathcal{K})$. The following assertions are equivalent.*

(i) *P is an orthogonal projection.*
(ii) *$I - P$ is an orthogonal projection.*
(iii) *$\mathrm{Ran}(P)$ is a regular subspace and, with respect to $\mathcal{K} = \mathrm{Ran}(P) [+] \mathrm{Ran}(P)^\perp$, $P(x + y) = x$ for arbitrary $x \in \mathrm{Ran}(P)$ and $y \in \mathrm{Ran}(P)^\perp$.*

Proof. (i)⇔(ii). If P is an orthogonal projection then $(I - P)^\sharp = I - P^\sharp = I - P$ and $(I - P)^2 = I - 2P + P^2 = I - 2P + P = I - P$, hence $I - P$ is an orthogonal projection as well. The converse implication follows from here since $P = I - (I - P)$.

(i)⇒(iii). If P is an orthogonal projection then $Px = x$ for all $x \in \mathrm{Ran}(P)$. Then, for any sequence $(x_n)_n$ of vectors from $\mathrm{Ran}(P)$ such that $x_n \xrightarrow[n\to\infty]{} x$ for some $x \in \mathcal{K}$, since P is bounded we have $Px_n \xrightarrow[n\to\infty]{} Px$ but, since $x_n = Px_n$ for all $n \in \mathbb{N}$, it follows that $Px = x$ hence $x \in \mathrm{Ran}(P)$. This shows that $\mathrm{Ran}(P)$ is closed.

Since $0 = P - P^2 = P(I - P)$ it follows that $\mathrm{Ran}(I - P) \subseteq \mathrm{Ker}(P)$. Conversely, if $x \in \mathrm{Ker}(P)$ then $(I - P)x = x$ hence $x \in \mathrm{Ran}(I - P)$. Thus, $\mathrm{Ker}(P) = \mathrm{Ran}(I - P)$. Since $P = P^\sharp$ we have $\mathrm{Ran}(P) = \mathrm{Clos}\,\mathrm{Ran}(P) = \mathrm{Ker}(P)^\perp = \mathrm{Ran}(I - P)^\perp$. But, for all $x \in \mathcal{K}$ we have $x = Px + (I - P)x$, hence $\mathcal{K} = \mathrm{Ran}(P) + \mathrm{Ran}(P)^\perp$, which implies that $\mathrm{Ran}(P)$ is a regular subspace. Consequently, its orthogonal companion $\mathrm{Ran}(P)^\perp = \mathrm{Ker}(I - P)$ is regular as well and then, for any $x \in \mathrm{Ran}(P)$ and $y \in \mathrm{Ran}(P)^\perp$ we have $P(x + y) = Px + Py = x$.

(iii)⇒(i). Let $z \in \mathcal{K}$ be arbitrary. Then $z = x + y$ with uniquely determined vectors $x \in \mathrm{Ran}(P)$ and $y \in \mathrm{Ran}(P)^\perp$. Then $Pz = P(x + y) = x$ and $P^2 z = Px = x = Pz$, hence $P^2 = P$. On the other hand, for arbitrary $z_1, z_2 \in \mathcal{K}$, representing $z_j = x_j + y_j$ with $x_j \in \mathrm{Ran}(P)$ and $y_j \in \mathrm{Ran}(P)$, $j = 1, 2$, we have

$$[Pz_1, z_2] = [x_1, x_2 + y_2] = [x_1, x_2] = [x_1 + y_1, x_2] = [z_1, Pz_2],$$

4.2 SOME CLASSES OF BOUNDED OPERATORS

hence $P^\sharp = P$. ∎

Sometimes we need nonorthogonal projections.

Lemma 4.2.4. *Let $E \in \mathcal{B}(\mathcal{K})$ be a projection, that is, $E^2 = E$.*

(i) *$E^\sharp$ is also a projection.*

(ii) *For any fundamental symmetry J of $\mathcal{K}$, the operator $E^\sharp J$ maps the subspace $E\mathcal{K}$ injectively onto the subspace $E^\sharp \mathcal{K}$.*

(iii) *The subspaces $E\mathcal{K}$ and $E^\sharp \mathcal{K}$ are in strong duality.*

(iv) *If E is selfadjoint then the subspace $E\mathcal{K}$ is regular and $(E\mathcal{K})^\perp = (I - E)\mathcal{K}$.*

Proof. (i) $(E^\sharp)^2 = (E^2)^\sharp = E^\sharp$.

(ii) Since E and $E^\sharp$ are projections it follows that the ranges $E\mathcal{K}$ and $E^\sharp \mathcal{K}$ are closed. Let J be a fundamental symmetry of $\mathcal{K}$ and denote by $\|\cdot\|$ the corresponding fundamental norm. Let $y \in E^\sharp \mathcal{K}$ be such that $\langle E^\sharp J E\mathcal{K}, y\rangle_J = \{0\}$. Then $E^* E J y = 0$ and hence $E J y = 0$. Since $y = E^\sharp y$ it follows that $0 = E J E^\sharp y = E E^* J y$ and hence $J y = J E^\sharp y = E^* J y = 0$, therefore $y = 0$. This shows that $E^\sharp J E \mathcal{K}$ is dense in $E^\sharp \mathcal{K}$.

It remains to show that the linear manifold $E^\sharp J E \mathcal{K}$ is closed. Assuming the contrary, let $(x_n)_{n\in\mathbb{N}}$ be a sequence in $E\mathcal{K}$ such that $\|x_n\| = 1$, $n \in \mathbb{N}$, and $\|E^\sharp J x_n\| \xrightarrow[n\to\infty]{} 0$. Then

$$\|Ex_n\|^2 = \|E^* E x_n\| = \|E^* x_n\| = \|J E^* x_n\| = \|E^\sharp J x_n\| \xrightarrow[n\to\infty]{} 0,$$

and hence $x_n = E x_n \xrightarrow[n\to\infty]{} 0$, a contradiction.

(iii) We have $(E\mathcal{K})^\perp = (I - E^\sharp)\mathcal{K}$ and $(E^\sharp \mathcal{K})^\perp = (I - E)\mathcal{K}$ and hence $E\mathcal{K} \cap (E^\sharp \mathcal{K})^\perp = \{0\}$ and $E\mathcal{K} + (E^\sharp \mathcal{K})^\perp = \mathcal{K}$.

(iv) Follows from (iii) and Remark 3.3.2. ∎

Corollary 4.2.5. *Let T be a closed and densely defined operator in $\mathcal{K}$. If σ is a spectral set of T then $\overline{\sigma} = \{\overline{\lambda} \mid \lambda \in \sigma\}$ is also a spectral set of $T^\sharp$ and the corresponding Riesz projections satisfy $E(\overline{\sigma}; T^\sharp) = E(\sigma; T)^\sharp$.*

Proof. Let C_σ be a positively oriented Jordan contour in $\rho(T)$ which separates the spectral set σ with respect to $\sigma(T) \setminus \sigma$. Then $C_{\overline{\sigma}} = \{\overline{\lambda} \mid \lambda \in \sigma\}$ is a negatively oriented closed Jordan contour separating the spectral set $\overline{\sigma}$ of $T^\sharp$ with respect to $\sigma(T^\sharp) \setminus \overline{\sigma}$. Therefore, using Cauchy contour integration we have

$$E(\sigma; T)^\sharp = \Big(\frac{1}{2\pi\mathrm{i}} \int_{C_\sigma} (\zeta I - T)^{-1} \mathrm{d}\zeta\Big)^\sharp = \frac{-1}{2\pi\mathrm{i}} \int_{C_{\overline{\sigma}}} (\overline{\zeta} I - T^\sharp)^{-1} \mathrm{d}\overline{\zeta} = E(\overline{\sigma}; T^\sharp). \; \blacksquare$$

Recall that an operator $V \in \mathcal{B}(\mathcal{K}_1, \mathcal{K}_2)$ is isometric if $[Vx, Vy]_{\mathcal{K}_2} = [x, y]_{\mathcal{K}_1}$. It is clear that V is isometric if and only if $V^\sharp V = I_1$. A bounded operator whose adjoint is isometric is called *coisometric* or a *coisometry*. In particular, a bounded operator is unitary if and only if it is both isometric and coisometric. We first record some simple geometric properties of bounded unitary operators.

Proposition 4.2.6. *Let V be a bounded unitary operator from the Kreĭn space $\mathcal{K}_1$ into the Kreĭn space $\mathcal{K}_2$ and let $\mathcal{L}$ be a linear manifold in $\mathcal{K}_1$.*

(a) $V\mathcal{L}^\perp = (V\mathcal{L})^\perp$ *and* $V\mathcal{L}^0 = (V\mathcal{L})^0$.

(b) $\mathcal{L}$ *is closed if and only if $V\mathcal{L}$ is closed.*

(c) $\mathcal{L}$ *is a nondegenerate (regular, pseudo-regular) subspace in $\mathcal{K}_1$ if and only if $V\mathcal{L}$ is a nondegenerate (regular, pseudo-regular, respectively) subspace in $\mathcal{K}_2$.*

(d) $\mathcal{L}$ *is a positive (strictly positive, uniformly positive) subspace in $\mathcal{K}_1$ if and only if $V\mathcal{L}$ is a positive (strictly positive, uniformly positive, respectively) subspace in $\mathcal{K}_2$.*

(e) $\mathcal{L}$ *is a maximal positive (strictly positive, uniformly positive) subspace in $\mathcal{K}_1$ if and only if $V\mathcal{L}$ is a maximal positive (strictly positive, uniformly positive, respectively) subspace in $\mathcal{K}_2$.*

Proof. (a) $V\mathcal{L}^\perp \subseteq (V\mathcal{L})^\perp$ holds since V is isometric. Conversely, assuming that $y \in (V\mathcal{L})^\perp$ then
$$0 = [y, Vx] = [V^\sharp y, x], \quad x \in \mathcal{L},$$
that is, $V^\sharp y \in \mathcal{L}^\perp$ and hence $y = VV^\sharp x \in V\mathcal{L}^\perp$. We have proven that $V\mathcal{L}^\perp = (V\mathcal{L})^\perp$. Then
$$V\mathcal{L}^0 = V(\mathcal{L} \cap \mathcal{L}^\perp) = V\mathcal{L} \cap V\mathcal{L}^\perp = V\mathcal{L} \cap (V\mathcal{L})^\perp = (V\mathcal{L})^0.$$

(b) This is a direct consequence of the bounded invertibility of V.

(c) From the second relation in (a) it follows that $\mathcal{L}$ is nondegenerate if and only if $V\mathcal{L}$ is nondegenerate. Then take into account that
$$(V\mathcal{L})^\perp + V\mathcal{L} = V\mathcal{L}^\perp + V\mathcal{L} = V(\mathcal{L}^\perp + \mathcal{L}),$$
combined with Theorem 3.1.1 and Theorem 3.2.1.

(d) V preserves positivity since it is isometric and it also preserves strict positivity since it is isometric and injective. On the other hand, if $\|\cdot\|_2$ is a fundamental norm on $\mathcal{K}_2$ then
$$\|x\|_1 = \|Vx\|_2, \quad x \in \mathcal{K}_1,$$
is a fundamental norm on $\mathcal{K}_1$ since V is an isometric isomorphism, hence V also preserves the uniform positivity of subspaces.

(e) This statement follows from (d), the first statement in (a), and the characterisation of maximalitiy in Corollary 2.1.7. ∎

From Corollary 3.1.6, for two Kreĭn spaces $\mathcal{K}_1$ and $\mathcal{K}_2$, it follows that there exists a bounded unitary operator U (isometric isomorphisms) from $\mathcal{K}_1$ into $\mathcal{K}_2$ if and only if $\kappa^\pm(\mathcal{K}_1) = \kappa^\pm(\mathcal{K}_2)$. A consequence of Theorem 5.3.1 will be that this statement remains true without the boundedness condition on the unitary operator U.

The next proposition shows that bounded isometric operators in Kreĭn spaces behave, from a certain point of view, like isometric operators in Hilbert spaces.

4.2 SOME CLASSES OF BOUNDED OPERATORS

Proposition 4.2.7. *Let V be a bounded isometric operator from the Kreĭn space $\mathcal{K}_1$ into the Kreĭn space $\mathcal{K}_2$.*
  (a) *V maps subspaces of $\mathcal{K}_1$ into subspaces of $\mathcal{K}_2$.*
  (b) *$\mathrm{Ran}(V)$ is a regular subspace of $\mathcal{K}_2$.*

Proof. (a) If $\mathcal{L}$ is a subspace of $\mathcal{K}_1$ then clearly $V\mathcal{L}$ is a linear manifold of $\mathcal{K}_2$. It remains to prove that it is closed. To see this, let $(x_n)_n$ be a sequence in $\mathcal{L}$ such that $Vx_n \xrightarrow[n\to\infty]{} y$ for some $y \in \mathcal{K}_2$. Since $V^\sharp$ is bounded and $V^\sharp V = I_1$ we have $x_n = V^\sharp V x_n \xrightarrow[n\to\infty]{} V^\sharp y$ and, since $\mathcal{L}$ is closed, it follows that $V^\sharp y \in \mathcal{L}$. Then, since V is bounded it follows that $Vx_n \xrightarrow[n\to\infty]{} VV^\sharp y$. By uniqueness of the limit, it follows that $y = VV^\sharp y$, hence $y \in V\mathcal{L}$.

(b) We assume that $\mathcal{K}_1$ is nontrivial. Let $\mathcal{K}_1 = \mathcal{K}_1^+[+]\mathcal{K}_1^-$ be a fundamental decomposition of $\mathcal{K}_1$, and let $\|\cdot\|_1$ denote the corresponding fundamental norm in $\mathcal{K}_1$ and fix a fundamental norm $\|\cdot\|_2$ in $\mathcal{K}_2$. Then

$$[x,x] = \pm\|x\|_1^2, \quad x \in \mathcal{K}_1^\pm,$$

hence

$$[Vx,Vx] = [x,x] = \|x\|_1^2 \geqslant \frac{1}{\|V\|^2}\|Vx\|_2^2, \quad x \in \mathcal{K}_1^+,$$

that is, $V\mathcal{K}_1^+$ is a uniformly positive linear manifold in $\mathcal{K}_2$. Similarly we prove that $V\mathcal{K}_1^-$ is uniformly negative. From (a) we know that both $V\mathcal{K}_1^+$ and $V\mathcal{K}_1^-$ are subspaces. Then $\mathrm{Ran}(V) = V\mathcal{K}_1^+[+]V\mathcal{K}_1^-$ and hence it is a regular subspace of $\mathcal{K}_2$. ∎

As a consequence of Proposition 4.2.7 it follows that if $V \in \mathcal{B}(\mathcal{K}_1,\mathcal{K}_2)$ is isometric then it can be viewed as a bounded unitary operator $V\colon \mathcal{K}_1 \to V\mathcal{K}_1$, when viewing the regular subspace $V\mathcal{K}_1$ as a Kreĭn space. With this interpretation, we have the following consequence of Proposition 4.2.6.

Corollary 4.2.8. *Let V be a bounded isometric operator from the Kreĭn space $\mathcal{K}_1$ into the Kreĭn space $\mathcal{K}_2$ and let $\mathcal{L}$ be a linear manifold in $\mathcal{K}_1$.*
  (a) *$V\mathcal{L}^\perp \subseteq (V\mathcal{L})^\perp$ and $V\mathcal{L}^0 = (V\mathcal{L})^0$.*
  (b) *$\mathcal{L}$ is closed if and only if $V\mathcal{L}$ is closed.*
  (c) *$\mathcal{L}$ is a nondegenerate (regular, pseudo-regular) subspace in $\mathcal{K}_1$ if and only if $V\mathcal{L}$ is a nondegenerate (regular, pseudo-regular, respectively) subspace in $\mathcal{K}_2$.*
  (d) *$\mathcal{L}$ is a positive (strictly positive, uniformly positive) subspace in $\mathcal{K}_1$ if and only if $V\mathcal{L}$ is a positive (strictly positive, uniformly positive, respectively) subspace in $\mathcal{K}_2$.*

An operator $S \in \mathcal{B}(\mathcal{K}_1,\mathcal{K}_2)$ is a *partial isometry* if $\mathrm{Ker}(S)$ and $\mathrm{Ran}(S)$ are regular subspaces and $S|\,\mathrm{Ker}(S)^\perp$ is an isometry. In this situation, the regular subspace $\mathrm{Ker}(S)^\perp$ is called the *initial space* and $\mathrm{Ran}(S)$ is called the *final space*. It is clear from Proposition 4.2.7 that any isometry and any coisometry are partial isometries.

Remark 4.2.9. Let $S \in \mathcal{B}(\mathcal{K}_1, \mathcal{K}_2)$ be a partial isometry. Then:

(a) $[Sx, Sy] = [x, y]$ for all $x, y \in \operatorname{Ker}(S)^\perp$.

(b) $S^\sharp$ is a partial isometry as well, with initial space $\operatorname{Ran}(S) = \operatorname{Ker}(S^\sharp)^\perp$ and final space $\operatorname{Ker}(S)^\perp = \operatorname{Ran}(S^\sharp)$.

(c) Letting P denote the orthogonal projection onto the regular subspace $\operatorname{Ker}(S)^\perp$ we have $S^\sharp S = P$ and $\operatorname{Ker}(S) = \operatorname{Ker}(P)$. Similarly, letting Q be the orthogonal projection onto $\operatorname{Ran}(S)$ we have $SS^\sharp = Q$ and $\operatorname{Ker}(S^\sharp) = \operatorname{Ker}(Q)$. ∎

Proposition 4.2.10. *Let $S \in \mathcal{B}(\mathcal{K}_1, \mathcal{K}_2)$. The following assertions are equivalent.*

(i) *S is a partial isometry.*

(i)' $SS^\sharp S = S$.

(ii) $S^\sharp S S^\sharp = S^\sharp$.

(ii)' *$S^\sharp$ is a partial isometry.*

(iii) *$S^\sharp S$ is a projection and $\operatorname{Ker}(S^\sharp S) = \operatorname{Ker}(S)$.*

(iii)' *$SS^\sharp$ is a projection and $\operatorname{Ker}(SS^\sharp) = \operatorname{Ker}(S^\sharp)$.*

Proof. (i)⇒(ii). Assume that S is a partial isometry. If $x \in \operatorname{Ker}(S)$ then $SS^\sharp Sx = 0 = Sx$. If $x \in \operatorname{Ker}(S)^\perp$ then, with notation as in Remark 4.2.9 (c), $SS^\sharp Sx = SPx = Sx$. Hence $SS^\sharp S = S$.

(ii)⇒(i). Assume that $SS^\sharp S = S$ and set $P = S^\sharp S$. Then $P = P^\sharp$ and $P^2 - P = S^\sharp(SS^\sharp S - S) = 0$, hence P is an orthogonal projection in $\mathcal{K}_1$. Similarly, since $S^\sharp SS^\sharp = S^\sharp$, it follows that $Q = SS^\sharp$ is an orthogonal projection in $\mathcal{K}_2$. By Proposition 4.2.3 $\operatorname{Ran}(P)$ and $\operatorname{Ran}(Q)$ are regular subspaces of $\mathcal{K}_1$ and of $\mathcal{K}_2$, respectively, and then

$$S \operatorname{Ran}(P) = SS^\sharp S \operatorname{Ran}(P) \subseteq SS^\sharp \mathcal{K}_2 = \operatorname{Ran}(Q),$$

and

$$S^\sharp \operatorname{Ran}(Q) = S^\sharp SS^\sharp \operatorname{Ran}(Q) \subseteq S^\sharp S \mathcal{K}_1 = \operatorname{Ran}(P).$$

Consequently, $\operatorname{Ran}(S) = \operatorname{Ran}(Q)$ and $S | \operatorname{Ran}(P)$ is an isometry.

In order to finish the proof, it remains to show that $\operatorname{Ker}(S) = \operatorname{Ran}(P)^\perp$. Indeed, for arbitrary $x \in \mathcal{K}_1$, if $Sx = 0$ then $Px = S^\sharp Sx = 0$ and, consequently, $x \in \operatorname{Ran}(P)^\perp$. Conversely, if $x \in \operatorname{Ran}(P)^\perp$ then $S^\sharp Sx = Px = 0$ and then $Sx = SS^\sharp Sx = 0$.

(iii)⇒(ii). Assume that $S^\sharp S$ is a projection and that $\operatorname{Ker}(S^\sharp S) = \operatorname{Ker}(S)$. Then $P = S^\sharp S$ is an orthogonal projection onto the regular subspace $\operatorname{Ker}(S)^\perp$. If $x \in \operatorname{Ker}(S)$ then $SS^\sharp Sx = 0 = Sx$. If $x \in \operatorname{Ker}(S)^\perp$ then $SS^\sharp Sx = SPx = Sx$. Hence $SS^\sharp S = S$. ∎

4.3 Continuity of Isometric Operators

Let $\mathcal{K}_1$ and $\mathcal{K}_2$ be Kreĭn spaces and $V \colon \operatorname{Dom}(V)(\subseteq \mathcal{K}_1) \to \mathcal{K}_2$ be a linear operator. We recall that V is called isometric if $[Vx, Vy] = [x, y]$, $x, y \in \operatorname{Dom}(V)$. In general, isometric operators in Kreĭn spaces are not necessarily continuous with respect to the strong topology.

4.3 CONTINUITY OF ISOMETRIC OPERATORS

Proposition 4.3.1. *Let V be an isometric operator with domain $\mathrm{Dom}(V)$ in $\mathcal{K}_1$ and range $\mathrm{Ran}(V)$ in $\mathcal{K}_2$ and assume that there exists a decomposition $\mathrm{Dom}(V) = \mathcal{D}_+ \dotplus \mathcal{D}_-$ with $\mathcal{D}_+$ a uniformly positive linear manifold, $\mathcal{D}_-$ a uniformly negative linear manifold and $\mathcal{D}_+ \perp \mathcal{D}_-$. Then V is bounded if and only if $V\mathcal{D}_+$ is uniformly positive and $V\mathcal{D}_-$ is uniformly negative.*

Proof. Assume that $V \neq 0$ is bounded and let $\|\cdot\|$ denote fundamental norms in both Kreĭn spaces $\mathcal{K}_1$ and $\mathcal{K}_2$. Let $\alpha > 0$ be such that
$$[x,x] \geqslant \alpha \|x\|^2, \quad x \in \mathcal{D}_+.$$
Then
$$[Vx, Vx] = [x,x] \geqslant \alpha \|x\|^2 \geqslant \frac{\alpha}{\|V\|^2} \|Vx\|^2, \quad x \in \mathcal{D}_+,$$
that is, $V\mathcal{D}_+$ is a uniformly positive linear manifold in $\mathcal{K}_2$. Similarly we prove that $V\mathcal{D}_-$ is uniformly negative.

Conversely, assume that both of $V\mathcal{D}_+$ and $V\mathcal{D}_-$ are uniformly definite. Then, the subspace $\mathrm{Clos}(V\mathcal{D}_+)$ is uniformly positive, $\mathrm{Clos}(V\mathcal{D}_-)$ is uniformly negative and, in addition, we have $\mathrm{Clos}(V\mathcal{D}_+) \perp \mathrm{Clos}(V\mathcal{D}_-)$. By Theorem 2.2.9 there exist fundamental decompositions $\mathcal{K}_i = \mathcal{K}_i^+ [+] \mathcal{K}_i^-$, $i = 1, 2$ such that
$$\mathcal{D}_\pm \subseteq \mathcal{K}_1^\pm, \quad \mathrm{Clos}(V\mathcal{D}_\pm) \subseteq \mathcal{K}_2^\pm. \tag{3.1}$$

Denoting by $\|\cdot\|$ the corresponding fundamental norms we have, for some $\alpha > 0$,
$$\|Vx\|^2 \leqslant \frac{1}{\alpha}[Vx, Vx] = \frac{1}{\alpha}[x,x] = \frac{1}{\alpha}\|x\|^2, \quad x \in \mathcal{D}_+,$$
and hence $V|\mathcal{D}_+$ is bounded. Similarly we prove that $V|\mathcal{D}_-$ is bounded. If $x \in \mathrm{Dom}(V)$ then $x = x_+ + x_-$ for some $x_\pm \in \mathcal{D}_\pm$ and hence, by (3.1)
$$\|Vx\|^2 \leqslant \|Vx_+\|^2 + \|Vx_-\|^2 \leqslant \frac{1}{\alpha}\left(\|x_+\|^2 + \|x_-\|^2\right) = \frac{1}{\alpha}\|x\|^2.$$

This shows that V is bounded. ∎

Corollary 4.3.2. *If the domain $\mathrm{Dom}(V)$ of the isometric operator V is regular then V is bounded if and only if its range $\mathrm{Ran}(V)$ is regular.*

Proof. If V is bounded it follows from the proof of Proposition 4.3.1 that $\mathrm{Clos}\,\mathrm{Ran}(V)$ is a regular subspace of $\mathcal{K}_2$. Considering V as a bounded isometry from the Kreĭn space $\mathcal{K}_1$ into the Kreĭn space $\mathcal{K}_2$ it follows that $V^\sharp V = I_1$, in particular V has a bounded left inverse and hence $\mathrm{Ran}(V)$ is closed.

For the converse implication use Proposition 4.3.1 and Theorem 3.1.1. ∎

Proposition 4.3.3. *Let V be a densely defined isometry with dense range, and assume that there exists a fundamental symmetry J_1 in $\mathcal{K}_1$ such that $J_1 \mathrm{Dom}(V) \subseteq \mathrm{Dom}(V)$. If either $J_1^+ \mathrm{Dom}(V)$ or $J_1^- \mathrm{Dom}(V)$ is closed, then V is bounded.*

Proof. Since both the domain and the range of V are dense, it follows that V is injective and that $V^\sharp$ exists and, since V is isometric, it follows that $V^{-1} \subseteq V^\sharp$, in particular, V is closable. From $J_1 \operatorname{Dom}(V) \subseteq \operatorname{Dom}(V)$ the following decomposition follows

$$\operatorname{Dom}(V) = J_1^+ \operatorname{Dom}(V) \dot{+} J_1^- \operatorname{Dom}(V), \qquad (3.2)$$

and $J_1^+ \operatorname{Dom}(V)$ is uniformly positive and $J_1^- \operatorname{Dom}(V)$ is uniformly negative. If, to make a choice, the linear manifold $J_1^+ \operatorname{Dom}(V)$ is closed, since V is closable and $J_1^+ \operatorname{Dom}(V) \subseteq \operatorname{Dom}(V)$ it follows that $V|J_1^+ \operatorname{Dom}(V)$ is bounded. We claim that $V J_1^+ \operatorname{Dom}(V)$ is a maximal uniformly positive subspace.

Indeed, let $\|\cdot\|$ denote the fundamental norm associated to J_1 as well as an arbitrary fundamental norm on $\mathcal{K}_2$. For any $x \in \operatorname{Dom}(V)$ we have

$$[VJ_1^+ x, V J_1^+ x] = [J_1^+ x, J_1^+ x] = \|J_1^+ x\|^2 \geqslant \frac{1}{\|VJ_1^+\|^2} \|VJ_1^+ x\|^2,$$

and hence $V J_1^+ \operatorname{Dom}(V)$ is uniformly positive. Since $J_1^+ \operatorname{Dom}(V)$ is closed and the operator $V|J_1^+ \operatorname{Dom}(V)$ is isometric and bounded, it follows that $V J_1^+ \operatorname{Dom}(V)$ is also closed, hence it is a uniformly positive subspace of $\mathcal{K}_2$.

On the other hand, from (3.2),

$$\operatorname{Ran}(V) = V J_1^+ \operatorname{Dom}(V) + V J_1^- \operatorname{Dom}(V),$$

and since V is isometric we have $V J_1^+ \operatorname{Dom}(V) \perp V J_1^- \operatorname{Dom}(V)$ and $V J_1^- \operatorname{Dom}(V)$ is negative. Applying Corollary 2.2.10 we conclude that $V J_1^+ \operatorname{Dom}(V)$ is maximal uniformly positive and, in addition, the closure of $V J_1^- \operatorname{Dom}(V)$ coincides with the subspace $(V J_1^+ \operatorname{Dom}(V))^\perp$ and is maximal uniformly negative.

Finally, we apply Proposition 4.3.1 to conclude that V is bounded. ∎

Proposition 4.3.4. *Let the domain* $\operatorname{Dom}(V)$ *of the isometric operator V be closed and nondegenerate. If the closure of* $\operatorname{Ran}(V)$ *is also nondegenerate, then V is bounded.*

Proof. According to the Closed Graph Principle, it is sufficient to prove that V is closed. To see this, let $(x_n)_{n \in \mathbb{N}}$ be a sequence of vectors in $\operatorname{Dom}(V)$ such that $\lim_{n \to \infty} x_n = x$ and $\lim_{n \to \infty} V x_n = y$. Since $\operatorname{Dom}(V)$ is closed, we obtain $x \in \operatorname{Dom}(V)$ and, since it is nondegenerate, V is injective. Therefore,

$$[y, z] = \lim_{n \to \infty} [V x_n, z] = \lim_{n \to \infty} [x_n, V^{-1} z] = [Vx, z], \quad z \in \operatorname{Ran}(V),$$

hence $Vx - y$ belongs to the isotropic subspace of the closure of $\operatorname{Ran}(V)^0 = \{0\}$, and hence, $y = Vx$. ∎

We conclude the discussion on the continuity of isometric operators in Kreĭn spaces by presenting an example of an everywhere defined, unbounded, and isometric operator.

4.3 CONTINUITY OF ISOMETRIC OPERATORS

Example 4.3.5. Let $\mathcal{K}$ be a separable infinite dimensional Kreĭn space with $\kappa^+(\mathcal{K}) = \kappa^-(\mathcal{K})$ infinite. For example, take $\mathcal{K} = \ell^2 \oplus \ell^2$ with indefinite inner product defined by the symmetry $J(x \oplus y) = x \oplus -y$. Consider a fundamental decomposition $\mathcal{K} = \mathcal{K}^+[+]\mathcal{K}^-$ and the decompositions $\mathcal{K}^+ = \mathcal{K}_1^+[+]\mathcal{K}_2^+$ and $\mathcal{K}^- = \mathcal{K}_1^-[+]\mathcal{K}_2^-$ where both subspaces $\mathcal{K}_1^+$ and $\mathcal{K}_2^+$ are infinite dimensional and $\dim \mathcal{K}_1^- = 1$. According to Example 3.2.10, there exists a strictly positive subspace $\mathcal{L} \subseteq \mathcal{K}_1 = \mathcal{K}_1^+[+]\mathcal{K}_1^-$ which is complete under the norm $\mathcal{L} \ni x \mapsto [x,x]^{1/2}$ but it is not uniformly positive. Let $U_1 \colon \mathcal{K}_1^+ \to \mathcal{L}$ be an isometry. By Proposition 4.3.1, U_1 is unbounded.

Let $U_2 \colon \mathcal{K}^- \to \mathcal{K}_2^-$ be a bounded isometric operator. Define $U \colon \mathcal{K} \to \mathcal{K}$ by

$$Ux = U_1 x_1^+ + x_2^+ + U_2 x^-,$$

where, for an arbitrary vector $x \in \mathcal{K}$, we used the representation $x = x_1^+ + x_2^+ + x^-$ according to the decomposition

$$\mathcal{K} = \mathcal{K}_1^+[+]\mathcal{K}_2^+[+]\mathcal{K}^-.$$

The operator U is everywhere defined, isometric and unbounded. ∎

In the following we specialise to isometric operators in Pontryagin spaces.

Theorem 4.3.6. *Let V be an isometric operator from a Kreĭn space $\mathcal{K}_1$, that is, $\mathrm{Dom}(V) \subseteq \mathcal{K}_1$, to a Pontryagin space $\mathcal{K}_2$. If V is closable then it is bounded.*

Proof. Denote by $\|\cdot\|$ fixed fundamental norms in both $\mathcal{K}_1$ and $\mathcal{K}_2$ and by $\langle \cdot, \cdot \rangle$ the corresponding positive definite inner products. Assuming that V is not continuous, it follows that there exists a sequence $(x_n)_{n \in \mathbb{N}} \subset \mathrm{Dom}(V)$ such that

$$\lim_{n \to \infty} \|x_n\| = 0, \quad \|V x_n\| = 1, \quad n \in \mathbb{N},$$

and hence

$$\lim_{n \to \infty} [V x_n, V x_n] = \lim_{n \to \infty} [x_n, x_n] = 0.$$

Considered as a closable operator from the Hilbert space $(\mathrm{Clos}\,\mathrm{Dom}(V), \langle \cdot, \cdot \rangle)$ to the Hilbert space $(\mathrm{Clos}\,\mathrm{Ran}(V), \langle \cdot, \cdot \rangle)$, V has a densely defined adjoint V^*. Since

$$\lim_{n \to \infty} \langle V x_n, y \rangle = \lim_{n \to \infty} \langle x_n, V^* y \rangle, \quad y \in \mathrm{Dom}(V^*),$$

and the sequence $(V x_n)_{n \in \mathbb{N}}$ is bounded, we get

$$\lim_{n \to \infty} \langle V x_n, y \rangle = 0, \quad y \in \mathrm{Clos}\,\mathrm{Dom}(V^*) = \mathrm{Clos}\,\mathrm{Ran}(V).$$

Notice that

$$\langle V x_n, y \rangle = 0, \quad y \in \overline{\mathrm{Ran}(V)}^\perp,$$

and hence
$$\lim_{n\to\infty} \langle Vx_n, y \rangle = 0, \quad y \in \mathcal{K}_2.$$

Since the weak topologies associated to the inner products $[\cdot,\cdot]$ and $\langle\cdot,\cdot\rangle$ coincide, we find that
$$\lim_{n\to\infty} [Vx_n, y] = 0, \quad y \in \mathcal{K}_2,$$
and finally, Theorem 1.4.11 yields $\lim_{n\to\infty} \|Vx_n\| = 0$, a contradiction. ∎

Corollary 4.3.7. *Let $\mathcal{K}_1$, $\mathcal{K}_2$, and V be as in Theorem 4.3.6. If $\operatorname{Clos}\operatorname{Ran}(V)$ is nondegenerate then V is continuous.*

Proof. According to Theorem 4.3.6 it is sufficient to prove that V is closable. Let $(x_n)_{n\in\mathbb{N}} \subset \operatorname{Dom}(V)$ be such that
$$\lim_{n\to\infty} x_n = 0, \quad \lim_{n\to\infty} Vx_n = y \in \overline{\operatorname{Ran}(V)}.$$
Then
$$[Vx, y] = \lim_{n\to\infty} [Vx, Vx_n] = \lim_{n\to\infty} [x, x_n] = 0, \quad x \in \operatorname{Dom}(V),$$
and then taking into account that $\overline{\operatorname{Ran}(V)}$ is nondegenerate, we get $y = 0$. ∎

Corollary 4.3.8. *Let $\mathcal{K}_1$ and $\mathcal{K}_2$ be Pontryagin spaces such that $\kappa = \kappa^-(\mathcal{K}_1) = \kappa^-(\mathcal{K}_2)$ is finite ($\kappa = \kappa^+(\mathcal{K}_1) = \kappa^+(\mathcal{K}_2) < \infty$) and let $V \colon \operatorname{Dom}(V)(\subseteq \mathcal{K}_1) \to \mathcal{K}_2$ be an isometric operator. If $\operatorname{Dom}(V)$ contains a negative (respectively, strictly positive) subspace of dimension κ, then both V and V^{-1} are continuous.*

We conclude this section with an example of an injective isometric operator in a Pontryagin space of type Π_1 which is not continuous.

Example 4.3.9. Let $\mathcal{H}$ be an infinite dimensional Pontryagin space with $\kappa(\mathcal{H}) = 1$, let e be a nonzero neutral vector in $\mathcal{H}$ and consider the subspace $\mathcal{L} = e^\perp$. Then e spans $\mathcal{L}^0$, the isotropic subspace of $\mathcal{L}$ and hence the decomposition $\mathcal{L} = \mathcal{L}_1[+]\mathbb{C}e$ holds. Since $\mathcal{L}$ is infinite dimensional, there exists an unbounded linear functional φ on $\mathcal{L}$. We define the linear operator V
$$\mathcal{L} \ni x \mapsto Vx = \varphi(x)e - x \in \mathcal{H}.$$
Then V is isometric and unbounded. Therefore, the restriction $V_0 = V|\mathcal{L}_1$ is isometric, unbounded and injective. Extend V_0 to an operator $\widetilde{V}$ on $\mathcal{L}$ by
$$\widetilde{V}x = \begin{cases} V_0 x, & x \in \mathcal{L}_1, \\ \lambda x, & x \in \mathbb{C}e, \end{cases}$$
where λ is a nonzero complex number. Then $\widetilde{V}$ is isometric, injective and unbounded. ∎

4.4 Dissipative Operators

Let $(\mathcal{X}, [\cdot, \cdot])$ be a space with (indefinite) inner product and $A\colon \text{Dom}(A)(\subseteq \mathcal{X}) \to \mathcal{X}$ a linear operator. A is called *dissipative* if

$$\text{Im}[Ax, x] = \frac{[Ax, x] - [x, Ax]}{2\mathrm{i}} \geqslant 0, \quad x \in \text{Dom}(A). \tag{4.1}$$

The dissipative operator A is called *maximal dissipative* if for any dissipative operator $B\colon \text{Dom}(B)(\subseteq \mathcal{X}) \to \mathcal{X}$ such that $A \subseteq B$ it follows that $A = B$. From Zorn's Lemma we see that any dissipative operator A admits a maximal dissipative extension $\widetilde{A}$ within the same space $\mathcal{X}$. The notion of accretive operator, discussed in Section 2.3 only for Hilbert spaces, can be extended word for word to this indefinite setting. Then A is dissipative if and only if the operator $-\mathrm{i}A$ is accretive.

As for accretive operators, cf. Section 2.3, there is a close connection between dissipative operators and semidefinite subspaces. To see this, let $\mathcal{X}_\mathrm{g} = \mathcal{X} \times \mathcal{X}$ be the space endowed with the following indefinite inner product

$$[(x_1, y_1), (x_2, y_2)] = \mathrm{i}[x_1, y_2] - \mathrm{i}[y_1, x_2], \quad x_1, y_1, x_2, y_2 \in \mathcal{X}. \tag{4.2}$$

We call $(\mathcal{X}_\mathrm{g}, [\cdot, \cdot])$ the *graph space*. For an arbitrary linear operator $A\colon \mathcal{D}(A) \subseteq \mathcal{X} \to \mathcal{X}$ we consider its graph $G(A) = \{(x, Ax) \mid x \in \mathcal{D}(A)\}$ as a linear manifold in the graph space $\mathcal{X}_\mathrm{g}$. Since

$$[(x, Ax), (x, Ax)] = 2\,\text{Im}[Ax, x], \quad x \in \text{Dom}(A),$$

it follows that the operator A is dissipative if and only if its graph $G(A)$ is a positive subspace in the graph space $(\mathcal{X}_\mathrm{g}, [\cdot, \cdot])$.

The above construction can be made even more explicit in the case of a Kreĭn space $(\mathcal{K}, [\cdot, \cdot])$. Let J be a fundamental symmetry on $\mathcal{K}$, let $J = J^+ - J^-$ be the corresponding Jordan decomposition and $\mathcal{K} = \mathcal{K}^+[+]\mathcal{K}^-$ the associated fundamental decomposition. Then let $\mathcal{K}_\mathrm{g} = \mathcal{K} \oplus \mathcal{K}$ as a Hilbert space. Moreover, the operator

$$J_\mathrm{g} = \begin{bmatrix} 0 & \mathrm{i}J \\ -\mathrm{i}J & 0 \end{bmatrix} \tag{4.3}$$

is a symmetry in such a way that it induces the indefinite inner product defined at (4.2) and hence $(\mathcal{K}_\mathrm{g}, [\cdot, \cdot])$ is a Kreĭn space. Considering the subspaces

$$\mathcal{K}_\mathrm{g}^\pm = G(\pm \mathrm{i}J) = \{(x, \pm \mathrm{i}Jx) \mid x \in \mathcal{K}\}, \tag{4.4}$$

then $\mathcal{K}_\mathrm{g} = \mathcal{K}_\mathrm{g}^+[+]\mathcal{K}_\mathrm{g}^-$ is the fundamental decomposition associated to J_g.

Lemma 4.4.1. *Let A be a linear operator in the Kreĭn space $\mathcal{K}$ and J a fundamental symmetry on $\mathcal{K}$.*

(a) A is dissipative in $\mathcal{K}$ if and only if JA (equivalently, AJ) is dissipative in the Hilbert space $(\mathcal{K}, \langle \cdot, \cdot \rangle_J)$.

(b) If the operator A is densely defined and $\mathcal{L}$ is a positive subspace in the Kreĭn space $\mathcal{K}_g$ defined before and such that $G(A) \subseteq \mathcal{L}$, then there exists a dissipative operator $\widetilde{A}$ in $\mathcal{K}$ such that $A \subseteq \widetilde{A}$ and $G(\widetilde{A}) = \mathcal{L}$.

(c) If A is densely defined and dissipative then it is closable and its closure is dissipative. In particular, if A is maximal dissipative then it is closed.

Proof. (a) The first assertion is clear, since $\mathrm{Im}[Ax, x] = \mathrm{Im}\langle JAx, x\rangle_J$ for all $x \in \mathrm{Dom}(A) = \mathrm{Dom}(JA)$. As for the latter assertion, note that $\mathrm{Dom}(AJ) = J\,\mathrm{Dom}(A)$ and that for $x = Jy$ with $y \in \mathrm{Dom}(A)$ we have $\langle AJx, x\rangle_J = [Ay, y]$.

(b) Let $\mathcal{L}$ be a positive subspace in $\mathcal{K}_g$ such that $G(A) \subseteq \mathcal{L}$ and consider the notation as in (4.3) and (4.4). By the angular operator representation $\mathcal{L} = \{x + K_\mathcal{L} x \mid x \in J_g^+ \mathcal{L}\}$. Since $J_g^+ G(A) \subseteq J_g^+ \mathcal{L}$ and $\mathrm{Dom}(A)$ is dense in $\mathcal{K}$, it follows that any vector $z \in J_g^+ \mathcal{L}$ can be represented uniquely $z = (x, \mathrm{i} Jx)$. Thus $\mathcal{L} = G(\widetilde{A})$ for some operator $\widetilde{A}$ that extends A and is dissipative, since its graph $\mathcal{L}$ is a positive subspace in $\mathcal{K}_g$.

(c) This is a consequence of (a), (b), and the existence of maximal extensions of positive subspaces in a Kreĭn space, cf. Lemma 2.1.3. ∎

Proposition 4.4.2. *Any dissipative closable operator A in a Kreĭn space $\mathcal{K}$ admits a maximal dissipative extension $\widetilde{A}$ which is densely defined, and hence, closed, in $\mathcal{K}$.*

Proof. From the previous lemma we can assume that A is closed. Then $G(A)$ is a closed positive subspace in the Kreĭn space $\mathcal{K}_g$. We claim that there exists a maximal positive subspace $\mathcal{L}$ in $\mathcal{K}_g$ such that $\mathcal{L} \supseteq G(A)$ and $\mathcal{L}^0 = G(A)^0$. Indeed, to see this, we represent $G(A) = \{x \oplus Kx \mid x \in J_g^+ G(A)\}$ where $K \colon J_g^+ G(A) \to \mathcal{K}_g^-$ is a contraction. Then we take $\mathcal{L} = G(A)[+](\mathcal{K}_g^+ \ominus J_g^+ G(A))$ and it is easy to see that it fulfills all the requirements.

There exists a maximal dissipative operator $\widetilde{A}$ such that $\mathcal{L} = G(\widetilde{A})$. Indeed, assume $(0, y) \in \mathcal{L}$. By the definition of the inner product $[\cdot, \cdot]$ in the Kreĭn space $\mathcal{K}_g$, the vector $(0, y)$ is neutral. Since $\mathcal{L}$ is positive it follows that $(0, y) \in \mathcal{L}^0 = G(A)^0$, and hence $y = 0$. This shows that there exists $\widetilde{A}$ an operator in $\mathcal{K}$ such that $\mathcal{L} = G(\widetilde{A})$. By construction, $\widetilde{A}$ is maximal dissipative and $\widetilde{A} \supseteq A$.

We prove now that $\widetilde{A}$ is densely defined in $\mathcal{K}$. To see this, let $y \in \mathcal{K}$ be such that $y \perp \mathrm{Dom}(\widetilde{A})$. Then $(0, y) \perp G(\widetilde{A})$ in $\mathcal{K}_g$, that is, $(0, y)$ is a neutral vector in $G(\widetilde{A})$. Thus, $(0, y) \in (G(\widetilde{A})^\perp)^0 = G(\widetilde{A})^0$ and hence $y = 0$. This shows that $\widetilde{A}$ is densely defined in $\mathcal{K}$, by Lemma 4.4.1, item (c). ∎

Corollary 4.4.3. *Assume that A is a closed dissipative operator in a Kreĭn space. Then A is maximal dissipative if and only if A is densely defined and the operator $-A^\sharp$ is dissipative.*

4.4 DISSIPATIVE OPERATORS

Proof. Assume that A is maximal dissipative. By Proposition 4.4.2 the operator A is densely defined, and hence, the adjoint $A^\sharp$ exists. Then $G(-A^\sharp) = \{-A^\sharp y \oplus y \mid y \in \text{Dom}(A^\sharp)\}$ and it coincides with its orthogonal companion $G(A)^\perp$ in $\mathcal{K}_g$, hence the subspace $G(-A^\sharp)$ is maximal positive in $\mathcal{K}_g$, equivalently, $-A^\sharp$ is dissipative. ∎

We recall that, given an operator $T \in \mathcal{B}(\mathcal{H})$, $\mathcal{H}$ a Hilbert space, a complex number $\lambda \in \mathbb{C}$ is *of regular type* with respect to T if the operator $\lambda I - T$ is injective and has closed range, equivalently, the operator $(\lambda I - T)^{-1} \colon \text{Ran}(\lambda I - T) \to \mathcal{H}$ exists as a bounded operator.

Lemma 4.4.4. *Let T be a closed dissipative operator in a Hilbert space $\mathcal{H}$.*

(1) *The lower half plane $\mathbb{C}^- = \{\lambda \in \mathbb{C} \mid \text{Im}\,\lambda < 0\}$ consists of points of regular type with respect to T and*

$$\|(T - \lambda I)^{-1}\| \leqslant \frac{1}{|\text{Im}\,\lambda|}, \quad \text{Im}\,\lambda < 0. \tag{4.5}$$

(2) *The following assertions are equivalent.*
  (a) T *is maximal dissipative;*
  (b) $\mathbb{C}^- \cap \rho(T) \neq \emptyset$;
  (c) $\mathbb{C}^- \subseteq \rho(T)$.

Proof. (1) Let $\lambda \in \mathbb{C}$, $\lambda = \xi + i\eta$, with $\xi, \eta \in \mathbb{R}$. Then, for arbitrary $x \in \mathcal{H}$ we have

$$\|(T - \lambda I)x\|^2 = \langle Tx, Tx \rangle - 2\xi\,\text{Re}\langle Tx, x \rangle - 2\eta\,\text{Im}\langle Tx, x \rangle + (\xi^2 + \eta^2)\langle x, x \rangle$$
$$= \|(T - \xi I)x\|^2 - 2\eta\,\text{Im}\langle Tx, x \rangle + \eta^2\|x\|^2.$$

If $\eta = \text{Im}\,\lambda < 0$ then

$$\|(T - \lambda I)x\|^2 \geqslant \eta^2\|x\|^2, \quad x \in \text{Dom}(T),$$

hence λ is a point of regular type with respect to T and (4.5) holds.

(2) (b)⇒(a). Let $\lambda \in \mathbb{C}^- \cap \rho(T)$. If T is not maximal dissipative then there exists a maximal dissipative operator $\widetilde{T} \supseteq T$ such that $\widetilde{T} \neq T$. Then $(\widetilde{T} - \lambda I)\,\text{Dom}(\widetilde{T}) \supseteq (T - \lambda I)\,\text{Dom}(T) = \mathcal{H}$. Since $\text{Dom}(\widetilde{T} - \lambda I) \setminus \text{Dom}(T - \lambda I) \neq \emptyset$ it follows that there exists $x \in \text{Dom}(\widetilde{T})$, $x \neq 0$, such that $(\widetilde{T} - \lambda I)x = 0$, that is λ is an eigenvalue for $\widetilde{T}$, a contradiction with assertion (1).

(a)⇒(c). Assume that T is a maximal dissipative operator and $\lambda \in \mathbb{C}^-$. We already know that the operator $T - \lambda I$ is injective and has closed range, from (1). Let $x \in \mathcal{H} \ominus \text{Ran}(T - \lambda I)$, $x \neq 0$. Then

$$0 = \langle x, (T - \lambda I)y \rangle = \langle x, Ty \rangle - \overline{\lambda}\langle x, y \rangle.$$

This implies

$$[(x, \overline{\lambda}x), (y, Ty)] = i(\langle x, Ty \rangle - \langle \overline{\lambda}x, y \rangle) = 0,$$

that is, the vector $(x, \bar\lambda x)$ in $\mathcal{K}_g$ is orthogonal to $\operatorname{Ran}(T - \lambda I)$. Since the vector $(x, \bar\lambda x)$ is positive in the Kreĭn space $(\mathcal{K}_g, [\cdot, \cdot])$, this contradicts the maximality of T, in view of Lemma 4.4.1.

(b)⇒(c). This implication is obvious. ∎

Theorem 4.4.5. *Let A be a closed dissipative operator in the Kreĭn space $\mathcal{K}$ such that there exists a maximal uniformly positive subspace $\mathcal{L}$ in $\mathcal{K}$ with $\mathcal{L} \subset \operatorname{Dom}(A)$. Let $\mathcal{K} = \mathcal{K}^+[+]\mathcal{K}^-$ be the fundamental decomposition of K with $\mathcal{L} = \mathcal{K}^+$, J the corresponding fundamental symmetry with its Jordan decomposition $J = J^+ - J^-$, and $\|\cdot\|$ the corresponding fundamental norm. Then the following assertions are equivalent.*

(a) A *is maximal dissipative.*
(b) $-J^-A|\mathcal{K}^-$ *is maximal dissipative in the Hilbert space* $(\mathcal{K}^-, \langle \cdot, \cdot \rangle_J)$.
(c) $\{\lambda \mid \operatorname{Im}\lambda > 2\|AJ^+\|\} \cap \rho(A) \ne \emptyset$.
(d) $\{\lambda \mid \operatorname{Im}\lambda > 2\|AJ^+\|\} \subseteq \rho(A)$.

In addition, under any of these four conditions, we have

$$\|(A - \lambda I)^{-1}\| = O\left(\frac{1}{\operatorname{Im}\lambda}\right) \quad (\operatorname{Im}\lambda \to \infty).$$

Proof. Taking into account that $\mathcal{K}^+ \subseteq \operatorname{Dom}(A)$, the operator A has the following matrix representation

$$A = \begin{bmatrix} A_{11} & A_{12} \\ A_{21} & A_{22} \end{bmatrix} : \mathcal{K}^+ \dotplus (\operatorname{Dom}(A) \cap \mathcal{K}^-) \to \mathcal{K}^+[+]\mathcal{K}^-. \qquad (4.6)$$

As in Lemma 4.4.1, the operator

$$JA = \begin{bmatrix} A_{11} & A_{12} \\ -A_{21} & -A_{22} \end{bmatrix}$$

is dissipative in the Hilbert space $(\mathcal{K}, \langle \cdot, \cdot \rangle_J)$. Therefore, for an arbitrary vector x in $\operatorname{Dom}(A) \cap \mathcal{K}^- = \operatorname{Dom}(A_{22})$, we have $\operatorname{Im}\langle JAx, x \rangle_J = \operatorname{Im}\langle -A_{22}x, x \rangle_J \geqslant 0$ and hence $-A_{22} = -J^-A|\mathcal{K}^-$ is closed and dissipative in the Hilbert space $(\mathcal{K}^-, \langle \cdot, \cdot \rangle_J)$.

(a)⇒(b). Assume that A is a maximal dissipative operator. By Lemma 4.4.4, in order to show that $-A_{22} = J^-A|\mathcal{K}^-$ is maximal dissipative in the Hilbert space $(\mathcal{K}^-, \langle \cdot, \cdot \rangle_J)$, we have to prove that $\rho(A_{22}) \cap \mathbb{C}^+ \ne \emptyset$. Since JA is a maximal dissipative operator in the Hilbert space $(\mathcal{K}, \langle \cdot, \cdot \rangle_J)$, we already know from Lemma 4.4.4 that $\mathbb{C}^- \subseteq \rho(JA)$, hence for $\lambda \in \mathbb{C}^+$ the operator $JA + \lambda I$ is boundedly invertible in $\mathcal{K}$ and

$$JAJ^- + \lambda I = (I - JAJ^+(JA + \lambda I)^{-1})(JA + \lambda I). \qquad (4.7)$$

In addition, if $\lambda \in \mathbb{C}^+$, again by Lemma 4.4.4 we have

$$\|JAJ^+(JA + \lambda I)^{-1}\| < 1,$$

4.4 Dissipative Operators

therefore, $(I - JAJ^+(JA + \lambda I)^{-1})$ is boundedly invertible and hence, by (4.7) we get $\lambda \in \rho(JAJ^-)$. Taking into account that

$$JAJ^- + \lambda I = \begin{bmatrix} \lambda I_+ & A_{12} \\ 0 & \lambda I_- - A_{22} \end{bmatrix}$$

and hence $(-A_{22} + \lambda I_-)^{-1} = J^-(JAJ^- + \lambda I)^{-1}|\mathcal{K}^-$, we get $\lambda \in \rho(A_{22})$.

(b)⇒(a). Assuming that the operator $-A_{22}$ is maximal dissipative in the Hilbert space $(\mathcal{K}^-, \langle \cdot, \cdot \rangle_J)$, let $\widetilde{A}$ be a maximal dissipative extension of A. Since $\mathcal{K}^+ \subseteq \mathrm{Dom}(A)$ it follows that $\widetilde{A}_{22} = J^-\widetilde{A}|(\mathrm{Dom}(A) \cap \mathcal{K}^-) \supseteq A_{22}$. Then $-\widetilde{A}_{22}$ is a maximal dissipative extension of $-A_{22}$, hence $A_{22} = \widetilde{A}_{22}$, which proves that $A = \widetilde{A}$, that is, A is maximal dissipative.

(b)⇒(d). If $-A_{22}$ is maximal dissipative in $(\mathcal{K}^-, \langle \cdot, \cdot \rangle_J)$, by Lemma 4.4.4 we have $\mathbb{C}^+ \subseteq \rho(A_{22})$. From (b)⇒(a) and Lemma 4.4.4 we have $\mathbb{C}^- \subseteq \rho(AJ)$ and $\mathrm{Im}\,\lambda > 2\|AJ^+\|$. Then $\|2AJ^+(AJ + \lambda I)^{-1}\| < 1$ and hence the operator $A - \lambda I = -(I - 2AJ^+(AJ + \lambda I)^{-1})(AJ + \lambda I)$ is boundedly invertible, that is, $\lambda \in \rho(A)$.

(d)⇒(c). Obvious.

(c)⇒(a). Let $\mathrm{Im}\,\lambda > 2\|AJ^+\|$ and $\lambda \in \rho(A)$. The operator A cannot admit dissipative extensions $\widetilde{A}$ out of A itself since, otherwise, $\lambda \in \sigma_p(\widetilde{A})$. Taking into account that $\widetilde{A}J^+ = AJ^+$, this contradicts (a)⇒(d). ∎

We recall now some notions from the operator theory on Hilbert spaces to be used subsequently. Let $\mathcal{H}_1$ and $\mathcal{H}_2$ be Hilbert spaces and S a linear operator from $\mathcal{H}_1$ into itself. An operator T from $\mathcal{H}_1$ into $\mathcal{H}_2$ is *S-bounded* if there exists $\lambda_0 \in \rho(S)$ such that the operator $T(\lambda_0 I - S)^{-1}$ is bounded. If, for some $\lambda_0 \in \rho(S)$, the operator $T(\lambda_0 I - S)^{-1}$ is compact then T is called *S-compact*. Note that, if $\lambda_0, \lambda \in \rho(S)$ then

$$T(\lambda I - S)^{-1} - T(\lambda_0 I - S)^{-1} = (\lambda - \lambda_0) T(\lambda_0 I - S)^{-1}(\lambda I - S)^{-1}$$

and hence the above definitions actually do not depend on $\lambda_0 \in \rho(S)$.

Assume now that $T \in \mathcal{B}(\mathcal{H})$ for some Hilbert space $\mathcal{H}$. A complex number λ is called a *normal point* of T if λ is an isolated eigenvalue of finite algebraic multiplicity of T. We denote by $\widetilde{\sigma}_p(T)$ the set of normal points of T and let $\widetilde{\rho}(T) = \rho(T) \cup \widetilde{\sigma}_p(T)$.

Theorem 4.4.6. *Let A be a maximal dissipative operator in a Kreĭn space $\mathcal{K}$ such that there exists a maximal uniformly positive subspace $\mathcal{L} \subseteq \mathrm{Dom}(A)$. Then, with notation as in Theorem 4.4.5 and (4.6), we have the following.*

(1) A_{12} is A_{22}-bounded.
(2) If $A_{11} - A_{11}^*$ is compact then $\mathbb{C}^+ \subseteq \widetilde{\rho}(A)$ and A_{12} is A_{22}-compact.

Proof. (1) Let $\lambda \in \rho(A) \cap \mathbb{C}^+$. Then

$$AJ^- - \lambda I = (I - AJ^+(A - \lambda I)^{-1})(A - \lambda I).$$

Since $\|(A - \lambda I)^{-1}\| = O(\frac{1}{\operatorname{Im} \lambda})$ ($\operatorname{Im} \lambda \to \infty$), it follows that for $\operatorname{Im} \lambda$ sufficiently large we have $\lambda \in \rho(AJ^-)$, that is, $\mathbb{C}^+ \cap \rho(AJ^-) \neq \emptyset$. Since

$$AJ^- - \lambda I = \begin{bmatrix} -\lambda I_+ & A_{12} \\ 0 & A_{22} - \lambda I_- \end{bmatrix}$$

it follows that

$$(AJ^- - \lambda I)_{12}^{-1} = \frac{1}{\lambda} A_{12}(A_{22} - \lambda I_-)^{-1},$$

and hence A_{12} is A_{22}-bounded.

(2). Let $\lambda_0 \in \mathbb{C}^+$. Then

$$A_{11} - \lambda_0 I^+ = \left(\frac{1}{2}(A_{11} + A_{11}^*) - \lambda_0 I^+\right) + \frac{1}{2}(A_{11} - A_{11}^*). \tag{4.8}$$

On the right-hand side of (4.8) the first operator has a bounded inverse, while the second one is compact. Therefore, $A_{11} - \lambda_0 I^+$ is a Fredholm operator of index 0. Since A_{11} is bounded we have $\rho(A_{11}) \cap \mathbb{C}^+ \neq \emptyset$ and hence $\mathbb{C}^+ \subseteq \widetilde{\rho}(A_{11})$. In addition,

$$A - \lambda_0 I = \left(\begin{bmatrix} A_{11} - \lambda_0 I & 0 \\ A_{21} & I_- \end{bmatrix} + \begin{bmatrix} 0 & A_{12}(A_{22} - \lambda I_-)^{-1} \\ 0 & 0 \end{bmatrix}\right) \begin{bmatrix} I_+ & 0 \\ 0 & A_{22} - \lambda_0 I_- \end{bmatrix},$$

where, on the right-hand side, the first operator in parentheses is Fredholm of index 0, the second one is compact, while the factor operator is Fredholm of index 0. This implies that $A - \lambda_0 I$ is a Fredholm operator of index 0. From Lemma 4.4.4 we have $\mathbb{C}^+ \cap \rho(A) \neq \emptyset$, hence $\mathbb{C}^+ \subseteq \widetilde{\rho}(A)$. ∎

Corollary 4.4.7. *Let A be a closed and dissipative operator in a Pontryagins space $\mathcal{K}$ with $\kappa^+(\mathcal{K}) < \infty$. Then A is maximal dissipative if and only if $\mathbb{C}^+ \cap \rho(A) \neq \emptyset$. In this case, $\mathbb{C}^+ \subseteq \widetilde{\rho}(A)$.*

Proof. Assume that A is maximal dissipative. Then $\operatorname{Dom}(A)$ is dense in $\mathcal{K}$ and, by Lemma 1.4.10, there exists a maximal uniformly positive subspace $\mathcal{K}^+ \subseteq \operatorname{Dom}(A)$. By Lemma 4.4.4, we have $\mathbb{C}^+ \cap \rho(A) \neq \emptyset$. By Theorem 4.4.6, A_{12} is A_{22}-compact and both operators $A_{12}(A_{22} - \lambda_0 I_-)^{-1}$, for $\lambda_0 \in \rho(A_{22})$, and A_{11} have finite ranks, hence they are compact. Therefore, $\mathbb{C}^+ \subseteq \widetilde{\rho}(A) = \rho(A) \cup \{\infty\}$, the extended resolvent set.

Conversely, assume that $\mathbb{C}^+ \cap \rho(A) \neq \emptyset$. Let $\lambda_0 \in \mathbb{C}^+ \cap \rho(A)$. Taking into account that

$$AJ^- - \lambda_0 I = \left(I - AJ^+(A - \lambda I)^{-1}\right)(A - \lambda_0 I),$$

AJ^+ has finite rank and that $\mathbb{C}^- \cap \sigma_{\mathrm{p}}(AJ^-) = \emptyset$, it follows that $AJ^- - \lambda_0 I$ has a bounded inverse. Therefore, $\lambda_0 \in \rho(A)$ and then, by Theorem 4.4.5, we get that A is a maximal dissipative operator. ∎

4.5 Cayley Transformations

As in the Hilbert space, there is a close connection, induced by linear fractional transformations mapping the upper half plane conformally onto the unit disc, between symmetric operators and isometric operators. This relation can be put into the more general framework of dissipative versus contractive operators.

For complex numbers ε and ζ with

$$|\varepsilon| = 1, \quad \zeta \neq \overline{\zeta}, \tag{5.1}$$

consider the mapping φ

$$\varphi(\lambda) = \varepsilon \frac{\lambda - \overline{\zeta}}{\lambda - \zeta}, \quad \lambda \in \mathbb{C} \setminus \{\zeta\}, \tag{5.2}$$

which maps conformally a half plane (the upper half plane, if $\operatorname{Im}\zeta < 0$, and respectively the lower half plane, if $\operatorname{Im}\zeta > 0$) onto the unit circle of the complex plane. Then the inverse mapping $\psi = \varphi^{-1}$ is

$$\psi(\mu) = \frac{\zeta\mu - \overline{\zeta}\varepsilon}{\mu - \varepsilon}, \quad \mu \in \mathbb{C} \setminus \{\varepsilon\}. \tag{5.3}$$

Let $\mathcal{K}$ denote a Kreĭn space and let A be a linear operator in $\mathcal{K}$ such that $\operatorname{Ker}(A - \zeta I) = \{0\}$. Then one can define an operator U in $\mathcal{K}$ by

$$U = \varepsilon(A - \overline{\zeta}I)(A - \zeta I)^{-1}, \quad \operatorname{Dom}(U) = \operatorname{Ran}(A - \zeta I). \tag{5.4}$$

Conversely, if U is a linear operator in $\mathcal{K}$ such that $\operatorname{Ker}(U - \varepsilon I) = \{0\}$ then one can define an operator A in $\mathcal{K}$ by

$$A = (\zeta U - \overline{\zeta}\varepsilon I)(U - \varepsilon I)^{-1}, \quad \operatorname{Dom}(A) = \operatorname{Ran}(U - \varepsilon I). \tag{5.5}$$

Before giving names to these transformations, we first present their interplay.

Proposition 4.5.1. *Let the complex numbers ε and ζ satisfy (5.1). If the operator A has the property $\operatorname{Ker}(A - \zeta I) = 0$ then the operator U defined in (5.4) has the property $\operatorname{Ker}(U - \varepsilon I) = 0$ and conversely, if U has the property $\operatorname{Ker}(U - \varepsilon I) = 0$ then the operator A defined in (5.5) has the property $\operatorname{Ker}(A - \zeta I) = 0$. In addition, the transformations (5.5) and (5.4) are inverse one to each other.*

Proof. Let x be a vector in $\operatorname{Dom}(U) = \operatorname{Ran}(A - \zeta I)$ such that $Ux = \varepsilon x$. Then $x = (A - \zeta I)f$ for some vector $f \in \operatorname{Dom}(A)$ and $\varepsilon(A - \overline{\zeta}I)f = \varepsilon(A - \zeta I)f$, equivalently, $\varepsilon(\zeta - \overline{\zeta})f = 0$. Using (5.1) this implies $f = 0$ and hence $x = 0$. We have proven that $\operatorname{Ker}(U - \varepsilon I) = 0$.

Let now g be in $\mathrm{Dom}(A)$. Then $U(A - \zeta I)g = \varepsilon(A - \overline{\zeta}I)g = \varepsilon(A - \zeta I)g(\zeta - \overline{\zeta})g$ and hence $(U - \varepsilon I)(A - \zeta I)g = \varepsilon(\zeta - \overline{\zeta})g$, and thus we can represent the vector g by

$$g = \frac{\overline{\varepsilon}}{\zeta - \overline{\zeta}}(U - \varepsilon I)(A - \zeta I)g \in \mathrm{Ran}(U - \varepsilon I).$$

Applying the operator U on both sides of this identity we obtain

$$\begin{aligned}(\zeta U - \overline{\zeta}\varepsilon I)(U - \varepsilon I)^{-1}g &= \frac{\overline{\varepsilon}}{\zeta - \overline{\zeta}}(\zeta U - \overline{\zeta}I)(A - \zeta I)g \\ &= \frac{\overline{\varepsilon}}{\zeta - \overline{\zeta}}(\varepsilon\zeta(A - \overline{\zeta}I)g - \varepsilon\overline{\zeta}(A - \zeta I)g) \\ &= \frac{|\varepsilon|^2}{\zeta - \overline{\zeta}}(\zeta - \overline{\zeta})Ag = Ag.\end{aligned}$$

This proves that $A \subseteq (\zeta U - \overline{\zeta}\varepsilon I)(U - \varepsilon I)^{-1}$.

In order to prove the converse inclusion, let h be a vector in $\mathrm{Ran}(U - \varepsilon I)$. Then $h = (U - \varepsilon I)y$ for some $y \in \mathrm{Dom}(U) = \mathrm{Ran}(A - \zeta I)$, therefore $y = (A - \zeta I)g$ for some $g \in \mathrm{Dom}(A)$. Then $h = (U - \varepsilon I)y = (U - \varepsilon I)(A - \zeta I)g = \varepsilon(A - \overline{\zeta}I)g - \varepsilon(A - \zeta I)g = \varepsilon(\zeta - \overline{\zeta})g$, and hence $h \in \mathrm{Dom}(A)$.

The converse implication follows in a similar way. ∎

Let the complex numbers ε and ζ satisfy (5.1). The reciprocal relations (5.5) and (5.4) are called the *Cayley transformations*.

In the following we see how the vectors associated to eigenvalues behave with respect to the Cayley transform. The proofs of the next two lemmas are straightforward and are left to the reader.

Lemma 4.5.2. *Let U denote the Cayley transformation of the operator A and, for $\lambda \neq \zeta$, let $f \in \mathrm{Dom}(A^n)$ be such that, for some $n \in \mathbb{N}$,*

$$(A - \lambda I)^n f = 0, \quad (A - \lambda I)^{n-1}f \neq 0.$$

Define

$$\mu = \varepsilon\frac{\lambda - \overline{\zeta}}{\lambda - \zeta}, \quad \alpha = \frac{\lambda - \zeta}{\varepsilon(\overline{\zeta} - \zeta)}, \quad x = \alpha^n(A - \zeta I)^n f.$$

Then μ is an eigenvalue of U, $x \in \mathrm{Dom}(U^n)$, $(U - \mu I)^n x = 0$, $(U - \mu I)^{n-1}x \neq 0$, and

$$\mathrm{Lin}\{f, (A - \lambda I)f, \ldots, (A - \lambda I)^{n-1}f\} = \mathrm{Lin}\{x, (U - \mu I)x, \ldots, (U - \mu I)^{n-1}x\}.$$

Lemma 4.5.3. *Let A denote the Cayley transformation of the operator U as in (5.5) and, for $\mu \neq \varepsilon$, let $x \in \mathrm{Dom}(U^n)$ be such that, for some $n \in \mathbb{N}$,*

$$(U - \mu I)^n x = 0, \quad (U - \mu I)^{n-1}x \neq 0.$$

4.5 Cayley Transformations

Define

$$\lambda = \frac{\zeta\mu - \overline{\zeta}\varepsilon}{\mu - \varepsilon}, \quad \beta = \frac{\mu - \varepsilon}{\varepsilon(\overline{\zeta} - \zeta)}, \quad x = \beta^n(U - \varepsilon I)^n x.$$

Then λ is an eigenvalue of A, $f \in \mathrm{Dom}(A^n)$, $(A - \lambda I)^n f = 0$, $(A - \lambda I)^{n-1} f \neq 0$, and

$$\mathrm{Lin}\{f, (A - \lambda I)f, \ldots, (A - \lambda I)^{n-1} f\} = \mathrm{Lin}\{x, (U - \mu I)x, \ldots, (U - \mu I)^{n-1} x\}.$$

The following result is an immediate consequence of Lemmas 4.5.2 and 4.5.3.

Corollary 4.5.4. *Let ε, ζ, A, and U satisfy (5.1), (5.5), and hence (5.4). Consider the complex numbers λ, μ, $\lambda \neq \zeta$, and $\mu \neq \varepsilon$, such that*

$$\mu = \varepsilon \frac{\lambda - \overline{\zeta}}{\lambda - \zeta}, \quad \lambda = \frac{\zeta\mu - \overline{\zeta}\varepsilon}{\mu - \varepsilon}.$$

Then λ is an eigenvalue of A if and only if μ is an eigenvalue of U, in this case the lengths of the corresponding Jordan chains are the same and the root manifolds $\mathfrak{S}_\lambda(A)$ and $\mathfrak{S}_\mu(U)$ coincide.

Corollary 4.5.5. *Let ε, ζ, A, and U satisfy (5.1), (5.5), and hence (5.4). Consider $\mathcal{L}$ a finite dimensional subspace in $\mathcal{K}$. Then $\mathcal{L} \subseteq \mathrm{Dom}(A)$ and $A\mathcal{L} \subseteq \mathcal{L}$ if and only if $\mathcal{L} \subseteq \mathrm{Dom}(U)$ and $U\mathcal{L} \subseteq \mathcal{L}$.*

Proof. Assume that $\mathcal{L} \subseteq \mathrm{Dom}(A)$ and $A\mathcal{L} \subseteq \mathcal{L}$. Since $\mathcal{L}$ is finite dimensional it follows that it can be decomposed into a direct sum of cyclic subspaces of A, that is

$$\mathcal{L} = \mathcal{L}_1 \dotplus \mathcal{L}_2 \dotplus \cdots \dotplus \mathcal{L}_k,$$

where $\mathcal{L}_i = \mathrm{Lin}\{f_i, (A - \lambda_i)f_i, \ldots, (A - \lambda_i)^{n_i} f_i\}$ for some $f_i \in \mathcal{L}$, $i \in \{1, 2, \ldots, k\}$. Then use Lemma 4.5.2. The converse implication is similar, this time using Lemma 4.5.3. ∎

Up to now we have used only the linear properties of the space $\mathcal{K}$. In the next result we use also the topological properties of the Kreĭn spaces.

Proposition 4.5.6. *Let ε, ζ, A, and U satisfy (5.1), (5.5), and hence (5.4). Then A is a closed operator and $\zeta \notin \sigma_\mathrm{p}(A)$ if and only if U is a closed operator and $\varepsilon \notin \sigma_\mathrm{p}(U)$.*

Proof. Assume that A is a closed operator and $\zeta \notin \sigma_\mathrm{p}(A)$. We note that

$$U = \varepsilon I(\zeta - \overline{\zeta})(A - \zeta I)^{-1}$$

and hence the operator U is closed, since the inverse and the translation of a closed operator are closed. The fact that $\varepsilon \notin \sigma_\mathrm{p}(U)$ follows from Proposition 4.5.1. If U is a closed operator and $\varepsilon \notin \sigma_\mathrm{p}(U)$, we remark that

$$A = \zeta I + \varepsilon(\zeta - \overline{\zeta})(U - \varepsilon I)^{-1},$$

and this shows that A is closed and for $\zeta \notin \sigma_\mathrm{p}(A)$ we apply again Proposition 4.5.1. ∎

Proposition 4.5.7. *Let ε, ζ, A, and U satisfy (5.1), (5.5), and hence (5.4). Then the linear fractional transformation φ in (5.2) establishes a bijective correspondence between $\sigma_p(A)$ and $\sigma_p(U)$, between $\sigma_c(A) \setminus \{\zeta\}$ and $\sigma_c(U) \setminus \{\varepsilon\}$, between $\sigma_r(A) \setminus \{\zeta\}$ and $\sigma_r(U) \setminus \{\varepsilon\}$, and, consequently, between $\rho(A)$ and $\rho(U)$.*

Proof. The fact that the mapping φ is a bijective correspondence between $\sigma_p(A)$ and $\sigma_p(U)$ is contained in Corollary 4.5.4. Let now $\lambda \neq \zeta$ and $\mu \neq \varepsilon$ be such that

$$\mu = \varphi(\lambda) = \varepsilon \frac{\lambda - \overline{\zeta}}{\lambda - \zeta}.$$

Again by Corollary 4.5.4 it follows that $\lambda \notin \sigma_p(A)$ if and only if $\mu \notin \sigma_p(U)$ and hence the proof would be finished if we show that

$$\operatorname{Ran}(U - \mu I) = \operatorname{Ran}(A - \lambda I).$$

To see this, let x be an arbitrary vector in $\operatorname{Dom}(U)$. Then

$$(U - \mu I)x = \varepsilon\left((A - \overline{\zeta}I) - \frac{\lambda - \overline{\zeta}}{\lambda - \zeta}(A - \zeta I)\right)(A - \zeta I)^{-1}x$$

$$= \varepsilon \frac{\overline{\zeta} - \zeta}{\lambda - \zeta}(A - \lambda I)(A - \zeta I)^{-1}x,$$

and hence $\operatorname{Ran}(U - \mu I) \subseteq \operatorname{Ran}(A - \lambda I)$. Since $\operatorname{Dom}(U) = \operatorname{Ran}(A - \zeta I)$, from the above equality we also have

$$(A - \lambda I)f = \overline{\varepsilon}\frac{\lambda - \zeta}{\overline{\zeta} - \zeta}(U - \mu I)(A - \zeta I)f, \quad f \in \operatorname{Dom}(A),$$

and hence $\operatorname{Ran}(U - \mu I) \supseteq \operatorname{Ran}(A - \lambda I)$. ∎

We come now to the original problem of this section, the relation between symmetric operators and isometric operators, respectively, the relation between contractions and dissipative operators. Recall that a linear operator $T\colon \operatorname{Dom}(T)(\subseteq \mathcal{K}_1) \to \mathcal{K}_2$ is called a contraction if $[Tx, Tx] \leq [x, x]$ for all $x \in \operatorname{Dom}(T)$.

Lemma 4.5.8. *Let ε, ζ, A, and U satisfy (5.1), (5.5), and hence (5.4). If $\zeta \notin \sigma_p(A)$, equivalently, $\varepsilon \notin \sigma_p(U)$, then we have the following.*

(i) A is a symmetric operator if and only if U is isometric.

(ii) If, in addition, $\zeta \in \mathbb{C}^+$, that is, $\operatorname{Im}\zeta > 0$, then A is dissipative if and only if U is a contraction.

Proof. Let $x, y \in \operatorname{Dom}(U)$ and $f, g \in \operatorname{Dom}(A)$ be such that $x = (A - \zeta I)f$, and $y = (A - \zeta I)g$. Then

$$[Ux, Uy] - [x, y] = [\varepsilon(A - \overline{\zeta}I)f, \varepsilon(A - \overline{\zeta}I)g] - [(A - \zeta I)f, (A - \zeta I)g]$$
$$= [Af, Ag] - \zeta[Af, g] - \overline{\zeta}[f, Ag] + |\zeta|^2[f, g] - [Af, Ag] + \overline{\zeta}[Af, g]$$
$$+ \zeta[f, Ag] - |\zeta|^2[f, g] = (\zeta - \overline{\zeta})([Af, g] - [f, Ag]).$$

4.5 Cayley Transformations

Taking into account Proposition 4.5.1, assertion (i) follows.

Assume now that $x = y$. Then

$$[Ux, Ux] - [x, x] = -4 \operatorname{Im} \zeta \, \operatorname{Im}[Af, f].$$

Since $\operatorname{Im} \zeta > 0$ this shows that U is contractive if and only if A is dissipative. ∎

Recall that a unitary operator U in a Kreĭn space $\mathcal{K}$ is, by definition, a densely defined operator, injective and with dense range, such that $U^\sharp = U^{-1}$, in particular it is a closed isometric operator.

Theorem 4.5.9. *Let ε, ζ, A, and U satisfy (5.1), (5.5), and hence (5.4). Then the operator A is selfadjoint and $\zeta \in \sigma_c(A) \cup \rho(A)$ ($\zeta \in \rho(A)$) if and only if the operator U is unitary (bounded unitary) and $\varepsilon \notin \sigma_p(U)$.*

Proof. Let A be a selfadjoint operator and $\zeta \in \sigma_c(A) \cup \rho(A)$. Then $\operatorname{Dom}(U) = \operatorname{Ran}(A - \zeta I)$ is dense in $\mathcal{K}$ and hence $U^\sharp$ exists. By Lemma 4.5.8, U is isometric and hence $U^{-1} \subseteq U^\sharp$. To prove the converse inclusion let y be an arbitrary vector in $\operatorname{Dom}(U^\sharp)$. Then there exists $y^\sharp \in \mathcal{K}$ such that

$$[Ux, y] = [x, y^\sharp], \quad x \in \operatorname{Dom}(U).$$

From (5.5) we obtain

$$[\varepsilon(A - \overline{\zeta}I)f, g] = [(A - \zeta I)f, y^\sharp], \quad f \in \operatorname{Dom}(A),$$

or, equivalently,

$$[Af, \overline{\varepsilon}y - y^\sharp] = [f, \overline{\varepsilon}\zeta y - \overline{\zeta}y^\sharp], \quad f \in \operatorname{Dom}(A).$$

Since A is selfadjoint, from above and (5.4) we get

$$\overline{\varepsilon}y - y^\sharp = (U - \varepsilon I)x, \quad \overline{\varepsilon}\zeta y - \overline{\zeta}y^\sharp = (\zeta U - \varepsilon\overline{\zeta}I)x,$$

for some vector $x \in \operatorname{Dom}(U)$. Solving the system with respect to y and $y^\sharp$ we find $y = \varepsilon Ux$ and hence $y \in \operatorname{Ran}(U) = \operatorname{Dom}(U^{-1})$. Finally we get $\varepsilon \notin \sigma_p(U)$ as in Proposition 4.5.1.

Conversely, let U be a unitary operator and $\varepsilon \notin \sigma_p(U)$. Since $|\varepsilon| = 1$, by Proposition 4.1.11 we obtain $\varepsilon \notin \sigma_r(U)$ and hence $\operatorname{Dom}(A) = \operatorname{Ran}(U - \varepsilon I)$ is dense in $\mathcal{K}$. From Lemma 4.5.8 A is symmetric and hence $A \subseteq A^\sharp$. In order to prove the converse inclusion let g be an arbitrary vector in $\operatorname{Dom}(A^\sharp)$. Then there exists $g^\sharp \in \mathcal{K}$ such that

$$[Af, g] = [f, g^\sharp], \quad f \in \operatorname{Dom}(A).$$

By means of (5.4) this is equivalent with

$$[(\zeta U - \varepsilon \overline{\zeta}I)x, y] = [(U - \varepsilon I)x, g^\sharp], \quad x \in \operatorname{Dom}(U),$$

which is equivalent with

$$[Ux, \overline{\zeta}g - g^\sharp] = [x, \overline{\varepsilon}\zeta g - \overline{\varepsilon}g^\sharp], \quad x \in \mathrm{Dom}(U).$$

Since U is unitary, that is, $U^\sharp = U^{-1}$, and observing that $\mathrm{Ran}(U) = \mathrm{Ran}(A - \overline{\zeta}I)$, from above we obtain

$$\overline{\zeta}g - g^\sharp = (A - \overline{\zeta}I)h, \quad \overline{\varepsilon}(\zeta g - g^\sharp) = \overline{\varepsilon}(A - \zeta I)h.$$

Solving this system with respect to g and $g^\sharp$ we find $g = -h \in \mathrm{Dom}(A)$. Finally, $\zeta \notin \sigma_\mathrm{p}(A)$ follows as in Proposition 4.5.1, and since $\mathrm{Dom}(U) = \mathrm{Ran}(A - \zeta I)$ is dense in $\mathcal{K}$ it follows that $\zeta \notin \sigma_\mathrm{r}(A)$ and hence $\zeta \in \sigma_\mathrm{c}(A) \cup \rho(A)$. The fact that $\zeta \in \rho(A)$ corresponds to a bounded unitary operator U follows from the Closed Graph Theorem. ∎

In a similar fashion and using Lemma 4.5.8 one proves the following.

Theorem 4.5.10. *Let ε, ζ, A, and U satisfy* (5.1), (5.5), *and hence* (5.4). *In addition, assume that $\zeta \in \mathbb{C}^+$. Then the operator A is maximal dissipative and $\zeta \in \rho(A)$ if and only if the operator $U \in \mathcal{B}(\mathcal{K})$ is a double contraction, that is, both U and $U^\sharp$ are contractions, and $\varepsilon \notin \sigma_\mathrm{p}(U)$.*

4.6 Notes

The material presented in Section 4.1 and Section 4.3, as well as the history of the results, can be found in J. Bognár [18] and T. Ya. Azizov and I. S. Iokhvidov [10]. Only Proposition 4.3.1 is from our joint article with T. Constantinescu [32] and Theorem 4.1.3 is from our joint article with A. Dijksma [43]. The material presented in Section 4.4 follows the presentation in T. Ya. Azizov and I. S. Iokhvidov [10], where a careful citation of the original sources can be found. The Cayley transformation approach goes back to I. S. Iokhvidov and M. G. Kreĭn [77].

Chapter 5

Selfadjoint Projections and Unitary Operators

The notion of orthogonal projection in a Hilbert space has a natural generalisation in a Kreĭn space, with the important difference that this leads to a class of unbounded operators. In order to emphasise the difference, we call them selfadjoint projections, while we call orthogonal projections those selfadjoint projections that are bounded and have been considered already in Chapter 4. From the geometry point of view, this comes from the difference between nondegenerate subspaces and regular subspaces.

A closely related class is that of symmetries which can be unbounded as well. Our main concern is to obtain canonical forms. We also investigate the properties of monotone nets of selfadjoint projections in a Kreĭn space, in connection with the problem of characterising the nondegeneracy or regularity of the corresponding limit subspaces.

The class of unitary operators is also more general in Kreĭn spaces than in Hilbert spaces. As before, the main difference relies on the idea that the unitary operators on a genuine Kreĭn space may be unbounded. However, some canonical representations of these operators with respect to fundamental decompositions exist. As an application we present a dichotomic characterisation of dense operator ranges.

5.1 Selfadjoint Projections

A linear, possibly unbounded, operator P in the Kreĭn space $\mathcal{K}$ is called a *projection* if it is idempotent, $P^2 = P$, that is, $\operatorname{Ran}(P) \subseteq \operatorname{Dom}(P)$ and $P^2 x = Px$ for all $x \in \operatorname{Dom}(P)$. If this holds then $I - P$ is also a projection. We are interested mainly in selfadjoint projections, that is, those densely defined projections P such that $P = P^\sharp$, see Section 4.1.

Proposition 5.1.1. *A subspace $\mathcal{L}$ of the Kreĭn space $\mathcal{K}$ is the range of a (bounded) selfadjoint projection if and only if it is nondegenerate (respectively, regular). This correspondence between nondegenerate (regular) subspaces and (respectively, bounded) selfadjoint projections is bijective.*

Proof. Let $\mathcal{L}$ be a nondegenerate subspace, hence $\mathcal{L}^0 = \mathcal{L} \cap \mathcal{L}^\perp = 0$ and $\mathcal{L} + \mathcal{L}^\perp$ is dense in $\mathcal{K}$, see Remark 1.4.9 (c). Define a linear operator P in $\mathcal{K}$ as follows: $\operatorname{Dom}(P) = \mathcal{L} \dotplus \mathcal{L}^\perp$ and
$$P(x_1 + x_2) = x_1, \quad x_1 \in \mathcal{L},\ x_2 \in \mathcal{L}^\perp.$$

Then P is correctly defined, $\operatorname{Ran}(P) = \mathcal{L}$ and $P^2 = P$. Moreover, for any $x_1, y_1 \in \mathcal{L}$ and $x_2, y_2 \in \mathcal{L}^\perp$ we have

$$[P(x_1 + x_2), y_1 + y_2] = [x_1, y_1 + y_2] = [x_1, y_1]$$
$$= [x_1 + x_2, y_1] = [x_1 + x_2, P(y_1 + y_2)].$$

This shows that $P \subseteq P^\sharp$. In order to prove the converse inclusion let y be an arbitrary vector in $\operatorname{Dom}(P^\sharp)$, that is,

$$[Px, y] = [x, P^\sharp y], \quad x \in \operatorname{Dom}(P).$$

Taking $x \in \mathcal{L}$ it follows that $y - P^\sharp y \in \mathcal{L}^\perp$ and then letting x in $\mathcal{L}^\perp$ it follows that $P^\sharp y \in \mathcal{L}$. Therefore

$$y = (y - P^\sharp y) + P^\sharp y \in \mathcal{L}^\perp \dotplus \mathcal{L} = \operatorname{Dom}(P).$$

Conversely, let P be a selfadjoint projection in $\mathcal{K}$. Then $I - P$ is also a selfadjoint projection in $\mathcal{K}$ and $\operatorname{Ran}(P) = \operatorname{Ker}(I - P)$, in particular $\operatorname{Ran}(P)$ is closed. In order to prove that $\operatorname{Ran}(P)$ is nondegenerate let $x \in \operatorname{Ran}(P)^0$. Then $x \in \operatorname{Ran}(P)$ and

$$0 = [x, Py] = [Px, y] = [x, y], \quad y \in \operatorname{Dom}(P),$$

and hence $x = 0$ follows, since $\operatorname{Dom}(P)$ is dense in $\mathcal{K}$.

From the above it is clear that the correspondence between nondegenerate subspaces and selfadjoint projections is bijective. Also, recall that a subspace $\mathcal{L}$ of $\mathcal{K}$ is regular if and only if $\mathcal{L} + \mathcal{L}^\perp = \mathcal{K}$. By the Closed Graph Principle and the above proved facts it follows that the regular subspaces are precisely the ranges of bounded selfadjoint projections. ∎

As a consequence of Proposition 5.1.1, given a nondegenerate subspace $\mathcal{L}$ of a Kreĭn space $\mathcal{K}$, we can talk, without any ambiguity, of the selfadjoint projection onto $\mathcal{L}$, meaning the selfadjoint projection P in $\mathcal{K}$ such that $\operatorname{Ran}(P) = \mathcal{L}$.

By definition, a linear operator A in the Kreĭn space is *positive* if $[Ax, x] \geq 0$, $x \in \operatorname{Dom}(A)$.

Remark 5.1.2. (a) A subspace $\mathcal{L}$ is the range of a (bounded) positive selfadjoint projection if and only if $\mathcal{L}$ is a strictly (respectively, uniformly) positive subspace.

(b) Using Example 2.2.11 and the pattern in the proof of Proposition 5.1.1, we can construct a positive, densely defined, and closed projection which is not selfadjoint.

(c) Let $\mathcal{L}$ be a nondegenerate subspace and let P be the corresponding selfadjoint projection onto $\mathcal{L}$. Then $I - P$ is the selfadjoint projection onto $\mathcal{L}^\perp$. Also, $\mathcal{L}$ is maximal strictly positive if and only if P is positive and $I - P$ is negative. ∎

In connection with selfadjoint projections it is useful to generalise one more class of operators from Hilbert spaces to Kreĭn spaces. A linear operator S in the Kreĭn space $\mathcal{K}$ is called a *symmetry* if it is selfadjoint and unitary, that is, $S = S^\sharp = S^{-1}$. Note that any symmetry is a densely defined and closed operator.

The proof of the next lemma is straightforward and we omit it.

5.1 Selfadjoint Projections

Lemma 5.1.3. (a) *A linear operator S in $\mathcal{K}$ is a symmetry if and only if the operator $P = \frac{1}{2}(S + I)$ is a selfadjoint projection. In this case $S = 2P - I$ and these relations establish a bijective correspondence between the class of symmetries and the class of selfadjoint projections.*

(b) *Let P be a selfadjoint projection and $S = 2P - I$. Denoting $\mathcal{L} = \operatorname{Ran}(P)$ then $\operatorname{Dom}(S) = \mathcal{L} \dotplus \mathcal{L}^\perp$ and S acts as follows:*

$$S(x_1 + x_2) = x_1 - x_2, \quad x_1 \in \mathcal{L}_1, \; x_2 \in \mathcal{L}^\perp.$$

Also we have

$$\mathcal{L} = \operatorname{Ker}(S - I), \quad \mathcal{L}^\perp = \operatorname{Ker}(S + I).$$

(c) *With notation as in item (b), S is a positive symmetry if and only if $\mathcal{L}$ is a maximal strictly positive subspace. In particular, S is a fundamental symmetry if and only if it is a bounded positive symmetry.*

As a consequence of Lemma 5.1.3, given a nondegenerate subspace $\mathcal{L}$ of a Kreĭn space $\mathcal{K}$, we can talk without any ambiguity about the symmetry S corresponding to $\mathcal{L}$, that is, the symmetry S in $\mathcal{K}$ such that $\operatorname{Ker}(I - S) = \mathcal{L}$.

Corollary 5.1.4. *The following assertions are equivalent.*
(a) *$\mathcal{K}$ is a Pontryagin space.*
(b) *Every selfadjoint projection in $\mathcal{K}$ is bounded.*
(c) *Every symmetry in $\mathcal{K}$ is bounded.*

Proof. This is a consequence of Poposition 5.1.1 and Proposition 3.1.8. ∎

Remark 5.1.5. Let P denote an unbounded selfadjoint projection in $\mathcal{K}$. From the previous corollary it follows that necessarily both $\kappa^+(\mathcal{K})$ and $\kappa^-(\mathcal{K})$ are infinite. Then the point spectrum $\sigma_{\mathrm{p}}(P) = \{0, 1\}$, the continuous spectrum $\sigma_{\mathrm{c}}(P) = \mathbb{C} \setminus \{0, 1\}$, and the residual spectrum $\sigma_{\mathrm{r}}(P) = \emptyset$. Letting $S = 2P - I$ denote the corresponding unbounded symmetry, then $\sigma_{\mathrm{p}}(S) = \{-1, 1\}$, $\sigma_{\mathrm{c}}(S) = \mathbb{C} \setminus \{-1, 1\}$, and $\sigma_{\mathrm{r}}(S) = \emptyset$.

In particular, if $\mathcal{K}$ is a genuine Kreĭn space and $\mathcal{L}$ is any strictly positive subspace which is not uniformly positive (this exists, by Proposition 3.1.8) then the spectrum of the positive selfadjoint projection corresponding to $\mathcal{L}$ covers the whole complex plane. ∎

We are now interested in the description of selfadjoint projections in terms of angular operators. The reader may take some time and recall the facts on angular operators as presented in Chapter 2.

Theorem 5.1.6. *Let $\mathcal{L}$ be a maximal strictly positive subspace of $\mathcal{K}$ and let K denote its angular operator with respect to a fixed fundamental decomposition $\mathcal{K} = \mathcal{K}^+[+]\mathcal{K}^-$. Denote*

$$\mathcal{D}_+ = \operatorname{Ran}(I_+ - K^*K) \subseteq \mathcal{K}^+, \quad \mathcal{D}_- = \operatorname{Ran}(I_- - KK^*) \subseteq \mathcal{K}^-.$$

Then the selfadjoint projection P onto $\mathcal{L}$ is the closure of the following operator

$$P_0 = \begin{bmatrix} (I_+ - K^*K)^{-1} & -K^*(I_- - KK^*)^{-1} \\ K(I_+ - K^*K)^{-1} & -KK^*(I_- - KK^*)^{-1} \end{bmatrix},$$

where the operator block matrix has to be understood with respect to the decompositions

$$\mathrm{Dom}(P_0) = \mathcal{D}_+ \dotplus \mathcal{D}_- \to \mathcal{K} = \mathcal{K}^+[+]\mathcal{K}^-.$$

Proof. Since $\mathcal{L}$ is maximal strictly positive then $\mathcal{L}^\perp$ is maximal strictly negative and hence both angular operators K and K^* are strict contractions (cf. Lemma 2.1.2), equivalently $I_+ - K^*K$ and $I_- - KK^*$ are injective. So the operator block matrix representation of P_0 makes sense.

Let z be arbitrary in $\mathrm{Dom}(P_0)$ and hence there exist $x \in \mathcal{K}^+$ and $y \in \mathcal{K}^-$ such that

$$z = (I_+ - K^*K)x + (I_- - KK^*)y.$$

Then, since

$$\mathcal{L} = \{s + Ks \mid s \in \mathcal{K}^+\} \quad \text{and} \quad \mathcal{L}^\perp = \{t + K^*t \mid t \in \mathcal{K}^-\},$$

we have

$$z = ((x - K^*y) + K(x - K^*y)) + ((y - Kx) + K^*(y - Kx)) \in \mathcal{L} + \mathcal{L}^\perp = \mathrm{Dom}(P),$$

and

$$P_0 z = (x - K^*y) + K(x - K^*y) = Pz,$$

and hence $P_0 \subseteq P$. Therefore, P_0 is closable and its closure is contained in P.

For the converse inclusion, let z be arbitrary in $\mathrm{Dom}(P)$. Then $z = x + y$ for some $x \in \mathcal{L}$ and $y \in \mathcal{L}^\perp$. This implies

$$z = x^+ + Kx^+ + y^- + K^*y^-,$$

for certain $x^+ \in \mathcal{K}^+$ and $y^- \in \mathcal{K}^-$. Consider the operator $A \in \mathcal{B}(\mathcal{K})$

$$A = \begin{bmatrix} I_+ & -K^* \\ -K & I_- \end{bmatrix} \quad \text{with respect to} \quad \mathcal{K} = \mathcal{K}^+[+]\mathcal{K}^-.$$

If J denotes the fundamental symmetry determined by the fundamental decomposition $\mathcal{K} = \mathcal{K}^+[+]\mathcal{K}^-$ then A is J-selfadjoint, that is, selfadjoint with respect to the positive definite inner product $\langle \cdot, \cdot \rangle_J$. Taking into account the Frobenius–Schur factorisation

$$A = \begin{bmatrix} I_+ & -K^* \\ 0 & I_- \end{bmatrix} \begin{bmatrix} I_+ - K^*K & 0 \\ 0 & I_- \end{bmatrix} \begin{bmatrix} I_+ & 0 \\ -K & I_- \end{bmatrix},$$

5.1 Selfadjoint Projections

since the extremal operators on the right side are invertible and $I_+ - K^*K$ is injective it follows that A has dense range in $\mathcal{K}$. Then there exist sequences $(x_n)_{n\in\mathbb{N}}$ in $\mathcal{K}^+$ and $(y_n)_{n\in\mathbb{N}}$ in $\mathcal{K}^-$ such that

$$x_n - K^*y_n \xrightarrow[n\to\infty]{} x^+, \quad y_n - Kx_n \xrightarrow[n\to\infty]{} y^-.$$

Consider the sequence $(z_n)_{n\in\mathbb{N}}$ in $\mathrm{Dom}(P_0)$ defined by

$$z_n = (I_+ - K^*K)x_n + (I_- - KK^*)y_n, \quad n \in \mathbb{N},$$

and note that

$$z_n = (x_n - K^*y_n) + K(x_n - K^*y_n) + (y_n - Kx_n) + K^*(y_n - Kx_n), \quad n \in \mathbb{N},$$

and hence

$$z_n \xrightarrow[n\to\infty]{} x_+ + Kx^+ + y^- + K^*y^- = z,$$

and

$$P_0 z_n = (x_n - K^*y_n) + K(x_n - K^*y_n) \xrightarrow[n\to\infty]{} x^+ + Kx^+ = x,$$

therefore the operator P is contained in the closure of the operator P_0, too. ∎

Corollary 5.1.7. *With notation as in Theorem 5.1.6, the symmetry S corresponding to the maximal strictly positive subspace $\mathcal{L}$, that is, $\mathrm{Ker}(I - S) = \mathcal{L}$, coincides with the closure of the operator*

$$S_0 = \begin{bmatrix} (I_+ + K^*K)(I_+ - K^*K)^{-1} & -2K^*(I_- - KK^*)^{-1} \\ 2K(I_+ - K^*K)^{-1} & -(I_- + KK^*)(I_- - KK^*)^{-1} \end{bmatrix}$$

where the operator block matrix has to be understood with respect to

$$\mathrm{Dom}(S_0) = \mathcal{D}_+ \dotplus \mathcal{D}_- \to \mathcal{K} = \mathcal{K}^+[+]\mathcal{K}^-.$$

Proof. It is sufficient to observe that $S_0 = 2P_0 - I$. Then the statement follows from Theorem 5.1.6 and Lemma 5.1.3. ∎

Remark 5.1.8. Let the subspace $\mathcal{L}$ be maximal uniformly positive. Then its angular operator K is a uniform contraction (cf. Lemma 2.1.2), equivalently, the bounded operators $I_+ - K^*K$ and $I_- - KK^*$ are boundedly invertible. If P denotes the bounded selfadjoint projection onto $\mathcal{L}$ and S denotes the symmetry with $\mathrm{Ker}(I - S) = \mathcal{L}$ then both P and S are positive and, as a consequence of Theorem 5.1.6 and Corollary 5.1.7, with respect to the fundamental decomposition $\mathcal{K} = \mathcal{K}^+[+]\mathcal{K}^-$, we have

$$P = \begin{bmatrix} (I_+ - K^*K)^{-1} & -K^*(I_- - KK^*)^{-1} \\ K(I_+ - K^*K)^{-1} & -KK^*(I_- - KK^*)^{-1} \end{bmatrix},$$

and

$$S = \begin{bmatrix} (I_+ + K^*K)(I_+ - K^*K)^{-1} & -2K^*(I_- - KK^*)^{-1} \\ 2K(I_+ - K^*K)^{-1} & -(I_- + KK^*)(I_- - KK^*)^{-1} \end{bmatrix}.$$

In particular, the latter formula gives a description (parametrisation) of all fundamental symmetries with respect to a fixed fundamental decomposition $\mathcal{K} = \mathcal{K}^+[+]\mathcal{K}^-$, the parameter K running through the open unit ball, with respect to the operator norm, of $\mathcal{B}(\mathcal{K}^+, \mathcal{K}^-)$, that is, $\|K\| < 1$. ∎

5.2 Monotone Nets of Selfadjoint Projections

In the case of Hilbert spaces, to a certain extent, operations with subspaces can be transposed onto operations with orthogonal projections. In the case of Kreĭn spaces there are important differences.

Remark 5.2.1. Let $\mathcal{L}$ and $\mathcal{M}$ be regular subspaces of a Kreĭn space $\mathcal{K}$ and denote by P and Q the corresponding bounded selfadjoint projections onto $\mathcal{L}$ and $\mathcal{M}$, respectively. If $PQ = QP$ then $\mathcal{L} \cap \mathcal{M}$ and $\mathcal{L} + \mathcal{M}$ are regular subspaces, the bounded selfadjoint projection onto $\mathcal{L} \cap \mathcal{M}$ is $P \wedge Q = PQ$, and the bounded selfadjoint projection onto $\mathcal{L} + \mathcal{M}$ is $P \vee Q = P + Q - PQ$.

If, in addition, P and Q are positive, equivalently $\mathcal{L}$ and $\mathcal{M}$ are uniformly positive subspaces, then $P \vee Q = P + Q - PQ = P + (I - P)Q$ is positive, equivalently $\mathcal{L} + \mathcal{M}$ is uniformly positive. ∎

The class of selfadjoint projections, equivalently the class of nondegenerate subspaces, has a natural order relation. Let E and F be selfadjoint projections (not necessarily bounded). By definition, E is dominated by F, denoted by $E \preceq F$, if $\operatorname{Ran}(E) \subseteq \operatorname{Ran}(F)$.

Remark 5.2.2. Let $(P_i)_{i \in \mathcal{J}}$ be a family of bounded selfadjoint projections that are mutually commutative. From Remark 5.2.1 it follows by induction that for any finite subset $\Lambda \subseteq \mathcal{J}$ the subspaces $\bigcap_{i \in \Lambda} P_i \mathcal{K}$ and $\bigvee_{i \in \Lambda} P_i \mathcal{K}$, which is, by definition, the closure of the linear span of $P_i \mathcal{K}$ when i is running in Λ, are regular. We denote by $P_\Lambda = \bigwedge_{i \in \Lambda} P_i$ and $P^\Lambda = \bigvee_{i \in \Lambda} P_i$, respectively, the corresponding selfadjoint projections. Denote $\mathcal{F}(\Lambda) = \{\Lambda \subseteq \mathcal{J} \mid \Lambda \text{ finite}\}$ and then observe that the net $(P_\Lambda)_{\Lambda \in \mathcal{F}(\mathcal{J})}$ is nonincreasing, in the sense that $\Lambda \subseteq \Delta$ implies $P_\Delta \preceq P_\Lambda$, while $(P^\Lambda)_{\Lambda \in \mathcal{F}(\mathcal{J})}$ is a nondecreasing net, in the sense that $\Lambda \subseteq \Delta$ implies $P^\Lambda \preceq P^\Delta$, and notice that

$$\bigcap_{i \in \mathcal{J}} P_i \mathcal{K} = \bigcap_{\Lambda \in \mathcal{F}(\mathcal{J})} P_\Lambda \mathcal{K}, \quad \bigcup_{i \in \mathcal{J}} P_i \mathcal{K} = \bigcup_{\Lambda \in \mathcal{F}(\mathcal{J})} P^\Lambda \mathcal{K}.$$

This shows that the problem of determining the structure of these subspaces (i.e. whether they are regular, nondegenerate, etc.) reduces to the case of nonincreasing and, respectively, nondecreasing nets of selfadjoint projections. ∎

5.2 Monotone Nets of Selfadjoint Projections

In the remaining part of this section we consider the problem of characterising the nondegeneracy and the regularity of the subspaces generated by monotone nets of selfadjoint projections. First we fix some definitions. Recall that, given a linear operator T acting between two vector spaces $\mathcal{X}$ and $\mathcal{Y}$, with domain $\mathrm{Dom}(T)$ a linear manifold in $\mathcal{X}$, we denote by $\mathcal{G}(T) = \{(x, Tx) \mid x \in \mathrm{Dom}(T)\} \subseteq \mathcal{X} \times \mathcal{Y}$ its graph.

Let $(C_i)_{i \in \mathcal{J}}$ be a net of operators in $\mathcal{K}$. Define two linear submanifolds $\mathcal{G}_\mathrm{s}((C_i)_{i \in \mathcal{J}})$ and $\mathcal{G}_\mathrm{w}((C_i)_{i \in I})$ in $\mathcal{K} \times \mathcal{K}$ as follows: a pair of vectors (ξ, η) belongs to $\mathcal{G}_\mathrm{s}((C_i)_{i \in \mathcal{J}})$ (respectively, to $\mathcal{G}_\mathrm{w}((C_i)_{i \in I})$) if there exists a net of vectors $(\xi_i)_{i \in \mathcal{J}}$ in $\mathcal{K}$ such that, for any $i \in \mathcal{J}$, $\xi_i \in \mathrm{Dom}(C_i)$ and

$$\xi_i \xrightarrow[i \in \mathcal{J}]{\mathrm{s}} \xi \quad \text{and} \quad C_i \xi_i \xrightarrow[i \in \mathcal{J}]{\mathrm{s}} \eta,$$

meaning that the limits hold with respect to the strong topology, respectively,

$$\xi_i \xrightarrow[i \in \mathcal{J}]{\mathrm{w}} \xi \quad \text{and} \quad C_i \xi_i \xrightarrow[i \in \mathcal{J}]{\mathrm{w}} \eta,$$

meaning that the limits hold with respect to the weak topology. In general, the linear manifolds $\mathcal{G}_\mathrm{s}((C_i)_{i \in \mathcal{J}})$ and $\mathcal{G}_\mathrm{w}((C_i)_{i \in \mathcal{J}})$ may not be graphs of operators. If there exists an operator C in $\mathcal{K}$ such that $\mathcal{G}_\mathrm{s}((C_i)_{i \in \mathcal{J}}) = \mathcal{G}(C)$ (respectively, $\mathcal{G}_\mathrm{w}((C_i)_{i \in I}) = \mathcal{G}(C)$) we say that the net $(C_i)_{i \in \mathcal{J}}$ *converges in the strong graph sense* (respectively, *in the weak graph sense*) to the operator C. Clearly, if either the strong limits or the weak limits exist, they are uniquely determined. Also, the strong graph convergence implies the weak graph convergence and, in this case, the two limits coincide.

Lemma 5.2.3. *Let $(P_i)_{i \in \mathcal{J}}$ be a nondecreasing net of selfadjoint projections in $\mathcal{K}$ and consider the subspace*

$$\mathcal{M} = \bigvee_{i \in I} \mathrm{Ran}(P_i).$$

Then

$$\mathcal{G}_\mathrm{s}((P_i)_{i \in \mathcal{J}}) = \mathcal{G}_\mathrm{w}((P_i)_{i \in \mathcal{J}}) = \{(x + y, x) \mid x \in \mathcal{M}, y \in \mathcal{M}^\perp\}.$$

Proof. The assertion will follow if we show the inclusions:

$$\mathcal{G}_\mathrm{s}((P_i)_{i \in \mathcal{J}}) \subseteq \mathcal{G}_\mathrm{w}((P_i)_{i \in \mathcal{J}}) \subseteq \{(x + y, x) \mid x \in \mathcal{M}, y \in \mathcal{M}^\perp\} \subseteq \mathcal{G}_\mathrm{s}((P_i)_{i \in \mathcal{J}}).$$

The first inclusion is obvious. For the second, let $(z, x) \in \mathcal{G}_\mathrm{w}((P_i)_{i \in \mathcal{J}})$. Then there exists a net of vectors $(z_i)_{i \in \mathcal{J}}$ such that $z_i \in \mathrm{Dom}(P_i)$, $i \in \mathcal{J}$, and the following weak convergences hold:

$$z_i \xrightarrow[i \in \mathcal{J}]{\mathrm{w}} z \quad \text{and} \quad P_i z_i \xrightarrow[i \in \mathcal{J}]{\mathrm{w}} x.$$

This implies $x \in \mathcal{M}$. If $j \in \mathcal{J}$ is fixed then for any $t \in \mathrm{Ran}(P_j)$ we have

$$[z_i - P_i z_i, t] \xrightarrow[i \in \mathcal{J}]{} [z - x, t].$$

Considering only those $i \geqslant j$ it follows that
$$[z_i - P_i z_i, t] = [(I - P_i)z_i, P_j t] = 0, \quad t \in \operatorname{Ran}(P_j),$$
and hence
$$[z - x, t] = 0, \quad t \in \operatorname{Ran}(P_j).$$
Since $j \in \mathcal{I}$ is arbitrary we get
$$z - x \in \bigcap_{j \in \mathcal{I}} \operatorname{Ran}(P_j)^\perp = \mathcal{M}^\perp,$$
that is, $z = x + y$ for some $y \in \mathcal{M}^\perp$.

In order to prove the last inclusion let $x \in \mathcal{M}$ and $y \in \mathcal{M}^\perp$. Then
$$y \in \bigcap_{i \in \mathcal{J}} \operatorname{Ran}(P_i)^\perp = \bigvee_{i \in \mathcal{J}} \operatorname{Ker}(P_i).$$
On the other hand, there exists a net of vectors $(x_i)_{i \in \mathcal{J}}$, $x_i \in \operatorname{Ran}(P_i)$, $i \in \mathcal{J}$, such that $x_i \xrightarrow[i \in \mathcal{J}]{\mathrm{s}} x$, hence the strong convergences
$$x_i + y \xrightarrow[i \in \mathcal{J}]{\mathrm{s}} x + y, \quad P_i(x_i + y) = P_i x_i = x_i \xrightarrow[i \in \mathcal{J}]{\mathrm{s}} x,$$
hold, that is, $(x + y, x) \in \mathcal{G}_\mathrm{s}((P_i)_{i \in \mathcal{J}})$. ∎

Theorem 5.2.4. *Let $(P_i)_{i \in \mathcal{J}}$ be a nondecreasing net of selfadjoint projections in $\mathcal{K}$ and consider the subspace $\mathcal{M}$ as in Lemma 5.2.3. The following assertions are equivalent.*

(i) *The subspace $\mathcal{M}$ is nondegenerate.*
(ii) *The net $(P_i)_{i \in \mathcal{J}}$ converges in the strong graph sense.*
(iii) *The net $(P_i)_{i \in \mathcal{J}}$ converges in the weak graph sense.*

Moreover, the limits in (ii) *and* (iii) *coincide with the selfadjoint projection onto the nondegenerate subspace $\mathcal{M}$.*

Proof. (i)⇒(ii). If $\mathcal{M}$ is nondegenerate let P be the selfadjoint projection onto $\mathcal{M}$. Then, the graph of P is
$$\mathcal{G}(P) = \{(x + y, x) \mid x \in \mathcal{M}, \ y \in \mathcal{M}^\perp\},$$
hence, by Lemma 5.2.3 the net $(P_i)_{i \in \mathcal{J}}$ converges in the strong graph sense to P.

(ii)⇒(iii). Obvious.

(iii)⇒(i). If the net $(P_i)_{i \in \mathcal{J}}$ converges in the weak graph sense then, by Lemma 5.2.3, the linear manifold $\{(x + y, x) \mid x \in \mathcal{M}, \ y \in \mathcal{M}^\perp\}$ is the graph of an operator, that is, $x + y = 0$, $x \in \mathcal{M}$ and $y \in \mathcal{M}^\perp$ implies $x = 0$, hence $\mathcal{M}^0 = \mathcal{M} \cap \mathcal{M}^\perp = \{0\}$, that is, $\mathcal{M}$ is a nondegenerate subspace. ∎

5.2 Monotone Nets of Selfadjoint Projections

Remark 5.2.5. As a direct consequence of Theorem 5.2.4 it follows that if $(P_i)_{i \in \mathcal{J}}$ is a nondecreasing net of bounded selfadjoint projections in $\mathcal{K}$ then $\bigvee_{i \in \mathcal{J}} P_i \mathcal{K}$ is regular if and only if $(P_i)_{i \in \mathcal{J}}$ converges in the weak (equivalently, in the strong) graph sense to a bounded operator $P \in \mathcal{B}(\mathcal{K})$. In this case P is the selfadjoint projection onto $\bigvee_{i \in \mathcal{J}} P_i \mathcal{K}$. ∎

The graph topologies are very weak topologies and hence, it is of interest to know whether the convergence of the net of selfadjoint projections can be obtained in stronger topologies.

Proposition 5.2.6. *Let $(P_i)_{i \in \mathcal{J}}$ be a nondecreasing net of bounded selfadjoint projections in $\mathcal{K}$. Then the following statements are equivalent.*

  (i) *The net $(P_i)_{i \in \mathcal{J}}$ is uniformly bounded.*
  (ii) *The net $(P_i)_{i \in \mathcal{J}}$ converges in the strong operator topology.*
  (iii) *The net $(P_i)_{i \in \mathcal{J}}$ converges in the weak operator topology.*

Moreover, if these statements hold then the subspace $\bigvee_{i \in \mathcal{J}} P_i \mathcal{K}$ is regular.

Proof. (i)⇒(ii). Let $\|\cdot\|$ denote a fundamental norm on $\mathcal{K}$ and assume

$$M = \sup_{i \in \mathcal{J}} \|P_i\| < \infty.$$

We first show that the net $(P_i)_{i \in \mathcal{J}}$ is convergent in the strong graph sense, that is, the linear manifold $\mathcal{G}_s((P_i)_{i \in \mathcal{J}})$ is the graph of an operator. To this end, let $(x_i)_{i \in \mathcal{J}}$ be a net of vectors in $\mathcal{K}$ such that the limits exist

$$\|x_i\| \xrightarrow[i \in \mathcal{J}]{} 0, \quad \|P_i x_i - y\| \to 0,$$

for a certain $y \in \mathcal{K}$. Then for arbitrary $i \in \mathcal{J}$,

$$\|y\| \leqslant \|y - P_i x_i\| + \|P_i x_i\| \leqslant \|y - P_i x_i\| + M\|x_i\|$$

and hence, passing to the limit following $i \in \mathcal{J}$, we get $y = 0$. Then, by Theorem 5.2.4 it follows that the subspace $\bigvee_{i \in \mathcal{J}} P_i \mathcal{K}$ is nondegenerate, in particular the linear manifold

$$\mathcal{D} = \bigcup_{i \in \mathcal{J}} P_i \mathcal{K} + \bigcap_{i \in \mathcal{J}} (I - P_i) \mathcal{K}$$

is dense in $\mathcal{K}$. Since for all $x \in \mathcal{D}$ the net $(P_i x)_{i \in \mathcal{J}}$ of vectors in $\mathcal{K}$ is strongly convergent and taking into account the uniform boundedness of $(P_i)_{i \in I}$ the strong operator convergence follows.

(ii)⇒(iii). Obvious.

(iii)⇒(i). Taking into account that the weak topology on $\mathcal{K}$ is equivalently defined by the inner product $\langle \cdot, \cdot \rangle_J$, this implication is a direct consequence of the Uniform Boundedness Principle in Hilbert spaces. ∎

Corollary 5.2.7. *Assume that $(P_i)_{i \in \mathcal{J}}$ is a nondecreasing net of bounded selfadjoint projections in $\mathcal{K}$ such that the subspace $\mathcal{M} = \bigvee_{i \in \mathcal{J}} P_i \mathcal{K}$ is regular and either $\kappa_+(\mathcal{M})$ or $\kappa_-(\mathcal{M})$ is finite. Then the net $(P_i)_{i \in I}$ is uniformly bounded.*

Proof. Assume that the subspace $\mathcal{M}$ is regular and that $\kappa_-(\mathcal{M})$ is finite. Without restricting the generality we can suppose $\mathcal{M} = \mathcal{K}$, in particular the subspace $\mathcal{K}$ is of Pontryagin type. Consider the dense linear manifold $\mathcal{D} = \bigcup_{i \in \mathcal{J}} P_i \mathcal{K}$. According to Lemma 1.4.10, there exists a fundamental decomposition $\mathcal{K} = \mathcal{K}^-[+]\mathcal{K}^+$ such that $\mathcal{K}^- \subseteq \mathcal{D}$. Since $\mathcal{K}^-$ is of finite dimension, it follows that there exists i_0 such for all $i \geqslant i_0$ we have $\mathcal{K}^- \subseteq P_i \mathcal{K}$, hence, letting J denote the fundamental symmetry corresponding to the fundamental decomposition $\mathcal{K} = \mathcal{K}^-[+]\mathcal{K}^+$, we have $J^- P_i = P_i J^-$ and then, since $J = I - 2J^-$, it follows that $JP_i = P_i J$. Consequently, with respect to the decomposition $\mathcal{K} = \mathcal{K}^-[+]\mathcal{K}^+$ the operator P_i is diagonal

$$P_i = \begin{bmatrix} I_- & 0 \\ 0 & P_i^+ \end{bmatrix}, \quad i \geqslant i_0,$$

where $(P_i^+)_{i \in \mathcal{J}}$ is a net of selfadjoint projections in the Hilbert space $\mathcal{K}^+$. Then the net $(P_i)_{i \in \mathcal{J}}$ is uniformly bounded. ∎

We conclude this section with an example of a nondecreasing sequence of bounded selfadjoint projections which is not uniformly bounded but the corresponding regular subspaces span a regular subspace. This shows that the graph convergences cannot be replaced, in general, by stronger convergences.

Example 5.2.8. Let $\mathcal{H}$ be a Hilbert space with orthonormal basis $(g_k)_{k \geqslant 1}$ and consider $\mathcal{K}$ the Kreĭn space and the fundamental symmetry J as in Example 4.1.6. Define the sequences of vectors

$$e_k = g_k \oplus \frac{k}{k+1} g_k, \quad f_k = \frac{k}{k+1} g_k \oplus g_k, \quad k \in \mathbb{N}.$$

Consider the regular subspaces

$$\mathcal{L}_k = \begin{cases} \operatorname{Lin}\{e_1\}, & k = 1, \\ \operatorname{Lin}\{e_1, \ldots, e_k, f_1, \ldots, f_{k-1}\}, & k \geqslant 2. \end{cases}$$

Let P_k denote the bounded selfadjoint projection onto $\mathcal{L}_k$. Since the subspaces $\mathcal{L}_k$, $k \geqslant 1$, span the whole space $\mathcal{K}$, by Remark 5.2.5 it follows that $P_k \xrightarrow[i \in \mathcal{J}]{} I$ in the strong graph sense. On the other hand,

$$P_k(g_k \oplus -g_k) = (k+1)e_k, \quad k \in \mathbb{N},$$

and hence

$$\|P_k\| \geqslant \frac{\|P_k(g_k \oplus -g_k)\|}{\|g_k \oplus -g_k\|} = \sqrt{k^2 + k + \frac{1}{2}} \xrightarrow[k \to \infty]{} \infty,$$

where $\|\cdot\|$ denotes the fundamental norm corresponding to the fundamental symmetry J. ∎

5.3 Unitary Operators

A closed densely defined operator U from the Kreĭn space $\mathcal{K}_1$ into the Kreĭn space $\mathcal{K}_2$ is *unitary* if $U^\sharp = U^{-1}$. Also, recall that an operator S in the Kreĭn space $\mathcal{K}$ is a symmetry if it is unitary and selfadjoint, that is, $S^\sharp = S = S^{-1}$. We emphasise now the close connection between unitary operators and positive symmetries.

Theorem 5.3.1. *Let J_i be fundamental symmetries on the Kreĭn spaces $\mathcal{K}_i$, $i = 1, 2$. Then any unitary operator U from $\mathcal{K}_1$ into $\mathcal{K}_2$ can be factored*

$$U = WS_1 = S_2 W, \tag{3.1}$$

where W, S_1 and S_2 are uniquely determined such that (3.1) holds and, in addition,

(α) $W\colon \mathcal{K}_1 \to \mathcal{K}_2$ is unitary and (J_1, J_2)-unitary (that is, unitary with respect to the corresponding Hilbert spaces $(\mathcal{K}_i, \langle \cdot, \cdot \rangle_{J_i})$, $i = 1, 2$),

(β) S_i are positive symmetries, $i = 1, 2$.

Proof. Consider U as a closed and densely defined operator from the Hilbert space $(\mathcal{K}_1, \langle \cdot, \cdot \rangle_{J_1})$ into the Hilbert space $(\mathcal{K}_2, \langle \cdot, \cdot \rangle_{J_2})$. Then U admits the polar decompositions

$$U = VA_1 = A_2 V, \tag{3.2}$$

where

$$A_1 = (U^*U)^{1/2}, \quad A_2 = (UU^*)^{1/2},$$

and V is (J_1, J_2)-unitary, in particular an everywhere defined bounded and boundedly invertible operator. Define

$$S_1 = J_1 A_1, \quad S_2 = J_2 A_2.$$

Since A_1 is J_1-positive and J_1-selfadjoint (that is, with respect to the inner product $\langle \cdot, \cdot \rangle_{J_1}$) it follows that S_1 is positive and selfadjoint in the Kreĭn space $\mathcal{K}_1$. Further, U unitary means

$$U^{-1} = U^\sharp = J_1 U^* J_2,$$

and hence U^* is also unitary (the involutions $*$ and $\sharp$ commute), that is,

$$U^{*-1} = U^{*\sharp} = J_1 U J_2,$$

and hence we have

$$(A_1^{-1})^2 = (U^*U)^{-1} = U^{-1}U^{*-1} = (J_1 U^* J_2)(J_2 U J_1)$$
$$= J_1(U^*U)J_1 = J_1 A_1^2 J_1 = (J_1 A_1 J_1)^2. \tag{3.3}$$

Since A_1^{-1} and $J_1 A_1 J_1$ are J_1-positive and J_1-selfadjoint, using the uniqueness of the square root, from (3.3) it follows that

$$A_1^{-1} = J_1 A_1 J_1,$$

and hence A_1 is unitary in the Kreĭn space $\mathcal{K}_1$. Then

$$S_1^\sharp = S_1 = J_1 A_1 = A_1^{-1} J_1 = S_1^{-1},$$

which shows that S_1 is also unitary and hence S_1 is a positive symmetry. Similarly we show that S_2 is a positive symmetry.

Let now $x, y \in \operatorname{Dom}(A_1)$ be arbitrary vectors. Then

$$[A_1 x, A_1 y] = [x, y] = [Ux, Uy] = [V A_1 x, V A_1 y] \tag{3.4}$$

and taking into account that $\operatorname{Ran}(A_1)$ is dense in $\mathcal{K}_1$ and that V is bounded and boundedly invertible, (3.4) shows that V is unitary. In addition,

$$V J_1 = V^\sharp J_1 = (J_1 V^* J_2)^{-1} J_1 = J_2 V^{*-1} = J_2 V$$

and hence, defining

$$W = V J_1 = J_2 V,$$

from (3.2) we obtain the factorisations as in (3.1). Then W is unitary and (J_1, J_2)-unitary, since V has the same properties.

Finally, the uniqueness of the factorisations (3.1) with the properties (α) and (β) are obtained by tracking back the same way and taking into account the uniqueness of the left and, respectively, right polar decompositions of closed and densely defined operators in Hilbert spaces. ∎

As a first consequence of Theorem 5.3.1 we can characterise the domains, as well as the ranges, of unitary operators.

Corollary 5.3.2. *A linear manifold $\mathcal{D}$ of the Kreĭn space $\mathcal{K}$ is the domain (equivalently, the range) of some unitary operator, if and only if $\mathcal{D} = \mathcal{L} + \mathcal{L}^\perp$ for some maximal strictly positive subspace $\mathcal{L}$ in $\mathcal{K}$.*

Proof. Indeed, by Theorem 5.3.1, $\mathcal{D}$ is the domain of some unitary operator if and only if it is the domain of some positive symmetry S in $\mathcal{K}$, that is, $\mathcal{D} = \mathcal{L}^+ + \mathcal{L}^\perp$, where $\mathcal{L} = \operatorname{Ker}(S - I)$ is a maximal strictly positive subspace. ∎

Any unitary operator is closed, densely defined, with dense range, and isometric. We have seen in Proposition 4.1.10 that a densely defined isometry and with dense range is closable and its closure has the same features, but it is not clear whether these imply unitarity. The answer is, in general, no.

Example 5.3.3. We have seen in Example 2.2.11 that in any Kreĭn space $\mathcal{K}$ with index of indefiniteness $\kappa(\mathcal{K}) = \min\{\kappa_-(\mathcal{K}), \kappa_+(\mathcal{K})\}$ infinite, there exists an alternating pair $(\mathcal{M}, \mathcal{N})$ of subspaces which is not maximal but such that $\mathcal{M} + \mathcal{N}$ is dense in $\mathcal{K}$, hence nondegenerate. Let V be the operator with domain $\mathcal{M} + \mathcal{N}$ and defined by $V(x + y) = x - y$ for all $x \in \mathcal{M}$ and $y \in \mathcal{N}$. Since $\mathcal{M} \perp \mathcal{N}$ and $\mathcal{M} + \mathcal{N}$ is nondegenerate we have $\mathcal{M} \cap \mathcal{N} = \{0\}$, hence the definition of V is correct. It is easy to see that V is an isometry.

5.3 Unitary Operators

V is a closed operator. Indeed, if $(x_n)_n$ is a sequence in $\mathcal{M}$ and $(y_n)_n$ is a sequence in $\mathcal{N}$ such that $x_n + y_n \xrightarrow[n\to\infty]{} h$ and $x_n - y_n \xrightarrow[n\to\infty]{} k$, then

$$x_n = ((x_n + y_n) + (x_n - y_n))/2 \xrightarrow[n\to\infty]{} (h+k)/2$$

and

$$y_n = ((x_n + y_n) - (x_n - y_n))/2 \xrightarrow[n\to\infty]{} (h-k)/2.$$

Since $\mathcal{M}$ is closed we have $(h+k)/2 \in \mathcal{M}$ and similarly, since $\mathcal{N}$ is closed it follows that $(h-k)/2 \in \mathcal{N}$, hence $h \in \mathcal{M} + \mathcal{N} = \mathrm{Dom}(V)$ and $k = Vh$.

So, V is a closed densely defined isometry and with dense range. However, by Corollary 5.3.2, V is not unitary, since the alternating pair $(\mathcal{M}, \mathcal{N})$ is not maximal. ∎

In the following we focus on canonical forms (certain representations with respect to fundamental decompositions) of unitary operators. We first characterise the action of unitary operators on fundamental decompositions.

Lemma 5.3.4. *Let $\mathcal{K}_1$, $\mathcal{K}_2$ be Kreĭn spaces and U a unitary operator from $\mathcal{K}_1$ into $\mathcal{K}_2$. For any fundamental decomposition $\mathcal{K}_1 = \mathcal{K}_1^-[+]\mathcal{K}_1^+$, the linear manifold $\mathrm{Dom}(U) \cap \mathcal{K}_1^+$ is dense in $\mathcal{K}_1^+$ (respectively, the linear manifold $\mathrm{Dom}(U) \cap \mathcal{K}_1^-$ is dense in $\mathcal{K}_1^-$) and $U(\mathrm{Dom}(U) \cap \mathcal{K}_1^+)$ is a maximal strictly positive subspace in $\mathcal{K}_2$ (respectively, $U(\mathrm{Dom}(U) \cap \mathcal{K}_1^-)$ is a maximal strictly negative subspace in $\mathcal{K}_2$).*

Proof. Applying Theorem 5.3.1 and Proposition 4.2.6 it follows that, without restricting the generality, we can assume $\mathcal{K}_1 = \mathcal{K}_2 = \mathcal{K}$ and that the unitary operator U can be replaced by a positive symmetry S in $\mathcal{K}$. Further, consider the maximal strictly positive subspace $\mathcal{L} = \mathrm{Ker}(S - I)$ and its angular operator K with respect to the fundamental decomposition $\mathcal{K} = \mathcal{K}^-[+]\mathcal{K}^+$. The operator K is a strict contraction. Then S is the closure of the operator S_0 as in Corollary 5.1.7 and hence

$$\mathrm{Dom}(S) \cap \mathcal{K}^+ \supseteq \mathrm{Dom}(S_0) \cap \mathcal{K}^+ = \mathcal{D}_+ = \mathrm{Ran}(I_+ - K^*K)$$

which implies that $\mathrm{Dom}(S) \cap \mathcal{K}^+$ is dense in $\mathcal{K}^+$.

From the operator block matrix representation of S_0 as in Corollary 5.1.7 it follows that

$$S_0 \mathcal{D}_+ = \{x + 2K(I_+ + K^*K)^{-1}x \mid x \in \mathcal{K}^+\}. \tag{3.5}$$

Since K is a strict contraction we have

$$\|(I_+ - K^*K)x\| > 0, \quad x \in \mathcal{K}^+ \setminus \{0\}, \tag{3.6}$$

and this is equivalent with

$$2\|Kx\| < \|(I_+ + K^*K)x\|, \quad x \in \mathcal{K}^+ \setminus \{0\}, \tag{3.7}$$

which shows that the operator $2K(I_+ + K^*K)^{-1}$ is strictly contractive. Applying Corollary 2.1.7, from (3.5) we obtain that $S_0\mathcal{D}_+$ is a maximal strictly positive subspace. On the other hand, $S(\mathrm{Dom}(S) \cap \mathcal{K}^+) \supseteq S_0\mathcal{D}_+$ and since $S(\mathrm{Dom}(S) \cap \mathcal{K}^+)$ is a positive linear manifold, from the maximality of $S_0\mathcal{D}_+$ it follows that

$$S(\mathrm{Dom}(S) \cap \mathcal{K}^+) = S_0\mathcal{D}_+,$$

and hence $S(\mathrm{Dom}(S) \cap \mathcal{K}^+)$ is a maximal strictly positive subspace in $\mathcal{K}$. ∎

Lemma 5.3.5. *Let S be a maximal strictly positive subspace of the Kreĭn space $\mathcal{K}$ and denote by T the angular operator of S with respect to the fundamental decomposition $\mathcal{K} = \mathcal{K}^-[+]\mathcal{K}^+$. Then there exists a unique positive symmetry S in $\mathcal{K}$ such that $S(\mathrm{Dom}(S) \cap \mathcal{K}^+) = S$ and it coincides with the closure of the operator S_0 defined by*

$$S_0 = \begin{bmatrix} (I_+ - T^*T)^{-1/2} & -T^*(I_- - TT^*)^{-1/2} \\ T(I_+ - T^*T)^{-1/2} & -(I_- - TT^*)^{-1/2} \end{bmatrix} : \begin{matrix} \mathcal{D}_+ \\ + \\ \mathcal{D}_- \end{matrix} \to \begin{matrix} \mathcal{K}^+ \\ [+] \\ \mathcal{K}^- \end{matrix},$$

where $\mathcal{D}_+ = \mathrm{Ran}((I_+ - T^*T)^{1/2}) \subseteq \mathcal{K}^+$ and $\mathcal{D}_- = \mathrm{Ran}((I_- - TT^*)^{1/2}) \subseteq \mathcal{K}^-$.

Proof. Let S be a positive symmetry in $\mathcal{K}$, $\mathcal{L} = \mathrm{Ker}(S - I)$ and K the angular operator of $\mathcal{L}$ with respect to the fundamental decomposition $\mathcal{K} = \mathcal{K}^-[+]\mathcal{K}^+$. By Lemma 5.3.4 $S(\mathrm{Dom}(S) \cap \mathcal{K}^+)$ is a maximal strictly positive subspace in $\mathcal{K}$ and it was shown in the course of its proof that $2K(I_+ + K^*K)^{-1}$ is its angular operator. It follows that $S(\mathrm{Dom}(S) \cap \mathcal{K}^+) = S$ if and only if

$$T = 2K(I_+ + K^*K)^{-1}. \tag{3.8}$$

In the following we show that the equation (3.8) has a unique solution K and this solution is a strict contraction.

Let us first remark that from (3.8) it follows that

$$|T| = 2|K|(I_+ + |K|^2)^{-1}, \tag{3.9}$$

where, as usual, $|T| = (T^*T)^{1/2}$ denotes the absolute value of the operator T. Consider the continuous function $\varphi \colon [0, 1] \to [0, 1]$

$$\varphi(k) = \frac{2k}{1 + k^2}, \quad 0 \leqslant k \leqslant 1.$$

By the functional calculus with continuous functions for selfadjoint operators in a Hilbert space, (3.9) yields $|T| = \varphi(|K|)$. The function φ is invertible, its inverse

$$\varphi^{-1}(t) = \begin{cases} \dfrac{1 - \sqrt{1 - t^2}}{t}, & 0 < t \leqslant 1, \\ 0, & t = 0, \end{cases}$$

5.3 Unitary Operators

is continuous and hence, again by functional calculus with continuous functions, from (3.9) we obtain $|K| = \varphi^{-1}(|T|)$. Consider now

$$T = X|T|, \quad K = Y|K|,$$

the left polar decompositions of T and K, respectively. From (3.8) we get $\mathrm{Ker}(T) = \mathrm{Ker}(K)$ and $\mathrm{Ran}(T) = \mathrm{Ran}(K)$ and hence, from (3.8), (3.9), and the uniqueness of the polar decomposition, we obtain $X = Y$. We have proven that $K = X\varphi^{-1}(|T|)$ is the unique solution of (3.8). In addition, K is strictly contractive since (3.6) is equivalent with (3.7).

It remains to compute the entries of S_0 in terms of T. To this end, from (3.8) we obtain

$$I_+ - T^*T = I_+ - 4K^*K(I_+ + K^*K)^{-1} = (I_+ - K^*K)^2(I_+ + K^*K)^{-2},$$

and using the uniqueness of the square root

$$(I_+ - T^*T)^{1/2} = (I_+ - K^*K)(I_+ + K^*K)^{-1},$$

and hence

$$(I_+ - T^*T)^{-1/2} = (I_+ + K^*K)(I_+ - K^*K)^{-1}. \tag{3.10}$$

Then we obtain

$$T(I_+ - T^*T)^{-1/2} = 2K(I_+ - K^*K)^{-1}. \tag{3.11}$$

Similarly we have

$$(I_- - TT^*)^{-1/2} = (I_- + KK^*)(I_- - KK^*)^{-1}, \tag{3.12}$$

and

$$T^*(I_- - TT^*)^{-1/2} = 2K^*(I_- - KK^*)^{-1}. \tag{3.13}$$

From (3.10)–(3.13) and Corollary 5.1.7 we obtain the required representation of S_0 and that S is the closure of S_0. ∎

Theorem 5.3.6. *Let $\mathcal{K}_1$ and $\mathcal{K}_2$ be Kreĭn spaces and fix fundamental decompositions $\mathcal{K}_i = \mathcal{K}_i^-[+]\mathcal{K}_i^+$, $i = 1, 2$. Let $\mathcal{L}$ be a maximal strictly positive subspace in $\mathcal{K}_2$, consider T its angular operator, and denote*

$$\mathcal{D}_+ = \mathrm{Ran}((I_+ - T^*T)^{1/2}) \subseteq \mathcal{K}_2^+, \quad \mathcal{D}_- = \mathrm{Ran}((I_- - TT^*)^{1/2}) \subseteq \mathcal{K}_2^-.$$

The following statements are equivalent.
  (i) *The unitary operator U from $\mathcal{K}_1$ into $\mathcal{K}_2$ satisfies $U(\mathrm{Dom}(U) \cap \mathcal{K}_1^+) = \mathcal{L}$.*
  (ii) *U is the closure of the linear operator U_0 defined by*

$$U_0 = \begin{bmatrix} (I_+ - T^*T)^{-1/2}V_+ & T^*(I_- - TT^*)^{-1/2}V_- \\ T(I_+ - T^*T)^{-1/2}V_+ & (I_- - TT^*)^{-1/2}V_- \end{bmatrix} : \begin{array}{c} V_+^{-1}\mathcal{D}_+ \\ + \\ V_-^{-1}\mathcal{D}_- \end{array} \to \begin{array}{c} \mathcal{K}_2^+ \\ [+] \\ \mathcal{K}_2^- \end{array}$$

where $V_\pm \in \mathcal{B}(\mathcal{K}_1^\pm, \mathcal{K}_2^\pm)$ are Hilbert space unitary operators.

Proof. Let U be a unitary operator from $\mathcal{K}_1$ into $\mathcal{K}_2$ and let $U = S_2 W$ be the representation as in Theorem 5.3.1. We denote by J_i the fundamental symmetries corresponding to the fundamental decomposition $\mathcal{K} = \mathcal{K}_i^-[+]\mathcal{K}_i^+$. Since W is a unitary and (J_1, J_2)-unitary operator (where J_i is the fundamental symmetry associated to the fundamental decompositions $\mathcal{K}_i = \mathcal{K}_i^-[+]\mathcal{K}_i^+$, $i = 1, 2$) it follows that with respect to these fundamental decompositions we have

$$W = \begin{bmatrix} W_+ & 0 \\ 0 & W_- \end{bmatrix},$$

where $W_\pm \in \mathcal{B}(\mathcal{K}_1^\pm, \mathcal{K}_2^\pm)$ are Hilbert space unitary operators. Define $V_\pm = \pm W_\pm$. Then note that $U(\mathrm{Dom}(U) \cap \mathcal{K}_1^+) = \mathcal{L}$ if and only if $S_2(\mathrm{Dom}(S_2) \cap \mathcal{K}_2^+) = \mathcal{L}$ and hence the equivalence of the statements (i) and (ii) follows from Lemma 5.3.5. ∎

When specialising to bounded unitary operators we have to consider maximal uniformly definite subspaces.

Corollary 5.3.7. *Let $\mathcal{K}_1$ and $\mathcal{K}_2$ be Kreĭn spaces and fix fundamental decompositions $\mathcal{K}_i = \mathcal{K}_i^-[+]\mathcal{K}_i^+$, $i = 1, 2$. Let $\mathcal{L}$ be a maximal uniformly positive subspace in $\mathcal{K}_2$, consider T its angular operator, and let $U \in \mathcal{B}(\mathcal{K}_1, \mathcal{K}_2)$ be a unitary operator. The following assertions are equivalent.*

(i) *U maps $\mathcal{K}_1$ onto $\mathcal{L}$.*

(ii) *With respect to the specified fundamental decompositions, U has the following operator block matrix representation*

$$U = \begin{bmatrix} (I_+ - T^*T)^{-1/2}V_+ & T^*(I_- - TT^*)^{-1/2}V_- \\ T(I_+ - T^*T)^{-1/2}V_+ & (I_- - TT^*)^{-1/2}V_- \end{bmatrix}, \qquad (3.14)$$

where $V_\pm$ are Hilbert space unitary operators in $\mathcal{K}^\pm$.

Theorem 5.3.6 has a dual counterpart corresponding to the first decomposition in Theorem 5.3.1. We first prove a lemma.

Lemma 5.3.8. *Let $\mathcal{K}$ be a Kreĭn space, S a symmetry on $\mathcal{K}$, $\mathcal{L}$ a maximal strictly positive subspace of $\mathcal{K}$ and $\mathcal{K} = \mathcal{K}^-[+]\mathcal{K}^+$ a fundamental decomposition. Then the closure of $S(\mathrm{Dom}(S) \cap \mathcal{L})$ coincides with $\mathcal{K}^+$ if and only if $S(\mathrm{Dom}(S) \cap \mathcal{K}^+) = \mathcal{L}$.*

Proof. If $S(\mathrm{Dom}(S) \cap \mathcal{K}^+) = \mathcal{L}$ then $\mathcal{L} \subseteq \mathrm{Ran}(S) = \mathrm{Dom}(S)$ and using Lemma 5.3.4 we have

$$\mathrm{Clos}\, S(\mathrm{Dom}(S) \cap \mathcal{L}) = \mathrm{Clos}\, S\mathcal{L} = \mathrm{Clos}\,(\mathrm{Dom}(S) \cap \mathcal{K}^+) = \mathcal{K}^+.$$

Conversely, assume that $S(\mathrm{Dom}(S) \cap \mathcal{K}^+) = \mathcal{L}$ holds. Then

$$\mathrm{Clos}\,(\mathrm{Dom}(S) \cap \mathcal{L}) \subseteq S(\mathcal{K}^+ \cap \mathrm{Dom}(S)), \qquad (3.15)$$

5.3 Unitary Operators

and taking into account that $S(\mathrm{Dom}(S) \cap \mathcal{K}^+)$ is closed it follows that

$$\mathrm{Clos}\,(\mathrm{Dom}(S) \cap \mathcal{L}) \subseteq S(\mathrm{Dom}(S) \cap \mathcal{K}^+) \subseteq \mathrm{Ran}(S) = \mathrm{Dom}(S)$$

and hence $\mathrm{Clos}\,(\mathrm{Dom}(S) \cap \mathcal{L}) \subseteq \mathrm{Dom}(S) \cap \mathcal{L}$, that is, $\mathrm{Dom}(S) \cap \mathcal{L}$ is closed.

Let $x \in S(\mathrm{Dom}(S) \cap \mathcal{K}^+)$ be a vector such that $x \perp \mathrm{Dom}(S) \cap \mathcal{L}$. Since S is isometric this yields $Sx \perp \mathrm{Clos}\,S(\mathrm{Dom}(S) \cap \mathcal{L})$ and hence $Sx \in \mathcal{K}^- \cap \mathcal{K}^+ = \{0\}$. We have thus proven that (3.15) holds with equality and then, since both $\mathcal{L}$ and $S(\mathrm{Dom}(S) \cap \mathcal{K}^+)$ are maximal strictly positive subspaces, we conclude that $\mathcal{L} = S(\mathcal{K}^+ \cap \mathrm{Dom}(S))$. ∎

Theorem 5.3.9. *Let $\mathcal{K}_1$ and $\mathcal{K}_2$ be Kreĭn spaces and fix fundamental decompositions $\mathcal{K}_i = \mathcal{K}_i^-[+]\mathcal{K}_i^+$, $i = 1, 2$. Let $\mathcal{L}$ be a maximal strictly positive subspace in $\mathcal{K}_1$, consider T its angular operator and denote*

$$\mathcal{D}_+ = \mathrm{Ran}((I_+ - T^*T)^{1/2}) \subseteq \mathcal{K}_1^+, \quad \mathcal{D}_- = \mathrm{Ran}((I_- - TT^*)^{1/2}) \subseteq \mathcal{K}_1^-.$$

The following statements are equivalent.

(i) *The unitary operator U from $\mathcal{K}_1$ into $\mathcal{K}_2$ satisfies $\mathrm{Clos}\,U(\mathrm{Dom}(U) \cap \mathcal{L}) = \mathcal{K}_2^+$.*

(ii) *U is the closure of the linear operator U_0 defined by*

$$U_0 = \begin{bmatrix} V_+(I_+ - T^*T)^{-1/2} & -V_+T^*(I_- - TT^*)^{-1/2} \\ -V_-T(I_+ - T^*T)^{-1/2} & V_-(I_- - TT^*)^{-1/2} \end{bmatrix} : \begin{matrix} \mathcal{D}_+ \\ \dot{+} \\ \mathcal{D}_- \end{matrix} \to \begin{matrix} \mathcal{K}_2^+ \\ [+] \\ \mathcal{K}_2^- \end{matrix}$$

where $V_\pm \in \mathcal{B}(\mathcal{K}_1^\pm, \mathcal{K}_2^\pm)$ are Hilbert space unitary operators.

Proof. Consider the factorisation $U = WS_1$ as in Theorem 5.3.1. Then we have $\mathrm{Clos}\,U(\mathrm{Dom}(U) \cap \mathcal{L}) = W\,\mathrm{Clos}\,S_1(\mathrm{Dom}(S_1) \cap \mathcal{L})$ and hence, by Lemma 5.3.8, the statement (i) is equivalent with $S_1(\mathcal{K}_1 \cap \mathrm{Dom}(S_1)) = \mathcal{L}$. Define $V_\pm$ as in the proof of Theorem 5.3.6. The equivalence of (i) and (ii) follows now from Lemma 5.3.5. ∎

Again, when specialising to bounded unitary operators we have to consider their action on uniformly definite subspaces.

Corollary 5.3.10. *Let $\mathcal{K}_1$ and $\mathcal{K}_2$ be Kreĭn spaces and fix fundamental decompositions $\mathcal{K}_i = \mathcal{K}_i^-[+]\mathcal{K}_i^+$, $i = 1, 2$. Let $\mathcal{L}$ be a maximal uniformly positive subspace in $\mathcal{K}_1$, consider T its angular operator, and let $U \in \mathcal{B}(\mathcal{K}_1, \mathcal{K}_2)$ be a unitary operator. The following assertions are equivalent.*

(i) *U maps $\mathcal{L}$ onto $\mathcal{K}_2^+$.*

(ii) *With respect to the specified fundamental decompositions, U has the following operator block matrix representation*

$$U = \begin{bmatrix} V_+(I_+ - T^*T)^{-1/2} & -V_+T^*(I_- - TT^*)^{-1/2} \\ -V_-T(I_+ - T^*T)^{-1/2} & V_-(I_- - TT^*)^{-1/2} \end{bmatrix}, \quad (3.16)$$

where $V_\pm$ are Hilbert space operators in $\mathcal{K}^\pm$.

5.4 Dense Operator Ranges

A linear manifold $\mathcal{M}$ of some Kreĭn space $\mathcal{K}$ is called an *operator range* if $\mathcal{M} = \mathrm{Ran}(T)$ for some operator $T \in \mathcal{B}(\mathcal{H}, \mathcal{K})$, where $\mathcal{H}$ is a Kreĭn space. It is clear that in this definition of operator ranges, only the Hilbert space structures of the spaces matter, so this is actually a notion from the operator theory in Hilbert spaces and, at this level of generality, the indefiniteness of the ambient space $\mathcal{K}$ plays no role. Also, an operator range can be defined as the range of some closed operator, but it does not make a more general definition since, using the graph norm on the domain, any closed operator can be regarded as a bounded operator from a different initial space.

We first show that operator ranges in Kreĭn spaces share some of the features of subspaces, for example, decomposability.

Proposition 5.4.1. *Any operator range $\mathcal{M}$ in a Kreĭn space $\mathcal{K}$ is decomposable, more precisely, $\mathcal{M} = \mathcal{M}_- \dot{+} \mathcal{M}^0 \dot{+} \mathcal{M}_+$, for some mutually orthogonal strictly strictly positive/negative linear submanifolds $\mathcal{M}_\pm$ of $\mathcal{K}$.*

Proof. Let $\mathcal{M} = \mathrm{Ran}(T)$ for some operator $T \in \mathcal{B}(\mathcal{H}, \mathcal{K})$ and consider the selfadjoint operator $A = T^\sharp T \in \mathcal{B}(\mathcal{H})$. Without restricting the generality we can assume that $\mathcal{H}$ is a Hilbert space and hence A is a bounded selfadjoint operator in the Hilbert space $\mathcal{H}$. Consider the polar decomposition $A = S_A |A|$ and the spectral decomposition

$$\mathcal{H} = \mathcal{H}_- \oplus \mathrm{Ker}(A) \oplus \mathcal{H}_+,$$

where $\mathcal{H}_\pm$ are the spectral subspaces of A corresponding to $(0, +\infty)$ and to $(-\infty, 0)$, respectively. It is easy to show that $\mathcal{M}^0 = T \mathrm{Ker}(A)$ and hence, letting

$$\mathcal{M}_\pm = T|A|^{-1/2}(\mathrm{Ran}(|A|^{1/2}) \cap \mathcal{H}_\pm),$$

it follows that $\mathcal{M} = \mathcal{M}_- \dot{+} \mathcal{M}^0 \dot{+} \mathcal{M}_+$. In addition, $\mathcal{M}_\pm$ are strictly positive/negative linear manifolds and mutually orthogonal. ∎

The main result of this section is a dichotomic characterisation of dense operator ranges. We first prove three lemmas.

Lemma 5.4.2. *Assume that the operator range $\mathcal{M}$ is contained into the domain of some unbounded unitary operator. Then $\mathcal{M}$ does not contain any maximal uniformly definite subspaces.*

Proof. According to Corollary 5.3.2, if the operator range $\mathcal{M}$ is included in the domain of some unbounded unitary operator it follows that $\mathcal{M} \subseteq \mathcal{L} + \mathcal{L}^\perp$, for some maximal strictly positive but not uniformly positive subspace $\mathcal{L}$. Let us assume that $\mathcal{M}$ contains some maximal uniformly positive subspace $\mathcal{P}$. Then $\mathcal{K} = \mathcal{P}[+]\mathcal{P}^\perp$ is a fundamental decomposition of $\mathcal{K}$. Letting K denote the angular operator of $\mathcal{L}$ with respect to this fundamental decomposition then K is a strict contraction of norm 1. Since $\mathcal{P} \subseteq \mathcal{L} + \mathcal{L}^\perp$ it follows that any $x \in \mathcal{P}$ has a unique representation $x = (f + K^*g) + (Kf + g)$ for

5.4 Dense Operator Ranges

some $f \in \mathcal{P}$ and $g \in \mathcal{P}^\perp$. This yields $g = -Kf$ and so $x = (I - K^*K)f$. But, since $\|K\| = 1$ it follows that $\mathrm{Ran}(I - K^*K) \neq \mathcal{P}$, a contradiction. ∎

Note that Lemma 5.4.2 can be viewed also as a consequence of Proposition 4.3.3, but we preferred a proof which is more elementary.

Lemma 5.4.3. *Let $\mathcal{M} = \mathcal{M}_- + \mathcal{M}_+$ be an operator range such that the linear manifolds $\mathcal{M}_\pm \subseteq \mathcal{K}_\pm$ for some fundamental decomposition $\mathcal{K} = \mathcal{K}^-[+]\mathcal{K}^+$. Then $\mathcal{M}$ contains a maximal uniformly definite subspace if and only if either $\mathcal{M}_- = \mathcal{K}^-$ or $\mathcal{M}_+ = \mathcal{K}^+$.*

Proof. To make a choice, let $\mathcal{P} \subseteq \mathcal{M}$ be some maximal uniformly positive subspace and let K denote the angular operator of $\mathcal{P}$ with respect to the fundamental decomposition $\mathcal{K} = \mathcal{K}_-[+]\mathcal{K}_+$. Then, for any $x \in \mathcal{K}^+$ we have $x + Kx \in \mathcal{P} \subseteq \mathcal{M} = \mathcal{M}_- + \mathcal{M}_+$ and, since $\mathcal{M}_\pm \subseteq \mathcal{K}^\pm$, it follows that $x \in \mathcal{M}_+$. Thus $\mathcal{M}_+ = \mathcal{K}^+$. ∎

The next lemma is essentially a result of Hilbert space operator theory, but we state it within our Kreĭn space context. In addition, note that an angular operator is always a Hilbert space contraction and a strict contraction of norm 1 determines exactly those strict contractions that are not uniform.

Lemma 5.4.4. *Let $\mathcal{K} = \mathcal{K}^+[+]\mathcal{K}^-$ be a fundamental decomposition and assume that $K \in \mathcal{B}(\mathcal{K}^+, \mathcal{K}^-)$ is a Hilbert space strict contraction of norm 1. Then there exists a Hilbert space strict contraction $L \in \mathcal{B}(\mathcal{K}^+, \mathcal{K}^-)$ of norm 1 such that*

$$\begin{bmatrix} I_+ & K^* \\ K & I_- \end{bmatrix} \leqslant \begin{bmatrix} I_+ & L^* \\ L & I_- \end{bmatrix}^2.$$

Proof. As in the proof of Lemma 5.3.5 there exists a strict Hilbert space contraction $L \in \mathcal{B}(\mathcal{K}^+, \mathcal{K}^-)$ of norm 1, the unique solution of the equation $K = 2L(I_+ + L^*L)^{-1}$. Then note that

$$\begin{bmatrix} I_+ & K^* \\ K & I_- \end{bmatrix} = \begin{bmatrix} I_+ & L^* \\ L & I_- \end{bmatrix}^2 - \begin{bmatrix} I_+ & 2L^*(I_- + LL^*)^{-1} \\ 2L(I_+ + L^*L)^{-1} & I_- \end{bmatrix}$$

$$= \begin{bmatrix} I_+ & 2L^*(I_- + LL^*)^{-1} \\ 2L(I_+ + L^*L)^{-1} & I_- \end{bmatrix} \begin{bmatrix} L^*L & 0 \\ 0 & LL^* \end{bmatrix} \geqslant 0,$$

where, in the second row, we took into account that the two positive operator block matrices commute and hence their product is a positive operator. ∎

Theorem 5.4.5. *Let $\mathcal{M}$ be a dense operator range in the Kreĭn space $\mathcal{K}$. Then one and only one of the following cases holds true.*

Case 1. *$\mathcal{M} \subseteq \mathcal{L} + \mathcal{L}^\perp$ for some maximal strictly positive subspace $\mathcal{L}$ of $\mathcal{K}$, which is not uniformly positive.*

Case 2. *$\mathcal{M}$ contains a maximal uniformly definite subspace.*

Proof. Since $\mathcal{M}$ is dense it follows that it is nondegenerate. In view of Proposition 5.4.1 $\mathcal{M} = \mathcal{M}_- + \mathcal{M}_+$ for some strictly positive/negative linear manifolds $\mathcal{M}_\pm$. Then their closures $\text{Clos}\,\mathcal{M}_\pm$ are strictly positive/negative subspaces. We distinguish two possibilities.

(a) *Neither $\mathcal{M}_-$ nor $\mathcal{M}_+$ are uniform.* According to Theorem 2.2.9 there exists a maximal strictly positive subspace $\mathcal{L}$ such that $\mathcal{M}_+ \subseteq \mathcal{P}$ and $\mathcal{M}_- \subseteq \mathcal{P}^\perp$ and hence Case 1 holds.

(b) *One of the linear manifolds $\mathcal{M}_\pm$ is uniform.* To make a choice, assume that $\mathcal{M}_+$ is uniformly positive. Then its closure $\text{Clos}\,\mathcal{M}_+$ is a uniformly positive subspace, $\text{Clos}\,\mathcal{M}_-$ is negative, $\text{Clos}\,\mathcal{M}_- \perp \text{Clos}\,\mathcal{M}_+$ and $\text{Clos}\,\mathcal{M}_- + \text{Clos}\,\mathcal{M}_+$ is dense in $\mathcal{K}$. In view of Corollary 2.2.10 it follows that $\text{Clos}\,\mathcal{M}_+$ is a maximal uniformly positive subspace and hence $\mathcal{M}_- \subseteq \mathcal{M}_+^\perp$ is uniformly negative. Thus, in case (b) both linear manifolds $\mathcal{M}_\pm$ are uniformly definite. Moreover, if one of the linear manifolds $\mathcal{M}_\pm$ is closed then the preceding argument proves that $\mathcal{M}$ contains a maximal definite subspace and hence, Case 2 holds.

Let us assume now that neither $\mathcal{M}_-$ nor $\mathcal{M}_+$ is closed. In view of Lemma 5.4.3 $\mathcal{M}$ does not contain any maximal uniformly definite subspace. Applying Theorem 2.2.9 we get a fundamental decomposition $\mathcal{K} = \mathcal{K}^+[+]\mathcal{K}^-$ such that $\mathcal{M}_\pm \subseteq \mathcal{K}^\pm$ and let J be the corresponding fundamental symmetry. Let $\mathcal{M} = \text{Ran}(T)$ for some bounded operator T. Since $J\mathcal{M} = \mathcal{M}$ it follows that $\mathcal{M}_\pm = \text{Ran}(J^\pm T)$ and hence, by using the polar decomposition and scaling, if necessary, there exist Hilbert space positive selfadjoint contractions $A_\pm \in \mathcal{B}(\mathcal{K}^\pm)$ such that $\mathcal{M}_\pm = \text{Ran}(A_\pm)$. In addition, since their ranges are dense but not closed, both $A_\pm$ are injective but there exist two sequences of orthonormal spectral vectors $(f_n^\pm)_{n\geqslant 1}$ in $\mathcal{K}^\pm$ such that $A_\pm f_n^\pm \xrightarrow[n\to\infty]{} 0$.

Define a partial isometry $V \in \mathcal{B}(\mathcal{K}^+, \mathcal{K}^-)$ such that $V f_n^+ = f_n^-$ for each $n \geqslant 1$ and then define the strict contraction $K \in \mathcal{B}(\mathcal{K}^+, \mathcal{K}^-)$ of norm 1 by

$$K = (I_- - A_-^2)^{1/2} V (I_+ - A_+^2)^{1/2}.$$

With this definition we have

$$\begin{bmatrix} A_+^2 & 0 \\ 0 & 0 \end{bmatrix} \leqslant \begin{bmatrix} I_+ - K^*K & 0 \\ 0 & I_- \end{bmatrix},$$

and hence

$$\begin{bmatrix} A_+^2 & 0 \\ 0 & 0 \end{bmatrix} = \begin{bmatrix} I_+ & K^* \\ 0 & I_- \end{bmatrix} \begin{bmatrix} A_+^2 & 0 \\ 0 & 0 \end{bmatrix} \begin{bmatrix} I_+ & 0 \\ K & I_- \end{bmatrix}$$
$$\leqslant \begin{bmatrix} I_+ & K^* \\ 0 & I_- \end{bmatrix} \begin{bmatrix} I_+ - K^*K & 0 \\ 0 & I_- \end{bmatrix} \begin{bmatrix} I_+ & 0 \\ K & I_- \end{bmatrix} = \begin{bmatrix} I_+ & K^* \\ K & I_- \end{bmatrix}.$$

This and Lemma 5.4.4 imply

$$\begin{bmatrix} A_+^2 & 0 \\ 0 & 0 \end{bmatrix} \leqslant \begin{bmatrix} I_+ & L^* \\ L & I_- \end{bmatrix}^2, \tag{4.1}$$

where $L \in \mathcal{B}(\mathcal{K}^+, \mathcal{K}^-)$ is the unique strict contraction of norm 1 that verifies the equation $K = 2L(I_+ + L^*L)^{-1}$. In a similar way we prove

$$\begin{bmatrix} 0 & 0 \\ 0 & A_-^2 \end{bmatrix} \leqslant \begin{bmatrix} I_+ & L^* \\ L & I_- \end{bmatrix}^2. \tag{4.2}$$

Finally, let $\mathcal{L} = \{x + Lx \mid x \in \mathcal{K}^+\}$ be the maximal strictly positive but not uniformly positive subspace associated to the angular operator L. From (4.1) and (4.2) it follows that $\mathcal{M} = \mathcal{M}_- + \mathcal{M}_+ \subseteq \mathcal{L} + \mathcal{L}^\perp$, that is, Case 1 holds. ∎

As an application of Theorem 5.4.5 we will add one more geometric characterisation of uniqueness of Kreĭn spaces induced by selfadjoint operators in Theorem 6.1.9.

5.5 Notes

The material in this chapter follows essentially our article [59]; only Theorem 5.4.5 is from M. A. Dritschel's article [50], with a slightly different proof. We have been inspired by the work of M. Tomita [142] in approaching unbounded unitary operators. The bounded cases of Theorem 5.3.6 and Theorem 5.3.9 as in Corollary 5.3.7 and Corollary 5.3.10 were obtained by M. G. Kreĭn and Yu. L. Shmulyan in [101].

Chapter 6

Techniques of Induced Kreĭn Spaces

An induced Kreĭn space is the analogue of renorming a Hilbert space and it is a very powerful tool in operator theory on Kreĭn spaces. We already considered the abstract case in Section 1.5 but here we are interested in a special situation, when the Kreĭn space is induced by a bounded selfadjoint operator. This allows us to clarify also the uniqueness question that we left open before. Of significant importance is the possibility of lifting operators to induced spaces. A first application of the technique of induced Kreĭn spaces is to operator valued holomorphic functions with a symmetry property with respect to complex conjugation and their representations as compressed resolvents of bounded selfadjoint operators in Kreĭn spaces, where an essential role is played by the Nevanlinna kernel. Then we see an application to linearisations of selfadjoint operator pencils. We also see a theorem in which an operator valued holomorphic function in a neighbourhood of 0 is represented as a compressed resolvent of a bounded unitary operator in a Kreĭn space, where an essential role is played by the Carathéodory kernel. The last application concerns elementary rotations, which are special types of unitary liftings, and an explicit form of the canonical unitary dilation, which leads to the Schur kernels.

6.1 Kreĭn Spaces Induced by Selfadjoint Operators

If $A \in \mathcal{B}(\mathcal{H})$ is a selfadjoint operator in the Kreĭn space $(\mathcal{H}, [\cdot, \cdot]_\mathcal{H})$ we can define a new inner product $[\cdot, \cdot]_A$ on $\mathcal{H}$ by

$$[x, y]_A = [Ax, y]_\mathcal{H}, \quad x, y \in \mathcal{H}. \tag{1.1}$$

In this section we investigate the existence and the properties of Kreĭn spaces induced by this kind of inner product space. This concept is a special case of that of a Kreĭn space induced by an inner product space as in Section 1.5. More precisely, a pair $(\mathcal{K}; \Pi)$ is called a *Kreĭn space induced by A* if

(i) $(\mathcal{K}; [\cdot, \cdot]_\mathcal{K})$ is a Kreĭn space,
(ii) $\Pi \colon \mathcal{H} \to \mathcal{K}$ is a bounded linear operator,
(iii) $\Pi \mathcal{H}$ is dense in $\mathcal{K}$,
(iv) for all $x, y \in \mathcal{H}$ we have $[\Pi x, \Pi y]_\mathcal{K} = [Ax, y]_\mathcal{H}$, equivalently, $A = \Pi^\sharp \Pi$.

We refer to Π as a *canonical map*.

6.1 Kreĭn Spaces Induced by Selfadjoint Operators

Remark 6.1.1. (a) The requirement that $(\mathcal{H}; [\cdot, \cdot]_\mathcal{H})$ is a Kreĭn space can be replaced, without loss of generality, with that of being a Hilbert space. Indeed, letting J be a fundamental symmetry on $\mathcal{K}$, we consider the Hilbert space $(\mathcal{H}; \langle \cdot, \cdot \rangle_J)$ and the bounded selfadjoint operator $D = JA$ on it. Then $[x, y]_D = \langle Dx, y \rangle_J = [JDx, y]_\mathcal{H} = [Ax, y]_\mathcal{H}$ for all $x, y \in \mathcal{H}$ and hence $(\mathcal{K}; \Pi)$ is a Kreĭn space induced by A if and only if it is a Kreĭn space induced by D.

(b) In Section 1.5 of Chapter 1 we considered a more general notion of a Kreĭn space induced by an inner product space $(\mathcal{X}; [\cdot, \cdot])$ and it is necessary to see how the two concepts are related. So, in view of the statement at item (a), let $A \in \mathcal{B}(\mathcal{H})$ a selfadjoint operator in the Hilbert space $(\mathcal{H}, \langle \cdot, \cdot \rangle_\mathcal{H})$ be given and assume that there exists $(\mathcal{K}; \Pi)$ a Kreĭn space induced by the inner product space $(\mathcal{H}; [\cdot, \cdot]_A)$ in the sense of the definition given in Section 1.5, with notation as in (1.1), that is, $(\mathcal{K}, [\cdot, \cdot]_\mathcal{K})$ is a Kreĭn space and $\Pi : \mathcal{H} \to \mathcal{K}$ is a linear operator such that $\operatorname{Ran}(\Pi) = \Pi\mathcal{H}$ is dense in $\mathcal{K}$ and such that

$$[\Pi x, \Pi y]_\mathcal{K} = \langle Ax, y \rangle_\mathcal{H}, \quad x, y \in \mathcal{H}, \tag{1.2}$$

holds. Since A is bounded its kernel $\operatorname{Ker}(A)$ is a subspace in $\mathcal{H}$, that is, a closed linear manifold in $\mathcal{H}$ and, from (1.2) it follows that Π leaves $\operatorname{Ker}(A)$ invariant hence, without loss of generality, by replacing $\mathcal{H}$ with $\mathcal{H} \ominus \operatorname{Ker}(A)$, we can assume that $\operatorname{Ker}(A) = \{0\}$. This implies that the inner product space $(\mathcal{H}; [\cdot, \cdot]_A)$ defined as in (1.1) is nondegenerate and then, from (1.2) it follows that Π is injective. Consequently, there exists the inverse operator $\Pi^{-1} : \Pi\mathcal{H} \to \mathcal{H}$ with dense domain and full range. On the other hand, for any $z \in \operatorname{Ran}(\Pi) = \mathcal{H}$, letting $y = \Pi^{-1} z$ in (1.2) we obtain

$$[\Pi x, z]_\mathcal{K} = \langle Ax, \Pi^{-1} z \rangle_\mathcal{H} = \langle x, A\Pi^{-1} z \rangle_\mathcal{H}.$$

This shows that Π has a densely defined adjoint, that is $A\Pi^{-1}$, and hence it is closable. Then, Π is everywhere defined, it is closed and hence, by the Closed Graph Principle, it is bounded. In conclusion, for the special case of a bounded selfadjoint operator A, the continuity of the canonical operator Π comes for free and hence, the two definitions agree.

(c) Let $(\mathcal{X}; [\cdot, \cdot])$ be an inner product space and assume that on $\mathcal{X}$ there is a quadratic norm $\|\cdot\|$, that is, associated to a positive definite inner product $\langle \cdot, \cdot \rangle$, making the inner product $[\cdot, \cdot]$ jointly continuous. Then, letting $\mathcal{H}$ be the Hilbert space completion of $\mathcal{X}$ with respect to the norm $\|\cdot\|$, the inner product $[\cdot, \cdot]$ can be uniquely extended to $\mathcal{H}$. Then, as in Remark 1.3.3, we can define the Gram operator $G \in \mathcal{B}(\mathcal{H})$, $G = G^*$, such that $[x, y] = \langle Gx, y \rangle$ for all $x, y \in \mathcal{H}$. If $(\mathcal{K}; \Pi)$ is a Kreĭn space induced by G then $(\mathcal{K}; \Pi|\mathcal{X})$ is a Kreĭn space induced by the inner product space $(\mathcal{X}; [\cdot, \cdot])$, with definition as in Section 1.5. ∎

Two Kreĭn spaces $(\mathcal{K}_i, \Pi_i)$, $i = 1, 2$, induced by the same bounded selfadjoint operator A, are called *unitary equivalent* if there exists a bounded unitary operator $U : \mathcal{K}_1 \to \mathcal{K}_2$ such that

$$U\Pi_1 x = \Pi_2 x, \quad x \in \mathcal{H}. \tag{1.3}$$

In view of Remark 6.1.1.(b) and the results in Section 1.5, we have some general characterisations of the existence of Kreĭn spaces induced by selfadjoint operators A in terms of the underlying inner product spaces $(\mathcal{H}; [\cdot, \cdot]_A)$ defined as in (1.1). However, we recall that the uniqueness question was not considered in Section 1.5 and it is only now that we are in a position to properly deal with it. On the other hand, this time we ask for criteria of existence of induced Kreĭn spaces in terms of the operator A.

Here we have the existence of Kreĭn spaces induced by bounded selfadjoint operators.

Proposition 6.1.2. *For any selfadjoint operator $A \in \mathcal{B}(\mathcal{H})$ in the Kreĭn space $(\mathcal{H}; [\cdot, \cdot]_\mathcal{H})$ there exists $(\mathcal{K}; \Pi)$ a Kreĭn space induced by A.*

Proof. Let J be a fundamental symmetry on the Kreĭn space $\mathcal{H}$ and consider the bounded selfadjoint operator $D = JA$ onto the Hilbert space $(\mathcal{H}, \langle \cdot, \cdot \rangle_J)$. Let E denote the spectral measure of D. Then the Hilbert space $\mathcal{H}$ can be decomposed as

$$\mathcal{H} = \mathcal{D}_- \oplus \mathrm{Ker}(A) \oplus \mathcal{D}_+, \tag{1.4}$$

where $\mathcal{D}_- = E(-\infty, 0)\mathcal{H}$ and $\mathcal{D}_+ = E(0, \infty)\mathcal{H}$. We consider the nondegenerate inner product space $(\hat{\mathcal{K}}, [\cdot, \cdot]_A)$ where

$$\hat{\mathcal{K}} = \mathcal{D}_- \dotplus \mathcal{D}_+.$$

Note that $\mathcal{D}_-$ is orthogonal to $\mathcal{D}_+$ with respect to the inner product $[\cdot, \cdot]_A$ and then we use Proposition 1.5.2 to embed densely and isometrically the inner product space $(\hat{\mathcal{K}}, [\cdot, \cdot]_A)$ into a Kreĭn space $(\mathcal{K}, [\cdot, \cdot]_A)$. Then define $\Pi \colon \mathcal{H} \to \mathcal{K}$ by

$$\mathcal{H} \ni x_- + x_0 + x_+ \mapsto x_- + x_+ \in \mathcal{K}.$$

Clearly $(\mathcal{K}; \Pi)$ is a Kreĭn space induced by A. ∎

Next we consider the situation when the induced Kreĭn space is actually a Pontryagin type space. Recall that $\kappa_-(A)$ and $\kappa_+(A)$ denote the negative and, respectively, the positive algebraic ranks of the inner product space $(\mathcal{H}; [\cdot, \cdot]_A)$ defined as in (1.1).

Corollary 6.1.3. *Let $A \in \mathcal{B}(\mathcal{K})$ be a selfadjoint operator for some Kreĭn space $(\mathcal{H}; [\cdot, \cdot]_\mathcal{H})$ such that either $\kappa_-(A) < \infty$ or $\kappa_+(A) < \infty$. Then there exists and it is unique, up to a unitary equivalence, a Kreĭn space of Pontryagin type $(\mathcal{K}; [\cdot, \cdot]_\mathcal{K})$ induced by A, more precisely, either $\kappa_-(\mathcal{K}) = \kappa_-(A) < \infty$ or $\kappa_+(\mathcal{K}) = \kappa_+(A) < \infty$.*

Proof. In view of Remark 6.1.1, this is a consequence of Corollary 1.5.4. ∎

Remark 6.1.4. We observe that the strong topology of the Kreĭn space induced by the selfadjoint operator A, as defined in Proposition 6.1.2, corresponds to the completion of $\hat{\mathcal{K}}$ with respect to the norm $\hat{\mathcal{K}} \ni x \mapsto \||JA|^{1/2} x\|$, where $\|\cdot\|$ denotes the fundamental norm corresponding to the fundamental symmetry J. ∎

6.1 KREĬN SPACES INDUCED BY SELFADJOINT OPERATORS

Remark 6.1.5. The Kreĭn space induced by the selfadjoint operator A and constructed in the proof of Proposition 6.1.2 does not depend on the fundamental symmetry J, up to a unitary equivalence. Indeed, let G be a second fundamental symmetry on the Kreĭn space $\mathcal{H}$. Then $JA = (JG)(GA)$ and the bounded operator JG is the adjoint of the identity operator $(\mathcal{H}, \langle \cdot, \cdot \rangle_J) \to (\mathcal{H}, \langle \cdot, \cdot \rangle_G)$. This shows that if $GA = (GA)^+ - (GA)^-$ is the Jordan decomposition of GA, in the Hilbert space $(\mathcal{H}, \langle \cdot, \cdot \rangle_G)$, then $JA = (JG)(GA)^+ - (JG)(GA)^-$ is the Jordan decomposition of JA in the Hilbert space $(\mathcal{H}, \langle \cdot, \cdot \rangle_J)$. Then observe that JG gives rise to a unitary equivalence of the two induced Kreĭn spaces obtained using the fundametal symmetry J and, respectively, the fundamental symmetry G. ■

By means of Remark 6.1.5, we can denote by $(\mathcal{K}_A; \Pi_A)$, without any ambiguity, the Kreĭn space induced by the selfadjoint operator A and constructed as in the proof of Proposition 6.1.2. This also enables us to define the *geometric positive/negative ranks* of the selfadjoint operator A by

$$\kappa^+(A) = \kappa^+(\mathcal{K}_A), \quad \kappa^-(A) = \kappa^-(\mathcal{K}_A). \tag{1.5}$$

The construction of the induced Kreĭn space $(\mathcal{K}_A; \Pi_A)$, where A is a selfadjoint operator in the Kreĭn space $\mathcal{H}$, has the disadvantage that it is obtained by a completion procedure and hence, some of the vectors in $\mathcal{K}_A$ can be outside of $\mathcal{H}$. In the following we present a different construction in which the induced Kreĭn space is actually a subspace of $\mathcal{H}$, the strong topology of this induced Kreĭn space is inherited from the strong topology of $\mathcal{H}$, but the cost is a more involved canonical mapping Π.

Example 6.1.6. Let A be a bounded selfadjoint operator in the Kreĭn space $\mathcal{H}$. We fix a fundamental symmetry J on $\mathcal{H}$ and consider the polar decomposition

$$JA = S_{JA}|JA| \tag{1.6}$$

of the bounded selfadjoint operator JA on the Hilbert space $(\mathcal{H}, \langle \cdot, \cdot \rangle_J)$. The operator S_{JA} is a selfadjoint partial isometry and we consider the subspace $\mathcal{H}_A = \text{Clos}(\text{Ran}(JA))$ that is invariant under S_{JA}. The restriction of S_{JA} to $\mathcal{H}_A$ is a symmetry and let us define the inner product $[\cdot, \cdot]_{S_{JA}}$ by

$$[x, y]_{S_{JA}} = \langle S_{JA} x, y \rangle_J, \quad x, y \in \mathcal{H}_A. \tag{1.7}$$

We consider the Kreĭn space $(\mathcal{H}_A, [\cdot, \cdot]_{S_{JA}})$. Since $\text{Ran}(|JA|^{1/2}) \subseteq \mathcal{H}_A$ we can define the operator $\pi_A \colon \mathcal{H} \to \mathcal{H}_A$ by

$$\pi_A x = |JA|^{1/2} x, \quad x \in \mathcal{H}. \tag{1.8}$$

It is easy to see that $(\mathcal{H}_A, \pi_A)$ is a Kreĭn space induced by A.

We prove now that the induced Kreĭn spaces $(\mathcal{H}_A; \pi_A)$ and $(\mathcal{K}_A; \Pi_A)$ are unitary equivalent. Recalling now the notation in the proof of Proposition 6.1.2, we define the operator U with $\mathrm{Dom}(U) = \mathcal{D}_- \dotplus \mathcal{D}_+$ where, letting $JA = (JA)_+ - (JA)_-$ be the Jordan decomposition of the operator JA, we have $\mathcal{D}_\pm = \mathrm{Clos}\,\mathrm{Ran}(JA_\pm)$, defined by

$$Ux = |JA|^{1/2}x, \quad x \in \mathcal{H}. \tag{1.9}$$

It follows that for all $x, y \in \mathcal{H}$ we have

$$[Ux, Uy]_{S_{JA}} = \langle S_{JA}|JA|^{1/2}x, |JA|^{1/2}y\rangle_J = [Ax, y]_\mathcal{H} = [x, y]_A,$$

which proves that U is isometric. By the construction of the Kreĭn space $\mathcal{K}_A$ as in Proposition 6.1.2, the closures of $\mathcal{D}_\pm$ are maximal uniformly definite subspaces in $\mathcal{K}_A$. We notice that $U\mathcal{D}_\pm \subseteq \mathcal{H}_A^\pm$, where $\mathcal{H}_A = \mathcal{H}_A^+[+]\mathcal{H}_A^-$ is the fundamental decomposition associated to the fundamental symmetry S_{JA}. Then we apply Proposition 4.3.1 to conclude that U is uniquely extended to a bounded unitary operator $U \in \mathcal{B}(\mathcal{K}_A, \mathcal{H}_A)$. Using the definition of U it follows that $U\Pi_A x = \pi_A x$, for all $x \in \mathcal{H}$.

From Remark 6.1.5 and the fact proved before, it follows that the definition of the induced Kreĭn space $(\mathcal{H}_A; \pi_A)$ does not depend on the fundamental symmetry J, in particular, this justifies the notation. ∎

The idea of a Kreĭn space induced by a bounded selfadjoint operator A on $\mathcal{H}$, in view of axiom (iv) of the definition, is to factor $A = \Pi^\sharp \Pi$, for some bounded operator $\Pi \colon \mathcal{H} \to \mathcal{K}$ and some Kreĭn space $\mathcal{K}$, such that Π has dense range. In the case of a Hilbert space, this kind of factorisation is an equivalent characterisation of positivity. The analogue of this fact for Kreĭn spaces is related to the positive/negative geometric ranks $\kappa^\pm(A)$, as defined at (1.5). On the other hand, let us observe that, as a by-product of the unitary equivalence of the induced Kreĭn spaces $(\mathcal{K}_A; \Pi_A)$ and $(\mathcal{H}_A; \pi_A)$ as in Example 6.1.6, it follows that

$$\kappa^+(A) = \dim(E_{JA}(0, +\infty)\mathcal{H}), \quad \kappa^-(A) = \dim(E_{JA}(-\infty, 0)\mathcal{H}), \tag{1.10}$$

where J is any fundamental symmetry of $\mathcal{H}$ and E_{JA} is the spectral measure of JA.

Proposition 6.1.7. *Let A be a bounded selfadjoint operator in a Kreĭn space $\mathcal{H}$ and let $\mathcal{K}$ be another Kreĭn space. Then, $A = T^\sharp T$, for some operator $T \in \mathcal{B}(\mathcal{H}, \mathcal{K})$, if and only if $\kappa^\pm(A) \leqslant \kappa^\pm(\mathcal{K})$.*

Proof. As in Remark 6.1.1.(a), without loss of generality we can assume that $\mathcal{H}$ is a Hilbert space and let E denote the spectral measure of A. Then, as in Example 6.1.6, $\mathcal{H}_A = \mathcal{H}_A^+[+]\mathcal{H}_A^-$, where $\mathcal{H}_A^+ = E(0, +\infty)\mathcal{H}$ and $\mathcal{H}_A^- = E(-\infty, 0)\mathcal{H}$, is a fundamental decomposition of the Kreĭn space $(\mathcal{H}_A; [\cdot, \cdot]_A)$.

Let $A = T^\sharp T$ for some operator $T \in \mathcal{B}(\mathcal{H}, \mathcal{K})$. Then,

$$[x, y]_A = \langle Ax, y\rangle_\mathcal{H} = \langle T^\sharp Tx, y\rangle_\mathcal{H} = [Tx, Ty]_\mathcal{K}, \quad x, y \in \mathcal{H},$$

hence T is isometric with respect to inner products $[\cdot, \cdot]_A$ and $[\cdot, \cdot]_\mathcal{K}$ and, consequently, injective on any strictly definite subspace of $\mathcal{H}_A$. In particular, $T\mathcal{H}_A^+$ is a uniformly positive

6.1 Kreĭn Spaces Induced by Selfadjoint Operators

linear manifold in $\mathcal{K}$ and $T\mathcal{H}_A^-$ is a uniformly negative manifold in $\mathcal{K}$. By Theorem 2.2.9 there exists a fundamental decomposition $\mathcal{K} = \mathcal{K}^+[+]\mathcal{K}^-$ such that $T\mathcal{H}_A^\pm \subseteq \mathcal{K}^\pm$. Since T is injective on $\mathcal{H}_A^\pm$ it follows that $\kappa^\pm(A) = \dim(\mathcal{H}_A^\pm) \subseteq \dim(\mathcal{K}^\pm) = \kappa^\pm(\mathcal{K})$.

Conversely, assume that $\kappa^\pm(A) \leqslant \kappa^\pm(\mathcal{K})$ and let $\mathcal{K} = \mathcal{K}^+[+]\mathcal{K}^-$ be an arbitrary fundamental decomposition of $\mathcal{K}$. Then, there exist bounded isometric operators $V_\pm\colon \mathcal{H}_A^\pm \to \mathcal{K}^\pm$. Recalling that the norm topology of $\mathcal{H}_A$ is the same as the topology induced by the norm of $\mathcal{H}$, it follows that, in view of the decomposition $\mathcal{H} = \mathcal{H}_A^+ \oplus \operatorname{Ker}(A) \oplus \mathcal{H}_A^-$, letting the operator $T\colon \mathcal{H} \to \mathcal{K}$ be defined by

$$Tx = (x_+ + x_0 + x_-) = V_+x_+ + V_-x_-,$$

for arbitrary $x = x_+ + x_0 + x_- \in \mathcal{H}$, with $x_\pm \in \mathcal{H}_A^\pm$ and $x_0 \in \operatorname{Ker}(A)$, T is bounded and then, with obvious notation,

$$\begin{aligned}
\langle T^\sharp Tx, y\rangle_\mathcal{H} &= [Tx, Ty]_\mathcal{K} = [V_+x_+ + V_-x_-, V_+y_+ + V_-y_-]_\mathcal{K}\\
&= [V_+x_+, V_+y_+]_\mathcal{K} + [V_-x_-, V_-y_-]_\mathcal{K}\\
&= [x_+, y_+]_A + [x_-, y_-]_A = [x_+ + x_-, y_+ + y_-]_A\\
&= \langle A(x_+ + x_-), y_+ + y_-\rangle_\mathcal{H} = \langle Ax, y\rangle_\mathcal{H}, \quad x, y \in \mathcal{H},
\end{aligned}$$

hence $A = T^\sharp T$. ∎

Remark 6.1.8. With notation as in Proposition 6.1.7, $A = T^\sharp T$ for some operator $T \in \mathcal{B}(\mathcal{H})$ if and only if $\kappa^\pm(A) \leqslant \kappa^\pm(\mathcal{H})$. In particular, if $\mathcal{H}$ is a separable Kreĭn space with infinite ranks $\kappa^\pm(\mathcal{H})$, it follows that any selfadjoint operator $A \in \mathcal{B}(\mathcal{H})$ can be factored as $A = T^\sharp T$, for some $T \in \mathcal{B}(\mathcal{H})$. ∎

We are now in a position to approach the uniqueness, modulo unitary equivalence, of the Kreĭn spaces induced by selfadjoint operators.

Theorem 6.1.9. *Let A be a bounded selfadjoint operator in the Kreĭn space $\mathcal{H}$. The following statements are equivalent.*

(i) *The Kreĭn space induced by A is unique, modulo unitary equivalence.*
(ii) *For some (equivalently, for any) fundamental symmetry J of $\mathcal{H}$, there exists an $\varepsilon > 0$ such that either $(0, \varepsilon) \subset \rho(JA)$ or $(-\varepsilon, 0) \subset \rho(JA)$.*
(iii) *For some (equivalently, for any) Kreĭn space $(\mathcal{K}; \Pi)$ induced by A, the linear manifold $\Pi\mathcal{H}$ contains a maximal uniformly definite subspace of $\mathcal{K}$.*
(iv) *For some (equivalently, for any) Kreĭn space $(\mathcal{K}; \Pi)$ induced by A, the linear manifold $\Pi \operatorname{Dom}(A)$ is not contained in the domain of any unbounded unitary operator in $\mathcal{K}$.*

Proof. (i)⇒(ii). Let us first notice that, without restricting the generality, we can assume that $\mathcal{H}$ is a Hilbert space. Indeed, as in Remark 6.1.1 (a), the statement (ii) is a topological property of the spectrum of JA which holds either for any fundamental symmetry J of $\mathcal{H}$ or for no fundamental symmetry and hence, we can replace the Kreĭn space

$\mathcal{H}$ with the Hilbert space $(\mathcal{H}, \langle\cdot,\cdot\rangle_J)$, for some arbitrary fundamental symmetry J, and then, instead of A, we consider the operator JA.

So, $(\mathcal{H}; \langle\cdot,\cdot\rangle_\mathcal{H})$ is a Hilbert space, $A = A^* \in \mathcal{B}(\mathcal{H})$, $A = A_+ - A_-$ is the Jordan decomposition, and let E denote its spectral measure. Also, as in Remark 6.1.1.(b), without restricting the generality, we can assume that $\operatorname{Ker}(A) = \{0\}$. Consequently, by Remark 6.1.4, $\mathcal{K}_A$ is just the completion of $\mathcal{H}$ under the norm $\|\cdot\|_A = \||A|^{1/2} \cdot \|_\mathcal{H}$.

Let us assume that the statement (ii) does not hold. Then there exists a strictly decreasing sequence of values $(\mu_n)_{n\geqslant 1} \subseteq \sigma(A)$, $0 < \mu_n < 1$ such that $\mu_n \xrightarrow[n\to\infty]{} 0$, and there exists a decreasing sequence of values $(\nu_n)_{n\geqslant 1} \subseteq \sigma(-A)$, $0 < \nu_n < 1$, such that $\nu_n \xrightarrow[n\to\infty]{} 0$. Then there exist sequences of vectors $(e_n)_{n\geqslant 1}$ and $(f_n)_{n\geqslant 1}$ such that

$$e_n \in E((\mu_{n+1}, \mu_n])\mathcal{H}, \quad f_n \in E([-\nu_n, -\nu_{n+1}))\mathcal{H}, \quad n \geqslant 1, \qquad (1.11)$$

$$\langle Ae_i, e_j\rangle_\mathcal{H} = \delta_{ij}, \quad \langle Af_i, f_j\rangle_\mathcal{H} = -\delta_{ij}, \quad i,j \geqslant 1. \qquad (1.12)$$

As a consequence, we have $\|e_n\|_A = 1 = \|f_n\|_A$, for all $n \geqslant 1$, and

$$\langle Ae_i, f_j\rangle_\mathcal{H} = 0, \quad i,j \geqslant 1. \qquad (1.13)$$

Define the sequence $(\lambda_n)_{n\geqslant 1}$ by

$$\lambda_n = \min\{\sqrt{1-\mu_n^2}, \sqrt{1-\nu_n^2}\}.$$

Then $0 < \lambda_n < 1$, $\lambda_n \xrightarrow[n\to\infty]{} 1$.

Let us consider now the sequence $(\mathcal{S}_n)_{n\geqslant 1}$, of subspaces of the Kreĭn space $\mathcal{K}_A$, defined by

$$\mathcal{S}_n = \mathbb{C}e_n \dotplus \mathbb{C}f_n, \quad n \geqslant 1,$$

and then define the isometric operators $U_n \in \mathcal{B}(\mathcal{S}_n)$

$$U_n = \frac{1}{\sqrt{1-\lambda_n^2}} \begin{bmatrix} 1 & -\lambda_n \\ \lambda_n & -1 \end{bmatrix}, \quad n \geqslant 1.$$

To see that U_n is isometric either perform a direct calculation or use Corollary 5.3.7. We observe that

$$\|U_n\| \geqslant \|U_n e_n\|_A = \frac{1}{\sqrt{1-\lambda_n^2}}\|e_n + \lambda_n f_n\|_A = \frac{\sqrt{1+\lambda_n^2}}{\sqrt{1-\lambda_n^2}} \xrightarrow[n\to\infty]{} \infty. \qquad (1.14)$$

Further, we define the linear manifold $\mathcal{D}_0$ in $\mathcal{K}_A$ by

$$\mathcal{D}_0 = \bigcup_{k\geqslant 1}(\mathcal{S}_1 + \cdots + \mathcal{S}_k)$$

and notice that the closure of $\mathcal{D}_0$ is a regular subspace in $\mathcal{K}_A$.

6.1 KREĬN SPACES INDUCED BY SELFADJOINT OPERATORS

Recalling the proof of Proposition 6.1.2, with $\mathcal{D}_+ = E(0,+\infty)\mathcal{H} = \operatorname{Clos}\operatorname{Ran}(A_+)$ in $\mathcal{H}$ and $\mathcal{D}_- = E(-\infty,0)\mathcal{H} = \operatorname{Clos}\operatorname{Ran}(A_-)$ in $\mathcal{H}$, since we assumed $\operatorname{Ker}(A) = \{0\}$, the linear manifold $\mathcal{H} = \mathcal{D}$, where $\mathcal{D} = \mathcal{D}_+ \dotplus \mathcal{D}_- = \operatorname{Ran}(\Pi_A)$ is dense in $\mathcal{K}_A$. By construction, $\mathcal{D}_0 \subseteq \mathcal{D}$ and the following decomposition holds

$$\mathcal{D} = \mathcal{D}_0 \dotplus (\mathcal{D} \cap \mathcal{D}_0^\perp).$$

Then define a linear operator U in $\mathcal{K}_A$, with domain $\mathcal{D}_0$ and the same range, given by $U|\mathcal{S}_n = U_n$, $n \geqslant 1$ and $U|(\mathcal{D}\cap\mathcal{D}_0^\perp) = I|(\mathcal{D}\cap\mathcal{D}_0^\perp)$. The operator U is isometric in $\mathcal{K}_A$, it has dense range as well as dense domain, and it is unbounded, by (1.14). Then define the operator $\Pi\colon \mathcal{H} \to \mathcal{K}_A$ by $\Pi = U\Pi_A$. We claim that $(\mathcal{K}_A, \Pi)$ is a Kreĭn space induced by A.

Indeed, $\Pi\mathcal{H} = U\Pi_A\mathcal{H} \supseteq \mathcal{D}$ which is dense in $\mathcal{K}_A$. Further,

$$[\Pi x, \Pi y] = [U\Pi_A x, U\Pi_A y] = [\Pi_A x, \Pi_A y] = [Ax, y], \quad x, y \in \mathcal{H}.$$

In addition, Π is bounded, as Remark 6.1.1 shows. This concludes the proof of the claim. Since U is unbounded it follows that $(\mathcal{K}_A; \Pi_A)$ is not unitary equivalent with $(\mathcal{K}_A; \Pi)$.

(ii)⇒(iii). As before, we can assume, without restricting the generality, that $\mathcal{H}$ is a Hilbert space and $\operatorname{Ker}(A) = \{0\}$. Denoting $\mathcal{H}_\pm = \operatorname{Clos}\operatorname{Ran}(A_\pm)$, the following decomposition holds

$$\mathcal{H} = \mathcal{H}_+ \oplus \mathcal{H}_-.$$

The operators $A_\pm$ induce bounded selfadjoint operators in the Hilbert spaces $\mathcal{H}_\pm$, respectively. As in Remark 6.1.4 it follows that the strong topology of $\mathcal{K}$ is determined by the norms $\mathcal{H}_\pm \ni x \mapsto \|A_\pm^{1/2} x\|$.

To make a choice, let us assume that there exists $\varepsilon > 0$ such that $(-\varepsilon, 0) \subseteq \rho(A)$, equivalently A_- has closed range, hence $\operatorname{Ran}(A_-) = \mathcal{H}_-$ is complete with respect to the norm $\|A_-^{1/2} \cdot \|$. By the definition of the Kreĭn space $\mathcal{K}_A$, this implies that $\mathcal{H}_-$ is a maximal uniformly negative subspace of $\mathcal{K}_A$.

In the case it is assumed that $(0, \varepsilon) \subseteq \rho(A)$, in a similar way we prove that $\mathcal{H}_+$ is a maximal uniformly positive subspace of $\mathcal{K}_A$.

(iii)⇒(i). Let $(\mathcal{K}_i; \Pi_i)$, $i = 1, 2$, be Kreĭn spaces induced by A. The equation $U\Pi_1 x = \Pi_2 x$, $x \in \mathcal{H}$, uniquely determines an isometric operator U densely defined in $\mathcal{K}_1$ and with dense range in $\mathcal{K}_2$. If $\Pi_1\mathcal{H}$ contains a maximal uniformly definite subspace then by Proposition 4.3.3 it follows that U has a unique extension to a bounded unitary operator and hence the two Kreĭn spaces induced by A are unitary equivalent. Moreover, since bounded unitary operators map maximal uniformly definite subspaces into maximal uniformly definite subspaces, it follows that if (ii) holds for some Kreĭn space induced by A then it holds for any other Kreĭn space induced by A.

(iii)⇔(iv). This is a consequence of Theorem 5.4.5. ∎

In order to exploit the full power of induced Kreĭn spaces we need to know which linear operators can be lifted to the corresponding induced Kreĭn spaces. Since we do not have, in general, uniqueness modulo unitary equivalence of induced Kreĭn spaces, the lifting theorem refers only to spaces of type $(\mathcal{K}_A; \Pi_A)$, and their unitary equivalence class.

Theorem 6.1.10. *Let $\mathcal{H}_1$ and $\mathcal{H}_2$ be Kreĭn spaces and let $A \in \mathcal{B}(\mathcal{H}_1)$, $A = A^\sharp$, $B \in \mathcal{B}(\mathcal{H}_2)$, $B = B^\sharp$, $T_1 \in \mathcal{B}(\mathcal{H}_1, \mathcal{H}_2)$, and $T_2 \in \mathcal{B}(\mathcal{H}_2, \mathcal{H}_1)$ be such that*

$$[T_1 x, y]_B = [x, T_2 y]_A, \quad x \in \mathcal{H}_1,\ y \in \mathcal{H}_2,$$

or, equivalently, $T_2^\sharp A = B T_1$. Then there exist uniquely determined operators $\widetilde{T}_1 \in \mathcal{B}(\mathcal{K}_A, \mathcal{K}_B)$ and $\widetilde{T}_2 \in \mathcal{B}(\mathcal{K}_B, \mathcal{K}_A)$ such that

$$\widetilde{T}_1 \Pi_A = \Pi_B T_1, \quad \widetilde{T}_2 \Pi_B = \Pi_A T_2,$$

and

$$[\widetilde{T}_1 x, y]_B = [x, \widetilde{T}_2 y]_A, \quad x \in \mathcal{K}_A,\ y \in \mathcal{K}_B.$$

Proof. Fix fundamental symmetries J_1 and J_2 on $\mathcal{H}_1$ and $\mathcal{H}_2$, respectively. Then the assumption $T_2^\sharp A = BT_1$ reads

$$T_2^* J_1 A = J_2 B T_1. \tag{1.15}$$

We consider the selfadjoint operators $J_1 A$ and $J_2 B$ on the Hilbert spaces $(\mathcal{H}_1, \langle \cdot, \cdot \rangle_{J_1})$ and $(\mathcal{H}_2, \langle \cdot, \cdot \rangle_{J_2})$, respectively, and let

$$J_1 A = S_{J_1 A} |J_1 A|, \quad J_2 B = S_{J_2 B} |J_2 B|,$$

be their polar decompositions. We claim that for any vector $x \in \mathcal{H}_1$ the following inequality holds

$$\||J_2 B|^{1/2} T_1 x\| \leqslant \|T_2 S_{J_2 B} T_1\|^{1/2} \cdot \||J_1 A|^{1/2} x\|. \tag{1.16}$$

Fix $x \in \mathcal{H}_1$ arbitrary and let $n \leqslant 1$ be a natural number. From the relation (1.15) and its dual relation $T_1^* B^* J_2 = A^* J_1 T_2$ it follows that, for all $n \geqslant 1$,

$$\langle J_1 A (T_2 S_{J_2 B} T_1)^{2^n} x, (T_2 S_{J_2 B} T_1)^{2^n} x \rangle_{J_1} = \langle J_1 A x, (T_2 S_{J_2 B} T_1)^{2^{n+1}} x \rangle_{J_1}. \tag{1.17}$$

By repeatedly using the Schwarz Inequality and (1.17) we obtain, for all $n \geqslant 1$,

$$\||J_2 B|^{1/2} T_1 x\|^2 \leqslant \||J_1 A|^{1/2} x\|^{1 + \frac{1}{2} + \cdots + \frac{1}{2^n}} \langle J_1 A x, (T_2 S_{J_2 B} T_1)^{2^{n+1}} x \rangle_{J_1}^{1/2^{n+1}},$$

and hence

$$\||J_2 B|^{1/2} T_1 x\|^2 \leqslant \||J_1 A|^{1/2} x\|^{1 + \frac{1}{2} + \cdots + \frac{1}{2^{n+1}}} \||J_1 A|^{1/2}\|^{1/2^{n+1}}$$
$$\cdot \|(T_2 S_{J_2 B} T_1)\| \|x\|^{1/2^{n+1}}. \tag{1.18}$$

Letting $n \to \infty$ in (1.18) we obtain the inequality (1.16).

From the assumption $T_2^\sharp A = BT_1$ it follows that $T_1 \operatorname{Ker}(A) \subseteq \operatorname{Ker}(B)$ and hence T_1 factors to an operator $\hat{T}_1 \colon \mathcal{H}_1 \ominus \operatorname{Ker}(A) \to \mathcal{H}_2 \ominus \operatorname{Ker}(B)$. Using the inequality (1.16) and Remark 6.1.4 it follows that $\hat{T}_1$ extends by continuity to an operator $\widetilde{T}_1 \in \mathcal{B}(\mathcal{K}_A, \mathcal{K}_B)$ such that $\widetilde{T}_1 \Pi_A = \Pi_B T_1$. Similarly, there exists a uniquely determined operator $\widetilde{T}_2 \in \mathcal{B}(\mathcal{K}_B, \mathcal{K}_A)$ such that $\widetilde{T}_2 \Pi_B = \Pi_A T_2$. The last statement follows from the assumption $T_2^\sharp A = BT_1$ and the continuity of $\widetilde{T}_1$ and $\widetilde{T}_2$. ∎

The lifting theorem has some consequences for the spectral behaviour. In the next corollary, if σ is a complex set then $\overline{\sigma} = \{\overline{\lambda} \mid \lambda \in \sigma\}$.

Corollary 6.1.11. *With notation and assumptions as in Theorem 6.1.10, assume that $\mathcal{H}_1 = \mathcal{H}_2$ and $A = B$. Then, $\rho(T_1) \cap \overline{\rho(T_2)} \subseteq \rho(\widetilde{T}_1) \cap \overline{\rho(\widetilde{T}_2)}$.*

Proof. If $\mathcal{H}_1 = \mathcal{H}_2 = \mathcal{H}$ and $A = B$ hold then Theorem 6.1.10 says that, considering the class

$$\mathcal{Z}_A = \{T \in \mathcal{B}(\mathcal{H}) \mid \text{ there exists } S \in \mathcal{B}(\mathcal{H}),\ AT = S^\sharp A\},$$

one can uniquely define a mapping

$$\mathcal{Z}_A \ni T \mapsto \widetilde{T} \in \mathcal{B}(\mathcal{K}_A),$$

such that

$$\Pi_A T = \widetilde{T} \Pi_A, \quad T \in \mathcal{Z}_A.$$

Moreover, the class of operators $\mathcal{Z}_A$ is clearly stable under multiplication and the same Theorem 6.1.10 implies that the mapping $\sim$ is unital, that is, $\widetilde{I}$ is the identity operator on $\mathcal{K}_A$, and it is also multiplicative, that is,

$$\widetilde{XY} = \widetilde{X}\widetilde{Y}, \quad X, Y \in \mathcal{Z}_A.$$

If $\lambda \in \rho(T_1) \cap (T_2)$ then, from the assumption $AT_1 = T_2^\sharp A$ it follows that

$$A(T_1 - \lambda I)^{-1} = (T_2^\sharp - \lambda I)^{-1} A,$$

in particular $T_1 - \lambda I \in \mathcal{Z}_A$ and, using the mapping $\sim$ mentioned before, it follows that $\widetilde{T} - \lambda \widetilde{I}$ is invertible, hence $\lambda \in \rho(\widetilde{T}_1)$. Similarly we prove that $\overline{\lambda} \in \rho(\widetilde{T}_2^\sharp)$. ∎

6.2 Nevanlinna Type Representations

In this section we show that certain classes of operator valued holomorphic functions that possess symmetry properties with respect to complex conjugation admit realisations as generalised resolvents of selfadjoint bounded operators in Kreĭn spaces. This result, although it is not the most general of this type, establishes a deep connection between operator valued holomorphic functions and operator theory in Kreĭn spaces, on the one hand, and on the other hand, it motivates once more the interest in the spectral theory of selfadjoint operators to be considered in the next chapters.

For $r > 0$ we consider the open unit disc $\mathbb{D}_r = \{z \in \mathbb{C} \mid |z| < r\}$ as well as the complement of its closure in the complex plain $\mathbb{D}_r^c = \{z \in \mathbb{C} \mid |z| > r\}$.

Theorem 6.2.1. *Let $(\mathcal{H}; [\cdot,\cdot]_{\mathcal{H}})$ be a Kreĭn space and $r > 0$. Assume that we are given a function $Q\colon \operatorname{Clos}\mathbb{D}_r^c \to \mathcal{B}(\mathcal{H})$ which is holomorphic on an open neighbourhood of $\operatorname{Clos}\mathbb{D}_r^c$ and at infinity, and is symmetric with respect to the complex conjugation, that is,*

$$Q(\bar{z}) = Q(z)^{\sharp}, \quad |z| \geqslant r.$$

Then there exists a triple $(\mathcal{K}; A; \Gamma)$, where $\mathcal{K}$ is a Kreĭn space, $A \in \mathcal{B}(\mathcal{K})$ is selfadjoint, $A^{\sharp} = A$, with spectrum $\sigma(A) \subseteq \operatorname{Clos}\mathbb{D}_r$, and $\Gamma \in \mathcal{B}(\mathcal{H},\mathcal{K})$ such that

$$Q(z) = Q(\infty) + \Gamma^{\sharp}(A - zI)^{-1}\Gamma, \quad z \in \mathbb{D}_r^c. \tag{2.1}$$

In addition, $\mathcal{K}$ can be chosen minimal in the sense that

$$\mathcal{K} = \operatorname{Clos}\operatorname{Lin}\{(A - zI)^{-1}\Gamma\mathcal{H} \mid z \in \mathbb{D}_r^c\}.$$

Proof. We first observe that, without loss of generality, we can assume that $\mathcal{H}$ is a Hilbert space. Indeed, letting J denote a fixed fundamental symmetry and $\langle \cdot,\cdot \rangle_J$ denote the corresponding positive definite inner product, we replace Q by the function JQ and observe that it has the symmetry property $JQ(\bar{z}) = JQ(z)^{\sharp} = J(JQ(z)^*J) = Q(z)^{\sharp}J = (JQ(z))^{\sharp}$. Then, once we prove the statement for the Hilbert space $(\mathcal{H}; \langle \cdot,\cdot \rangle_{\mathcal{H}})$, we only have to replace the operator Γ with the operator ΓJ. So, in the following, we let $(\mathcal{H}; \langle \cdot,\cdot \rangle_{\mathcal{H}})$ be a Hilbert space.

We consider the Nevanlinna kernel $N_Q\colon \mathbb{D}_r^c \times \mathbb{D}_r^c \to \mathcal{B}(\mathcal{H})$ corresponding to Q defined by

$$N_Q(\zeta, z) = \begin{cases} \frac{Q(z) - Q(\zeta)^*}{z - \bar{\zeta}}, & z \neq \bar{\zeta}, \\ Q'(z), & z = \bar{\zeta}. \end{cases} \tag{2.2}$$

Since Q is symmetric with respect to complex conjugation, the kernel N_Q is Hermitian in the sense that $N_Q(\zeta, z) = N_Q(z, \zeta)^*$ for all $z, \zeta \in \mathbb{D}_r^c$, and the function $N_Q(\zeta, \cdot)$ is holomorphic on an open neighbourhood of the closure of $\mathbb{D}_r^c$ and null at infinity, for all $|\zeta| > r$.

Step 1. *For all z, ζ with $|z|, |\zeta| > r$ we have*

$$N_Q(\zeta, z) = -\frac{1}{2\pi \mathrm{i}} \int_{|u|=r} \frac{Q(u)}{(u-z)(u-\bar{\zeta})} \mathrm{d}u, \tag{2.3}$$

where the circle $\{z \mid |z| = r\}$ is positively oriented.

6.2 NEVANLINNA TYPE REPRESENTATIONS

Indeed, for $z \neq \bar{\zeta}$ we write the Cauchy formula at infinity for $Q(z)$ and $Q(\zeta)$ and get

$$N_Q(\zeta, z) = \frac{1}{z - \bar{\zeta}}\left(-\frac{1}{2\pi i}\int_{|u|=r}\frac{Q(u)}{u-z}du + Q(\infty) + \frac{1}{2\pi i}\int_{|u|=r}\frac{Q(u)}{u-\bar{\zeta}}du - Q(\infty)\right)$$

$$= -\frac{1}{2\pi i}\int_{|u|=r}\frac{Q(u)}{z - \bar{\zeta}}\left(\frac{1}{u-z} - \frac{1}{u-\bar{\zeta}}\right)du$$

$$= -\frac{1}{2\pi i}\int_{|u|=r}\frac{Q(u)}{(u-z)(u-\bar{\zeta})}du.$$

If $z = \bar{\zeta}$ the formula (2.3) reads

$$Q'(z) = -\frac{1}{2\pi i}\int_{|u|=r}\frac{Q(u)}{(u-z)^2}du,$$

which is a consequence of (2.3) for $z \neq \bar{\zeta}$, by passing to the limit as $\bar{\zeta} \to z$ on both sides and using the uniform continuity on compact sets with respect to $\bar{\zeta}$. The claim at Step 1 is proven.

We consider the linear space

$$\mathcal{F}_0(\mathbb{D}_r^c; \mathcal{H}) = \{f: \mathbb{D}_r^c \to \mathcal{H} \mid \operatorname{supp}(f) \text{ is finite}\}.$$

It is convenient to use the canonical basis $\{\delta_z\}_{z \in \mathbb{D}_r^c}$ of the linear space $\mathcal{F}_0(\mathbb{D}_r^c; \mathbb{C})$, where

$$\delta_z(\zeta) = \begin{cases} 1, & \zeta = z, \\ 0, & \zeta \neq z. \end{cases}$$

Then for all $f \in \mathcal{F}_0(\mathbb{D}_r^c; \mathcal{H})$ we have the representation

$$f = \sum_{|z|>r} \delta_z f_z,$$

where $f_z = f(z)$ for all $|z| > r$ and we interpret $\delta_z h \in \mathcal{F}_0(\mathbb{D}_r^c; \mathcal{H})$ for arbitrary $h \in \mathcal{H}$.

On $\mathcal{F}_0(\mathbb{D}_r^c; \mathcal{H})$ we define the inner product $[\cdot, \cdot]$ as follows: if $f, g \in \mathcal{F}_0(\mathbb{D}_r^c; \mathcal{H})$ have representations

$$f = \sum_{|z|>r} \delta_z f_z, \quad g = \sum_{|\zeta|>r} \delta_\zeta g_\zeta,$$

then

$$[f, g] = \sum_{|z|,|\zeta|>r} \langle N_Q(\zeta, z) f_z, g_\zeta \rangle_{\mathcal{H}}. \qquad (2.4)$$

Also, on $\mathcal{F}_0(\mathbb{D}_r^c;\mathcal{H})$ we consider the inner product $\langle\cdot,\cdot\rangle_\mathcal{G}$ defined by

$$\Big\langle \sum_{|z|>r}\delta_z f_z, \sum_{|\zeta|>r}\delta_\zeta g_\zeta \Big\rangle_\mathcal{G} = \frac{r^2}{2\pi}\int_0^{2\pi}\Big\langle \sum_{|z|>r}\frac{f_z}{re^{i\theta}-z}, \sum_{|\zeta|>r}\frac{g_\zeta}{re^{i\theta}-\zeta}\Big\rangle_\mathcal{H} d\theta$$

$$= \frac{r^2}{2\pi}\sum_{|z|>r,\,|\zeta|>r}\int_0^{2\pi}\frac{\langle f_z, g_\zeta\rangle_\mathcal{H}}{(re^{i\theta}-z)(re^{-i\theta}-\overline{\zeta})}d\theta,$$

and it is easy to see that it is positive definite. Let $\|\cdot\|_\mathcal{G}$ denote the associated norm, that is,

$$\Big\|\sum_{|z|>r}\delta_z f_z\Big\|_\mathcal{G} = \frac{r}{\sqrt{2\pi}}\Big(\int_0^{2\pi}\Big\|\sum_{|z|>r}\frac{f_z}{re^{i\theta}-z}\Big\|_\mathcal{H}^2 d\theta\Big)^{1/2}.$$

Step 2. *For all $f,g\in\mathcal{F}_0(\mathbb{D}_r^c;\mathcal{H})$ we have*

$$|[f,g]| \leqslant \frac{1}{r}\max_{|v|=r}\|Q(v)\|_\mathcal{H}\,\|f\|_\mathcal{G}\,\|g\|_\mathcal{G}. \tag{2.5}$$

Indeed, taking into account the results at Step 1, we have

$$|[f,g]| = \Big|\sum_{|z|,|\zeta|>r}\langle N_Q(\zeta,z)f_z, g_\zeta\rangle_\mathcal{H}\Big| = \Big|\sum_{|z|,|\zeta|>r}\frac{1}{2\pi i}\int_{|u|=r}\Big\langle Q(u)\frac{f_z}{u-z},\frac{g_\zeta}{\overline{u}-\zeta}\Big\rangle_\mathcal{H} du\Big|$$

$$= \Big|\sum_{|z|,|\zeta|>r}\int_0^{2\pi}\Big\langle Q(re^{i\theta})\frac{f_z}{re^{i\theta}-z},\frac{g_\zeta}{re^{-i\theta}-\zeta}\Big\rangle_\mathcal{H} re^{i\theta}d\theta\Big|$$

$$\leqslant \frac{1}{2\pi}\int_0^{2\pi}\Big|\Big\langle rQ(re^{i\theta})\sum_{|z|>r}\frac{f_z}{re^{i\theta}-z},\sum_{|\zeta|>r}\frac{g_\zeta}{re^{-i\theta}-\zeta}\Big\rangle_\mathcal{H}\Big| d\theta$$

$$\leqslant \frac{1}{2\pi}\int_0^{2\pi}\Big\|re^{i\theta}Q(re^{i\theta})\sum_{|z|>r}\frac{f_z}{re^{i\theta}-z}\Big\|_\mathcal{H}\Big\|\sum_{|\zeta|>r}\frac{g_\zeta}{re^{-i\theta}-\zeta}\Big\|_\mathcal{H} d\theta$$

$$\leqslant \max_{|u|=r}\|Q(u)\|\frac{r}{2\pi}\int_0^{2\pi}\Big\|\sum_{|z|>r}\frac{f_z}{re^{i\theta}-z}\Big\|_\mathcal{H}\Big\|\sum_{|\zeta|>r}\frac{g_\zeta}{re^{-i\theta}-\zeta}\Big\|_\mathcal{H} d\theta$$

and then, by using the Cauchy–Bunyakovsky Inequality, we get

$$\leqslant \frac{1}{r}\max_{|u|=r}\|Q(u)\|$$

$$\times\Big(\frac{r^2}{2\pi}\int_0^{2\pi}\sum_{|z|>r}\Big\|\frac{f_z}{re^{i\theta}-z}\Big\|_\mathcal{H}^2 d\theta\Big)^{1/2}\Big(\frac{r^2}{2\pi}\int_0^{2\pi}\Big\|\sum_{|\zeta|>r}\frac{g_\zeta}{re^{-i\theta}-\zeta}\Big\|_\mathcal{H} d\theta\Big)^{1/2}$$

$$= \frac{1}{r}\max_{|u|=r}\|Q(u)\|\,\|f\|_\mathcal{G}\,\|g\|_\mathcal{G}.$$

6.2 Nevanlinna Type Representations

This proves the inequality (2.5) and completes the proof of Step 2.

Further, let $\mathcal{G}$ denote the Hilbert space completion of the linear space $\mathcal{F}_0(\mathbb{D}_r^c;\mathcal{H})$ with respect to the norm $\|\cdot\|_\mathcal{G}$. In view of the inequality (2.5), the inner product $[\cdot,\cdot]$ is uniquely extended to $\mathcal{G}$. Then, for arbitrary $h \in \mathcal{H}$ we define $\Gamma_0 h \in \mathcal{G}$ by

$$\Gamma_0 h = \lim_{z\to\infty}(-z\delta_z h). \tag{2.6}$$

Step 3. *The operator Γ_0 is correctly defined and $\Gamma_0 \in \mathcal{B}(\mathcal{H},\mathcal{G})$.*

Indeed, for all $z,\zeta \in \mathbb{C}$ with $|z|,|\zeta| > r$ we have

$$\|\zeta\delta_\zeta h - z\delta_z h\|_\mathcal{G}^2 = \frac{r^2}{2\pi}\int_0^{2\pi}\left\|\frac{\zeta h}{re^{i\theta}-\zeta} - \frac{zh}{re^{i\theta}-z}\right\|_\mathcal{H}^2 d\theta$$

$$= \frac{r^2}{2\pi}\int_0^{2\pi}\|h\|_\mathcal{G}^2\left|\frac{1}{\frac{re^{i\theta}}{\zeta}-1} - \frac{1}{\frac{re^{i\theta}}{z}-1}\right|^2 d\theta.$$

Using Lebesgue's Theorem of Dominated Convergence, for any sequence $(z_n)_{n\geq 1}$ in $\mathbb{C}$ such that $z_n \xrightarrow[n\to\infty]{} \infty$ the sequence $(-z_n\delta_{z_n}h)_{n\geq 1}$ in $\mathcal{F}_0(\mathbb{D}_r^c;\mathcal{H})$ is a Cauchy sequence with respect to the norm $\|\cdot\|_\mathcal{G}$, hence the definition of Γ_0 is correct. Clearly, the operator Γ_0 is linear. If $h \in \mathcal{H}$ is fixed then

$$\|\Gamma_0 h\|_\mathcal{G}^2 = \lim_{z\to\infty}\frac{r^2}{2\pi}\int_0^{2\pi}\left\|\frac{-zh}{re^{i\theta}-z}\right\|_\mathcal{H}^2 d\theta = \lim_{z\to\infty}\frac{r^2\|h\|_\mathcal{H}^2}{2\pi}\int_0^{2\pi}\left|\frac{-z}{re^{i\theta}-z}\right|^2 d\theta$$

$$= \frac{r^2}{2\pi}\|h\|_\mathcal{H}^2\int_0^{2\pi}d\theta = r^2\|h\|_\mathcal{H}^2.$$

Thus, Γ_0 can be uniquely extended to a bounded linear operator $\Gamma_0\colon \mathcal{H} \to \mathcal{G}$. This ends the proof of Step 3.

Step 4. *For all $h,k \in \mathcal{H}$ we have*

$$[\delta_z h, \Gamma_0 k] = \langle(Q(z)-Q(\infty))h,k\rangle_\mathcal{H}.$$

Indeed,

$$[\delta_z h, \Gamma_0 k] = \lim_{\zeta\to\infty}[\delta_z h, -\zeta k\delta_\zeta] = \lim_{\zeta\to\infty}\langle N_Q(z,\overline{\zeta})(-\overline{\zeta})h,k\rangle_\mathcal{H}$$

$$= \lim_{\zeta\to\infty}\langle(Q(z)-Q(\overline{\zeta}))h,k\rangle_\mathcal{H} - \lim_{\zeta\to\infty}\langle N_Q(\zeta,z)zh,k\rangle_\mathcal{H}$$

$$= \langle(Q(z)-Q(\infty))h,k\rangle_\mathcal{H},$$

hence the desired formula holds. The claim at Step 4 is proven.

Let us define the linear manifold $\mathcal{D}$ by

$$\mathcal{D} = \{f \in \mathcal{F}_0(\mathbb{D}_r^c;\mathcal{H}) \mid \sum_{|z|>r} f_z = 0\}. \tag{2.7}$$

Step 5. $\mathcal{D}$ *is dense in* $\mathcal{F}_0(\mathbb{D}_r^c;\mathcal{H})$ *with respect to the norm* $\|\cdot\|_\mathcal{G}$.

To see this, we first observe that for all $h \in \mathcal{H}$ and all $\zeta \in \mathbb{C}$ with $|\zeta| > r$ we have

$$\|\delta_\zeta h\|_\mathcal{G}^2 = \frac{r^2}{2\pi}\int_0^{2\pi}\left\|\frac{h}{re^{\mathrm{i}\theta}-\zeta}\right\|_\mathcal{H}^2 \mathrm{d}\theta = \frac{r^2\|h\|_\mathcal{H}^2}{2\pi}\int_0^{2\pi}\left|\frac{1}{re^{\mathrm{i}\theta}-\zeta}\right|^2 \mathrm{d}\theta,$$

hence, Lebesgue's Theorem of Dominated Convergence implies $\|\delta_\zeta h\|_\mathcal{G} \xrightarrow[\zeta\to\infty]{} 0$.

Then, let $f \in \mathcal{F}_0(\mathbb{D}_r^c;\mathcal{H})$ be arbitrary and consider $g_\zeta \in \mathcal{F}_0(\mathbb{D}_r^c;\mathcal{H})$ defined by

$$g_\zeta = f - \delta_\zeta\Big(\sum_{|z|>r} f_z\Big), \quad |\zeta| > r.$$

Clearly $g_\zeta \in \mathcal{D}$ and from what we have proven before we obtain that $g_\zeta \xrightarrow[\zeta\to\infty]{} f$. The statement at Step 5 is proven.

Define the linear operator $A_0\colon \mathcal{D} \to \mathcal{F}_0(\mathbb{D}_r^c;\mathcal{H})$ as follows: for any $f \in \mathcal{D}$ defined as in (2.7), with representation $f = \sum_{|z|>r}\delta_z f_z$ and $\sum_{|z|>r} f_z = 0$, we let

$$A_0 f = \sum_{|z|>r} z\delta_z f_z. \tag{2.8}$$

Step 6. *The operator* A_0 *is Hermitian, that is,* $[A_0 f, g] = [f, A_0 g]$ *for all* $f, g \in \mathcal{D}$, *and bounded with respect to the norm* $\|\cdot\|_\mathcal{G}$, *more precisely, its operator norm* $\|A_0\| = r$.

Indeed, we first observe that, for all $z,\zeta \in \mathbb{D}_r^c$ we have

$$zN_Q(\zeta,z) = \big(Q(z)-Q(\overline{\zeta})\big) + \overline{\zeta}N_Q(\zeta,z).$$

Then, for arbitrary $f, g \in \mathcal{D}$,

$$[A_0 f, g] = \Big[\sum_{|z|>r} z\delta_z f_z, \sum_{|\zeta|>r}\delta_\zeta g_\zeta\Big] = \sum_{|z|>r,|\zeta|>r}\langle N_Q(\zeta,z)zf_z, g_\zeta\rangle_\mathcal{H}$$

$$= \sum_{|z|,|\zeta|>r}\langle(Q(z)-Q(\overline{\zeta}))f_z, g_\zeta\rangle_\mathcal{H} + \sum_{|z|,|\zeta|>r}\langle N_Q(\zeta,z)f_z,\zeta g_\zeta\rangle_\mathcal{H}$$

$$= \Big\langle\sum_{|z|>r|}Q(z)f_z, \sum_{|\zeta|>r|}g_\zeta\Big\rangle_\mathcal{H} + \Big\langle\sum_{|z|>r|}f_z, \sum_{|\zeta|>r|}Q(\zeta)g_\zeta\Big\rangle_\mathcal{H}$$

$$+ \sum_{|z|,|\zeta|>r}\langle N_Q(\zeta,z)f_z,\zeta g_\zeta\rangle_\mathcal{H},$$

and then, taking into account that $\sum_{|z|>r} f_z = \sum_{|\zeta|>r} g_\zeta = 0$, we get

$$[A_0 f, g] = \sum_{|z|>r,|\zeta|>r}\langle N_Q(\zeta,z)f_z,\zeta g_\zeta\rangle_\mathcal{H} = [f, A_0 g].$$

6.2 Nevanlinna Type Representations

On the other hand, for arbitrary $f \in \mathcal{D}$ we have

$$\|A_0 f\|_{\mathcal{G}}^2 = \frac{r^2}{2\pi}\int_0^{2\pi}\Big\|\sum_{|z|>r}\frac{zf_z}{re^{i\theta}-z}\Big\|_{\mathcal{H}}^2 d\theta$$

and, taking into account that $\sum_{|z|>r} f_z = 0$,

$$= \frac{r^2}{2\pi}\int_0^{2\pi}\Big\|\sum_{|z|>r}\Big(\frac{zf_z}{re^{i\theta}-z}+f_z\Big)\Big\|_{\mathcal{H}}^2 d\theta$$

$$= \frac{r^4}{2\pi}\int_0^{2\pi}\Big\|\sum_{|z|>r}\frac{f_z}{re^{i\theta}-z}\Big\|_{\mathcal{H}}^2 d\theta = r^2\|f\|_{\mathcal{G}}^2.$$

The statement at Step 6 is proven.

On the grounds of Step 5 and Step 6, the operator A_0 extends uniquely to a bounded and selfadjoint operator, denoted again by $A_0 \in \mathcal{B}(\mathcal{G})$, $A_0 = A_0^\sharp$, and with operator norm $\|A_0\| = r$.

Step 7. *For arbitrary $f \in \mathcal{F}_0(\mathbb{D}_r^c; \mathcal{H})$ we have*

$$A_0 f = \sum_{|z|>r} z\delta_z f_z + \Gamma_0 \sum_{|z|>r} f_z. \qquad (2.9)$$

For arbitrary $h \in \mathcal{H}$, as seen during the proof of Step 5, we have $\|\delta_\zeta h\|_{\mathcal{G}}^2 \xrightarrow[\zeta\to\infty]{} 0$, hence, for any $h \in \mathcal{H}$ and any $|z| > r$ we have, with respect to the norm $\|\cdot\|_{\mathcal{G}}$,

$$\lim_{\zeta\to\infty}(\delta_z h - \delta_\zeta h) = \delta_z h.$$

Since, for any $z, \zeta \in \mathbb{D}_r^c$ we have $(\delta_z h - \delta_\zeta h) \in \mathcal{D}$, recalling the definition of Γ_0 as in (2.6), it follows that

$$\lim_{\zeta\to\infty} A_0(\delta_z h - \delta_\zeta h) = \lim_{\zeta\to\infty}(z\delta_z h - \zeta\delta_\zeta h) = z\delta_z h + \Gamma_0 h.$$

By linearity, the density of $\mathcal{D}$ in $\mathcal{F}_0(\mathbb{D}_r^c); \mathcal{H})$ and the continuity of A_0 we obtain the formula (2.9). The claim at Step 7 is proven.

Since the inner product space $(\mathcal{G}; [\cdot,\cdot])$ has a Hilbert space norm $\|\cdot\|_{\mathcal{G}}$ that makes the inner product $[\cdot,\cdot]$ jointly continuous, let $G \in \mathcal{B}(\mathcal{G})$ denote its Gramian, that is, a selfadjoint bounded operator with respect to the Hilbert space $(\mathcal{G}; \langle\cdot,\cdot\rangle_{\mathcal{G}})$ such that

$$[a,b] = \langle Ga,b\rangle_{\mathcal{G}}, \quad a,b \in \mathcal{G}. \qquad (2.10)$$

Let $(\mathcal{K}_G; \Pi_G)$ denote the Kreĭn space induced by G, defined as in the proof of Proposition 6.1.2. Let $\Gamma = \Pi_G \Gamma_0 \in \mathcal{B}(\mathcal{H}, \mathcal{K}_G)$. We are now in a position to finish the proof.

Step 8. *There exists uniquely a selfadjoint operator* $A \in \mathcal{B}(\mathcal{K}_G)$ *such that* $\Pi_G A_0 = A\Pi_G$,
$$Q(z) = Q(\infty) + \Gamma^\sharp (A - zI)^{-1}\Gamma, \quad |z| > r,$$
and $\mathcal{K}_G = \text{Clos Lin}\{(A - zI)^{-1}\Gamma\mathcal{H} \mid |z| > r\}$.

As before, we consider the Kreĭn space $(\mathcal{K}_G; \Pi_G)$ induced by the Gram operator G and, since the operator $A_0 \in \mathcal{B}(\mathcal{G})$ is selfadjoint with respect to the inner product space $(\mathcal{G}; [\cdot,\cdot])$, we can apply Theorem 6.1.10 in order to obtain the unique selfadjoint operator $A \in \mathcal{B}(\mathcal{K}_G)$ such that $\Pi_G A_0 = A\Pi_G$.

From Step 7 we have
$$(A_0 - zI)\delta_z h = \Gamma_0 h, \quad h \in \mathcal{H}, |z| > r,$$
hence
$$\Pi_G \Gamma_0 h = \Pi_G (A_0 - zI)\delta_z h = (A - zI)\Pi_G \delta_z h, \quad h \in \mathcal{H}, |z| > r. \quad (2.11)$$

At Step 6 we proved that the operator norm $\|A_0\| = r$ hence $\sigma(A_0) \subseteq \text{Clos}\,\mathbb{D}_r$ and then, by Corollary 6.1.11, it follows that $\sigma(A) \subseteq \text{Clos}\,\mathbb{D}_r$ as well. Hence, for any $z \in \mathbb{C}$ with $|z| > r$ the operator $A - zI$ is boundedly invertible and from (2.11) we have
$$\Pi_G \delta_z h = (A - zI)^{-1}\Pi_G \Gamma_0 h, \quad h \in \mathcal{H}.$$

Then, taking into account Step 4, for all $h, k \in \mathcal{H}$ we have
$$\begin{aligned}
\langle (Q(z) - Q(\infty))h, k\rangle_\mathcal{H} &= [\delta_z h, \Gamma_0 k] = \langle G\delta_z h, \Gamma_0 k\rangle_\mathcal{G}\\
&= \langle \Pi_G^\sharp \Pi_G \delta_z h, \Gamma_0 k\rangle_\mathcal{G}\\
&= [\Pi_G \delta_z h, \Pi_G \Gamma_0 K]_{\mathcal{K}_G}\\
&= [(A - zI)^{-1}\Pi_G \Gamma_0 h, \Pi_G \Gamma_0 k]_{\mathcal{K}_G}\\
&= [(A - zI)^{-1}\Gamma h, \Gamma k]_{\mathcal{K}_G}
\end{aligned}$$
hence
$$Q(z) = Q(\infty) + \Gamma^\sharp(A - zI)^{-1}\Gamma, \quad |z| > r.$$

Finally, by (2.11), it follows that $\text{Lin}\{(A-zI)^{-1}\Gamma\mathcal{H} \mid |z| > r\}$ contains $\Pi_G \mathcal{F}_0(\mathbb{D}_r^c; \mathcal{H})$ which is dense in $\mathcal{K}_G$. We take $\mathcal{K} = \mathcal{K}_G$ and observe that the triple $(\mathcal{K}; A; \Gamma)$ has all the required properties. ∎

Remark 6.2.2. (a) With notation and assumptions as in Theorem 6.2.1, let $(\mathcal{K}; A; \Gamma)$ be a triple as in the statement. Then, for all $|z|, |\zeta| > r$ and all $h, k \in \mathcal{H}$,
$$[(A - zI)^{-1}\Gamma h, (A - \zeta I)^{-1}\Gamma k]_\mathcal{K} = [N_Q(\zeta, z)h, k]_\mathcal{H}. \quad (2.12)$$
Indeed, if $z \neq \overline{\zeta}$, then
$$[(A - zI)^{-1}\Gamma h, (A - \zeta I)^{-1}\Gamma k]_\mathcal{K} = [\Gamma^\sharp (A - \overline{\zeta})^{-1}(A - zI)^{-1}\Gamma h, k]_\mathcal{K}$$

6.2 NEVANLINNA TYPE REPRESENTATIONS

which, by Hilbert's Resolvent Formula, equals

$$= \frac{1}{z-\bar\zeta}([\Gamma^\sharp(A-zI)\Gamma h, k]_\mathcal{K} - [\Gamma^\sharp(A-\bar\zeta I)\Gamma h, k]_\mathcal{K})$$

$$= \left[\frac{Q(z)-Q(\zeta)^\sharp}{z-\bar\zeta}h, k\right]_\mathcal{K} = [N_Q(\zeta, z)h, k]_\mathcal{K}.$$

For $z = \bar\zeta$, we write (2.12) for a $z \neq \bar\zeta$ and then let $\bar\zeta \to z$, taking into account the continuity with respect to ζ on both sides.

The formula (2.12) shows that the appearance of the Nevanlinna kernel N_Q in the definition of the inner product $[\cdot, \cdot]$ as in (2.4) is quite natural.

(b) The minimality property $\mathcal{K} = \operatorname{Clos} \operatorname{Lin}\{(A-zI)^{-1}\Gamma\mathcal{H} \mid z \in \mathbb{D}_r^c\}$ obtained in Theorem 6.2.1 can be used in order to obtain a "weak" uniqueness property, in the following sense: for any other triple $(\mathcal{K}'; A'; \Gamma')$ with the same properties as in Theorem 6.2.1 and the minimality property $\mathcal{K}' = \operatorname{Clos} \operatorname{Lin}\{(A'-zI)^{-1}\Gamma'\mathcal{H} \mid z \in \mathbb{D}_r^c\}$, there exists a linear operator $U\colon \operatorname{Dom}(U)(\subseteq \mathcal{K}) \to \mathcal{K}'$, that is densely defined, isometric, and with dense range, defined by

$$U\sum_{j=1}^n (A-z_jI)^{-1}\Gamma h_j = \sum_{j=1}^n (A'-z_jI)^{-1}\Gamma'h_j, \qquad (2.13)$$

for arbitrary $n \in \mathbb{N}$, $z_1, \ldots, z_n \in \mathbb{D}_r^c$, and $h_1, \ldots, h_n \in \mathcal{H}$. By (2.12) it follows that U is isometric and well defined, since $\operatorname{Lin}\{(A-zI)^{-1}\Gamma h \mid z \in \mathbb{D}_r^c, h \in \mathcal{H}\}$ is dense in $\mathcal{K}$ and hence nondegenerate. Then the closure of U has the same properties.

If uniqueness to a unitary equivalence is required, that is, the operator U is asked to be bounded and hence, has a unique extension to a bounded unitary operator $U\colon \mathcal{K} \to \mathcal{K}'$ then, in view of Theorem 6.1.9, additional conditions are necessary.

(c) With notation and assumptions as in item (b), assume that the operator U defined as in (2.13) is bounded, and let $U \in \mathcal{B}(\mathcal{K}, \mathcal{K}')$ be its extension to a bounded unitary operator. Then, by the Riesz–Dunford functional calculus, for $R > 0$ sufficiently large and all $h \in \mathcal{H}$, we have

$$\Gamma'h = -\frac{1}{2\pi\mathrm{i}}\int_{|z|=R}(A'-zI)^{-1}\Gamma'h\,\mathrm{d}z$$

$$= -\frac{1}{2\pi\mathrm{i}}\int_{|z|=R}U(A-zI)^{-1}\Gamma h\,\mathrm{d}z$$

$$= U\left(-\frac{1}{2\pi\mathrm{i}}\int_{|z|=R}(A-zI)^{-1}\Gamma h\,\mathrm{d}z\right) = U\Gamma h,$$

hence $U\Gamma = \Gamma'$. Similarly we show that $UA = A'U$. ∎

Given a nonempty set X and a Hilbert space $\mathcal{H}$, a mapping $\mathbf{k}\colon X \times X \to \mathcal{B}(\mathcal{H})$ is called an operator valued *kernel*. The adjoint kernel $\mathbf{k}^*$ is defined by $\mathbf{k}^*(x, y) = \mathbf{k}(y, x)^*$, for all $x, y \in X$. The kernel $\mathbf{k}$ is called *Hermitian* if $\mathbf{k}^* = \mathbf{k}$. Letting $\mathcal{F}_0(X;\mathcal{H})$ denote the vector space of all maps $f: X \to \mathcal{H}$ of finite support, as in the proof of Theorem 6.2.1, any $f \in \mathcal{F}_0(X;\mathcal{H})$ has a unique representation $f = \sum_{x \in X} \delta_x f_x$, where $f_x = f(x)$ for all $x \in X$. Then, we can define a pairing $[\cdot, \cdot]_\mathbf{k}$ on $\mathcal{F}_0(X;\mathcal{H})$ by

$$[f,g]_\mathbf{k} = \sum_{x,y \in X} \langle \mathbf{k}(y,x) f(x), g(y) \rangle_\mathcal{H}, \quad f, g \in \mathcal{F}_0(X;\mathcal{H}). \tag{2.14}$$

Since $\mathbf{k}$ is Hermitian, it follows that $[\cdot, \cdot]_\mathbf{k}$ is an inner product. Consequently, we can define the *ranks of positivity/negativity* $\kappa_\pm(\mathbf{k})$ as the ranks of positivity/negativity of the inner product space $(\mathcal{F}_0(X;\mathcal{H}); [\cdot, \cdot]_\mathbf{k})$, as in Section 1.1. In the literature, they are also called the *numbers of positive/negative squares* of the kernel $\mathbf{k}$.

Corollary 6.2.3. *With notation and assumptions as in Theorem 6.2.1, assume that the Nevanlinna kernel N_Q defined at (2.2) has a finite number of negative squares, $\kappa_-(N_Q) < \infty$. Then there exists a triple $(\mathcal{K}; A; \Gamma)$, where $\mathcal{K}$ is a Pontryagin space with $\kappa_-(\mathcal{K}) = \kappa_-(N_Q)$, $A^\sharp = A \in \mathcal{B}(\mathcal{K})$, with spectrum $\sigma(A) \subseteq \operatorname{Clos} \mathbb{D}_r$, and $\Gamma \in \mathcal{B}(\mathcal{H}, \mathcal{K})$ such that*

$$Q(z) = Q(\infty) + \Gamma^\sharp (A - zI)^{-1} \Gamma, \quad |z| > r.$$

In addition, $\mathcal{K}$ can be chosen such that $\mathcal{K} = \operatorname{Clos} \operatorname{Lin}\{(A - zI)^{-1} \Gamma \mathcal{H} \mid z \in \mathbb{D}_r^c\}$ and it is unique, modulo unitary equivalence.

Proof. The inner product $[\cdot, \cdot]$ defined as in (2.4) coincides with the inner product $[\cdot, \cdot]_{N_Q}$ as defined in (2.14). Consequently, an inspection of the proof of Theorem 6.2.1 and Corollary 6.1.3 shows that $\kappa_-(\mathcal{K}) = \kappa_-(N_Q)$. ∎

Example 6.2.4. Let $\mathcal{H}$ be a Hilbert space and $Q \colon \operatorname{Dom}(Q) \to \mathcal{B}(\mathcal{H})$ be an operator valued function defined on a nonempty subset $\operatorname{Dom}(Q)$ in the upper half plane of the complex plane $\mathbb{C}_+ = \{z \in \mathbb{C} \mid \operatorname{Im} z > 0\}$. The operator valued *Nevanlinna kernel* associated to the function Q is defined by

$$N_Q(\zeta, z) = \frac{Q(z) - Q(\zeta)^*}{z - \overline{\zeta}}, \quad z, \zeta \in \operatorname{Dom}(Q), \tag{2.15}$$

and is Hermitian in the sense $N_Q(\zeta, z) = N_Q(z, \zeta)^*$.

Given a nonnegative integer number κ, with definition in (2.14), one defines the class $\mathcal{N}_\kappa(\mathcal{H})$ of all meromorphic functions Q in $\mathbb{C}_+$, such that, letting $\operatorname{Dom}(Q)$ denote the domain of holomorphy of Q, the associated Nevanlinna kernel $N_Q \colon \operatorname{Dom}(Q) \times \operatorname{Dom}(Q) \to \mathcal{B}(\mathcal{H})$ defined as in (2.15) has κ negative squares. Corollary 6.2.3 provides representations with generalised resolvents of bounded selfadjoint operators in Kreĭn spaces for those functions $Q \in \mathcal{N}_\kappa(\mathbb{C}_+; \mathcal{H})$ that admit holomorphic extensions to $\mathbb{D}_r^c$ for some $r > 0$ and at infinity, with the symmetry property $Q(\overline{\zeta}) = Q(\zeta)^*$ for all $\zeta \in \mathbb{D}_r^c$. ∎

6.3 Linearisation of Selfadjoint Operator Pencils

In this section we show that the Nevanlinna type representation obtained in Theorem 6.2.1 contains, as a particular case, the method of linearisation of selfadjoint operator pencils.

Let $(\mathcal{H}; \langle \cdot, \cdot \rangle_\mathcal{H})$ denote a Hilbert space and let $L_0, L_1, \ldots, L_{n-1} \in \mathcal{B}(\mathcal{H})$ be selfadjoint operators, for some $n \in \mathbb{N}$. Define the *operator pencil* L by

$$L(z) = z^n I + z^{n-1} L_{n-1} + \cdots + z L_1 + L_0, \quad z \in \mathbb{C}. \tag{3.1}$$

By definition, the *spectrum* of the operator pencil L is the set $\sigma(L) = \{z \in \mathbb{C} \mid L(z)$ is not boundedly invertible$\}$ while its *resolvent set* is $\rho(L) = \mathbb{C} \setminus \sigma(L)$.

Lemma 6.3.1. *The spectrum of any selfadjoint operator pencil L is a nonempty compact subset.*

Proof. Using the familiar fact that if $T \in \mathcal{B}(\mathcal{H})$, $\|T\| < 1$, then the operator $I - T$ is boundedly invertible and taking into account that, for z sufficiently large, we have

$$\|z^{n-1} L_{n-1} + \cdots + z L_1 + L_0\| \leq |z|^{n-1} \|L_{n-1}\| + \cdots + |z| \|L_1\| + \|L_0\| < |z|^n,$$

it follows that $\sigma(L)$ is bounded. On the other hand,

$$\rho(L) = \{z \in \mathbb{C} \mid L(z) \in \mathrm{GL}(\mathcal{H})\},$$

where $\mathrm{GL}(\mathcal{H})$ denotes the open set of all boundedly invertible operators in $\mathcal{B}(\mathcal{H})$, hence, since $z \mapsto L(z) \in \mathcal{B}(\mathcal{H})$ is continuous, it follows that $\rho(L)$ is open, hence its complement $\sigma(L)$ is closed. Since $\rho \ni z \mapsto L(z)^{-1} \in \mathcal{B}(\mathcal{H})$ is holomorphic and null at infinity, it follows by Liouville's Theorem that $\rho(L)$ cannot be the whole complex plane, hence $\sigma(L) \neq \emptyset$. ∎

By the previous lemma, there exists $r > 0$ such that

$$Q(z) = -L(z)^{-1}, \quad |z| \geq r, \tag{3.2}$$

is an operator valued function that is holomorphic on a neighbourhood of the closure of $\mathbb{D}_r^c$, null at infinity, and satisfies the symmetry property $Q(\bar{z}) = Q(z)^*$. As an application of Theorem 6.2.1, it follows that there exists a triple $(\mathcal{K}; A; \Gamma)$ as in the statement and such that the function Q has the representation (2.1). However, for $n > 1$, we have

$$\lim_{z \to \infty} z Q(z) = \lim_{z \to \infty} -z L(z)^{-1} = 0, \tag{3.3}$$

which implies that $\Gamma^\sharp \Gamma = 0$. This means, see Proposition 4.2.2, that $\Gamma \mathcal{H}$ is a neutral subspace of $\mathcal{K}$ and hence $\Gamma \mathcal{H}$ has no information on the topology of $\mathcal{K}$. Therefore, we have to describe the representation (2.1) in more precise terms.

We consider the Hilbert space $\mathcal{H}^n = \mathcal{H} \oplus \cdots \oplus \mathcal{H}$, the orthogonal direct sum of n copies of $\mathcal{H}$, with inner product $\langle \cdot, \cdot \rangle_{\mathcal{H}^n}$, for $f = (f_j)_{j=1}^n$ and $g = (g_j)_{j=1}^n$ arbitrary vectors in $\mathcal{H}$, defined by

$$\langle f, g \rangle_{\mathcal{H}^n} = \langle f_1, g_1 \rangle_{\mathcal{H}} + \cdots + \langle f_n, g_n \rangle_{\mathcal{H}}.$$

Also, for any $f \in \mathcal{H}$ and $|z| \geqslant r$ we consider the vector $f^{(n)}(z) \in \mathcal{H}^n$ defined by

$$f^{(n)}(z) = (z^{n-1} L(z)^{-1} f, z^{n-2} L(z)^{-1} f, \ldots, L(z)^{-1} f). \tag{3.4}$$

Lemma 6.3.2. *With notation as before, letting N_Q denote the Nevanlinna kernel corresponding to the operator valued function Q as in (2.2), we have*

$$\langle N_Q(z, \zeta) f, g \rangle_{\mathcal{H}} = \langle G f^{(n)}(z), g^{(n)}(\zeta) \rangle_{\mathcal{H}^n}, \quad f, g \in \mathcal{H}, \; |z|, |\zeta| \geqslant r, \tag{3.5}$$

where $G \in \mathcal{B}(\mathcal{H}^n)$ denotes the selfadjoint operator defined by

$$G = \begin{bmatrix} 0 & 0 & \cdots & 0 & I \\ 0 & 0 & \cdots & I & L_{n-1} \\ \vdots & & & & \\ 0 & I & \cdots & L_3 & L_2 \\ I & L_{n-1} & \cdots & L_2 & L_1 \end{bmatrix}. \tag{3.6}$$

Proof. To see this, we first observe that, from (3.4) and (3.6) we have

$$G f^{(n)}(z) = \begin{bmatrix} L(z)^{-1} f \\ z L(z)^{-1} f + L_{n-1} L(z)^{-1} f \\ \vdots \\ z^{n-2} L(z)^{-1} f + z^{n-1} L_{n-1} L(z)^{-1} f + \cdots + L_2 L(z)^{-1} f \\ z^{n-1} L(z)^{-1} f + z^{n-2} L_{n-1} L(z)^{-1} f + \cdots + L_1 L(z)^{-1} f \end{bmatrix}. \tag{3.7}$$

Then, we observe that

$$\begin{aligned} L(z) - L(\bar{\zeta}) &= z^n I + z^{n-1} L_{n-1} + \cdots + L_0 - \bar{\zeta}^n I - \bar{\zeta}^{n-1} L_{n-1} - \cdots - \bar{\zeta} L_1 - L_0 \\ &= (z - \bar{\zeta})(z^{n-1} + z^{n-2} \bar{\zeta} + \cdots + z \bar{\zeta}^{n-2} + \bar{\zeta}^{n-1}) I \\ &\quad + (z - \bar{\zeta})(z^{n-2} + z^{n-3} \bar{\zeta} + \cdots + z \bar{\zeta}^{n-3} + \bar{\zeta}^{n-2}) + \cdots \\ &\quad + (z - \bar{\zeta}) L_1. \end{aligned} \tag{3.8}$$

From (2.15) we have, for all $z \neq \bar{\zeta}$, $|z|, |\zeta| \geqslant r$,

$$\begin{aligned} \langle N_Q(z, \zeta) f, g \rangle_{\mathcal{H}} &= \frac{\langle L(\bar{\zeta})^{-1} f - L(z)^{-1} f, g \rangle_{\mathcal{H}}}{z - \bar{\zeta}} \\ &= \frac{\langle f, L(\zeta)^{-1} f \rangle_{\mathcal{H}} - \langle L(z)^{-1} f, g \rangle_{\mathcal{H}}}{z - \bar{\zeta}} \end{aligned}$$

6.3 Linearisation of Selfadjoint Operator Pencils

letting $h = L(z)^{-1}f$ and $k = L(\zeta)^{-1}g$, equivalently, $f = L(z)h$ and $g = L(\zeta)k$,

$$= \frac{\langle L(z)h, k\rangle_{\mathcal{H}} - \langle h, L(\zeta)k\rangle}{z - \overline{\zeta}}$$

$$= \frac{\langle (L(z) - L(\overline{\zeta}))h, k\rangle_{\mathcal{H}}}{z - \overline{\zeta}}$$

then, taking into account (3.8),

$$= \langle (z^{n-1} + z^{n-2}\overline{\zeta} + \cdots + z\overline{\zeta}^{n-2} + \overline{\zeta}^{n-1})h, k\rangle_{\mathcal{H}}$$
$$+ \langle (z^{n-2} + z^{n-3}\overline{\zeta} + \cdots + z\overline{\zeta}^{n-3} + \overline{\zeta}^{n-2})L_{n-1}h, k\rangle_{\mathcal{H}}$$
$$+ \cdots + \langle L_1 h, k\rangle_{\mathcal{H}}$$
$$= \langle (z^{n-1} + z^{n-2}\overline{\zeta} + \cdots + \overline{\zeta}^{n-1})L(z)^{-1}f, L(\zeta)^{-1}g\rangle_{\mathcal{H}}$$
$$+ \langle (z^{n-2} + z^{n-3}\overline{\zeta} + \cdots + \overline{\zeta}^{n-2})L_{n-1}L(z)^{-1}f, L(\zeta)^{-1}g\rangle_{\mathcal{H}}$$
$$+ \cdots + \langle L_1 L(z)^{-1}f, L(\zeta)^{-1}g\rangle_{\mathcal{H}}$$

and, finally, rearranging the terms and using (3.7),

$$= \langle Gf^{(n)}(z), g^{(n)}(\zeta)\rangle_{\mathcal{H}^n},$$

hence (3.5) is proven for $z \neq \overline{\zeta}$. For $z = \overline{\zeta}$ it follows by letting $\overline{\zeta} \to z$ in the formula just proven and taking into account the continuity and differentiability properties. ∎

We consider now the *companion operator* of the operator pencil

$$L = \begin{bmatrix} -L_{n-1} & -L_{n-2} & \cdots & -L_1 & -L_0 \\ I & 0 & \cdots & 0 & 0 \\ \vdots & & & & \\ 0 & 0 & \cdots & I & 0 \end{bmatrix}, \quad (3.9)$$

which is a bounded operator on the Hilbert space $\mathcal{H}^n$.

Lemma 6.3.3. *With notation as before, we have*

$$\langle GLf, g\rangle_{\mathcal{H}^n} = \langle Gf, Lg\rangle_{\mathcal{H}^n}, \quad f, g \in \mathcal{H}^n. \quad (3.10)$$

This symmetry property of the operator L can be proven directly by performing matrix multiplication in order to show that $GL = L^*G$, which is equivalent with (3.10). The method of linearisation of selfadjoint operator pencils uses the operators G and the companion L and then Theorem 6.1.10 to lift L to a bounded selfadjoint operator $\widetilde{L}$ on a Kreĭn space and make a connection between the spectral properties of the operator pencil

L and those of $\widetilde{L}$. In the following, we show that the Nevanlinna kernel representation as in Theorem 6.2.1 provides the same representation.

With notation as in the proof of Theorem 6.2.1, the linear space $\mathcal{F}_0(\mathbb{D}_r^c; \mathcal{H})$ can be naturally identified with the vector space of all vectors $f^{(n)}(z)$, where z runs in $\mathbb{D}_r^c$ and $f \in \mathcal{H}$, more precisely, we identify $\delta_z f$ with $f^{(n)}(z)$. With this identification, the definition of the operator Γ_0 as in (2.6) yields

$$\Gamma_0 h = \lim_{z \to \infty}(z\delta_z h) = (-h, 0, \ldots, 0), \quad h \in \mathcal{H}. \tag{3.11}$$

Then, taking into account the definition of the companion L we have

$$Lf^{(n)}(z) = zf^{(n)}(z) - (f, 0, \ldots, 0), \quad f \in \mathcal{H}, \ |z| > r, \tag{3.12}$$

hence, taking into account the formula (2.9) it follows that, modulo the identification of $\mathcal{F}_0(\mathbb{D}_r^c; \mathcal{H})$ as before, L is identified with the operator A_0, the selfadjoint bounded operator obtained as in the proof of Theorem 6.2.1, by extension of the operator defined at (2.8).

Remark 6.3.4. If instead of $-L(z)^{-1}$ we consider the operator valued function $Q_0(z) = -z^{n-1}L(z)^{-1}$, then

$$Q_0(\infty) = 0, \quad \lim_{z \to \infty}(-zQ_0(z)) = I. \tag{3.13}$$

When applying Theorem 6.2.1 and obtaining the triple $(\mathcal{K}; A; \Gamma)$, it can be shown that in the representation (2.1) of Q_0, the operator A is again the lifting of the companion operator L defined as in (3.9) and, by (3.13), Γ is actually an isometric embedding of $\mathcal{H}$ in the Kreĭn space $\mathcal{K}$, hence

$$L(z)^{-1} = z^{1-n} P_{\mathcal{H}}(zI - A)^{-1}|\mathcal{H}, \quad |z| > r. \tag{3.14}$$

However, the price for obtaining a better representation as in (3.14) is a more involved Kreĭn space $\mathcal{K}$. ∎

6.4 Carathéodory Type Representations

The following theorem is a generalised resolvent representation for operator valued functions that are holomorphic in a neighbourhood of the origin. It refers to the Carathéodory kernel instead of the Nevanlinna kernel and, in a certain sense, is more general than Theorem 6.2.1. For a subset S of the complex plane we denote $S^* = \{z^* \mid z \in S \setminus \{0\}\}$, where $z^* = 1/\bar{z}$.

Theorem 6.4.1. *Let $\mathcal{H}$ be a Kreĭn space and $F: \mathbb{D}_r \to \mathcal{B}(\mathcal{H})$ a function holomorphic on an open neighbourhood of the closure of $\mathbb{D}_r$, for some $0 < r < 1$. Then, there exists a triple $(\mathcal{K}; U; \Gamma)$, where $\mathcal{K}$ is a Kreĭn space, $U \in \mathcal{B}(\mathcal{K})$ is a unitary operator, and $\Gamma \in \mathcal{B}(\mathcal{H}, \mathcal{K})$, such that*

$$F(z) = \frac{F(0) - F(0)^\sharp}{2} + \Gamma^\sharp(U + zI)(U - zI)^{-1}\Gamma, \quad z \in \mathbb{D}_r. \tag{4.1}$$

6.4 CARATHÉODORY TYPE REPRESENTATIONS

In addition, $\mathcal{K}$ can be chosen with the following minimality property

$$\mathcal{K} = \text{Clos Lin}\{(U - zI)^{-1}\Gamma\mathcal{H} \mid z \in \mathbb{D}_r \cup \mathbb{D}_r^*\}. \tag{4.2}$$

Proof. Let the indefinite inner product on $\mathcal{H}$ be denoted by $[\cdot,\cdot]_\mathcal{H}$ and let $\langle \cdot,\cdot \rangle_\mathcal{H}$ be a positive inner product on $\mathcal{H}$ corresponding to a fixed fundamental symmetry, and then let $\|\cdot\|_\mathcal{H}$ denote the associated Hilbert space norm. Extend F to the closure of $\mathbb{D}_r^* = \mathbb{D}_{r^{-1}}^c$ by letting

$$F(z) = -F(z^*)^\sharp, \quad |z| \geq 1/r. \tag{4.3}$$

Then F is locally holomorphic on some open neighbourhood of the closure of $\mathbb{D}_r \cup \mathbb{D}_{r^{-1}}^c$ and at infinity.

We consider the following Hilbert spaces of sequences in $\mathcal{H}$

$$\mathcal{G}_+ = \left\{ f_+ = (f_n)_{n=0}^\infty \mid f_n \in \mathcal{H} \text{ for all } n \geq 0, \ \sum_{n=0}^\infty \frac{1}{r^{2n}} \|f_n\|_\mathcal{H}^2 < \infty \right\}, \tag{4.4}$$

and

$$\mathcal{G}_- = \left\{ f_- = (f_{-n})_{n=1}^\infty \mid f_{-n} \in \mathcal{H} \text{ for all } n \geq 1, \ \sum_{n=1}^\infty \frac{1}{r^{2n}} \|f_{-n}\|_\mathcal{H}^2 < \infty \right\}, \tag{4.5}$$

with positive definite inner products

$$\langle f_+, g_+ \rangle_\mathcal{G} = \sum_{n=0}^\infty \frac{1}{r^{2n}} \langle f_n, g_n \rangle_\mathcal{H}, \quad \langle f_-, g_- \rangle_\mathcal{G} = \sum_{n=1}^\infty \frac{1}{r^{2n}} \langle f_{-n}, g_{-n} \rangle_\mathcal{H}. \tag{4.6}$$

We identify in a natural way each $f_+ = (f_n)_{n=0}^\infty \in \mathcal{G}_+$ with the $\mathcal{H}$-valued function

$$f_+(z) = \sum_{n=0}^\infty z^n f_n, \tag{4.7}$$

holomorphic on $\mathbb{D}_{1/r}$. By analogy with the scalar Hardy space $H^2(\mathbb{D})$, we observe that for almost all $\theta \in [0, 2\pi)$ the nontangential limits

$$\lim_{z \to e^{i\theta}/r} f_+(z) = f_+(e^{i\theta}/r) = \sum_{n=0}^\infty \frac{e^{in\theta}}{r^n} f_n \tag{4.8}$$

exist and the series converges in the mean with respect to θ, as well. Then, a similar argument as for the Hardy space $H^2(\mathbb{D})$ shows that

$$\|f_+\|_\mathcal{G} = \frac{1}{\sqrt{2\pi}} \left(\int_0^{2\pi} \left\| f_+\left(\frac{e^{i\theta}}{r}\right) \right\|_\mathcal{H}^2 d\theta \right)^{1/2}. \tag{4.9}$$

In addition, for any $f_+ = (f_n)_{n=0}^\infty \in \mathcal{G}_+$ we have

$$\frac{1}{2\pi i} \int_{|t|=\frac{1}{r}} t^n f_+(t) dt = \frac{1}{2\pi r^{n+1}} \int_0^{2\pi} e^{i(n+1)\theta} f_+\left(\frac{e^{i\theta}}{r}\right) d\theta = 0, \quad n \geq 0, \qquad (4.10)$$

where the integral converges weakly.

Similarly, we identify in a natural way $f_- = (f_{-n})_{n=1}^\infty \in \mathcal{G}_-$ with the $\mathcal{H}$-valued function

$$f_-(z) = \sum_{n=1}^\infty z^{-n} f_{-n}, \qquad (4.11)$$

holomorphic on $\mathbb{D}_r^c$ such that for almost all $\theta \in [0, 2\pi)$ the nontangential limits

$$\lim_{z \to r e^{i\theta}} f_-(z) = f_-(re^{i\theta}) = \sum_{n=1}^\infty r^{-n} e^{-in\theta} f_{-n}, \qquad (4.12)$$

exist and the series converges in the mean with respect to θ, as well. Then, we have

$$\|f_-\|_\mathcal{G} = \frac{1}{\sqrt{2\pi}} \left(\int_0^{2\pi} \|f_-(re^{i\theta})\|_\mathcal{H}^2 d\theta \right)^{1/2}. \qquad (4.13)$$

Consider the Hilbert space $\mathcal{G} = \mathcal{G}_+ \oplus \mathcal{G}_-$ with the quadratic norm $\|\cdot\|_\mathcal{G}$ obtained from the norms in (4.6). Elements $f \in \mathcal{G}$ can be viewed as sequences $f = (f_n)_{n=-\infty}^\infty$ with entries in $\mathcal{H}$, identified with pairs $f = (f_+, f_-)$ where $f_+ = (f_n)_{n=0}^\infty$ and $f_- = (f_{-n})_{n=1}^\infty$, and then

$$\|f\|_\mathcal{G}^2 = \|f_+\|_\mathcal{G}^2 + \|f_-\|_\mathcal{G}^2, \qquad (4.14)$$

and as functions $f(z) = f_+(z) + f_-(z)$ holomorphic on the annulus

$$A_r = \{z \in \mathbb{C} \mid r < |z| < 1/r\}. \qquad (4.15)$$

Let

$$A_r^c = \mathbb{C} \setminus A_r = \{z \in \mathbb{C} \mid \text{either } |z| \leq r \text{ or } |z| \geq 1/r\},$$

and for $z, \zeta \in A_r^c$ consider the *Carathéodory kernel* associated to the function F and defined by

$$C_F(\zeta, z) = \begin{cases} \frac{F(z) + F(\zeta)^\sharp}{2(1 - z\bar\zeta)}, & \text{if } z \neq \bar\zeta^{-1}, \\ \frac{1}{2} z F'(z), & \text{if } z = \bar\zeta^{-1}. \end{cases} \qquad (4.16)$$

Observe that, from (4.3) it follows that the Carathéodory kernel is Hermitian in the sense that $C_F(z, \zeta)^\sharp = C_F(\zeta, z)$ for all $z, \zeta \in A_r^c$, that for any fixed $\zeta \in A_r^c$ the vector valued function $C_F(\zeta, \cdot)$ is locally holomorphic on A_r^c, and that it is uniformly bounded, in the sense that $\sup_{z, \zeta \in A_r^c} \|C_F(\zeta, z)\|_\mathcal{H} < \infty$. In addition, observe that

$$z\bar\zeta C_F(\zeta, z) = C_F(\zeta, z) - \frac{F(z) + F(\zeta)^\sharp}{2}, \quad z, \zeta \in A_r^c. \qquad (4.17)$$

6.4 CARATHÉODORY TYPE REPRESENTATIONS

For $z\bar{\zeta} \neq 1$ this comes from the definition, while for $z\bar{\zeta} = 1$ we can use the formula for $z\bar{\zeta} \neq 1$ and let $\bar{\zeta} \to z$ in view of the continuity properties.

On $\mathcal{G}$ we introduce an inner product denoted by $[\cdot, \cdot]_\mathcal{G}$ and defined, for arbitrary $f = (f_+, f_-)$ and $g = (g_+, g_-)$ in $\mathcal{G}$, by

$$[f, g]_\mathcal{G} = \frac{1}{4\pi^2} \int_{|s|=\frac{1}{r}} \int_{|t|=\frac{1}{r}} [C_F(t,s) f_+(s), g_+(t)]_\mathcal{H} \, \mathrm{d}s \overline{\mathrm{d}t}$$
$$- \frac{1}{4\pi^2} \int_{|s|=r} \int_{|t|=\frac{1}{r}} [C_F(t,s) f_-(s), g_+(t)]_\mathcal{H} \, \mathrm{d}s \overline{\mathrm{d}t}$$
$$- \frac{1}{4\pi^2} \int_{|s|=\frac{1}{r}} \int_{|t|=r} [C_F(t,s) f_+(s), g_-(t)]_\mathcal{H} \, \mathrm{d}s \overline{\mathrm{d}t}$$
$$+ \frac{1}{4\pi^2} \int_{|s|=r} \int_{|t|=r} [C_F(t,s) f_-(s), g_-(t)]_\mathcal{H} \, \mathrm{d}s \overline{\mathrm{d}t}. \quad (4.18)$$

Step 1. *The inner product $[\cdot, \cdot]_\mathcal{G}$ is jointly continuous with respect to the norm $\|\cdot\|_\mathcal{G}$.*

To see this, let $f, g \in \mathcal{G}$. In view of the definition (4.18), we have to estimate the absolute values of four double integrals. For the first double integral we have

$$\left| \int_{|s|=\frac{1}{r}} \int_{|t|=\frac{1}{r}} [C_F(t,s) f_+(s), g_+(t)]_\mathcal{H} \, \mathrm{d}s \overline{\mathrm{d}t} \right|$$
$$= \left| \int_0^{2\pi} \int_0^{2\pi} \frac{1}{r^2} [C_F(\frac{e^{i\sigma}}{r}, \frac{e^{i\theta}}{r}) f_+(\frac{e^{i\sigma}}{r}), g_+(\frac{e^{i\theta}}{r})]_\mathcal{H} \, \mathrm{d}\sigma \mathrm{d}\theta \right|$$
$$\leqslant \frac{1}{r^2} \int_0^{2\pi} \int_0^{2\pi} \left| [C_F(\frac{e^{i\sigma}}{r}, \frac{e^{i\theta}}{r}) f_+(\frac{e^{i\sigma}}{r}), g_+(\frac{e^{i\theta}}{r})]_\mathcal{H} \right| \mathrm{d}\sigma \mathrm{d}\theta$$

and, letting $M = \sup_{0 \leqslant \sigma, \theta < 2\pi} \|C_F(\frac{e^{i\sigma}}{r}, \frac{e^{i\theta}}{r})\| < \infty$,

$$\leqslant \frac{M}{r^2} \int_0^{2\pi} \int_0^{2\pi} \|f_+(\frac{e^{i\sigma}}{r})\|_\mathcal{H} \|g_+(\frac{e^{i\theta}}{r})\|_\mathcal{H} \, \mathrm{d}\sigma \mathrm{d}\theta$$
$$\leqslant \frac{M}{r^2} \int_0^{2\pi} \|f_+(\frac{e^{i\sigma}}{r})\|_\mathcal{H} \, \mathrm{d}\sigma \int_0^{2\pi} \|g_+(\frac{e^{i\theta}}{r})\|_\mathcal{H} \, \mathrm{d}\theta$$

then, using the Cauchy–Bunyakovsky Inequality

$$\leqslant 2\pi \frac{M}{r^2} \left(\int_0^{2\pi} \|f_+(\frac{e^{i\sigma}}{r})\|_\mathcal{H}^2 \mathrm{d}\sigma \right)^{1/2} \left(\int_0^{2\pi} \|g_+(\frac{e^{i\theta}}{r})\|_\mathcal{H}^2 \mathrm{d}\theta \right)^{1/2}$$
$$= 2\pi \frac{M}{r^2} \|f_+\|_\mathcal{G} \|g_+\|_\mathcal{G}. \quad (4.19)$$

In a similar way we obtain the following estimations for the second, the third, and the fourth double integrals in the definition (4.18), for some constant $C \geqslant 2\pi M/r^2$ independent of f and g,

$$\left|\int_{|s|=r}\int_{|t|=\frac{1}{r}} [C_F(t,s)f_-(s), g_+(t)]_{\mathcal{H}} \mathrm{d}s \overline{\mathrm{d}t}\right| \leqslant C\|f_-\|_{\mathcal{G}} \|g_+\|_{\mathcal{G}}, \tag{4.20}$$

$$\left|\int_{|s|=\frac{1}{r}}\int_{|t|=r} [C_F(t,s)f_+(s), g_-(t)]_{\mathcal{H}} \mathrm{d}s \overline{\mathrm{d}t}\right| \leqslant C\|f_+\|_{\mathcal{G}} \|g_-\|_{\mathcal{G}}, \tag{4.21}$$

$$\left|\int_{|s|=r}\int_{|t|=r} [C_F(t,s)f_-(s), g_-(t)]_{\mathcal{H}} \mathrm{d}s \overline{\mathrm{d}t}\right| \leqslant C\|f_-\|_{\mathcal{G}} \|g_-\|_{\mathcal{G}}. \tag{4.22}$$

In view of the definition of the Hilbert space $(\mathcal{G}; \|\cdot\|_{\mathcal{G}})$, see (4.14), we observe that the estimations (4.19)–(4.22) prove that the inner product $[\cdot,\cdot]_{\mathcal{G}}$ is jointly continuous with respect to the norm $\|\cdot\|_{\mathcal{G}}$. The claim at Step 1 is proven.

We define the linear operator $U_0 \colon \mathcal{G} \to \mathcal{G}$, when viewing elements $f \in \mathcal{G}$ as $\mathcal{H}$-valued functions holomorphic on the annulus A_r, by

$$(U_0 f)(z) = zf(z), \quad r < |z| < \frac{1}{r}. \tag{4.23}$$

Equivalently, in terms of the representation $f = (f_+, f_-) \in \mathcal{G}$, we have

$$(U_0 f)_+(z) = zf_+(z) + f_{-1}, \quad (U_0 f)_-(z) = zf_-(z) - f_{-1}, \quad r < |z| < \frac{1}{r}. \tag{4.24}$$

Clearly, by (4.23), U_0 is invertible, with its inverse $U_0^{-1} \colon \mathcal{G} \to \mathcal{G}$, defined by

$$(U_0^{-1} f)(z) = \frac{1}{z} f(z), \quad r < |z| < \frac{1}{r}, \tag{4.25}$$

when viewing $f \in \mathcal{G}$ as an $\mathcal{H}$-valued function holomorphic on the annulus A_r, equivalently, in terms of the representation $f = (f_+, f_-) \in \mathcal{G}$,

$$(U_0^{-1} f)_+(z) = \frac{f_+(z) - f(0)}{z}, \quad (U_0^{-1} f)_-(z) = \frac{f_-(z) + f(0)}{z}, \quad r < |z| < \frac{1}{r}. \tag{4.26}$$

Step 2. *Both U_0 and U_0^{-1} are bounded operators in the Hilbert space $(\mathcal{G}; \langle\cdot,\cdot\rangle_{\mathcal{G}})$, with operator norm $\|U_0\| \leqslant \frac{1}{r}$, and isometric with respect to the inner product $[\cdot,\cdot]_{\mathcal{G}}$.*

Indeed, letting $f = (f_n)_{n=-\infty}^{\infty} \in \mathcal{G}$ arbitrary, from (4.23) it follows

$$\|U_0 f\|_{\mathcal{G}}^2 = \|f_{-1}\|_{\mathcal{H}}^2 + \sum_{n=1}^{\infty} \frac{1}{r^{2n}} \|f_{n-1}\|_{\mathcal{H}}^2 + \sum_{n=1}^{\infty} \frac{1}{r^{2n}} \|f_{-n-1}\|_{\mathcal{H}}^2$$

$$\leqslant \frac{1}{r^2} \|f\|_{\mathcal{H}}^2,$$

6.4 CARATHÉODORY TYPE REPRESENTATIONS

which proves that U_0 is bounded and its operator norm is $\leqslant \frac{1}{r}$. As for U_0^{-1}, either perform a similar estimation or use the Closed Graph Principle in order to conclude that it is bounded as well.

In the following we prove that U_0 is isometric with respect to the inner product $[\cdot, \cdot]_\mathcal{G}$, so let $f = (f_+, f_-)$ and $g = (g_+, g_-)$ be two arbitrary elements in $\mathcal{G} = \mathcal{G}_+ \oplus \mathcal{G}_-$. Then, in view of (4.24) and (4.18), and taking into account (4.17), we have

$$
\begin{aligned}
[U_0 f_+, U_0 g_+]_\mathcal{G} &= \frac{1}{4\pi^2} \int_{|s|=\frac{1}{r}} \int_{|t|=\frac{1}{r}} [C_F(t,s)(sf_+(s) + f_{-1}), tg_+(t) + g_{-1}]_\mathcal{H} \mathrm{d}s\overline{\mathrm{d}t} \\
&= \frac{1}{4\pi^2} \int_{|s|=\frac{1}{r}} \int_{|t|=\frac{1}{r}} [C_F(t,s)f_+(s), g_+(t)]_\mathcal{H} \mathrm{d}s\overline{\mathrm{d}t} \\
&\quad - \frac{1}{4\pi^2} \int_{|s|=\frac{1}{r}} \int_{|t|=\frac{1}{r}} \frac{1}{2}[F(s)f_+(s), g_+(t)]_\mathcal{H} \mathrm{d}s\overline{\mathrm{d}t} \\
&\quad - \frac{1}{4\pi^2} \int_{|s|=\frac{1}{r}} \int_{|t|=\frac{1}{r}} \frac{1}{2}[f_+(s), F(t)g_+(t)]_\mathcal{H} \mathrm{d}s\overline{\mathrm{d}t} \\
&\quad + \frac{1}{4\pi^2} \int_{|s|=\frac{1}{r}} \int_{|t|=\frac{1}{r}} [C_F(t,s)f_{-1}, tg_+(t)]_\mathcal{H} \mathrm{d}s\overline{\mathrm{d}t} \\
&\quad + \frac{1}{4\pi^2} \int_{|s|=\frac{1}{r}} \int_{|t|=\frac{1}{r}} [C_F(t,s)sf_+(s), g_{-1}]_\mathcal{H} \mathrm{d}s\overline{\mathrm{d}t} \\
&\quad + \frac{1}{4\pi^2} \int_{|s|=\frac{1}{r}} \int_{|t|=\frac{1}{r}} [C_F(t,s)f_{-1}, g_{-1}]_\mathcal{H} \mathrm{d}s\overline{\mathrm{d}t}. \quad (4.27)
\end{aligned}
$$

We observe that, due to (4.10), the second and the third terms from the right-hand side of (4.27) vanish.

By Cauchy's Integral Formula, for any $|t| > \frac{1}{r}$ we have

$$\frac{1}{2\pi \mathrm{i}} \int_{|s|=\frac{1}{r}} C_F(t,s)\mathrm{d}s = -\frac{1}{2\overline{t}}(F(\infty) + F(t)^\sharp) = \frac{F(0)^\sharp - F(t)^\sharp}{2\overline{t}}. \quad (4.28)$$

From here, we obtain that (4.28) holds for any $|t| = \frac{1}{r}$ by taking limits on both sides and using uniform continuity on each compact subset. Consequently,

$$
\begin{aligned}
&\frac{1}{4\pi^2} \int_{|s|=\frac{1}{r}} \int_{|t|=\frac{1}{r}} [C_F(t,s)f_{-1}, tg_+(t)]_\mathcal{H} \mathrm{d}s\overline{\mathrm{d}t} \\
&= -\frac{1}{2\pi \mathrm{i}} \int_{|t|=\frac{1}{r}} \left[\frac{F(0)^\sharp - F(t)^\sharp}{2\overline{t}} f_{-1}, tg_+(t)\right]_\mathcal{H} \overline{\mathrm{d}t} \\
&= \left[f_{-1}, \frac{1}{2\pi \mathrm{i}} \int_{|t|=\frac{1}{r}} \frac{F(0)}{2} g_+(t)\mathrm{d}t\right]_\mathcal{H} - \left[f_{-1}, \frac{1}{2\pi \mathrm{i}} \int_{|t|=\frac{1}{r}} F(t)g_+(t)\mathrm{d}t\right]_\mathcal{H}
\end{aligned}
$$

and then, since the first integral from the last row vanishes, due to (4.10),

$$= -\left[f_{-1}, \frac{1}{2\pi i}\int_{|t|=\frac{1}{r}} F(t)g_+(t)dt\right]_{\mathcal{H}}. \tag{4.29}$$

Then, performing similar calculations for the last three terms in (4.27), we get

$$[U_0 f_+, U_0 g_+]_{\mathcal{G}} = \frac{1}{4\pi^2}\int_{|s|=\frac{1}{r}}\int_{|t|=\frac{1}{r}} [C_F(t,s)f_+(s), g_+(t)]_{\mathcal{H}} ds\overline{dt}$$

$$- \left[f_{-1}, \frac{1}{2\pi i}\int_{|t|=\frac{1}{r}} \frac{F(t)}{2}g_+(t)dt\right]_{\mathcal{H}}$$

$$- \left[\frac{1}{2\pi i}\int_{|s|=\frac{1}{r}} \frac{F(s)}{2}f_+(s)ds, g_{-1}\right]_{\mathcal{H}}$$

$$+ \left[\frac{F(0)^{\sharp} + F(0)}{2}f_{-1}, g_{-1}\right]_{\mathcal{H}}. \tag{4.30}$$

Similarly to the calculations we performed in order to obtain (4.30), we get

$$[U_0 f_-, U_0 g_+]_{\mathcal{G}} = -\frac{1}{4\pi^2}\int_{|s|=r}\int_{|t|=\frac{1}{r}} [C_F(t,s)(sf_-(s) - f_{-1}), tg_+(t) + g_{-1}]_{\mathcal{H}} ds\overline{dt}$$

$$= -\frac{1}{4\pi^2}\int_{|s|=r}\int_{|t|=\frac{1}{r}} [C_F(t,s)f_{-1}, g_+(t)]_{\mathcal{H}} ds\overline{dt}$$

$$+ \left[f_{-1}, \frac{1}{2\pi i}\int_{|t|=\frac{1}{r}} \frac{F(t)}{2}g_+(t)dt\right]_{\mathcal{H}}$$

$$+ \left[\frac{1}{2\pi i}\int_{|s|=r} \frac{F(s)}{2}f_-(s)ds, g_{-1}\right]_{\mathcal{H}} - \left[\frac{F(0)}{2}f_{-1}, g_{-1}\right]_{\mathcal{H}}, \tag{4.31}$$

$$[U_0 f_+, U_0 g_-]_{\mathcal{G}} = -\frac{4}{\pi^2}\int_{|s|=\frac{1}{r}}\int_{|t|=r} [C_F(t,s)(sf_+(s) + f_{-1}), tg_-(t) - g_{-1}]_{\mathcal{H}} ds\overline{dt}$$

$$= -\frac{1}{4\pi^2}\int_{|s|=\frac{1}{r}}\int_{|t|=r} [C_F(t,s)f_+(s), g_-(t)]_{\mathcal{H}} ds\overline{dt}$$

$$+ \left[f_{-1}, \frac{1}{2\pi i}\int_{|t|=r} \frac{F(t)}{2}g_-(t)dt\right]_{\mathcal{H}}$$

$$+ \left[\frac{1}{2\pi i}\int_{|s|=\frac{1}{r}} \frac{F(s)}{2}f_+(s)ds, g_{-1}\right]_{\mathcal{H}} - \left[\frac{F(0)^{\sharp}}{2}f_{-1}, g_{-1}\right]_{\mathcal{H}}, \tag{4.32}$$

6.4 CARATHÉODORY TYPE REPRESENTATIONS

$$[U_0 f_-, U_0 g_-]_{\mathcal{G}} = \frac{4}{\pi^2} \int_{|s|=r} \int_{|t|=r} [C_F(t,s)(sf_-(s) - f_{-1}), tg_-(t) - g_{-1}]_{\mathcal{H}} \mathrm{d}s\overline{\mathrm{d}t}$$

$$= \frac{4}{\pi^2} \int_{|s|=r} \int_{|t|=r} [C_F(t,s)f_-(s), g_-(t)]_{\mathcal{H}} \mathrm{d}s\overline{\mathrm{d}t}$$

$$- \left[\frac{1}{2\pi\mathrm{i}} \int_{|s|=r} \frac{F(s)}{2} f_-(s)\mathrm{d}s, g_{-1}\right]_{\mathcal{H}}$$

$$- \left[f_{-1}, \frac{1}{2\pi\mathrm{i}} \int_{|t|=r} \frac{F(t)}{2} g_-(t)\mathrm{d}t\right]_{\mathcal{H}}. \quad (4.33)$$

Adding side by side (4.30), (4.32), (4.31), and (4.33), and taking into account the definition of the inner product $[\cdot, \cdot]_{\mathcal{G}}$ as in (4.18), we obtain

$$[U_0 f, U_0 g]_{\mathcal{G}} = \frac{1}{4\pi^2} \int_{|s|=\frac{1}{r}} \int_{|t|=\frac{1}{r}} [C_F(t,s)f_+(s), g_+(t)]_{\mathcal{H}} \mathrm{d}s\overline{\mathrm{d}t}$$

$$- \frac{1}{4\pi^2} \int_{|s|=r} \int_{|t|=\frac{1}{r}} [C_F(t,s)f_-(s), g_+(t)]_{\mathcal{H}} \mathrm{d}s\overline{\mathrm{d}t}$$

$$- \frac{1}{4\pi^2} \int_{|s|=\frac{1}{r}} \int_{|t|=r} [C_F(t,s)f_+(s), g_-(t)]_{\mathcal{H}} \mathrm{d}s\overline{\mathrm{d}t}$$

$$+ \frac{1}{4\pi^2} \int_{|s|=r} \int_{|t|=r} [C_F(t,s)f_-(s), g_-(t)]_{\mathcal{H}} \mathrm{d}s\overline{\mathrm{d}t}$$

$$= [f, g]_{\mathcal{G}},$$

which completes the proof that U_0 is isometric with respect to the inner product $[\cdot, \cdot]_{\mathcal{G}}$. Since U_0 is invertible, this shows that its inverse U_0^{-1} is isometric as well. Thus, all statements at Step 2 are proven.

We define now a linear operator $\Gamma_0: \mathcal{H} \to \mathcal{G}$ by

$$(\Gamma_0 h) = \frac{1}{z} h, \quad h \in \mathcal{H}, \ r < |z| < \frac{1}{r}. \quad (4.34)$$

Step 3. *The linear operator Γ_0 is bounded and, for all $h \in \mathcal{H}$ and all $g = (g_+, g_-) \in \mathcal{G}$, we have*

$$[\Gamma_0 h, g]_{\mathcal{G}} = \left[h, -\frac{1}{2\pi\mathrm{i}} \int_{|t|=\frac{1}{r}} \frac{F(t)}{2} g_+(t)\mathrm{d}t + \frac{1}{2\pi\mathrm{i}} \int_{|t|=r} \frac{F(0)^\sharp + F(t)}{2} g_-(t)\mathrm{d}t\right]_{\mathcal{H}}. \quad (4.35)$$

Indeed, we first observe that $\mathrm{Ran}(\Gamma_0) \subset \mathcal{G}_-$ and then, for arbitrary $h \in \mathcal{H}$, we have

$$\|\Gamma_0 h\|_{\mathcal{G}}^2 = \|\Gamma_0 h\|_{\mathcal{G}_-}^2 = \frac{\|h\|_{\mathcal{H}}^2}{r^2},$$

which proves the boundedness of Γ_0.

In order to prove (4.35), let $h \in \mathcal{H}$ be fixed, but arbitrary. We first observe that, for any $|t| \leq r$, by the Cauchy Integral Formula at 0 applied to $C_F(t, \cdot)$, we have

$$\frac{1}{2\pi i} \int_{|s|=r} \frac{C_F(t,s)}{s} h \, \mathrm{d}s = C_F(t,0)h = \frac{F(0) + F(t)^\sharp}{2} h. \tag{4.36}$$

On the other hand, we prove that, for any $|t| \geq \frac{1}{r}$ we have

$$\frac{1}{2\pi i} \int_{|s|=r} \frac{C_F(t,s)}{s} h \, \mathrm{d}s = \frac{F(t)^\sharp}{2} h. \tag{4.37}$$

Indeed, for $|t| > \frac{1}{r}$ this follows from the Cauchy Integral Formula at infinity applied to $C_F(t, \cdot)$. For $|t| = \frac{1}{r}$, we approximate t by a sequence of complex numbers $(t_n)_n$ such that $|t_n| > \frac{1}{r}$ for all n and $t_n \xrightarrow[n\to\infty]{} t$. Since (4.37) holds for each t_n, $n \geq 1$, we then pass to the limit following $n \to \infty$ and take into account that $C_F(t_n, \cdot) \xrightarrow[n\to\infty]{} C_F(t, \cdot)$ uniformly on any compact subset, in particular on $\partial \mathbb{D}_r$, and that $F(t_n)h \xrightarrow[n\to\infty]{} F(t)h$.

Then, for arbitrary $g = (g_+, g_-) \in \mathcal{G}$, taking into account that $\Gamma_0 h \in \mathcal{G}_-$ and the definition of the inner product $[\cdot, \cdot]_\mathcal{G}$ in (4.18), we have

$$[\Gamma_0 h, g]_\mathcal{H} = -\frac{1}{4\pi^2} \int_{|s|=r} \int_{|t|=\frac{1}{r}} [\frac{C_F(t,s)}{s} h, g_+(t)]_\mathcal{H} \mathrm{d}s \overline{\mathrm{d}t}$$
$$+ \frac{1}{4\pi^2} \int_{|s|=r} \int_{|t|=r} [\frac{C_F(t,s)}{s} h, g_-(t)]_\mathcal{H} \mathrm{d}s \overline{\mathrm{d}t}$$

and then, taking into account (4.37) and (4.36),

$$= \frac{1}{2\pi i} \int_{|t|=\frac{1}{r}} [\frac{F(t)^\sharp}{2} h, g_+(t)]_\mathcal{H} \overline{\mathrm{d}t} - \frac{1}{2\pi i} \int_{|t|=r} [\frac{F(0) + F(t)^\sharp}{2} h, g_-(t)]_\mathcal{H} \overline{\mathrm{d}t}$$
$$= \Big[h, -\frac{1}{2\pi i} \int_{|t|=\frac{1}{r}} \frac{F(t)}{2} g_+(t) \mathrm{d}t + \frac{1}{2\pi i} \int_{|t|=r} \frac{F(0)^\sharp + F(t)}{2} g_-(t) \mathrm{d}t\Big]_\mathcal{H},$$

which concludes the proof of (4.35). The claim at Step 3 is proven.

In view of (4.35), we define the linear operator $\Gamma_1 \colon \mathcal{G} \to \mathcal{H}$ by

$$\Gamma_1 f = -\frac{1}{2\pi i} \int_{|t|=\frac{1}{r}} \frac{F(t)}{2} f_+(t) \mathrm{d}t + \frac{1}{2\pi i} \int_{|t|=r} \frac{F(0)^\sharp + F(t)}{2} f_-(t) \mathrm{d}t. \tag{4.38}$$

Step 4. *The operator* $\Gamma_1 \colon \mathcal{G} \to \mathcal{H}$ *is bounded and we have*

$$[\Gamma_0 h, g]_\mathcal{G} = [h, \Gamma_1 g]_\mathcal{H}, \quad h \in \mathcal{H}, \ g \in \mathcal{G}. \tag{4.39}$$

6.4 Carathéodory Type Representations

Indeed, (4.39) is a direct consequence of the definition of Γ_1 as in (4.38) and the formula (4.35). In order to prove the boundedness of Γ_1, for arbitrary $f = (f_+, f_-) \in \mathcal{G}$, we have

$$\|\Gamma_1 f\|_{\mathcal{H}} = \left\| -\frac{1}{2\pi i} \int_{|t|=\frac{1}{r}} \frac{F(t)}{2} f_+(t) dt + \frac{1}{2\pi i} \int_{|t|=r} \frac{F(0)^\sharp + F(t)}{2} f_-(t) dt \right\|_{\mathcal{H}}$$

and, making the change of variables $t = \frac{e^{i\theta}}{r}$ in the first integral and, respectively, $t = re^{i\theta}$ in the second integral,

$$\leqslant \frac{1}{2\pi r} \int_0^{2\pi} \left\| \frac{F(e^{i\theta}/r)}{2} \right\| \|f_+(\frac{e^{i\theta}}{r})\|_{\mathcal{H}} d\theta$$
$$+ \frac{r}{2\pi} \int_0^{2\pi} \left\| \frac{F(0)^\sharp + F(re^{i\theta})}{2} \right\| \|f_-(re^{i\theta})\|_{\mathcal{H}} d\theta$$

then, applying the Cauchy–Bunyakovsky Inequality to each integral, for some constant $C > 0$,

$$\leqslant C \left(\frac{1}{\sqrt{2\pi}} \int_0^{2\pi} \|f_+(\frac{e^{i\theta}}{r})\|_{\mathcal{H}}^2 d\theta \right)^{1/2} + C \left(\frac{1}{\sqrt{2\pi}} \int_0^{2\pi} \|f_-(re^{i\theta})\|_{\mathcal{H}}^2 d\theta \right)^{1/2}$$

whence, taking into account (4.9) and (4.13),

$$= C(\|f_+\|_{\mathcal{G}} + \|f_-\|_{\mathcal{G}}) = C\|f\|_{\mathcal{G}},$$

which proves the boundedness of Γ_1.

Step 5. *For every $|\zeta| < r$ we have*

$$F(\zeta) = \frac{F(0) - F(0)^\sharp}{2} + \Gamma_1(U_0 + \zeta I)(U_0 - \zeta I)^{-1} \Gamma. \tag{4.40}$$

Indeed, fix $|\zeta| < r$ and observe that, for every $r < |z| < \frac{1}{r}$, we have

$$((U_0 - \zeta I)f)(z) = (z - \zeta)f(z),$$

hence $U_0 - \zeta I$ is boundedly invertible. Then, fixing $f \in \mathcal{G}$, for every $r < |z| < \frac{1}{r}$, we have

$$((U_0 + \zeta I)(U_0 - \zeta I)^{-1} \Gamma_0 f)(z) = \frac{z + \zeta}{z(z - \zeta)} f,$$

whence, taking into account that,

$$\frac{z + \zeta}{z(z - \zeta)} f = -\frac{1}{z} f + 2 \sum_{n=1}^{\infty} \frac{\zeta^{n-1}}{z^n} f,$$

which, in view of the identification of the space $\mathcal{G}_-$ as in (4.11), means that the $\mathcal{H}$-valued function $A_r \ni z \mapsto \frac{z+\zeta}{z(z-\zeta)} f$ belongs to $\mathcal{G}_-$, we have

$$\Gamma_1((U_0 + \zeta I)(U_0 - \zeta I)^{-1}\Gamma_0 f)(z) = \frac{1}{2\pi\mathrm{i}} \int_{|s|=r} \frac{F(0)^\sharp + F(s)}{2} \frac{s+\zeta}{s-\zeta} f \mathrm{d}s$$

$$= \frac{1}{2\pi\mathrm{i}} \int_{|s|=r} \frac{F(0)^\sharp + F(s)}{2} \left(-\frac{1}{s} + \frac{2}{s-\zeta}\right) f \mathrm{d}s$$

then, applying the Cauchy Integral Formula at 0 and at ζ,

$$= -\frac{F(0)^\sharp + F(0)}{2} f + \left(F(0)^\sharp + F(\zeta)\right) f$$

$$= F(\zeta)f + \frac{F(0)^\sharp - F(0)}{2} f,$$

which concludes the proof of (4.40) and of Step 5.

We are now in a position to finish the proof.

Step 6. *There exists a triple* $(\mathcal{K}; U; \Gamma)$, *where* $\mathcal{K}$ *is a Kreĭn space,* $U \in \mathcal{B}(\mathcal{K})$ *is a unitary operator, and* $\Gamma \in \mathcal{B}(\mathcal{H}, \mathcal{K})$, *such that* (4.1) *and* (4.2) *hold.*

Indeed, by Step 1, the inner product $[\cdot, \cdot]_\mathcal{G}$ is bounded on $\mathcal{G}$ hence there exists its Gram operator $G \in \mathcal{B}(\mathcal{G})$, $G = G^*$, such that $[f, g]_\mathcal{G} = \langle Gf, g \rangle_\mathcal{G}$ for all $f, g \in \mathcal{G}$. We consider the induced Kreĭn space $(\mathcal{K}_G; \Pi_G)$ and let $\mathcal{K} = \mathcal{K}_G$. In view of Step 4, by Theorem 6.1.10, there exists a uniquely determined bounded operator $\Gamma \colon \mathcal{H} \to \mathcal{K}$ such that $\Gamma = \Pi_G \Gamma_0$ and $\Gamma^\sharp \Pi_G = \Gamma_1$.

On the other hand, since U_0^{-1} is isometric with respect to the inner product $[\cdot, \cdot]_\mathcal{G}$, we have

$$[U_0 f, g]_\mathcal{G} = [U_0^{-1} U_0 f, U_0^{-1} g]_\mathcal{G} = [f, U_0^{-1} g]_\mathcal{G},$$

hence, again by Theorem 6.1.10, there exists uniquely a bounded operator $U \in \mathcal{B}(\mathcal{K})$ such that $\Pi_G U_0 = U \Pi_G$ and $\Pi_G U_0^{-1} = U^\sharp \Pi_G$, hence U is boundedly invertible and $U^{-1} = U^\sharp$, that is, U is a bounded unitary operator in the Kreĭn space $\mathcal{K}$. On the other hand, we proved at Step 2 that the operator norm of U_0 is at most $\frac{1}{r}$ hence $\sigma(U_0) \subseteq \mathrm{Clos}\,\mathbb{D}_{1/r}$ and then, in view of Corollary 6.1.11, $\sigma(U) \subseteq \mathrm{Clos}\,\mathbb{D}_{1/r}$. Since U is unitary, its spectrum lies symmetrically with respect to the transformation $\lambda \mapsto \lambda^* = 1/\overline{\lambda}$, hence $\mathbb{D}_r \subseteq \rho(U)$. Consequently, for any $\zeta \in \mathbb{D}_r$ and any $h, k \in \mathcal{H}$, we have

$$[\Gamma_1(U_0 + \zeta I)(U_0 - \zeta I)^{-1}\Gamma_0 h, k]_\mathcal{G} = [(U_0 + \zeta I)(U_0 - \zeta I)^{-1}\Gamma_0 h, \Gamma_0 k]_\mathcal{G}$$

$$= [\Pi_G(U_0 + \zeta I)(U_0 - \zeta I)^{-1}\Gamma_0 h, \Pi_G \Gamma_0 k]_\mathcal{K}$$

$$= [(U + \zeta I)(U - \zeta I)^{-1}\Gamma h, \Gamma k]_\mathcal{K}$$

$$= [\Gamma^\sharp (U + \zeta I)(U - \zeta I)^{-1}\Gamma h, k]_\mathcal{K},$$

6.4 CARATHÉODORY TYPE REPRESENTATIONS

which, in view of Step 5, proves the formula (4.1).

Also, by construction, the set $\{(U_0-\zeta I)^{-1}\Gamma_0 h \mid h \in \mathcal{H}$ and either $|\zeta| < r$ or $|\zeta| > \frac{1}{r}\}$ is total in $\mathcal{G}$ and hence, taking into account that, for each $h \in \mathcal{H}$ and complex number ζ with either $|\zeta| < r$ or $|\zeta| > \frac{1}{r}$, we have

$$\Pi_\mathcal{G}(U_0 - \zeta I)^{-1}\Gamma_0 h = (U - \zeta I)^{-1} h,$$

which proves (4.2) as well. ∎

Remark 6.4.2. (a) The proof presented for Theorem 6.4.1 is, in a certain way, similar to that of Theorem 6.2.1. The Hilbert spaces $\mathcal{G}$ and the selfadjoint operators G used to induce the Kreĭn spaces $\mathcal{K}$ as in Theorem 6.2.1 and Theorem 6.4.1 are only formally different. One of the key differences is that, in the proof of Theorem 6.2.1 one starts with the vector space of formal sums $\sum_{|z|>r} \delta_z f_z$ on which an appropriate topology is defined that provides the completion to the Hilbert space $\mathcal{G}$, while in the case of Theorem 6.4.1 one starts directly with the Hilbert space $\mathcal{G}$ obtained as the direct sum of two vector valued Hardy spaces H^2. In the latter case, one could start as well with the vector space of finite sums $\sum_{z \in A_r^c} \delta_z f_z$, indexed on the $A_r^c = \{z \in \mathbb{C} \mid |z| < r$ or $|z| > \frac{1}{r}\}$ and define the indefinite inner product by

$$\Big[\sum_{z \in A_r^c} \delta_z f_z, \sum_{\zeta \in A_r^c} \delta_\zeta f_\zeta\Big] = \sum_{z,\zeta \in A_r^c} [C_F(\zeta, z) f_z, g_\zeta]_\mathcal{H},$$

and then an appropriate Hilbert space topology should be defined.

(b) With notation and assumptions as in Theorem 6.4.1, we observe that, for any complex numbers $z, \zeta \in A_r^c$ and any $h, k \in \mathcal{H}$, we have

$$[(U - zI)^{-1}\Gamma h, (U - \zeta I)^{-1}\Gamma k]_\mathcal{K} = [C_F(\zeta, z) h, k]_\mathcal{H}. \tag{4.41}$$

To see this, we first assume $z\bar\zeta \neq 1$. Then

$$[C_F(\zeta, z) h, k]_\mathcal{H} = \Big[\frac{F(z) + F(\zeta)}{2(1 - z\bar\zeta)} h, k\Big]_\mathcal{H}$$

$$= \frac{\Gamma^\sharp((U + zI)(U - zI)^{-1} + (U^\sharp + \bar\zeta I)(U^\sharp - \bar\zeta I)^{-1})\Gamma}{2(1 - z\bar\zeta)}. \tag{4.42}$$

A simple calculation shows that

$$\frac{1}{2(1 - z\bar\zeta)}\Big(\frac{\lambda + z}{\lambda - z} + \frac{\lambda^{-1} + \bar\zeta}{\lambda^{-1} - \bar\zeta}\Big) = \frac{1}{(\lambda - z)(\lambda^{-1} - \bar\zeta)}, \tag{4.43}$$

for all complex numbers λ such that $r - \varepsilon < |\lambda| < 1/(r - \varepsilon)$, where $0 < \varepsilon$ is chosen in such a way that both z and $\bar\zeta$ do not belong to any of the rings centred at 0 and of radii

$r - \varepsilon$ and $1/(r - \varepsilon)$. Then, from (4.43), by holomorphic functional calculus for U we have

$$\frac{1}{2(1 - z\overline{\zeta})} \left((U + zI)(U - zI)^{-1} + (U^\sharp + \overline{\zeta}I)(U^\sharp - \overline{\zeta}I)^{-1} \right) = (U^{-1} - \overline{\zeta}I)^{-1}(U - zI)^{-1},$$

whence, taking into account that $U^\sharp = U^{-1}$, (4.42) and (4.43), we get (4.41). If $z\overline{\zeta} = 1$ then we use the case $z\overline{\zeta} \neq 1$ just proven and pass to the limit $\overline{\zeta} \to 1/z$ using the continuity and differentiability properties. So, (4.41) is proven for all z and ζ in the complement of the ring A_r.

The formula (4.41) shows that the appearance of the Carathéodory kernel C_F in the definition of the inner product $[\cdot, \cdot]_{\mathcal{G}}$ as in (4.18) is natural.

(c) Formula (4.41) also shows that, assuming that the triple $(\widetilde{\mathcal{K}}; \widetilde{U}; \widetilde{\Gamma})$ has the same properties as the triple $(\mathcal{K}; U, \Gamma)$ in Theorem 6.4.1, including the minimality condition, there exists an isometric operator V from $\mathcal{K}$ to $\mathcal{K}'$, densely defined and with dense range, defined by

$$V(U - zI)^{-1}\Gamma h = (U' - zI)^{-1}\Gamma' h, \qquad (4.44)$$

for all $h \in \mathcal{H}$ and all $z \in A_r^c$, hence a weak type of uniqueness of minimal triples holds, compare with Remark 6.2.2. Strong uniqueness, that is, the existence of a bounded unitary operator V as before, depends on the same type of spectral condition on the Gram operator G as in Theorem 6.1.9. ∎

An application of Theorem 6.4.1 is referring to the existence of unitary dilations for arbitrary bounded operators in Kreĭn spaces.

Corollary 6.4.3. *Let $\mathcal{H}$ be a Kreĭn space and $T \in \mathcal{B}(\mathcal{H})$. Then there exists a Kreĭn space $\mathcal{K}$ such that $\mathcal{H}$ is a regular subspace of it, and a unitary operator $U \in \mathcal{B}(\mathcal{K})$ such that*

$$T^n = P_{\mathcal{H}} T^n | \mathcal{H}, \quad T^{\sharp n} = P_{\mathcal{H}} U^{-n} | \mathcal{H}, \quad n \geq 0.$$

In addition, $\mathcal{K}$ can be chosen with the following minimality property

$$\mathcal{K} = \operatorname{Clos} \operatorname{Lin}\{U^n \mathcal{H} \mid n \in \mathbb{Z}\}.$$

Proof. Let $r > 0$ such that $r < \min\{1, |\sigma(T)|^{-1}\}$ and $F_T \colon \mathbb{D}_r \to \mathcal{B}(\mathcal{H})$ be defined by

$$F_T(z) = (I + zT)(I - zT)^{-1}, \quad z \in \mathbb{D}_r, \qquad (4.45)$$

and observe that $F_T(0) = I$. Then, by Theorem 6.4.1, we obtain a Kreĭn space $\mathcal{K}$, a unitary operator $V \in \mathcal{B}(\mathcal{K})$ and $\Gamma \in \mathcal{B}(\mathcal{H}, \mathcal{K})$ such that

$$F_T(z) = \Gamma^\sharp (V + zI)(V - zI)^{-1}\Gamma = \Gamma^\sharp (I + zV^{-1})(I - zV^{-1})^{-1}\Gamma, \quad z \in \mathbb{D}_r. \quad (4.46)$$

Since $I = F_T(0) = \Gamma^\sharp \Gamma$, it follows that Γ is a bounded isometry and hence, we can identify $\mathcal{H}$ with the regular subspace $\Gamma\mathcal{H}$ of $\mathcal{K}$. With this identification and letting $U = V^{-1}$, (4.46) becomes

$$F_T(z) = P_{\mathcal{H}}(I + zU)(I - zU)^{-1}|\mathcal{H}, \quad z \in \mathbb{D}_r, \qquad (4.47)$$

whence, expanding $(I - zU)^{-1}$ in a Neumann series, performing the same operation for $(I - zT)^{-1}$ in (4.45), and identifying the coefficients corresponding to z^n from both sides, we obtain $T^n = P_{\mathcal{H}} U^n | \mathcal{H}$ for all $n = 0, 1, 2, \ldots$. Taking adjoints, we obtain $T^{\sharp n} = P_{\mathcal{H}} U^{-n} | \mathcal{H}$ for all $n = 0, 1, 2, \ldots$.

The minimality property follows by the observation that

$$\mathrm{Lin}\{U^n \mathcal{H} \mid n \in \mathbb{Z}\} = \mathrm{Lin}\{(I - zU^{-1})^{-1} \mathcal{H} \mid z \in \mathbb{D}_r \cup \mathbb{D}_r^*\}$$
$$= \mathrm{Lin}\{(V - zI)\Gamma\mathcal{H} \mid z \in \mathbb{D}_r \cup \mathbb{D}_r^*\}. \blacksquare$$

Example 6.4.4. Given a function $F \colon \mathrm{Dom}(F) \to \mathcal{B}(\mathcal{H})$, for some Hilbert space $\mathcal{H}$, where $\mathrm{Dom}(F)$ is a subset of the open unit disc of the complex plane $\mathbb{D}$, the operator valued *Carathéodory kernel* associated to the function F is defined by

$$C_F(\zeta, z) = \frac{F(z) + F(\zeta)^*}{1 - z\overline{\zeta}}, \quad z, \zeta \in \mathrm{Dom}(F), \tag{4.48}$$

and is Hermitian in the sense $C_F(\zeta, z) = C_F(z, \zeta)^*$.

Given a nonnegative integer number κ, one defines the class $\mathcal{C}_\kappa(\mathcal{H})$ of all meromorphic functions F in $\mathbb{D}$, such that, letting $\mathrm{Dom}(F)$ denote the domain of holomorphy of F, the associated Carathéodory kernel $C_F \colon \mathrm{Dom}(F) \times \mathrm{Dom}(F) \to \mathcal{B}(\mathcal{H})$ defined as in (4.48) has κ negative squares, with definition as in (2.14).

Let $\varphi \colon \mathbb{D} \to \mathbb{C}_+$ be the fractional linear transformation as in Chapter 4, (5.2), for $\mathrm{Im}\,\zeta > 0$ and $|\varepsilon| = 1$. Then

$$F = \mathrm{i} Q \circ \varphi, \quad Q \in \mathcal{N}_\kappa(\mathcal{H}),$$

establishes an injective mapping between $\mathcal{N}_\kappa(\mathcal{H})$ and $\mathcal{C}_\kappa(\mathcal{H})$. $\blacksquare$

Example 6.4.5. Letting $\Theta \colon \mathrm{Dom}(\Theta) \to \mathcal{B}(\mathcal{H})$, where $\mathcal{H}$ is a Hilbert space, be defined on the nonempty subset $\mathrm{Dom}(\Theta)$ of $\mathbb{D}$, then the operator valued *Schur kernel* is defined by

$$S_\Theta(\zeta, z) = \frac{1 - \Theta(\zeta)^* \Theta(z)}{1 - z\overline{\zeta}}, \quad z, \zeta \in \mathbb{D}. \tag{4.49}$$

If $\mathcal{H}$ has finite dimension then for any $\alpha \in \mathbb{C}$ with $\mathrm{Re}\,\alpha > 0$ and $F \in \mathcal{C}_\kappa(\mathcal{H})$ the function Θ defined by

$$\Theta(z) = (F(z) - \overline{\alpha}I)(F(z) + \alpha I)^{-1}$$

belongs to the class $\mathcal{S}_\kappa(\mathcal{H})$, with obvious definition. $\blacksquare$

6.5 Elementary Rotations

Let $\mathcal{K}_1$ and $\mathcal{K}_2$ be Kreĭn spaces and let T be an operator in $\mathcal{B}(\mathcal{K}_1, \mathcal{K}_2)$. An *elementary rotation* or *Julia operator* of the operator T is a triple $(U; \mathcal{K}_1', \mathcal{K}_2')$ with the following properties.

(i) $\mathcal{K}_1'$ and $\mathcal{K}_2'$ are Kreĭn spaces.
(ii) $U \in \mathcal{B}(\mathcal{K}_1[+]\mathcal{K}_1', \mathcal{K}_2[+]\mathcal{K}_2')$ is a unitary operator lifting the operator T, that is, $P_{\mathcal{K}_2} U|\mathcal{K}_1 = T$.
(iii) The following mutually equivalent minimality conditions hold:
$$\mathcal{K}_2 \vee U\mathcal{K}_1 = \mathcal{K}_2[+]\mathcal{K}_2', \quad \mathcal{K}_1 \vee U^\sharp \mathcal{K}_2 = \mathcal{K}_1[+]\mathcal{K}_1'.$$

Remark 6.5.1. Let $T \in \mathcal{B}(\mathcal{K}_1, \mathcal{K}_2)$ for some Kreĭn spaces $\mathcal{K}_1, \mathcal{K}_2, \mathcal{K}_1'$, and $\mathcal{K}_2'$. Let $U \in \mathcal{B}(\mathcal{K}_1[+]\mathcal{K}_1', \mathcal{K}_2[+]\mathcal{K}_2')$ be a bounded linear operator such that $P_{\mathcal{K}_2} U|\mathcal{K}_1 = T$, that is, U is a lifting of T, hence, with respect to the decompositions $\mathcal{K}_1[+]\mathcal{K}_1'$ and $\mathcal{K}_2[+]\mathcal{K}_2'$, U has the operator block matrix representation

$$U = \begin{bmatrix} T & X \\ Y & Z \end{bmatrix}. \tag{5.1}$$

(a) U is unitary if and only if the following identities hold:
$$I_1 - T^\sharp T = Y^\sharp Y, \quad I_2 - T T^\sharp = X X^\sharp, \tag{5.2}$$
$$T Y^\sharp = -X Z^\sharp, \quad T^\sharp X = -Y^\sharp Z, \tag{5.3}$$
$$I_1' - Y Y^\sharp = Z Z^\sharp, \quad I_2' - X X^\sharp = Z^\sharp Z. \tag{5.4}$$

(b) $\mathcal{K}_2 \vee U\mathcal{K}_1 = \mathcal{K}_2[+]\mathcal{K}_2'$ if and only if $Y \in \mathcal{B}(\mathcal{K}_1, \mathcal{K}_2')$ has dense range. On the other hand, $\mathcal{K}_1 \vee U^\sharp \mathcal{K}_2 = \mathcal{K}_1[+]\mathcal{K}_1'$ if and only if $X^\sharp \in \mathcal{B}(\mathcal{K}_2, \mathcal{K}_1')$ has dense range. ∎

In order to prove the existence of elementary rotations of some arbitrary operator T, Remark 6.5.1 shows that this is connected with induced Kreĭn spaces of the operators $I - T^\sharp T$ and $I - T T^\sharp$ and hence it gives motivation for the following definition.

For an arbitrary operator T in $\mathcal{B}(\mathcal{K}_1, \mathcal{K}_2)$, an operator $D \in \mathcal{B}(\mathcal{K}_1, \mathcal{D})$, for some Kreĭn space $\mathcal{D}$, is called a *defect operator* of T if the pair $(\mathcal{D}; D)$ is a Kreĭn space induced by $I_1 - T^\sharp T$, that is, $I_1 - T^\sharp T = D^\sharp D$ and D has dense range.

Remark 6.5.2. Defect operators always exist. Indeed, let J_1 and J_2 be fundamental symmetries on $\mathcal{H}_1$ and $\mathcal{H}_2$, respectively. We consider the selfadjoint operator $J_1 - T^* J_2 T$ in the Hilbert space $(\mathcal{H}_1, \langle \cdot, \cdot \rangle_{J_1})$ and

$$J_T = \operatorname{sgn}(J_1 - T^* J_2 T), \quad D_T = |J_1 - T^* J_2 T|^{1/2}.$$

Consider $\mathcal{D}_T = \operatorname{Clos} \operatorname{Ran}(D_T)$ and endow it with the (indefinite) inner product induced by the symmetry J_T. Then $D_T \in \mathcal{B}(\mathcal{H}_1, \mathcal{D}_T)$ is a defect operator of T. To see this, observe that we have the following

$$D_T^\sharp D_T = J_1 D_T^* J_T D_T = J_1 D_T J_T D_T$$
$$= J_1(J_1 - T^* J_2 T) = I - J_1 T^* J_2 T = I - T^\sharp T.$$

Actually, recalling Proposition 6.1.2, we observe that the construction of the defect operator D_T and the underlying Kreĭn space $\mathcal{D}_T$ is exactly, with notation as in Section 6.1, the Kreĭn space $(\mathcal{K}_{I-T^\sharp T}; \Pi_{I-T^\sharp T})$ induced by $I - T^\sharp T$. ∎

6.5 ELEMENTARY ROTATIONS

With this definition, it follows that if $(U; \mathcal{K}'_1; \mathcal{K}'_2)$ is an elementary rotation of T, with operator block matrix as in (5.1), then Y is a defect operator of T and $X^\sharp$ is a defect operator of $T^\sharp$. Since we know that defect operators of bounded selfadjoint operators always exist, see Proposition 6.1.2, we use this as a starting point for proving the existence of elementary rotations.

Theorem 6.5.3. *Let $T \in \mathcal{B}(\mathcal{K}_1, \mathcal{K}_2)$ and $D \in \mathcal{B}(\mathcal{K}_1, \mathcal{D})$ be a defect operator of T. Then there exists $(U; \mathcal{K}'_1; \mathcal{K}'_2)$, an elementary rotation of T, with $\mathcal{K}'_2 = \mathcal{D}$ and $P_{\mathcal{K}'_1} U | \mathcal{K}_1 = D$.*

Proof. Let $V \colon \mathcal{K}_1 \to \mathcal{K}_2[+]\mathcal{D}$ be defined by

$$V = \begin{bmatrix} T \\ D \end{bmatrix} \colon \mathcal{K}_1 \longrightarrow \begin{matrix} \mathcal{K}_2 \\ [+] \\ \mathcal{D} \end{matrix} . \tag{5.5}$$

Then, since $D^\sharp D = I_1 - T^\sharp T$,

$$V^\sharp V = T^\sharp T + D^\sharp D = T^\sharp T + I_1 - T^\sharp T = I_1,$$

hence V is a bounded isometry and, consequently, $\operatorname{Ran}(V)$ and its orthogonal companion $\operatorname{Ran}(V)^\perp = (\mathcal{K}_2[+]\mathcal{D})[-]\operatorname{Ran}(V)$ are regular subspaces of the Kreĭn space $\mathcal{K}_2[+]\mathcal{D}$.

Letting the Kreĭn space $\mathcal{K}'_1 = \operatorname{Ran}(V)^\perp$, denote by $W \colon \mathcal{K}'_1 \to \mathcal{K}_2[+]\mathcal{D}$ the embedding operator which clearly is a bounded isometry, and then define

$$U = [V \ W] \colon \begin{matrix} \mathcal{K}_1 \\ [+] \\ \mathcal{K}'_1 \end{matrix} \longrightarrow \mathcal{K}_2[+]\mathcal{D}. \tag{5.6}$$

We prove that U is an elementary rotation of T.

Firstly, $P_{\mathcal{K}_2} U | \mathcal{K}_1 = P_{\mathcal{K}_2} V | \mathcal{K}_1 = T$. Then, since $\operatorname{Ran}(V)$, the final space of V, is orthogonal to $\mathcal{K}'_1 = \operatorname{Ran}(V)^\perp$, the initial space of $W^\sharp$, it follows that $W^\sharp V = 0$, hence $V^\sharp W = 0$. Then

$$U^\sharp U = \begin{bmatrix} V^\sharp V & V^\sharp W \\ W^\sharp V & W^\sharp W \end{bmatrix} = \begin{bmatrix} I_1 & 0 \\ 0 & I_2 \end{bmatrix},$$

hence U is isometric.

On the other hand, $VV^\sharp$ is the orthogonal projection on $\operatorname{Ran}(V)$ and $WW^\sharp$ is the orthogonal projection on $\operatorname{Ran}(V)^\perp$, hence $UU^\sharp = VV^\sharp + WW^\sharp$ is the identity operator on $\mathcal{K}_2[+]\mathcal{D}$ and, consequently, U is a unitary operator.

Finally, since D is a defect operator of T it follows that D has dense range in $\mathcal{D} = \mathcal{K}'_2$. In view of Remark 6.5.1 (b) it follows that U satisfies the minimality condition $\mathcal{K}_2 \vee U\mathcal{K}_1 = \mathcal{K}_2[+]\mathcal{D}$, hence $(U; \mathcal{K}'_1; \mathcal{K}'_2)$ is an elementary rotation of T. ∎

As we observed before, see Remark 6.5.1, the off-diagonal entries of elementary rotations are a defect operator and, respectively, an adjoint of a defect operator, which are special cases of induced Kreĭn spaces, but we did not say much about the lower diagonal entry. Now we focus on finding a more explicit form of elementary rotations. Our approach is to use the induced Kreĭn space of type $(\mathcal{H}_A; \pi_A)$, as in Example 6.1.6, in order to model the defect operators.

We define the *defect spaces* $\mathcal{D}_T = \mathcal{H}_{I-T^\sharp T}$ and $\mathcal{D}_{T^\sharp} = \mathcal{H}_{I-TT^\sharp}$, where the induced Kreĭn spaces $\mathcal{H}_{I-T^\sharp T}$ and $\mathcal{H}_{I-TT^\sharp}$ are defined as in Example 6.1.6. To be more specific, we fix some fundamental symmetries J_1 and J_2, respectively, and consider the *defect operators* of T

$$D_T = |J_1 - T^*J_2T|^{1/2}, \quad D_{T^\sharp} = |J_2 - TJ_1T^*|^{1/2}, \tag{5.7}$$

where the moduli are calculated with respect to the Hilbert spaces $(\mathcal{K}_1, \langle \cdot, \cdot \rangle_{J_1})$ and to $(\mathcal{K}_2, \langle \cdot, \cdot \rangle_{J_2})$, respectively. We also define the *sign operators* of the defects of T by

$$J_T = \operatorname{sgn}(J_1 - T^*J_2T), \quad J_{T^\sharp} = \operatorname{sgn}(J_2 - TJ_1T^*), \tag{5.8}$$

which are obtained by the Borelian functional calculus and the signum function

$$\operatorname{sgn}(t) = \begin{cases} 1, & t > 0, \\ 0, & t = 0, \\ -1, & t < 0. \end{cases}$$

Further, considering the closure $\mathcal{D}_T = \operatorname{Clos} \operatorname{Ran}(D_T)$ in $\mathcal{K}_1$, the Kreĭn space structure of $\mathcal{D}_T$ is defined as follows: the strong topology is inherited from the strong topology of $\mathcal{K}_1$ and the indefinite inner product is determined by the symmetry $J_T | \mathcal{D}_T$. Similarly one defines the Kreĭn space $\mathcal{D}_{T^\sharp}$, using the operators D_T and $J_{T^\sharp}$.

Lemma 6.5.4. *With the previous notation, there exist uniquely determined operators* $L_T \in \mathcal{B}(\mathcal{D}_T, \mathcal{D}_{T^\sharp})$ *and* $L_{T^\sharp} \in \mathcal{B}(\mathcal{D}_{T^\sharp}, \mathcal{D}_T)$ *such that:*

$$D_{T^\sharp} L_T = TJ_1 D_T | \mathcal{D}_T, \quad D_T L_{T^\sharp} = T^*J_2 D_{T^\sharp} | \mathcal{D}_{T^\sharp}. \tag{5.9}$$

Proof. Let us first note that

$$TJ_1(J_1 - T^*J_2T) = (J_2 - TJ_1T^*)J_2T.$$

Taking into account the factorisations

$$J_1 - T^*J_2T = J_T D_T^2, \quad J_2 - TJ_1T^* = J_{T^\sharp} D_{T^\sharp}^2,$$

it follows that the formally defined operator $L_T = D_{T^\sharp}^{-1} TJ_1 D_T$ is at least densely defined in $\mathcal{D}_T$. It remains to prove that the operator L_T is bounded or, equivalently, that for some nonnegative μ we have

$$TJ_1 D_T^2 J_1 T^* \leq \mu D_{T^\sharp}^2. \tag{5.10}$$

To this end we introduce a new inner product $\langle \cdot, \cdot \rangle$ on $\mathcal{K}_2$

$$\langle x, y \rangle = \langle D_{T^\sharp}^2 x, y \rangle_{J_2}, \quad x, y \in \mathcal{K}_2,$$

6.5 ELEMENTARY ROTATIONS

and we first show that the operator $S = J_{T^\sharp} J_2 T J_T J_1 T^*$ is $\langle \cdot, \cdot \rangle$ symmetric. Indeed, we have

$$D_{T^\sharp}^2 S = (J_2 - T J_1 T^*) J_2 T J_T J_1 T^*$$
$$= T J_1 (J_1 - T^* J_2 T) J_T J_1 T^* = T J_1 D_T^2 J_1 T^*, \tag{5.11}$$

and analogously

$$S^* D_{T^\sharp}^2 = T J_1 D_T^2 J_1 T^*. \tag{5.12}$$

From (5.11) and (5.12) the claim is proved. Then, in view of Theorem 6.1.10 it follows that the operator S can be lifted to the Kreĭn space induced by the selfadjoint operator $J_2 D_{T^\sharp}^2$, which, by Remark 6.1.4, is equivalent with

$$D_{T^\sharp}^4 S \leqslant \nu D_{T^\sharp}^4, \tag{5.13}$$

for some nonnegative ν. Applying the Schwarz Inequality with respect to the positive semidefinite inner product $\langle \cdot, \cdot \rangle$ and taking into account (5.11), from (5.13) we obtain (5.10).

The statement concerning the operator $L_{T^\sharp}$ follows similarly. ∎

We refer to L_T and $L_{T^\sharp}$ as the *link operators* associated to T.

Remark 6.5.5. If the operator T intertwines some fundamental symmetries J_1 and J_2, that is, $T J_1 = J_2 T$, then $L_T = J_2 T$ and $L_{T^\sharp} = J_1 T^*$. Moreover, in this case the equations (5.9) become $D_{T^\sharp} = T D_T$ and $D_T T^* = T^* D_{T^*}$, respectively. ∎

Corollary 6.5.6. *The link operators associated with an operator T are adjoints one to the other:*

$$L_{T^\sharp} = L_T^\sharp. \tag{5.14}$$

Proof. From (5.9) we infer:

$$D_{T^\sharp} L_T J_T D_T = T J_1 D_T J_T D_T$$
$$= T J_1 (J_1 - T^* J_2 T) = (J_2 - T J_1 T^*) J_2 T$$
$$= D_{T^\sharp} J_{T^\sharp} D_{T^\sharp} J_2 T.$$

Since $D_{T^\sharp}$ is injective on $\mathcal{D}_{T^\sharp}$ this implies $L_T J_T D_T = J_{T^\sharp} D_{T^\sharp} J_2 T$ or, equivalently,

$$D_T J_T L_{T^\sharp} J_{T^\sharp} = T^* J_2 D_{T^\sharp}. \tag{5.15}$$

By the uniqueness of L_T as in Lemma 6.5.4 and (5.15) we get $L_{T^\sharp} = J_T L_T^* J_{T^\sharp} | \mathcal{D}_{T^\sharp}$, which is equivalent with (5.14). ∎

Corollary 6.5.7. *With notation as before, the following formulae hold.*

$$(J_T - D_T J_1 D_T) | \mathcal{D}_T = L_T^* J_{T^\sharp} L_T, \tag{5.16}$$

$$(J_{T^\sharp} - D_{T^\sharp} J_2 D_{T^\sharp}) | \mathcal{D}_{T^\sharp} = L_{T^\sharp}^* J_T L_{T^\sharp}. \tag{5.17}$$

Proof. In view of Corollary 6.5.6, (5.16) is equivalent with

$$L_{T^\sharp}L_T = (I - J_T D_T J_1 D_T)|\mathcal{D}_T. \tag{5.18}$$

To prove (5.18), take $x, y \in \mathcal{K}_1$ and, using (5.9), compute

$$\langle L_{T^\sharp}L_T D_T x, D_T y\rangle_{J_1} = \langle D_T L_{T^\sharp} D_T x, y\rangle_{J_1}$$
$$= \langle T^* J_2 D_{T^\sharp} L_T D_T x, y\rangle_{J_1} = \langle T^* J_2 T J_1 D_T^2 x, y\rangle_{J_1}$$

and then

$$\langle (I - J_T D_T J_1 D_T) D_T x, D_T y\rangle_{J_1} = \langle D_T^2 x, y\rangle_{J_1} - \langle J_T D_T^2 J_1 D_T^2 x, y\rangle_{J_1}$$
$$= \langle D_T^2 x, y\rangle_{J_1} - \langle (J_1 - T^* J_2 T) J_1 D_T^2 x, y\rangle_{J_1}$$
$$= \langle T^* J_2 T J_1 D_T^2 x, y\rangle_{J_1}.$$

These prove (5.18), while (5.17) can be proved similarly. ∎

Theorem 6.5.8. *For any operator $T \in \mathcal{B}(\mathcal{K}_1, \mathcal{K}_2)$, using the previous notation, the triple $(R(T); \mathcal{D}_{T^\sharp}, \mathcal{D}_T)$, where*

$$R(T)\colon \mathcal{K}_1[+]\mathcal{D}_{T^\sharp} \to \mathcal{K}_2[+]\mathcal{D}_T,$$
$$R(T) = \begin{bmatrix} T & D_{T^\sharp} \\ D_T & -J_T L_{T^\sharp} \end{bmatrix} \tag{5.19}$$

is an elementary rotation of T.

Proof. Clearly, the operator $R(T)$ is a lifting of T. To prove that $R(T)$ is unitary we use (5.9) and (5.17), by means of a direct calculation that we leave to the reader. Finally, taking into account the matrix representation of $R(T)$ it follows that

$$\mathcal{K}_2 \vee R(T)\mathcal{K}_1 = (\mathcal{K}_2[+]0) \vee (0[+]\operatorname{Ran}(D_T)) = \mathcal{K}_2[+]\mathcal{D}_T.$$

Thus, all three axioms in the definition of an elementary rotation are fulfilled. ∎

Corollary 6.5.9. *For any operator $T \in \mathcal{B}(\mathcal{K}_1, \mathcal{K}_2)$ we have*

$$\kappa^\pm(I_1 - T^\sharp T) + \kappa^\pm(\mathcal{K}_2) = \kappa^\pm(I_2 - TT^\sharp) + \kappa^\pm(\mathcal{K}_1).$$

Proof. From Theorem 6.5.8 the Kreĭn spaces $\mathcal{K}_1[+]\mathcal{D}_{T^\sharp}$ and $\mathcal{K}_2[+]\mathcal{D}_T$ are unitary equivalent. Therefore, their ranks of negativity and, respectively, of positivity are equal, that is, the desired result. ∎

The triple $(R(T); \mathcal{D}_{T^\sharp}, \mathcal{D}_T)$ is called the *canonical elementary rotation* of T. This is explicitly described in terms of T. Of course, the problem of uniqueness of elementary rotations must be settled. We first clarify what kind of uniqueness we can expect.

6.5 ELEMENTARY ROTATIONS

Two elementary rotations $(U;\mathcal{K}_1',\mathcal{K}_2')$ and $(V;\mathcal{H}_1',\mathcal{H}_2')$ of the same operator $T \in \mathcal{B}(\mathcal{K}_1,\mathcal{K}_2)$ are called *unitary equivalent* if there exist unitary operators $\Phi_1\colon \mathcal{K}_1' \to \mathcal{H}_1'$, $\Phi_2\colon \mathcal{K}_2' \to \mathcal{H}_2'$ such that

$$\begin{bmatrix} I_2 & 0 \\ 0 & \Phi_2 \end{bmatrix} U = V \begin{bmatrix} I_1 & 0 \\ 0 & \Phi_1 \end{bmatrix}.$$

The uniqueness of elementary rotations will be characterised in terms of certain spectral properties. For an arbitrary operator $T \in \mathcal{B}(\mathcal{K}_1,\mathcal{K}_2)$ we introduce the spectral properties $(\alpha)_-$ and $(\alpha)_+$ as follows.

$(\alpha)_-$ *There exists $\varepsilon > 0$ such that for some fundamental symmetries J_1 and J_2 we have $(-\varepsilon, 0) \cap \sigma(J_1 - T^*J_2T) = \emptyset$.*

$(\alpha)_+$ *There exists $\varepsilon > 0$ such that for some fundamental symmetries J_1 and J_2 we have $(0, \varepsilon) \cap \sigma(J_1 - T^*J_2T) = \emptyset$.*

As in the proof of Theorem 6.1.9 it is easy to show that if one of the properties $(\alpha)_-$ or $(\alpha)_+$ holds for some fundamental symmetries then it holds for any other fundamental symmetries (of course, the ε may be different). Moreover, these properties share a certain symmetry. More precisely, let us introduce the following dual properties.

$(\alpha)_-^\sharp$ *There exists $\varepsilon > 0$ such that for some fundamental symmetries J_1 and J_2 we have $(-\varepsilon, 0) \cap \sigma(J_2 - TJ_1T^*) = \emptyset$.*

$(\alpha)_+^\sharp$ *There exists $\varepsilon > 0$ such that for some fundamental symmetries J_1 and J_2 we have $(0, \varepsilon) \cap \sigma(J_2 - TJ_1T^*) = \emptyset$.*

Lemma 6.5.10. *The operator T has the property $(\alpha)_-$ (the property $(\alpha)_+$) if and only if it has the property $(\alpha)_-^\sharp$ (respectively, the property $(\alpha)_+^\sharp$).*

Proof. We consider the Hilbert spaces $|\mathcal{K}_i| = (\mathcal{K}_i, \langle \cdot, \cdot \rangle_{J_i})$ and the selfadjoint operator $H \in \mathcal{B}(|\mathcal{K}_1| \oplus |\mathcal{K}_2|)$

$$H = \begin{bmatrix} J_1 & T^* \\ T & J_2 \end{bmatrix},$$

which can be factored, following the two dual Frobenius–Schur factorisations, as

$$H = \begin{bmatrix} I_1 & 0 \\ TJ_1 & I_2 \end{bmatrix} \begin{bmatrix} J_1 & 0 \\ 0 & J_2 - TJ_1T^* \end{bmatrix} \begin{bmatrix} I_1 & J_1T^* \\ 0 & I_2 \end{bmatrix}, \tag{5.20}$$

and

$$H = \begin{bmatrix} I_1 & T^*J_2 \\ 0 & I_2 \end{bmatrix} \begin{bmatrix} J_1 - T^*J_2T & 0 \\ 0 & J_2 \end{bmatrix} \begin{bmatrix} I_1 & 0 \\ J_2T & I_2 \end{bmatrix}. \tag{5.21}$$

Remark that the first and the third operators appearing on the right sides in (5.20) and (5.21) are boundedly invertible. The factorisation (5.20) implies that T has the property $(\alpha)_\pm$ if and only if H has a spectral gap on the right/left of 0 which, by the other Schur factorisation (5.21), is equivalent with the fact that T has the property $(\alpha)_\pm^\sharp$. ∎

Theorem 6.5.11. *Let T be an operator in $\mathcal{B}(\mathcal{K}_1, \mathcal{K}_2)$. The following statements are equivalent.*

(i) *The operator T has unique elementary rotation, modulo unitary equivalence.*
(ii) *The operator T has either the property $(\alpha)_-$ or the property $(\alpha)_+$.*
(ii)$^\sharp$ *The operator T has either the property $(\alpha)_-^\sharp$ or the property $(\alpha)_+^\sharp$.*
(iii) *There exists an (equivalently, for all) induced Kreĭn space $(\mathcal{D}, D)$ of $I - T^\sharp T$ such that $\operatorname{Ran}(D)$ contains either a maximal uniformly positive subspace or a maximal uniformly negative subspace.*
(iii)$^\sharp$ *There exists an (equivalently, for all) induced Kreĭn space $(\mathcal{D}_\sharp, D_\sharp)$ of $I - TT^\sharp$ such that $\operatorname{Ran}(D_\sharp)$ contains either a maximal uniformly positive subspace or a maximal uniformly negative subspace.*

Proof. (i)$\Rightarrow$(ii). Assume that the operator T has neither the property $(\alpha)_-$ nor the property $(\alpha)_+$. By Theorem 6.1.9 there exists an isometry $V \colon \operatorname{Ran}(D_T)(\subseteq \mathcal{D}_T) \to \mathcal{D}_T$ with dense range, unbounded, and such that the operator $B = VD_T \in \mathcal{B}(\mathcal{K}_1, \mathcal{D}_T)$. Since V is isometric it follows that

$$J_1 - T^* J_2 T = B^* J_T B,$$

and hence, the operator $T_c = [T \ B]^t \in \mathcal{B}(\mathcal{K}_1, \mathcal{K}_2[+]\mathcal{D}_T)$ is isometric, in particular $\operatorname{Ran}(T_c)$ is a regular subspace of $\mathcal{K}_2[+]\mathcal{D}_T$. Let $\mathcal{K}_1'$ denote the orthogonal complement of $\operatorname{Ran}(T_c)$, that is,

$$\mathcal{K}_2[+]\mathcal{D}_T = \operatorname{Ran}(T)[+]\mathcal{K}_1'.$$

If R denotes the inclusion $\mathcal{K}_1' \hookrightarrow \mathcal{K}_2[+]\mathcal{D}_T$ then

$$\begin{bmatrix} J_2 & 0 \\ 0 & J_T \end{bmatrix} - T_c J_1 T_c^* = RJ_1'R^*, \tag{5.22}$$

where J_1' is a fixed fundamental symmetry on $\mathcal{K}_1'$. Define $U \in \mathcal{B}(\mathcal{K}_1[+]\mathcal{K}_1', \mathcal{K}_2[+]\mathcal{D}_T)$ by $U = [T_c \ R]$ and we claim that $(U; \mathcal{K}_1', \mathcal{D}_T)$ is an elementary rotation of T. Indeed, U is clearly a lifting of T and it is isometric since

$$R^* \begin{bmatrix} J_2 & 0 \\ 0 & J_T \end{bmatrix} R = J_1', \quad R^* \begin{bmatrix} J_2 & 0 \\ 0 & J_T \end{bmatrix} T_c = 0.$$

Using (5.22) it follows that $U^\sharp$ is also isometric. The analogue of the minimality condition holds since B has dense range and U is unitary.

It remains to remark that the elementary rotations $(U; \mathcal{K}_1', \mathcal{D}_T)$ and $(R(T); \mathcal{D}_{T^\sharp}, \mathcal{D}_T)$ are not unitary equivalent since the operator V is unbounded.

(ii)$\Leftrightarrow$(ii)$^\sharp$. This was proven in Lemma 6.5.10.
(ii)$\Leftrightarrow$(iii). This is a consequence of Theorem 6.1.9.
(iii)$\Leftrightarrow$(iii)$^\sharp$. This equivalence is a consequence of the previous two equivalences.

6.5 Elementary Rotations

(ii)⇒(i). Assume that T has either the property $(\alpha)_-$ or the property $(\alpha)_+$ and let $(U; \mathcal{K}_1', \mathcal{K}_2')$ be an elementary rotation of T. Let U be represented as in (5.1). Fix fundamental symmetries J_1, J_2, J_1', and J_2' on $\mathcal{K}_1, \mathcal{K}_2, \mathcal{K}_1'$, and $\mathcal{K}_2'$, respectively. We consider the canonical elementary rotation $R(T)$ and we will prove that U is unitary equivalent with $R(T)$.

To this end, notice first that by (5.2) and Theorem 6.1.9 we have

$$Y = \Phi_2 D_T, \qquad (5.23)$$

where $\Phi_2 \colon \mathcal{D}_T \to \mathcal{K}_1'$ is a bounded unitary operator. Similarly we have

$$X = D_{T^\sharp} \Phi_1, \qquad (5.24)$$

for a bounded unitary operator $\Phi_1 \colon \mathcal{K}_1' \to \mathcal{D}_{T^\sharp}$. Using the first equation in (5.3) and the representations of X and Y as in (5.23) and (5.24), respectively, we obtain

$$T^* J_2 D_{T^\sharp} \Phi_1 = -D_T J_T \Phi_2^* Z.$$

In view of the uniqueness property of the link operator $L_{T^\sharp}$ this implies

$$Z = -\Phi_2 L_{T^\sharp} J_{T^\sharp} \Phi_1. \qquad (5.25)$$

From equations (5.23), (5.24), and (5.25) we conclude that U and $R(T)$ are unitary equivalent. ∎

Given $T \in \mathcal{B}(\mathcal{K}_1, \mathcal{K}_2)$, an operator $W \in \mathcal{B}(\mathcal{K}_1[+]\mathcal{K}_1', \mathcal{K}_2[+]\mathcal{K}_2')$ having the following operator block matrix representation

$$W = \begin{bmatrix} T & X \\ Y & Z \end{bmatrix} : \begin{matrix} \mathcal{K}_1 \\ [+] \\ \mathcal{K}_1' \end{matrix} \longrightarrow \begin{matrix} \mathcal{K}_2 \\ [+] \\ \mathcal{K}_2' \end{matrix},$$

is called a *node* or *colligation*. The $\mathcal{B}(\mathcal{K}_1', \mathcal{K}_2')$-valued function

$$\Theta_W(\lambda) = Z + \lambda Y(I - \lambda T)^{-1} X, \quad \lambda \in \rho(T)^{-1} \cup \{0\} \qquad (5.26)$$

is called a *characteristic function*, and note that it is always holomorphic in a neighbourhood of 0.

Proposition 6.5.12. *With notation as in (5.26), if W is isometric then, whenever $\overline{\lambda}, \mu \in \rho(T)^{-1} \cup \{0\}$, we have*

$$I - \Theta_W(\lambda)^\sharp \Theta_W(\mu) = (1 - \overline{\lambda}\mu) X^\sharp (I - \overline{\lambda}T^\sharp)^{-1}(I - \mu T)^{-1} X.$$

Proof. The operator W is isometric if and only if the following identities hold:

$$Y^\sharp Y = I - T^\sharp T, \quad T^\sharp X = -Y^\sharp Z, \quad I - Z^\sharp Z = X^\sharp X. \qquad (5.27)$$

From (5.26) we have

$$I - \Theta_W(\lambda)^\sharp \Theta_W(\mu) = I - Z^\sharp Z$$
$$+ \overline{\lambda} X^\sharp (I - \overline{\lambda} T^\sharp)^{-1} Y^\sharp Z + \mu Z^\sharp Y (I - \mu T)^{-1} X$$
$$- \overline{\lambda}\mu X^\sharp (I - \overline{\lambda} T^\sharp)^{-1} Y^\sharp Y (I - \mu T)^{-1} X.$$

Taking into account of (5.27) this shows that

$$I - \Theta_W(\lambda)^\sharp \Theta_W(\mu) = X^\sharp A(\lambda,\mu) X, \qquad (5.28)$$

where

$$A(\lambda,\mu) = I - \overline{\lambda}(I - \overline{\lambda} T^\sharp)^{-1} T^\sharp - \mu T (I - \mu T)^{-1}$$
$$- \overline{\lambda}\mu (I - \overline{\lambda} T^\sharp)^{-1}(I - T^\sharp T)(I - \mu T)^{-1}.$$

A straightforward calculation shows that

$$A(\lambda,\mu) = (1 - \overline{\lambda}\mu)(I - \overline{\lambda} T^\sharp)^{-1}(I - \mu T)^{-1}. \qquad (5.29)$$

From (5.28) and (5.29) we get the desired formula. ∎

Given Kreĭn spaces $\mathcal{K}_1$, $\mathcal{K}_2$ and a $\mathcal{B}(\mathcal{K}_1,\mathcal{K}_2)$-valued function Θ defined on some set $\Lambda \subseteq \mathbb{C}$, the *Schur kernel* S_Θ associated to Θ is defined by

$$S_\Theta(\lambda,\mu) = \frac{I - \Theta(\lambda)^\sharp \Theta(\mu)}{1 - \overline{\lambda}\mu}, \quad \lambda,\mu \in \Lambda,\ \overline{\lambda}\mu \neq 1. \qquad (5.30)$$

In this respect, Proposition 6.5.12 says that elementary rotations are connected to Schur kernels. For example, letting $R(T)$ be the elementary rotation as in (5.19), then the characteristic function Θ_T

$$\Theta_T(\lambda) = -J_T L_{T^\sharp} + \lambda D_T (I - \lambda T)^{-1} D_{T^\sharp}$$

has its Schur kernel

$$S_{\Theta_T}(\lambda,\mu) = D_{T^\sharp}^\sharp (I - \overline{\lambda} T^\sharp)^{-1}(I - \mu T)^{-1} D_{T^\sharp}.$$

6.6 Isometric and Unitary Dilations

As a consequence of the existence of elementary rotations we can provide a more constructive proof for the existence of unitary dilations, compare with Corollary 6.4.3. More precisely, we prove that to any elementary rotation $(U;\mathcal{K}_1';\mathcal{K}_2')$ of a given operator $T \in \mathcal{B}(\mathcal{H})$, for some Kreĭn space $(\mathcal{H};[\cdot,\cdot]_\mathcal{H})$, we can associate in a canonical way a minimal isometric dilation and a minimal unitary dilation.

6.6 Isometric and Unitary Dilations

A pair $(V;\mathcal{K})$ is called an *isometric dilation* of T if $\mathcal{K}$ is a Kreĭn space containing $\mathcal{H}$ as a regular subspace and $V \in \mathcal{B}(\mathcal{K})$ is an isometry such that

$$T^n = P_{\mathcal{H}} V^n | \mathcal{H}, \quad n = 1, 2, 3, \ldots.$$

An isometric dilation $(V;\mathcal{K})$ is called *minimal* if $\mathcal{K} = \bigvee_{n=0}^{\infty} V^n \mathcal{H}$.

Let $D \in \mathcal{B}(\mathcal{H}, \mathcal{D})$ be a defect operator for T and define the Kreĭn space $\ell_{\mathbb{N}}^2(\mathcal{D})$ as follows. Fix J a fundamental symmetry on $\mathcal{H}$ and let $\|\cdot\|$ denote its associated fundamental norm. Let

$$\ell_{\mathbb{N}}^2(\mathcal{D}) = \{(x_n)_{n=1}^{\infty} \mid x_n \in \mathcal{D} \text{ for all } n \geq 1 \text{ and } \sum_{n=1}^{\infty} \|x_n\|^2 < \infty\}. \tag{6.1}$$

Then $\ell_{\mathbb{N}}^2(\mathcal{D})$ is a Hilbert space with inner product

$$\langle x, y \rangle_{\ell^2} = \sum_{n=1}^{\infty} \langle x_n, y_n \rangle_J, \quad x = (x_n)_{n=1}^{\infty}, y = (y_n)_{n=1}^{\infty} \in \ell_{\mathbb{N}}^2(\mathcal{D}). \tag{6.2}$$

On $\ell_{\mathbb{N}}^2(\mathcal{D})$ define the inner product

$$[x, y]_{\ell^2} = \sum_{n=1}^{\infty} [x_n, y_n]_{\mathcal{D}}, \quad x = (x_n)_{n=1}^{\infty}, y = (y_n)_{n=1}^{\infty} \in \ell_{\mathbb{N}}^2(\mathcal{D}), \tag{6.3}$$

with the observation that the series converges with respect to the norm $\|\cdot\|_{\ell^2}$ associated to the inner product $\langle \cdot, \cdot \rangle_{\ell^2}$ as in (6.2).

We leave to the reader to verify the following statement.

Lemma 6.6.1. *With notation as before, $\ell_{\mathbb{N}}^2(\mathcal{D})$ is a Kreĭn space with inner product defined as in (6.3) and in such a way that the norm $\|\cdot\|_{\ell^2}$ is a fundamental norm.*

Let the Kreĭn space $\mathcal{K}$ be defined by

$$\mathcal{K} = \mathcal{H}[+]\ell_{\mathbb{N}}^2(\mathcal{D}), \tag{6.4}$$

and the operator $V \in \mathcal{B}(\mathcal{K})$ be defined by the following operator block matrix

$$V = \begin{bmatrix} T & 0 & 0 & 0 & \cdots \\ D & 0 & 0 & 0 & \cdots \\ 0 & I & 0 & 0 & \cdots \\ 0 & 0 & I & 0 & \cdots \\ \vdots & & & & \end{bmatrix}, \tag{6.5}$$

more precisely, for any $h \in \mathcal{H}$ and $x = (x_n)_{n=1}^{\infty} \in \ell_{\mathbb{N}}^2(\mathcal{D})$,

$$V(h, x) = (Th, Dh, x_1, x_2, x_3, \ldots).$$

We leave to the reader to verify the following statement.

Lemma 6.6.2. *With notation as in* (6.4) *and* (6.5), $(V;\mathcal{K})$ *is a minimal isometric dilation of* T.

Recall that, a pair $(U;\mathcal{K})$ is a unitary dilation of T if $\mathcal{K}$ is a Kreĭn space containing $\mathcal{H}$ as a regular subspace, $U \in \mathcal{B}(\mathcal{K})$ is a unitary operator and

$$T^n = P_{\mathcal{H}} U^n |\mathcal{H}, \quad T^{\sharp n} = P_{\mathcal{H}} U^{\sharp n} |\mathcal{H}, \quad n = 1, 2, 3, \ldots, \tag{6.6}$$

and it is minimal if $\bigvee_{n \in \mathbb{Z}} U^n \mathcal{H} = \mathcal{K}$.

By Theorem 6.5.3, we can consider elementary rotations of T. Having in mind Remark 6.5.1(a), letting $(R; \mathcal{D}; \mathcal{D}_\sharp)$ be an elementary rotation of T, then the unitary operator R has the operator block matrix representation

$$R = \begin{bmatrix} T & D_\sharp \\ D & L \end{bmatrix}, \tag{6.7}$$

where $D \in \mathcal{B}(\mathcal{H}, \mathcal{D})$ is a defect operator of T, $D_\sharp^\sharp \in \mathcal{B}(\mathcal{H}, \mathcal{D}_\sharp)$ is a defect operator of $T^\sharp$, and $L \in \mathcal{B}(\mathcal{D}_\sharp, \mathcal{D})$.

With notation as in (6.1)–(6.3), define the Kreĭn space

$$\mathcal{K} = \ell^2_{-\mathbb{N}}(\mathcal{D}_\sharp)[+]\mathcal{H}[+]\ell^2_{\mathbb{N}}(\mathcal{D}), \tag{6.8}$$

and the operator $U \in \mathcal{B}(\mathcal{K})$ by the operator block matrix

$$U = \begin{bmatrix} & & \vdots & & & & & \\ \cdots & I & 0 & 0 & 0 & 0 & 0 & \cdots \\ \cdots & 0 & I & 0 & 0 & 0 & 0 & \cdots \\ \cdots & 0 & 0 & D_\sharp & T & 0 & 0 & 0 & \cdots \\ \cdots & 0 & 0 & L & D & 0 & 0 & 0 & \cdots \\ \cdots & 0 & 0 & 0 & 0 & I & 0 & \cdots \\ \cdots & 0 & 0 & 0 & 0 & 0 & I & 0 & \cdots \\ & & & \vdots & & & & \end{bmatrix}, \tag{6.9}$$

where the entry T is on the $(0,0)$ position, when viewing $\mathcal{K}$ as in (6.8) with entries indexed on $\mathbb{Z}$. More precisely, letting an arbitrary vector $x \in \mathcal{K}$ be represented, in accordance with (6.8), as

$$x = (\ldots, x_{-3}, x_{-2}, x_{-1}, h, x_1, x_2, x_3, \ldots)$$

with $x_{-n} \in \mathcal{D}_\sharp$ and $x_n \in \mathcal{D}$, for all $n \geq 1$, and $h \in \mathcal{H}$ viewed on the 0 position, then

$$Ux = (\ldots, x_{-3}, x_{-2}, D_\sharp x_{-1} + Th, Lx_{-1} + Dh, x_1, x_2, x_3, \ldots)$$

when viewing $D_\sharp x_{-1} + Th$ on the 0 position.

Theorem 6.6.3. *With notation as in* (6.8) *and* (6.9), *the pair* $(U;\mathcal{K})$ *is a minimal unitary dilation of* T.

In view of the properties of the elementary rotation R with representation as in (6.7), the proof is straightforward and we omit it.

6.7 Notes

The idea of a Kreĭn space induced either by inner product spaces or by selfadjoint operators is old and can be seen implicitly in many articles. The formalisation in Section 6.1 was firstly used in our joint article with T. Constantinescu [32], see also [30] and [31]. Theorem 6.1.9 on uniqueness was essentially proven, in this setting, in [32], but related theorems have been obtained by T. Hara [72], M. A. Dritschel [50], T. Constantinescu and the author [31], and B. Ćurgus and H. Langer [41]. Theorem 6.1.10 has been proven in an equivalent setting by A. Dijksma, H. Langer, and H. S. V. de Snoo in [44], but the idea of the proof goes back to M. G. Kreĭn [91], W. T. Reid [124], P. D. Lax [108], and J. Dieudonnée [42].

The realisation of operator valued holomorphic functions with internal symmetries by generalised resolvents of Hilbert space operators is a classical topic related to early investigations of C. Carathéodory [26], I. Schur [133], and R. Nevanlinna [111]. The generalisation to indefinite inner product spaces is mostly due to M. G. Kreĭn and H. Langer [95, 96, 97, 99]. Theorem 6.2.1 and Theorem 6.4.1 in this general formulation were obtained by A. Dijksma, H. Langer, and H. S. V. de Snoo in [45] and the connections with selfadjoint operator pencils and Corollary 6.4.3 on existence of unitary dilations are from that article as well.

Elementary rotations, also called Julia operators, are a concept coming from dilation theory, for example, C. Foiaş and B. Sz.-Nagy [141]. For the indefinite setting, the canonical form is from Gr. Arsene, T. Constantinescu and the author [7] while the more general approach with defect operators is from M. A. Dritschel [48, 49] and M. A. Dritschel and J. Rovnyak [52]. Theorem 6.5.11 on uniqueness of elementary rotation follows [31], but related or equivalent forms were obtained in [72] and [50] as well, or more detailed versions as in [41]. For the explicit descriptions of isometric and unitary dilations we followed our joint article with T. Constantinescu [29], M. A. Dritschel [48], and M. A. Dritschel and J. Rovnyak [52, 53].

Chapter 7

Plus/Minus-Operators

Plus-operators and minus-operators have no counterparts in the operator theory on Hilbert spaces. Under very general hypothesis one can obtain interesting results on the bounds associated to plus/minus-operators. At the general level one can view these objects also as spaces with two inner products. Due to the interplay of these two classes, most of the results need a proof only for one of them, and then the corresponding result for the other class can be easily obtained by some changes of sign.

Strong minus-operators are collinear with contractions, a class that is of special interest and will be studied from different perspectives during the following chapters. Of course, the most relevant cases correspond to the Kreĭn space setting where of particular interest are the uniform minus-operators. To make things even more complicated, in a genuine Kreĭn space setting the class of minus-operators is not closed under the involution. This makes it natural to introduce double minus-operators. We see that uniform minus-operators possess extensions to doubly uniform minus-operators.

7.1 Spaces with Two Inner Products

We fix an inner product space $(\mathcal{X}, [\cdot, \cdot])$ which is supposed to be *indefinite* throughout this section. To make this assumption clearer, let us recall that this means that there exist vectors $x, y \in \mathcal{X}$ such that $[x, x] > 0$ and $[y, y] < 0$. In particular, $\dim(\mathcal{X}) \geqslant 2$. As we shall see, this assumption is essential for the correctness of most of the results. Also, recall (e.g. see Lemma 1.1.1) that this assumption implies that $\mathcal{X}$ always contains nontrivial neutral vectors.

Proposition 7.1.1. *If $\omega(\cdot, \cdot)$ is a second inner product on $\mathcal{X}$ such that*

$$\text{whenever } x \in \mathcal{X} \text{ is such that } [x, x] = 0 \text{ we have } \omega(x, x) \leqslant 0, \tag{1.1}$$

then,

(i) $\sup\limits_{[x,x]>0} \frac{\omega(x,x)}{[x,x]} =: \nu_\omega < +\infty,$

(ii) $\omega(x, x) \leqslant \nu_\omega [x, x]$ *for all* $x \in \mathcal{X}.$

Conversely, if for some real number ν condition (ii) *holds, then ω has the property* (1.1).

7.1 Spaces with Two Inner Products

Proof. We first prove that for any vectors $x, y \in \mathcal{X}$ such that $[x, x] < 0$ and $[y, y] > 0$, the following inequality holds

$$\frac{\omega(x, x)}{[x, x]} \geqslant \frac{\omega(y, y)}{[y, y]}. \tag{1.2}$$

Indeed, if (1.2) is false, then we can find two vectors $x, y \in \mathcal{X}$ such that $[x, x] = -1$, $[y, y] = 1$ and $-\omega(x, x) < \omega(y, y)$. Taking $z = x + \lambda y$ with $|\lambda| = 1$, we obtain $[z, z] = 2\operatorname{Re}\overline{\lambda}[x, y]$ and $\omega(z, z) > 2\operatorname{Re}\overline{\lambda}\omega(x, y)$. Since there exist at least two solutions λ such that $[z, z] = 0$, and for at least one of these solutions we have $\omega(z, z) > 0$, we get a contradiction with (1.1).

The indefiniteness of $(\mathcal{X}, [\cdot, \cdot])$ and the inequality (1.2) proves (i). As for the statement (ii), if x is positive this follows from the definition of ν_ω, if x is neutral it follows from (1.1), and if x is negative it follows from (1.2). ∎

Simply by replacing ω with $-\omega$, from Proposition 7.1.1 we get the following result.

Corollary 7.1.2. *If $\omega(\cdot, \cdot)$ is a second inner product on $\mathcal{X}$ such that*

$$\text{whenever } x \in \mathcal{X} \text{ is such that } [x, x] = 0 \text{ we have } \omega(x, x) \geqslant 0, \quad x \in \mathcal{X}, \tag{1.3}$$

then,

(i) $\inf\limits_{[x,x]>0} \frac{\omega(x,x)}{[x,x]} = \mu_\omega > -\infty$,

(ii) $\omega(x, x) \geqslant \mu_\omega [x, x], \quad x \in \mathcal{X}$.

Conversely, if the condition (ii) holds for some real μ, then ω has the property (1.3).

Corollary 7.1.3. *Assume that the second inner product ω on $\mathcal{X}$ satisfies the following property*

$$[x, x] = 0 \text{ implies } \omega(x, x) = 0, \quad x \in \mathcal{X}. \tag{1.4}$$

Then there exists $\mu \in \mathbb{R}$ such that

$$\omega(x, y) = \mu[x, y], \quad x \in \mathcal{X}. \tag{1.5}$$

Proof. By Proposition 7.1.1 and Corollary 7.1.2 there exist real numbers ν and μ such that

$$\mu[x, x] \leqslant \omega(x, x) \leqslant \nu[x, x], \quad x \in \mathcal{X}.$$

Since $(\mathcal{X}, [\cdot, \cdot])$ is indefinite, we obtain $\mu = \nu$ and then (1.5) follows from the polarisation formula (1.4) in Chapter 1. ∎

Concerning the hypothesis on the indefiniteness of $(\mathcal{X}, [\cdot, \cdot])$ in Proposition 7.1.1, it is useful to record the following fact.

Proposition 7.1.4. *Assume that $\kappa_-(\mathcal{X}) > 0$ ($\kappa_+(\mathcal{X}) > 0$) and let a second inner product ω be defined on $\mathcal{X}$. If $[y, y] < 0$ implies $\omega(y, y) \leqslant 0$ (respectively, $[y, y] > 0$ implies $\omega(y, y) \leqslant 0$) then $[x, x] = 0$ implies $\omega(x, x) \leqslant 0$.*

Proof. Suppose that there exists $x \in \mathcal{X}$ such that $[x,x] = 0$ and $\omega(x,x) = 1$. If $\kappa_-(\mathcal{X}) > 0$ then there exists $y \in \mathcal{X}$ such that $[y,y] = -1$ and $\operatorname{Re}[x,y] \geq 0$. Set $y_t = x + ty$, $t \in \mathbb{R}$. Then
$$[y_t, y_t] = 2t\operatorname{Re}[x,y] - t^2,$$
$$\omega(y_t, y_t) = 1 + 2t\operatorname{Re}\omega(x,y) + t^2,$$
hence, for $t < 0$ and of modulus sufficiently small, we get $[y_t, y_t] < 0$ and $\omega(y_t, y_t) > 0$, a contradiction. ∎

Remark 7.1.5. Assume that the second inner product ω on $\mathcal{X}$ has the following property
$$\text{whenever } x \in \mathcal{X} \setminus \{0\} \text{ is such that } [x,x] = 0, \text{ we have } \omega(x,x) \neq 0. \tag{1.6}$$
Then one and only one of the following statements holds:

(i) $[x,x] = 0$, $x \neq 0$ implies $\omega(x,x) > 0$, $x \in \mathcal{X}$;
(ii) $[x,x] = 0$, $x \neq 0$ implies $\omega(x,x) < 0$, $x \in \mathcal{X}$.

If instead of (1.6) we assume that ω has the property
$$\text{whenever } x \in \mathcal{X} \setminus \{0\} \text{ is such that } [x,x] = 0, \text{ we have } \omega(x,x) \geq 0, \tag{1.7}$$
then one and only one of the following statements holds:

(iii) $[x,x] < 0$ implies $\omega(x,x) \geq 0$;
(iv) $[x,x] > 0$ implies $\omega(x,x) \geq 0$. ∎

Remark 7.1.6. Assume that ω has the property (1.6). Then for any vectors $x, y \in \mathcal{X}$ with $[x,x] < 0$ and $[y,y] > 0$, we have
$$\frac{\omega(x,x)}{[x,x]} < \frac{\omega(y,y)}{[y,y]}.$$
In particular, one of the following statements holds:

(v) $[x,x] < 0$ implies $\omega(x,x) > 0$;
(vi) $[x,x] > 0$ implies $\omega(x,x) > 0$. ∎

7.2 Minus-Operators

Let $\mathcal{X}$ and $\mathcal{X}'$ be indefinite inner product spaces and let $T\colon \mathcal{X} \to \mathcal{X}'$ be a linear operator. Associated to the operator T there is the set
$$M(T) = \{\mu \geq 0 \mid \mu[x,x] - [Tx,Tx] \geq 0, \text{ for all } x \in \mathcal{X}\}. \tag{2.1}$$

7.2 MINUS-OPERATORS

It can be easily proven that if the set $M(T)$ is not empty then it is convex and closed in $[0,+\infty)$. In this case we define the bounds $\|T\|_\pm$

$$\|T\|_+^2 = \inf M(T), \quad \|T\|_-^2 = \sup M(T). \tag{2.2}$$

Two other bounds can be associated to a general operator $T\colon \mathcal{X} \to \mathcal{X}'$,

$$\mu_+(T) = \sup_{[x,x]=1} [Tx,Tx], \quad \mu_-(T) = \inf_{[x,x]=-1} -[Tx,Tx]. \tag{2.3}$$

These definitions make sense in the case that $\mathcal{X}$ is indefinite only; we let $\mu_+(T) = -\infty$ if $\mathcal{X}$ is negative semidefinite and, respectively, $\mu_-(T) = +\infty$ if $\mathcal{X}$ is positive semidefinite. T is called a *minus-operator* if there exists $\mu \geqslant 0$ such that

$$[Tx, Tx] \leqslant \mu[x,x], \quad x \in \mathcal{X}, \tag{2.4}$$

that is, the set $M(T)$ is nonvoid. Similarly, T is called a *plus-operator* if there exists $\mu \geqslant 0$ such that

$$[Tx, Tx] \geqslant \mu[x,x], \quad x \in \mathcal{X}.$$

By changing the inner product $[\cdot,\cdot]$ into $-[\cdot,\cdot]$ on $\mathcal{X}$ and on $\mathcal{X}'$, respectively, minus-operators can be transformed into plus-operators and vice versa. Thus, each result about minus-operators has a precise transcription for plus-operators and hence, we will state the results mainly for minus-operators.

If both $\mathcal{X}$ and $\mathcal{X}'$ are positive definite then $\mu_+(T) = \|T\| \leqslant +\infty$ and $\mu_-(T) = +\infty$. If both $\mathcal{X}$ and $\mathcal{X}'$ are negative definite then $\mu_+(T) = -\infty$ and $\mu_-(T) = \gamma(T)$ (here $\gamma(T)$ denotes the *minimum modulus* of the operator T, with respect to the seminorms associated to $-[\cdot,\cdot]$).

A distinct situation corresponds to the case when the inner product space $(\mathcal{X},[\cdot,\cdot])$ is indefinite, that is, it contains positive vectors as well as negative vectors. In this case we can use the results in the previous section.

Proposition 7.2.1. *Let $T\colon \mathcal{X} \to \mathcal{X}'$ be a linear operator and assume that $\mathcal{X}$ is indefinite. Then T is a minus-operator if and only if $[Tx,Tx] \leqslant 0$ for all $x \in \mathcal{X}$ with $[x,x] \leqslant 0$.*

Proof. If T is a minus-operator then (2.4) holds for some $\mu \geqslant 0$ and hence, if $[x,x] \leqslant 0$ then $[Tx,Tx] \leqslant 0$.

For the converse implication, let us assume that $[Tx, Tx] \leqslant 0$ for all $x \in \mathcal{X}$ with $[x,x] \leqslant 0$. We consider the inner product $\omega_T(x,y) = [Tx,Ty]$, $x,y \in \mathcal{X}$, and observe that the assumptions of Proposition 7.1.1 are fulfilled for ω_T hence

$$\sup_{[x,x]>0} \frac{[Tx,Tx]}{[x,x]} = \nu_T < +\infty,$$

and $[Tx,Tx] \leqslant \nu_T[x,x]$ for all $x \in \mathcal{X}$. We distinguish two cases. If $\nu_T \geqslant 0$ we let $\mu = \nu_T$. If $\nu_T < 0$ then $[Tx,Tx] < 0$ whenever $[x,x] > 0$. Hence, in view of the hypothesis, we have $[Tx,Tx] \leqslant 0$ for all $x \in \mathcal{X}$, and we let $\mu = 0$. ∎

Another characterisation of minus-operators can be obtained by using the bounds $\mu_+(T)$ and $\mu_-(T)$ defined as in (2.3).

Proposition 7.2.2. (a) *If T is a minus-operator then* $\max\{0, \mu_+(T)\} \leqslant \mu_-(T)$.

(b) *Assume that $\mathcal{X}$ is nondegenerate and decomposable. Then $T\colon \mathcal{X} \to \mathcal{X}'$ is a minus-operator if and only if* $\max\{0, \mu_+(T)\} \leqslant \mu_-(T)$. *In this case, if $\mathcal{X}$ is not positive definite we have $\mu_+(T) \geqslant 0$ and a nonnegative number μ satisfies (2.4) if and only if* $\max\{0, \mu_+(T)\} \leqslant \mu \leqslant \mu_-(T)$. *Moreover, with notation as in (2.2), we have*

$$\|T\|_+^2 = \max\{0, \mu_+(T)\}, \quad \|T\|_-^2 = \mu_-(T).$$

Proof. (a) If $\mu \geqslant 0$ satisfies (2.4), then clearly we have $\max\{0, \mu_+(T)\} \leqslant \mu \leqslant \mu_-(T)$, in particular, $\max\{0, \mu_+(T)\} \leqslant \mu_-(T)$.

(b) Assume now that $\mathcal{X}$ is nondegenerate and decomposable. Let us assume that $\max\{0, \mu_+(T)\} \leqslant \mu_-(T)$ and consider a number $\mu \geqslant 0$ such that $\max\{0, \mu_+(T)\} \leqslant \mu \leqslant \mu_-(T)$. We first assume that $\mathcal{X}$ is indefinite and let $x, y \in \mathcal{X}$ be such that $[x, x] > 0$ and $[y, y] < 0$. On the grounds of (2.3) we have

$$\frac{[Tx, Tx]}{[x, x]} \leqslant \mu \leqslant \frac{[Ty, Ty]}{[y, y]}.$$

From here it follows that (2.4) holds for all definite vectors $x \in \mathcal{X}$. Assume now that $z \in \mathcal{X}$ is a neutral vector in $\mathcal{X}$. Represent $z = z_+ + z_-$ with z_+ strictly positive and z_- strictly negative, see Lemma 1.1.1, and let $\varepsilon > 0$ be arbitrary. Then $z_\varepsilon = z_+ + (1+\varepsilon)z_-$ is a negative vector and hence $[Tz_\varepsilon, Tz_\varepsilon] \leqslant \mu[z_\varepsilon, z_\varepsilon] \leqslant 0$. But $Tz_\varepsilon = Tz_+ + \varepsilon Tz_-$ and hence

$$0 \geqslant [Tz_\varepsilon, Tz_\varepsilon] = [Tz, Tz] + \varepsilon \operatorname{Re}[Tz, Tz_-] + \varepsilon^2 [Tz_-, Tz_-].$$

Letting ε tend down to zero it follows that $[Tz, Tz] \leqslant 0$. Thus (2.4) holds for all $x \in \mathcal{X}$.

The remaining part of the proof is straightforward. ∎

Proposition 7.2.3. *Let $T\colon \mathcal{X}_1 \to \mathcal{X}_2$ and $S\colon \mathcal{X}_2 \to \mathcal{X}_3$ be minus-operators. Then $ST\colon \mathcal{X}_1 \to \mathcal{X}_3$ is a minus-operator and*

$$\|ST\|_+ \leqslant \|S\|_+ \cdot \|T\|_+, \quad \|ST\|_- \geqslant \|S\|_- \cdot \|T\|_-.$$

Proof. Let μ and ν be nonnegative numbers such that $\mu[x, x] - [Tx, Tx] \geqslant 0$ and $\nu[y, y] - [Sy, Sy] \geqslant 0$ for all $x \in \mathcal{X}_1$ and $y \in \mathcal{X}_2$. Then, for an arbitrary vector $x \in \mathcal{X}$, we have

$$\mu\nu[x, x] - [STx, STx] = \nu(\mu[x, x] - [Tx, Tx]) + (\nu[Tx, Tx] - [STx, STx]) \geqslant 0.$$

The remaining statements follow from this inequality and the definition at (2.2). ∎

The operator $T\colon \mathcal{X} \to \mathcal{X}'$ is called a *strong minus-operator* if T is a minus-operator and $\mu_+(T) > 0$. By Proposition 7.2.2 it follows that if T is a strong minus-operator then, for all $x \in \mathcal{X}$, we have $[Tx, Tx] \leqslant \mu_+(T)[x, x]$.

7.2 MINUS-OPERATORS

Remark 7.2.4. (a) If T is a strong minus-operator and $x \in \mathcal{X}$, then $[Tx, Tx] < 0$ whenever $[x, x] < 0$.
(b) If T is a minus-operator and $\mu_+(T) = 0$ then $\mathrm{Ran}(T)$ is a negative linear manifold.
(c) If T is a minus-operator such that, for any $x \in \mathcal{X}$ with $[x, x] < 0$ we have $[Tx, Tx] = 0$, then $\mu_+(T) = 0$. ∎

As a consequence of Corollary 7.1.3 we have the following result.

Lemma 7.2.5. *Assume that the operator* $T: \mathcal{X} \to \mathcal{X}'$ *satisfies at least one of the following conditions:*

(i) $[x, x] = 0$ *implies* $[Tx, Tx] = 0$;
(ii) *if* $[x, x] \geqslant 0$ *then* $[Tx, Tx] = 0$ *and, if* $[x, x] \leqslant 0$ *then* $[Tx, Tx] \leqslant 0$.

Then $[Tx, Ty] = \mu_+(T)[x, y]$, *for all* $x, y \in \mathcal{X}$.

Corollary 7.2.6. *Under the assumptions of Lemma 7.2.5 the following alternative holds: either* $\mu_+(T) = 0$ *and* $\mathrm{Ran}(T)$ *is neutral, or* $\mu_+(T) > 0$ *and* $\mu_+(T)^{-\frac{1}{2}}T$ *is isometric.*

We also have the following result.

Proposition 7.2.7. *Let* $T: \mathcal{X} \to \mathcal{X}'$ *be a linear operator and assume that* $\mathcal{X}$ *is indefinite. The following assertions are equivalent.*

(i) T *maps bijectively* $\{x \in \mathcal{X} \mid [x, x] < 0\}$ *onto* $\{y \in \mathcal{X}' \mid [y, y] < 0\}$.
(ii) T *maps bijectively* $\{x \in \mathcal{X} \mid [x, x] \leqslant 0\}$ *onto* $\{y \in \mathcal{X}' \mid [y, y] \leqslant 0\}$.
(iii) T *is an invertible minus-operator,* $\mu_+(T) > 0$ *and* $\mu_+(T)^{-1/2}T$ *is isometric.*
(i)' T *maps bijectively* $\{x \in \mathcal{X} \mid [x, x] > 0\}$ *onto* $\{y \in \mathcal{X}' \mid [y, y] > 0\}$.
(ii)' T *maps bijectively* $\{x \in \mathcal{X} \mid [x, x] \geqslant 0\}$ *onto* $\{y \in \mathcal{X}' \mid [y, y] \geqslant 0\}$.
(iii)' T *is an invertible plus-operator,* $\mu_-(T) > 0$, *and* $\mu_-(T)^{-1/2}T$ *is isometric.*

Proof. (i)⇒(iii). First note that by Lemma 1.1.3 we have $T\mathcal{X} = \mathcal{X}'$. Then, if $x \in \mathrm{Ker}(T)$ we can write $x = x_1 + x_2$ with $[x_k, x_k] < 0$, $k = 1, 2$, and hence $Tx_1 = -Tx_2$. Since T is injective on the cone $\{z \in \mathcal{X} \mid [z, z] < 0\}$, it follows that $x_1 = x_2$ and $x = 0$. Thus T is invertible.

Let $x \in \mathcal{X}$ be neutral. We want to prove that $[Tx, Tx] \leqslant 0$. We claim that there exists $\varepsilon > 0$ and $u \in \mathcal{X}$ such that $[x + tu, x + tu] < 0$ for all $0 < t < \varepsilon$. Indeed, we first choose $u \in \mathcal{X}$ such that $[x, u] = -1$. Then, if $t > 0$ we have

$$[x + tu, x + tu] = -2t + t^2[u, u] = t(t[u, u] - 2).$$

If $[u, u] \leqslant 0$ we let ε be an arbitrary positive number. If $[u, u] > 0$ we choose $\varepsilon = 2/[u, u]$, and the claim is proven.

Further, for $0 < t < \varepsilon$ we have

$$0 < [T(x + tu), T(x + tu)] = [Tx, Tx] + 2t\,\mathrm{Re}[Tx, Tu] + t^2[Tu, Tu].$$

Since $T(x+tu) = Tx + tTu$, letting t tend down to 0 we get $[Tx, Tx] \leqslant 0$. By Proposition 7.1.1 this implies that T is a minus-operator.

Assume that $\mu_+(T) = 0$. Then $T\mathcal{X} = \mathcal{X}'$ is a negative linear manifold and hence the cone $\{x \in \mathcal{X} \mid [x,x] < 0\}$ is a linear manifold, which contradicts the indefiniteness of the space $\mathcal{X}$. Therefore, $\mu_+(T) > 0$. Since T is invertible it follows now that $[x,x] > 0$ implies $[Tx, Tx] > 0$. We apply Lemma 7.2.5 to conclude that $\mu_+(T)^{-1/2}T$ is isometric.

(ii)⇒(iii). As before, we prove that T is an invertible minus-operator and $\mu_+(T) > 0$. In particular, if $[x,x] < 0$ then $[Tx, Tx] < 0$.

We claim that if $[x,x] = 0$ then $[Tx, Tx] = 0$. Indeed, assume that $[Tx, Tx] \neq 0$. Then $[Tx, Tx] > 0$ because, otherwise, the assumption (ii) would be contradicted. But, using the same approximation argument as before, we conclude that $[Tx, Tx] \leqslant 0$. Thus, the only possibility is $[Tx, Tx] = 0$. By Corollary 7.1.3 we obtain that $\mu_+(T)^{-1/2}T$ is isometric.

The remainder of the proof is only a matter of logic. Clearly, (iii) implies (i) and (ii), hence all these three assertions are equivalent. In the same way, or using the parallelism via the anti-space, we have the equivalence of (i)', (ii)', and (iii)'. Finally, remark that (iii) and (iii)' are obviously equivalent. ∎

From now on we deal with minus-operators acting between Kreĭn spaces. If $\mathcal{K}_1$ and $\mathcal{K}_2$ are Kreĭn spaces and $T \in \mathcal{B}(\mathcal{K}_1, \mathcal{K}_2)$, let J_i be fundamental symmetries on $\mathcal{K}_i$, $i = 1, 2$. Then T is a minus-operator if and only if $\mu J_1 - T^* J_2 T \geqslant 0$ for some $\mu \geqslant 0$.

Assume now that $T \in \mathcal{B}(\mathcal{K}_1, \mathcal{K}_2)$ is a minus-operator. If, in addition, $T^\sharp$ is also a minus-operator, then T is called a *double minus-operator*. Let us note, from the following example, that in the case when the underlying Kreĭn space is indefinite there exist minus-operators which are not double minus-operators.

Example 7.2.8. Let the Kreĭn space $\mathcal{K} = \mathcal{K}^+[+]\mathcal{K}^-$ be such that $\dim(\mathcal{K}^-) = \infty$ and let J denote the corresponding fundamental symmetry. Let $V \in \mathcal{B}(\mathcal{K}^-)$ be isometric ($V^*V = I_-$) and nonunitary ($VV^* \neq I_-$), for example a unilateral shift. With respect to the above fundamental decomposition, define

$$T = \begin{bmatrix} I_+ & 0 \\ 0 & V \end{bmatrix}.$$

Then T is a strong minus-operator, since it is isometric and hence a contraction, but, for arbitrary $\mu \geqslant 0$, we have

$$\mu J - TJT^* = \begin{bmatrix} (\mu-1)I_+ & 0 \\ 0 & -\mu I_- + VV^* \end{bmatrix},$$

which is never positive. Therefore, T is not a double minus-operator. ∎

An operator $T \in \mathcal{B}(\mathcal{K}_1, \mathcal{K}_2)$ is called a *doubly strong minus-operator* if both T and $T^\sharp$ are strong minus-operators.

7.3 Uniform Minus-Operators

Theorem 7.2.9. *Let $T \in \mathcal{B}(\mathcal{K}_1, \mathcal{K}_2)$ be a doubly strong minus-operator. Then, with definitions as in (2.2), we have $\|T\|_\pm = \|T^\sharp\|_\pm$.*

Proof. Let $\mathcal{K}_i = \mathcal{K}_i^+[+]\mathcal{K}_i^-$, $i = 1, 2$, be fundamental decompositions that induce underlying Hilbert space structures. Let $\mathcal{K} = \mathcal{K}_1 \oplus \mathcal{K}_2$, first considered as a Hilbert space, and then letting $\mathcal{K}^+ = \mathcal{K}_1^+ \oplus \mathcal{K}_2^-$ and $\mathcal{K}^- = \mathcal{K}_1^- \oplus \mathcal{K}_2^+$, we take $\mathcal{K} = \mathcal{K}^+[+]\mathcal{K}^-$ as a fundamental decomposition. In particular, the indefinite inner product on $\mathcal{K}$ is

$$[x_1 \oplus y_1, x_2 \oplus y_2] = [x_1, x_2] - [y_1, y_2], \quad x_1, y_1 \in \mathcal{K}_1, \; x_2, y_2 \in \mathcal{K}_2.$$

In this way, we have completely specified the Kreĭn space structure on $\mathcal{K}$.

Without any loss of generality it is sufficient to prove that, if T is a contraction then $T^\sharp$ is a contraction; indeed, this follows easily by rescaling T adequately. If T is a contraction then its graph

$$\mathcal{G}(T) = \{x \oplus Tx \mid x \in \mathcal{K}_1\} \subseteq \mathcal{K}$$

is a positive subspace. Note that $\mathcal{G}(T^\sharp) = \mathcal{G}(T)^\perp$. In the following we prove that $\mathcal{G}(T)$ is a maximal positive subspace in $\mathcal{K}$. To this end, let $x \in \mathcal{K}_1 \oplus \mathcal{K}_2^-$ be such that $x \perp \mathcal{G}(T)$. Then $x = x_1^+ \oplus x_2^- \in \mathcal{G}(T^\sharp)$, that is, $x_1^+ = T^\sharp x_2^-$. Since $T^\sharp$ is a strong minus-operator it follows that there exists $\mu > 0$ such that $\mu[y, y] - [T^\sharp y, T^\sharp y] \geq 0$, in particular, $\mu[x_2^-, x_2^-] \geq [x_1^+, x_1^+]$. Then $x_1^+ = 0$, $x_2^- = 0$, and hence $x = 0$. This proves that $\mathcal{G}(T)$ is a maximal positive subspace and then $\mathcal{G}(T^\sharp) = \mathcal{G}(T)^\perp$ is a negative subspace, that is, $T^\sharp$ is a contraction. ∎

More information on strong minus-operators will be revealed in the next chapter which is dedicated to the geometry of contractive operators in Kreĭn spaces and their adjoints. Note that if T is a strong minus-operator then it is colinear with a contractive operator, more precisely $\mu_+(T)^{-1/2}T$ is a contractive operator.

7.3 Uniform Minus-Operators

Given a Kreĭn space $\mathcal{K}$, an operator $A \in \mathcal{B}(\mathcal{K})$ is called *uniformly positive* if it is positive, that is, $[Ah, h] \geq 0$ for all $h \in \mathcal{K}$, and boundedly invertible. It is easy to see that this is equivalent to saying that, for any fundamental norm (equivalently, for some fundamental norm) $\|\cdot\|$ of $\mathcal{K}$, and some $\delta > 0$, we have $[Ah, h] \geq \delta \|h\|^2$, for all $h \in \mathcal{K}$. In the case when $\mathcal{K}$ is a Hilbert space, it is convenient to denote $A > 0$ for A uniformly positive.

Let $\mathcal{K}_1$ and $\mathcal{K}_2$ be Kreĭn spaces and let $T \in \mathcal{B}(\mathcal{K}_1, \mathcal{K}_2)$. T is called a *uniform minus-operator* if there exists $\mu > 0$ such that the selfadjoint operator $\mu I - T^\sharp T$ is uniformly positive. To any uniform minus-operator we associate the set

$$M^{\mathrm{u}}(T) = \{\mu > 0 \mid \mu I - T^\sharp T \text{ is uniformly positive}\}. \tag{3.1}$$

Any uniform minus-operator is a strong minus-operator as well, in particular, it makes sense to consider the bounds $\|T\|_\pm$, see (2.2).

Remark 7.3.1. Let J_i denote fundamental symmetries on $\mathcal{K}_i$, $i = 1, 2$. Then $T \in \mathcal{B}(\mathcal{K}_1, \mathcal{K}_2)$ is a uniform minus-operator if and only if, for some $\mu > 0$ we have $\mu J_1 - T^* J_2 T > 0$, that is, for some $\alpha > 0$ we have $\langle (\mu J_1 - T^* J_2 T)h, h \rangle_{J_1} \geqslant \delta \|h\|^2$, for all $h \in \mathcal{K}_1$, where $\|\cdot\|$ denotes the fundamental norm on $\mathcal{K}_1$ corresponding to J_1. ∎

Lemma 7.3.2. *If $T \in \mathcal{B}(\mathcal{K}_1, \mathcal{K}_2)$ is a uniform minus-operator then $0 \leqslant \|T\|_+ < \|T\|_- \leqslant \infty$ and $M^u(T) = (\|T\|_+, \|T\|_-)$.*

Proof. Fix J_i, $i = 1, 2$, fundamental symmetries on $\mathcal{K}_i$, $i = 1, 2$, respectively. Fix $\mu \in M^u(T)$. Then there exists $\varepsilon > 0$ such that $\mu J_1 - T^* J_2 T \geqslant \varepsilon I$. For arbitrary $\lambda \in \mathbb{R}$ with $|\lambda - \mu| \leqslant \frac{\varepsilon}{2}$ we have

$$\lambda J_1 - T^* J_2 T = (\lambda - \mu)J + (\mu J_1 - T^* J_2 T) \geqslant \frac{\varepsilon}{2} I.$$

This shows that the set $M^u(T)$ is open.

Let $\mu_1, \mu_2 \in M^u(T)$ and μ be such that $\mu_1 < \mu < \mu_2$. For some $0 < t < 1$ we have $\mu = t\mu_1 + (1-t)\mu_2$. In addition,

$$\mu_i J_1 - T^* J_2 T \geqslant \varepsilon_i I, \quad i = 1, 2.$$

Then

$$\mu J_1 - T^* J_2 T = t(\mu_1 J_1 - T^* J_2 T) + (1-t)(\mu_2 J_1 - T^* J_2 T) \geqslant (t\varepsilon_1 + (1-t)\varepsilon_2)I,$$

which shows that $\mu \in M^u(T)$. Thus $M^u(T)$ is convex.

Clearly, the closure of $M^u(T)$ is $M(T)$, see (2.1) and the statements that follow. ∎

Remark 7.3.3. Let $\mathcal{K}$ be a Kreĭn space. Then $\mathcal{K}$ is positive definite, that is, it is a Hilbert space, if and only if all $T \in \mathcal{B}(\mathcal{K})$ are uniform minus-operators. In this case, $\|T\|_+ = \|T\|$ and $\|T\|_- = +\infty$.

On the other hand, $\mathcal{K}$ is negative definite, that is, it is the antispace of a Hilbert space, if and only if all boundedly invertible operators $T \in \mathcal{B}(\mathcal{K})$ are uniform minus-operators and, in this case, $\|T\|_+ = 0$ and $\|T\|_- = \|T^{-1}\|^{-1}$ hold. ∎

The next result gives a characterisation of uniform minus-operators T in terms of the bounds $\mu_\pm(T)$.

Theorem 7.3.4. *Let $T \in \mathcal{B}(\mathcal{K}_1, \mathcal{K}_2)$. Then T is a uniform minus-operator if and only if*

$$\max(0, \mu_+(T)) < \mu_-(T).$$

In this case, $\mu_+(T) = \|T\|_+$ and $\mu_-(T) = \|T\|_-$.

Proof. Assume that T is a uniform minus-operator. Then there exist $\mu > 0$ and $\varepsilon > 0$ such that

$$\mu[x,x] - [Tx, Tx] \geqslant \varepsilon \|x\|^2, \quad x \in \mathcal{K}_1. \tag{3.2}$$

7.3 UNIFORM MINUS-OPERATORS

From Proposition 7.2.2 we have $\max\{0, \mu_+(T)\} \leqslant \mu_-(T)$ and $\max\{0, \mu_+(T)\} \leqslant \mu \leqslant \mu_-(T)$.

We prove now that $\max\{0, \mu_+(T)\} < \mu < \mu_-(T)$. To see this, assume that $\mu = \mu_-(T)$. Then there exists a sequence $(x_n) \subset \mathcal{K}_1$ such that $[x_n, x_n] = -1$ for all $n \in \mathbb{N}$ and $[Tx_n, Tx_n] \to -\mu_-(T)$ as $n \to \infty$, and hence

$$\mu[x_n, x_n] - [Tx_n, Tx_n] \to -\mu_-(T) + \mu_-(T) = 0.$$

Using (3.2) we get from here that $x_n \xrightarrow[n \to \infty]{} 0$, a contradiction with $[x_n, x_n] = -1$ for all $n \in \mathbb{N}$. In a similar fashion one proves that $\mu \neq \mu_+(T)$. Thus, our claim is proved and hence $\max\{0, \mu_+(T)\} < \mu_-(T)$.

Conversely, let us assume that $\max\{0, \mu_+(T)\} < \mu_-(T)$. By Proposition 7.2.2 it follows that the operator T is a minus-operator. According to the same Proposition 7.2.2, for any $\mu > 0$ such that $\max\{0, \mu_+(T)\} < \mu < \mu_-(T)$, if J_i are fundamental symmetries on $\mathcal{K}_i$, $i = 1, 2$, we have $\mu J_1 - T^* J_2 T \geqslant 0$.

In order to finish the proof it remains to prove that the operator $\mu J_1 - T^* J_2 T$ is boundedly invertible. To see this, let $(x_n)_n$ be a sequence of vectors in $\mathcal{K}_1$ such that $(\mu J_1 - T^* J_2 T) x_n \xrightarrow[n \to \infty]{} 0$. Then, either there exists a subsequence $(x_{k_n})_n$ consisting only of positive vectors, or there exists a subsequence $(x_{k_n})_n$ consisting only of negative vectors. We analyse separately these two cases.

In the first case let μ_+ be such that

$$\max\{0, \mu_+(T)\} < \mu_+ < \mu < \mu_-(T).$$

Then

$$\langle (\mu_+ J_1 - T^* J_2 T) x_{k_n}, x_{k_n} \rangle_{J_1} = \mu_+ [x_{k_n}, x_{k_n}] - [Tx_{k_n}, Tx_{k_n}]$$
$$\leqslant [x_{k_n}, x_{k_n}] - [Tx_{k_n}, Tx_{k_n}] \to 0 \quad (n \to \infty).$$

Since $\mu_+ J_1 - T^* J_2 T \geqslant 0$ and $\mu J_1 - T^* J_2 T \geqslant 0$ we get from here

$$(\mu_+ J_1 - T^* J_2 T) x_{k_n} \xrightarrow[n \to \infty]{} 0, \quad (\mu J_1 - T^* J_2 T) x_{k_n} \xrightarrow[n \to \infty]{} 0,$$

which by subtraction yield $(\mu - \mu_+) J_1 x_{k_n} \xrightarrow[n \to \infty]{} 0$. Since J_1 is boundedly invertible it follows that $x_{k_n} \xrightarrow[n \to \infty]{} 0$.

In the latter case we consider μ_- such that

$$\max\{0, \mu_+(T)\} < \mu < \mu_- < \mu_-(T)$$

and reasoning as above we also conclude that $x_{k_n} \xrightarrow[n \to \infty]{} 0$.

We thus have proved that whenever the sequence $(x_n)_n$ of vectors in $\mathcal{K}_1$ is chosen such that $(\mu J_1 - T^* J_2 T) x_n \xrightarrow[n \to \infty]{} 0$, there exists a subsequence $(x_{k_n})_n$ converging to

0. Since $\mu J_1 - T^* J_2 T \geqslant 0$, this implies that the operator $\mu J_1 - T^* J_2 T$ is boundedly invertible, that is, the operator T is a uniform minus-operator. ∎

The next result gives a characterisation of uniform minus-operators within the class of strong minus-operators.

Theorem 7.3.5. *Let $T \in \mathcal{B}(\mathcal{K}_1, \mathcal{K}_2)$. Then T is a uniform minus-operator if and only if T is a strong minus-operator and there exists $\gamma > 0$ such that for some fundamental norm $\|\cdot\|$ on $\mathcal{K}_2$ we have*

$$[Tx, Tx] \leqslant -\gamma \|Tx\|^2, \quad x \in \mathcal{K}_1, \ [x, x] \leqslant 0. \tag{3.3}$$

Proof. Let J_i be fundamental symmetries on $\mathcal{K}_i$ and their Jordan decompositions $J_i = J_i^+ - J_i^-$, $i = 1, 2$. We easily exclude the case $T = 0$. Assume that T is a uniform minus-operator. As noted before, T is a strong minus-operator. Let $\mu > 0$ and $\varepsilon > 0$ be such that $\mu J_1 - T^* J_2 T \geqslant \varepsilon I_1$. For any $x \in \mathcal{K}_1$ such that $[x, x] \leqslant 0$ we have

$$[Tx, Tx] \leqslant -\mu[x, x] + [Tx, Tx] \leqslant -\varepsilon \|x\|^2 \leqslant \frac{-\varepsilon}{\|T\|^2} \|Tx\|^2,$$

that is, (3.3) holds with $\gamma = \varepsilon / \|T\|^2$.

Conversely, assume that T is a strong minus-operator and (3.3) holds. Then we have $\max\{0, \mu_+(T)\} \leqslant \mu_-(T)$. If $\mu_+(T) \leqslant 0$, since T is a strong minus-operator it follows that $\max\{0, \mu_+(T)\} < \mu_-(T)$.

So, let us assume that $\mu_+(T) > 0$. For $\varepsilon > 0$ consider $S_\varepsilon = J_2^+ + (1-\varepsilon)J_2^- \in \mathcal{B}(\mathcal{K}_2)$ and define $T_\varepsilon = S_\varepsilon T$. Remark that for any $x \in \mathcal{K}_1$ with $[x, x] \geqslant 0$ we have $\|J_2^+ Tx\|^2 \leqslant \|J_2^- Tx\|^2$. Therefore,

$$[T_\varepsilon x, T_\varepsilon x] = \|J_2^+ Tx\|^2 - (1-\varepsilon)^2 \|J_2^- Tx\|^2 = [Tx, Tx] + (2\varepsilon - \varepsilon^2)\|J_2^- Tx\|^2$$
$$\leqslant (-\gamma + 2\varepsilon - \varepsilon^2)\|Tx\|^2 \leqslant (-\gamma + 2\varepsilon - \varepsilon^2)\|T_\varepsilon x\|^2.$$

For $\varepsilon > 0$ and sufficiently small we have $-\gamma + 2\varepsilon + \varepsilon^2 < 0$ and hence T_ε maps negative vectors into negative vectors and, by Proposition 7.2.1, this shows that T_ε is a minus-operator. Taking into account that for all $x \in \mathcal{K}_1$ we have $[T_\varepsilon x, T_\varepsilon x] \geqslant [Tx, Tx]$, we get

$$\mu_+(T) \leqslant \mu_+(T_\varepsilon) \leqslant \mu_-(T_\varepsilon) \leqslant \mu_-(T). \tag{3.4}$$

We now remark that for $x \in \mathcal{K}_1$ with $[x, x] = -1$ we have

$$[T_\varepsilon x, T_\varepsilon x] = \|J_2^+ Tx\|^2 - (1-\varepsilon)^2 \|J_2^- Tx\|^2 \geqslant (1-\varepsilon)^2 [Tx, Tx],$$

and hence

$$\mu_-(T_\varepsilon) < (1-\varepsilon)\mu_-(T) < \mu_-(T),$$

which, taking into account (3.4), yields $\mu_+(T) < \mu_-(T)$. Since $\mu_+(T) > 0$ was supposed, this shows that $\max\{0, \mu_+(T)\} < \mu_-(T)$, and hence, by Theorem 7.3.4, T is a uniform minus-operator. ∎

7.3 Uniform Minus-Operators

Given $T \in \mathcal{B}(\mathcal{K}_1, \mathcal{K}_2)$, we call T a *doubly uniform minus-operator* if both T and its adjoint $T^\sharp$ are uniform minus-operators. As in Example 7.2.8, it can be shown that in genuine Kreĭn spaces there exist uniform minus-operators which are not doubly uniform minus-operators

Proposition 7.3.6. *Let $T \in \mathcal{B}(\mathcal{K}_1, \mathcal{K}_2)$ be a uniform minus-operator. Then T is a doubly uniform minus-operator if and only if $T^\sharp$ is a strong minus-operator.*

Proof. If T is a uniform minus-operator then there exists $\alpha > 0$ such that the operator $\alpha^2 I - T^\sharp T$ is positive and boundedly invertible. If, in addition, $T^\sharp$ is a strong minus-operator, then by Theorem 7.2.9 it follows that $\alpha^2 I - TT^\sharp$ is positive. Thus, in order to prove that $T^\sharp$ is a uniform minus-operator it remains to prove that the operator $\alpha^2 I - TT^\sharp$ is boundedly invertible. To this end, we use a trick based on the Frobenius–Schur factorisations.

In the Kreĭn space $\mathcal{K}_1[+]\mathcal{K}_2$ we consider the bounded selfadjoint operator H defined by
$$H = \begin{bmatrix} I & \frac{1}{\alpha}T \\ \frac{1}{\alpha}T^\sharp & I \end{bmatrix},$$
and note that the following dual factorisations hold
$$H = \begin{bmatrix} I & 0 \\ \frac{1}{\alpha}T^\sharp & I \end{bmatrix} \begin{bmatrix} I & 0 \\ 0 & I - \frac{1}{\alpha^2}T^\sharp T \end{bmatrix} \begin{bmatrix} I & \frac{1}{\alpha}T \\ 0 & I \end{bmatrix},$$
$$H = \begin{bmatrix} I & \frac{1}{\alpha}T \\ 0 & I \end{bmatrix} \begin{bmatrix} I - \frac{1}{\alpha^2}TT^\sharp & 0 \\ 0 & I \end{bmatrix} \begin{bmatrix} I & 0 \\ \frac{1}{\alpha}T^\sharp & I \end{bmatrix}.$$
Since all the upper or lower tringular operator block matrices appearing in the above formulae are boundedly invertible, it follows that the operators $\alpha^2 I - T^\sharp T$ and $\alpha^2 I - TT^\sharp$ are simultaneously boundedly invertible. ∎

Lemma 7.3.7. *Let $T \in \mathcal{B}(\mathcal{K}_1, \mathcal{K}_2)$ and $S \in \mathcal{B}(\mathcal{K}_2, \mathcal{K}_3)$ be strong minus-operators. If either T or S is a uniform minus-operator and if T is boundedly invertible, then ST is a uniform minus-operator. If both T and S are doubly uniform minus-operators, in order for the operator ST to be also a doubly uniform minus-operator it is sufficient that either T or S be a uniform minus-operator.*

Proof. Fixing fundamental symmetries J_i on $\mathcal{K}_i$, $i = 1, 2, 3$, the first assertion follows from the identity
$$\mu\nu J_1 - T^* S^* J_3 ST = \nu(\mu J_1 - T^* J_2 T) + T^*(\nu J_2 - S^* J_3 S)T.$$
The latter statement follows from Proposition 7.3.6. ∎

Theorem 7.3.8. *Given Kreĭn spaces $\mathcal{K}_1$ and $\mathcal{K}_2$, the set of all uniform minus-operators, as well as the set of all doubly uniform minus-operators, in $\mathcal{B}(\mathcal{K}_1, \mathcal{K}_2)$ are open with respect to the uniform operator topology on $\mathcal{B}(\mathcal{K}_1, \mathcal{K}_2)$. Moreover, the closure of the first set coincides with the set of all strong minus-operators, while the closure of the latter set coincides with the set of all doubly strong minus-operators, in $\mathcal{B}(\mathcal{K}_1, \mathcal{K}_2)$.*

Proof. Fix J_i fundamental symmetries on $\mathcal{K}_i$, $i = 1, 2$. Then T is a uniform minus-operator if and only if there exists $\mu > 0$ and $\varepsilon > 0$ such that $\mu J_1 - T^* J_2 T \geqslant \varepsilon I_1$. Since the mapping $T \mapsto \mu J_1 - T^* J_2 T$ is continuous with respect to the uniform topology on $\mathcal{B}(\mathcal{K}_1, \mathcal{K}_2)$ it follows that the set of uniform (doubly uniform) minus-operators is open.

Let $T \in \mathcal{B}(\mathcal{K}_1, \mathcal{K}_2)$ be a strong (doubly strong) minus-operator and let $0 < \varepsilon < 1$. Then, by Lemma 7.3.7 the operator $T_\varepsilon = T(\sqrt{1+\varepsilon} J_1^+ + \sqrt{1-\varepsilon} J_1^-)$ is a uniform (doubly uniform) minus-operator and $\|T_\varepsilon - T\| \to 0$ as $\varepsilon \downarrow 0$. ∎

7.4 Extensions of Uniform Minus-Operators

The following theorem shows that the "gap" between uniform minus-operators and, respectively, doubly uniform minus-operators can be characterised in terms of extensions by anti-Hilbert spaces. A similar statement holds for plus-operators, of course, with the difference that the extensions are performed by Hilbert spaces.

Theorem 7.4.1. *Let T be a uniform plus-operator (uniform minus-operator) in a Kreĭn space $\mathcal{K}$ and assume that $\|T\|_- < \mu < \|T\|_+$ (respectively, $\|T\|_+ < \mu < \|T\|_-$). Then there exists a Hilbert space (respectively, anti-Hilbert space) $\mathcal{H}$ and a doubly uniform plus-operator (respectively, doubly uniform minus-operator) $\widetilde{T} \in \mathcal{B}(\mathcal{K}[+]\mathcal{H})$ such that $\widetilde{T}|\mathcal{K} = T$ and $\|\widetilde{T}\|_- < \mu < \|\widetilde{T}\|_+$ (respectively, $\|\widetilde{T}\|_+ < \mu < \|\widetilde{T}\|_-$).*

Proof. We prove the statement only for plus-operators, in order to avoid the use of anti-Hilbert spaces. Without restricting the generality, we can assume $\mu = 1$, that is, the uniform plus-operator $T \in \mathcal{B}(\mathcal{K})$ is actually a *uniformly expansive operator*, that is, for some fundamental norm $\|\cdot\|$ on $\mathcal{K}$ there exists $\delta > 0$ such that

$$[Tx, Tx] \geqslant [x, x] + \delta \|x\|^2, \quad x \in \mathcal{K}, \tag{4.1}$$

or, equivalently, letting J denote the corresponding fundamental symmetry, we have $J - T^*JT \geqslant \delta I$. Denote $A = T^\sharp T - I$ and note that A is a bounded uniformly positive operator (that is, A is bounded, positive, and boundedly invertible) and hence, see Proposition 4.1.7, there exists $\mathcal{K} = \mathcal{K}^+[+]\mathcal{K}^-$, a fundamental decomposition reducing A. In particular, $T^\sharp T \mathcal{K}^\pm \subseteq \mathcal{K}^\pm$ and hence $T\mathcal{K}^+ \perp T\mathcal{K}^-$. By (4.1) it follows that $T\mathcal{K}^+$ is a uniformly positive subspace. Let

$$T = \begin{bmatrix} T_{11} & T_{12} \\ T_{21} & T_{22} \end{bmatrix} : \mathcal{K}^+[+]\mathcal{K}^- \to \mathcal{K}^+[+]\mathcal{K}^-, \tag{4.2}$$

be the operator block matrix representation of T. We divide the remainder of this proof in three steps.

Step 1. *Without restricting the generality we can assume that $\mathcal{K}^+$ is invariant under T (that is, $T_{12} = 0$) and $T\mathcal{K}^+ \perp \mathcal{K}^-$ (that is, $T_{11}^* T_{12} = 0$).*

Indeed, we have

$$T\mathcal{K}^+ = \left\{ \begin{bmatrix} T_{11}x \\ T_{21}x \end{bmatrix} \mid x \in \mathcal{K}^+ \right\} = \left\{ \begin{bmatrix} y \\ T_{21}T_{11}^{-1}y \end{bmatrix} \mid y \in T_{11}\mathcal{K}^+ \right\},$$

7.4 Extensions of Uniform Minus-Operators

where T_{11} is one-to-one and has closed range, in particular there exists the bounded operator $T_{11}^{-1}: T_{11}\mathcal{K}^+ \to \mathcal{K}^+$ and $\|T_{21}T_{11}^{-1}\| < 1$. Consider the operator $\Gamma = T_{21}T_{11}^{-1} \in \mathcal{B}(T_{11}\mathcal{K}^+, \mathcal{K}^-)$. Letting $\Gamma|(\mathcal{K}^+ \ominus T_{11}\mathcal{K}^+) = 0$ we extend Γ to an operator, still denoted by $\Gamma: \mathcal{K}^+ \to \mathcal{K}^-$, $\|\Gamma\| < 1$. We consider a bounded unitary operator $U \in \mathcal{B}(\mathcal{K})$ such that $U\mathcal{K}^+ = \mathcal{G}(\Gamma)$, more precisely, with respect to the fundamental decomposition $\mathcal{K} = \mathcal{K}^+[+]\mathcal{K}^-$, U can be chosen as

$$U = \begin{bmatrix} (I_+ - \Gamma^*\Gamma)^{-1/2} & \Gamma^*(I_+ - \Gamma^*\Gamma)^{-1/2} \\ \Gamma(I_+ - \Gamma^*\Gamma)^{-1/2} & (I_- \Gamma\Gamma^*)^{-1/2} \end{bmatrix}.$$

Then the operator $W = U^{-1}T$ is uniformly expansive and $W\mathcal{K}^+ = U^{-1}T\mathcal{K}^+ \subseteq \mathcal{K}^+$. In addition, $[W\mathcal{K}^+, W\mathcal{K}^-] = [T\mathcal{K}^+, T\mathcal{K}^-] = 0$, that is, $W\mathcal{K}^+ \perp W\mathcal{K}^-$.

On the other hand, let $\widetilde{\mathcal{K}} = \mathcal{K}^+[+]\mathcal{K}^-[+]\mathcal{H}$, where $\mathcal{H}$ is a Hilbert space such that $\widetilde{W} \in \mathcal{B}(\widetilde{\mathcal{K}})$ is double uniformly expansive, that is, both W and $W^\sharp$ are uniformly expansive, see (4.1), and $\widetilde{W}|\mathcal{K} = W$. We consider a unitary operator $\widetilde{U} \in \mathcal{B}(\widetilde{\mathcal{K}})$ such that $\widetilde{U}|\mathcal{K} = U$ and $\widetilde{U}|\mathcal{K} = I_\mathcal{H}$, and let $\widetilde{T} = \widetilde{U}\widetilde{W}$. Then $\widetilde{T}$ is double uniformly expansive, $\widetilde{T}|\mathcal{K} = \widetilde{U}\widetilde{W}|\mathcal{K} = \widetilde{U}W = UW = T$.

These show that we can replace the operator T with the operator W, which has all the required properties, and hence the claim is proved.

As a consequence, we can represent the operator T by

$$T = \begin{bmatrix} T_{11} & 0 \\ T_{21} & T_{22} \end{bmatrix}: \mathcal{K}^+[+]\mathcal{K}^- \to \mathcal{K}^+[+]\mathcal{K}^-, \tag{4.3}$$

and such that $T_{11}^*T_{12} = 0$. Let P denote the Hilbert space orthogonal projection of $\mathcal{K}^+$ onto $\mathcal{K}^+ \ominus T_{11}\mathcal{K}^+$. Then $PT_{12} = T_{12}$ and $(I - P)T_{12} = 0$. For $\beta > 1$ define

$$T_{13} = (\beta I_+ + T_{22}T_{12}^*)^{1/2}P: \mathcal{K}^+ \to \mathcal{K}^+,$$
$$T_{23} = T_{22}T_{12}^*(\beta I_+ + T_{12}T_{12}^*)^{-1/2}P: \mathcal{K}^+ \to \mathcal{K}^+,$$
$$T_{33} = T_{11}^*: \mathcal{K}^+ \to \mathcal{K}^+.$$

Using these operators define $\widetilde{T}$ by

$$\widetilde{T} = \begin{bmatrix} T_{11} & T_{12} & T_{13} \\ 0 & T_{22} & T_{23} \\ 0 & 0 & T_{33} \end{bmatrix}: \mathcal{K}^+[+]\mathcal{K}^-[+]\mathcal{K}^+ \to \mathcal{K}^+[+]\mathcal{K}^-[+]\mathcal{K}^+. \tag{4.4}$$

Step 2. $\widetilde{T}$ *is a uniformly expansive operator.*

To see this, note first that the operator $\widetilde{T}$ acts in the Kreĭn space $\mathcal{K}[+]\mathcal{K}^+$ and hence we can choose the fundamental symmetry

$$\widetilde{J} = \begin{bmatrix} I_+ & 0 & 0 \\ 0 & -I_- & 0 \\ 0 & 0 & I_+ \end{bmatrix}.$$

Then
$$\widetilde{J} - \widetilde{T}^*\widetilde{J}\widetilde{T} = \begin{bmatrix} I_+ - T_{11}^*T_{11} & 0 & 0 \\ 0 & A_{11} & A_{12} \\ 0 & A_{12}^* & A_{22} \end{bmatrix},$$
where
$$A_{11} = -I_- - T_{12}^*T_{12} + T_{22}^*T_{22},$$
$$A_{12} = -T_{12}^*T_{13} + T_{22}^*T_{23},$$
$$A_{22} = I_+ - T_{13}^*T_{13} + T_{23}^*T_{23} - T_{33}^*T_{33}.$$

Consider the operators
$$D = (\beta I_- + T_{12}^*T_{12})^{1/2}, \quad D_* = (\beta I_+ + T_{12}T_{12}^*)^{1/2} \qquad (4.5)$$
and note that both D and D_* are boundedly invertible in $\mathcal{K}^-$ and in $\mathcal{K}^+$, respectively. An argument similar to that provided during the proof of Lemma 2.2.1 shows that
$$T_{12}D = D_*T_{12}, \quad D_*P = PD_*. \qquad (4.6)$$

Taking into account that T is uniformly expansive, that is,
$$J - T^*JT = \begin{bmatrix} I_+ - T_{11}^*T_{11} & 0 \\ 0 & -I_- - T_{12}^*T_{12} + T_{22}^*T_{22} \end{bmatrix} \leqslant -\varepsilon I,$$
we choose β such that $-I_- - T_{12}^*T_{12} + T_{22}^*T_{22} < (\beta - 1)I_-$, equivalently, $T_{22}^*T_{22} < \beta I_- + T_{12}^*T_{12}$. Then $T_{22} = \Gamma D$, where $\Gamma \colon \mathcal{K}^- \to \mathcal{K}^-$, $\|\Gamma\| < 1$. Further, let us note that
$$\begin{bmatrix} A_{11} & A_{12} \\ A_{12}^* & A_{22} \end{bmatrix} \leqslant \begin{bmatrix} (1-\beta)I_- + A_{11} & A_{12} \\ A_{12}^* & A_{22} \end{bmatrix}, \qquad (4.7)$$
and
$$(1-\beta)I_- + A_{11} = -\beta I_- - T_{12}^*T_{12} + T_{22}^*T_{22} = -D(I - \Gamma^*\Gamma)D,$$
$$A_{12} = -D(I_- - \Gamma^*\Gamma)T_{12}^*P,$$
$$A_{22} = I_+ - \beta P - PT_{12}(I_- - \Gamma^*\Gamma)T_{12}^*P - T_{11}T_{11}^*.$$

All these show that we have the following factorisation
$$\begin{bmatrix} (1-\beta)I_- + A_{11} & A_{12} \\ A_{12}^* & A_{22} \end{bmatrix} = \begin{bmatrix} DD_\Gamma & 0 \\ 0 & I_+ \end{bmatrix}$$
$$\times \begin{bmatrix} -I_- & -D_\Gamma T_{12}^*P \\ -PT_{12}D_\Gamma & I_+ - \beta P - PT_2(I_- - \Gamma^*\Gamma)T_{12}^*P - T_{11}T_{11}^* \end{bmatrix}$$
$$\times \begin{bmatrix} D_\Gamma D & 0 \\ 0 & I_+ \end{bmatrix}.$$

7.4 Extensions of Uniform Minus-Operators

By performing a Frobenius–Schur factorisation we get

$$I_+ - \beta P - PT_{12}(I_- - \Gamma^*\Gamma)T_{12}^*P - T_{11}T_{11}^* + PT_{12}(I_- - \Gamma^*\Gamma)T_{12}^*P$$
$$= \begin{bmatrix} (1-\beta)P & 0 \\ 0 & (I-P)(I_+ - T_{11}T_{11}^*)(I-P) \end{bmatrix} \leqslant -\varepsilon I_+.$$

This shows that the operator block matrices in (4.7) are uniformly negative and hence the operator $\widetilde{J} - \widetilde{T}^*\widetilde{J}\widetilde{T}$ is uniformly negative, that is, $\widetilde{T}$ is uniformly expansive.

Step 3. *The operator $\widetilde{T}^\sharp$ is uniformly expansive.*

In view of Proposition 7.3.6 and Step 2, it is sufficient to prove that $\widetilde{T}^\sharp$ is expansive. To see this we write the operator block matrix representation of the operator $\widetilde{T} - \widetilde{T}\widetilde{J}\widetilde{T}^*$ with respect to the same decomposition as in (4.4)

$$\widetilde{T} - \widetilde{T}\widetilde{J}\widetilde{T}^* = \begin{bmatrix} B_{11} & B_{12} & B_{13} \\ B_{12}^* & B_{22} & B_{23} \\ B_{13}^* & B_{23}^* & B_{33} \end{bmatrix}.$$

Using the representation (4.4), the operators D and D_* introduced at (4.5), and taking into account (4.6), we have:

$$B_{13} = -T_{13}T_{33}^* = -D_*PT_{11} = 0,$$
$$B_{23} = -T_{23}T_{33}^* = -\Gamma T_{12}^*PT_{11} = 0,$$
$$B_{12} = T_{13}T_{23}^* - T_{12}T_{22}^* = D_*PD_*^{-1}T_{12}T_{22}^* - T_{12}T_{22}^* = T_{12}T_{22}^* - T_{12}T_{22}^* = 0.$$

These show that the operator block matrix representation of the selfadjoint operator $\widetilde{T} - \widetilde{T}\widetilde{J}\widetilde{T}^*$ is diagonal and, in order to prove that it is uniformly negative, it remains only to prove that all its diagonal entries are uniformly negative. Indeed,

$$B_{11} = I_+ - T_{11}T_{11}^* + T_{12}T_{12}^* - T_{13}T_{13}^*$$
$$= I_+ - T_{11}T_{11}^* - (\beta I_+ + T_{12}T_{12}^*)P$$
$$= I_+ - T_{11}T_{11}^* + T_{12}T_{12}^* - \beta P - T_{12}T_{12}^*$$
$$= (1-\beta)P + (I_+ - P)(I_+ - T_{11}T_{11}^*)(I_+ - P) < 0,$$

and

$$B_{22} = -I_- + T_{22}T_{22}^* - T_{23}T_{23}^*$$
$$= -I_- + T_{22}T_{22}^* - T_{22}T_{12}^*(\beta I_+ + T_{12}T_{12}^*)^{-1}T_{22}^*$$
$$= -I_- + T_{22}T_{22}^* - T_{22}T_{22}^* + \beta(\beta I_+ + T_{12}T_{12}^*)$$
$$= -I_- + \beta(\beta I_+ + T_{12}T12^*)^{-1} < 0.$$

Since, clearly, $I_+ - T_{33}T_{33}^* < 0$, it follows that $\widetilde{T} - \widetilde{T}\widetilde{J}\widetilde{T}^* < 0$. ∎

7.5 Notes

Except for Section 7.4, I learned most of the material of this chapter from M. G. Kreĭn and Yu. L. Shmulyan [100, 101] and the lecture notes of T. Ando [4].

Chapter 8

Geometry of Contractive Operators

In this chapter we present contractive operators in Kreĭn spaces, a special class of strong minus-operators, from a geometric point of view. More precisely, we will consider the action of a contraction on different subspaces, with special attention to negative and maximal negative subspaces. As in the case of strict minus-operators, contractions in Kreĭn spaces may have adjoints that are not contractive, hence the concept of a double contraction shows up.

In contradistinction to the case of Hilbert spaces, contractive operators in Kreĭn spaces may be unbounded, hence criteria of boundedness are of special interest. Our presentation will then be focused only on bounded contractions.

An interplay between contractions in Kreĭn spaces and contractions in Hilbert spaces is provided by the scattering transform. This provides the opportunity to obtain some parametrisations, in terms of operator balls, of double contractions in Kreĭn spaces, and hence plenty of examples. Also, corresponding to the action of contractions on maximal negative subspaces, one can associate a certain linear fractional transformation on operator balls, via the angular operator representation, and then one can explore their properties in an even more geometric fashion.

8.1 Contractions in Kreĭn Spaces

Let $\mathcal{H}_1$ and $\mathcal{H}_2$ be Kreĭn spaces. A linear operator $T\colon \operatorname{Dom}(T)(\subseteq \mathcal{H}_1) \to \mathcal{H}_2$ is called *contractive* if $[Tx, Tx] \leqslant [x,x]$ for all $x \in \operatorname{Dom}(T)$. In general, contractive operators between Kreĭn spaces are not necessarily bounded. In the next section we will give criteria of boundedness for contractive operators.

Throughout this section, all linear operators will be everywhere defined and bounded. Clearly, if T is contractive then it maps negative vectors into negative vectors and hence it is a minus-operator. Conversely, if T is a strong minus-operator, that is, T maps negative vectors into negative vectors and $\mu_+(T) > 0$ (see Proposition 7.1.1), then the operator $\mu_+(T)^{-1/2}T$ is contractive. Thus, everything we prove here for contractions can be easily applied to bounded strong minus-operators in Kreĭn spaces.

The bounded operator T is called *expansive* if $[Tx, Tx] \geqslant [x,x]$ for all $x \in \mathcal{H}_1$. Clearly, the operator T is contractive if and only if the operator T acting between the Kreĭn spaces $(\mathcal{H}_1, -[\cdot,\cdot])$ and $(\mathcal{H}_2, -[\cdot,\cdot])$ is expansive. For this reason every result we prove here for contractions has a counterpart for expansive operators. Also, as before, the statements obtained for expansive operators in this way can be translated for bounded strong plus-operators in Kreĭn spaces as well.

Lemma 8.1.1. *If T is contractive then $\mathrm{Ker}(T)$ is a uniformly positive subspace and $\mathrm{Ker}(T^\sharp) = \mathrm{Ker}(TT^\sharp)$.*

Proof. Since T is contractive, clearly $\mathrm{Ker}(T)$ is a positive subspace. Fix fundamental symmetries J_1 and J_2 on $\mathcal{H}_1$ and $\mathcal{H}_2$, respectively. On the Hilbert space $(\mathcal{H}_1, \langle \cdot, \cdot \rangle_{J_1})$ we consider the positive operator

$$D = J_1 - T^* J_2 T. \tag{1.1}$$

If $D = 0$ then T is isometric and $\mathrm{Ker}(T) = \{0\}$ is, by definition, uniformly positive. Assume $D \neq 0$. If $x \in \mathrm{Ker}(T)$ we have $Dx = J_1 x$ and hence

$$\|x\|^2 = \|Dx\|^2 \leqslant \|D\| \langle Dx, x \rangle_{J_1} = \|D\| [x, x],$$

which proves the uniform positivity of $\mathrm{Ker}(T)$.

Clearly, $\mathrm{Ker}(TT^\sharp) \supseteq \mathrm{Ker}(T^\sharp)$. Assume that $x \in \mathrm{Ker}(TT^\sharp) \setminus \mathrm{Ker}(T^\sharp)$. Then $T^\sharp x \in \mathrm{Ker}(T)$ which is uniformly positive, in particular $0 < [T^\sharp x, T^\sharp x] = [TT^\sharp x, x] = 0$, a contradiction. ∎

Lemma 8.1.2. *Let $T\colon \mathcal{H}_1 \to \mathcal{H}_2$ be a contraction. Fix fundamental symmetries J_1 and J_2 on $\mathcal{H}_1$ and $\mathcal{H}_2$, respectively, as well as the associated fundamental norms $\|\cdot\|$. Then, for all $x \in \mathcal{H}_1$ with $[x, x] \geqslant 0$, we have the inequality*

$$\|Tx\| \geqslant \frac{\|x\|}{\|T\| + (1 + \|T\|^2)^{1/2}}.$$

Proof. Let D be the positive (with respect to the positive definite inner product $\langle \cdot, \cdot \rangle_{J_1}$) operator, as in (1.1). For all $x \in \mathcal{H}_1$ we have

$$\|x\| - \|T\| \|Tx\| \leqslant \|Dx\| \leqslant \|D\|^{1/2} \langle Dx, x \rangle_{J_1}^{1/2} \leqslant (1 + \|T\|^2)^{1/2} \langle Dx, x \rangle_{J_1}^{1/2}.$$

On the other hand, if $[x, x] \leqslant 0$ then

$$\langle Dx, x \rangle_{J_1} = [x, x] - [Tx, Tx] \leqslant -[Tx, Tx] \leqslant \|Tx\|^2.$$

Therefore,

$$\|x\| - \|T\| \|Tx\| \leqslant (1 + \|T\|^2)^{1/2} \|Tx\|,$$

which is equivalent with the required inequality. ∎

Corollary 8.1.3. *Let $T\colon \mathcal{H}_1 \to \mathcal{H}_2$ be contractive. If $\mathcal{L}$ is a negative (strictly negative, uniformly negative) subspace in $\mathcal{H}_1$, then $T\mathcal{L}$ is a negative (respectively, strictly negative, uniformly negative) subspace of $\mathcal{H}_2$ and $T|\mathcal{L}\colon \mathcal{L} \to T\mathcal{L}$ is boundedly invertible.*

8.1 CONTRACTIONS IN KREĬN SPACES

Proof. If $\mathcal{L}$ is a negative subspace in $\mathcal{H}_1$, since $[Tx, Tx] \leqslant [x, x]$ for all $x \in \mathcal{H}_1$, hence for $x \in \mathcal{L}$, it follows that the linear manifold $T\mathcal{L}$ is negative. The inequality in Lemma 8.1.2 shows that the operator $T|\mathcal{L}$ is bounded from below and hence its image $T\mathcal{L}$ is closed and $T|\mathcal{L} \colon \mathcal{L} \to T\mathcal{L}$ is boundedly invertible. If, in addition, $\mathcal{L}$ is a strictly negative subspace of $\mathcal{H}_1$ then the same is true for $T\mathcal{L}$.

Assume that $\mathcal{L}$ is uniformly negative and fix fundamental norms $\|\cdot\|$ on $\mathcal{H}_1$ and $\mathcal{H}_2$. Then, $[x, x] \leqslant -\alpha \|x\|^2$ for some $\alpha > 0$ and all $x \in \mathcal{L}$. Consequently,

$$[Tx, Tx] \leqslant [x, x] \leqslant -\alpha \|x\|^2 \leqslant -\alpha \|T\|^2 \|Tx\|^2, \quad x \in \mathcal{L},$$

and hence the subspace $T\mathcal{L}$ is also uniformly negative. ∎

As in the case of strong minus-operators, in genuine Kreĭn spaces the adjoint of a contraction need not be a contraction, cf. Example 7.2.8. A contraction with the property that its adjoint is a contraction as well is called a *double contraction*. The next theorem gives a characterisation of doubly contractive operators in terms of the action of the operator on maximal negative subspaces.

Theorem 8.1.4. *Let $T \in \mathcal{B}(\mathcal{H}_1, \mathcal{H}_2)$ be a contraction. The following assertions are equivalent.*

(1) $T^\sharp$ *is contractive.*

(2) $\alpha T^\sharp$ *is contractive for some $\alpha \in \mathbb{R} \setminus \{0\}$ (equivalently, $T^\sharp$ is a strong minus-operator).*

(3) *For every maximal negative subspace $\mathcal{L} \subseteq \mathcal{H}_1$ the image $T\mathcal{L}$ is a maximal negative subspace in $\mathcal{H}_2$.*

(4) *For some maximal negative subspace $\mathcal{L} \subseteq \mathcal{H}_1$ the image $T\mathcal{L}$ is a maximal negative subspace in $\mathcal{H}_2$.*

Proof. (1)⇒(2). Evident.

(2)⇒(3). Let $\mathcal{L}$ be a maximal negative subspace of $\mathcal{H}_1$ and fix a fundamental decomposition $\mathcal{H}_2 = \mathcal{H}_2^-[+]\mathcal{H}_2^+$. By assumption, $\alpha T^\sharp$ is contractive, for some $\alpha \neq 0$. Let $x \in \mathcal{H}_2^- \cap (T\mathcal{L})^\perp$ be arbitrary. Then $T^\sharp x \perp \mathcal{L}$ and, since $\mathcal{L}$ is maximal negative, its orthogonal companion is maximal positive, therefore,

$$0 \leqslant |\alpha|^2 [T^\sharp x, T^\sharp x] \leqslant [x, x].$$

Thus, $\mathcal{H}_2^- \cap (T\mathcal{L})^\perp = 0$. Since, by Corollary 8.1.3, the subspace $T\mathcal{L}$ is negative, this proves that it is actually maximal negative (e.g. see Corollary 2.1.6).

(3)⇒(4). Evident.

(4)⇒(3). Assume that $\mathcal{L}_0$ is a maximal negative subspace of $\mathcal{H}_1$ such that $T\mathcal{L}_0$ is a maximal negative subspace in $\mathcal{H}_2$ and let $\mathcal{L}_1$ be an arbitrary maximal negative subspace in $\mathcal{H}_1$. Fix fundamental decompositions $\mathcal{H}_k = \mathcal{H}_k^-[+]\mathcal{H}_k^+$, $k = 1, 2$, and the corresponding fundamental norms. Using the angular operator representation, we get $\mathcal{L}_i = \{x + K_i x \mid x \in \mathcal{H}_1^-\}$, for some operators $K_i \in \mathcal{B}(\mathcal{H}_1^-, \mathcal{H}_1^+)$, $\|K_i\| \leqslant 1$, $i = 0, 1$. Let $K_t =$

$tK_0 + (1-t)K_1$, for $0 \leqslant t \leqslant 1$, and remark that $\|K_t\| \leqslant 1$. Therefore, the subspaces $\mathcal{L}_t = \{x + K_t x \mid x \in \mathcal{H}_1^-\}$ are maximal negative in $\mathcal{H}_1$.

With respect to the fundamental decompositions $\mathcal{H}_k = \mathcal{H}_k^- [+] \mathcal{H}_k^+$, $k = 1, 2$, we consider the operator block matrix representation of the operator T

$$T = \begin{bmatrix} T_{11} & T_{12} \\ T_{21} & T_{22} \end{bmatrix}. \tag{1.2}$$

For every t in the interval $[0, 1]$ and $x \in \mathcal{H}_1^-$, the vector $z = x + K_t x$ is in the maximal negative subspace $\mathcal{L}_t$. Therefore we have

$$\sqrt{2} \, \|T_{21} K_t x + T_{22} x\| \geqslant \|T(K_t x + x)\|,$$

and, by applying Lemma 8.1.2 to the negative vector z, we get

$$\|T(K_t x + x)\| \geqslant \frac{\|x + K_t x\|}{\|T\| + \sqrt{1 + \|T\|^2}} \geqslant \frac{\|x\|}{\|T\| + \sqrt{1 + \|T\|^2}}.$$

Combining these two inequalities we obtain that, for some $\delta > 0$,

$$\|T_{21} K_t x + T_{22} x\| \geqslant \delta \|x\|, \quad x \in \mathcal{H}_1^-,$$

which shows that for all t in the interval $[0, 1]$ the operator $T_{21} K_t + T_{22} \colon \mathcal{H}_1^- \to \mathcal{H}_2^-$ is injective and has closed range. In particular, these operators are semi-Fredholm and, from the local stability of the index of semi-Fredholm operators, we get

$$\mathrm{ind}(T_{21} K_0 + T_{22}) = \mathrm{ind}(T_{21} K_1 + T_{22}).$$

But, since $T\mathcal{L}_0$ is, by assumption, maximal positive, the index on the left-hand side is 0. Therefore, the operator $T_{21} K_1 + T_{22}$ is a semi-Fredholm and injective operator of index 0 and hence it is surjective. Remark that

$$T\mathcal{L} = \{(T_{11} K_t x + T_{12} x) + (T_{21} K_t x + T_{22} x) \mid x \in \mathcal{H}_1^-\},$$

and hence, by Corollary 2.1.6, it follows that the negative subspace $T\mathcal{L}$ is maximal.

(3)$\Rightarrow$(1). Assume that for every maximal negative subspace $\mathcal{L}$ in $\mathcal{H}_1$ the subspace $T\mathcal{L}$ is maximal negative. We first prove that

$$[T^\sharp y, T^\sharp y] \leqslant [y, y], \quad y \in \mathcal{H}_2, \ [y, y] < 0. \tag{1.3}$$

To see this, let $y \in \mathcal{H}_2$ be a strictly negative vector. If $[T^\sharp y, T^\sharp y] \geqslant 0$ then there exists a maximal positive subspace $\mathcal{M}$ in $\mathcal{H}_2$ such that $T^\sharp y \in \mathcal{M}$, see Lemma 2.1.3. Then $T\mathcal{M}^\perp$ is a maximal negative subspace and $y \in T\mathcal{M}^\perp$, which contradicts the negativity of the vector y. Therefore $T^\sharp y$ is also a strictly negative vector.

Let $\mathcal{L}$ be a maximal uniformly negative subspace of $\mathcal{H}_1$ such that $T^\sharp y \in \mathcal{L}$. By assumption $T\mathcal{L}$ is maximal negative and, by Corollary 8.1.3, it is maximal uniformly

8.1 CONTRACTIONS IN KREĬN SPACES

negative. Therefore, if P denotes the bounded selfadjoint projection onto $T\mathcal{L}$, then $I - P$ is a positive operator and hence

$$[y, y] = [Py, y] + [(I - P)y, y] \geqslant [Py, y]. \tag{1.4}$$

Since $\mathrm{Ran}(P) = T\mathcal{L}$, there exists $x \in \mathcal{L}$ such that $Py = Tx$ and hence

$$[x, T^\sharp y] = [Tx, y] = [Py, y] = [Py, Py] = [Tx, Tx]. \tag{1.5}$$

On the other hand, since x and $T^\sharp y$ are vectors in the same negative subspace $\mathcal{L}$, by (1.5) and the Schwarz Inequality we have

$$-[Tx, Tx] = -[x, T^\sharp y] = |[x, T^\sharp y]| \leqslant |[x, x]|^{1/2} |[T^\sharp y, T^\sharp y]|^{1/2}.$$

Therefore, since $[Tx, Tx] \leqslant [x, x]$ and hence $|[x, x]|^{1/2} \leqslant |[Tx, Tx]|^{1/2}$, from here we get

$$-[Tx, Tx] \leqslant -[T^\sharp y, T^\sharp y]. \tag{1.6}$$

Using (1.4), (1.6), and $[Tx, Tx] \leqslant [x, x]$, we obtain (1.3).

Finally, taking into account (1.3) and Proposition 7.2.1 it follows that $T^\sharp$ is a contraction. ∎

Corollary 8.1.5. *Assume that* $\kappa_-(\mathcal{H}_1) = \kappa_-(\mathcal{H}_2) = \kappa < \infty$. *Then every contraction* $T \in \mathcal{B}(\mathcal{H}_1, \mathcal{H}_2)$ *is a double contraction.*

Proof. Indeed, let $\mathcal{L}$ be a maximal negative subspace in $\mathcal{H}_1$ and hence $\dim(\mathcal{L}) = \kappa$. By Corollary 8.1.3 it follows that $T\mathcal{L}$ is a negative subspace and $\dim(T\mathcal{L}) = \dim(\mathcal{L}) = \kappa$ and hence $T\mathcal{L}$ is maximal within the class of negative subspaces of $\mathcal{H}_2$. By Theorem 8.1.4 $T^\sharp$ is also a contraction. ∎

Remark 8.1.6. There is a simpler proof of Corollary 8.1.5 based on the Frobenius–Schur factorisations as in the proof of Proposition 7.3.6. This is a particular case of the following "index formula"

$$\kappa_-(I - T^\sharp T) + \kappa_-(\mathcal{H}_2) = \kappa_-(I - TT^\sharp) + \kappa_-(\mathcal{H}_1), \tag{1.7}$$

which will be proven in a more general context in Remark 12.1.9. ∎

Corollary 8.1.7. *Let* $T \in \mathcal{B}(\mathcal{H}_1, \mathcal{H}_2)$ *be a bounded contraction represented as in* (1.2). *Then* T *is doubly contractive if and only if the operator* $T_{22} \in \mathcal{B}(\mathcal{H}_1^-, \mathcal{H}_2^-)$ *is boundedly invertible. Moreover, in this case* $\|T_{12} T_{22}^{-1}\| < 1$ *and* $\|T_{22}^{-1} T_{21}\| < 1$ *hold too.*

Proof. By Theorem 8.1.4, T is doubly contractive if and only if for some maximal negative subspace $\mathcal{L}$ in $\mathcal{H}_1$ the negative subspace $T\mathcal{L}$ is maximal. Take $\mathcal{L} = \mathcal{H}_1^-$ and note that $T\mathcal{H}_1^- = \{T_{12}x + T_{22}x \mid x \in \mathcal{H}_1^-\}$. By Corollary 2.1.7, the negative subspace $T\mathcal{H}_1^-$ is maximal if and only if T_{22} is boundedly invertible.

To prove the second part of the statement, note that $T_{12} T_{22}^{-1}$ is the angular operator of the maximal uniformly negative subspace $T\mathcal{H}_1^-$ and hence $\|T_{12} T_{22}^{-1}\| < 1$. Similarly, the operator $-T_{21}^* T_{22}^{*-1}$ is the angular operator of the maximal uniformly negative subspace $T^\sharp \mathcal{H}_2^-$ and hence $\|T_{21}^* T_{22}^{*-1}\| = \|T_{22}^{-1} T_{21}\| < 1$. ∎

Corollary 8.1.8. *Let $T \in \mathcal{B}(\mathcal{H}_1, \mathcal{H}_2)$ be a bounded double contraction represented as in* (1.2). *Then for every contraction $K \in \mathcal{B}(\mathcal{H}_1^-, \mathcal{H}_1^+)$ the operator $T_{21}K + T_{22} \in \mathcal{B}(\mathcal{H}_1^-, \mathcal{H}_2^-)$ is boundedly invertible.*

Proof. If $K \in \mathcal{B}(\mathcal{H}_1^-, \mathcal{H}_1^+)$, $\|K\| \leqslant 1$, let $\mathcal{M} = \{x + Tx \mid x \in \mathcal{H}_1^-\}$ be the maximal positive subspace with angular operator K. By Theorem 8.1.4, $T\mathcal{M}$ is a maximal negative subspace in $\mathcal{H}_2$. As in the proof of the implication (3)⇒(4) in Theorem 8.1.4, it follows that the operator $T_{21}K + T_{22} \in \mathcal{B}(\mathcal{H}_1^-, \mathcal{H}_2^-)$ is boundedly invertible. ∎

8.2 Boundedness of Contractions in Kreĭn Spaces

Unlike the case of Hilbert space contractions, a densely defined contraction in a Kreĭn space is not necessarily continuous; recall that even isometric operators may be unbounded, see Section 4.3. Thus, criteria for the boundedness of densely defined contractive operators are of interest.

Theorem 8.2.1. *Let $\mathcal{H}_i$, $i = 1, 2$ be Kreĭn spaces and let $T \colon \mathrm{Dom}(T)(\subseteq \mathcal{H}_1) \to \mathcal{H}_2$ be a densely defined contraction such that $\mathrm{Dom}(T)$ contains a maximal uniformly negative subspace $\mathcal{L}$ of $\mathcal{H}_1$ with the property that $T\mathcal{L}$ is a maximal uniformly negative subspace of $\mathcal{H}_2$. Then T is bounded and it extends uniquely to a double contraction $T \in \mathcal{B}(\mathcal{H}_1, \mathcal{H}_2)$.*

Proof. We exclude from the beginning the case when $\kappa_-(\mathcal{H}_1) = 0$, in which case the statement is trivial. Denote $\mathcal{H}_1^- = \mathcal{L}$ and then, letting $\mathcal{H}_1^+ = \mathcal{L}^\perp$, $\mathcal{H}_1 = \mathcal{H}_1^+[+]\mathcal{H}_1^-$ is a fundamental decomposition of $\mathcal{H}_1$. Since $\mathcal{H}_1^- \subseteq \mathrm{Dom}(T)$, it follows that $\mathrm{Dom}(T)$ splits

$$\mathrm{Dom}(T) = \mathcal{H}_1^- \dotplus (\mathrm{Dom}(T) \cap \mathcal{H}_1^+). \tag{2.1}$$

By assumption, $T\mathcal{L} = T\mathcal{H}_1^-$ hence, we can choose the fundamental decomposition $\mathcal{H}_2 = \mathcal{H}_2^+[+]\mathcal{H}_2^-$ with $\mathcal{H}_2^- = T\mathcal{H}_1^-$. Then, with respect to this decomposition and (2.1), T is represented by the operator block matrix

$$T = \begin{bmatrix} T_{11} & T_{12} \\ 0 & T_{22} \end{bmatrix}. \tag{2.2}$$

Since T is contractive it follows that

$$\|T_{22}f\|^2 = -[Tf, Tf] \geqslant -[f, f] = \|f\|^2, \quad f \in \mathcal{H}_1^-,$$

and hence $T_{22}\colon \mathcal{H}_1^- \to \mathcal{H}_2^-$ is boundedly invertible. By the Open Mapping Principle T_{22} is bounded.

We prove now that T_{21} is bounded. To see this, we first prove that

$$\|T_{21}g\| < \|T_{22}g\|, \quad g \in \mathrm{Dom}(T) \cap \mathcal{H}_1^-, \ 0 < \|g\| < 1. \tag{2.3}$$

8.3 THE ADJOINT OF A CONTRACTION

Indeed, assuming the converse is true, there exists $g \in \mathrm{Dom}(T) \cap \mathcal{H}_1^-$ such that $< 0 \|g\| < 1$ and $\|T_{21}g\| \geqslant \|T_{22}g\|$. Since $T_{21}g = P_{\mathcal{H}_2^-} Tg$ and T_{22} is boundedly invertible there exists $h = T_{22}^{-1} T_{21} g$ such that $T_{21}g = T_{22}h$. Then

$$\|T_{22}g\| \leqslant \|T_{21}g\| = \|T_{22}h\| \leqslant \|T_{22}\| \cdot \|h\|,$$

and hence $\|h\| \geqslant 1$. Therefore,

$$T(h-g) = \begin{bmatrix} T_{11}g \\ T_{21}g - T_{22}h \end{bmatrix} = \begin{bmatrix} T_{11}g \\ 0 \end{bmatrix},$$

and hence

$$\|T_{11}g\|^2 = [T(h-g), T(h-g)] \leqslant [h-g, h-g] = -\|h\|^2 + \|g\|^2 < 1 - 1 = 0,$$

hence $T_{11} = 0$, which contradicts

$$-\|T_{11}g\|^2 = [Tg, Tg] \leqslant [g, g] = -\|g\|^2 < 0.$$

It follows that (2.3) is true, which shows that T_{21} is bounded.

Let now $h \in \mathcal{H}_1^+$ be arbitrary. Then

$$\|T_{11}h\|^2 - \|T_{21}h\|^2 = [Th, Th] \leqslant [h, h] = \|h\|^2,$$

and hence

$$\|T_{11}h\|^2 \leqslant (1 + \|T_{21}\|^2) \|h\|,$$

which shows that T_{11} is also bounded. We have thus proven that all the operator entries in (2.2) are bounded. Thus, T is bounded and extends by continuity to a contraction $T \in \mathcal{B}(\mathcal{H}_1, \mathcal{H}_2)$. By Theorem 8.1.4, $T^\sharp$ is also a contraction. ∎

Corollary 8.2.2. *Any densely defined contraction* $T \colon \mathrm{Dom}(T)(\subseteq \mathcal{H}_1) \to \mathcal{H}_2$ *with* $\kappa_-(\mathcal{H}_1) = \kappa_-(\mathcal{H}_2) < \infty$ *is necessarily continuous.*

Proof. Since $\mathrm{Dom}(T)$ is a dense linear manifold of the Pontryagin space $\mathcal{H}_1$ with $\kappa_-(\mathcal{H}_1) < \infty$, by Lemma 1.4.10 it follows that there exists a maximal uniformly negative subspace $\mathcal{H}_1^- \subseteq \mathrm{Dom}(T)$. Since $\mathcal{H}_1^-$ is finite dimensional and T is contractive, it follows that $T\mathcal{H}_1^-$ is a uniformly negative subspace in $\mathcal{H}$, with $\dim(T\mathcal{H}_1^-) = \kappa_-(\mathcal{H}_1) = \kappa_-(\mathcal{H}_2)$, and then $T\mathcal{H}_1^-$ is a maximal uniformly negative subspace of $\mathcal{H}_2$. We can apply Theorem 8.2.1 to conclude that T is bounded. ∎

8.3 The Adjoint of a Contraction

In this section we prove that, given a bounded contraction between Kreĭn spaces, one can decompose its range in such a way as to discard a uniformly negative subspace to leave the remaining operator a double contraction. The reader may compare this with Theorem 7.4.1 where an extension to a double contraction was obtained. We first investigate some consequences of such a situation.

Proposition 8.3.1. *Let $\mathcal{H}_1$ and $\mathcal{H}_2$ be two Kreĭn spaces and let $T \in \mathcal{B}(\mathcal{H}_1, \mathcal{H}_2)$ be a contraction. Assume that there exists a regular subspace $\mathcal{R} \subseteq \mathcal{H}_2$, such that $\mathcal{R}^\perp$ is uniformly negative and $T^\sharp|\mathcal{R} \colon \mathcal{R} \to \mathcal{H}_2$ is doubly contractive. If $\mathcal{R}_-$ is an $\mathcal{R}$-maximal uniformly negative subspace, then the subspace $\mathcal{H}_2^- = \mathcal{R}_-[+]\mathcal{R}^\perp$ is $\mathcal{H}_2$-maximal uniformly negative and, denoting*
$$\mathcal{L} = \{h \in \mathcal{H}_2^- \mid T^\sharp h \perp T^\sharp \mathcal{R}_-\},$$
we have $\mathcal{H}_2^- = \mathcal{R}_- \dotplus \mathcal{L}$.

Proof. If $\mathcal{R}_-$ is an $\mathcal{R}$-maximal uniformly negative subspace, then the subspace $\mathcal{H}_2^- = \mathcal{R}_-[+]\mathcal{R}^\perp$ is clearly an $\mathcal{H}_2$-maximal uniformly negative subspace. Assuming that $T^\sharp|\mathcal{R}$ is a double contraction then, by Theorem 8.1.4, $T^\sharp \mathcal{R}_-$ is an $\mathcal{H}_1$-maximal uniformly negative subspace. Let $\mathcal{H}_1 = \mathcal{H}_1^+[+]\mathcal{H}_1^-$ be the fundamental decomposition with $\mathcal{H}_1^- = T^\sharp \mathcal{R}_-$.

Let $f \in \mathcal{H}_2^- = \mathcal{R}_-[+]\mathcal{R}^\perp$ be arbitrary. Then $T^\sharp f = h_- + h_+$ with $h_\pm \in \mathcal{H}_1^\pm$. Since $\mathcal{H}_1^- = T^\sharp \mathcal{R}_-$, we have $h_- = T^\sharp g$ for some $g \in \mathcal{R}_-$, hence $h_+ = T^\sharp x$ with $x = f - g \in \mathcal{H}_2^-$. Since $[T^\sharp x, T^\sharp \mathcal{R}_-] = [h_+, \mathcal{H}_1^-] = \{0\}$ then $x \in \mathcal{L}$, by the definition of the subspace $\mathcal{L}$. This implies that $f = g + x \in \mathcal{R}_- + \mathcal{L}$. Thus, we have proven $\mathcal{H}_2^- \subseteq \mathcal{R}_- + \mathcal{L}$. Since the converse inclusion is obvious it follows that $\mathcal{H}_2^- = \mathcal{R}_- + \mathcal{L}$.

Let $f \in \mathcal{R}_- \cap \mathcal{L}$. Then
$$T^\sharp f \in T^\sharp(\mathcal{R}_- \cap \mathcal{L}) = T^\sharp \mathcal{R}_- \cap T^\sharp \mathcal{L} \subseteq \mathcal{H}_1^- \cap \mathcal{H}_1^+ = \{0\},$$
and hence $T^\sharp f = 0$. Since $T^\sharp|\mathcal{R}_-$ is injective this implies $f = 0$. Thus $\mathcal{R}_- \cap \mathcal{L} = \{0\}$. ∎

Let $\mathcal{H}_1$ and $\mathcal{H}_2$ be Kreĭn spaces and $T \in \mathcal{B}(\mathcal{H}_1, \mathcal{H}_2)$. Recall that a *defect operator* of the operator T is, by definition, an operator $D \in \mathcal{B}(\mathcal{H}_1, \mathcal{D})$, for some Kreĭn space $\mathcal{D}$, such that $I_1 - T^\sharp T = D^\sharp D$ and D has dense range. Defect operators always exist, see Remark 6.5.2. Recalling the concept of a Kreĭn space induced by a bounded selfadjoint operator as in Section 6.1, it follows that D is a defect operator of T if and only if $(\mathcal{D}; D)$ is a Kreĭn space induced by $I - T^\sharp T$. In particular, defect operators are not unique, modulo unitary equivalence, and a characterisation of uniqueness can be obtained by applying Theorem 6.1.9 to $A = I - T^\sharp T$.

Lemma 8.3.2. *Let $T \in \mathcal{B}(\mathcal{H}_1, \mathcal{H}_2)$ be a contraction, $D \in \mathcal{B}(\mathcal{H}_2, \mathcal{D})$ a defect operator of $T^\sharp$, $\mathcal{H}_2^-$ and $\mathcal{D}_-$ maximal uniformly negative subspaces of $\mathcal{H}_2$ and $\mathcal{D}$, respectively. Then there exist two subspaces $\mathcal{R}_-$ and $\mathcal{L}$ of $\mathcal{H}_2$ such that*
$$\mathcal{H}_2^- = \mathcal{R}_- \dotplus \mathcal{L},$$
$T^\sharp$ maps $\mathcal{R}$ injectively onto a maximal uniformly negative subspace of $\mathcal{H}_1$, $T^\sharp \mathcal{L} \perp T^\sharp \mathcal{R}_-$, and $D\mathcal{R}_- \perp \mathcal{D}_-$.

Proof. Consider the extension $R \in \mathcal{B}(\mathcal{H}_1[+]\mathcal{D}, \mathcal{H}_2)$ given in the row matrix representation by
$$R = \begin{bmatrix} T & D^\sharp \end{bmatrix}. \tag{3.1}$$

8.3 THE ADJOINT OF A CONTRACTION

A straightforward calculation shows that R is a double contraction. Consider the fundamental decomposition $\mathcal{D} = \mathcal{D}_+[+]\mathcal{D}_-$ and let $D_- = P_{\mathcal{H}_1[+]\mathcal{D}_-} D \in \mathcal{B}(\mathcal{H}_2, \mathcal{D}_-)$, where $P_{\mathcal{H}_1[+]\mathcal{D}_-}$ denotes the selfadjoint projection onto the regular subspace $\mathcal{H}_1[+]\mathcal{D}_-$. Since the orthogonal complement of $\mathcal{H}_1[+]\mathcal{D}_-$ in $\mathcal{H}_1[+]\mathcal{D}$ is the uniformly positive subspace $0[+]\mathcal{D}_+$, it follows that the operator

$$R_- = P_{\mathcal{H}_1[+]\mathcal{D}_-} R = \begin{bmatrix} T & D_-^\sharp \end{bmatrix} \in \mathcal{B}(\mathcal{H}_1[+]\mathcal{D}_-, \mathcal{H}_2) \qquad (3.2)$$

is a double contraction, too. Therefore, $R_-^\sharp \mathcal{H}_2^-$ is a maximal uniformly negative subspace in $\mathcal{H}_1[+]\mathcal{D}_-$. Let $\mathcal{H}_1 = \mathcal{H}_1^+[+]\mathcal{H}_1^-$ be a fundamental decomposition of $\mathcal{H}_1$. Then $\mathcal{H}_1[+]\mathcal{D}_- = \mathcal{H}^+[+](\mathcal{H}_1^-[+]\mathcal{D}_-)$ is a fundamental decomposition of $\mathcal{H}_1[+]\mathcal{D}_-$. By the angular operator representation, the uniformly negative subspace $R_-^\sharp \mathcal{H}_2^-$ can be described as

$$R_-^\sharp \mathcal{H}_2^- = \{f + Kf \mid f \in \mathcal{H}_1[+]\mathcal{D}_-\},$$

for some operator $K \in \mathcal{B}(\mathcal{H}_1^-[+]\mathcal{D}_-, \mathcal{H}_1^+)$, $\|K\| < 1$. Then

$$\mathcal{H}_1 \cap R_-^\sharp \mathcal{H}_2^- = \{f + Kf \mid f \in \mathcal{H}_1^+\}$$

is a maximal uniformly negative subspace of $\mathcal{H}_1$. Let

$$\mathcal{G} = \mathcal{H}_1 \cap (\mathcal{H}_1 \cap R_-^\sharp \mathcal{H}_2^-)^\perp = \mathcal{H}_1 \cap (R_-^\sharp \mathcal{H}_2^-)^\perp.$$

Since $R_-^\sharp | \mathcal{H}_2^-$ is injective and $\mathcal{G} \subseteq R_-^\sharp \mathcal{H}_2^-$, it follows that there exists a uniquely determined subspace $\mathcal{R}_- \subseteq \mathcal{H}_2^-$ such that $R_-^\sharp \mathcal{R}_- = \mathcal{G}$.

Note that

$$T^\sharp | \mathcal{R}_- = R_-^\sharp | \mathcal{R}_-. \qquad (3.3)$$

Therefore, $T^\sharp | \mathcal{R}_-$ is injective and $T^\sharp \mathcal{R}_- = \mathcal{G}$. Also, for arbitrary $f \in \mathcal{R}_-$, from (3.3) it follows that $D_- f = 0$ and hence $D_- \mathcal{R}_- \perp \mathcal{D}_-$. Define

$$\mathcal{L} = \{h \in \mathcal{H}_2^- \mid T^\sharp h \perp T^\sharp \mathcal{R}_-\}.$$

As in the proof of Proposition 8.3.1 we can show that $\mathcal{H}_2^- = \mathcal{R}_- \dotplus \mathcal{L}$ and, by the definition of $\mathcal{L}$, we have $T^\sharp \mathcal{L} \perp T^\sharp \mathcal{R}_-$. ∎

Theorem 8.3.3. *Let $T \in \mathcal{B}(\mathcal{H}_1, \mathcal{H}_2)$ be a contraction. Then there exists a regular subspace $\mathcal{R} \in \mathcal{H}_2$ such that $\mathcal{R}^\perp$ is uniformly negative and $T^\sharp | \mathcal{R}$ is a double contraction. In addition, $\mathcal{R}$ is maximal with these two properties.*

Proof. Let $D \in \mathcal{B}(\mathcal{H}_2, \mathcal{D})$ be a defect operator of $T^\sharp$, $\mathcal{D}_-$ a $\mathcal{D}$-maximal uniformly negative subspace, $D_- = P_{\mathcal{D}_-} D$, and the double contraction R as in (3.1). Therefore, $D^\sharp$ is a contraction and hence $D_-^\sharp \mathcal{D}_- = D^\sharp \mathcal{D}_-$ is a uniformly negative subspace in $\mathcal{H}_2$. Let $\mathcal{H}_2 = \mathcal{H}_2^+[+]\mathcal{H}_2^-$ be a fundamental decomposition such that $D_-^\sharp \mathcal{D}_- \subseteq \mathcal{H}_2^-$. Let

$\mathcal{R}_-$ and $\mathcal{L}$ be subspaces of $\mathcal{H}_2^-$ as in Lemma 8.3.2, and denote $\mathcal{R} = \mathcal{R}_-[+]\mathcal{H}_2^+$. By construction, $\mathcal{R}^\perp \subseteq \mathcal{H}_2^-$ is a uniformly negative subspace.

For arbitrary $f \in \mathcal{R}$ we have

$$[f, f] - [T^\sharp f, T^\sharp f] = [(I_2 - TT^\sharp)f, f] = [Df, Df].$$

But $D\mathcal{R}_- \perp \mathcal{D}_-$ and $D_-^\sharp \mathcal{D}_- \subseteq \mathcal{H}_2^-$ implies

$$D_-\mathcal{R} = D_-(\mathcal{R}_-[+]\mathcal{H}_2) = \{0\}.$$

Denoting $D_+ = D - D_-$ it follows that $[Df, Df] = [D_+f, D_+f] \geqslant 0$, hence $T^\sharp|\mathcal{R}$ is a contraction. As $\mathcal{R}_-$ is mapped by $T^\sharp|\mathcal{R}$ into a maximal uniformly negative subspace of $\mathcal{H}_1$, by Theorem 8.1.4 it follows that $T^\sharp|\mathcal{R}$ is a double contraction.

The maximality of the space $\mathcal{R}$ is a consequence of Theorem 8.1.4. ∎

As a consequence of Theorem 8.3.3 we have a new characterisation of double contractions within the set of contractions (compare with Theorem 8.1.4).

Theorem 8.3.4. *Let $T \in \mathcal{B}(\mathcal{H}_1, \mathcal{H}_2)$ be a contraction. Then T is doubly contractive if and only if $T^\sharp$ maps some maximal negative subspace of $\mathcal{H}_2$ injectively onto a negative subspace of $\mathcal{H}_1$.*

Proof. The first implication follows from Corollary 8.1.3.

Conversely, assume that $\mathcal{N}$ is a maximal negative subspace of $\mathcal{H}_2$ such that $T^\sharp|\mathcal{N}$ is injective and $T^\sharp \mathcal{N}$ is a negative subspace. Let $D \in \mathcal{B}(\mathcal{H}_2, \mathcal{D})$ be a defect operator for $T^\sharp$. As in the proof of Theorem 8.3.3, fix two fundamental decompositions $\mathcal{D} = \mathcal{D}_+[+]\mathcal{D}_-$ and $\mathcal{H}_i = \mathcal{H}_i^+[+]\mathcal{H}_i^-$, $i = 1, 2$, such that $\mathcal{H}_2^- \supseteq D^\sharp \mathcal{D}_-$. We consider again the operator $D_- = P_{\mathcal{D}_-}D$ and the double contraction $R_- \in \mathcal{B}(\mathcal{H}_1[+]\mathcal{D}_-, \mathcal{H}_2)$ as in (3.2). Then $D_-^\sharp \mathcal{D}_- = D^\sharp \mathcal{D}_- \subseteq \mathcal{H}_2^-$ and hence $D_-|\mathcal{H}_2^+ = 0$. By Corollary 8.1.3 and Theorem 8.1.4, it follows that $R_-^\sharp$ maps $\mathcal{N}$ injectively onto a maximal negative subspace of $\mathcal{H}_1[+]\mathcal{D}_-$. By the angular operator representation, let

$$R_-^\sharp \mathcal{N} = \{f + Kf \mid f \in \mathcal{H}_1^-[+]\mathcal{D}_-\},$$

for some $K \in \mathcal{B}(\mathcal{H}_1^-[+]\mathcal{D}_-, \mathcal{H}_1^+)$ with $\|K\| \leqslant 1$. Then

$$\mathcal{M} = \mathcal{H}_1 \cap R_-^\sharp \mathcal{N} = \{f + Kf \mid f \in \mathcal{H}_1^-\}$$

is a maximal negative subspace of $\mathcal{H}_1$.

Since $\mathcal{M} \supseteq R_-^\sharp \mathcal{N}$, it follows that there exists a subspace $\mathcal{L} \subseteq \mathcal{N}$ such that $R_-^\sharp \mathcal{L} = \mathcal{M}$ and then, taking into account that $\mathcal{M} \subseteq \mathcal{H}_1$, we get

$$T^\sharp|\mathcal{L} = R_-^\sharp|\mathcal{L}. \tag{3.4}$$

But $T^\sharp|\mathcal{N}$ maps $\mathcal{N}$ injectively onto $R_-^\sharp \mathcal{N}$, hence $T^\sharp|\mathcal{L}$ maps $\mathcal{L}$ injectively onto $\mathcal{M}$, which is maximal negative. Since $\mathcal{L} \subseteq \mathcal{N}$, $T^\sharp|\mathcal{N}$ is injective, and its range is a negative subspace, we get $\mathcal{L} = \mathcal{N}$. Therefore, from (3.4) we get now $D_-|\mathcal{N} = 0$ and, taking into account that $\mathcal{H}_2 = \mathcal{H}_2^+ \dotplus \mathcal{N}$, we get $D_- = 0$. Thus, $\mathcal{D}$ is a Hilbert space and, since

$$[f,f] - [T^\sharp f, T^\sharp f] = [(I_2 - TT^\sharp)f, f] = [Df, Df] \geqslant 0,$$

it follows that $T^\sharp$ is contractive. ∎

8.4 The Scattering Transform

The scattering transform, or the Potapov–Ginzburg transform, has been considered in Section 3.5 already, in the context of linear relations. Here we reconsider it in the special case of contractions, with the advantage of explaining its geometric roots and interpretation.

Let $\mathcal{K}_1$ and $\mathcal{K}_2$ be Kreĭn spaces, $T \colon \mathcal{K}_1 \to \mathcal{K}_2$ a bounded contraction, and fix fundamental symmetries $J_k = J_k^+ - J_k^-$ as well as the corresponding fundamental decompositions

$$\mathcal{K}_k = \mathcal{K}_k^+[+]\mathcal{K}_k^-, \quad k = 1, 2. \tag{4.1}$$

Define the Kreĭn space

$$\mathcal{K} = \mathcal{K}_1[+] - \mathcal{K}_2, \tag{4.2}$$

that is, $\mathcal{K}$ is the orthogonal direct sum of $\mathcal{K}_1$ and $\mathcal{K}_2$, with inner product defined by

$$[x_1 + x_2, y_1 + y_2] = [x_1, y_1] - [x_2, y_2], \quad x_1, y_1 \in \mathcal{K}_1, \ x_2, y_2 \in \mathcal{K}_2. \tag{4.3}$$

Consider now the Hilbert spaces

$$\mathcal{H}_1 = \mathcal{K}_1^+ \oplus \mathcal{K}_2^-, \quad \mathcal{H}_2 = \mathcal{K}_1^- \oplus \mathcal{K}_2^+, \tag{4.4}$$

and note that, identifying these canonically with subspaces of $\mathcal{K}$, $\mathcal{K}^+ = \mathcal{H}_1$ is a maximal uniformly positive subspace, $\mathcal{K}^- = \mathcal{H}_2$ is a maximal uniformly negative subspace, and

$$\mathcal{K} = \mathcal{K}^+[+]\mathcal{K}^- = (\mathcal{K}_1^+ \oplus \mathcal{K}_2^-)[+](\mathcal{K}_1^- \oplus \mathcal{K}_2^+) \tag{4.5}$$

is a fundamental decomposition of the Kreĭn space $\mathcal{K}$. Consequently, the graph of T can be viewed as a subspace of $\mathcal{K}$

$$\mathcal{G}(T) = \{x + Tx \mid x \in \mathcal{K}_1\} \subset \mathcal{K}.$$

By (4.3) and taking into account that T is contractive we have

$$[x + Tx, x + Tx] = [x, x] - [Tx, Tx] \geqslant 0, \quad x \in \mathcal{K}_1,$$

hence $\mathcal{G}(T)$ is a positive subspace of $\mathcal{K}$. Therefore, by the angular operator representation of positive subspaces, there exists a Hilbert space contraction

$$S \colon \mathrm{Dom}(S) = P_{\mathcal{H}_1}\mathcal{G}(T) \to \mathcal{H}_2, \tag{4.6}$$

uniquely determined such that

$$\mathcal{G}(T) = \{f + Sf \mid f \in \mathrm{Dom}(S)\}. \tag{4.7}$$

The operator $S = S(T)$ is called the *scattering transform*, or the *Potapov–Ginzburg transform*, of the contraction T. In terms of the scattering transform Π for linear relations defined as in Section 3.5, we have $\mathcal{G}(S) = \Pi(\mathcal{G}(T))$, with appropriate identifications of operators with linear relations by means of their graphs.

In order to give a more explicit form of the scattering transform of a contraction T, let us consider the matrix representation of T with respect to the fundamental decompositions as in (4.1)

$$T = \begin{bmatrix} T_{11} & T_{12} \\ T_{21} & T_{22} \end{bmatrix}, \tag{4.8}$$

and consider the operators

$$J_1^+ + J_2^- T = \begin{bmatrix} I_1^+ & 0 \\ T_{21} & T_{22} \end{bmatrix} : \mathcal{K}_1 \to \mathcal{H}_1, \tag{4.9}$$

$$J_1^- + J_2^+ T = \begin{bmatrix} T_{11} & T_{12} \\ 0 & I_1^- \end{bmatrix} : \mathcal{K}_1 \to \mathcal{H}_2. \tag{4.10}$$

Lemma 8.4.1. *With notation as before, if $T \in \mathcal{B}(\mathcal{K}_1, \mathcal{K}_2)$ is a contraction then the scattering transform S of T, defined as in (4.6) and (4.7), is*

$$S = (J_2^+ T + J_1^-)(J_1^+ + J_2^- T)^{-1}, \quad \mathrm{Dom}(S) = \mathrm{Ran}(J_1^+ + J_2^- T).$$

Proof. Since T is contractive, by Corollary 8.1.7 the operator T_{22} is injective and has closed range. Then, taking into account (4.9), the same is true for the operator $J_1^+ + J_2^- T$ and hence, there exists $(J_1^+ + J_2^- T)^{-1}$, its inverse on the subspace $\mathrm{Ran}(J_1^+ + J_2^- T)$.

On the other hand, for arbitrary $x = x^+ + x^-$ in $\mathcal{K}_1 = \mathcal{K}_1^+[+]\mathcal{K}_1^-$, and taking into account of the operator block matrix representation (4.8), we have

$$\begin{aligned} x + Tx &= \begin{bmatrix} I_1^+ & 0 \\ T_{21} & T_{22} \end{bmatrix} \begin{bmatrix} x^+ \\ x^- \end{bmatrix} + \begin{bmatrix} T_{11} & T_{12} \\ 0 & I_1^- \end{bmatrix} \begin{bmatrix} x^+ \\ x^- \end{bmatrix} \\ &= (J_1^+ + J_2^- T)x + (J_2^+ T + J_1^-)x \\ &= f + (J_2^+ T + J_1^-)(J_1^+ + J_2^- T)^{-1} f, \end{aligned} \tag{4.11}$$

where $f = (J_1^+ + J_2^- T)x$ is a generic vector in the subspace $\mathrm{Ran}(J_1^+ + J_2^- T)$. Finally, it remains to recall the definition of the scattering operator S of T as in (4.7). ∎

Proposition 8.4.2. *If T is a contraction and S denotes its scattering transform, then the operator T is doubly contractive if and only if $\mathrm{Dom}(S) = \mathcal{H}_1 (= \mathcal{K}_1^+ \oplus \mathcal{K}_2^-)$.*

8.4 The Scattering Transform

Proof. Since T is contractive, its graph $\mathcal{G}(T)$, viewed as a subspace of the Kreĭn space $\mathcal{K}$ as in (4.2), is a positive subspace. But, from (4.3) we obtain that

$$\mathcal{G}(T)^\perp = \{T^\sharp y + y \mid y \in \mathcal{K}_2\} = \mathcal{G}(T^\sharp).$$

Thus, $T^\sharp$ is contractive if and only if $\mathcal{G}(T)^\perp$ is a negative subspace, and then, by Corollary 2.1.7 we see that this holds if and only if the subspace $\mathcal{G}(T)$ is maximal positive in $\mathcal{K}$. But, the subspace $\mathcal{G}(T)$ is maximal positive if and only if its angular operator, that is, the scattering operator S, is defined everywhere on $\mathcal{K}^+$, which we identify with the Hilbert space $\mathcal{H}_1$. ∎

Corollary 8.4.3. *Let T be a double contraction with operator block matrix representation (4.8) with respect to the fundamental decomposition (4.1). Then, with respect to the decompositions (4.4), the scattering transform S has the operator block matrix representation*

$$S = \begin{bmatrix} T_{11} - T_{12}T_{22}^{-1}T_{21} & T_{12}T_{22}^{-1} \\ -T_{22}^{-1}T_{21} & T_{22}^{-1} \end{bmatrix}.$$

Proof. By Corollary 8.1.7, the entry T_{22} is boundedly invertible, hence the operator $J_1^+ + J_2^- T$ is boundedly invertible and

$$(J_1^+ + J_2^- T)^{-1} = \begin{bmatrix} I_1^+ & 0 \\ -T_{22}^{-1}T_{21} & T_{22}^{-1} \end{bmatrix}.$$

Then use this formula in combination with Lemma 8.4.1 and (4.10), and get the required operator block matrix representation of the scattering operator S. ∎

Corollary 8.4.4. *With the previous notation, let T be a double contraction and S its scattering operator. Then*

$$T = (J_2^+ S + J_2^-)(J_1^+ + J_1^- S)^{-1}, \tag{4.12}$$

the scattering operator of $T^\sharp$ is S^* and hence, in particular,

$$T^\sharp = (J_1^+ S^* + J_1^-)(J_2^+ + J_2^- S^*)^{-1}, \tag{4.13}$$

and the following formulae hold

$$J_1 - T^* J_2 T = (J_1^+ + J_2^- T)^* (1 - S^* S)(J_1^+ + J_2^- T), \tag{4.14}$$

$$J_2 - T J_1 T^* = (J_2^+ + J_1^- T^\sharp)^* (1 - SS^*)(J_2^+ + J_1^- T^\sharp). \tag{4.15}$$

Proof. We use the operator block matrix representations of T as in (4.8) and of S as in Corollary 8.4.3. The calculations are straightforward and are left to the reader. ∎

Corollary 8.4.5. *Consider Kreĭn spaces $\mathcal{K}_1$ and $\mathcal{K}_2$ and define the Hilbert spaces as in (4.4). Let $S\colon \mathcal{H}_1 \to \mathcal{H}_2$ be a Hilbert space contraction. Then, S is the scattering operator of some double contraction $T\colon \mathcal{K}_1 \to \mathcal{K}_2$ if and only if the entry $S_{22} = P_{\mathcal{K}_1^-} S | \mathcal{K}_2^-$ is boundedly invertible in $\mathcal{B}(\mathcal{K}_2^-, \mathcal{K}_1^-)$. In this case, if S has the operator bock matrix representation*

$$S = \begin{bmatrix} S_{11} & S_{12} \\ S_{21} & S_{22} \end{bmatrix},$$

then there exists a unique double contraction $T\colon \mathcal{K}_1 \to \mathcal{K}_2$ such that $S = S(T)$, and this is given by

$$T = \begin{bmatrix} S_{11} - S_{12} S_2^{-1} S_{21} & S_{12} S_{22}^{-1} \\ -S_{22}^{-1} S_{21} & S_{22}^{-1} \end{bmatrix}. \qquad (4.16)$$

Proof. If S is the scattering operator of some double contraction T, by Corollary 8.4.3 the entry S_{22} is boundedly invertible.

Conversely, if S_{22} is boundedly invertible we can define the operator T as in (4.16). Then we prove that the formulae (4.14) and (4.15) hold. Since S is a Hilbert space contraction we have $I - S^*S \geqslant 0$ and $I - SS^* \geqslant 0$ and hence $J_1 - T^* J_2 T \geqslant 0$, that is, T is a Kreĭn space contraction, and $J_2 - T J_1 T^* \geqslant 0$, that is, $T^\sharp$ is also a Kreĭn space contraction. Calculating the matrix representation of the scattering transform $S(T)$ as in Corollary 8.4.3, we obtain that $S = S(T)$. ∎

The bijective correspondence induced by the scattering transform, between the class of doubly contractive operators T and the class of Hilbert space operators S with the entry S_{22} boundedly invertible, enables us to produce a yet more explicit description of doubly contractive operators, if we take advantage of the parametric description of operator 2×2 matrix representations of Hilbert space contractions obtained in Section 2.2. More precisely, with notation as in Section 2.2, we have the following.

Theorem 8.4.6. *Let the Kreĭn spaces $\mathcal{K}_i$ and their fundamental decompositions $\mathcal{K}_i = \mathcal{K}_i^-[+]\mathcal{K}_i^+$, $i = 1, 2$, be fixed. Then, the formula*

$$T = \begin{bmatrix} -\Gamma_1 C^{-1} \Gamma_2 + D_{\Gamma_1^*} \Gamma D_{\Gamma_1} & \Gamma_1 D_C C^{-1} \\ C^{-1} D_{C^*} \Gamma \Gamma_2 & C^{-1} \end{bmatrix}, \qquad (4.17)$$

establishes a bijective correspondence between the set of all doubly contractive operators $T\colon \mathcal{K}_1 = \mathcal{K}_1^+[+]\mathcal{K}_1^- \to \mathcal{K}_2 = \mathcal{K}_2^+[+]\mathcal{K}_2^-$ and the set of all quadruples $(C, \Gamma_1, \Gamma_2, \Gamma)$ subject to the following conditions:

$$\begin{cases} C\colon \mathcal{K}_2^- \to \mathcal{K}_2^+, \text{ is a boundedly invertible Hilbert space contraction,} \\ \Gamma_1\colon \mathcal{D}_C \to \mathcal{K}_1^- \text{ is a Hilbert space contraction,} \\ \Gamma_2\colon \mathcal{K}_1^+ \to \mathcal{D}_{C^*} \text{ is a Hilbert space contraction,} \\ \Gamma\colon \mathcal{D}_{\Gamma_2} \to \mathcal{D}_{\Gamma_1^*} \text{ is a Hilbert space contraction.} \end{cases} \qquad (4.18)$$

Proof. To prove this, using Theorem 2.2.7 we first parametrise the set of all Hilbert space contractions S with bounded invertible entry S_{22}, by the formula

$$S = \begin{bmatrix} -\Gamma_1 C^* \Gamma_2 + D_{\Gamma_1^*} \Gamma D_{\Gamma_2} & \Gamma_1 D_C \\ D_{C^*} \Gamma_2 & C \end{bmatrix} : \begin{matrix} \mathcal{K}_1^+ \\ \oplus \\ \mathcal{K}_2^- \end{matrix} \to \begin{matrix} \mathcal{K}_1^+ \\ \oplus \\ \mathcal{K}_2^+ \end{matrix},$$

where the parameters are determined by the conditions (4.18). Then we apply Corollary 8.4.5 to get the matrix representation as in (4.18). ∎

We can use the parametrisation formula (4.17), for example in order to produce examples of double contractions in Kreĭn spaces with different patterns.

8.5 Linear Fractional Transformations

Let $\mathcal{H}$ be a Kreĭn space and consider a doubly contractive operator $T \in \mathcal{B}(\mathcal{H})$. Fix a fundamental norm $\|\cdot\|$, the associated fundamental decomposition $\mathcal{H} = \mathcal{H}^+[+]\mathcal{H}^-$ and, with respect to this, consider the operator 2×2 block matrix representation of T

$$T = \begin{bmatrix} T_{11} & T_{12} \\ T_{21} & T_{22} \end{bmatrix}. \tag{5.1}$$

We denote the open unit ball of $\mathcal{B}(\mathcal{H}^-, \mathcal{H}^+)$ by $\mathcal{B}_0(\mathcal{H}^-, \mathcal{H}^+)$ and let $\mathcal{B}_1(\mathcal{H}^-, \mathcal{H}^+)$ be the closed unit ball of $\mathcal{B}(\mathcal{H}^-, \mathcal{H}^+)$, more precisely,

$$\mathcal{B}_0(\mathcal{H}^-, \mathcal{H}^+) = \{K \in \mathcal{B}(\mathcal{H}^-, \mathcal{H}^+) \mid \|K\| < 1\}, \tag{5.2}$$
$$\mathcal{B}_1(\mathcal{H}^-, \mathcal{H}^+) = \{K \in \mathcal{B}(\mathcal{H}^-, \mathcal{H}^+) \mid \|K\| \leq 1\}. \tag{5.3}$$

As a consequence of Corollary 8.1.8, the linear fractional mapping $\Phi_T \colon \mathcal{B}_1(\mathcal{H}^-, \mathcal{H}^+) \to \mathcal{B}_1(\mathcal{H}^-, \mathcal{H}^+)$ defined by

$$\Phi_T(K) = (T_{11}K + T_{12})(T_{21}K + T_{22})^{-1}, \quad K \in \mathcal{B}_1(\mathcal{H}^-, \mathcal{H}^+) \tag{5.4}$$

is correctly defined.

The linear fractional transformation Φ_T has a geometric interpretation as follows. If $K \in \mathcal{B}_1(\mathcal{H}^-, \mathcal{H}^+)$ then its graph $\mathcal{G}(K)$ is a maximal negative subspace in $\mathcal{H}$. By Theorem 8.1.4, $T\mathcal{G}(K)$ is a maximal negative subspace in $\mathcal{H}$ and it is easy to see that $\Phi_T(K)$ is nothing else but the angular operator of $T\mathcal{G}(K)$. In addition, by Lemma 8.1.2 the open unit ball $\mathcal{B}_0(\mathcal{H}^-, \mathcal{H}^+)$ is invariant under Φ_T.

We first record some composition properties of the linear fractional mapping Φ_T, that follow easily from this geometric interpretation.

Lemma 8.5.1. *Let T and S be two double contractions on $\mathcal{H}$. Then $\Phi_T \circ \Phi_S = \Phi_{TS}$. Moreover, for every $\zeta \in \mathbb{C}$, $|\zeta| = 1$, the linear fractional transformation $\Phi_{\zeta I}$ is the identity mapping on $\mathcal{B}_1(\mathcal{H}^-, \mathcal{H}^+)$, in particular, $\Phi_{\zeta T} = \Phi_T$.*

We now investigate the injectivity of the linear fractional transformation Φ_T.

Theorem 8.5.2. *If $T \in \mathcal{B}(\mathcal{H})$ is a double contraction then the following assertions are equivalent.*

(1) Φ_T *is injective on* $\mathcal{B}_1(\mathcal{H}^-, \mathcal{H}^+)$.

(2) Φ_T *is injective on* $\mathcal{B}_0(\mathcal{H}^-, \mathcal{H}^+)$.

(3) T *is injective on* $\mathcal{H}$.

Proof. (2)$\Rightarrow$(3). Since T is a double contraction, by Corollary 8.1.7 it follows that $T_{22} \in \mathcal{B}(\mathcal{H}^-)$ is boundedly invertible. Performing a Frobenius–Schur factorisation in (5.1) we have

$$T = \begin{bmatrix} I_+ & T_{12}T_{22}^{-1} \\ 0 & I_- \end{bmatrix} \begin{bmatrix} T_{11} - T_{12}T_{22}^{-1}T_{21} & 0 \\ 0 & T_{22} \end{bmatrix} \begin{bmatrix} I_+ & 0 \\ T_{22}^{-1}T_{21} & I_- \end{bmatrix},$$

and hence T is injective on $\mathcal{H}$ if and only if the operator $T_{11} - T_{12}T_{22}^{-1}T_{21}$ is injective on $\mathcal{H}^+$. In particular, assuming that T is not injective on $\mathcal{H}$, it follows that $(T_{11} - T_{12}T_{22}^{-1}T_{21})K = 0$ for some $K \in \mathcal{B}_0(\mathcal{H}^-, \mathcal{H}^+)$, $K \neq 0$. Therefore, $T_{12}T_{22}^{-1} = (A_{11}K + A_{12})(A_{21}K + A_{22})^{-1}$ and hence $\Phi_T(0) = \Phi_T(K)$, contradicting the assumption.

(3)$\Rightarrow$(1). Let $K, L \in \mathcal{B}_1(\mathcal{H}^-, \mathcal{H}^+)$ be such that $\Phi_T(K) = \Phi_T(L)$. In view of the geometric interpretation of the linear fractional transformation, it follows that $T\mathcal{G}(K) = T\mathcal{G}(L)$. Since T is injective, this implies $\mathcal{G}(K) = \mathcal{G}(L)$ and, in view of the uniqueness of the angular operator, from here we get $K = L$. Thus, the fractional linear mapping Φ_T is injective on $\mathcal{B}_1(\mathcal{H}^-, \mathcal{H}^+)$. ∎

Corollary 8.5.3. *Let the double contraction T be noninjective. Then, for any $K \in \mathcal{B}_0(\mathcal{H}^-, \mathcal{H}^+)$ there exists $L \in \mathcal{B}_0(\mathcal{H}^-, \mathcal{H}^+)$ with $L \neq K$ and such that $\Phi_T(K) = \Phi_T(L)$.*

Proof. Let $K \in \mathcal{B}_0(\mathcal{H}^-, \mathcal{H}^+)$ be arbitrary and let $\mathcal{M} = \mathcal{G}(K)$ be the maximal uniformly negative subspace of angular operator K. By Lemma 5.1.3, there exists a positive symmetry $S \in \mathcal{B}(\mathcal{H})$ such that $S\mathcal{H}^- = \mathcal{M}$, that is, $\Phi_S(0) = K$.

Note that the linear fractional mapping $\Phi_{TS} = \Phi_T \circ \Phi_S$ is not injective and hence, as in the proof of Theorem 8.5.2, there exists a nontrivial $N \in \mathcal{B}_0(\mathcal{H}^-, \mathcal{H}^+)$ such that $\Phi_{TS}(0) = \Phi_{TS}(N)$. Let $L = \Phi_S(N)$. Then

$$\Phi_T(L) = \Phi_{TS}(N) = \Phi_{TS}(0) = \Phi_T(K).$$

Since S is injective and hence, so is Φ_S, it follows that $L \neq K$. ∎

We characterise now the range of the linear fractional mapping as an operator ball.

8.5 Linear Fractional Transformations

Theorem 8.5.4. *Let T be a double contraction on $\mathcal{H}$. Then $\Phi_T(\mathcal{B}_1(\mathcal{H}^-, \mathcal{H}^+))$ consists of all contractions $L \in \mathcal{B}_1(\mathcal{H}^-, \mathcal{H}^+)$ such that*

$$L = (T_{12}T_{22}^* - T_{11}T_{21}^*)(T_{21}T_{21}^* - T_{22}T_{22}^*)^{-1}$$
$$+ \left(T_{11}T_{11}^* - T_{12}T_{12}^* - (T_{12}T_{22}^* - T_{11}T_{21}^*)(T_{21}T_{21}^* - T_{22}T_{22}^*)^{-1}(T_{21}T_{11}^* - T_{22}T_{12}^*)\right)^{1/2}$$
$$\times N(T_{21}T_{21}^* - T_{22}T_{22}^*)^{-1/2},$$

for some $N \in \mathcal{B}_1(\mathcal{H}^-, \mathcal{H}^+)$.

Proof. Let $K \in \mathcal{B}_1(\mathcal{H}^-, \mathcal{H}^+)$ and denote $L = \Phi_T(K)$. This is equivalent with $\mathcal{G}(L) = T\mathcal{G}(K)$ and hence, with $T\mathcal{G}(K) \perp \mathcal{G}(L)^\perp$. Therefore, $\mathcal{G}(K) \perp T^\sharp(\mathcal{G}(L)^\perp)$, in particular the subspace $T^\sharp(\mathcal{G}(L)^\perp)$ is positive. Taking into account that $\mathcal{G}(L)^\perp = \{y + L^*y \mid y \in \mathcal{H}^+\}$, it follows that

$$[I_+ \ \ -L] TT^\sharp \begin{bmatrix} I_+ \\ L^* \end{bmatrix} \geq 0. \tag{5.5}$$

With respect to the fundamental decomposition $\mathcal{H} = \mathcal{H}^+[+]\mathcal{H}^-$, we consider the operator block matrix representation of the double contraction $TT^\sharp$

$$TT^\sharp = \begin{bmatrix} T_{11}T_{11}^* - T_{12}T_{12}^* & T_{12}T_{22}^* - T_{11}T_{12}^* \\ T_{21}T_{11}^* - T_{22}T_{12}^* & T_{22}T_{22}^* - T_{21}T_{21}^* \end{bmatrix} = \begin{bmatrix} B_{11} & B_{12} \\ B_{21} & B_{22} \end{bmatrix}. \tag{5.6}$$

Performing a Frobenius–Schur factorisation in (5.6) and using (5.5), we get

$$B_{11} - B_{12}B_{22}^{-1}B_{21} - (L - B_{12}B_{22}^{-1})B_{22}(L^* - B_{22}^{-1}B_{21}) \geq 0,$$

or, equivalently,

$$B_{11} - B_{12}B_{22}^{-1}B_{21} \geq LB_{22}^{1/2} - B_{12}B_{22}^{1/2}(B_{22}L^* - B_{22}^{-1/2}B_{21}).$$

Therefore, there exists $N \in \mathcal{B}_1(\mathcal{H}^-, \mathcal{H}^+)$ such that

$$LB_{22}^{1/2} - B_{12}B_{22}^{-1/2} = (B_{22}L^* - B_{22}^{-1/2}B_{21})N,$$

or, equivalently,

$$L = B_{12}B_{22}^{-1} + (B_{11} - B_{12}B_{22}^{-1}B_{21})^{1/2} N B_{22}^{1/2}. \tag{5.7}$$

Identifying the corresponding corners in (5.6) and taking account of (5.7), we get the desired formula. ∎

We are now in a position to characterise two extreme cases, that is, when the range of the linear fractional transformation reduces to a single element and, respectively, when the linear fractional mapping is surjective.

Corollary 8.5.5. *Let T be a double contraction on $\mathcal{H}$ with the block-operator matrix representation as in (5.1). The following assertions are equivalent:*

(1) $\Phi_T(\mathcal{B}_1(\mathcal{H}^-,\mathcal{H}^+))$ *reduces to a unique element.*

(1)' $\Phi_{T^\sharp}(\mathcal{B}_1(\mathcal{H}^-,\mathcal{H}^+))$ *reduces to a unique element.*

(2) $T^\sharp T$ *is negative.*

(2)' $TT^\sharp$ *is negative.*

(3) *The range of T is a negative subspace.*

(3)' *The range of $T^\sharp$ is a negative subspace.*

(4) $\mathrm{Ker}(T)$ *is a maximal negative subspace.*

(4)' $\mathrm{Ker}(T^\sharp)$ *is a maximal negative subspace.*

(5) $T_{11} = T_{12} T_{22}^{-1} T_{21}$.

(5)' $T_{11}^* = T_{21}^* T_{22}^{*-1} T_{12}^*$.

(6) *There exist bounded unitary operators U_1 and U_2 such that, the operator block matrix representation of $U_1 T U_2$, with respect to the fundamental decomposition $\mathcal{H} = \mathcal{H}^+[+]\mathcal{H}^-$, is of the form* $\begin{bmatrix} 0 & 0 \\ 0 & * \end{bmatrix}$.

Proof. (1)⇔(2)'. As in the proof of Theorem 8.5.4 we consider the operator block matrix representation of (5.6) and then, from (5.7) it follows that the range of Φ_T reduces to a single element if and only if

$$B_{11} - B_{12} B_{22}^{-1} B_{21} = 0. \tag{5.8}$$

Note that, a by-product of the proof of Theorem 8.5.4 is the inequality

$$B_{11} - B_{12} B_{22}^{-1} B_{21} \geqslant 0.$$

On the other hand, since T is a double contraction, from Corollary 8.1.7 it follows that

$$B_{22} = T_{22}(I_{22} - T_{22}^{-1} T_{21} T_{21}^* T_{22}^{*-1}) T_{22}^* \geqslant 0.$$

Therefore, performing a Frobenius–Schur factorisation on the operator block matrix representation of $JTT^\sharp$, it follows that $JTT^\sharp \geqslant 0$ if and only if $B_{11} - B_{12} B_{22}^{-1} B_{21} \leqslant 0$. Thus, $TT^\sharp$ is negative if and only if (5.8) holds.

(2)'⇔(3)'. Obvious.

(3)'⇔(4). Suppose that the range of $T^\sharp$ is negative. Since $T^\sharp$ is a double contraction, from Theorem 8.1.4 it follows that $T^\sharp \mathcal{H}^-$ is a maximal negative subspace, hence the negative linear manifold $T^\sharp \mathcal{H}$ is a maximal negative subspace of $\mathcal{H}$. Therefore, its orthogonal companion $\mathrm{Ker}(T)$ is a maximal positive subspace.

Conversely, if $\mathrm{Ker}(T)$ is a maximal positive subspace then $\mathrm{Clos}(T^\sharp \mathcal{H}) = \mathrm{Ker}(T)^\perp$ is a maximal negative subspace.

8.5 Linear Fractional Transformations

(1)⇔(5). As in the proof of Theorem 8.5.4, the range of Φ_T is reduced to a single element if and only if

$$(T_{11} - T_{12}T_{22}^{-1}T_{21})K = 0, \quad K \in \mathcal{B}_1(\mathcal{H}^-, \mathcal{H}^+),$$

and this is clearly equivalent with $T_{11} = T_{12}T_{22}^{-1}T_{21}$.

So far, we have established the equivalence of the assertions (1), (2)′, (3)′, (4), and (5). Therefore, the equivalence of (1)′, (2), (3), (4)′, and (5)′ holds. Since (5) and (5)′ are clearly equivalent, it remains only to prove that (1) and (6) are equivalent.

(6)⇒(1). Let the bounded unitary operators U_1 and U_2 be as in (6). Then the double contraction $S = U_1 T U_2$ is such that $SS^\sharp$ is a negative operator and hence, by the equivalence of (2) and (1), the range of $\Phi_{U_1 T U_2}$ reduces to a single element. Since Φ_{U_1} and Φ_{U_2} are bijective, it follows that the range of $\Phi_T = \Phi_{U_1^\sharp} \circ \Phi_{U_1 T U_2} \circ \Phi_{U_2^\sharp}$ reduces to a single element.

(1)⇒(6). If the range of Φ_T reduces to a single element, as in the proof of (3)′ ⇔ (4) it follows that $T\mathcal{H}$ is a maximal negative subspace and $\operatorname{Ker}(T)$ is a maximal positive subspace. Then there exist unitary operators U_1 and U_2 such that $U_1 T\mathcal{H} = \mathcal{H}^-$ and $U_2 \mathcal{H}^- = \operatorname{Ker}(T)$. Then the operator block matrix representation of $U_1 T U_2$, with respect to the fundamental decomposition $\mathcal{H} = \mathcal{H}^+ [+] \mathcal{H}^-$, is of the form $\begin{bmatrix} 0 & 0 \\ 0 & * \end{bmatrix}$. ∎

Corollary 8.5.6. *Let $\mathcal{H}$ be an indefinite Kreĭn space and let $T \in \mathcal{B}(\mathcal{H})$ be a doubly contractive operator. The following assertions are equivalent.*

(1) $\Phi_T(\mathcal{B}_1(\mathcal{H}^-, \mathcal{H}^+)) = \mathcal{B}_1(\mathcal{H}^-, \mathcal{H}^+)$.

(2) $\Phi_T(\mathcal{B}_0(\mathcal{H}^-, \mathcal{H}^+)) = \mathcal{B}_0(\mathcal{H}^-, \mathcal{H}^+)$.

(3) $T^\sharp$ *is isometric.*

Proof. (1)⇒(3). Assume that $\Phi_T(\mathcal{B}_1(\mathcal{H}^-, \mathcal{H}^+)) = \mathcal{B}_1(\mathcal{H}^-, \mathcal{H}^+)$ and let $x \in \mathcal{H}$ be a strictly negative vector. Since there exists a maximal negative subspace $\mathcal{M} \ni x$, as in the proof of Theorem 8.5.4 we obtain

$$[T^\sharp x, T^\sharp x] \geq 0.$$

Therefore, by Proposition 7.2.1, there exists $\gamma \geq 0$ such that

$$[T^\sharp y, T^\sharp y] \geq \gamma [y, y], \text{ for all } y \in \mathcal{H},$$

and hence, since $T^\sharp$ is contractive we get $[T^\sharp y, T^\sharp y] = [y, y]$ for all $y \in \mathcal{H}$, that is, $T^\sharp$ is isometric.

(3)⇒(1). If $T^\sharp$ is isometric, that is, $TT^\sharp = I$, it follows that $\Phi_T \circ \Phi_{T^\sharp}$ is the identity mapping, therefore Φ_T is surjective.

The proof of (1)′⇔ (3) follows similarly. ∎

Corollary 8.5.7. *Let $\mathcal{H}$ be an indefinite Kreĭn space and let $T \in \mathcal{B}(\mathcal{H})$ be a doubly contractive operator. The following assertions are equivalent.*

(1) *Φ_T is bijective on $\mathcal{B}_1(\mathcal{H}^-, \mathcal{H}^+)$.*

(2) *Φ_T is bijective on $\mathcal{B}_0(\mathcal{H}^-, \mathcal{H}^+)$.*

(3) *T is unitary.*

Proof. It follows from Theorem 8.5.2 and Corollary 8.5.6 that (1) and (2) are equivalent, and that (1) is equivalent with T injective and $T^\sharp$ isometric. But T injective implies that $T^\sharp$ has dense range and hence $T^\sharp$ is unitary. ∎

8.6 Notes

The material in this chapter, except Sections 8.2 and 8.3, is based mainly on the investigation of M. G. Kreĭn and Yu. P. Shmulyan [100, 101, 102, 103], as well as the presentation of T. Ando [4]. The results in Section 8.2 were obtained by M. A. Dritschel and J. Rovnyak [52]. The scattering transform for the general case of linear relations was introduced by Yu. L. Shmulyan [135], by generalisation of the investigations of the finite dimensional case of V. P. Potapov [122] and the infinite dimensional operator case of Yu. P. Ginzburg [65].

Chapter 9

Invariant Maximal Semidefinite Subspaces

This chapter is dedicated to one of the most important problems in operator theory on indefinite inner product spaces, namely, the existence of maximal semidefinite subspaces that are invariant under certain operators, or certain collections of operators. There are, basically, two main approaches for invariant maximal semidefinite subspace problems, spectral and geometric.

The spectral methods are limited, for the moment, due to the lack of power of spectral functions that we will see in Chapter 11. What we can do now is to first impose certain conditions on bounded operators in such a way that their spectra are conveniently separated and then, via Riesz–Dunford functional calculus, we can obtain invariant maximal semidefinite spaces as spectral subspaces. Then we use various perturbation and approximation methods in order to enlarge the class of operators to which we can apply this method. The Cayley transform proves to be extremely efficient in translating results on invariant subspaces to classes of unbounded operators as well.

The second approach makes use of nonlinear analysis methods in the form of fixed point theorems. In order to ease the presentation, we carefully prove all the fixed point theorems that we use, except the Knaster–Kuratowsky–Mazurkiewich Theorem which, because of its complexity, falls beyond our grasp. In view of the applications of the following chapter, we also explore the existence of invariant maximal semidefinite subspaces in the more general G-spaces, that is, subspaces of Kreĭn spaces, see Section 2.4.

Finally, we approach the existence of certain maximal semidefinite subspaces that are jointly invariant under groups of unitary operators or under Boolean algebras of bounded selfadjoint projections, in terms of fundamental reducibility. Some related questions of stability of fundamental reducibility bring us back to spectral methods.

9.1 Questions and Discussions

We are, approximately, at equal distance between the beginning and the end of our journey and maybe it is time for a discussion about what we have got so far and what will come from now on. Most of the results we have seen up to now refer to the geometry and topology of indefinite inner product spaces and their operator theory, and we have just a few results on spectral theory. This is understandable, to a certain extent, but our expectations are higher. However, we have enough knowledge to approach now, probably, the most important question in operator theory on indefinite inner product spaces, namely, that of invariant maximal semidefinite subspaces. Let us consider, in order to emphasise the ideas only, one of the simplest and most tractable classes of operators, at least when comparing with Hilbert space theory, that of bounded unitary operators. If $U \in \mathcal{B}(\mathcal{K})$ is a unitary operator, for some Kreĭn space $\mathcal{K}$, we know, for example recall Sections 4.3 and 5.3, that U maps maximal semidefinite subspaces into subspaces of the same type, so the question of whether there exists a maximal semidefinite subspace in $\mathcal{L}$ in $\mathcal{K}$ that is *invariant* under U, that is, $U\mathcal{L} \subseteq \mathcal{L}$, makes sense. It is not clear why we ask directly for so much, instead of being less ambitious and asking for the existence of a nontrivial invariant subspace, as in the Hilbert case. There are many reasons, especially relating to applications, why this is the right question that we should answer, and it will become clearer during the next chapter.

Anyway, the question of whether a unitary operator in a Kreĭn space $\mathcal{K}$ has invariant maximal semidefinite subspaces can be put in an explicit and computationally appealing form of the fractional linear transformation Φ_U, that makes sense either as a selfmap in $\mathcal{B}_1(\mathcal{K}^-, \mathcal{K}^+)$, or as a selfmap in $\mathcal{B}_1(\mathcal{K}^+, \mathcal{K}^-)$, cf. the definition at (5.4) in Section 8.5. To make a choice, let us assume that we are interested in invariant maximal positive subspaces $\mathcal{L}$, with the remark that, taking the orthogonal companion $\mathcal{L}^\perp$ and recalling that $(U\mathcal{L})^\perp = U\mathcal{L}^\perp$, this is equivalent to the question of existence of invariant maximal negative subspaces.

Let then $\mathcal{K} = \mathcal{K}^+[+]\mathcal{K}^-$ be a fixed fundamental decomposition of $\mathcal{K}$, with respect to which we consider the operator block matrix representation of the bounded unitary operator U

$$U = \begin{bmatrix} U_{11} & U_{12} \\ U_{21} & U_{22} \end{bmatrix}, \tag{1.1}$$

and then consider its fractional linear transformation

$$\Phi_U(K) = (U_{21} + U_{22}K)(U_{11} + U_{12}K)^{-1}, \quad K \in \mathcal{B}_1(\mathcal{K}^+, \mathcal{K}^-). \tag{1.2}$$

Please note that this definition does not correspond exactly to that in (5.4) because there it was referring to maximal negative subspaces. As in Section 8.5, the fractional linear transformation Φ_U has the geometric interpretation of mapping the angular operator K of the maximal positive subspace $\mathcal{L}$ into the angular operator $\Phi_U(K)$ of the maximal positive subspace $U\mathcal{L}$. This immediately shows that, due to the maximality of the positive subspace $U\mathcal{L}$, we have $U\mathcal{L} \subseteq \mathcal{L}$ if and only if $U\mathcal{L} = \mathcal{L}$, and that, in terms of the map Φ_U, this is equivalent to the operator equation $\Phi_U(K) = K$, that is,

$$(U_{21} + U_{22}K)(U_{11} + U_{12}K)^{-1} = K. \tag{1.3}$$

This is a classical "fixed point problem" and in nonlinear analysis there are many strong fixed point theorems available nowadays, so let us evaluate our invariant subspace problem from this perspective. We are looking for the solution K of the equation (1.3) in the closed unit ball $\mathcal{B}_1(\mathcal{K}^+, \mathcal{K}^-)$, which is convex. So, taking into account the infinite dimensional feature of our problem and the available topological fixed point theorems, we have

9.1 Questions and Discussions

to look for an appropriate topology to make compact the convex set $\mathcal{B}_1(\mathcal{K}^+, \mathcal{K}^-)$. According to the Alaoglu Theorem, it is the weak operator topology that makes $\mathcal{B}_1(\mathcal{K}^+, \mathcal{K}^-)$ a compact set. The next step is to see whether Φ_U is continuous with respect to the weak operator topology: here we find a very serious obstruction. In order to understand this obstruction, recall that the weak operator topology does not make the product of operators jointly continuous. Let us put this remark in a simpler manner, by noting that equation (1.3) is equivalent with an "algebraic Riccati equation"

$$U_{21} + U_{22}K = KU_{11} + KU_{12}K, \tag{1.4}$$

and that it is exactly the latter term on the right-hand side that causes the trouble. An additional assumption on the operator U that makes all terms in (1.4) continuous with respect to the weak operator topology is the assumption that U_{12} should be a compact operator.

Recall (1.1) to see that U_{12} is just the upper right corner in the matrix representation of U with respect to the fixed fundamental decomposition $\mathcal{K} = \mathcal{K}^+[+]\mathcal{K}^-$. To see what this compactness condition means in geometrical terms, recall Corollary 5.3.7: it means exactly that the angular operator of the maximal uniformly positive subspace $U\mathcal{K}^+$ is compact. Let us also note, from the same Corollary 5.3.7, that the compactness of the corner U_{12} is equivalent to the compactness of the corner U_{21} so, searching for invariant maximal negative subspaces is perfectly equivalent from the point of view of this technical condition.

On the other hand, there is a natural question that can be raised in connection to this "compact corner" condition: to what extent does the compactness of the corner $U_{21} = P_{\mathcal{K}^+}U|\mathcal{K}^-$ depend on the fundamental decomposition $\mathcal{K} = \mathcal{K}^+[+]\mathcal{K}^-$? This is a general question that can be raised for, not necessarily unitary operators, any bounded operator in a Kreĭn space. The answer is that this compact corner condition depends on the fundamental symmetry with respect to which we consider the corner. To see this, let $\mathcal{K} = \mathcal{L}[+]\mathcal{L}^\perp$ be another fundamental decomposition of $\mathcal{K}$, where $\mathcal{L}$ is a maximal uniformly positive subspace. By Remark 5.1.8, the bounded selfadjoint projection $P_\mathcal{L}$ onto $\mathcal{L}$ has the operator block matrix representation

$$P_\mathcal{L} = \begin{bmatrix} (I_+ - K^*K)^{-1} & -(I_+ - K^*K)^{-1}K^* \\ (I_- - KK^*)^{-1}K & -KK^*(I_- - KK^*)^{-1} \end{bmatrix}$$
$$= \begin{bmatrix} (I_+ - K^*K)^{-1} & 0 \\ 0 & (I_- - KK^*)^{-1} \end{bmatrix} \begin{bmatrix} I_+ & -K^* \\ K & -KK^* \end{bmatrix}.$$

Then, the bounded selfadjoint projection $P_{\mathcal{L}^\perp}$ onto the maximal negative subspace $\mathcal{L}^\perp$ has the operator block matrix representation

$$P_{\mathcal{L}^\perp} = I - P_\mathcal{L} = \begin{bmatrix} -K^*K(I_+ - K^*K)^{-1} & (I_+ - K^*K)^{-1}K^* \\ -(I_- - KK^*)^{-1}K & (I_- - KK^*)^{-1} \end{bmatrix}$$
$$= \begin{bmatrix} -K^*K & K^* \\ -K & I_- \end{bmatrix} \begin{bmatrix} (I_+ - K^*K)^{-1} & 0 \\ 0 & (I_- - KK^*)^{-1} \end{bmatrix}.$$

Then note that, since the diagonal operator block matrices in the factorisations from before are invertible, in order to investigate the compactness of the operator $P_{\mathcal{L}^\perp} U P_\mathcal{L}$ we only have to consider the operator

$$\begin{bmatrix} I_+ & -K^* \\ K & -KK^* \end{bmatrix} \begin{bmatrix} U_{11} & U_{12} \\ U_{21} & U_{22} \end{bmatrix} \begin{bmatrix} -K^*K & K^* \\ -K & I_- \end{bmatrix}. \tag{1.5}$$

Calculating explicitly the operator block matrix in (1.5) we get

$$P_\mathcal{L} U|\mathcal{L}^\perp = (U_{11}K^* - K^*U_{22}) + (U_{12} - K^*U_{21}K^*).$$

Proposition 9.1.1. *Let $U \in \mathcal{B}(\mathcal{K})$ have the operator block matrix representation (1.1) with respect to a fixed fundamental decomposition $\mathcal{K} = \mathcal{K}^+[+]\mathcal{K}^-$ and let $\mathcal{L}$ be a maximal uniformly definite subspace of $\mathcal{K}$, and hence $\mathcal{K} = \mathcal{L}[+]\mathcal{L}^\perp$ is a fundamental decomposition of $\mathcal{K}$. Denote by K the angular operator of $\mathcal{L}$ with respect to the fixed fundamental decomposition. Then, assuming that both U_{12} and U_{21} are compact, the operator $P_\mathcal{L} U P_{\mathcal{L}^\perp}$ is compact if and only if $U_{11}K^* - K^*U_{22}$ is compact.*

Taking into account that, for example in the case when U is unitary, both operators U_{11} and U_{22} are boundedly invertible, it follows easily from Proposition 9.1.1 that, unless K is compact, the operator $P_\mathcal{L} U P_{\mathcal{L}^\perp}$ may not be compact. Thus, the compact corner condition *depends* on the fundamental decomposition.

A second approach to the problem of invariant maximal semidefinite subspaces can be imagined by taking into account that once the spectrum of a bounded operator in a Banach space is separated, then the Riesz–Dunford functional calculus can be used in order to obtain spectral idempotents. In order to explain this, let us enlarge a bit the class of operators by employing expansive operators, that is, operators $T \in \mathcal{B}(\mathcal{K})$ with the property that $[Tx, Tx] \geqslant [x, x]$ for all $x \in \mathcal{B}(\mathcal{K})$. Any expansive operator maps positive vectors to vectors of the same type so the question of whether there exist maximal positive subspaces $\mathcal{L}$ in $\mathcal{K}$ that are invariant under the expansive operator T makes sense. However, Theorem 8.1.4 (actually, its translation for expansive operators) tells us that we necessarily need some extra conditions. First, we ask for T to be uniformly expansive, that is, for some $\delta > 0$, we should have $[Tx, Tx] \geqslant [x, x] + \delta \|x\|^2$ for all $x \in \mathcal{K}$, and then that the same property should have $T^\sharp$ as well, that is, T should be a double uniformly expansive operator. These two conditions prove to be sufficient to nicely separate the spectrum of T and then, the Riesz spectral projections produce the required invariant maximal semidefinite subspaces. Then we consider expansive operators that can be approximated by double uniformly expansive operators and we want to obtain invariant maximal semidefinite subspaces as "limits" of maximal semidefinite subspaces that are invariant under the perturbed operators. In order to make precise what kind of "limits" we have to employ, we go back to the angular operator representation and note that we should deal with weak limits. What is funny (or maybe not) is that, in order to use this perturbation approach, we meet again the compact corner condition! So why is this "funny"? Because we try to change the trek to get where we want by avoiding an obstruction and all that happens is that we meet the same obstruction. Seems familiar?

9.2 Spectral Methods

Questions about "invariant" maximal semidefinite subspaces can be raised in connection with unbounded operators as well, but here one should first make precise what "invariant" means. So, let A be a densely defined operator in a Kreĭn space $\mathcal{K}$. Since $\mathrm{Dom}(A)$ is only dense in $\mathcal{K}$, most of the maximal semidefinite subspaces may not be entirely contained in it or, even worse, they may intersect $\mathrm{Dom}(A)$ only at 0. There are at least two choices that may be considered. First, one can say that $\mathcal{L}$ is invariant under A if $\mathcal{L} \subseteq \mathrm{Dom}(A)$ and $A\mathcal{L} \subseteq \mathcal{L}$. The second possible definition might be to ask less, namely, that $\mathcal{L} \cap \mathrm{Dom}(A)$ be only dense in $\mathcal{L}$ and that $A(\mathcal{L} \cap \mathrm{Dom}(A)) \subseteq \mathcal{L}$.

In order to approach problems on invariant maximal semidefinite subspaces under unbounded operators, the most useful tool seems to be the Cayley transform, see Section 4.5. Thus, by Theorem 4.5.9, letting A be a, possibly unbounded, selfadjoint operator in $\mathcal{K}$ and $\zeta \in \rho(A)$, the operator $U = \varepsilon(A - \overline{\zeta}I)(A - \zeta I)^{-1}$ is a bounded unitary operator in $\mathcal{K}$. What we have to consider carefully is the behaviour of maximal semidefinite subspaces with respect to invariance under A and respectively U. Note that, in Corollary 4.5.5 we only considered finite dimensional subspaces, so this kind of question has to be investigated now in a broader generality.

Coming back to a bounded unitary operator U, recall that maximal uniformly definite subspaces correspond to fundamental decompositions, cf. Lemma 2.1.8. Then, a bit of algebraic calculation shows that U has an invariant maximal uniformly definite subspace if and only if its operator block matrix representation with respect to some fundamental decomposition is diagonal. Thus, it makes sense to call such a unitary operator U *fundamentally reducible*. Note that fundamentally reducible unitary operators are extremely important: such an operator is unitary in the Hilbert space sense, hence its spectrum is contained in the unit circle and it has the nicest spectral theory we may dream of. So, equivalent characterisations of fundamentally reducible unitary operators are of interest. These operators are investigated from yet another point of view, namely, that of stability under small perturbations. It turns out that this stability is related to a separation property of the spectrum, as in Section 9.5.

9.2 Spectral Methods

Recall that a bounded operator T in a Kreĭn space $\mathcal{K}$ is a *plus-operator* if there exists $\mu \geq 0$ such that
$$[Tx, Tx] \geq \mu[x, x], \quad \text{for all } x \in \mathcal{K}, \tag{2.1}$$
equivalently, the set
$$P(T) = \{\mu \geq 0 \mid [Tx, Tx] - \mu[x, x] \geq 0, \text{ for all } x \in \mathcal{K}\} \tag{2.2}$$
is nonvoid. The results presented in Chapter 7 for minus-operators can be translated to plus-operators by changing the sign of the indefinite inner product on $\mathcal{K}$. Thus, if $P(T)$ is nonvoid then it is always a convex and closed subset in $\mathbb{R}_+$, and hence an interval.

Moreover, letting

$$\mu_+(T) = \sup_{[x,x]=1} [Tx, Tx], \quad \mu_-(T) = \inf_{[x,x]=-1} -[Tx, Tx], \qquad (2.3)$$

where we define $\mu_+(T) = -\infty$ if $\mathcal{K}$ is negative definite and, respectively, $\mu_-(T) = +\infty$ if $\mathcal{K}$ is positive definite, then a particular case of the counter part of Proposition 7.2.2, when translated for plus-operators, is the following result.

Proposition 9.2.1. (a) *If T is a plus-operator then* $\max\{0, \mu_-(T)\} \leqslant \mu_+(T)$.
(b) *T is a plus-operator if and only if* $\max\{0, \mu_-(T)\} \leqslant \mu_+(T)$. *In this case, if $\mathcal{K}$ is not positive definite we have $\mu_-(T) \geqslant 0$, and a positive number μ satisfies (2.1) if and only $\max\{0, \mu_-(T)\} \leqslant \mu \leqslant \mu_+(T)$. Moreover, the set $P(T)$ coincides with the closed interval* $[\max\{0, \mu_+(T)\}, \mu_-(T)]$.

The operator $T \in \mathcal{B}(\mathcal{K})$ is a *strong plus-operator* if the set $P(T)$ contains strictly positive numbers, equivalently, $\mu_+(T) > 0$. A strong plus-operator is called a *uniform plus-operator* if for some $\mu > 0$ in $P(T)$ the operator $\mu I - T^\sharp T$ is boundedly invertible. A uniform plus-operator T such that its adjoint $T^\sharp$ is a uniform plus-operator is called a *doubly uniform plus-operator*. By Proposition 7.3.6, a uniform plus-operator T is a doubly uniform plus-operator if and only if $T^\sharp$ is a strong plus-operator.

Recall that, given an operator T in a Hilbert space, a complex number λ is called a *of regular type* for T if the operator $\lambda I - T$ is injective and has closed range.

Lemma 9.2.2. *If T is a uniform plus-operator then the ring*

$$C = \{\lambda \in \mathbb{C} \mid \sqrt{\max\{0, \mu_-(T)\}} < |\lambda| < \sqrt{\mu_+(T)}\} \qquad (2.4)$$

contains only points of regular type for T.

If T is a doubly uniform plus-operator then the ring C is contained in $\rho(T)$, the resolvent set of T.

Proof. Let λ be an arbitrary complex number in the set C. By replacing T with $|\lambda|^{-1/2}T$, without loss of generality we can assume that T is a uniformly expansive operator, that is, for some $\delta > 0$ and all $x \in \mathcal{K}$ we have

$$[Tx, Tx] \geqslant [x, x] + \delta \|x\|^2, \qquad (2.5)$$

where $\|\cdot\|$ denotes a fundamental norm on $\mathcal{K}$. We have to prove that the unit circle $\mathbb{T}$ consists only of points of regular type for T. To this end, let $\mu \in \mathbb{T}$ and assume, by contradiction, that μ is not a point of regular type for T, that is, there exists a sequence of vectors $(x_n)_n$ in $\mathcal{K}$ with $\|x_n\| = 1$ for all $n \geqslant 1$ and $(T - \mu I)x_n \xrightarrow[n \to \infty]{} 0$. On the other hand, since T is supposed to be uniformly expansive, (2.5) holds hence,

$$\delta \|x\|^2 \leqslant [Tx_n, Tx_n] - [x_n, x_n]$$
$$= \mu[x_n, (T - \mu I)x_n] + \overline{\mu}[(T - \mu I)x_n, x_n] + [(T - \mu I)x_n, (T - \mu I)x_n],$$

9.2 Spectral Methods

which, letting $n \to \infty$, yields a contradiction since the right-hand side expression converges to 0.

In order to prove the assertion on doubly uniform plus-operators we first note, as before, that without loss of generality we can assume that T is a doubly uniform expansion and what we have to prove is that $\mathbb{T} \subset \rho(T)$. To this end, letting $\mu \in \mathbb{T}$ be arbitrary, then μ is a point of regular type for T and $\overline{\mu}$ is a point of regular type for $T^\sharp$ hence, by definition, the operators $T - \mu I$ and $T^\sharp - \overline{\mu} I = (T - \mu I)^\sharp$ are injective and have closed ranges. Taking into account that $\operatorname{Ker}(T - \mu I)^\perp = \operatorname{Clos} \operatorname{Ran}(T - \mu I)^\sharp$ we get that $T - \mu I$ is boundedly invertible, hence $\mu \in \rho(T)$. ∎

Before applying the previous lemma to invariant maximal semidefinite subspaces of doubly uniform plus-operators we record a spectral characterisation of doubly uniform plus-operators within the class of uniform plus-operators.

Corollary 9.2.3. *Let $T \in \mathcal{B}(\mathcal{K})$ be a uniform plus-operator. If the ring C, defined in (2.4), contains a circle Γ centred at zero and such that $\Gamma \subset \rho(T)$ then T is a doubly uniform plus-operator.*

Proof. As noted during the proof of Lemma 9.2.2, without loss of generality we can assume that T is uniformly expansive and that $\mathbb{T} \subset \rho(T)$. Then $\mathbb{T} \subset \rho(T^\sharp)$ and we have the factorisation

$$TT^\sharp - I = (T - I)(T^\sharp - I)^{-1}(T^\sharp T - I)(T - I)^{-1}(T^\sharp - I),$$

hence, for arbitrary $x \in \mathcal{K}$ we have

$$\begin{aligned}[][T^\sharp x, T^\sharp x] - [x, x] &= [(TT^\sharp - I)x, x] \\ &= [(T^\sharp T - I)(T - I)^{-1}(T^\sharp - I)x, ((T - I)^{-1}(T^\sharp - I)x] \\ &\geq \delta \|(T - I)^{-1}(T^\sharp - I)x\|^2 \\ &\geq \frac{\delta}{\|(T^\sharp - I)^{-1}(T - I)\|^2} \|x\|^2,\end{aligned}$$

where $\delta > 0$ is chosen as in (2.5). This proves that $T^\sharp$ is uniformly expansive as well. ∎

Here is one of the most general results on the existence of invariant maximal semidefinite subspaces based on spectral separation methods. Recall that, by $|\sigma(T)|$ we denote the spectral radius of the operator T.

Theorem 9.2.4. *Let V be a doubly uniform plus-operator in a Kreĭn space $\mathcal{K}$.*

(1) There exist uniquely determined maximal uniformly positive/negative subspaces $\mathcal{L}^\pm$, such that $V\mathcal{L}^+ = \mathcal{L}^+$, $V\mathcal{L}^- \subseteq \mathcal{L}^-$, and

$$|\sigma(V|\mathcal{L}^+)| \geq \sqrt{\mu_+(V)}, \quad |\sigma(V|\mathcal{L}^-)| \leq \sqrt{\max\{0, \mu_-(V)\}}. \tag{2.6}$$

(2) $\mathcal{L}^\pm$ is the only pair of maximal positive/negative subspaces in $\mathcal{K}$ invariant under V.

(3) *For any pair $\mathcal{M}$, $\mathcal{N}$ of positive/negative subspaces in $\mathcal{K}$ and invariant under V we have $\mathcal{M} \subseteq \mathcal{L}^+$ and $\mathcal{N} \subseteq \mathcal{L}^-$.*

Proof. (1) Let V be a doubly uniform plus-operator in $\mathcal{K}$. By Lemma 9.2.2 the ring $C = \{\lambda \in \mathbb{C} \mid \sqrt{\max\{0, \mu_-(V)\}} < |\lambda| < \sqrt{\mu_+(V)}\}$ is contained in $\rho(V)$, the resolvent set of V. Without restricting the generality we can assume, as in the proof of Lemma 9.2.2, that V is a uniformly expansive operator in $\mathcal{K}$ and hence, that the unit circle $\mathbb{T}$ is contained in the ring C that splits $\sigma(V)$ into two components.

Consider the spectral projections

$$P_- = \frac{1}{2\pi i} \int_{\mathbb{T}} (\lambda I - V)^{-1} d\lambda, \quad P_+ = I - P_-. \tag{2.7}$$

Let $\mathcal{L}^{\pm} = P_{\pm}\mathcal{K}$. Using polar coordinates $\lambda = e^{i\theta}$ we get

$$P_- = \frac{1}{2\pi} \int_0^{2\pi} e^{i\theta} (e^{i\theta} I - V)^{-1} d\theta. \tag{2.8}$$

Since V is expansive, that is, for any fixed fundamental symmetry J, we have $J - V^*JV \leqslant 0$, it follows that

$$J - e^{i\theta} J (e^{i\theta} I - V)^{-1} - e^{-i\theta} (e^{-i\theta} I - V^*)^{-1} J$$
$$= (e^{-i\theta} I - V^*)^{-1} (V^*JV - J)(e^{i\theta} I - V)^{-1} \geqslant 0.$$

Integrating this inequality we get

$$J - JP_- - P_-^* J \geqslant 0. \tag{2.9}$$

Applying $P_-^* \cdot P_-$ to (2.9) we get

$$P_-^* J P_- - P_-^* J P_- - P_-^* J P_- = -P_-^* J P_- \geqslant 0,$$

that is, $[P_- x, P_- x] \leqslant 0$ for all $x \in \mathcal{K}$, and hence its range $\mathcal{L}^- = P_- \mathcal{K}$ is a negative subspace.

Similarly, applying $(I - P_-^*) \cdot (I - P_-)$ to (2.9) we get $(I - P_-^*)J(I - P_-) \geqslant 0$, that is, $[(I - P_-)x, (I - P_-)x] \geqslant 0$, for all $x \in \mathcal{K}$, and hence the range $\mathcal{L}^+ = P_+\mathcal{K}$ is a positive subspace. In addition, since $P_+ + P_- = I$, that is, $\mathcal{K} = \mathcal{L}^+ + \mathcal{L}^-$, the subspaces $\mathcal{L}^{\pm}$ are maximal semidefinite (see Corollary 2.1.6).

Clearly, $\mathcal{L}^+$ is invariant under V and, since $\mathcal{L}^+$ is positive and V is uniformly expansive, by Theorem 8.1.4 we obtain $V\mathcal{L}^+ = \mathcal{L}^+$. Let $y \in \mathcal{L}^+$ be an arbitrary vector. Then $y = Vx$ for a certain $x \in \mathcal{L}^+$ and hence, for some $\varepsilon > 0$,

$$[y, y] = [Vx, Vx] \geqslant [x, x] + \varepsilon \|x\|^2 \geqslant \frac{\varepsilon}{\|V\|^2} \|y\|^2,$$

that is, $\mathcal{L}^+$ is uniformly positive.

9.2 Spectral Methods

Let now $x \in \mathcal{L}^-$ be arbitrary. Then $Vx \in \mathcal{L}^-$ and

$$-[x, x] \geq -[Vx, Vx] + \varepsilon \|x\|^2 \geq \varepsilon \|x\|^2,$$

that is, $[x, x] \leq -\varepsilon \|x\|^2$, hence $\mathcal{L}^-$ is uniformly negative. Moreover, the inequalities in (2.6) are consequences of the properties of the spectral projections.

(2) Assume now that the two subspaces $\mathcal{L}_\pm$ are invariant under V and maximal positive/negative in $\mathcal{K}$. Since V is double uniformly expansive, by Theorem 8.1.4 we get $V\mathcal{L}_+ = \mathcal{L}_+$. As before, it follows that $\mathcal{L}_\pm$ are uniformly positive/negative subspaces. Now, since $V|\mathcal{L}_+$ is a uniformly expansive and surjective operator in the Hilbert space $(\mathcal{L}_+, [\cdot, \cdot])$, it follows that $|\sigma(V|\mathcal{L}_+)| > 1$. Similarly, since $V|\mathcal{L}_-$ is a uniformly contractive operator in the Hilbert space $(\mathcal{L}_-, -[\cdot, \cdot])$, it follows that $|\sigma(V|\mathcal{L}_-)| < 1$. We now use these facts for the operator λV, where

$$\frac{1}{\sqrt{\mu_+(V)}} < |\lambda| < \frac{1}{\max\{0, \mu_-(V)\}},$$

taking into account that λV is yet a double uniformly expansive operator and $\sigma(\lambda V) = \lambda \sigma(V)$, to conclude,

$$|\sigma(V|\mathcal{L}_+)| \geq \sqrt{\mu_+(V)}, \quad |\sigma(V|\mathcal{L}_-)| \leq \sqrt{\max\{0, \mu_-(V)\}}.$$

Using again the properties of the spectral projections P_+ and P_-, we conclude that $\mathcal{L}_\pm \subseteq \mathcal{L}^\pm$. By maximality we then get $\mathcal{L}_\pm = \mathcal{L}^\pm$.

(3) This is a consequence of the fact that the subspaces $\mathcal{L}^\pm$ are spectral subspaces and the properties of Riesz–Dunford functional calculus. ∎

Corollary 9.2.5. *Let V be a uniform plus-operator and $\mathcal{L}_+$ a positive subspace such that $V\mathcal{L}_+ = \mathcal{L}_+$. Then $\mathcal{L}_+$ is a uniformly positive subspace and there exists a maximal uniformly positive subspace $\mathcal{L}^+ \supseteq \mathcal{L}_+$ and such that $V\mathcal{L}^+ \subseteq \mathcal{L}^+$.*

Proof. By Theorem 7.4.1 there exists a Hilbert space $\mathcal{H}$ and a double uniformly expansive operator $\widetilde{V} \in \mathcal{B}(\mathcal{K}[+]\mathcal{H})$ such that $\widetilde{V} \supseteq V$. Then $\widetilde{V}\mathcal{L}_+ = V\mathcal{L}_+ = \mathcal{L}_+$ and hence, by Theorem 9.2.4, there exists a maximal uniformly positive subspace $\widetilde{\mathcal{L}}_+$ in the Kreĭn space $\mathcal{K}[+]\mathcal{H}$, such that $\widetilde{V}\widetilde{\mathcal{L}}_+ = \mathcal{L}_+$. Then $\mathcal{L}^+ = \widetilde{\mathcal{L}}^+ \cap \mathcal{K} \supseteq \mathcal{L}_+$ is a maximal uniformly positive subspace in $\mathcal{K}$, and invariant under V. ∎

We can now obtain the existence of maximal semidefinite subspaces invariant under certain operators that have the compact corner condition, as discussed before. The idea is to make a small perturbation in such a way that the previous theorem becomes applicable.

Theorem 9.2.6. *Let $V \in \mathcal{B}(\mathcal{K})$ be a strong plus-operator such that there exists a fundamental decomposition $\mathcal{K} = \mathcal{K}^+[+]\mathcal{K}^-$ with respect to which the operator $P_{\mathcal{K}^+}V|\mathcal{K}^-$ ($= V_{12}$ in the matrix representation (2.10)) is compact. Then there exists a maximal positive subspace $\mathcal{L} \subseteq \mathcal{K}$ that is invariant under the operator V.*

Proof. Let $\mathcal{K} = \mathcal{K}^+[+]\mathcal{K}^-$ be the fundamental decomposition with respect to which the operator $P_{\mathcal{K}^+}V|\mathcal{K}^- = V_{12}$ is compact. We consider J the associated fundamental symmetry and $\|\cdot\|$ the fundamental norm, respectively. Then V has an operator block matrix

$$V = \begin{bmatrix} V_{11} & V_{12} \\ V_{21} & V_{22} \end{bmatrix}, \qquad (2.10)$$

with respect to the fundamental decomposition $\mathcal{K} = \mathcal{K}^+[+]\mathcal{K}^-$.

As mentioned before, without loss of generality, we can assume that V is expansive. For any $0 < \varepsilon < 1$ consider the operator $S_\varepsilon \in \mathcal{B}(\mathcal{K})$

$$S_\varepsilon = \begin{bmatrix} \sqrt{1+\varepsilon} & 0 \\ 0 & \sqrt{1-\varepsilon} \end{bmatrix} \text{ with respect to } \mathcal{K} = \mathcal{K}^+[+]\mathcal{K}^-.$$

Then

$$J - S_\varepsilon^* J S_\varepsilon = -\varepsilon I,$$

and hence S_ε is a uniformly expansive operator (actually it is a double uniformly expansive operator, but we do not need this). Consider the operator $V_\varepsilon = VS_\varepsilon$. Then

$$J - V_\varepsilon^* J V_\varepsilon = J - S_\varepsilon^* V^* J V S_\varepsilon + S_\varepsilon^* J S_\varepsilon - S_\varepsilon^* J S_\varepsilon$$
$$= (J - S_\varepsilon^* J S_\varepsilon) + S_\varepsilon^*(J - V^* J V)S_\varepsilon \leqslant -\varepsilon I,$$

and hence V_ε is uniformly expansive. By Corollary 9.2.5, for any $0 < \varepsilon < 1$, there exists a maximal positive subspace $\mathcal{L}_\varepsilon$ in $\mathcal{K}$, invariant under V_ε. Let K_ε be the angular operator of $\mathcal{L}_\varepsilon$. K_ε are contractions in $\mathcal{B}(\mathcal{K}^+, \mathcal{K}^-)$ and satisfy the algebraic Riccati equation

$$K_\varepsilon V_{11,\varepsilon} + K_\varepsilon V_{12,\varepsilon} K_\varepsilon = V_{21,\varepsilon} + V_{22,\varepsilon} K_\varepsilon. \qquad (2.11)$$

By Alaoglu's Theorem, there exists a sequence $(\varepsilon_n)_{n \geqslant 1}$, with $\varepsilon_n > 0$ and $\varepsilon_n \xrightarrow[n \to \infty]{} 0$ such that, letting $K_n = K_{\varepsilon_n}$, the sequence $(K_n)_{n \geqslant 1}$ converges to some contraction $K \in \mathcal{B}(\mathcal{K}^+, \mathcal{K}^-)$ in the weak operator topology.

Note that as $\varepsilon \to 0$ we have $V_\varepsilon \to V$ uniformly. This happens also for the sequence $V_n = V_{\varepsilon_n}$. Since V_{21} is a compact operator, we can pass to the weak operator limit in (2.11) for $\varepsilon_n \xrightarrow[n \to \infty]{} 0$ and get

$$KV_{11} + KV_{12}K = V_{21} + V_{22}K,$$

that is, the maximal positive subspace $\mathcal{L} = \{Kx + x \mid x \in \mathcal{K}^+\}$ is invariant under the operator V. ∎

We derive from the previous theorem some important consequences.

Corollary 9.2.7. *Let $\mathcal{K}$ be a Pontryagin space with $\kappa_+(\mathcal{K}) < \infty$. Then any plus-operator $V \in \mathcal{B}(\mathcal{K})$ has an invariant maximal positive subspace.*

9.2 SPECTRAL METHODS

Proof. Let $\mathcal{K} = \mathcal{K}^+[+]\mathcal{K}^-$ be an arbitrary fundamental decomposition of the Pontryagin space $\mathcal{K}$, V a plus-operator on $\mathcal{K}$ with matrix representation as in (2.10). Since $\dim(\mathcal{K}^+) = \kappa^+(\mathcal{K}) < \infty$ it follows that the operator entry V_{12} in (2.10) has finite rank, hence it is compact. Then we apply Theorem 9.2.6. ∎

Corollary 9.2.8. *Let $\mathcal{K}$ be a Pontryagin space. Then any unitary operator $U \in \mathcal{B}(\mathcal{K})$ has an alternating pair $(\mathcal{M}, \mathcal{N})$ of invariant maximal positive/negative subspaces.*

Proof. Let $U \in \mathcal{B}(\mathcal{K})$ be a unitary operator. To make a choice, let us assume that $\kappa^+(\mathcal{K}) < \infty$. By Corollary 9.2.7, U has an invariant maximal positive subspace $\mathcal{M}$. Then $U\mathcal{M} = \mathcal{M}$, for example see Corollary 8.5.7. This shows that $\mathcal{M}$ is invariant under the operator $U^{-1} = U^\sharp$ and hence, the orthogonal companion $\mathcal{N} = \mathcal{M}^\perp$ is invariant under U.

If $\kappa^-(\mathcal{K}) < \infty$ then a unitary operator U on $\mathcal{K}$ remains unitary on $(\mathcal{K}; -[\cdot,\cdot])$ and we apply the above argument. ∎

We pass now to invariant maximal semidefinite subspaces obtained by the Cayley transformation. We use notation as in Section 4.5.

Theorem 9.2.9. *Let A be a maximal dissipative operator in the Kreĭn space $\mathcal{K}$ such that $\mathrm{Dom}(A)$ contains a maximal uniformly positive subspace. Then, there exists $\zeta \in \rho(A) \cap \mathbb{C}^+$ such that, for any maximal positive subspaces $\mathcal{L} \subseteq \mathcal{K}$, denoting $U = (A - \bar{\zeta}I)(A - \zeta I)^{-1}$, the following assertions are equivalent:*
(a) $\mathcal{L} \subseteq \mathrm{Dom}(A)$ *and* $A\mathcal{L} \subseteq \mathcal{L}$;
(b) $U\mathcal{L} \subseteq \mathcal{L}$.

Proof. (a)⇒(b). Let $\mathcal{K} = \mathcal{K}^+[+]\mathcal{K}^-$ be a fundamental decomposition of $\mathcal{K}$ such that $\mathcal{K}^+$ is a maximal uniformly positive subspace contained in the domain of A and let J be the corresponding fundamental symmetry. With respect to this fundamental decomposition we consider the operator block matrix representation of the operator U

$$U = \begin{bmatrix} U_{11} & U_{12} \\ U_{21} & U_{22} \end{bmatrix}.$$

Recall that by Theorem 4.5.10 the operator U is doubly expansive and bounded.

On the other hand, since A is maximal dissipative, from Proposition 4.4.2 it follows that A is densely defined. Since $\mathcal{K}^+ \subseteq \mathrm{Dom}(A)$, it follows that $\mathrm{Dom}(A) = \mathcal{K}^+ \dotplus (\mathrm{Dom}(A) \cap \mathcal{K}^-)$ and then that $\mathrm{Dom}(A) \cap \mathcal{K}^-$ is dense in $\mathcal{K}^-$. Therefore, A has the operator block matrix representation

$$A = \begin{bmatrix} A_{11} & A_{12} \\ A_{21} & A_{22} \end{bmatrix} : \begin{matrix} \mathcal{K}^+ \\ \dotplus \\ \mathrm{Dom}(A) \cap \mathcal{K}^- \end{matrix} \to \begin{matrix} \mathcal{K}^+ \\ [+] \\ \mathcal{K}^- \end{matrix},$$

with the observation that the operators A_{21} and A_{22} are defined on $\mathrm{Dom}(A) \cap \mathcal{K}^-$, as in the proof of Lemma 4.4.4. This also shows that, for all $\zeta \in \mathbb{C}$ with $\mathrm{Im}\,\zeta > 2\|AJ^+\|$, we have $\zeta \in \rho(A)$. Thus, $\{\zeta \in \mathbb{C} \mid \operatorname{Im} \zeta > 2\|AJ^+\|\}$ contains only points of regular type for the operator $A|\mathcal{L}\colon \mathcal{L} \to \mathcal{L}$. Since $A|\mathcal{L}$ is a bounded operator, this means $(A - \zeta I)^{-1}\mathcal{L} = \mathcal{L}$, and hence

$$U\mathcal{L} = (A - \overline{\zeta}I)(A - \zeta I)^{-1}\mathcal{L} \subseteq \mathcal{L}.$$

(b)$\Rightarrow$(a). We divide the proof of this implication into three steps.

Step 1. *If $\zeta \in \mathbb{C}$ and $\operatorname{Im} \zeta > 2\|AJ^+\|$, then $1 \in \rho(U_{11})$.*

Indeed, without loss of generality we can assume $\zeta \in \rho(A_{11})$. Then

$$\bigl(U_{11} - (I_+ - A_{12})(A_{22} - \zeta I_-)^{-1}A_{21}\bigr)(A_{11} - \zeta I_+)^{-1} = (\zeta - \overline{\zeta})(A_{11} - \zeta I_+)^{-1},$$

and hence $\operatorname{Ran}(U_{11} - I_+) = \mathcal{K}^+$. Since U is doubly expansive this implies $0 \in \rho(U_{11})$, $\|U_{11}^{-1}\| \leq 1$, and $I_+ - U_{11}^{-1} = U_{11}^{-1}(U_{11} - I_+)$. Therefore, $\operatorname{Ran}(I_+ - U_{11}^{-1}) = \mathcal{K}^+$, that is, $1 \in \rho(U_{11}^{-1})$, equivalently, $1 \in \rho(U_{11})$.

Step 2. *There exists $\zeta \in \mathbb{C}$ with $\operatorname{Im} \zeta > 2\|AJ^+\|$ such that*

$$\|A_{12}(A_{22} - \zeta I_-)^{-1}\| < 1. \tag{2.12}$$

To see this, we write down the inequality $\operatorname{Im}[Ay, y] \geq 0$ for $y = y_+ + y_-$, where $y_+ \in \mathcal{K}^+$ and $y_- \in \mathcal{K}^- \cap \operatorname{Dom}(A)$, and get

$$\operatorname{Im}[A_{11}y_+, y_+] + \operatorname{Im}[A_{12}y_-, y_-] + \operatorname{Im}[A_{21}y_+, y_-] + \operatorname{Im}[A_{22}y_-, y_-] \geq 0. \tag{2.13}$$

Let $\zeta = \xi + i\eta$, $\xi, \eta \in \mathbb{R}$, with $\eta > 2\|AJ^+\|$, and $x_- \in \mathcal{K}^-$ with $\|x_-\| = 1$. If $A_{12}(A_{22} - \zeta I_-)^{-1}x_- = 0$, then (2.12) holds.

Let us assume that $A_{12}(A_{22} - \zeta I_-)^{-1}x_- \neq 0$ and denote

$$y_- = \sqrt{\eta}(A_{22} - \zeta I_-)^{-1}x_-, \quad y_+ = i\frac{A_{12}(A_{22} - \zeta I_-)^{-1}x_-}{\|A_{12}(A_{22} - \zeta I_-)^{-1}x_-\|}.$$

Note that $\|y_+\| = 1$. From (2.13) we have

$$\operatorname{Im}[A_{11}y_+, y_-] - \sqrt{\eta}\|A_{12}(A_{22} - \zeta I_-)^{-1}x_-\| + \sqrt{\eta}\operatorname{Im}[A_{21}y_+, (A_{22} - \zeta I_-)^{-1}x_-]$$
$$+ \eta \operatorname{Im}[A_{22}(A_{22} - \zeta I_-)^{-1}x_-, (A_{22} - \zeta I_-)^{-1}x_-] \geq 0,$$

whence

$$\|A_{12}(A_{22} - \zeta I_-)^{-1}x_-\|$$
$$\leq \frac{1}{\sqrt{\eta}}\operatorname{Im}\langle A_{11}y_+, y_+\rangle - \operatorname{Im}\langle A_{21}y_+, (A_{22} - \zeta I_-)^{-1}x_-\rangle|$$
$$- \sqrt{\eta}\operatorname{Im}\langle A_{22}(A_{22} - \zeta I_-)^{-1}x_-, (A_{22} - \zeta I_-)^{-1}x_-\rangle|$$
$$\leq \frac{1}{\sqrt{\eta}}\|A_{11}\| + \|A_{21}\|\|(A_{22} - \zeta I_-)^{-1}\|$$
$$+ \sqrt{\eta}\|A_{22}\|\|(A_{22} - \zeta I_-)^{-1}\|\|A_{22}(A_{22} - \zeta I_-)^{-1}\|. \tag{2.14}$$

9.2 Spectral Methods

The operator $-A_{22}$ is dissipative in the Hilbert space $(\mathcal{K}^-, \langle \cdot, \cdot \rangle)$, hence

$$\|(A_{22} - \zeta I_-)^{-1}\| \leqslant \frac{1}{\operatorname{Im} \zeta} = \frac{1}{\eta},$$

and the operator $V = (A_{22} - \overline{\zeta} I_-)(A_{22} - \zeta I_-)^{-1}$ is a contraction, that is $\|V\| \leqslant 1$, and hence

$$\|A_{22}(A_{22} - \zeta I)^{-1}\| \leqslant \|(A_{22} - \overline{\zeta} I_-)(A_{22} - \zeta I_-)^{-1}\| + |\overline{\zeta}| \|(A_{22} - \zeta I_-)^{-1}\|$$

$$\leqslant 1 + \frac{|\overline{\zeta}|}{\eta} \leqslant 1 + \sqrt{2}, \quad (2.15)$$

provided that $|\xi| = |\operatorname{Re} \zeta| \leqslant \operatorname{Im} \xi = \eta$. Thus, for $\operatorname{Im} \zeta > 2\|AJ^+\|$ and $\operatorname{Im} \xi \geqslant |\operatorname{Re} \zeta|$, from (2.14) and (2.15) we have

$$\|A_{12}(A_{22} - \zeta I_-)^{-1} x_-\| \leqslant \frac{\|A_{11}\|}{\sqrt{\eta}} + \frac{\|A_{21}\|}{\eta} + \frac{1 + \sqrt{2}}{\sqrt{\eta}}.$$

Since x_- is arbitrary with $\|x_-\| = 1$ we get

$$\|A_{12}(A_{22} - \zeta I_-)^{-1}\| \leqslant \frac{\|A_{11}\|}{\sqrt{\eta}} + \frac{\|A_{21}\|}{\eta} + \frac{1 + \sqrt{2}}{\sqrt{\eta}} < 1,$$

provided $\eta = \operatorname{Im} \xi$ is sufficiently large. Thus, the claim made at Step 2 is proved.

Let $\mathcal{L}$ be a maximal positive subspace such that $U\mathcal{L} \subseteq \mathcal{L}$, and let $K \in \mathcal{B}(\mathcal{K}^+, \mathcal{K}^-)$, $\|K\| \leqslant 1$ be the angular operator of $\mathcal{L}$. Then $(U - I)\mathcal{L} \subseteq \mathcal{L}$, in particular $(U - I)\mathcal{L}$ is a positive subspace.

Step 3. $(U - I)\mathcal{L} = \mathcal{L}$.

Note that it is sufficient to prove that $(U - I)\mathcal{L}$ is a maximal positive subspace. But, taking into account that

$$\begin{bmatrix} U_{11} - I_+ & U_{12} \\ U_{21} & U_{22} - I_- \end{bmatrix} \begin{bmatrix} I_+ \\ K \end{bmatrix} = \begin{bmatrix} U_{11} - I_+ + U_{12}K \\ U_{21} + (U_{22} - I_-)K \end{bmatrix},$$

it follows that it is sufficient to prove that $(U_{11} - I_+ + U_{12}K)\mathcal{K}^+ = \mathcal{K}^+$.

Note that

$$U_{12} = -(U_{11} - I_+)A_{12}(A_{22} - \zeta I_-)^{-1}$$

and then

$$U_{11} - I_+ + U_{12}K = (U_{11} - I_+)(I_+ - A_{12}(A_{22} - \zeta I_-)^{-1}K).$$

Taking into account the assertions proven at Step 1 and Step 2, it follows that the operator $U_{11} - I_+ + U_{12}K$ has a bounded inverse, hence $(U_{11} - I_+ + U_{12}K)\mathcal{K}^+ = \mathcal{K}^+$. Thus, the claim made at Step 3 is proven.

Finally, Step 3 shows that $\mathcal{L} \subseteq \text{Dom}(A)$ and $A\mathcal{L} \subseteq \mathcal{L}$. ∎

Theorem 9.2.9 provides the necessary link to apply Theorem 9.2.6 in order to obtain maximal semidefinite subspaces that are invariant under certain unbounded operators.

Corollary 9.2.10. *Let A be a maximal dissipative operator in a Kreĭn space $\mathcal{K}$, subject to the following assumptions.*

(α) $\text{Dom}(A)$ contains a maximal uniformly positive subspace.

(β) There exists $\zeta \in \rho(A) \cap \mathbb{C}^+$ and a fundamental decomposition $\mathcal{K} = \mathcal{K}^+[+]\mathcal{K}^-$ such that the operator $P_{\mathcal{K}^+}(A - \overline{\zeta}I)(A - \zeta I)^{-1}|\mathcal{K}^-$ is compact.

Then, there exists $\mathcal{L}$ a maximal positive subspace of $\mathcal{K}$ such that $\mathcal{L} \subseteq \text{Dom}(A)$ and $A\mathcal{L} \subseteq \mathcal{L}$.

We can specialise even more to Pontryagin spaces.

Corollary 9.2.11. *Let A be a maximal dissipative operator in a Pontryagin space with $\kappa_+(\mathcal{K}) < \infty$. Then there exists a maximal positive subspace $\mathcal{L}$ in $\mathcal{K}$ such that $\mathcal{L} \subseteq \text{Dom}(A)$ and $A\mathcal{L} \subseteq \mathcal{L}$.*

In the case of selfadjoint operators on Pontryagin spaces one can draw a stronger conclusion.

Corollary 9.2.12. *Let A be a selfadjoint operator in a Pontryagin space $\mathcal{K}$. Then, there exists a pair $\mathcal{L}_\pm$ of maximal positive/negative subspaces of $\mathcal{K}$ such that $A(\text{Dom}(A) \cap \mathcal{L}_\pm) \subseteq \mathcal{L}_\pm$ and either $\mathcal{L}_+ \subseteq \text{Dom}(A)$ or $\mathcal{L}_- \subseteq \text{Dom}(A)$.*

9.3 Fixed Point Approach

A set $\{s_1, \ldots, s_n\}$ of vectors in the Euclidian space $\mathbb{R}^{n-1}$, for some natural number $n \geq 3$, is called an $(n-1)$-*simplex* if there exists no affine manifold $\mathcal{A}$ properly contained in $\mathbb{R}^{n-1}$ such that $\mathcal{A} \supset \{s_1, \ldots, s_n\}$. The *standard* $(n-1)$-*simplex* is $\{e_0, e_1, \ldots, e_{n-1}\}$, where $e_0 = 0$ and $e_1, \ldots, e_{n-1}$ is the canonical basis of $\mathbb{R}^{n-1}$.

We will use the following result of B. Knaster, C. Kuratowski and S. Mazurkiewicz [89], known by the name of *KKM Theorem*. Given a subset $\mathcal{A}$ of a vector space $\mathcal{X}$ we denote by $\text{Conv}(\mathcal{A})$ the *convex hull* of $\mathcal{A}$, that is, the collection of all convex combinations of vectors in $\mathcal{A}$, more precisely, all vectors $\sum_{j=1}^{N} \alpha_j a_j$, where $N \in \mathbb{N}$ and $\alpha_j \geq 0$ with $\sum_{j=1}^{N} \alpha_j = 1$.

Theorem 9.3.1 (Knaster–Kuratowski–Mazurkiewicz). *Let $\{s_1, \ldots, s_n\} \subset \mathbb{R}^{n-1}$ be an $(n-1)$-simplex and let $V_1, \ldots, V_n$ be relatively open subsets of $\text{Conv}\{s_1, \ldots, s_n\}$ such that, for any proper subset $\sigma \subset \{1, 2, \ldots, n\}$ we have*

$$\left(\bigcap_{k \in \sigma} V_k\right) \cap \text{Conv}\{s_k \mid k \in \sigma\} = \emptyset.$$

9.3 FIXED POINT APPROACH

Then
$$\text{Conv}\{s_1,\ldots,s_n\} \setminus \bigcup_{k=1}^{n} V_k \neq \emptyset.$$

Theorem 9.3.1 has the following equivalent form.

Theorem 9.3.2. *Let $\{s_1,\ldots,s_n\} \subset \mathbb{R}^{n-1}$ be an $(n-1)$-simplex and let $K_1,\ldots,K_n$ be compact subsets in $\mathbb{R}^{n-1}$ such that, for any proper subset $\sigma \in \{1,2,\ldots,n\}$ we have*
$$\text{Conv}\{s_k \mid k \in \sigma\} \subset \bigcup_{k \in \sigma} K_k.$$

Then $\bigcap_{j=1}^{n} K_j \neq \emptyset$.

In the following we apply the KKM Theorem to fixed point theorems in linear topological spaces.

Theorem 9.3.3. *Let $\mathcal{L}$ be a Hausdorff linear topological space, $\mathcal{X}$ a convex and compact subset of $\mathcal{L}$, and K a closed subset of $\mathcal{X} \times \mathcal{X}$, subject to the following conditions.*

(i) $(x,x) \in K$, *for all* $x \in \mathcal{X}$.
(ii) *For all $x \in \mathcal{X}$ the set $\{y \in \mathcal{X} \mid (x,y) \notin K\}$ is either convex or void.*

Then there exists $x_0 \in \mathcal{X}$ such that for all $y \in \mathcal{X}$ we have $(x_0, y) \in K$.

Proof. For arbitrary $y \in \mathcal{X}$ we consider the y-section $K_y = \{x \in \mathcal{X} \mid (x,y) \in K\}$. Then K_y is a nonempty (since, by (i) always $y \in K_y$) and a compact subset of $\mathcal{X}$. Note that the conclusion will follow if we prove that $\bigcap_{y \in \mathcal{X}} K_y \neq \emptyset$. Since $\mathcal{X}$ is compact, this is equivalent with proving that for any finite set of points $\{y_1,\ldots,y_n\} \in \mathcal{X}$ and any natural number n, we have
$$\bigcap_{i=1}^{n} K_{y_i} \neq \emptyset. \tag{3.1}$$

To prove (3.1) we first show that for any proper subset σ of $\{1,\ldots,n\}$ we have
$$\bigcap_{k \in \sigma} V_{y_k} \cap \text{Conv}\{y_k \mid k \in \sigma\} = \emptyset, \tag{3.2}$$

where $V_y = \mathcal{X} \setminus K_y$, $y \in \mathcal{X}$. Indeed, if $x \in \bigcap_{k \in \sigma} V_{y_k} \cap \text{Conv}\{y_k \mid k \in \sigma\}$ then $y_k \in \{y \in \mathcal{X} \mid (x,y) \notin K\}$, and taking into account assumption (ii), it follows that
$$x \in \text{Conv}\{y_k \mid k \in \sigma\} \subseteq \{y \in \mathcal{X} \mid (x,y) \notin K\},$$

and hence, $(x,x) \notin K$, a contradiction with the assumption (i). Thus (3.2) is proved.

Let then $\{s_1,\ldots,s_n\}$ be the standard $(n-1)$-simplex in $\mathbb{R}^{n-1}$ and, since $\mathcal{X}$ is convex we can consider the linear mapping $g\colon \mathrm{Conv}\{s_1,\ldots,s_n\} \to \mathcal{X}$,

$$g\Big(\sum_{i=1}^n \alpha_i s_i\Big) = \sum_{i=1}^n \alpha_i y_i, \quad \alpha_1,\ldots,\alpha_n \in \mathbb{R}.$$

Clearly g is continuous and, taking into account of (3.2), the sets $U_k = g^{-1}(V_k)$, $k = 1,\ldots,n$, satisfy the conditions of Theorem 9.3.1. Therefore,

$$\mathrm{Conv}\{y_1,\ldots,y_n\} \not\subseteq \bigcup_{i=1}^n V_{y_k},$$

which is equivalent with (3.1). ∎

Theorem 9.3.4. *Let $\mathcal{L}$ be a Hausdorff locally convex space, $\mathcal{X}$ a convex and compact subset of $\mathcal{L}$ and $F\colon \mathcal{X} \times \mathcal{X} \to \mathcal{L}$ a jointly continuous mapping subject to the following conditions.*

(a) *F is affine in the second variable, that is, for all $x, y_1, y_2 \in \mathcal{X}$ and $0 \leqslant \alpha \leqslant 1$ we have*
$$F(x, \alpha y_1 + (1-\alpha) y_2) = \alpha F(x, y_1) + (1-\alpha) F(x, y_2).$$

(b) *For any $x \in \mathcal{X}$ there exists $y \in \mathcal{X}$ such that $F(x,y) = 0$.*

Then there exists $x_0 \in \mathcal{X}$ such that $F(x_0, x_0) = 0$.

Proof. Let $\{p_\nu\}_{\nu \in \mathcal{N}}$ be a family of seminorms generating the Hausdorff locally convex topology of $\mathcal{L}$. Consider the family of subsets

$$R_\nu = \{x \in \mathcal{X} \mid p_\nu(F(x,x)) = 0\}, \quad \nu \in \mathcal{N}.$$

Let us fix $\nu_1,\ldots,\nu_n \in \mathcal{N}$ and consider the subset of $\mathcal{X} \times \mathcal{X}$

$$K = \Big\{(x,y) \in \mathcal{X} \times \mathcal{X} \mid \sum_{k=1}^n p_{\nu_k}(F(x,x)) \leqslant \sum_{k=1}^n p_{\nu_k}(F(x,y))\Big\}.$$

We remark that K satisfies the conditions of Theorem 9.3.3. Indeed, since F is jointly continuous, the set K is closed in $\mathcal{X} \times \mathcal{X}$. The condition (i) is clearly satisfied. If $x \in \mathcal{X}$ is arbitrary, then the set

$$\{y \in \mathcal{X} \mid (x,y) \notin \mathcal{X}\} = \Big\{(x,y) \in \mathcal{X} \times \mathcal{X} \mid \sum_{k=1}^n p_{\nu_k}(F(x,x)) > \sum_{k=1}^n p_{\nu_k}(F(x,y))\Big\}$$

is convex since F is affine in the second variable. Thus, condition (ii) is also satisfied. Applying Theorem 9.3.3 we conclude that there exists $x_0 \in \mathcal{X}$ such that for all $y \in \mathcal{X}$ we have

$$\sum_{k=1}^n p_{\nu_k}(F(x_0, x_0)) \leqslant \sum_{k=1}^n p_{\nu_k}(F(x_0, y)). \tag{3.3}$$

9.3 FIXED POINT APPROACH

Due to the assumption (b), we get $y_0 \in \mathcal{X}$ such that $F(x_0, y_0) = 0$. Therefore, from the inequality (3.3) we obtain

$$p_{\nu_k}(F(x_0, y_0)) = 0, \quad k = 1, 2, \ldots, n,$$

or, equivalently

$$\bigcap_{k=1}^{n} F_{\nu_k} \neq \emptyset.$$

Since this holds for any $n \in \mathbb{N}$ and any seminorms $p_{\nu_1}, \ldots, p_{\nu_n}$, and taking into account that $\mathcal{X}$ is compact, we conclude that $\bigcap_{\nu \in \mathcal{N}} F_\nu \neq \emptyset$. Finally, since $\mathcal{L}$ is Hausdorff separated we conclude that there exists $x_0 \in \mathcal{X}$ such that $F(x_0, x_0) = 0$. ∎

Corollary 9.3.5. *If $\mathcal{L}$ is a Hausdorff locally convex space, $\mathcal{X}$ is a compact and convex subset in $\mathcal{L}$ and $f \colon \mathcal{X} \to \mathcal{X}$ is continuous, then there exists $x \in \mathcal{X}$ such that $f(x) = x$.*

Proof. We apply the previous theorem for $F(x, y) = f(x) - y$, $x, y \in X$. ∎

The following is a multi-valued variant of Theorem 9.3.4. We omit the proof since it follows in a very similar fashion to the proof of Theorem 9.3.4.

Theorem 9.3.6. *Let $\mathcal{X}$ be a nonempty, convex and compact subset of a Hausdorff topological space, and let $\Phi \colon \mathcal{X} \to \mathcal{P}(\mathcal{X})$ be a set valued mapping, subject to the following conditions.*

(a) For any $x \in \mathcal{X}$, the set $\Phi(x)$ is nonempty, convex and compact.
(b) The graph of Φ, that is, the set $\{(x, y) \mid y \in \Phi(x)\}$, is closed in the product topology of $\mathcal{X} \times \mathcal{X}$.

Then there exists $x_0 \in \mathcal{X}$ such that $x_0 \in \Phi(x_0)$.

Theorem 9.3.6 enables us to prove more general results on invariant maximal semidefinite subspaces than those in Section 9.2, by allowing us to replace the ambiental Kreĭn spaces with certain G-spaces, as well as by manipulating maximal extensions of certain semidefinite subspaces that are invariant under some sets of operators. We first recall the setting of generalised angular operators in G-spaces as in Section 2.5. Let $(\mathcal{H}, [\cdot, \cdot])$ be a G-space and let $G \in \mathcal{B}(\mathcal{H})$ be the Gram operator of $[\cdot, \cdot]$ with respect to the positive definite inner product $\langle \cdot, \cdot \rangle$ that turns $\mathcal{H}$ into a Hilbert space, that is, G is selfadjoint with respect to the positive definite inner product $\langle \cdot, \cdot \rangle$ and $[x, y] = \langle Gx, y \rangle$ for all $x, y \in \mathcal{H}$. Consider $G = G_+ - G_-$ the Jordan decomposition of G, let $\mathcal{H}_+$ denote the spectral subspace corresponding to the positive semiaxis $[0, +\infty)$ and $\mathcal{H}_-$ be the spectral subspace corresponding to the negative semiaxis $(-\infty, 0)$. Then

$$\mathcal{H} = \mathcal{H}_+ \oplus \mathcal{H}_- \tag{3.4}$$

and, if $x = x_+ + x_-$ and $y = y_+ + y_-$ are the corresponding representations of arbitrary vectors $x, y \in \mathcal{H}$, then

$$[x, y] = \langle G_+ x_+, x_- \rangle - \langle G_- x_-, y_- \rangle.$$

The notions of positivity, negativity, neutrality, etc. with respect to the indefinite inner product $(\mathcal{H}, [\cdot, \cdot])$ are similar to those in Kreĭn spaces. In the following we fix the decomposition (3.4). We state the following theorem for invariant positive subspaces: the reader is invited to make the transcription for invariant negative subspaces.

Theorem 9.3.7. *Assume that the operator G_- has closed range and let $\mathcal{V} \subset \mathcal{B}(\mathcal{H})$ be a finite set of bounded operators subject to the following conditions.*

(i) *For any positive subspace $\mathcal{M}$ of $\mathcal{H}$, the subspace $\bigvee_{V \in \mathcal{V}} V\mathcal{M}$ is also positive.*
(ii) *The operators $G_+^{1/2} P_{\mathcal{H}_+} V P_{\mathcal{H}_-}$ are compact, for all $V \in \mathcal{V}$.*

Then, for any positive subspace $\mathcal{L}_0$ in $\mathcal{H}$ such that $\bigvee_{V \in \mathcal{L}} V\mathcal{L}_0 = \mathcal{L}_0$, there exists a maximal positive subspace $\mathcal{L} \subseteq \mathcal{H}$, invariant under all the operators $V \in \mathcal{V}$, and such that $\mathcal{L}_0 \subseteq \mathcal{L}$.

Proof. We use Theorem 2.5.2 as well as a few by-products of its proof. Since $\mathcal{L}_0$ is a positive subspace of $\mathcal{H}$ and G_- is closed, the angular operator $K_{\mathcal{L}_0}$ can be considered. More precisely, an application of (5.7) from Section 2.5 yields

$$K_{\mathcal{L}_0} = G_-^{-1/2} X_0 G_+^{1/2} | P_+ \mathcal{L}_0, \qquad (3.5)$$

for a uniquely determined contraction $X_0 \colon \operatorname{Clos} \operatorname{Ran}(G_+^{1/2} P_+ | \mathcal{L}_0) \to \mathcal{H}_-$. We are looking for a maximal positive subspace $\mathcal{L}$, in terms of generalised angular operators, that is, a contraction $X \colon \mathcal{H}_+ \ominus \operatorname{Ker}(G) \to \mathcal{H}_-$ such that $K_\mathcal{L} = G_-^{-1/2} X G_+^{1/2}$, or equivalently,

$$\mathcal{L} = \{x + G_-^{-1/2} X G_+^{1/2} x \mid x \in \mathcal{H}_+\}.$$

Then $\mathcal{L}_0 \subseteq \mathcal{L}$ if and only if $X_0 \subseteq X$.

On the other hand, let $\mathcal{M}$ be another maximal positive subspace of $\mathcal{H}$, with angular operator $K_\mathcal{M} = G_-^{-1/2} Y G_+^{1/2}$, where $Y \colon \mathcal{H}_+ \ominus \operatorname{Ker}(G) \to \mathcal{H}_-$ is a uniquely determined contraction. Let us consider the matrix representations

$$V = \begin{bmatrix} V_{11} & V_{12} \\ V_{21} & V_{22} \end{bmatrix},$$

with respect to the decomposition $\mathcal{H} = \mathcal{H}_+ \oplus \mathcal{H}_-$. Then $V\mathcal{L} \subseteq \mathcal{M}$ if and only if

$$G_-^{-1/2} Y G_+^{1/2} V_{11} + G_-^{-1/2} Y G_+^{1/2} V_{12} G_-^{-1/2} X G_+^{1/2} = V_{21} + V_{22} G_-^{-1/2} X G_+^{1/2}. \qquad (3.6)$$

Thus, in order to apply Theorem 9.3.6 we are led to consider the following setting. Let $\mathcal{B}(\mathcal{H}_+ \ominus \operatorname{Ker}(G), \mathcal{H}_-)$ be endowed with the weak operator topology, consider its subset

$$\mathcal{X} = \{X \in \mathcal{B}(\mathcal{H}_+ \ominus \operatorname{Ker}(G), \mathcal{H}_-) \mid \|X\| \leqslant 1, \ X_0 \subseteq X\},$$

and for each $X \in \mathcal{X}$ let $\Phi(X) \subseteq \mathcal{P}(\mathcal{X})$ be the collection of all $Y \in \mathcal{X}$ such that the equation (3.6) holds for all $V \in \mathcal{V}$. We verify that the assumptions of Theorem 9.3.6 are fulfilled.

9.4 Fundamental Reducibility

Firstly we apply Lemma 2.2.4 to see that $\mathcal{X}$ is actually an operator ball, in particular, it is convex and compact with respect to the weak operator topology. Let $X \in \mathcal{X}$ be arbitrary, that is, the maximal positive subspace $\mathcal{L}$ with angular operator $K_L = G_-^{-1/2} X G_+^{1/2}$ is an extension of $\mathcal{L}_0$. By assumption, the linear manifold $\bigvee_{V \in \mathcal{V}} V \mathcal{L}$ is positive and hence, there exists a maximal positive subspace $\mathcal{M} \supseteq \bigvee_{V \in \mathcal{V}} V \mathcal{L}$. Let $K_\mathcal{M}$ be its angular operator, that is, $K_\mathcal{M} = G_-^{-1/2} Y G_+^{1/2}$, where $Y \colon \mathcal{H}_+ \ominus \operatorname{Ker}(G) \to \mathcal{H}_-$ is a uniquely determined contraction. As explained above, the inclusion $V \mathcal{L} \subseteq \mathcal{M}$ means that Y verifies the equation (3.6). In addition, by assumption,

$$\mathcal{L}_0 = \bigvee_{V \in \mathcal{V}} V \mathcal{L}_0 \subseteq \bigvee_{V \in \mathcal{V}} V \mathcal{L} \subseteq \mathcal{M},$$

that is, $\mathcal{L}_0 \subseteq \mathcal{M}$, or equivalently $X_0 \subseteq Y$, and hence $Y \in \mathcal{X}$. In other words, we have proven that for any $X \in \mathcal{X}$ the set $\Phi(X)$ is nonempty.

In order to prove that the graph of Φ is closed with respect to the product of the weak operator topology on $\mathcal{X} \times \mathcal{X}$, we use the definition of the mapping Φ by means of the operator equations (3.6) to see that this is equivalent with the joint continuity of the mapping $\Psi \colon \mathcal{X} \times \mathcal{X} \to \mathcal{B}(\mathcal{H}_+ \ominus \operatorname{Ker}(G), \mathcal{H}_-)$

$$\Psi(X,Y) = G_-^{-1/2} Y G_+^{1/2} V_{11} + G_-^{-1/2} Y G_+^{1/2} V_{12} G_-^{-1/2} X G_+^{1/2} - V_{21} - V_{22} G_-^{-1/2} X G_+^{1/2},$$

with respect to the weak operator topology. Then we observe that only the second term in the expression of Ψ needs special attention, and for this we employ the assumption that all the operators $G_+^{1/2} V_{12}$, $V \in \mathcal{V}$, are compact, to see that it is jointly weak operator continuous.

Thus, Φ satifies the hypotheses of Theorem 9.3.6 and hence, there exists $X \in \mathcal{X}$ such that $X \in \Phi(X)$, that is, there exists a maximal positive subspace $\mathcal{L}$ invariant under all operators $V \in \mathcal{V}$, and such that $\mathcal{L}_0 \subseteq \mathcal{L}$. ∎

9.4 Fundamental Reducibility

Given $\mathcal{K}$ a Kreĭn space, an operator $A \in \mathcal{B}(\mathcal{K})$ is *fundamentally reducible* if there exists a fundamental decomposition $\mathcal{K} = \mathcal{K}^-[+]\mathcal{K}^+$ that reduces A, that is, $A\mathcal{K}^\pm \subseteq \mathcal{K}^\pm$. Equivalently, this means that, with respect to the decomposition $\mathcal{K} = \mathcal{K}^-[+]\mathcal{K}^+$, the operator A has a diagonal operator block matrix representation.

A subset $\mathcal{A} \subset \mathcal{B}(\mathcal{K})$, where $\mathcal{K}$ is a Kreĭn space, is called *jointly fundamentally reducible* if there exists a fundamental decomposition $\mathcal{K} = \mathcal{K}^+[+]\mathcal{K}^-$ with the property $A\mathcal{K}^\pm \subseteq \mathcal{K}^\pm$ for all $A \in \mathcal{A}$, equivalently, there exists a fundamental symmetry J on $\mathcal{K}$ such that $JA = AJ$ for all $A \in \mathcal{A}$.

Theorem 9.4.1. *Let $\mathcal{U}$ be a group of unitary operators on a Kreĭn space $\mathcal{K}$, that is jointly fundamentally reducible. Then, for any pair $(\mathcal{L}_+, \mathcal{L}_-)$ of orthogonal positive/negative subspaces that are invariant under each operator $U \in \mathcal{U}$, there exists a pair of orthogonal* maximal positive/negative subspaces $(\widetilde{\mathcal{L}}_+, \widetilde{\mathcal{L}}_-)$, that are invariant under each operator in $\mathcal{U}$, and such that $\mathcal{L}_+ \subseteq \widetilde{\mathcal{L}}_+$ and $\mathcal{L}_- \subseteq \widetilde{\mathcal{L}}_-$. In addition, if the pair $(\mathcal{L}_+, \mathcal{L}_-)$ consists of strictly definite (uniformly definite) subspaces, then the pair $(\widetilde{\mathcal{L}}_+, \widetilde{\mathcal{L}}_-)$ can be determined such that it consists of maximal strictly definite (maximal uniformly definite) subspaces.

Proof. Fix $\mathcal{K} = \mathcal{K}^+[+]\mathcal{K}^-$ a fundamental decomposition reducing all the unitary operators $U \in \mathcal{U}$. By the angular operator representation, there exist unique subspaces $\mathrm{Dom}(T)$ and $\mathrm{Dom}(S)$ and contractions $T \colon \mathrm{Dom}(T)(\subseteq \mathcal{K}^+) \to \mathcal{K}^-$ and $S \colon \mathrm{Dom}(S)(\subseteq \mathcal{K}^-) \to \mathcal{K}^+$ such that

$$\mathcal{L}_+ = \{x + Tx \mid x \in J^+\mathcal{L}_+ = \mathrm{Dom}(T)\}, \quad \mathcal{L}_- = \{Sy + y \mid y \in J^-\mathcal{L}_- = \mathrm{Dom}(S)\},$$

and
$$\langle Tx, y \rangle = \langle x, Sy \rangle, \quad x \in \mathrm{Dom}(T),\ y \in \mathrm{Dom}(S).$$

Denote $T_0 = P_{\mathrm{Dom}(S)} T = S^* | \mathrm{Dom}(T)$. As in Lemmas 2.2.4 and 2.2.6, there exist uniquely determined contractions $\Gamma_1 \colon \mathcal{K}^+ \ominus \mathrm{Dom}(T) \to \mathcal{D}_{T_0^*}$ and $\Gamma_2 \colon \mathcal{D}_{T_0} \to \mathcal{K}^- \ominus \mathrm{Dom}(S)$ such that

$$T = \begin{bmatrix} T_0 \\ \Gamma_2 D_{T_0} \end{bmatrix}, \quad S^* = [T_0 \quad D_{T_0^*} \Gamma_1]. \tag{4.1}$$

By assumption, every operator $U \in \mathcal{U}$ has the diagonal matrix representation

$$U = \begin{bmatrix} U^+ & 0 \\ 0 & U^- \end{bmatrix}, \tag{4.2}$$

with respect to the fundamental decomposition $\mathcal{K} = \mathcal{K}^+[+]\mathcal{K}^-$.

It is readily seen that the invariance of $\mathcal{L}_+$ under U is equivalent with

$$U^+ \mathrm{Dom}(T) = \mathrm{Dom}(T), \quad U^- T = T U^+ | \mathrm{Dom}(T). \tag{4.3}$$

In particular, the decomposition $\mathcal{K}^+ = \mathrm{Dom}(T) \oplus (\mathcal{K}^+ \ominus \mathrm{Dom}(T))$ reduces U^+, that is, U^+ is diagonal with respect to it,

$$U^+ = \begin{bmatrix} U_1^+ & 0 \\ 0 & U_2^+ \end{bmatrix}. \tag{4.4}$$

Similarly, the invariance of $\mathcal{L}_-$ under U is equivalent with

$$U^- \mathrm{Dom}(S) = \mathrm{Dom}(S), \quad U^+ S = S U^- | \mathrm{Dom}(S), \tag{4.5}$$

hence, with respect to the decomposition $\mathcal{K}^- = \mathrm{Dom}(S) \oplus (\mathcal{K}^- \ominus \mathrm{Dom}(S))$, U^- has the diagonal representation

$$U^- = \begin{bmatrix} U_1^- & 0 \\ 0 & U_2^- \end{bmatrix}. \tag{4.6}$$

9.4 Fundamental Reducibility

Now, the latter condition in (4.3) and the representation of T as in (4.1) imply

$$U_1^- T_0 = T_0 U_1^+, \quad U_2^- \Gamma_2 D_{T_0} = \Gamma_2 D_{T_0} U_1^+, \tag{4.7}$$

and, consequently,

$$D_{T_0} U_1^+ = U_1^+ D_{T_0}, \quad U_1^- D_{T_0^*} = D_{T_0^*} U_1^-, \quad U_2^- \Gamma_2 D_{T_0} = \Gamma_2 U_1^+ D_{T_0}. \tag{4.8}$$

Thus,

$$U_1^+ \mathcal{D}_{T_0} = \mathcal{D}_{T_0}, \quad U_2^- \Gamma_2 = \Gamma_2 U_1^+ |\mathcal{D}_{T_0}, \quad D_{\Gamma_2^*} U_2^- = U_2^- D_{\Gamma_2^*}. \tag{4.9}$$

Similarly, the latter condition in (4.5) and the representation of S as in (4.1) imply

$$T_0^* U_1^- = U_1^+ T_0^*, \quad U_2^+ \Gamma_1^* D_{T_0^*} = \Gamma_1^* D_{T_0^*} U_1^-, \tag{4.10}$$

and, consequently,

$$U_2^+ \Gamma_1^* D_{T_0^*} = \Gamma_1^* U_1^- D_{T_0^*}. \tag{4.11}$$

Thus, $U_1^- \mathcal{D}_{T_0^*} = \mathcal{D}_{T_0^*}$ and $U_2^+ \Gamma_1^* = \Gamma_1^* U_1^- |\mathcal{D}_{T_0^*}$, and hence

$$D_{\Gamma_1} U_2^+ = U_2^+ D_{\Gamma_1}. \tag{4.12}$$

Since U are all unitary operators and hence $U\mathcal{L}^\perp = (U\mathcal{L})^\perp$ for any subspace $\mathcal{L}$, according to Theorem 2.2.7 we are searching for an operator

$$\widetilde{T} = \begin{bmatrix} T & D_{T^*} \Gamma_1 \\ \Gamma_2 D_T & -\Gamma_2 T^* \Gamma_1 + D_{\Gamma_2^*} \Gamma D_{\Gamma_1} \end{bmatrix}, \tag{4.13}$$

where $\Gamma \colon \mathcal{D}_{\Gamma_1} \to \mathcal{D}_{\Gamma_2^*}$ is a contraction, and such that the subspace $G(\widetilde{T}) = \{x + \widetilde{T}x \mid x \in \mathcal{K}^+\}$ is invariant under all the operators $U \in \mathcal{U}$. Then U leaves invariant $G(\widetilde{T})$ if and only if $\widetilde{T} U^+ = U^- \widetilde{T}$. Corresponding to the decomposition

$$\mathcal{K} = \mathrm{Dom}(T) \oplus (\mathcal{K}^+ \ominus \mathrm{Dom}(T)) \oplus \mathrm{Dom}(S) \oplus (\mathcal{K}^- \ominus \mathrm{Dom}(S)),$$

the unitary operator U has the diagonal matrix representation, cf. (4.2), (4.4), and (4.6). Using (4.13), $\widetilde{T} U^+ = U^- \widetilde{T}$ is equivalent with

$$\begin{bmatrix} T_0 U_1^+ & D_{T_0^*} \Gamma_1 U_2^+ \\ \Gamma_2 D_{T_0} U_1^+ & -\Gamma_2 T_0^* \Gamma_1 U_2^+ + D_{\Gamma_2^*} \Gamma D_{\Gamma_1} U_2^+ \end{bmatrix}$$
$$= \begin{bmatrix} U_1^- T_0 & U_1^- D_{T_0^*} \Gamma_1 \\ U_2^- \Gamma_2 D_{T_0} & -U_2^- \Gamma_2 T_0^* \Gamma_1 + U_2^- D_{\Gamma_2^*} \Gamma D_{\Gamma_1} \end{bmatrix}. \tag{4.14}$$

Identifying the corresponding entries in (4.14) and taking into account (4.7), (4.8), and (4.10), this is equivalent with

$$-\Gamma_2 T_0^* \Gamma_1 U_2^+ + D_{\Gamma_2^*} \Gamma D_{\Gamma_1} U_2^+ = -U_2^- \Gamma_2 T_0^* \Gamma_1 + U_2^- D_{\Gamma_2^*} \Gamma D_{\Gamma_1}. \tag{4.15}$$

On the other hand, from (4.8) and (4.12) we have

$$\Gamma_2 T_0^* \Gamma_1 U_2^* = \Gamma_2 T_0^* P_{\mathcal{D}_{T_0^*}} U_1^- \Gamma_1 = \Gamma_2 U_1^+ P_{\mathcal{D}_{T_0}} T_0^* \Gamma_1$$
$$= U_2^- \Gamma_2 T_0^* \Gamma_1,$$

hence, $G(\widetilde{T})$ is invariant under U if and only if

$$D_{\Gamma_2^*} \Gamma D_{\Gamma_1} U_2^+ = U_2^- D_{\Gamma_2^*} \Gamma D_{\Gamma_1}. \tag{4.16}$$

Taking into account (4.9) and (4.12), it turns out that (4.16) is equivalent with

$$U_2^- P_{\mathcal{D}_{\Gamma_2^*}} \Gamma = \Gamma U_2^+ P_{\mathcal{D}_{\Gamma_1}}. \tag{4.17}$$

Thus, the existence of maximal dual pairs of subspaces that are invariant under all operators $U \in \mathcal{U}$ is reduced to the existence of a contraction $\Gamma \colon \mathcal{D}_{\Gamma_1} \to \mathcal{D}_{\Gamma_2^*}$ such that (4.17) holds for all $U \in \mathcal{U}$. Clearly, $\Gamma = 0$ is a solution. ∎

Remark 9.4.2. Inspecting the proof of Theorem 9.4.1, it follows that the formula (4.13), with the constraint (4.17), provides a description of all maximal dual pairs of subspaces invariant under a group $\mathcal{U}$ of fundamentally reducible operators.

Lemma 9.4.3. *Let U be a bounded unitary operator on a Kreĭn space $\mathcal{K}$. Then U is fundamentally reducible if and only if it is similar to a unitary operator on a Hilbert space, that is, there exist a Hilbert space $\mathcal{H}$ and $S \in \mathcal{B}(\mathcal{K}, \mathcal{H})$ boundedly invertible such that SUS^{-1} is unitary in $\mathcal{H}$.*

Proof. If U is fundamentally reducible, we choose the fundamental decomposition of $\mathcal{K} = \mathcal{K}^+[+]\mathcal{K}^-$ that reduces U and hence, with respect to the underlying Hilbert space structure on $\mathcal{K}$, which we denote by $\mathcal{H}$, U is unitary. Then $S = I$.

Conversely, let $\mathcal{H}$ be a Hilbert space and let $S \in \mathcal{B}(\mathcal{K}, \mathcal{H})$ be a boundedly invertible operator such that $V = SUS^{-1}$ is unitary on $\mathcal{H}$. Fix an arbitrary fundamental decomposition $\mathcal{K} = \mathcal{K}^+[+]\mathcal{K}^-$ of $\mathcal{K}$, J the associated fundamental symmetry, and let $(\mathcal{K}; \langle \cdot, \cdot \rangle_J)$ be the associated Hilbert space.

In order to keep the notation simple, let us first note that, without restricting the generality, we can assume that $\mathcal{H} = (\mathcal{K}; \langle \cdot, \cdot \rangle_J)$ and $S \in \mathcal{B}(\mathcal{H})$ is a positive boundedly invertible operator. Indeed, to see this we perform the polar decomposition $S = W|S|$ and replace S with its absolute value $|S|$.

With the assumption as above, let us then note that V commutes with $G = S^{-1}JS^{-1}$. To see this we take into account that $V^{-1} = V^*$ and that $U^{-1} = JU^*J$ and get

$$V^{-1} = (SUS^{-1})^* = S^{-1}U^*S = S^{-1}JU^{-1}JS = S^{-1}JS^{-1}V^{-1}SJS,$$

that is, $V = GVG^{-1}$, and hence $VG = GV$.

Since G is a boundedly invertible selfadjoint operator in the Hilbert space $(\mathcal{K}; \langle \cdot, \cdot \rangle_J)$ we consider its spectral measure E_G as well as its spectral subspaces $E_G(-\infty, 0)\mathcal{K}$ and

9.4 Fundamental Reducibility

$E_G(0, +\infty)\mathcal{K}$, which are invariant under V, and hence $\mathcal{L}^- = S^{-1}E_G(-\infty, 0)\mathcal{K}$ and $\mathcal{L}^+ = S^{-1}(0, +\infty)\mathcal{K}$ are invariant under $U = S^{-1}VS$. It remains to verify that $\mathcal{L}^+$ is orthogonal on $\mathcal{L}^-$ with respect to the inner product $[\cdot, \cdot]$, that $\mathcal{L}^+ \dot{+} \mathcal{L}^- = \mathcal{K}$, and that $\mathcal{L}^\pm$ is positive/negative. Thus, by Corollary 2.1.6, it follows that $\mathcal{K} = \mathcal{L}^+[+]\mathcal{L}^-$ is a fundamental decomposition of $\mathcal{K}$ that reduces U. ∎

A group $\mathcal{U}$ of bounded unitary operators on a Kreĭn space $\mathcal{K}$ is called *uniformly bounded* if $\sup_{U \in \mathcal{U}} \|U\| < \infty$ with respect to some, and hence, to any, fundamental norm $\|\cdot\|$ on $\mathcal{K}$. A unitary operator $U \in \mathcal{B}(\mathcal{K})$ is called *power bounded* if the group $\{U^n \mid n \in \mathbb{Z}\}$ is uniformly bounded. Sometimes, power bounded operators are called *stable*. Since $U^{-n} = U^{\sharp n}$ for all $n \in \mathbb{Z}$ and $\|U^\sharp\| = \|U\|$ for any fundamental norm $\|\cdot\|$ on $\mathcal{K}$, in the definition of a power bounded unitary operator it is sufficient to ask that $\{U^n \mid n \geqslant 0\}$ is uniformly bounded.

It is easy to see that a group $\mathcal{U}$ of joint fundamentally reducible bounded unitary operators on a Kreĭn space is necessarily uniformly bounded. The converse implication is closely related to the assumption of amenability of the group. Although we will only apply amenability to groups, we consider now amenability for semigroups: one reason is that there is no additional complication in the group case and, in addition, for a group the left multiplication is usually defined differently due to functorial reasons that fall outside of our interest.

Given an abstract semigroup $\mathcal{G}$, let $\ell^\infty(\mathcal{G})$ denote the Banach algebra of all bounded complex functions $\varphi \colon \mathcal{G} \to \mathbb{C}$, with uniform norm (the sup-norm). A function $\varphi \in \ell^\infty(\mathcal{G})$ is called a *positive function* if $\varphi(g) \geqslant 0$ for all $g \in \mathcal{G}$. For any $g \in \mathcal{G}$ there exists the left shift $L_g \colon \ell^\infty(\mathcal{G}) \to \ell^\infty(\mathcal{G})$ defined by $(L_g\varphi)(h) = \varphi(gh)$ for all $h \in \mathcal{G}$. A *left invariant mean* on $\mathcal{G}$ is, by definition, a linear mapping $\mu \colon \ell^\infty(\mathcal{G}) \to \mathbb{C}$ subject to the following properties.

(LIM1) μ is unital, that is, $\mu(1) = 1$, where we denote by $1 \in \ell^\infty(\mathcal{G})$ the function constantly equal to 1.

(LIM2) μ is positive, that is, $\mu(\varphi) \geqslant 0$ for all positive $\varphi \in \ell^\infty(\mathcal{G})$ (that is, $\varphi(g) \geqslant 0$ for all $g \in \mathcal{G}$).

(LIM3) μ is left invariant, that is, $\mu(L_g\varphi) = \mu(\varphi)$ for all $\varphi \in \ell^\infty(\mathcal{G})$ and all $g \in \mathcal{G}$.

A left invariant mean is given the acronym *LIM*, or a *Banach LIM* (erroneously, sometimes the term Banach limit is used, especially when the semigroup is $\mathbb{N}$).

Axiom (LIM2) implies that μ is symmetric, in the sense that $\mu(\varphi^*) = \overline{\mu(\varphi)}$ for all $\varphi \in \ell^\infty(\mathcal{G})$, where $\varphi^*(g) = \overline{\varphi(g)}$ is the natural involution on $\ell^\infty(\mathcal{G})$. In particular, if $\varphi \in \ell^\infty(\mathcal{G})$ is real valued then $\mu(\varphi) \in \mathbb{R}$. Then, it is easy to see that axioms (LIM1) and (LIM2), together, make a statement equivalent with the following: for any real valued function $\varphi \in \ell^\infty(\mathcal{G})$ we have

$$\inf_{g \in \mathcal{G}} \mu(g) \leqslant \mu(\varphi) \leqslant \sup_{g \in \mathcal{G}} \mu(g). \tag{4.18}$$

A semigroup that has a left invariant mean is called *amenable*.

Theorem 9.4.4. *Let $\mathcal{U}$ be an amenable group of bounded unitary operators on a Kreĭn space $\mathcal{K}$. If $\mathcal{U}$ is uniformly bounded then it is jointly fundamentally reducible.*

Proof. Let J be a fundamental symmetry of $\mathcal{K}$ and $\|\cdot\|$ the corresponding fundamental norm. Denote
$$c = \sup_{U \in \mathcal{U}} \|U\| < \infty,$$
and let μ be an invariant mean on the amenable group $\mathcal{U}$. For arbitrary $x, y \in \mathcal{K}$ consider the mapping $\Phi_{x,y} : \mathcal{U} \to \mathbb{C}$ defined in the following way
$$\mathcal{U} \ni U \mapsto \Phi_{x,y}(U) = \langle Ux, Uy \rangle_J \in \mathbb{C}.$$
Then
$$\sup_{U \in \mathcal{U}} |\Phi_{x,y}(U)| = \sup_{U \in \mathcal{U}} |\langle Ux, Uy \rangle_J| \leqslant c^2 \|x\| \|y\|,$$
hence $\Phi_{x,y} \in \ell^\infty(\mathcal{U})$. This proves that it makes sense to consider the mapping
$$\mathcal{K} \times \mathcal{K} \ni (x, y) \mapsto \mu(\Phi_{x,y}) \in \mathbb{C}, \tag{4.19}$$
and a moment of thought shows that this is a positive definite inner product on $\mathcal{K}$.

We claim that the new inner product $\mu(\Phi_{\cdot,\cdot})$ defined at (4.19) is equivalent with the positive definite inner product $\langle \cdot, \cdot \rangle_J$. Indeed, taking into account the inequalities in (4.18) it follows that for any $x \in \mathcal{K}$,
$$\mu(\Phi_{x,x}) \leqslant \sup\{\Phi_{x,x}(U) \mid U \in \mathcal{U}\} \leqslant c^2 \|x\|^2,$$
and
$$\mu(\Phi_{x,x}) \geqslant \inf\{\Phi_{x,x}(U) \mid U \in \mathcal{U}\} \geqslant \frac{1}{c^2} \|x\|^2.$$
On the other hand, from the left invariance of the mean μ we have
$$\mu(\langle VUx, VUy \rangle_J) = \mu(\langle Ux, Uy \rangle_J),$$
for all $U, V \in \mathcal{U}$ and all $x, y \in \mathcal{K}$, hence $(\mathcal{K}, \mu(\Phi_{\cdot,\cdot}))$ is a Hilbert space with respect to which $\mathcal{U}$ is a group of unitary operators. Then, taking into account the equivalence of the fundamental norm $\|\cdot\|$ with that induced by the inner product $\mu(\Phi_{\cdot,\cdot})$, see (4.19), by Lemma 9.4.3 it follows that $\mathcal{U}$ is jointly fundamentally reducible. ∎

In order to substantiate the statement in Theorem 9.4.4 we also prove the following.

Theorem 9.4.5. *Any Abelian group is amenable.*

In order to prove this theorem, we first reformulate the definition of amenability of a semigroup in a more convenient way.

9.4 Fundamental Reducibility

Lemma 9.4.6. *Let $\mathcal{G}$ be a semigroup. The following assertions are equivalent.*

(i) $\mathcal{G}$ *is amenable.*

(ii) *For all $N \in \mathbb{N}$, all real valued functions $\varphi_1, \ldots, \varphi_N \in \ell^\infty(\mathcal{G})$, and all $g_1, \ldots, g_N \in \mathcal{G}$, we have*

$$\inf\{\sum_{k=1}^{N}(\varphi_k(h) - \varphi_k(g_k h)) \mid h \in \mathcal{G}\} \leqslant 0.$$

Proof. Let us consider

$$\mathcal{L} := \text{Clos Lin}_\mathbb{R}\{\varphi - L_g\varphi \mid \varphi \in \ell^\infty_\mathbb{R}(\mathcal{G}),\ g \in \mathcal{G}\}$$

as a real linear subspace of $\ell^\infty_\mathbb{R}(\mathcal{G})$. Then, clearly, the assertion (ii) is equivalent with

(iii) $\inf\{\varphi(g) \mid g \in \mathcal{G}\} \leqslant 0$, for all $\varphi \in \mathcal{L}$.

So, in the following we prove that the assertions (i) and (iii) are equivalent.

(i)$\Rightarrow$(iii). Let μ be a LIM on $\mathcal{G}$. Then $\mu|\mathcal{L} = 0$ hence, for any $\varphi \in \mathcal{L}$, by (4.18) we have

$$0 = \mu(\varphi) \geqslant \inf_{g \in \mathcal{G}}\varphi(g).$$

(iii)$\Rightarrow$(i). Consider the open convex subset K of the real linear space $\ell^\infty_\mathbb{R}(\mathcal{G})$

$$K = \{\varphi \in \ell^\infty_\mathbb{R}(\mathcal{G}) \mid \inf_{g \in \mathcal{G}}\varphi(g) > 0\},$$

and observe that it is disjoint to the closed subspace $\mathcal{L}$. Thus, we can apply the Hahn–Banach Theorem in order to get a bounded real linear functional $\mu \colon \ell^\infty_\mathbb{R}(\mathcal{G}) \to \mathbb{R}$ such that

$$\mu|\mathcal{L} = 0 \text{ and } \mu(\varphi) > 0, \text{ for all } \varphi \in K.$$

By scaling, without loss of generality we can assume that $\mu(1) = 1$. Extending μ to a bounded complex linear functional μ on $\ell^\infty(\mathcal{G}) = \ell^\infty_\mathbb{R}(\mathcal{G}) + i\ell^\infty_\mathbb{R}(\mathcal{G})$ by

$$\mu(\varphi + i\psi) = \mu(\varphi) + i\mu(\psi), \quad \varphi, \psi \in \ell^\infty_\mathbb{R}(\mathcal{G}),$$

we get a LIM μ on $\mathcal{G}$. ∎

Proof of Theorem 9.4.5. Fix $N \in \mathbb{N}$, $\varphi_1, \ldots, \varphi_N \in \ell^\infty_\mathbb{R}(\mathcal{G})$, and $g_1, \ldots, g_N \in \mathcal{G}$. For any $p \in \mathbb{N}$ let

$$\Lambda_p := \{(\lambda_1, \ldots, \lambda_N) \in \mathbb{N}^p \mid \lambda_k \leqslant p,\ k = 1, \ldots, N\}$$

and observe that $\text{card}(\Lambda_p) = p^N$. Define

$$g(\lambda) = g_1^{\lambda_1} \cdots g_N^{\lambda_N}, \text{ for all } \lambda = (\lambda_1, \ldots, \lambda_N) \in \Lambda_p$$

and observe that, due to the assumption that $\mathcal{G}$ is Abelian, all terms in the sum

$$\sum_{j=1}^{N}\sum_{\lambda \in \Lambda_p}(\varphi_j(g(\lambda)) - \varphi_j(g_j g(\lambda)))$$

cancel, except, possibly, those with some λ_k equal to either 1 or p. Since, for each $k = 1, \ldots, N$ we have

$$\operatorname{card}(\{\lambda \in \Lambda_p \mid \lambda_k = p\}) = \operatorname{card}(\{\lambda \in \Lambda_p \mid \lambda_k = 1\}) = p^{N-1}$$

it follows that

$$\sum_{j=1}^{N} \sum_{\lambda \in \Lambda_p} (\varphi_j(g(\lambda)) - \varphi_j(g_j g(\lambda))) \leqslant \sum_{j=1}^{N} 2p^{N-1} \|\varphi_j\|_\infty = Mp^{N-1}, \qquad (4.20)$$

where $M = 2\sum_{j=1}^{N} \|\varphi_j\|_\infty < \infty$. Thus,

$$\inf\{\sum_{j=1}^{N} (\varphi_j(h) - \varphi_j(g_j h)) \mid h \in \mathcal{G}\} \leqslant \frac{1}{p^N} \sum_{j=1}^{N} \sum_{\lambda \in \Lambda_p} (\varphi_j(g(\lambda)) - \varphi_j(g_j g(\lambda)))$$

hence, by (4.20),

$$\leqslant \frac{M}{p} \xrightarrow[p \to \infty]{} 0,$$

which implies

$$\inf\{\sum_{j=1}^{N} (\varphi_j(h) - \varphi_j(g_j h)) \mid h \in \mathcal{G}\} \leqslant 0.$$

By Lemma 9.4.6 it follows that G is amenable. ∎

Corollary 9.4.7. *A commutative group of bounded unitary operators on a Kreĭn space is jointly fundamentally reducible if and only if it is uniformly bounded.*

Proof. Indeed, we use Theorem 9.4.5 and then apply Theorem 9.4.4. ∎

We end this section with a consequence for joint fundamentally reducible selfadjoint operators.

Proposition 9.4.8. *Let $\mathfrak{B}$ be a Boolean algebra of selfadjoint projections on a Kreĭn space $\mathcal{K}$. Then $\mathfrak{B}$ is jointly fundamentally reducible if and only if it is uniformly bounded.*

Proof. Clearly, if $\mathfrak{B}$ is a jointly fundamentally reducible Boolean algebra of selfadjoint projections, then there exists a fundamental symmetry J in $\mathcal{K}$ commuting with all operators $P \in \mathfrak{B}$. In particular, with respect to the Hilbert space $(\mathcal{K}, \langle \cdot, \cdot \rangle_J)$, all the operators $P \in \mathfrak{B}$ are selfadjoint projections, hence of norm 1 (except the null projection).

Conversely, assume that the Boolean algebra $\mathfrak{B}$ of selfadjoint projections is uniformly bounded. Note that whenever P is a selfadjoint projection in $\mathcal{K}$, the operator $U = 1 - 2P$ is a symmetry, that is, $U^\sharp = U = U^{-1}$. We consider $\mathcal{U} = \{1 - 2P \mid P \in \mathfrak{B}\}$. Since $\mathfrak{B}$ is a Boolean algebra it follows that $\mathcal{U}$ is a commutative group of unitary operators. Thus, we can apply Corollary 9.4.7 and conclude that $\mathcal{U}$ is fundamentally reducible. This means that there exists a fundamental symmetry J commuting with all the operators $U \in \mathcal{U}$, and hence, with all the selfadjoint projections $P \in \mathfrak{B}$. ∎

9.5 Strong Stability

In connection with fundamental reducibility, power boundedness, and invariant maximal semidefinite subspaces, there is one more concept that we consider. A bounded unitary operator $U \in \mathcal{B}(\mathcal{K})$ is called *strongly stable* if for any (equivalently, there exists a) fundamental norm $\|\cdot\|$ on $\mathcal{K}$, there exists $\varepsilon > 0$ such that, any unitary operator $V \in \mathcal{B}(\mathcal{K})$ with $\|U - V\| < \varepsilon$ is fundamentally reducible. Of course, this definition implies that a strongly stable unitary operator should be fundamentally reducible as well, in particular its spectrum lies entirely on the unit circle. We present a characterisation of a strongly stable unitary operator from the spectral point of view.

We need a lemma on stability of certain bounded selfadjoint projections under small perturbations that may be interesting by itself.

Lemma 9.5.1. *Let $\mathcal{L}$ be a maximal uniformly positive subspace in the Kreĭn space $\mathcal{K}$ and denote by P the bounded selfadjoint projection onto $\mathcal{L}$. If $Q \in \mathcal{B}(\mathcal{K})$ is another selfadjoint projection in $\mathcal{K}$ such that, with respect to some fundamental norm $\|\cdot\|$, we have $\|P - Q\|$ sufficiently small, then $Q\mathcal{K}$ is a maximal uniformly positive subspace as well.*

Proof. Since $\mathcal{L} = P\mathcal{K}$ is a maximal uniformly positive subspace in $\mathcal{K}$ and hence, without loss of generality, we can assume that $P\mathcal{K} = \mathcal{K}^+$, where $\mathcal{K} = \mathcal{K}^-[+]\mathcal{K}^+$ is a fundamental decomposition associated to a fundamental $\|\cdot\|$. Then, letting $J = J^+ - J^-$ be the Jordan decomposition of the associated fundamental symmetry J, we have $P = J^+$. Let us assume that $\delta = \|J^+ - Q\| < 1$. Then, for any $x \in \mathrm{Ran}(Q) = Q\mathcal{K}$ we have $Qx = x = J^+x + J^-x$ and hence

$$\|J^-x\|^2 = \|(J^+ - Q)x\|^2$$
$$\leqslant \delta^2\|x\|^2 = \delta^2\|J^+\|^2 + \delta^2\|J^-x\|^2,$$

hence

$$\|J^-x\| \leqslant \frac{\delta}{\sqrt{1-\delta^2}}\|J^+x\|.$$

This shows that the linear operator $T\colon J^+x \mapsto J^-x$ is correctly defined and a Hilbert space contraction $T\colon J^+\mathrm{Ran}(Q) \to \mathcal{K}^-$. Then $\mathrm{Ran}(Q)$ is the graph of T, hence it is a positive subspace, actually a uniformly positive subspace.

In order to see that $\mathrm{Ran}(Q)$ is a maximal positive subspace, by Corollary 2.1.7, we have to prove that $J^+\mathrm{Ran}(Q) = \mathcal{K}^+$. To see this, assume, by contradiction, that $\mathcal{K}^+ \setminus J^+\mathrm{Ran}(Q) \neq \emptyset$. Since $J^+\mathrm{Ran}(Q)$ is closed, we then can find a vector x in $\mathcal{K}^+$ that is orthogonal on $J^+\mathrm{Ran}(Q)$. Then

$$\|J^+x - Qx\|^2 = \|x - Qx\|^2 = \|x - J^+Qx - J^-Qx\|^2$$
$$= \|x\|^2 + \|J^+Qx\|^2 + \|J^-Qx\|^2 \geqslant \|x\|^2,$$

and hence $\|P - Q\| = \|J^+ - Q\| \geq 1$, a contradiction. ∎

Recall that, given an operator $T \in \mathcal{B}(\mathcal{X})$, for some complex Banach space $\mathcal{X}$, a nonempty closed subset K of $\sigma(T)$ is called a *spectral set* of T if there exist open nonempty sets U and V in $\mathbb{C}$ such that $K \subset U$, $\sigma(T) \setminus K \subseteq V$, and $U \cap V = \emptyset$.

Theorem 9.5.2. *A bounded unitary operator $U \in \mathcal{B}(\mathcal{K})$ is strongly stable if and only if its spectrum $\sigma(U)$ can be split into two disjoint spectral sets $\sigma_\pm$ such that the corresponding spectral subspaces $E(\sigma_\pm; U)\mathcal{K}$ are (actually, maximal) uniformly positive/negative subspaces.*

Proof. Assume that U is a strongly stable unitary operator, let $\mathcal{K} = \mathcal{K}^-[+]\mathcal{K}^+$ be a fundamental decomposition that reduces U, and let $U_\pm = P_{\mathcal{K}^\pm}U|\mathcal{K}^\pm$ be the corresponding compressions. Note that $U_\pm \in \mathcal{B}(\mathcal{K}^\pm)$ are Hilbert space unitary operators and that $\sigma(U) = \sigma(U_-) \cup \sigma(U_+) \subseteq \mathbb{T}$. In particular, U is a Hilbert space unitary operator as well, and let E denote its spectral measure.

We claim that $\sigma(U_-) \cap \sigma(U_+) = \emptyset$. Indeed, assume, by contradiction, that there exists $\theta_0 \in [0, 2\pi)$ such that $e^{i\theta_0} \in \sigma(U_-) \cap \sigma(U_+)$, let $0 \leq \theta_1 < \theta_0 < \theta_2 \leq 2\pi$, and let $\Delta_{\theta_1,\theta_2} = \{e^{i\theta} \mid \theta_1 \leq \theta \leq \theta_2\}$ be a closed arc on the unit circle $\mathbb{T}$. We prove that for any ε there exists a unitary operator $V \in \mathcal{B}(\mathcal{K})$ with $\|U - V\| < \varepsilon$ that is not fundamentally reducible. To see this we define another bounded unitary operator W according to the decomposition
$$\mathcal{K} = E(\Delta_{\theta_1,\theta_2})\mathcal{K}[+](I - E(\Delta_{\theta_1,\theta_2}))\mathcal{K}$$
by
$$W = e^{i\theta_0}E(\Delta_{\theta_1,\theta_2})\mathcal{K} + U(I - E(\Delta_{\theta_1,\theta_2})).$$
Then
$$U - W = (U - e^{i\theta_0})E(\Delta_{\theta_1,\theta_2}) = \int_{\theta_1}^{\theta_2}(e^{it} - e^{i\theta_0})dE(t)$$
and hence
$$\|U - W\| < 2(\theta_2 - \theta_1) < \frac{\varepsilon}{2}, \tag{5.1}$$
for $\theta_2 - \theta_1 < \varepsilon/4$. Further, by the choice of θ_0 the regular subspace $E(\Delta_{\theta_1,\theta_2})\mathcal{K}$ is indefinite and hence, there exist two orthogonal vectors $e, f \in E(\Delta_{\theta_1,\theta_2})\mathcal{K}$ such that $[e, e] = 1$ and $[f, f] = -1$. We first define the operator V on the two dimensional regular subspace $\mathcal{M} = \text{Lin}\{e, f\} = \mathbb{C}e[+]\mathbb{C}f$ by
$$V = \frac{e^{i\theta_0}}{\sqrt{1 - \gamma^2}}\begin{bmatrix} 1 & \gamma \\ \gamma & 1 \end{bmatrix},$$
for some $0 < \gamma < 1$, and note, for example by Corollary 5.3.7, that V is unitary on $\mathcal{M}$ and that
$$\|V^n\| \geq \frac{1}{(1 - \gamma^2)^{n/2}} \xrightarrow[n \to \infty]{} +\infty,$$

9.5 Strong Stability

hence V is not power bounded, equivalently, V is not fundamentally reducible on $\mathcal{M}$. We then extend V to the whole space $\mathcal{K}$ by letting V be the multiplication with $e^{i\theta_0}$ on $E(\Delta_{\theta_1,\theta_2})\mathcal{K} \cap \mathcal{M}^\perp$ and then letting V be U on $(I - E(\Delta_{\theta_1,\theta_2}))\mathcal{K}$. Thus, on the one hand V is a bounded unitary operator on $\mathcal{K}$ that is not fundamentally reducible and, on the other hand,

$$\|W - V\| = \frac{1}{\sqrt{1-\gamma^2}} \left\| \begin{bmatrix} 0 & \gamma \\ \gamma & 0 \end{bmatrix} \right\| = \frac{\gamma}{\sqrt{1-\gamma^2}} < \frac{\varepsilon}{2}, \quad (5.2)$$

for $0 < \gamma$ and sufficiently small. From (5.1) and (5.2) we get a contradiction with the assumption that U is strongly stable, and hence the claim is proved.

If we let $\sigma_\pm = \sigma(U_\pm)$ we get two spectral sets of U that are disjoint and $E(\sigma_\pm; U)\mathcal{K} = \mathcal{K}^\pm$ are maximal uniformly positive/negative subspaces of $\mathcal{K}$.

Conversely, assume that U is a bounded unitary operator and let $\sigma_\pm$ be the two disjoint spectral sets of U such that the corresponding spectral subspaces $E(\sigma_\pm; U)\mathcal{K}$ are uniformly positive/negative subspaces. Then $E(\sigma_-; U)\mathcal{K} \dotplus E(\sigma_+; U)\mathcal{K} = \mathcal{K}$ and hence, for example, by Corollary 2.1.6, it follows that these subspaces are maximal in each of their corresponding class. On the other hand, since the maximal uniformly positive subspace $E(\sigma_+; U)\mathcal{K}$ is invariant under the unitary operator U, then it actually reduces U and hence U is fundamentally reducible, in particular the spectrum of U is contained in the unit circle $\mathbb{T}$.

Further, since $\sigma_\pm$ are on the unit circle they are symmetric with respect to the transformation $\lambda \mapsto \lambda^* = 1/\bar{\lambda}$. Thus, since $\sigma_\pm$ are spectral sets of U, we can choose two disjoint closed Jordan contours $\Gamma_\pm$ surrounding $\sigma_\pm$, respectively, and such that each one is symmetric with respect to the transformation $\lambda \mapsto \lambda^*$. By the Riesz–Dunford functional calculus we have

$$E(\sigma_\pm; U) = \frac{1}{2\pi i} \int_{\Gamma_\pm} (\zeta I - U)^{-1} d\zeta,$$

and taking into account that U is unitary it follows that $E(\sigma_\pm; U) = E(\sigma_\pm; U)^\sharp$. In addition, $E(\sigma_+; U)E(\sigma_-; U) = 0$, hence their ranges $E(\sigma_\pm; U)\mathcal{K}$ are orthogonal. Then $E(\sigma_-; U)\mathcal{K}[\dotplus]E(\sigma_+; U)\mathcal{K} = \mathcal{K}$ is actually a fundamental decomposition of $\mathcal{K}$. Thus, we can choose the fundamental decomposition $\mathcal{K} = \mathcal{K}^-[\dotplus]\mathcal{K}^+$ with $\mathcal{K}^\pm = E(\sigma_\pm; U)\mathcal{K}$ that reduces U, and let $\|\cdot\|$ denote the associated fundamental norm.

Let $V \in \mathcal{B}(\mathcal{K})$ be a unitary operator such that $\|U - V\| < \delta$ for $0 < \delta$ and sufficiently small. By the upper semicontinuity of the spectrum, it follows that $\sigma(V)$ is also split into two disjoint spectral sets $\sigma'_\pm$, each one surrounded by $\Gamma_\pm$, respectively, hence we can consider the spectral projections

$$E(\sigma'_\pm; V) = \frac{1}{2\pi i} \int_{\Gamma_\pm} (\zeta I - V)^{-1} d\zeta.$$

Since V is unitary and $\Gamma_\pm$ are symmetric with respect to the transformation $\lambda \mapsto \lambda^*$, it follows that both projections $E(\sigma'_\pm; V)$ are selfadjoint. In addition, $E(\sigma'_-; V)E(\sigma'_+; V) = 0$, hence the corresponding regular subspaces $E(\sigma'_\pm; V)\mathcal{K}$ are orthogonal.

We next show that, for any $\varepsilon > 0$ we can find $\delta > 0$ such that for any unitary operator $V \in \mathcal{B}(\mathcal{K})$ such that $\|U - V\| < \delta$ we have $\|E(\sigma_\pm; U) - E(\sigma'_\pm; V)\| < \varepsilon$. To this end, let $\zeta \in \Gamma_+$ be arbitrary. Then

$$\begin{aligned}(\zeta I - V)^{-1} - (\zeta I - U)^{-1} &= \big((\zeta I - U) - (V - U)\big) - (\zeta I - U)^{-1} \\ &= (\zeta I - U)^{-1}\big(I - (V - U)(\zeta I - U)^{-1}\big)^{-1} - (\zeta I - U)^{-1} \\ &= (\zeta I - U)^{-1}\big(\big((I - (V - U)(\zeta I - V)^{-1}\big) - I\big)\end{aligned}$$

and then, letting $\delta < 1/\sup_{\zeta \in \Gamma_+} \|(\zeta I - U)^{-1}\|$, we can expand $I - (V - U)(\zeta I - U)^{-1}$ in a Neumann series and get

$$\begin{aligned}&= (\zeta I - U)^{-1}\Big(\sum_{k=0}^{\infty}\big((V - U)(\zeta I - U)^{-1}\big)^k - I\Big) \\ &= (\zeta I - U)\sum_{k=1}^{\infty}\big((V - U)(\zeta I - U)^{-1}\big)^k.\end{aligned}$$

Integrating the previous identity on the closed Jordan curve Γ_+ we obtain

$$E(\sigma'_+; V) - E(\sigma_+; U)\| = \frac{1}{2\pi i}\int_{\Gamma_+}(\zeta I - V)^{-1}\sum_{k=1}^{\infty}\big((V - U)(\zeta I - V)^{-1}\big)^k d\zeta,$$

and hence

$$\begin{aligned}\|E(\sigma'_+; V) - E(\sigma_+; U)\| &\leq \frac{1}{2\pi}\int_{\Gamma_+}\|(\zeta I - U)^{-1}\|\sum_{k=1}^{\infty}\|V - U\|^k\|(\zeta I - U)^{-1}\|^k|d\zeta| \\ &= \frac{\|V - U\|}{2\pi}\int_{\Gamma_+}\|(\zeta I - U)^{-1}\|^2\sum_{k=0}^{\infty}\|V - U\|^k\|(\zeta I - U)^{-1}\|^k|d\zeta| \\ &= \frac{\|V - U\|}{2\pi}\int_{\Gamma_+}\frac{\|(\zeta I - U)^{-1}\|^2}{1 - \|V - U\|\|(\zeta I - U)^{-1}\|}|d\zeta|.\end{aligned}$$

Since we integrate a continuous function on a compact curve, the claim is proved for the projections $E(\sigma'_+; V)$. In a similar way we prove the claim for the projections $E(\sigma'_-; V)$.

Finally, let us observe that, by the particular choice of the fundamental decomposition, letting J be the associated fundamental symmetry and $J = J^+ - J^-$ be its Jordan decomposition, we actually have $J^\pm = E(\sigma_\pm; U)$. Applying Lemma 9.5.1 it follows that $E(\sigma'_\pm; V)$ are positive/negative bounded selfadjoint projections and hence, their ranges, the spectral subspaces $E(\sigma'_\pm; V)\mathcal{K}$, are maximal uniformly positive/negative subspaces. ∎

9.6 Notes

The article by L. S. Pontryagin [121] was the first to consider a problem on spaces with indefinite inner product spaces and Corollary 9.2.12 is its main theorem. In the original

9.6 NOTES

article the proof is different from many points of view, compared with the one we present here. The inspiration for that article was the result of S. L. Sobolev, which was kept secret for twenty years, cf. [136].

The results we have presented in Section 9.2 follow, to a certain extent, the presentation in T. Ya. Azizov and I. S. Iokhvidov [10] and hence the attribution of most of them can be found there.

Theorem 9.3.3 is from K. Fan [56], Theorem 9.3.4 was obtained by K. Fan [56] and I. L. Glicksberg [67], while Theorem 9.3.6 is from I. L. Glicksberg [67]. The history of these fixed point theorems is actually more complicated, but we do not want to enter this territory. Theorem 9.3.7 is from I. S. Iokhvidov [76], based on an idea of K. Fan.

The interest in jointly invariant maximal semidefinite subspaces originates with the seminal paper of R. S. Phillips [120], whose motivation is based on his previous investigations on systems of partial differential equations [118, 119]. The results on amenable groups are from J. Dixmier [47] but the basic idea comes from S. Banach's notion of LIM (left invariant mean) on $\mathbb{Z}$ and a theorem of B. Sz.-Nagy [139]. In our presentation we have used also the monograph of F. P. Greenleaf [71].

The notion of strong stability and the results in Section 9.5 are from M. G. Kreĭn [93]. Lemma 9.5.1 is essentially a result of B. Sz.-Nagy.

Chapter 10

Hankel Operators and Interpolation Problems

In this chapter we present some applications of the results on invariant maximal semidefinite subspaces, to Nehari type problems and interpolation problems for meromorphic functions. We first consider a generalised version of the Nehari problem that can be modelled by means of a problem of existence of invariant maximal semidefinite subspaces. We then apply the Nehari type problem to estimate bounds and singular values of different types of Hankel operators, and to a four block problem. The Nehari type problem can be reformulated as a problem of contractive intertwining dilations, whose utility is exemplified on problems of Carathéodory–Schur type for meromorphic functions.

A different approach to the interpolation problem is to consider a generalised interpolation problem in the sense of D. Sarason for which, again, the main ingredient is invariant maximal semidefinite subspaces. This is then applied to the bitangential Nevanlinna–Pick problem for matrix valued meromorphic functions.

10.1 A Generalised Nehari Problem

Let $\mathcal{G}_1$ be a Kreĭn space and let S_1 be a bounded operator in $\mathcal{G}_1$. Also, let $\mathcal{G}_2$ be another Kreĭn space such that $\mathcal{G}_2$ contains a space $\mathcal{H}_2$ as a *regular subspace* (that is, $\mathcal{H}_2$ is a subspace of $\mathcal{G}_2$ which is also a Kreĭn space with the induced indefinite inner product and the same strong topology). We also consider S_2 a bounded operator in $\mathcal{G}_2$ and assume that the subspace $\mathcal{G}_2 \cap \mathcal{H}_2^\perp$ is *invariant* under S_2. A bounded operator $\Gamma\colon \mathcal{G}_1 \to \mathcal{H}_2$ is called an (S_1, S_2)-*Hankel operator* if $\Gamma S_1 = P_{\mathcal{H}_2} S_2 \Gamma$.

With the above notation, let Γ be an (S_1, S_2)-Hankel operator, $\rho > 0$ and $\kappa \in \mathbb{N}\cup\{\infty\}$. By definition, the set $\mathrm{N}_\kappa(\Gamma; \rho)$ consists of those pairs $(M; \mathcal{E})$ subject to the following conditions.

(1) $\mathcal{E}$ is a subspace of $\mathcal{G}_1$, invariant under S_1, and of codimension at most κ.
(2) $M\colon \mathcal{E} \to \mathcal{G}_2$ is bounded, $[Mx, Mx] \leqslant \rho^2[x, x]$ for all $x \in \mathcal{E}$, and $MS_1|\mathcal{E} = S_2 M$.
(3) $\Gamma|\mathcal{E} = P_{\mathcal{H}_2} M$.

The *generalised Nehari type problem* that we consider here is to determine the elements of the set $\mathrm{N}_\kappa(\Gamma; \rho)$. First notice that

$$\mathrm{N}_\kappa(\Gamma; \rho) = \{\rho^{-1} M \mid M \in \mathrm{N}_\kappa(\rho^{-1}\Gamma; 1)\}. \tag{1.1}$$

As a conclusion, it is sufficient to determine the set $\mathrm{N}_\kappa(\Gamma; 1)$; in the following this set will be denoted by $\mathrm{N}_\kappa(\Gamma)$, for simplicity.

10.1 A Generalised Nehari Problem

Lemma 10.1.1. *If* $N_\kappa(\Gamma; \rho)$ *is nonvoid, then* $\kappa_-(\rho^2 I - \Gamma^\sharp \Gamma) \leqslant \kappa + \kappa_-(\mathcal{G}_2 \ominus \mathcal{H}_2)$.

Proof. Let $(M; \mathcal{E})$ be in $N_\kappa(\Gamma; \rho)$. Then for all $x \in \mathcal{E}$ we have

$$\rho^2[x, x] - [\Gamma x, \Gamma x] = \rho^2[x, x] - [P_{\mathcal{H}_2} Mx, P_{\mathcal{H}_2} Mx]$$
$$= \rho^2[x, x] - [Mx, Mx] + [P_{\mathcal{G}_2 \cap \mathcal{H}_2^\perp} Mx, P_{\mathcal{G}_2 \cap \mathcal{H}_2^\perp} Mx].$$

Taking into account that the quadratic form $\rho^2[x, x] - [Mx, Mx]$ is positive on $\mathcal{E}$ and that the codimension of $\mathcal{E}$ is at most κ, from here we obtain the desired inequality. ∎

Let us forget, for the moment, about the operators S_1 and S_2 and keep only the Kreĭn spaces $\mathcal{G}_1$ and $\mathcal{G}_2$, assume that $\mathcal{H}_2$ is a regular subspace of $\mathcal{G}_2$, and let the operator $\Gamma \in \mathcal{B}(\mathcal{G}_1, \mathcal{H}_2)$ be given. As Lemma 10.1.1 shows, if the negative rank of the space $\mathcal{G}_2 \cap \mathcal{H}_2^\perp$ is not null, some important difficulties may occur. The approach we follow makes it necessary to assume in addition that $\mathcal{G}_2 \cap \mathcal{H}_2^\perp$ is *uniformly positive*. In this case, and provided that there exist solutions of the problem $N_\kappa(\Gamma; \rho)$, from Lemma 10.1.1 it follows that

$$\kappa_-(\rho^2 I - \Gamma^\sharp \Gamma) \leqslant \kappa. \tag{1.2}$$

We now describe the basic construction. Let

$$\mathcal{K} = \mathcal{G}_1 \oplus \mathcal{G}_2 \tag{1.3}$$

on which we consider the indefinite inner product $[\cdot, \cdot]$ defined by

$$[x_1 + x_2, y_1 + y_2] = [x_1, y_1] - [x_2, y_2], \quad x_1, y_1 \in \mathcal{G}_1, \ x_2, y_2 \in \mathcal{G}_2.$$

Then $(\mathcal{K}, [\cdot, \cdot])$ becomes a Kreĭn space. Fix fundamental symmetries J_1 and J_2 on $\mathcal{G}_1$ and $\mathcal{H}_2$, respectively. On $\mathcal{K}$ we have the fixed fundamental symmetry J where, with respect to the decomposition

$$\mathcal{K} = \mathcal{G}_1 \oplus \mathcal{H}_2 \oplus (\mathcal{G}_2 \ominus \mathcal{H}_2),$$

the operator J has the representation

$$J = \begin{bmatrix} J_1 & 0 & 0 \\ 0 & -J_2 & 0 \\ 0 & 0 & -I \end{bmatrix}.$$

We consider the linear manifold $\mathcal{H}$ in $\mathcal{K}$

$$\mathcal{H} = \{x + \Gamma x \mid x \in \mathcal{G}_1\} \oplus (\mathcal{G}_2 \ominus \mathcal{H}_2). \tag{1.4}$$

Taking into account that $\mathcal{H}$ is the direct orthogonal sum of the graph of a bounded operator, hence a subspace, with another subspace, it follows that $\mathcal{H}$ itself is closed, that is, it is a subspace of $\mathcal{K}$. Endowing $\mathcal{H}$ with the indefinite inner product $[\cdot, \cdot]$, we thus obtain a G-space $(\mathcal{H}, [\cdot, \cdot])$ with the strong topology induced by the strong toplogy of $\mathcal{K}$. This implies that the Gram operator of $\mathcal{H}$ is $G = P_\mathcal{H} J | \mathcal{H}$.

It will be helpful to use the following decomposition of $\mathcal{H}$

$$\mathcal{H} = \mathcal{H}_0 \oplus (\mathcal{G}_2 \ominus \mathcal{H}_2), \tag{1.5}$$

where $\mathcal{H}_0 = \{x + \Gamma x \mid x \in \mathcal{G}_1\}$ is the graph of Γ. Letting $G_0 = P_{\mathcal{H}_0} J | \mathcal{H}_0$, with respect to the decomposition (1.5) we have

$$G = \begin{bmatrix} G_0 & 0 \\ 0 & -I \end{bmatrix}. \tag{1.6}$$

$\mathcal{H}_0$ is a subspace of $\mathcal{K}$, and of $\mathcal{H}$ as well. Note that $\kappa_-(\mathcal{H}_0) = \kappa_-(I - \Gamma^\sharp \Gamma)$. Indeed, for arbitrary $x \in \mathcal{G}_1$, we have

$$[x + \Gamma x, x + \Gamma x] = [x, x] - [\Gamma x, \Gamma x] = [(I - \Gamma^\sharp \Gamma)x, x].$$

Consider now the Jordan decomposition $G_0 = G_{0+} - G_{0-}$ of the Gram operator G_0 and let $\mathcal{H}_{0-} = \operatorname{Clos} \operatorname{Ran}(G_{0-})$ and $\mathcal{H}_{0+} = \mathcal{H}_0 \ominus \mathcal{H}_{0-}$. Then

$$\operatorname{rank} G_{0-} = \dim \mathcal{H}_{0-} = \kappa_-(\mathcal{H}_0) = \kappa_-(I - \Gamma^\sharp \Gamma). \tag{1.7}$$

Further, letting

$$G_+ = G_{0+}, \quad G_- = G_{0-} \oplus I_{\mathcal{G}_2 \ominus \mathcal{H}_2},$$

from (1.6) it follows that $G = G_+ - G_-$ is the Jordan decomposition of G and $\mathcal{H} = \mathcal{H}_+ \oplus \mathcal{H}_-$ is the corresponding spectral decomposition, where

$$\mathcal{H}_+ = \mathcal{H}_{0+}, \quad \mathcal{H}_- = \mathcal{H}_{0-} \oplus (\mathcal{G}_2 \ominus \mathcal{H}_2). \tag{1.8}$$

Although we can compute explicitly the operator G_0, and hence the operator G, it does not help us too much, unless we can calculate its Jordan decomposition. However, we next adapt to our setting the correspondence of $\mathcal{H}$-maximal positive subspaces with contractive dilations of Γ, even though, due to the indefiniteness of the spaces $\mathcal{G}_1$ and $\mathcal{H}_2$, we cannot simply use the angular operator method. On the other hand, we are able to perform this only in the case when Γ is a *quasi-contraction*, that is, $\kappa_-(I - \Gamma^\sharp \Gamma) < \infty$. Note that, in this case, the operator G_- has closed range and hence the results in Theorem 2.5.2 apply.

Lemma 10.1.2. *Assume that $\kappa_-(I - \Gamma^\sharp \Gamma) < \infty$ and let $\mathcal{L}$ be an $\mathcal{H}$-maximal positive subspace. Then*

$$M: \mathcal{E} = P_{\mathcal{G}_1} \mathcal{L} \ni P_{\mathcal{G}_1} f \mapsto P_{\mathcal{G}_2} f \in \mathcal{G}_2, \quad f \in \mathcal{L}, \tag{1.9}$$

is correctly defined, contractive and bounded, $\operatorname{codim}_{\mathcal{G}_1} \mathcal{E} = \kappa_-(I - \Gamma^\sharp \Gamma)$, and $\Gamma | \mathcal{E} = P_{\mathcal{H}_2} M$.

Conversely, if $M: \mathcal{E}(\subseteq \mathcal{G}_1) \to \mathcal{G}_2$ is a bounded contraction, $\operatorname{codim}_{\mathcal{G}_1} \mathcal{E} \leqslant \kappa_-(I - \Gamma^\sharp \Gamma)$ and $\Gamma | \mathcal{E} = P_{\mathcal{H}_2} M$, then

$$\mathcal{L} = \{x + Mx \mid x \in \mathcal{E}\} \tag{1.10}$$

10.1 A Generalised Nehari Problem

is an $\mathcal{H}$-maximal positive subspace and $\operatorname{codim}_{\mathcal{G}_1} \mathcal{E} = \kappa_-(I - \Gamma^\sharp \Gamma)$.

In addition, this correspondence is inverse to that given in (1.9) and establishes a bijective correspondence between the set of all pairs $(M; \mathcal{E})$ such that $M: \mathcal{E} \to \mathcal{G}_2$ is a bounded contraction, $\mathcal{E} \subseteq \mathcal{G}_1$ with $\operatorname{codim}_{\mathcal{G}_1} \mathcal{E} \leqslant \kappa$, and $\Gamma|\mathcal{E} = P_{\mathcal{H}_2}M$, and the set of all $\mathcal{H}$-maximal positive subspaces.

Proof. Let $\mathcal{L}$ be an $\mathcal{H}$-maximal positive subspace. We first note that, since $\mathcal{L} \subseteq \mathcal{H}$ and (1.5) holds, we have

$$\Gamma P_{\mathcal{G}_1} f = P_{\mathcal{H}_2} f, \quad f \in \mathcal{L}. \tag{1.11}$$

We first prove that $P_{\mathcal{G}_1}|\mathcal{L}$ is injective. To this end, let $f \in \mathcal{L}$ be such that $P_{\mathcal{G}_1} f = 0$. It follows that $P_{\mathcal{H}_2} f = 0$. Therefore, $f \in \mathcal{K} \ominus (\mathcal{G}_1 \oplus \mathcal{H}_2) = \mathcal{G}_2 \ominus \mathcal{H}_2$. This implies $f = 0$, since f is a positive vector and the subspace $\mathcal{G}_2 \ominus \mathcal{H}_2$ is negative as a subspace of $\mathcal{H}$.

As a consequence of the injectivity of the operator $P_{\mathcal{G}_1}|\mathcal{L}$, we get that the operator M as in (1.9) is correctly defined and (1.10) holds. For the moment, M is a closed operator and $\mathcal{E}$ is a linear manifold in $\mathcal{G}_1$. Since $\mathcal{L}$ is positive, we have

$$[Mx, Mx] \leqslant [x, x], \quad x \in \mathcal{E},$$

that is, M is contractive. In addition, for arbitrary $x \in \mathcal{E}$ we have $x = P_{\mathcal{G}_1} f$ for some $f \in \mathcal{L}$ and then

$$\Gamma x = \Gamma P_{\mathcal{G}_1} f = P_{\mathcal{H}_2} f = P_{\mathcal{H}_2} P_{\mathcal{G}_2} f = P_{\mathcal{H}_2} M x,$$

hence $P_{\mathcal{H}_2} M = \Gamma|\mathcal{E}$ holds.

We now prove that the codimension of the linear manifold $\mathcal{E}$ in $\mathcal{G}_1$ is exacly $\kappa_-(I - \Gamma^\sharp \Gamma)$. Since $\mathcal{L}$ is an $\mathcal{H}$-maximal positive subspace, by Theorem 2.5.2 there exists the generalised angular operator $K_\mathcal{L}: \mathcal{H}_+ \to \mathcal{H}_-$ such that

$$\mathcal{L} = \{x + K_\mathcal{L} x \mid x \in \mathcal{H}_+\}.$$

Taking into account (1.8) we get

$$P_{\mathcal{G}_1} \mathcal{L} + \mathcal{H}_{0-} \supseteq P_{\mathcal{G}_1}(\mathcal{L} + \mathcal{H}_{0-}) = \mathcal{G}_1. \tag{1.12}$$

We claim now that the operator $P_{\mathcal{G}_1}$ is injective when restricted to the subspace $\mathcal{L} + \mathcal{H}_{0-}$. Indeed, let $l \in \mathcal{L}$ and $h \in \mathcal{H}_{0-}$ be such that $P_{\mathcal{G}_1}(l + h) = 0$, equivalently $P_{\mathcal{G}_1} l = -P_{\mathcal{G}_2} h$. Taking into account (1.5) it follows that $l = (x + \Gamma x) + g_2$ for some $g_2 \in \mathcal{G}_2 \ominus \mathcal{H}_2$ and $x = -P_{\mathcal{G}_1} h$. But, by the construction of the space $\mathcal{H}_{0-}$ we have $h = x + \Gamma x$ where $x = -P_{\mathcal{G}_1} h$, and hence $l = -h + g_2$. Now remark that the subspaces $\mathcal{H}_{0-}$ and $\mathcal{G}_2 \ominus \mathcal{H}_2$ are negative subspaces and orthogonal with respect to the inner product $[\cdot, \cdot]$ of $\mathcal{K}$, hence the vector $l = -h + g_2$ is either negative or null. But l is positive, as any other vector in $\mathcal{L}$, and hence $l = 0$ and $h = 0$. The claim is proved.

Since $\mathcal{L}$ is a positive subspace and $\mathcal{H}_{0-}$ is a negative subspace, it follows that the sum $\mathcal{L} + \mathcal{H}_{0-}$ is direct and, taking into account that $P_{\mathcal{G}_1}$ is injective on $\mathcal{L} \dotplus \mathcal{H}_{0-}$, from (1.12) we get

$$\mathcal{E} \dotplus P_{\mathcal{G}_1} \mathcal{H}_{0-} = \mathcal{G}_1,$$

which proves that the codimension of $\mathcal{E}$ in $\mathcal{G}_1$ is exactly $\dim \mathcal{H}_{0-} = \kappa_-(I - \Gamma^\sharp \Gamma)$.

We now prove that $\mathcal{E}$ is closed. First, consider the subspace $\mathcal{H}'_+ = \text{Ker}(P_{\mathcal{H}_{0-}} K_\mathcal{L})$ $\subseteq \mathcal{H}_+$ and remark that $\text{codim}_{\mathcal{H}_+} \mathcal{H}'_+ \leqslant \dim \mathcal{H}_{0-} = \kappa$. Define the subspace of $\mathcal{L}$

$$\mathcal{L}' = \{x + K_\mathcal{L} x \mid x \in \mathcal{H}'_+\},$$

and note that, since $K_\mathcal{L} \mathcal{H}'_1 \subseteq \mathcal{G}_2 \ominus \mathcal{H}_2$ it follows that $P_{\mathcal{G}_1} \mathcal{L}' = P_{\mathcal{G}_1} \mathcal{H}'_+$. Since $\mathcal{H}'_+$ is a subspace of $\mathcal{H}_+$ it follows that

$$\mathcal{H}'_+ = \{f + \widetilde{\Gamma} f \mid f \in P_{\mathcal{G}_1} \mathcal{H}'_+\}.$$

Since $\mathcal{H}'_+$ is closed and $\widetilde{\Gamma}$ is bounded it follows that $P_{\mathcal{G}_1} \mathcal{H}'_+ = P_{\mathcal{G}_1} \mathcal{L}'$ is closed. Taking into account that $\text{codim}_\mathcal{E} P_{\mathcal{G}_1} \mathcal{L}' \leqslant \kappa < \infty$, it follows that the linear manifold $\mathcal{E}$ is closed as well. Since the operator M is closed and its domain $\mathcal{E}$ is a subspace, the Closed Graph Theorem implies that M is bounded.

Conversely, let $M \colon \mathcal{E}(\subseteq \mathcal{G}_1) \to \mathcal{G}_2$ be a bounded contraction, $\text{codim}_{\mathcal{G}_1} \mathcal{E} \leqslant \kappa$ and $\Gamma | \mathcal{E} = P_{\mathcal{H}_2} M$. Since M is contractive, we readily check that $\mathcal{L}$ is positive. To prove that $\mathcal{L}$ is a subspace of $\mathcal{H}$, let $f = x + Mx$, for some vector $x \in \mathcal{E}$. Then, for $f = x + Mx$ and $x \in \mathcal{E}$ we have

$$\Gamma P_{\mathcal{G}_1} f = \Gamma P_{\mathcal{G}_1}(x + Mx) = \Gamma x = P_{\mathcal{H}_2} Mx = P_{\mathcal{H}_2}(x + Mx) = P_{\mathcal{H}_2} f.$$

In view of the definition of $\mathcal{H}$ this implies that $\mathcal{L} \subseteq \mathcal{H}$.

From Theorem 2.5.2 it follows that there exists an $\mathcal{H}$-maximal positive subspace $\widetilde{\mathcal{L}} \supseteq \mathcal{L}$. Then, by Lemma 10.1.2 we get that $P_{\mathcal{G}_1} \widetilde{\mathcal{L}}$ is a subspace of $\mathcal{G}_1$ of codimension κ. Since $P_{\mathcal{G}_1} \widetilde{\mathcal{L}} \supseteq P_{\mathcal{G}_1} \mathcal{L} = \mathcal{E}$ is a subspace in $\mathcal{G}_1$ of codimension at most κ, it follows that $P_{\mathcal{G}_1} \widetilde{\mathcal{L}} = \mathcal{E}$, in particular that $\mathcal{L} = \widetilde{\mathcal{L}}$ is an $\mathcal{H}$-maximal positive subspace and $\text{codim}_{\mathcal{G}_1} \mathcal{E} = \kappa$. ∎

Coming back to the Nehari type problem formulated at the beginning of this section, consider the space $\mathcal{H}$ as in (1.4). With respect to the decomposition (1.3) of the Kreĭn space $\mathcal{K}$, we define

$$S = \begin{bmatrix} S_1 & 0 \\ 0 & S_2 \end{bmatrix}. \tag{1.13}$$

Clearly, by the definition of the space $\mathcal{H}$ and taking into account that $\Gamma S_1 = P_{\mathcal{H}_2} S_2 \Gamma$ it follows that $\mathcal{H}$ is invariant under S.

By means of Lemma 10.1.2 and in view of the facts described up to now, we conclude that if the necessary condition (1.2) holds then the set $\text{N}_\kappa(\Gamma; \rho)$ can be represented by means of a class of invariant subspaces.

Lemma 10.1.3. *Let $(M, \mathcal{E})$ be an element in $\text{N}_\kappa(\Gamma)$. Then the subspace $\mathcal{L}$ defined as in (1.10) is $\mathcal{H}$-maximal positive and invariant under S. The correspondence defined as in (1.10) between the set $\text{N}_\kappa(\Gamma)$ and the set of all $\mathcal{H}$-maximal positive subspaces invariant under S is bijective.*

10.1 A GENERALISED NEHARI PROBLEM

Proof. Most of the statements are already proven in Lemma 10.1.2. Only the invariance property must be checked. Let $(M; \mathcal{E})$ be in $N_\kappa(\Gamma; 1)$. If f is an arbitrary vector in $\mathcal{L}$ then $f = x + Mx$ for some $x \in \mathcal{E}$. Then

$$Sf = S(x + Mx) = S_1 x + S_2 Mx = Sx + MS_1 x \in \mathcal{L}.$$

Conversely, let $\mathcal{L}$ be an $\mathcal{H}$-maximal positive subspace invariant under S. If x is an arbitrary vector in $\mathcal{E}$, then $S(x + Mx) \in \mathcal{L}$ and hence, $S_1 x = y$ and $S_2 Mx = My$ for some $y \in \mathcal{E} = P_{\mathcal{G}_1}\mathcal{L}$. Therefore, $S_1 \mathcal{E} \subseteq \mathcal{E}$ and $S_2 M = MS_1|\mathcal{E}$. ∎

The main result of this section is the existence of elements in the class $N_\kappa(\Gamma; \rho)$, under certain assumptions.

Theorem 10.1.4. *Assume that the subspace $\mathcal{G}_2 \cap \mathcal{H}_2^\perp$ is positive definite, S_1 is expansive and S_2 is contractive. Let Γ be an (S_1, S_2)-Hankel operator and $\rho > 0$ be such that $\kappa_-(\rho^2 I - \Gamma^\sharp \Gamma) < \infty$ and $\kappa < \infty$. Then, the set $N_\kappa(\Gamma; \rho)$ is nonvoid if and only if $\kappa_-(\rho^2 I - \Gamma^\sharp \Gamma) \leqslant \kappa$.*

Proof. The necessity of the condition $\kappa_-(\rho^2 I - \Gamma^\sharp \Gamma) \leqslant \kappa$ follows from Lemma 10.1.1.

Conversely, as noted at the beginning of this section, it is sufficient to prove the result for $\rho = 1$. Taking into account the description of the solutions of the problem $N_\kappa(\Gamma)$ in Lemma 10.1.3, we verify that the hypotheses of Theorem 9.3.7 are fulfilled.

Firstly, since S_1 is expansive and S_2 is contractive, for any vector $x = x_1 + x_2$, $x_1 \in \mathcal{G}_1$ and $x_2 \in \mathcal{G}_2$, we have

$$[Sx, Sx] = [S_1 x_1, S_1 x_1] - [S_2 x_2, S_2 x_2] \geqslant [x_1, x_1] - [x_2, x_2] = [x, x],$$

that is, S is expansive and hence it maps positive vectors into positive vectors.

Since $\dim \mathcal{H}_{0-} = \kappa_-(I - \Gamma^\sharp \Gamma) < \infty$, it follows that the operator G_- has closed range, cf. (1.7). We now take into account (1.5) and get

$$P_{\mathcal{H}_+} S P_{\mathcal{H}_-} = P_{\mathcal{H}_{0-}} S(P_{\mathcal{H}_{0-}} + P_{\mathcal{G}_2 \ominus \mathcal{H}_2})$$
$$= P_{\mathcal{H}_{0-}} S P_{\mathcal{H}_{0-}} + P_{\mathcal{H}_{0+}} S P_{\mathcal{G}_2 \ominus \mathcal{H}_2}.$$

Since $\mathcal{G}_2 \ominus \mathcal{H}_2$ is invariant under S_2 and $\mathcal{H}_{0+} \subseteq \mathcal{G}_1 \oplus \mathcal{H}_2$, we have $P_{\mathcal{H}_{0+}} S P_{\mathcal{G}_2 \ominus \mathcal{H}_2} = 0$. Therefore,

$$P_{\mathcal{H}_+} S P_{\mathcal{H}_-} = S P_{\mathcal{H}_{0-}} S P_{\mathcal{H}_{0-}},$$

and hence $\operatorname{rank} P_{\mathcal{H}_+} S P_{\mathcal{H}_-} \leqslant \dim \mathcal{H}_{0-} = \kappa < \infty$, in particular, the operator $P_{\mathcal{H}_+} S P_{\mathcal{H}_-}$ is compact.

The assumptions of Theorem 9.3.7 are verifed and hence there exists an $\mathcal{H}$-maximal positive subspace invariant under S. In view of Lemma 10.1.3 this implies that there exists a solution $(M; \mathcal{E})$ of the problem $N_\kappa(\Gamma; 1)$. ∎

10.2 More or Less Classical Hankel Operators

In the follwing we present some applications of the generalised Nehari problem to Hankel operators on different types of spaces.

Minus-Operators. Let $\mathcal{G}_1$ and $\mathcal{G}_2$ be Kreĭn spaces and consider linear operators $S_1 \in \mathcal{B}(\mathcal{G}_1)$ and $S_2 \in \mathcal{B}(\mathcal{G}_2)$. An (S_1, S_2)-*multiplier* is, by definition, an operator $M \in \mathcal{B}(\mathcal{G}_1, \mathcal{G}_2)$ intertwining the operators S_1 and S_2, that is, $MS_1 = S_2M$. Assume, in addition, that $\mathcal{H}_2$ is a Kreĭn subspace of $\mathcal{G}_2$ such that $\mathcal{G}_2 \cap \mathcal{H}_2^\perp$ is invariant under S_2 and uniformly positive. If M is an (S_1, S_2)-multiplier then $\Gamma_M = P_{\mathcal{H}_2} M$ is an (S_1, S_2)-Hankel operator. In addition, if M is a minus-operator and $\mu \in \mathbb{R}$ then it follows from the positive definiteness of $\mathcal{G}_2 \cap \mathcal{H}_2^\perp$ that

$$\mu[x,x] - [Mx, Mx] = \mu[x,x] - [\Gamma_M x, \Gamma_M x] - [P_{\mathcal{G}_2 \cap \mathcal{H}_2^\perp} Mx, P_{\mathcal{G}_2 \cap \mathcal{H}_2^\perp} Mx]$$
$$\leqslant \mu[x,x] - [\Gamma_M x, \Gamma_M x],$$

hence the interval $[\mu_+(M), \mu_-(M)]$ is contained in the interval $[\mu_+(\Gamma_M), \mu_-(\Gamma_M)]$. This shows that, if $\Gamma \in \mathcal{B}(\mathcal{G}_1, \mathcal{H}_2)$ is an (S_1, S_2)-Hankel minus-operator, that is, Γ is a minus-operator and $\Gamma S_1 = P_{\mathcal{H}_2} S_2 \Gamma$, then

$$\bigcup_{MS_1 = S_2M,\ \Gamma = P_{\mathcal{H}_2} M} [\mu_+(M), \mu_-(M)] \subseteq [\mu_+(\Gamma), \mu_-(\Gamma)]. \tag{2.1}$$

All these considerations are more or less trivial consequences of the definitions. The interesting part of this discussion is that Theorem 10.1.4 shows that if Γ is a strong minus-operator then the inclusion converse to (2.1) holds, too.

Theorem 10.2.1. *Let $S_1 \in \mathcal{B}(\mathcal{G}_1)$ be expansive, $S_2 \in \mathcal{B}(\mathcal{G}_2)$ be contractive, and let $\Gamma \in \mathcal{B}(\mathcal{G}_1, \mathcal{H}_2)$ be an (S_1, S_2)-Hankel strong minus-operator. If the Kreĭn subspace $\mathcal{G}_2 \cap \mathcal{H}_2^\perp$ is positive definite then, for any $\mu \in [\mu_+(\Gamma), \mu_-(\Gamma)]$, there exists an (S_1, S_2)-multiplier M such that $\Gamma = P_{\mathcal{H}_2} M$ and $[Mx, Mx] \leqslant \mu[x,x]$ for all $x \in \mathcal{G}_1$.*

Singular Numbers of Generalised Hankel Operators. We consider Hilbert spaces $\mathcal{G}_1$ and $\mathcal{G}_2$ and the remainder of the notation is as in Section 10.1. Let Γ be an (S_1, S_2)-Hankel operator. Let $\{s_k(\Gamma)\}_{k \geqslant 0}$ be the sequence of the singular numbers of Γ and note that

$$s_k(\Gamma) = \min\{\rho > 0 \mid \kappa_-(\rho^2 I - \Gamma^* \Gamma) \leqslant k\}, \quad k \geqslant 0.$$

This observation and Theorem 10.1.4 imply the following characterisation of singular numbers of generalised Hankel operators.

Theorem 10.2.2. *Assume that S_1 is expansive and S_2 is contractive and let $\Gamma \in \mathcal{B}(\mathcal{G}_1, \mathcal{H}_2)$ be an (S_1, S_2)-Hankel operator. Then, for all intergers $k \geqslant 0$ we have*

$$s_k(\Gamma) = \min\{\|M\| \mid M \in \mathcal{B}(\mathcal{E}, \mathcal{G}_2),\ \mathrm{codim}_{\mathcal{G}_1} \mathcal{E} \leqslant k,\ S_1 \mathcal{E} \subseteq \mathcal{E},$$
$$MS_1|\mathcal{E} = S_2 M,\ \Gamma|\mathcal{E} = P_{\mathcal{H}_2} M\}.$$

10.2 More or Less Classical Hankel Operators

Hankel Operators on Weighted ℓ^2 Spaces. A sequence $u = (u_n)_{n \geq 0}$ of positive numbers is called a *weight sequence*, or simply a *weight*. Assume that the weight $(u_n)_n$ is nondecreasing, that is, $u_n \leq u_{n+1}$ for all $n \geq 0$, and that

$$\sup_{n \geq 0} \frac{u_{n+1}}{u_n} < \infty. \tag{2.2}$$

We consider the weighted space $\ell^2(u)$

$$\ell^2(u) = \{(x_n)_{n \geq 0} \subset \mathbb{C} \mid \sum_{n=0}^{\infty} u_n |x_n|^2 < \infty\}. \tag{2.3}$$

If v is another nondecreasing weight sequence, let $\Gamma \colon \ell^2(u) \to \ell^2(v)$ be a bounded operator. Γ is called a *Hankel operator* if, with respect to the standard bases of these spaces, its matrix $[\gamma_{j,k}]_{j,k \geq 0}$ has the *Hankel property*, that is, $\gamma_{j,k} = \gamma_{j+k}$ for all $j, k \geq 0$. Thus

$$\Gamma x = \Big(\sum_{j=0}^{\infty} \gamma_{j+k} x_j\Big)_{k=0}^{\infty}, \quad x = (x_j)_{j \geq 0} \in \ell^2(u). \tag{2.4}$$

Associated to the weight v is the bilateral sequence $w = \{w_j\}_{j \in \mathbb{Z}}$ defined as follows

$$w_j = \begin{cases} v_j, & j \leq 0, \\ v_0, & j > 0. \end{cases} \tag{2.5}$$

Consider the weighted space $\ell^2(w)$

$$\ell^2(w) = \{(x_j)_{j \in \mathbb{Z}} \subset \mathbb{C} \mid \sum_{j \in \mathbb{Z}} w_j |x_j|^2 < \infty\}, \tag{2.6}$$

as well as the following two subspaces

$$\mathcal{H}_2 = \{x \in \ell^2(w) \mid x_j = 0, \text{ for all } j > 0\}, \tag{2.7}$$

$$\ell^2(w) \ominus \mathcal{H}_2 = \{x \in \ell^2(w) \mid x_j = 0, \text{ for all } j \leq 0\}. \tag{2.8}$$

A bilateral sequence of complex numbers $a = (a_k)_{k \in \mathbb{Z}}$ is called an (u, w)-*multiplier* if the infinite matrix M_a defined by

$$M_a x = \Big(\sum_{j=0}^{\infty} a_{k-j} x_j\Big)_{k \in \mathbb{Z}}, \quad x = (x_j)_{j \geq 0} \in \ell^2(u), \tag{2.9}$$

defines a bounded operator $M_a \colon \ell^2(u) \to \ell^2(w)$. Let a be an (u, w)-multiplier. Then the operator $\Gamma_a = P_{\mathcal{H}_2} M_a \colon \ell^2(u) \to \mathcal{H}_2$ has, in the standard bases of the underlying ℓ^2 spaces, an infinite matrix with the Hankel property, more precisely $\gamma_{j,k} = a_{-j-k}$ for all $j, k \geq 0$.

Theorem 10.2.3. *Let u and v be nondecreasing weights such that u satifies the condition (2.2), and let w be the weight defined as in (2.5). If $\Gamma\colon \ell^2(u) \to \ell^2(v)$ is a bounded Hankel operator, then there exists an (u,w)-multiplier a such that $\Gamma = P_{\mathcal{H}_2}M_a$ and $\|\Gamma\| = \|M_a\|$, modulo the identification of $\ell^2(v)$ with the subspace $\mathcal{H}_2$ defined in (2.7).*

Proof. On the Hilbert space $\mathcal{G}_1 = \ell^2(u)$ we consider the canonical shift operator S_1 defined as follows: $S_1 e_n = e_{n+1}$ for all $n \geq 0$, where $\{e_n\}_{n \geq 0}$ is the standard basis of $\ell^2(u)$. The operator S_1 is bounded due to the condition (2.2) and it is expansive since the weight u is nondecreasing.

On the Hilbert space $\mathcal{G}_2 = \ell^2(w)$ we consider the canonical shift operator S_2 defined similarly: $S_2 f_n = f_{n+1}$ for all $n \in \mathbb{Z}$, where $\{f_n\}_{n \geq 0}$ is the standard basis of $\ell^2(w)$. The operator S_2 is contractive since the weight v is nondecreasing.

In the following we identify canonically the Hilbert space $\ell^2(v)$ with the subspace $\mathcal{H}_2$ as in (2.7). With this identification, it is easy to see that a bounded operator $\Gamma\colon \ell^2(u) \to \ell^2(v)$ is Hankel if and only if it is (S_1, S_2)-Hankel. On the other hand, a straightforward calculation shows that a bounded operator $M\colon \ell^2(u) \to \ell^2(w)$ is an (S_1, S_2)-multiplier if and only if there exists an (u,w)-multiplier a such that $M = M_a$. Thus, we can apply Theorem 10.2.1 for $k = 0$, and get the required conclusion. ∎

Without essential modifications, Theorem 10.2.3 can be stated for Hankel operators with matrix entries, or, more generally, with operator entries, when the underlying Hilbert space is assumed to be separable.

Singular Values of Four Block Operators. Let m, n be positive integer numbers and denote by $M_{n,m}$ the set of $n \times m$ matrices with complex entries, identified with $\mathcal{B}(\mathbb{C}^m, \mathbb{C}^n)$ as Banach spaces. We denote by $H^\infty(M_{n,m})$ the Banach space of all functions $F\colon \mathbb{D} \to M_{n,m}$ which are analytic and uniformly bounded in $\mathbb{D}$, with norm

$$\|F\|_\infty = \sup_{z \in \mathbb{D}} \|F(z)\| < \infty.$$

Equivalently, $H^\infty(M_{n,m})$ is identified as a Banach space with the projective tensor product $H^\infty \otimes M_{n,m}$, where H^∞ is the Hardy space on $\mathbb{D}$ and $M_{n,m}$ carries the uniform norm. Also, letting L^∞ denote the Banach space of functions $f\colon \mathbb{T} \to \mathbb{C}$ that are essentially bounded and $\|f\|_\infty = \operatorname{ess\,sup}_{\mathbb{T}}(f)$, recall that the radial limit provides a canonical isometric embedding $H^\infty \hookrightarrow L^\infty$ and then, by the tensor product representation, we have a canonical isometric embedding $H^\infty(M_{n,m}) \hookrightarrow L^\infty(M_{n,m})$.

Let $\alpha \in \mathbb{D}$. We consider the Möbius transformation

$$b_\alpha(z) = \frac{|\alpha|}{\alpha} \frac{\alpha - z}{1 - \overline{\alpha}z}, \quad z \in \mathbb{D},\ \alpha \in \mathbb{D} \setminus \{0\},$$

and $b_0 = z$, which maps conformally the unit disc into itself. A *Blaschke–Potapov cell* of order $q \leq n$ is, by definition, a square matrix of order n

$$B_\alpha(z) = \begin{bmatrix} I_r & 0 & 0 \\ 0 & b_\alpha(z)I_q & 0 \\ 0 & 0 & I_s \end{bmatrix},$$

10.2 More or Less Classical Hankel Operators

where $n = r + q + s$. A *Blaschke–Potapov product of finite order* is, by definition, a finite product of analytic functions, each one being unitary equivalent with a Blaschke–Potapov cell. The order of a Blaschke–Potapov product is the sum of the orders of all its factors. Since the functions b_α map $\partial \mathbb{D}$ into itself, a Blaschke–Potapov product is always of norm one and the corresponding multiplication operator is isometric $H^2(\mathbb{C}^n) \to H^2(\mathbb{C}^n)$.

Here and throughout this chapter, $H^2(\mathbb{C}^n)$ denotes the Hilbert space of analytic functions $f \colon \mathbb{D} \to \mathbb{C}^n$ with Taylor representation

$$f(z) = \sum_{k=0}^{\infty} a_k z^k, \quad |z| < 1,$$

such that the coefficient sequences $(a_k)_k$ of vectors in $\mathbb{C}^n$ are square summable

$$\|f\|_2^2 = \sum_{k=0}^{\infty} \|a_k\|^2 < \infty,$$

where $\mathbb{C}^n$ is viewed as a (finite dimensional) Hilbert space with the standard inner product. This is equivalent with $H^2(\mathbb{C}^n)$ identified with the Hilbert space tensor product $H^2 \otimes \mathbb{C}^n$, where H^2 denotes the Hardy space on $\mathbb{D}$.

Also, any element $\Phi \in H^\infty(M_{n,m})$ yields the operator $M_\Phi \colon H^2(\mathbb{C}^n) \to H^2(\mathbb{C}^m)$ of multiplication with Φ and the underlying map

$$H^\infty(M_{n,m}) \ni \Phi \mapsto M_\Phi \in \mathcal{B}(H^2(\mathbb{C}^n), H^2(\mathbb{C}^m))$$

is a linear isometry. Recalling that, by means of the radial limit, there is a canonical isometric embedding $H^2 \hookrightarrow L^2$, where L^2 denotes the Hilbert space of square integrable functions $f \colon \mathbb{T} \to \mathbb{C}$, by means of the tensor product representation we get the canonical isometric embedding $H^2(\mathbb{C}^n) \hookrightarrow L^2(\mathbb{C}^n)$ of Hilbert spaces.

Let l be a positive integer. By $H_l^\infty(M_{n,m})$ we denote the class of functions G that can be represented by $G = F\Psi^{-1}$, where $F \in H^\infty(M_{m,n})$ and $\Psi \in H^\infty(\mathbb{C}^n)$ is a Blaschke–Potapov product of order $\leq l$. If we impose the additional condition that no zero of the function G coincides with any zero of the Blaschke–Potapov product Ψ then this representation of functions in $\mathcal{S}_l(M_{m,n})$ or $H_l^\infty(M_{n,m})$ is unique, modulo unitary equivalence. Such a factorisation is called *right coprime*.

In the following we will use a consequence of the Beurling–Lax Theorem, cf. [17] and [109], and the representation of matrix valued inner functions.

Theorem 10.2.4. *A subspace $\mathcal{E} \in H^2(\mathbb{C}^n)$ is shift invariant and of codimension $l < \infty$ if and only if $\mathcal{E} = \Psi H^2(\mathbb{C}^n)$ where $\Psi \in H^\infty(\mathbb{C}^n)$ is a Blaschke–Potapov product of order l.*

The problem we are considering in this paragraph is concerning the "four block operator". Let Φ be a function in $L^\infty(M_{n,l})$, consider the corresponding multiplication operator $M_\Phi \colon L^2(\mathbb{C}^l) \to L^2(\mathbb{C}^n)$, and denote $H_-^2(\mathbb{C}^n) = L^2(\mathbb{C}^n) \ominus H^2(\mathbb{C}^n)$. Then the *classical Hankel operator* $\Gamma_\Phi \colon H^2(\mathbb{C}^l) \to H^2(\mathbb{C}^n)$ with symbol Φ is defined by

$$\Gamma_\Phi f = P_{H_-^2(\mathbb{C}^n)} M_\Phi f, \quad f \in H^2(\mathbb{C}^l). \tag{2.10}$$

The *four block operator* K is defined as follows

$$K: H^2(\mathbb{C}^l) \oplus L^2(\mathbb{C}^m) \to H^2_-(\mathbb{C}^n) \oplus L^2(\mathbb{C}^p),$$
$$K = \begin{bmatrix} \Gamma_\Phi & P_{H^2_-(\mathbb{C}^n)} R \\ Q & F \end{bmatrix}, \qquad (2.11)$$

where, R, Q, and F are multiplication operators with matrix valued symbols in L^∞ and of appropriate dimensions.

Theorem 10.2.5. *Let K be the four block operator defined in (2.11). Then, for any integer $k \geqslant 0$, the singular numbers of K can be calculated as follows*

$$s_k(K) = \min\left\{ \left\| \begin{bmatrix} M_\Psi & R \\ Q & F \end{bmatrix} \right\| \mid \Psi \in L^\infty(M_{n,l}) \text{ such that } \Phi - \Psi \in H^\infty_k(M_{n,l}) \right\}.$$

Proof. We consider the Hilbert spaces $\mathcal{G}_1 = H^2(\mathbb{C}^l) \oplus L^2(\mathbb{C}^m)$ and $\mathcal{G}_2 = L^2_-(\mathbb{C}^n) \oplus L^2(\mathbb{C}^p)$, as well as the subspace $\mathcal{H}_2 = H^2_-(\mathbb{C}^n) \oplus L^2(\mathbb{C}^p) \subset \mathcal{G}_2$. Let S_1 and S_2 denote the shift operators on $\mathcal{G}_1$ and $\mathcal{G}_2$, respectively, that is, the multiplication operators with the independent variable e^{it}. In particular, both S_1 and S_2 are isometric operators, and $\mathcal{G}_2 \ominus \mathcal{H}_2 = H^2(\mathbb{C}^n) \oplus 0$ is invariant under S_2. In addition, the four block operator is an (S_1, S_2)-Hankel operator and hence Theorem 10.2.2 can be applied. Thus, for any $k \geqslant 0$ we have

$$s_k(K) = \min\{\|M\| \mid M \in \mathcal{B}(\mathcal{E}, \mathcal{G}_2), \text{ codim}_{\mathcal{G}_1} \mathcal{E} \leqslant k, \ S_1 \mathcal{E} \subseteq \mathcal{E},$$
$$MS_1|\mathcal{E} = S_2 M, \ K|\mathcal{E} = P_{\mathcal{H}_2} M\}.$$

Let $\mathcal{E}$ be a shift invariant subspace of $\mathcal{G}_1$ and of codimension less than k, and let $M \in \mathcal{B}(\mathcal{E}, \mathcal{G}_2)$ be such that $MS_1|\mathcal{E} = S_2 M$ and $K = P_{\mathcal{H}_2} M$. A moment of thought and an application of Theorem 10.2.4 shows that there exists a Blaschke–Potapov product $B \in H^\infty(\mathbb{C}^l)$ such that

$$\mathcal{E} = BH^2(\mathbb{C}^l) \oplus L^2(\mathbb{C}^m).$$

Then, the equalities $K|\mathcal{E} = P_{\mathcal{H}_2} M$ and $MS_1|\mathcal{E} = S_2 M$ imply that there exists $\Psi \in L^\infty(M_{n,l})$ such that

$$M = \begin{bmatrix} M_\Psi & R \\ Q & F \end{bmatrix},$$

and $\Gamma_\Phi | BH^2(\mathbb{C}^l) = \Gamma_\Psi | BH^2(\mathbb{C}^l)$, or equivalently,

$$P_{H^2_-(\mathbb{C}^n)} \Phi BH^2(\mathbb{C}^l) = P_{H^2_-(\mathbb{C}^n)} \Psi BH^2(\mathbb{C}^l),$$

which is equivalent with $(\Phi - \Psi)B \in H^\infty(M_{n,l})$, and hence, with $\Phi - \Psi \in H^\infty_k(M_{n,l})$. ∎

Let us also record the particular case when R, Q, and F are null, which provides a formula to compute the singular values of the classical Hankel operators, with matrix entries.

Corollary 10.2.6. *Let $\Phi \in L^\infty(M_{n,l})$ and $k \geq 0$ an integer. Then*

$$s_k(\Gamma_\Phi) = \min\{\|\Psi\| \mid \Psi \in L^\infty(M_{n,l}), \text{ such that } \Phi - \Psi \in H_k^\infty(M_{n,l})\}.$$

Also, specialising this corollary to $k = 0$, we get a formula for the norm of the classical Hankel operators, cf. Z. Nehari [110].

Corollary 10.2.7. *For any matrix valued function $\Phi \in L^\infty(M_{n,l})$ we have*

$$\|\Gamma_\Phi\| = \text{dist}(\Phi, H^\infty(M_{n,l})),$$

and there exists $\Psi \in H^\infty(M_{n,l})$ such that $\|\Gamma_\Phi\| = \|\Phi - \Psi\|_\infty$.

10.3 Intertwining Dilations

In this section, we show that a certain contractive intertwining dilation problem is equivalent to the generalised Nehari type problem considered in the previous section.

Let $\mathcal{H}_1$ and $\mathcal{H}_2$ be Kreĭn spaces and consider two operators $T_i \in \mathcal{B}(\mathcal{H}_i)$, $i = 1, 2$, and assume that for $i = 1, 2$, there exist pairs $(V_i; \mathcal{G}_i)$, subject to the following conditions:

(a_i) $\mathcal{G}_i$ is a Kreĭn space extension of $\mathcal{H}_i$;
(b_i) $V_i \in \mathcal{B}(\mathcal{G}_i)$ is a dilation of T_i, that is, $P_{\mathcal{H}_i} V_i = T_i P_{\mathcal{H}_i}$.

As a consequence of assumption (b_i) it follows that $\mathcal{G}_i \cap \mathcal{H}_i^\perp$ is invariant under the operator V_i, $i = 1, 2$.

Let $A \in \mathcal{B}(\mathcal{H}_1, \mathcal{H}_2)$ be an operator intertwining the operators T_1 and T_2, that is, $AT_1 = T_2 A$. Denote by $\text{CID}_\kappa(A; T_1, T_2)$ the set of *contractive intertwining dilations* of A consisting of all pairs $(A_\infty, \mathcal{E})$ subject to the following conditions:

(1) $\mathcal{E}$ is a subspace of $\mathcal{G}_1$, of codimension at most κ, and invariant under V_1;
(2) $A_\infty \in \mathcal{B}(\mathcal{E}, \mathcal{G}_2)$ is a contraction, that is, $[A_\infty x, A_\infty x] \leq [x, x]$ for all $x \in \mathcal{E}$;
(3) $P_{\mathcal{H}_2} A_\infty = A P_{\mathcal{H}_1} | \mathcal{E}$;
(4) $A_\infty V_1 | \mathcal{E} = V_2 A_\infty$.

Simply by inspecting the definitions we obtain.

Lemma 10.3.1. *Let A, T_1, V_1, etc. be as above and denote $\Gamma = AP_{\mathcal{H}_1}: \mathcal{G}_1 \to \mathcal{H}_2$. Then Γ is a (V_1, V_2)-Hankel operator and*

$$\text{CID}_\kappa(A; V_1, V_2) = \text{N}_\kappa(\Gamma).$$

As a consequence of this equality and Theorem 10.1.4 we obtain the following result.

Theorem 10.3.2. *If both subspaces $\mathcal{G}_i \cap \mathcal{H}_i^\perp$, $i = 1, 2$, are uniformly positive, V_1 is expansive, V_2 is a contraction, and A is a quasi-contraction, that is, $\kappa_-(I - A^\sharp A) < \infty$, then the set $\text{CID}_\kappa(A; T_1, T_2)$ is nonempty if and only if $\kappa_-(I - A^\sharp A) \leq \kappa$.*

Conversely, under the assumptions of Theorem 10.3.2, it is easy to see that each set $N_\kappa(\Gamma)$ can be realised as a set $\text{CID}_\kappa(A; V_1, V_2)$. To see this, let Γ be an (S_1, S_2)-Hankel operator. Define $T_1 = V_1 = S_1$, $T_2 = P_{\mathcal{H}_2} S_2$, $V_2 = S_2$, and $A = \Gamma$. Then, it is readily checked that, under the conditions in Theorem 10.3.2, we have $N_\kappa(\Gamma) = \text{CID}_\kappa(A; V_1, V_2)$.

We illustrate the applicability of Theorem 10.3.2 to an interpolation problem of Carathéodory–Schur type for meromorphic matrix valued functions. By $\mathcal{S}_l(M_{n,m})$ we denote the *generalised Schur class* of functions G, that can be represented by $G = F\Psi^{-1}$, where $F \in H^\infty(M_{m,n})$, $\|F\|_\infty \leq 1$, and $\Psi \in H^\infty(\mathbb{C}^n)$ is a Blaschke–Potapov product of order $\leq l$.

With notation as in Section 10.2, let $C = (C_l)_{l=1}^k \subset \mathcal{B}(\mathbb{C}^m, \mathbb{C}^n)$ be a sequence of $n \times m$ complex matrices. For $\kappa \in \mathbb{N}$ define the set $\text{CS}_\kappa(C)$ consisting of all meromorphic matrix valued functions $G \in \mathcal{S}_\kappa(\mathbb{C}^m, \mathbb{C}^n)$ such that the first $k+1$ "Taylor coefficients" at 0 of G coincide, respectively, with $C_0, C_1, \ldots, C_k$. More precisely, if the function G has the representation $G = F\Psi^{-1}$, where $F \in H^\infty(M_{m,n})$ and $\Psi \in H^\infty(M_n)$ is a Blaschke–Potapov product of order $\leq \kappa$, then the first $k+1$ Taylor coefficients at 0 of G coincide, respectively, with the first Taylor coefficients of the analytic matrix valued function $(C_0 + zC_1 + \cdots + z^k C_k)\Psi(z)$. Note that, in the above definition, a function $G \in \text{CS}_\kappa(C)$ is not necessarily analytic at 0.

Associated to the data $C = (C_0, C_1, \ldots, C_k)$ there is the following lower triangular operator block matrix of Toeplitz type

$$T_C = \begin{bmatrix} C_0 & 0 & 0 & \cdots & 0 \\ C_1 & C_0 & 0 & \cdots & 0 \\ C_2 & C_1 & C_0 & \cdots & 0 \\ \vdots & & & & \\ C_k & C_{k-1} & C_{k-2} & \cdots & C_0 \end{bmatrix}. \tag{3.1}$$

Theorem 10.3.3. *Let κ be a finite positive integer. Then the set $\text{CS}_\kappa(C)$ is not empty if and only if $\kappa_-(I - T_C^* T_C)$ does not exceed κ.*

Proof. We consider the Hilbert space $\mathcal{G}_1 = H^2(\mathbb{C}^m)$ and the forward shift operator $V_1 \in \mathcal{B}(H^2(\mathbb{C}^m))$, $(V_1 g)(z) = zg(z)$, for all $g \in H^2(\mathbb{C}^m)$ and $z \in \mathbb{D}$. Let $\mathcal{H}_1$ be the subspace of $H^2(\mathbb{C}^m)$ of polynomials with coefficients in $\mathbb{C}^m$ of degree not exceeding k, and define $T_1 = P_{\mathcal{H}_1} V_1 | \mathcal{H}_1$. Since $\mathcal{G}_1 \ominus \mathcal{H}_1$ is invariant under V_1, it follows that V_1 is a dilation of T_1.

Similarly, denote $\mathcal{G}_2 = H^2(\mathbb{C}^n)$ and let V_2 be the forward shift operator on $H^2(\mathbb{C}^n)$. Denote by $\mathcal{H}_2$ the subspace of $H^2(\mathbb{C}^n)$ consisting of all polynomials with coefficients in $\mathbb{C}^n$ of degree not exceeding k, and let $T_2 = P_{\mathcal{H}_2} V_2 | \mathcal{H}_2$. Since $\mathcal{G}_2 \ominus \mathcal{H}_2$ is invariant under V_2 it follows that V_2 is a dilation of T_2.

Let us consider now the matrix valued polynomial $C(z) = C_0 + zC_1 + \cdots + z^k C_k$ and let us denote by $M_C \in \mathcal{B}(H^2(\mathbb{C}^m), H^2(\mathbb{C}^n))$ the multiplication operator with $C \in$

$H^\infty(\mathbb{C}^m, \mathbb{C}^n)$. Then define $A \in \mathcal{B}(\mathcal{H}_1, \mathcal{H}_2)$ by $A = P_{\mathcal{H}_1} M_C|\mathcal{H}_2$. Since M_C is a multiplication operator it intertwines the operators V_1 and V_2. Therefore,

$$AT_1 = P_{\mathcal{H}_2} M_C P_{\mathcal{H}_1} V_1|\mathcal{H}_1 = P_{\mathcal{H}_2} M_C V_1|\mathcal{H}_1 = P_{\mathcal{H}_2} V_2 M_C|\mathcal{H}_1 = T_2 A.$$

These relations show that the problem $\text{CID}_\kappa(A; T_1, T_2)$ makes sense. In the following we prove that the problem $\text{CID}_\kappa(A; T_1, T_2)$ has solutions if and only if so does $\text{CS}_\kappa(C)$.

Indeed, let $(A_\infty, \mathcal{E}) \in \text{CID}_\kappa(A; T_1, T_2)$. Since $\mathcal{E}$ is a shift invariant subspace of $H^2(\mathbb{C}^m)$ of codimension at most $\kappa < \infty$, from Theorem 10.2.4 it follows that $\mathcal{E} = \Psi H^2(\mathbb{C}^m)$ for some Blaschke–Potapov product $\Psi \in H^\infty(\mathbb{C}^m)$ of finite order. Taking into account that $A_\infty V_1|\mathcal{E} = V_2 A_\infty$ we get

$$A_\infty \Psi V_1 h = A_\infty V_1 \Psi h = V_2 A_\infty \Psi h, \quad h \in H^2(\mathbb{C}^m),$$

that is, letting $F = A_\infty \Psi$ we have $FV_1 = V_2 F$ and hence F is a multiplication operator with some function $F \in H^\infty(\mathbb{C}^m, \mathbb{C}^n)$. Since A_∞ is contractive, and the multiplication operator with the function Ψ is isometric, we have

$$\|Fh\| = \|A_\infty \Psi h\| \leqslant \|\Psi h\| = \|h\|, \quad h \in H^2(\mathbb{C}^m),$$

that is, F is contractive. These relations prove that A_∞ is the multiplication operator with a function in $\mathcal{S}_\kappa(\mathbb{C}^m, \mathbb{C}^n)$.

We now take into account that $P_{\mathcal{H}_2} A_\infty = A P_{\mathcal{H}_1}|\mathcal{E}$. For arbitrary $h \in H^2(\mathbb{C}^m)$ we have

$$P_{\mathcal{H}_2} M_F h = P_{\mathcal{H}_2} A_\infty \Psi h = A P_{\mathcal{H}_1} M_\Psi h = P_{\mathcal{H}_2} M_{C\Psi} h.$$

This proves that the first $k+1$ Taylor coefficients at 0 of G coincide, respectively, with the first Taylor coefficients of the analytic matrix valued function $(C_0 + zC_1 + \cdots + z^k C_k)\Psi(z)$ and hence we have a solution of the problem $\text{CS}_\kappa(C)$.

The converse implication, namely, that once we have a solution of the problem $\text{CS}_\kappa(C)$ we have also a solution of the problem $\text{CID}_\kappa(A; T_1, T_2)$, is straighforward and we omit the details.

We note now that we can identify $\mathcal{H}_1$ with a direct sum of $k+1$ copies of $\mathbb{C}^m$ and, similarly, $\mathcal{H}_2$ can be identified with the direct sum copies of $k+1$ copies of $\mathbb{C}^n$. With these identifications it is easy to see that the operator A coincides with the operator T_C as in (3.1). The proof is now concluded as an application of Theorem 10.3.2. ∎

10.4 Generalised Interpolation

A *phase function* is, by definition, a function $\varphi \in L^\infty(M_n)$, for some integer $n \geqslant 1$, such that $\varphi(e^{it})$ is isometric almost everywhere $t \in [0, 2\pi]$. If φ is a phase function then its norm in $L^\infty(M_n)$ is 1 and the corresponding multiplication operator on $L^2(\mathbb{C}^n)$ is isometric.

With the notation introduced in the previous section, the problem of *generalised interpolation* is formulated as follows. For given $K \in L^\infty(M_{m,n})$, phase functions $\theta \in L^\infty(M_m)$ and $\varphi \in L^\infty(\mathbb{C}^n)$, and some integer $l \geq 0$, we have the set

$$C_l(K, \theta, \varphi) = \{F \in K + \theta \mathcal{S}_l(M_{m,n})\varphi \mid \|F\|_\infty \leq 1\}, \tag{4.1}$$

and we are required to determine its elements. Here $\mathcal{S}_l(M_{m,n})$ denotes the generalised Schur class as in the previous section.

Since φ is a phase function, so is its matrix adjoint φ^*, and hence the corresponding multiplication operator is isometric, in particular $\varphi H^2(\mathbb{C}^n)$ is a closed subspace of $L^2(\mathbb{C}^n)$. Thus, we can consider the Kreĭn space $L^2(\mathbb{C}^m) \oplus \varphi^* H^2(\mathbb{C}^n)$, regarded as a subspace of the Kreĭn space $L^2(\mathbb{C}^m \oplus \mathbb{C}^n) = L^2(\mathbb{C}^m) \oplus L^2(\mathbb{C}^n)$, with the indefinite inner product $[\cdot, \cdot]$ induced by the indefinite inner product on $\mathbb{C}^m \oplus \mathbb{C}^n$, where the first component $\mathbb{C}^m$ is uniformly positive definite and the second component $\mathbb{C}^n$ is uniformly negative definite. Thus, the ambient Kreĭn space will be

$$\mathcal{K} = L^2(\mathbb{C}^m) \oplus \varphi^* H^2(\mathbb{C}^n), \tag{4.2}$$

with the induced structure of Kreĭn space as described before, more precisely, the indefinite inner product $[\cdot, \cdot]$ is defined by

$$[f_1 \oplus \varphi^* g_1, f_2 \oplus \varphi^* g_2] = \frac{1}{2\pi} \int_0^{2\pi} \langle f_1(e^{it}), f_2(e^{it}) \rangle \mathrm{d}t - \frac{1}{2\pi} \int_0^{2\pi} \langle g_1(e^{it}), g_2(e^{it}) \rangle \mathrm{d}t,$$

taking into account that $\varphi(e^{it})\varphi(e^{it})^* = I_n$, almost everywhere $t \in [0, 2\pi]$. On the Kreĭn space $L^2(\mathbb{C}^m \oplus \mathbb{C}^n)$ the bilateral shift operator $S = M_{e^{it}}$ is defined and it is unitary with respect to both the positive definite inner product $\langle \cdot, \cdot \rangle$ and the indefinite inner product $[\cdot, \cdot]$. Clearly, $\mathcal{K}$ is invariant under S.

We denote by $\mathcal{M}_{K,\theta,\varphi}$ the subspace of $\mathcal{K}$ generated by the graphs of multiplication operators with all the functions of type $K + \theta h \varphi$, with $h \in H^\infty(M_{m,n})$, more precisely,

$$\mathcal{M}_{K,\theta,\varphi} = \mathrm{Clos}\left\{(K + \theta h \varphi)\varphi^* g \oplus \varphi^* g \mid h \in H^\infty(M_{m,n}),\ g \in H^2(\mathbb{C}^n)\right\}. \tag{4.3}$$

Lemma 10.4.1. *The subspace $\mathcal{M}_{K,\theta,\varphi}$ is invariant under the shift operator S and*

$$\mathcal{M}_{K,\theta,\varphi} = \left\{(K\varphi^* g + \theta f) \oplus \varphi^* g \mid f \in H^2(\mathbb{C}^m),\ g \in H^2(\mathbb{C}^n)\right\}$$
$$= \left\{K\varphi^* g \oplus \varphi^* g \mid g \in H^2(\mathbb{C}^n)\right\} \dotplus \left\{\theta f \oplus 0 \mid f \in H^2(\mathbb{C}^m)\right\}.$$

Proof. Let $h \in H^\infty(M_{m,n})$ and $g \in H^2(\mathbb{C}^n)$ be arbitrary. Since $\varphi \varphi^* = I_n$ almost everywhere on $[0, 2\pi]$, we have

$$(K + \theta h \varphi)\varphi^* g = K \varphi^* g + \theta h g,$$

and taking into account that $H^\infty(M_{m,n}) H^2(\mathbb{C}^n) = H^2(\mathbb{C}^m)$ we denote $f = hg \in H^2(\mathbb{C}^m)$. Thus, the subspace $\mathcal{M}_{K,\theta,\varphi}$ is generated by functions $(K\varphi^* g + \theta f) \oplus \varphi^* g$, where $f \in H^2(\mathbb{C}^m)$ and $g \in H^2(\mathbb{C}^n)$. On the other hand, we note that the subspace

$$\mathcal{M}_1 = \left\{K\varphi^* g \oplus \varphi^* g \mid g \in H^2(\mathbb{C}^n)\right\}$$

10.4 GENERALISED INTERPOLATION

is closed, since it is the graph of a bounded operator. Clearly, the subspace

$$\mathcal{M}_2 = \{\theta f \oplus 0 \mid f \in H^2(\mathbb{C}^m)\}$$

is uniformly positive. Then it follows easily that $\mathcal{M}_1 \cap \mathcal{M}_2 = 0$ and the sum $\mathcal{M}_1 \dot{+} \mathcal{M}_2$ is closed.

The subspace $\mathcal{M}_{K,\theta,\varphi}$ is invariant under the shift S, since it is generated by graphs of multiplication operators. ∎

The orthogonal companion, in $\mathcal{K}$, of the subspace $\mathcal{M}_{K,\theta,\varphi}$ can be calculated. For this reason, consider the operator $\Gamma_{\theta,\varphi}(K)$ defined as follows:

$$L^2(\mathbb{C}^m) \ominus \theta H^2(\mathbb{C}^m) \ni f \mapsto \Gamma_{\theta,\varphi}(K)f = P_{\varphi^* H^2(\mathbb{C}^n)} K^* f \in H^2(\mathbb{C}^n). \qquad (4.4)$$

Lemma 10.4.2. *The orthogonal subspace of $\mathcal{M}_{K,\theta,\varphi}$, calculated in the Kreĭn space $\mathcal{K}$, see (4.2), has angular operator $\Gamma_{\theta,\varphi}(K)$, more precisely,*

$$\mathcal{K} \cap \mathcal{M}_{K,\theta,\varphi}^{\perp} = \{a \oplus \Gamma_{\theta,\varphi}(K)a \mid a \in L^2(\mathbb{C}^m) \ominus \theta H^2(\mathbb{C}^m)\}.$$

Proof. By Lemma 10.4.1, a function $a \oplus b$ belongs to $\mathcal{K} \cap \mathcal{M}_{K,\theta,\varphi}^{\perp}$ if and only if

$$\langle K\varphi^* g + \theta f, a \rangle = \langle \varphi^* g, b \rangle, \quad f \in H^2(\mathbb{C}^m), \ g \in H^2(\mathbb{C}^n).$$

Letting $g = 0$ and $f \in H^2(\mathbb{C}^m)$ be arbitrary, we get $a \in L^2(\mathbb{C}^m) \ominus \theta H^2(\mathbb{C}^m)$. Therefore $b = P_{\varphi^* H^2(\mathbb{C}^n)} K^* a$. This proves that the subspace $\mathcal{K} \cap \mathcal{M}_{K,\theta,\varphi}^{\perp}$ has the angular operator $\Gamma_{\theta,\varphi}(K)$. ∎

Recall that, for a given selfadjoint operator A in a Hilbert space $\mathcal{H}$, we denote by $\kappa^-(A)$ the dimension of the spectral subspace of A corresponding to the negative semi-axis. As a consequence of Lemma 10.4.2 we get the following result.

Corollary 10.4.3. $\kappa^-(\mathcal{M}_{K,\theta,\varphi}^{\perp}) = \kappa^-(I - \Gamma_{\theta,\varphi}(K)\Gamma_{\theta,\varphi}(K)^*).$

The main idea in this approach is to model the generalised interpolation problem by means of shift invariant subspaces in $\mathcal{M}_{K,\theta,\varphi}$.

Lemma 10.4.4. *With the notation as before, the angular operator induces a bijective correspondence between the set $C_l(K, \theta, \varphi)$ and the set of all shift invariant negative subspaces of $\mathcal{M}_{K,\theta,\varphi}$ such that $\kappa^-(\mathcal{K} \cap \mathcal{L}^{\perp}) \leqslant l$.*

Proof. Let $F \in C_l(K, \theta, \varphi)$, that is, $F = K + \theta G \Psi^{-1} \varphi$ for some $G \in H^\infty(M_{m,n})$ and some Blaschke–Potapov product $\Psi \in H^\infty(\mathbb{C}^n)$ of degree at most l, and $\|F\|_\infty \leqslant 1$. Taking into account that $\varphi\varphi^* = I_n$ almost everywhere on $[0, 2\pi]$, we consider the subspace

$$\mathcal{L} = \begin{bmatrix} F \\ I \end{bmatrix} \varphi^* \Psi H^2(\mathbb{C}^n)$$
$$= \{(K\varphi^* \Psi g + \theta G g) \oplus \varphi^* \Psi g \mid g \in H^2(\mathbb{C}^n)\} \subseteq \mathcal{M}_{K,\theta,\varphi}.$$

From $\|F\|_\infty \leq 1$ it follows that $\mathcal{L}$ is a negative subspace. Since Ψ is a Blaschke–Potapov product of order $\leq l$, it follows that the codimension of $\Psi H^2(\mathbb{C}^n)$ in $H^2(\mathbb{C}^n)$ is $\leq l$ and hence the codimension of $\varphi^*\Psi H^2(\mathbb{C}^n)$ in $\varphi^* H^2(\mathbb{C}^n)$ is less than or equal to l. Then the codimension of $\mathcal{L}$ in some $\mathcal{K}$-maximal negative subspace is at most l. Since $\mathcal{L}$ is the graph of a multiplication operator it follows that $\mathcal{L}$ is invariant under the shift S.

Conversely, let $\mathcal{L}$ be a shift invariant negative subspace of $\mathcal{M}_{K,\theta,\varphi}$ of codimension at most l, in some $\mathcal{K}$-maximal negative subspace. Then $\mathcal{L}$ is the graph of some contraction $F\colon P_{\varphi^* H^2(\mathbb{C}^n)}\mathcal{L} \to L^2(\mathbb{C}^m)$

$$\mathcal{L} = \{Fx \oplus x \mid x \in P_{\varphi^* H^2(\mathbb{C}^n)}\mathcal{L}\}.$$

Writing down the invariance of $\mathcal{L}$ under the shift operator S we get that the subspace $P_{\varphi^* H^2(\mathbb{C}^n)}\mathcal{L}$ is shift invariant and hence, by the Beurling–Lax Theorem, there exists an inner function $\Psi \in H^\infty(\mathbb{C}^n)$ such that $P_{\varphi^* H^2(\mathbb{C}^n)}\mathcal{L} = \varphi^*\Psi H^2(\mathbb{C}^n)$, and

$$SF\varphi^*\Psi g = FS\varphi^*\Psi g, \quad g \in H^2(\mathbb{C}^n). \tag{4.5}$$

Since the subspace $\mathcal{L}$ has codimension at most l in some $\mathcal{K}$-maximal negative subspace, it follows that the codimension of $\varphi^*\Psi H^2(\mathbb{C}^n)$ in $\varphi^* H^2(\mathbb{C}^n)$ is at most l and hence, by Theorem 10.2.4, Ψ is a Blaschke–Potapov factor of order at most l.

Further, taking into account that the commutant of the shift operator consists of multiplication operators, it follows that $\mathcal{L}$ is the graph of some multiplication operator, that is,

$$\mathcal{L} = \begin{bmatrix} F \\ I \end{bmatrix} \varphi^*\Psi H^2(\mathbb{C}^n),$$

where $F \in L^\infty(M_{m,n})$.

Now we take into account that $\mathcal{L}$ is a subspace of $\mathcal{M}_{K,\theta,\varphi}$ and, by means of Lemma 10.4.1, we get that $(F - K)\varphi^*\Psi = \theta G$ for some $G \in H^\infty(M_{m,n})$, that is, $F = K + \theta G\Psi^{-1}\varphi$. Since $\|F\|_\infty \leq 1$ this means that $F \in C_l(K,\theta,\varphi)$. ∎

Corollary 10.4.5. *If the set $C_l(K,\theta,\varphi)$ is nonempty then $\kappa^-(\mathcal{M}_{K,\theta,\varphi})^\perp \leq l$.*

Proof. Indeed, by Lemma 10.4.4, if $C_l(K,\theta,\varphi)$ is nonempty then there exists a negative subspace $\mathcal{L}$ of $\mathcal{M}_{K,\theta,\varphi}$ having the codimension at most l in some $\mathcal{K}$-maximal negative subspace. By Proposition 2.1.11 the codimension of $\mathcal{L}$ in some $\mathcal{K}$-maximal negative subspace is $\kappa^-(\mathcal{L}^\perp)$. Since $\mathcal{M}_{K,\theta,\varphi}^\perp \subseteq \mathcal{L}^\perp$, we have $\kappa^-(\mathcal{M}_{K,\theta,\varphi}^\perp) \leq \kappa^-(\mathcal{L}^\perp) \leq l$. ∎

In the following we assume that K is analytic and that φ and θ are Blaschke–Potapov products of finite order. A key observation that can be made in this case is the following.

Lemma 10.4.6. *If K is analytic and φ and θ are Blaschke–Potapov products of finite orders, then the subspace $\mathcal{M}_{K,\theta,\varphi}$ is pseudo-regular.*

10.4 GENERALISED INTERPOLATION

Proof. Indeed, we use the second representation as in Lemma 10.4.1 to decompose the subspace $\mathcal{M}_{K,\theta,\varphi}$ into a direct orthogonal sum

$$\mathcal{M}_{K,\theta,\varphi} = \{P_{L^2(\mathbb{C}^m)\ominus\theta H^2(\mathbb{C}^m)} K\varphi^* g \oplus \varphi^* g \mid g \in H^2(\mathbb{C}^n)\}$$
$$[+] \{\theta f \oplus 0 \mid f \in H^2(\mathbb{C}^m)\}. \quad (4.6)$$

Clearly, the latter subspace from the right-hand side of (4.6) is uniformly positive and, hence, it remains to prove that the first space is pseudo-regular. To see this, we first consider the decomposition

$$H^2(\mathbb{C}^n) = \varphi H^2(\mathbb{C}^n) \oplus \left(H^2(\mathbb{C}^n) \ominus \varphi H^2(\mathbb{C}^n) \right). \quad (4.7)$$

Since φ is a Blaschke product of finite order, the second subspace on the right-hand side is finite dimensional. Taking into account that $\varphi^*\varphi = I_m$ almost everywhere on $[0, 2\pi]$ and that K is analytic, it follows that

$$\{P_{L^2(\mathbb{C}^m)\ominus\theta H^2(\mathbb{C}^m)} K\varphi^* g \oplus \varphi^* g \mid g \in \varphi H^2(\mathbb{C}^n)\}$$
$$= \{P_{H^2(\mathbb{C}^m)\ominus\theta H^2(\mathbb{C}^m)} Kg \oplus g \mid g \in H^2(\mathbb{C}^n)\}. \quad (4.8)$$

Therefore, since the subspace $H^2(\mathbb{C}^m) \ominus \theta H^2(\mathbb{C}^m)$ is finite dimensional, the angular operator of the subspace in (4.8) is of finite rank and hence the subspace is pseudo-regular. ∎

Corollary 10.4.7. *Assume that K is analytic and that φ and θ are Blaschke–Potapov products of finite orders. If $\kappa^-(\mathcal{M}_{K,\theta,\varphi}^\perp) = l$ then the angular operator induces a bijective correspondence between the set $C_l(K,\theta,\varphi)$ and the set of all shift invariant $\mathcal{M}_{K,\theta,\varphi}$-maximal negative subspaces.*

Proof. This is a consequence of Lemma 10.4.4, Lemma 10.4.6 and Proposition 3.2.8. ∎

The main result on the generalised interpolation problem, which we can obtain using this angular operator approach, is a necessary and sufficient condition for the existence of elements in the set $C_l(K,\theta,\varphi)$.

Theorem 10.4.8. *Suppose that K is in $H^\infty(M_{m,n})$ and that the functions $\theta \in H^\infty(\mathbb{C}^m)$ and $\varphi \in H^\infty(\mathbb{C}^n)$ are Blaschke–Potapov products of finite orders. Then the set $C_l(K,\theta,\varphi)$ is nonempty if and only if $\kappa^-(I - \Gamma_{\theta,\varphi}(K)\Gamma_{\theta,\varphi}(K)^*) \leq l$.*

Proof. Indeed, if the set $C_l(K,\theta,\varphi)$ is nonempty then, on the grounds of Lemma 10.4.4, it follows that $\kappa^-(\mathcal{M}_{K,\theta,\varphi}^\perp) \leq l$ and then, taking into account Lemma 10.4.2, we obtain that $\kappa^-(I - \Gamma_{\theta,\varphi}(K)\Gamma_{\theta,\varphi}(K)^*) \leq l$.

Conversely, assume that $\kappa^-(I - \Gamma_{\theta,\varphi}(K)\Gamma_{\theta,\varphi}(K)^*) \leq l$ and hence, on the grounds of Lemma 10.4.2, $\kappa^-(\mathcal{M}_{K,\theta,\varphi}^\perp) \leq l$. In view of Lemma 10.4.4 we only have to prove that there exists at least one shift invariant $\mathcal{M}_{K,\theta,\varphi}$-maximal negative subspace. For this we show that the hypotheses of Theorem 9.3.7 are fulfilled. We consider the shift operator S on the indefinite inner product space $\mathcal{M}_{K,\theta,\varphi}$ and we verify that the following hold.

(1) *The operator S maps negative vectors into negative vectors.*

Indeed, this follows trivially since S is isometric on the Kreĭn space $\mathcal{K}$ and, in particular, on its subspace $\mathcal{M}_{K,\theta,\varphi}$.

(2) *Let $\mathcal{M} = \mathcal{M}_{K,\theta,\varphi}$, $G = P_\mathcal{M} J|\mathcal{M}$ its Gram operator and let $G = G_+ - G_-$ be the corresponding Jordan decomposition of G. Then G_+ has closed range.*

Indeed, we already proved in Lemma 10.4.6 that $\mathcal{M}_{K,\theta,\varphi}$ is pseudo-regular. With respect to the fundamental symmetry J on $\mathcal{K}$ we have $G = P_{\mathcal{M}_{K,\theta,\varphi}} J | \mathcal{M}_{K,\theta,\varphi}$. Thus, in order to prove that G_+ has closed range, it remains to observe that the corresponding spectral subspace $G_+ \mathcal{M}_{K,\theta,\varphi} = \operatorname{Clos} G_+ \mathcal{M}_{K,\theta,\varphi}$ is regular.

(3) *Let $\mathcal{M} = \mathcal{M}^+ [+] \mathcal{M}^-$ be the corresponding spectral decomposition associated to the Jordan decomposition $G = G_+ - G_-$. Then the operator $P_{\mathcal{M}^+} S | \mathcal{M}^-$ is of finite rank, in particular, it is compact.*

Indeed, we go back to the proof of Lemma 10.4.6 and note that from (4.6), (4.7), and (4.8) we get the following decomposition

$$\mathcal{M}_{K,\theta,\varphi} = \left(\{ P_{H^2(\mathbb{C}^m) \ominus \theta H^2(\mathbb{C}^m)} Kh \oplus h \mid h \in H^2(\mathbb{C}^n) \} \right.$$
$$\left. \dotplus \{ P_{L^2(\mathbb{C}^m) \ominus \theta H^2(\mathbb{C}^m)} K\varphi^* g \oplus \varphi^* g \mid g \in H^2(\mathbb{C}^n) \ominus \varphi H^2(\mathbb{C}^n) \} \right)$$
$$[+] \{ \theta f \oplus 0 \mid f \in H^2(\mathbb{C}^m) \}.$$

Taking into account that the subspace $\{\theta f \oplus 0 \mid f \in H^2(\mathbb{C}^m)\}$ is contained in $\mathcal{M}_+$ we obtain

$$\operatorname{rank}(P_{\mathcal{M}^+} S | \mathcal{M}^-) \leqslant \dim(H^2(\mathbb{C}^m) \ominus \theta H^2(\mathbb{C}^m)) + \dim(H^2(\mathbb{C}^n) \ominus \varphi H^2(\mathbb{C}^n)) < \infty$$

and the claim is proved.

Finally, we can now apply Theorem 9.3.7 to conclude that there exists a shift invariant $\mathcal{M}_{K,\theta,\varphi}$-maximal negative subspace and hence, in view of Corollary 10.4.7, that the set $\mathcal{C}_l(K, \theta, \varphi)$ is nonvoid. ∎

10.5 The Bitangential Nevanlinna–Pick Problem

In this section we apply the results on the generalised interpolation problem obtained during the previous section to the so-called bitangential Nevanlinna–Pick problem, more precisely, we will show that it falls under the assumptions of Theorem 10.4.8.

Let m, n be positive integers, $\mathbf{z} = \{z_j\}_{j=1}^N \cup \{z'_j\}_{j=1}^{N'}$ be a finite set of distinct complex numbers in the unit disc $\mathbb{D}$, $\mathbf{p} = \{p_j\}_{j=1}^N \cup \{p'_j\}_{j=1}^{N'}$ be a finite set of points in $\mathbb{C}^m$ and $\mathbf{q} = \{q_j\}_{j=1}^N \cup \{q'_j\}_{j=1}^{N'}$ be a finite set of points in $\mathbb{C}^n$. With notation as in the preceding section, and for some integer $l \geqslant 0$, we consider the set

$$\operatorname{NP}_l(\mathbf{z}, \mathbf{p}, \mathbf{q}) = \{ F \in \mathcal{S}_l(M_{m,n}) \mid F(z_j)^* p_j = q_j, \, j = 1, 2, \ldots, N,$$
$$F(z'_j) q'_j = p'_j, \, j = 1, 2, \ldots, N' \}.$$

10.5 The Bitangential Nevanlinna–Pick Problem

In order to make this definition precise, we should clarify the meaning of the interpolation constraints in the definition of the set $\mathrm{NP}_l(\mathbf{z}, \mathbf{p}, \mathbf{q})$. Thus, let F be a function in the generalised Schur class $\mathcal{S}_l(M_{m,n})$. Then $F = G\Psi^{-1}$ for some function $G \in H^\infty(M_{m,n})$ of norm $\leqslant 1$ and some Blaschke–Potapov product $\Psi \in H^\infty(\mathbb{C}^n)$ of order at most l. If $z \in \mathbb{D}$ and $u \in \mathbb{C}^m$ and $v \in \mathbb{C}^n$, then $F(z)u = v$ has the meaning $G(z)u = \Psi(z)v$ if z is a zero of some Blaschke–Potapov factor of Ψ.

The problem of finding and explicitly describing all the elements of the set $\mathrm{NP}_l(\mathbf{z}, \mathbf{p}, \mathbf{q})$ is called the *bitangential Nevanlinna–Pick problem*. We first show that the set $\mathrm{NP}_l(\mathbf{z}, \mathbf{p}, \mathbf{q})$ can be realised as a set of type $C_l(K, \theta, \varphi)$ as in (4.1).

Lemma 10.5.1. *There exist two Blaschke–Potapov functions of finite orders $\theta \in H^\infty(M_m)$ and $\varphi \in H^\infty(M_n)$ such that*

$$\mathrm{NP}_l(\mathbf{z}, \mathbf{p}, \mathbf{q}) = C_l(K, \theta, \varphi),$$

where $K \in H^\infty(M_{m,n})$ is any function satisfying the interpolation conditions

$$K(z_j)^* p_j = q_j, \ j = 1, 2, \ldots, N \ \text{and} \ K(z_j')q_j' = p_j', \quad j = 1, 2, \ldots, N'.$$

Proof. Indeed, consider the space

$$\mathcal{S} = \{f \in H^2(\mathbb{C}^m) \mid \langle p_j, f(z_j)\rangle_{\mathbb{C}^m} = 0, \ j = 1, 2, \ldots, N\}.$$

Then $\mathcal{S}$ is a closed subspace of $H^2(\mathbb{C}^m)$, invariant under the shift S and of finite codimension. According to Theorem 10.2.4, there exists a Blaschke–Potapov product of finite order $\theta \in H^\infty(\mathbb{C}^m)$ such that $\mathcal{S} = \theta H^\infty(\mathbb{C}^m)$.

Let

$$\mathcal{S}' = \{g \in H^2(\mathbb{C}^n) \mid \langle q_j', g(z_j)\rangle_{\mathbb{C}^n} = 0, \ j = 1, 2, \ldots, N'\}.$$

Similarly, $\mathcal{S}'$ is a closed subspace of $H^2(\mathbb{C}^n)$, invariant under the shift S and of finite codimension. According to Theorem 10.2.4, there exists a Blaschke–Potapov product of finite order $\psi \in H^\infty(\mathbb{C}^n)$ such that $\mathcal{S}' = \psi H^\infty(\mathbb{C}^m)$. We take $\varphi \in H^\infty(M_n)$ defined by $\varphi(e^{it}) = \psi^*(e^{-it})$, for all $t \in [0, 2\pi)$.

We now make the remark that it is always possible to find a function $K \in H^\infty(M_{m,n})$ satisfying the interpolation conditions $K(z_j)^* p_j = q_j, j = 1, 2, \ldots, N$ and $K(z_j')q_j' = p_j', \ j = 1, 2, \ldots, N'$. Indeed, let $\{F_j\}_{j=1}^N \in M_{m,n}$ and $\{F_j'\}_{j=1}^{N'} \in M_{m,n}$ be such that $F_j^* p_j = q_j, j = 1, 2, \ldots, N$ and $F_j'q_j' = p_j', \ j = 1, 2, \ldots, N'$. Then we can use Lagrange interpolation corresponding to complex points $\{z_j\}_{j=1}^N \cup \{z_j'\}_{j=1}^{N'}$ and values $\{F_j\}_{j=1}^N \cup \{F_j'\}_{j=1}^{N'}$ to determine such a function K which is a complex polynomial with matrix coefficients, in particular $K \in H^\infty(M_{m,n})$.

Let us now consider a function $F \in C_l(K, \theta, \varphi)$, that is, $F \in L^\infty(M_{m,n}), \|F\|_\infty \leqslant 1$, and $F = K + \theta H \Psi^{-1} \varphi$ for some $H \in H^\infty(M_{m,n})$ and some Blaschke–Potapov product $\Psi \in H^\infty(M_n)$ of order at most l. Since K satisfies the interpolation conditions, in order to prove that F satisfies the interpolation conditions it remains to prove that

$$(\theta H \Psi^{-1}\varphi)^*(z_j)p_j = 0, \text{ for } j = 1, 2, \ldots, N,$$

and that
$$(\theta H \Psi^{-1} \varphi)(z'_j) q'_j = 0, \text{ for } j = 1, 2, \ldots, N'.$$

These assertions follow easily from the properties of the functions θ and φ, taking into account the meaning of the interpolation constraints, in the case when either of the interpolation data z_i or z_j is a pole of F, as specified before.

Conversely, let $F \in \mathrm{NP}_l(\mathbf{z}, \mathbf{p}, \mathbf{q})$ be arbitrary. Since K satisfies the interpolation conditions it follows that the function $F - K$ satisfies the conditions
$$(F - K)^*(z_j) p_j = 0, \text{ for } j = 1, 2, \ldots, N,$$
and
$$(F - K)(z'_j) q'_j = 0, \text{ for } j = 1, 2, \ldots, N'.$$

Using the properties of the functions θ and φ it follows from here that $F - K = \theta G \varphi$ and taking into account that $F \in H^\infty(M_{m,n})$, we get that $G \in H^\infty(M_{m,n})$. ∎

As a consequence of Lemma 10.5.1 and Theorem 10.4.8, a necessary and sufficient condition for the solvability of the bitangential Nevanlinna–Pick problem for meromorphic functions can be obtained.

Theorem 10.5.2. *Let the functions K, θ and φ be associated to the data $(\mathbf{z}, \mathbf{p}, \mathbf{q})$ as in Lemma 10.5.1. There exist solutions of the bitangential Nevannlinna–Pick problem $\mathrm{NP}_l(\mathbf{z}, \mathbf{p}, \mathbf{q})$ if and only if $\kappa^-(I - \Gamma_{\theta,\varphi}(K)\Gamma_{\theta,\varphi}(K)^*) \leqslant l$.*

In the next proposition, in order to relate this criterion with the criterion of solvability of the classical Nevanlinna–Pick problem in terms of the Pick matrix, we consider the one-sided tangential Nevanlinna–Pick problem.

Proposition 10.5.3. *Assume that we are given the data $\mathbf{z}, \mathbf{p}, \mathbf{q}$, and $N' = 0$. Then the one-sided tangential Nevanlinna–Pick problem $\mathrm{NP}_l(\mathbf{z}, \mathbf{p}, \mathbf{q})$ has at least one solution, if and only if the Pick matrix $\Lambda_{\mathbf{z},\mathbf{p},\mathbf{q}}$ defined by*
$$\Lambda_{\mathbf{z},\mathbf{p},\mathbf{q}} = \left[\frac{\langle p_j, p_i \rangle_{\mathbb{C}^m} - \langle q_j, q_i \rangle_{\mathbb{C}^n}}{1 - \overline{z}_j z_i} \right]_{i,j=1}^N$$
has at most l negative eigenvalues, counted with their multiplicities.

Proof. If $N' = 0$ we necessarily have $\varphi = 1$, where the functions K, θ and φ are associated to the problem $\mathrm{NP}_l(\mathbf{z}, \mathbf{p}, \mathbf{q})$ as in Lemma 10.5.1. We consider the set $C_l(K, \theta) = C_l(K, \theta, 1)$ and the space $\mathcal{M}_{K,\theta} = \mathcal{M}_{K,\theta,1}$, see (4.3). The first fact we prove is that
$$\mathcal{M}_{K,\theta} = \{ f \in H^2(\mathbb{C}^{m,n}) \mid [p_j \oplus q_j, f(z_j)]_{\mathbb{C}^{m,n}} = 0, \ j = 1, 2, \ldots, N \}. \quad (5.1)$$

Indeed, since $\varphi = 1$ we have $\mathcal{M}_{K,\theta} \subseteq H^2(\mathbb{C}^{m,n}) = H^2(\mathbb{C}^m) \oplus H^2(\mathbb{C}^n)$. Thus, without restricting the generality, we will replace the Kreĭn space $\mathcal{K}$ as in (4.2) with its

10.6 NOTES   271

regular subspace $H^2(\mathbb{C}^m) \oplus H^2(\mathbb{C}^n)$. Further, we take into account the definition of the functions K and θ as in the proof of Lemma 10.5.1 and then (5.1) follows.

The Schur kernel is reproducing for the Hardy space H^2 and hence

$$[p_j \oplus q_j, f(z_j)]_{\mathbb{C}^{m,n}} = \left[\frac{1}{1-z\overline{z}_j}(p_j \oplus q_j), f(z)\right]_{H^2(\mathbb{C}^m)}.$$

This implies that $\mathcal{M}_{K,\theta}^\perp$ (calculated in the Kreĭn space $H^2(\mathbb{C}^m) \oplus H^2(\mathbb{C}^n)$) is spanned by the set of functions

$$\left\{\frac{1}{1-z\overline{z}_j}(p_j \oplus q_j)\right\}_{j=1}^N. \tag{5.2}$$

Let now $h \in \mathcal{M}_{K,\theta}^\perp$ be arbitrary,

$$h(z) = \sum_{j=1}^N c_j \frac{1}{1-z\overline{z}_j}(p_j \oplus q_j).$$

Then

$$[h,h]_{H^2(\mathbb{C}^{m,n})} = \sum_{i=1}^N \sum_{j=1}^N \overline{c}_i c_j \frac{1}{1-z_i\overline{z}_j}[(p_j \oplus q_j),(p_i \oplus q_i)]_{\mathbb{C}^{m,n}}$$

$$= \sum_{i=1}^N \sum_{j=1}^N \frac{\langle p_j, p_i\rangle_{\mathbb{C}^m} - \langle q_j, q_i\rangle_{\mathbb{C}^n}}{1-\overline{z}_j z_i}\overline{c}_i c_j.$$

This proves that the Gram operator of the indefinite inner product $[\cdot,\cdot]$ restricted to $\mathcal{M}_{K,\theta}^\perp$ is given by the Pick matrix $\Lambda_{\mathbf{z},\mathbf{p},\mathbf{q}}$, hence the negative signature of the subspace $\mathcal{M}_{K,\theta}^\perp$ coincides with the number of negative eigenvalues of the matrix $\Lambda_{\mathbf{z},\mathbf{p},\mathbf{q}}$. The statement now follows from Theorem 10.4.8, via the identifications as in Lemma 10.5.1. ∎

10.6 Notes

The vast theory of interpolation problems for holomorphic functions has its roots in the articles of C. Carathéodory [26], G. Pick [117], I. Schur [133], and R. Nevanlinna [111] (a reprint of the last three articles is available in [75]). An investigation of Z. Nehari [110] on Hankel operators was later observed to contain as special cases most of these interpolation problems (for a useful presentation of the theory of Hankel operators see S. C. Power [123]). An operator approach through commutant lifting (intertwining dilations) is due to D. Sarason [130] and B. Sz.-Nagy and C. Foiaş [140] which triggered a whole theory of harmonic analysis on Hilbert spaces as presented in [141]. Shift invariant subspaces were investigated by A. Beurling [17] and P. D. Lax [109]. For a pertinent and modern treatment of most of these interpolation problems and many of their generalisations in operator theoretic terms, see M. Rosenblum and J. Rovnyak [126] and C. Foiaş and A. E. Frazho [57].

The investigations of V. M. Adamyan, D. Z. Arov, and M. G. Kreĭn in [1, 2] opened new directions and connections with operator theory in Kreĭn spaces. From here on it is difficult to track all the contributions, so I only refer to those articles from which I learned myself. Firstly I mention the series of articles of M. G. Kreĭn and H. Langer [96, 97, 98, 99] in which these problems were approached through extensions of certain Hermitian (unbounded) operators in indefinite inner product spaces. Secondly, I mention the series of articles of J. A. Ball and J. W. Helton [11, 12, 13, 14, 15, 16] in which these problems were approached through a Beurling–Lax type representation in terms of geometric aspects of indefinite inner product spaces. Related results were obtained by R. Arocena, T. Ya. Azizov, A. Dijksma, S. A. M. Marcantognini [5, 6] as well.

The presentation of this chapter follows closely our results in [62] and [63], as well as the article of T. Constantinescu and the author [33]. These investigations were influenced by the approach of S. Treil and A. Volberg [143] to the Nehari type problems by invariant maximal semidefinite subspaces, as well as by the approach of J. A. Ball and J. W. Helton through Beurling–Lax type theorems.

Chapter 11

Spectral Theory for Selfadjoint Operators

Selfadjoint operators are one of the most intensively studied classes of operators in Kreĭn spaces due to their applications. We start with some general properties referring to eigenvalues and root manifolds for symmetric operators and then we specialise to Pontryagin spaces. Since genuine selfadjoint operators in general Kreĭn spaces may have a very complicated spectral structure, we first discuss the spectral theory of selfadjoint operators in finite dimensional Kreĭn spaces and show that a special form of their Jordan decomposition can be obtained.

Historically, one of the most tractable classes of selfadjoint operators on Kreĭn spaces proved to be the class of definitisable operators of H. Langer. We first consider their general properties and prove that this contains the class of selfadjoint operators in a Pontryagin space. The main result is the existence and uniqueness of the spectral function associated to a definitisable selfadjoint operator in a Kreĭn space. The proof we provide is rather long and technical but we think that the effort is worthwhile. By using Herglotz's Theorem on integral representation for holomorphic functions mapping the upper half plane to itself, for which we provide a detailed proof, we first obtain a certain integral representation for the resolvent map. Then, using the Stieltjes Inversion Formulae, which we carefully prove as well, the spectral function is obtained by improperly integrating the resolvent map on certain arcs, symmetric with respect to the real axis. We conclude with some general spectral properties of definitisable selfadjoint operators that are derived from the spectral function and then we specialise to positive operators with nontrivial resolvent sets.

11.1 Eigenvalues and Root Manifolds

Recall that if λ is an eigenvalue of an operator T in $\mathcal{K}$ then the *root subspace* of T and λ is, by definition,
$$\mathfrak{S}_\lambda(T) = \bigvee_{n \in \mathbb{N}} \operatorname{Ker}(\lambda I - T)^n.$$

Lemma 11.1.1. *Let A be a symmetric operator in a Kreĭn space $\mathcal{K}$ and λ and μ eigenvalues of A such that $\lambda \neq \overline{\mu}$. Then $\mathfrak{S}_\lambda \perp \mathfrak{S}_\mu$.*

Proof. It is sufficient to prove that for any $n, m \in \mathbb{N}$, $x \in \operatorname{Dom}(A^m)$ and $y \in \operatorname{Dom}(A^n)$ such that
$$(A - \lambda I)^m x = 0, \quad (A - \mu I)^n y = 0,$$

it follows that $x \perp y$. We proceed by induction following $m+n$. If $m+n=2$, that is $m=n=1$, we have

$$0 = [Ax,y] - [x,Ay] = (\lambda - \overline{\mu})[x,y],$$

therefore, since $\lambda - \overline{\mu} \neq 0$ we obtain $[x,y] = 0$. Passing to the general induction step, let us denote $x_1 = (A-\lambda I)x \in \operatorname{Ker}(A-\lambda I)^{m-1}$ and $y_1 = (A-\mu I)y \in \operatorname{Ker}(A-\mu I)^{n-1}$ and hence, by the induction hypothesis, we have $[x_1,y] = [x,y_1] = 0$. Remark that $Ax = \lambda x + x_1$ and $Ay = \mu y + y_1$ and then

$$0 = [Ax,y] - [x,Ay] = [\lambda x + x_1, y] - [x, \mu y + y_1] = (\lambda - \overline{\mu})[x,y],$$

therefore $[x,y] = 0$. ∎

Remark 11.1.2. From Lemma 11.1.1 it follows that if λ is a nonreal eigenvalue of a symmetric operator A then the corresponding root subspace $\mathfrak{S}_\lambda(A)$ is neutral. ∎

Lemma 11.1.3. *Any real eigenvalue λ of a closed symmetric operator A in $\mathcal{K}$, such that $\operatorname{Ker}(A-\lambda I)$ is nondegenerate, is semisimple.*

Proof. Since A is closed then $\operatorname{Ker}(A-\lambda I) = \operatorname{Ran}(A-\lambda I)^\perp$ is closed and $\operatorname{Ran}(A-\lambda I) \subset \operatorname{Ker}(A-\lambda I)^\perp$. Let $y \in \operatorname{Ker}(A-\lambda I)^2$. Then

$$(A-\lambda I)y \in \operatorname{Ker}(A-\lambda I) \cap \operatorname{Ran}(A-\lambda I) \subseteq \operatorname{Ker}(A-\lambda I) \cap \operatorname{Ker}(A-\lambda I)^\perp = \operatorname{Ker}(A-\lambda I)^0.$$

If $\operatorname{Ker}(A-\lambda I)$ is nondegenerate then $(A-\lambda I)y = 0$. We have thus proven that $\operatorname{Ker}(A-\lambda I) = \operatorname{Ker}(A-\lambda I)^2$, that is, the eigenvalue λ is semisimple. ∎

Some of the features of eigenvalues pointed out in the previous lemmas can be stated for isolated components of the spectrum of a selfadjoint operator.

Proposition 11.1.4. *Let A be a selfadjoint operator in the Kreĭn space $\mathcal{K}$.*

(i) *If σ is a spectral set of A such that $\sigma = \overline{\sigma}$ then the Riesz projection $E(\sigma;A)$ is selfadjoint, $E(\sigma;A) = E(\sigma;A)^\sharp$, in particular the spectral subspace $E(\sigma;A)\mathcal{K}$ is regular.*

(ii) *If σ is a spectral set of A such that $\sigma \cap \overline{\sigma} = \emptyset$ then $E(\sigma;A)^\sharp E(\sigma;A) = 0$, in particular, the spectral subspaces $E(\sigma;A)\mathcal{K}$ and $E(\overline{\sigma};A)\mathcal{K}$ are neutral and in strong duality.*

(iii) *If λ_0 is an isolated eigenvalue of A then $\overline{\lambda}_0$ is also an isolated eigenvalue and the lengths of the corresponding Jordan chains of λ_0 and $\overline{\lambda}_0$ coincide.*

Proof. (i) If σ is a spectral set of A such that $\sigma = \overline{\sigma}$ then the Riesz projection $E(\sigma;A)$ has the property $E(\sigma;A)^\sharp = E(\overline{\sigma};A) = E(\sigma;A)$ and then from Lemma 4.2.4 we conclude that the spectral subspace $E(\sigma;A)\mathcal{K}$ is regular.

(ii) If $\sigma \cap \overline{\sigma} = \emptyset$ then, by the Riesz–Dunford–Taylor functional calculus it follows that $E(\sigma;A)^\sharp E(\sigma;A) = E(\sigma \cap \overline{\sigma};A) = 0$ and then, by Lemma 4.2.4, we obtain that the spectral subspaces $E(\sigma;A)\mathcal{K}$ and $E(\overline{\sigma};A)\mathcal{K}$ are neutral and in strong duality.

11.1 Eigenvalues and Root Manifolds

(iii) Let λ_0 be an isolated eigenvalue of A. Proposition 5.1.1 implies that $\overline{\lambda}_0$ is also isolated in the spectrum of A. Considering the Laurent series about λ_0 of the resolvent map

$$(\lambda I - A)^{-1} = \sum_{-\infty}^{\infty}(\lambda - \lambda_0)^n A_n$$

then

$$(\lambda I - A)^{-1} = \sum_{-\infty}^{\infty}(\lambda - \overline{\lambda}_0)^n A_n^\sharp$$

is the Laurent expansion of the resolvent function of A in a neighbourhood of $\overline{\lambda}_0$. This shows that $\overline{\lambda}_0$ is also an eigenvalue of A and the dimensions of the root subspaces corresponding to λ_0 and $\overline{\lambda}_0$ do coincide.

Finally, let J be a fundamental symmetry of $\mathcal{K}$ and consider an orthonormal basis $\{e_j\}_{j\in\mathcal{J}}$ of the Hilbert space $(E(\{\lambda_0\}; A)\mathcal{K}, (\cdot,\cdot)_J)$. Then there exists a dual basis $\{f_j\}_{j\in\mathcal{J}}$ of $(E(\{\overline{\lambda}_0\}; A)\mathcal{K}, \langle\cdot,\cdot\rangle_J)$, that is $[e_k, f_j] = \delta_{kj}$, $k, j \in \mathcal{J}$. Then the matrix of $A|E(\{\overline{\lambda}_0\}; A)\mathcal{K}$ with respect to the basis $\{f_j\}_{j\in\mathcal{J}}$ is the Hermitian adjoint of the matrix of $A|E(\{\lambda_0\}; A)\mathcal{K}$ with respect to the basis $\{e_j\}_{j\in\mathcal{J}}$. This shows that the lengths of the Jordan chains corresponding to λ_0 and $\overline{\lambda}_0$ do coincide. ∎

In the following we confine our discussion to the case of Pontryagin spaces. Recall that $\kappa(\mathcal{K}) = \min\{\kappa^-(\mathcal{K}), \kappa^+(\mathcal{K})\}$ denotes the rank of indefiniteness of a Kreĭn space $\mathcal{K}$.

Lemma 11.1.5. *Let A be a closed symmetric operator in the Pontryagin space $\mathcal{K}$ and λ a real eigenvalue of A. Then the length of any Jordan chain associated to λ is $\leqslant 2\kappa(\mathcal{K}) + 1$ and the corresponding root subspace $\mathfrak{S}_\lambda(A)$ can be decomposed as*

$$\mathfrak{S}_\lambda(A) = \mathcal{S}[+]\mathcal{S}',$$

where $\mathcal{S}$ is a finite dimensional subspace invariant under A and $\mathcal{S}' \subseteq \mathrm{Ker}(A - \lambda I)$.

Proof. Without restricting the generality we can assume that $\lambda = 0$ (otherwise replace A by $A - \lambda I$) and that $\mathfrak{S}_0(A)$ is infinite dimensional (otherwise we simply take $\mathcal{S}' = 0$). We divide the proof into three steps.

Step 1. *For any $n > 1$ the dimension of $\mathrm{Ker}(A^n)/\mathrm{Ker}(A^{n-1})$ is finite.*

Indeed, let $\mathcal{R}_n$ denote a direct summand of $\mathrm{Ker}(A^{n-1})$ in $\mathrm{Ker}(A^n)$, that is,

$$\mathrm{Ker}(A^n) = \mathrm{Ker}(A^{n-1}) \dotplus \mathcal{R}_n, \qquad (1.1)$$

and notice that $A^{n-1}\mathcal{R}_n \perp \mathrm{Ker}(A)$. Since $A^{n-1}\mathcal{R}_n \subseteq A^{n-1}\mathrm{Ker}(A^n) \subseteq \mathrm{Ker}(A)$ it follows that the linear manifold $A^{n-1}\mathcal{R}_n$ is neutral, therefore $\dim A^{n-1}\mathcal{R}_n \leqslant \kappa(\mathcal{K}) < \infty$. Then, since $\mathrm{Ker}(A^{n-1}) \cap \mathcal{R}_n = \{0\}$ it follows that $A^{n-1}|\mathcal{R}_n$ is injective and hence $\mathcal{R}_n$ is finite dimensional.

Step 2. *The length of any Jordan chain corresponding to 0 is $\leqslant 2\kappa(\mathcal{K}) + 1$.*

Let n be such that $\operatorname{Ker}(A^{n-1}) \neq \operatorname{Ker}(A^n)$. Then for all $m \leqslant n$ we have $\operatorname{Ker}(A^{m-1}) \neq \operatorname{Ker}(A^m)$. Let $\mathcal{R}_n$ be a (finite dimensional) subspace such that (1.1) holds. Then $A\mathcal{R}_n \subseteq A\operatorname{Ker}(A^n) \subseteq \operatorname{Ker}(A^{n-1})$ and $A\mathcal{R}_n \cap \operatorname{Ker}(A^{n-2}) = \{0\}$ hold, therefore we can choose a subspace $\mathcal{R}_{n-1}$ such that $\operatorname{Ker}(A^{n-1}) = \operatorname{Ker}(A^{n-2}) \dotplus \mathcal{R}_{n-1}$ and $A\mathcal{R}_n \subseteq \mathcal{R}_{n-1}$. Pursuing this way we obtain a family of nontrivial finite dimensional subspaces $\{\mathcal{R}_m\}_{m=2}^n$ such that

$$\operatorname{Ker}(A^m) = \operatorname{Ker}(A^{m-1}) \dotplus \mathcal{R}_m, \quad A\mathcal{R}_m \subseteq \mathcal{R}_{m-1}, \quad 2 \leqslant m \leqslant n. \tag{1.2}$$

Further, consider the finite dimensional subspaces

$$\mathcal{Q}_k = A^k \mathcal{R}_{2k}, \quad k \leqslant \left\lfloor \frac{n}{2} \right\rfloor, \tag{1.3}$$

where $\lfloor \cdot \rfloor$ stands for the integer part. Since $A^k|\mathcal{R}_{2k}$ is injective it follows that

$$\dim \mathcal{Q}_k = \dim \mathcal{R}_{2k} \geqslant 1, \quad k \leqslant \left\lfloor \frac{n}{2} \right\rfloor.$$

We claim that

$$\mathcal{Q}_k \perp \mathcal{Q}_j, \quad k, j \leqslant \left\lfloor \frac{n}{2} \right\rfloor. \tag{1.4}$$

Indeed, let us consider two vectors $x \in \mathcal{R}_{2k}$ and $y \in \mathcal{R}_{2j}$. To make a choice, let us say $k \leqslant j$. Then, since $x \in \mathcal{R}_{2k} \subset \operatorname{Ker}(A^{2k})$ we have

$$[A^k x, A^j y] = [A^{2k} x, A^{j-k} y] = 0,$$

and hence the claim is proven.

Since $\mathcal{Q}_k \subseteq \mathcal{R}_{2k}$ we can introduce the direct sum

$$\mathcal{P}_n = \mathcal{Q}_1 \dotplus \mathcal{Q}_2 \dotplus \cdots \dotplus \mathcal{Q}_{\lfloor \frac{n}{2} \rfloor}. \tag{1.5}$$

By (1.4) the space $\mathcal{P}_n$ is neutral and hence

$$\left\lfloor \frac{n}{2} \right\rfloor \leqslant \dim \mathcal{P}_n \leqslant \kappa(\mathcal{K}).$$

This implies $n \leqslant 2\kappa(\mathcal{K}) + 1$.

Step 3. $\mathfrak{S}_0(A) = \mathcal{S}[+]\mathcal{S}'$, *where $\mathcal{S}$ is a finite dimensional subspace invariant under A and $\mathcal{S}' \subseteq \operatorname{Ker}(A)$.*

To make a choice we assume that $\kappa^-(\mathcal{K}) < \infty$. From Step 1 we have that there exists $n_0 \in \mathbb{N}$ such that $\mathfrak{S}_0(A) = \operatorname{Ker}(A^{n_0})$. Consider a fundamental decomposition of the space $\operatorname{Ker}(A)$

$$\operatorname{Ker}(A) = \mathcal{L}^-[+]\mathcal{L}^0[+]\mathcal{L}^+. \tag{1.6}$$

Both $\mathcal{L}^-$ and $\mathcal{L}^0$ have dimensions $\leqslant \kappa(\mathcal{K}) < \infty$. We let

$$\mathcal{S}' = \mathcal{L}^+ \subseteq \operatorname{Ker}(A), \text{ and } \mathcal{S} = \mathfrak{S}_0(A) \cap (\mathcal{L}^+)^\perp.$$

11.1 Eigenvalues and Root Manifolds

Then $S \cap S' = 0$ and the decomposition $\mathfrak{S}_0(A) = S[+]S'$ holds. If x is a vector in S then $Ax \in \mathfrak{S}_0(A)$ and

$$[Ax, y] = [x, Ay], \quad y \in \mathcal{L}^+ \subseteq \mathrm{Ker}(A),$$

and hence $Ax \in S$. Thus, we have proved that S is invariant under A. It remains to prove that its dimension is finite. To see this, let us consider the subspaces

$$\mathcal{L}_n = \{x \in S \mid A^n x = 0\}, \quad n \geq 1. \tag{1.7}$$

Considering the Steps 1 and 2 for S instead of $\mathfrak{S}_0(A)$ it can be proven that $\dim S/\mathcal{L}_1 < \infty$. Since $\mathcal{L}_1 \subseteq \mathrm{Ker}(A)$ and $\mathcal{L}_1 \cap S' = 0$, taking account of (1.6) it follows that the dimension of $\mathcal{L}_1$ is finite, hence the dimension of S is finite. ∎

Inspecting the proof of Lemma 11.1.5 it follows that, given a closed symmetric operator A in the Pontryagin space $\mathcal{K}$ and a real eigenvalue λ, one can speak about elementary divisors of A at λ. Every such elementary divisor has a finite order and the number of elementary divisors of A at λ of order greater than two is finite. Moreover, if there is given a decomposition of the root subspace $\mathfrak{S}_\lambda(A) = S[+]S'$ as in Lemma 11.1.5, the elementary divisors of A of order greater than two are precisely determined by the Jordan matrix of the finite rank operator $A|S$. Let m_λ be the number of elementary divisors of $A|S$ of order greater than two and let $d_1, d_2, \ldots, d_{m_\lambda}$ be their orders. Define

$$\rho_A(\lambda) = \sum_{k=1}^{m_\lambda} \left\lfloor \frac{d_k}{2} \right\rfloor. \tag{1.8}$$

If λ is a nonreal eigenvalue of A we define

$$\rho_A(\lambda) = \dim \mathfrak{S}_\lambda(A). \tag{1.9}$$

Theorem 11.1.6. *Let A be a closed symmetric operator in the Pontryagin space $\mathcal{K}$ and consider the functions ρ_A as in (1.8) and (1.9). If σ is a set of eigenvalues of A such that $\lambda \neq \overline{\lambda}$ and $\lambda \in \sigma$ implies $\overline{\lambda} \notin \sigma$, then*

$$\sum_{\lambda \in \sigma} \rho_A(\lambda) \leq \kappa(\mathcal{K}).$$

Proof. Let us first consider a real eigenvalue λ. We prove that there exists a neutral subspace of $\mathfrak{S}_\lambda(A)$ of dimension exactly $\rho_A(\lambda)$. To see this, we assume $\lambda = 0$ and consider the decomposition $\mathfrak{S}_0(A) = S[+]S'$, as in Lemma 11.1.5. Recalling some facts proven during the proof of Lemma 11.1.5, let q be the Riesz index of λ, that is, q is the least natural number such that $\mathfrak{S}_0(A) = \mathrm{Ker}(A^q)$. Let $\mathcal{L}_n, n \geq 1$ be the finite dimensional subspaces defined by (1.7) and consider the subspaces $\{\mathcal{R}_m\}_{m=2}^q$ with the properties (1.2) (with $q = n$). Then define the subspaces $\mathcal{Q}_k$, $k \lfloor \frac{q}{2} \rfloor$ as in (1.3) and the neutral subspace $\mathcal{P}_q$ of S as in (1.5). We claim that

$$\rho_A(0) = \dim \mathcal{P}_q. \tag{1.10}$$

To this end, let us denote
$$r_m = \dim \mathcal{P}_m, \quad m \leqslant q.$$
Then
$$\dim \mathcal{P}_q = r_2 + r_4 + \cdots + r_{2\lfloor \frac{q}{2} \rfloor}.$$
For arbitrary $p \leqslant q$ let s_p denote the number of elementary divisors of order p. We claim that
$$s_p = r_p - r_{p+1}, \quad 2 \leqslant p \leqslant q, \tag{1.11}$$
where $r_{q+1} = 0$. Indeed, consider the finite rank operator $A|\mathcal{S}$. Let $x_1^{(q)}, \ldots, x_{n_q}^{(q)}$ be a complete system of linearly independent vectors in $\mathcal{P}_q$. Then the cyclic subspaces
$$\operatorname{Lin}\{x_i^{(q)}, Ax_i^{(q)}, \ldots, A^{q-1}x_i^{(q)}\}, \quad i = 1, 2, \ldots, n_q,$$
provide a description of all elementary divisors of order q and hence $n_q = r_q = s_q$. Let then $x_1^{q-1}, \ldots, x_{n_{q-1}}^{q-1}$ be a maximal system of vectors in $\mathcal{R}_{q-1}$ linearly independent modulo $A\mathcal{R}_q$. The cyclic subspaces
$$\operatorname{Lin}\{x_i^{(q-1)}, Ax_i^{(q-1)}, \ldots, A^{q-1}x_i^{(q-1)}\}, \quad i = 1, 2, \ldots, n_q,$$
give a description of all elementary divisors of order $q-1$ and hence $n_{q-1} = r_{q-1} - r_q = s_{q-1}$. Continuing this way we prove (1.11) for all $p \geqslant 2$.

Further, from (1.11) it follows that for all $2 \leqslant p \leqslant q$ we have
$$r_p = s_p + r_{p+1} = s_p + s_{p+1} + r_{p+2} = \cdots = s_p + \cdots + s_q.$$
Writing down this for even p we obtain
$$\begin{aligned} r_2 &= s_2 + s_3 + & s_4 + s_5 + & & s_6 + s_7 + & \cdots & +s_q, \\ r_4 &= & s_4 + s_5 + & & s_6 + s_7 + & \cdots & +s_q, \\ r_6 &= & & & s_6 + s_7 + & \cdots & +s_q, \\ &\vdots & & & & & \\ r_{2\lfloor \frac{q}{2} \rfloor} &= & & & & \cdots & +s_q, \end{aligned}$$
which, by summation give
$$\sum_{k=1}^{2\lfloor \frac{q}{2} \rfloor} r_{2k} = \sum_{j=2}^{q} \left\lfloor \frac{j}{2} \right\rfloor s_j = \sum_{i=1}^{m_0} \left\lfloor \frac{d_i}{2} \right\rfloor = \rho_A(0). \tag{1.12}$$
Taking into account that
$$\dim \mathcal{P}_0 = \sum_{k=1}^{2\lfloor \frac{q}{2} \rfloor} r_{2k},$$

11.2 JORDAN CANONICAL FORMS

by (1.12) we conclude that (1.10) holds.

Finally, from (1.10) it follows that for any real eigenvalue λ of A there exists a neutral subspace $\mathcal{P}_\lambda$ contained in the root manifold $\mathfrak{S}_\lambda(A)$ such that

$$\dim \mathcal{P}_\lambda = \rho_A(\lambda).$$

The direct sum subspace

$$\mathfrak{S}_\sigma = \sum_{\lambda \in \mathbb{R} \cap \sigma} \mathfrak{S}_\lambda(A) \dotplus \sum_{\lambda \in \sigma \setminus \mathbb{R}} \mathfrak{S}_\lambda(A)$$

is neutral (by Lemma 11.1.1), therefore

$$\sum_{\lambda \in \sigma} \rho_A(\lambda) = \dim \mathfrak{S}_\sigma \leqslant \kappa(\mathcal{K}). \quad \blacksquare$$

Proposition 11.1.7. *Let A be a selfadjoint operator in a Pontryagin space $\mathcal{K}$.*

(i) *The spectrum of A is contained in the real line with the exception of a finite set of points.*

(ii) *The nonreal spectrum $\sigma_0(A) = \sigma(A) \setminus \mathbb{R}$ consists of a finite number of eigenvalues, and the dimension of the spectral subspace corresponding to $\sigma_0(A)$ is $\leqslant 2\kappa(\mathcal{K})$.*

(iii) *The residual spectrum of A is empty.*

Proof. We first prove that the continuous spectrum $\sigma_c(A)$ is real. Let $\zeta \neq \bar{\zeta}$ be in $\sigma_c(A)$. Then for an arbitrary ε with $|\varepsilon| = 1$ we consider U the Cayley transform of A corresponding to ζ and ε, see Section 4.5. By Theorem 4.5.9, U is unitary. In particular, U is closed and isometric and hence by Theorem 4.3.6 it follows that U is bounded. Applying again Theorem 4.5.9 it follows that $\zeta \in \rho(A)$, a contradiction.

From Theorem 11.1.6 it follows that the dimension of the spectral subspace corresponding to $\sigma_0(A)$ is $\leqslant 2\kappa(\mathcal{K})$.

Finally we prove that the residual spectrum $\sigma_r(A)$ is empty. Indeed, if $\zeta \in \sigma_r(A)$, by Proposition 4.1.11 it follows that $\zeta \neq \bar{\zeta}$ and $\bar{\zeta} \in \sigma_p(A)$. Since $\sigma_p(A) \setminus \mathbb{R}$ is finite it follows that $\bar{\zeta}$ is isolated in $\sigma(A)$ and hence by Proposition 11.1.4 it follows that $\zeta \in \sigma_p(A)$, a contradiction. $\blacksquare$

11.2 Jordan Canonical Forms

If $\mathcal{K}$ is a finite dimensional Kreĭn space then densely defined symmetric operators in $\mathcal{K}$ are necessarily bounded, everywhere defined selfadjoint operators. In this section it will be shown that selfadjoint operators in a finite dimensional Kreĭn space have Jordan canonical forms, modulo unitary transformations of the space, simultaneously with a canonical form of a certain fundamental symmetry.

Let n be a nonnegative integer and consider the finite dimensional complex vector space $\mathbb{C}^n$. We denote by $\langle \cdot, \cdot \rangle$ the canonical positive definite inner product on $\mathbb{C}^n$. A *sip* matrix (*standard involutory permutation matrix*) J is, by definition, a square matrix of the form

$$J = \begin{bmatrix} 0 & 0 & \cdots & 0 & 1 \\ 0 & 0 & \cdots & 1 & 0 \\ \vdots & & \ddots & & \vdots \\ 0 & 1 & \cdots & 0 & 0 \\ 1 & 0 & \cdots & 0 & 0 \end{bmatrix}. \tag{2.1}$$

Clearly, a sip matrix J (as well as $-J$) is always a symmetry, that is, $J = J^* = J^{-1}$, in particular, these operators determine nondegenerate indefinite inner products $[\cdot, \cdot]$ (respectively, $-[\cdot, \cdot]$) defined by

$$[x, y] = \langle Jx, y \rangle, \quad x, y \in \mathbb{C}^n, \tag{2.2}$$

that is, finite dimensional Kreĭn spaces on $\mathbb{C}^n$, where n is the size of J.

Let $\lambda \in \mathbb{R}$ and consider the *Jordan block* of size n

$$A = \begin{bmatrix} \lambda & 1 & \cdots & 0 \\ 0 & \lambda & \ddots & \\ & & \ddots & 1 \\ 0 & 0 & \cdots & \lambda \end{bmatrix}. \tag{2.3}$$

Then $A^*J = JA$ and hence A is a selfadjoint operator on both Kreĭn spaces $(\mathcal{K}, \pm[\cdot, \cdot])$. On the other hand, let $n = 2k$ and $\alpha \in \mathbb{C}$ be such that $\operatorname{Im} \alpha > 0$ and consider the $k \times k$ Jordan blocks

$$H = \begin{bmatrix} \alpha & 1 & \cdots & 0 \\ 0 & \alpha & \ddots & \\ & & \ddots & 1 \\ 0 & 0 & \cdots & \alpha \end{bmatrix}, \quad \overline{H} = \begin{bmatrix} \overline{\alpha} & 1 & & 0 \\ 0 & \overline{\alpha} & \ddots & \\ & & \ddots & 1 \\ 0 & 0 & \cdots & \overline{\alpha} \end{bmatrix}. \tag{2.4}$$

Denote $A = H \oplus \overline{H}$. Then again $A^*J = JA$, that is, A is selfadjoint in the Kreĭn spaces $(\mathcal{K}, \pm[\cdot, \cdot])$.

The following theorem shows that any selfadjoint operator in a finite dimensional Kreĭn space can be reduced, modulo a unitary transformation, to a direct sum of selfadjoint operators of the two types mentioned before.

Theorem 11.2.1. *Let A be a selfadjoint operator in the Kreĭn space $\mathcal{K}$ of dimension n. Then $\sigma(A) = \{\alpha_j, \overline{\alpha}_j\}_{j=1}^r \cup \{\lambda_j\}_{j=1}^k$, where $\operatorname{Im} \alpha_j > 0$, for all $j \in \{1, 2, \ldots, r\}$, and $\lambda_j \in \mathbb{R}$, for all $j \in \{1, \ldots, k\}$.*

Moreover, there exist a fundamental symmetry J and an orthonormal basis $\{f_i\}_{i=1}^n$ of the Hilbert space $(\mathcal{K}, \langle \cdot, \cdot \rangle_J)$, with respect to which A is represented as a direct sum

$$A = \bigoplus_{j=1}^r \bigoplus_{p=1}^{l_j} \left(H_j^{(p)} \oplus \overline{H}_j^{(p)} \right) \oplus \bigoplus_{j=1}^k \bigoplus_{p=1}^{m_j} A_j^{(p)}, \tag{2.5}$$

11.2 JORDAN CANONICAL FORMS

and J has the representation

$$J = \bigoplus_{j=1}^{r}\bigoplus_{p=1}^{l_j} \widetilde{J}_j^{(p)} \oplus \bigoplus_{j=1}^{k}\bigoplus_{p=1}^{m_j} \varepsilon_j^{(p)} J_j^{(p)}, \tag{2.6}$$

where we have the following

(i) $\{H_j^{(p)}\}_{p=1}^{l_j}$ *are Jordan blocks, respectively, of size* $t_j^{(p)}$, *corresponding to* α_j, *for all* $j \in \{1,2,\ldots,r\}$;

(ii) $\{\overline{H}_j^{(p)}\}_{p=1}^{l_j}$ *are Jordan blocks, respectively, of the same size* $t_j^{(p)}$, *corresponding to* $\bar{\alpha}_j$, *for all* $j \in \{1,2,\ldots,r\}$;

(iii) $\{A_j^{(p)}\}_{p=1}^{m_j}$ *are Jordan blocks, respectively, of size* $s_j^{(p)}$ *corresponding to* λ_j, *for all* $j \in \{1,2,\ldots,k\}$;

(iv) $\{\widetilde{J}_j^{(p)}\}_{p=1}^{l_j}$ *are sip matrices, respectively, of size* $2t_j^{(p)}$;

(v) $\{J_j^{(p)}\}_{p=1}^{m_j}$ *are sip matrices, respectively, of size* $s_j^{(p)}$;

(vi) *the numbers* $\varepsilon_j^{(p)}$ *are signs* ± 1.

The set of signs $\{\varepsilon_j^{(p)}\}$ *is uniquely determined by A up to a permutation corresponding to a permutation of equal Jordan blocks.*

Proof. Since the Kreĭn space $\mathcal{K}$ is finite dimensional it follows that the spectrum of A consists of a finite set of eigenvalues and hence, by symmetry with respect to the real axis, for example, see Proposition 11.1.4, we have $\sigma(A) = \{\alpha_j, \bar{\alpha}_j\}_{j=1}^r \cup \{\lambda_j\}_{j=1}^k$, where $\operatorname{Im}\alpha_j > 0$, for all $j \in \{1,2,\ldots,r\}$, and $\lambda_j \in \mathbb{R}$, for all $j \in \{1,\ldots,k\}$. From Proposition 11.1.4 it follows that

$$\mathcal{K} = \mathfrak{S}_{\lambda_1}[+]\cdots[+]\mathfrak{S}_{\lambda_k}[+](\mathfrak{S}_{\alpha_1}\dotplus\mathfrak{S}_{\bar{\alpha}_1})[+]\cdots[+](\mathfrak{S}_{\alpha_r}\dotplus\mathfrak{S}_{\bar{\alpha}_r}). \tag{2.7}$$

We distinguish the following two cases.

(a) *Assume that* $\sigma(A) = \{\alpha,\bar{\alpha}\}$ *with* $\operatorname{Im}\alpha > 0$. By Proposition 11.1.4 we have $\mathcal{K} = \mathfrak{S}_\alpha \dotplus \mathfrak{S}_{\bar{\alpha}}$ and the root subspaces $\mathfrak{S}_\alpha$ and $\mathfrak{S}_{\bar{\alpha}}$ are in duality (strong and weak are the same in this finite dimensional case). There exists an integer $s > 0$ such that $(A-\alpha I)^s|\mathfrak{S}_\alpha = 0$ and $(A-\alpha I)^{s-1}x_1 \neq 0$ for some nontrivial $x_1 \in \mathfrak{S}_\alpha$. Moreover, there exists a nontrivial vector $y_1 \in \mathfrak{S}_{\bar{\alpha}}$ such that

$$[(A-\alpha I)^{s-1}x_1, y_1] = 1. \tag{2.8}$$

Indeed, this follows, for example, by letting S denote a fundamenal symmetry of $\mathcal{K}$ such that $S\mathfrak{S}_\alpha = \mathfrak{S}_{\bar{\alpha}}$ (cf. Theorem 3.3.3) and defining

$$y_1 = \frac{S(A-\alpha I)^{s-1}x_1}{\|(A-\alpha I)^{s-1}x_1\|^2},$$

where $\|\cdot\|$ denotes the unitary norm associated to S.

Consider the vectors $x_2, \ldots, x_s \in \mathfrak{S}_\alpha$ and $y_2, \ldots, y_s \in \mathfrak{S}_{\overline{\alpha}}$ defined by

$$x_j = (A - \alpha I)^{j-1} x_1, \quad y_j = (A - \overline{\alpha} I)^{j-1} y_1, \quad j = 2, \ldots, s. \tag{2.9}$$

Clearly the vectors $x_2, \ldots, x_s$ make a Jordan chain corresponding to α, that is,

$$A x_j - \alpha x_j = x_{j+1} \quad (j = 1, \ldots, s-1), \quad A x_s = \alpha x_s.$$

We claim that $y_1, \ldots, y_s$ make a Jordan chain corresponding to $\overline{\alpha}$. To this end we first remark that

$$[x_1, y_s] = [x_1, (A - \overline{\alpha}I)^{s-1} y_1] = [(A - \alpha I)^{s-1} x_1, y_1] = 1, \tag{2.10}$$

therefore $y_s \neq 0$ and then $y_j \neq 0$ for all $j \geq 2$. In addition, for any vector $x \in \mathfrak{S}_\alpha$ we have

$$[x, (A - \overline{\alpha}I) y_s] = [(A - \alpha I)^s x, y_1] = 0.$$

Therefore $(A - \overline{\alpha}I) y_s \in \mathfrak{S}_\alpha^\perp \cap \mathfrak{S}_{\overline{\alpha}}$ and hence $(A - \overline{\alpha}I) y_s = 0$. Since, by the definition of the vectors $y_1, \ldots, y_s$, we have

$$A y_j - \overline{\alpha} y_j = y_{j+1} \quad (j = 1, \ldots, s-1), \quad A y_s = \overline{\alpha} y_s,$$

the claim follows.

We claim now that there exist uniquely determined complex numbers $\zeta_2, \ldots, \zeta_s$ such that defining

$$e_1 = x_1 + \sum_{j=2}^{s} \zeta_j x_j,$$

we have

$$[e_1, y_k] = 0, \quad 1 \leq k \leq s-1. \tag{2.11}$$

Indeed, letting $j + k = s + 1$ from (2.8) it follows that

$$[x_k, y_k] = [(A - \alpha I)^{j-1} x_1, (A - \overline{\alpha}I)^{k-1} y_1] = [(A - \alpha I)^{s-1} x_1, y_1] = 1. \tag{2.12}$$

Taking into account that $(A - \alpha I)^s x_1 = 0$ a similar calculations shows that

$$[x_j, y_k] = 0, \quad j + k > s + 1. \tag{2.13}$$

From (2.12) and (2.13) it follows that the equations (2.11) make an inhomogeneous linear system in the unknowns $\zeta_2, \ldots, \zeta_s$ such that the principal matrix is upper triangular with the diagonal constant 1 and hence it has a unique solution.

We define now the *Jordan chain* of vectors associated to α by

$$e_{j+1} = (A - \alpha I) e_j, \quad 1 \leq j \leq s-1,$$

11.2 JORDAN CANONICAL FORMS

and the Jordan chain of vectors associated to $\overline{\alpha}$

$$e_{s+j} = y_j, \quad 1 \leqslant j \leqslant s.$$

Then, considering the subspace $\mathcal{N} = \mathrm{Lin}\{e_j\}_{j=1}^{2s}$ it follows that

$$A|\mathcal{N} = H \oplus \overline{H},$$

where H and $\overline{H}$ are the Jordan blocks as in (2.4), each one of the same size s. We claim that

$$[x, y] = (\widetilde{J}x, y), \quad x, y \in \mathcal{N}. \tag{2.14}$$

Indeed, if $j + k \leqslant s$ then

$$[e_j, y_k] = [(A - \alpha I)^{j-1} e_1, (A - \overline{\alpha} I)^{j-1} y_k] = [e_1, y_{j+k-1}] = 0, \tag{2.15}$$

and if $j + k \geqslant s + 1$ we have

$$[e_j, y_k] = [(A - \alpha I)^{j+k-2} e_1, y_1] = [(A - \alpha I)^{j+k-2} x_1, y_1],$$

therefore,

$$[e_j, y_k] = \begin{cases} 1, & \text{for } j + k = s + 1, \\ 0, & \text{for } j + k > s + 1. \end{cases} \tag{2.16}$$

From (2.15) and (2.16) the claim (2.14) follows.

In view of (2.14), it follows that the subspace $\mathcal{N}$ is regular and, since it is also invariant under A, we can apply the same process to the selfadjoint operator $A|\mathcal{N}^\perp$ and, within a finite number of steps, the Jordan blocks corresponding to α and $\overline{\alpha}$ are exhausted.

(b) *Assume that* $\sigma(A) = \{\lambda\}$ *for some* $\lambda \in \mathbb{R}$. Let $s > 0$ be the integer number such that $(A - \lambda I)^s = 0$ and $(A - \lambda I)^{s-1} \neq 0$. If S is a fixed fundamental symmetry of $\mathcal{K}$ it follows that the operator H can be defined by

$$H = S(A - \lambda I)^{s-1}.$$

Then H is nontrivial and selfadjoint in the Hilbert space $(\mathcal{K}, \langle \cdot, \cdot \rangle_S)$ and hence there exists a real, nontrivial eigenvalue of H. As a consequence, there exists an eigenvector x_1 of H such that

$$(Hx_1, x_1)_S = \varepsilon,$$

where $\varepsilon \in \mathbb{R}$ and $|\varepsilon| = 1$. In particular,

$$[(A - \lambda I)^{s-1} x_1, x_1] = \varepsilon. \tag{2.17}$$

Then define $x_j = (A - \lambda I)^{j-1} x_1$, $j = 2, \ldots, s$. We claim that there exist uniquely determined complex numbers $\zeta_2, \ldots, \zeta_s$ such that defining the vectors $e_1, \ldots, e_s$ by

$$e_1 = x_1 + \sum_{j=2}^{s} \zeta_j x_j, \quad e_j = (A - \lambda I)^{j-1} x_1, \quad j = 2, \ldots, s,$$

we have
$$[e_1, e_k] = 0, \quad k = 1, \ldots, s-1. \tag{2.18}$$

To this end we first remark that, in view of (2.17), for $j + k = s + 1$ we have
$$[x_j, x_k] = [(A - \lambda I)^{j-1} x_1, (A - \lambda I)^{k-1} x_1] = [(A - \lambda I)^{s-1} x_1, x_1] = \varepsilon, \tag{2.19}$$

while for $j + k > s + 1$ we have
$$[x_j, x_k] = [(A - \lambda I)^{j+k-2} x_1, x_1] = 0. \tag{2.20}$$

Using (2.19) and (2.20) it follows that equations (2.18) can be written as the following nonlinear system in the unknowns $\zeta_2, \ldots, \zeta_s$
$$[x_1, x_k] + 2\varepsilon \zeta_{s-k+1} + \varphi_k(\zeta_2, \ldots, \zeta_{s-k}) = 0, \quad k = 1, s-1,$$

where $\varphi_k(\zeta_2, \ldots, \zeta_{s-k})$ are algebraic functions. This shows that this system has a unique solution to be determined successively in the order $\zeta_s, \ldots, \zeta_2$.

Now we notice that $e_1, \ldots, e_s$ is a maximal Jordan chain associated to the eigenvalue λ and hence, considering the subspace $\mathcal{N} = \text{Lin}\{e_1, \ldots, e_s\}$, it follows that $A|\mathcal{N}_1$ has the representation as the Jordan block as in (2.3) of size s, with respect to the basis $e_1, \ldots, e_s$. Moreover, in view of (2.17) and (2.18) it follows that with respect to the same basis we have
$$[x, y] = \varepsilon[Jx, y], \quad x, y \in \mathcal{N},$$

where J is the sip matrix of size s. In particular, this shows that the subspace $\mathcal{N}_1$ is regular and, since it is also invariant under A, we apply the same procedure to $A|\mathcal{N}_1^\perp$ and, within a finite number of steps, all the Jordan blocks of A, of the same size s, are exhausted. Let these blocks be $A_1, \ldots, A_p$, let $J_1, \ldots, J_p$ be the corresponding sip matrices, all of them of size s, and $\varepsilon_1, \ldots, \varepsilon_p$ be the corresponding signs. We consider the regular subspace $\mathcal{N}$, invariant under A, which is defined by
$$\mathcal{N} = \mathcal{N}_1[+] \cdots [+] \mathcal{N}_p.$$

The dimension of $\mathcal{N}$ is s times the number of Jordan blocks of A of size s, in particular it is uniquely determined by A. We remark that
$$[(A - \lambda I)^{s-1} x, y] = \sum_{j=1}^{p} \varepsilon_j \langle x_j^{(s)}, y_j^{(s)} \rangle, \tag{2.21}$$

where
$$x = \sum_{j=1}^{p} x_j, \quad y = \sum_{j=1}^{p} y_j, \quad x_j, y_j \in \mathcal{N}_j, \quad j = 1, \ldots, p,$$

and $x_j^{(s)}$ and $y_j^{(p)}$ are the $j \times s$ components of the vectors x_j and y_j, respectively, in the basis $\{e_k\}_{k=1}^{sp}$. This shows that the cardinal of the set $\{j \mid \varepsilon_j = +1\}$ (as well as the cardinal of the set $\{j \mid \varepsilon_j = -1\}$) depends only on the operator A, more precisely, in view of (2.21), it coincides with the number of positive squares of the inner product space $(\mathcal{N}, [(A - \lambda I)^{s-1} \cdot, \cdot])$. In particular, the sequence of signs $\varepsilon_1, \ldots, \varepsilon_p$ is unique, modulo a permutation corresponding to a permutation of the sip matrices J_j, $j = 1, \ldots, p$.

We continue the process as before and, within a finite number of steps, we exhaust all the Jordan blocks of A. In addition, the uniqueness of all the signs ε_j, modulo permutations of the corresponding Jordan blocks of the same size, is obtained.

Finally, in view of (2.7) and the cases (a) and (b), we obtain the existence of the fundamental symmetry J and the basis $\{e_j\}_{j=1}^n$, such that the representations (2.5) and (2.6) hold. In addition, the set of signs $\{\varepsilon_j^{(p)}\}$ is uniquely determined by A, up to a permutation corresponding to a permutation of the Jordan blocks corresponding to the same eigenvalue and of the same size. ∎

11.3 Definitisable Selfadjoint Operators

We already noticed in Proposition 4.1.5 the location (with respect to the real axis) of the point spectrum, the continuous spectrum and the residual spectrum of a selfadjoint operator in a Kreĭn space and, as a consequence, the symmetry of the spectrum with respect to the real axis was obtained. Also, in Example 4.1.6 it was pointed out that, apart from this symmetry, the spectrum of a selfadjoint operator can be fairly arbitrary.

For the beginning we focus on positive and selfadjoint operators. By Remark 5.1.5, it may happen that the spectrum of a positive and selfadjoint operator is the whole complex plane, in particular, the spectrum may not be contained in the real line. For bounded positive operators the situation is, fortunately, much better: according to Proposition 4.1.7, any bounded and positive operator in a Kreĭn space has real spectrum.

The result in Proposition 4.1.7 is valid also for unbounded positive and selfadjoint operators under the additional assumption that the resolvent set is nonempty. We will obtain this in the more general context of definitisable selfadjoint operators, see Lemma 11.3.3 from below.

Definition 11.3.1. A selfadjoint operator A in the Kreĭn space is called *definitisable* if the resolvent set $\rho(A)$ is nonempty and there exists a nontrivial real polynomial p such that

$$[p(A)x, x] \geqslant 0, \quad x \in \mathrm{Dom}(A^n), \tag{3.1}$$

where $n \in \mathbb{N}$ is the degree of the polynomial p. A nontrivial real polynomial p such that (3.1) holds is called *definitising*.

Remark 11.3.2. The requirement, in the definition of a definitisable selfadjoint operator A, that the resolvent set $\rho(A)$ is nonempty is extremely important. As mentioned before, Remark 5.1.5 shows that selfadjointness of the operator A does not imply that its resolvent set is nonempty. But there is more to be said about this. In general, we do not know whether $\mathrm{Dom}(p(A)) = \mathrm{Dom}(A^n)$, where $n = \deg(p) \geqslant 3$, is dense, assuming that

$\rho(A)$ is not empty. However, as a consequence of Lemma 11.3.3, for any definitisable selfadjoint operator A we have that $\text{Dom}(A^n)$ is dense in $\mathcal{K}$ for any $n \geqslant 1$. ∎

Recall that for a field $\mathbb{F}$ we denote by $\mathbb{F}[X]$ the algebra of polynomials in the undetermined variable X with coefficients in $\mathbb{F}$. If $p \in \mathbb{C}[X]$ is an arbitrary polynomial we denote by $Z(p) = \{\lambda \in \mathbb{C} \mid p(\lambda) = 0\}$ the set of roots of p. If $\lambda \in Z(p)$ then $k(\lambda)$ designates the order of multiplicity of the root λ with respect to p. If $\lambda \notin Z(p)$ then, by definition, $k(\lambda) = 0$. Also, for arbitrary $\lambda \in \sigma_p(A)$, $\nu_A(\lambda)$, called the *Riesz index*, denotes either the least natural number n such that $\text{Ker}((\lambda I - A)^n) = \text{Ker}((\lambda I - A)^{n+1})$, if it exists, or the symbol $+\infty$. Recall that $\widetilde{\sigma}(T)$ denotes the extended spectrum of the closed linear operator T, that is, $\widetilde{\sigma}(T) = \sigma(T)$ if T is bounded and $\widetilde{\sigma}(T) = \sigma(T) \cup \{\infty\}$ if T is unbounded.

Lemma 11.3.3. *If A is a definitisable selfadjoint operator then the nonreal spectrum of A, $\sigma_0(A) = \sigma(A) \setminus \mathbb{R}$, is a finite set of points, symmetric with respect to the real line. More precisely, for any definitising polynomial p we have $\sigma_0(A) \subseteq Z(p) \setminus \mathbb{R}$ and any $\lambda \in \sigma_0(A)$ is an eigenvalue of A such that $\nu(\lambda) \leqslant k(\lambda)$.*

Proof. Let p be a definitisable polynomial for A and fix $\lambda_0 \in \rho(A)$. Denote $n = \deg(p)$ and consider $z_0 \in \mathbb{C} \setminus \mathbb{R}$ such that $p(z_0) \neq 0$. There exists $\zeta_0 \in \mathbb{C} \setminus \mathbb{R}$ such that, considering the rational function

$$q(\lambda) = p(\lambda)(\lambda - \lambda_0)^{-n-1}(\lambda - \overline{\lambda}_0)^{-n-1}(\lambda - \zeta_0)(\lambda - \overline{\zeta}_0),$$

we have $q(z_0) \notin \mathbb{R}$. Notice that

$$Z(q) = Z(p) \cup \{\zeta_0, \overline{\zeta}_0\}.$$

By the Riesz–Dunford–Taylor functional calculus we get

$$[q(A)x, x] = [p(A)(A - \zeta_0)(A - \lambda_0 I)^{-n-1}x, (A - \zeta_0)(A - \lambda_0 I)^{-n-1}x] \geqslant 0, \quad x \in \mathcal{K},$$

and hence the operator $q(A) \in \mathcal{B}(\mathcal{K})$ is positive. By Proposition 4.1.7 it follows that $\sigma(q(A)) \subset \mathbb{R}$. From the Spectral Mapping Theorem of the Riesz–Dunford–Taylor functional calculus we get $q(\widetilde{\sigma}(A)) = \sigma(q(A)) \subseteq \mathbb{R}$ and hence $\zeta_0 \notin \sigma(A)$. This proves that $\sigma_0(A) \subseteq Z(p) \setminus \mathbb{R}$ and hence $\sigma_0(A)$ is a finite set of points. The symmetry of $\sigma_0(A)$ with respect to the real axis follows from Proposition 11.1.4.

Let λ be in $\sigma_0(A)$. Then λ is isolated in the spectrum of A and hence we can consider the Riesz projections $E(\lambda; A)$ and $E(\overline{\lambda}; A) = E(\lambda; A)^\sharp$. Then $E = E(\lambda; A) + E(\overline{\lambda}; A) = E(\{\lambda, \overline{\lambda}\}; A)$ is a bounded selfadjoint projection, $\text{Ran}(E) \subseteq \text{Dom}(A)$, $EA \subseteq AE$, and $A|E\mathcal{K}$ is bounded. Since $\sigma(A|E\mathcal{K}) \subseteq \{\lambda, \overline{\lambda}\}$, without restricting the generality we can assume $E = I$. Then $A \in \mathcal{B}(\mathcal{K})$ and $\sigma(A) = \{\lambda, \overline{\lambda}\}$. Moreover,

$$p(\zeta) = (\zeta - \lambda)^{k(\lambda)}(\zeta - \overline{\lambda})^{k(\lambda)} p_1(\zeta),$$

where p_1 is a nontrivial real polynomial such that $p_1(\lambda) \neq 0$ and $p_1(\overline{\lambda}) \neq 0$. Restricting the domain of the polynomial function p_1 to a suitable domain Ω that contains $\widetilde{\sigma}(A)$, it

11.3 Definitisable Selfadjoint Operators

follows that there exists a function r that is holomorphic on Ω and such that $p_1 = r^2$ and $r(\bar\lambda) = \overline{r(\lambda)}$ for all $\lambda \in \Omega$. Then

$$[(A - \bar\lambda)^{k(\lambda)} r(A)x, (A - \bar\lambda)^{k(\lambda)} r(A)x] = [p(A)x, x], \quad x \in \mathcal{K},$$

and, taking into account that $r(A)$ is boundedly invertible, it follows that the closure of $\mathrm{Ran}((A - \bar\lambda)^{k(\lambda)})$ is a positive subspace.

Let us assume now that $\mathrm{Ker}(A - \lambda I)^{k(\lambda)} \neq E(\lambda; A)\mathcal{K}$. By Proposition 11.1.4 we know that the subspaces $E(\lambda; A)\mathcal{K}$ and $E(\bar\lambda; A)\mathcal{K}$ are neutral and in strong duality. Therefore, by Theorem 3.3.3 there exists a fundamental symmetry J such that $E(\bar\lambda; A)\mathcal{K} = JE(\lambda; A)\mathcal{K}$. Then $J\mathrm{Ker}(A - \lambda I)^{k(\lambda)} \neq E(\bar\lambda; A)\mathcal{K}$ and we can introduce the regular subspace

$$\widehat{\mathcal{K}} = (E(\lambda; A)\mathcal{K} \ominus \mathrm{Ker}(A - \lambda I)^{k(\lambda)}) \oplus (E(\bar\lambda; A)\mathcal{K} \ominus J\mathrm{Ker}(A - \lambda I)^{k(\lambda)}).$$

The subspace $\widehat{\mathcal{K}}$ is indefinite and since

$$\mathcal{K} = (\mathrm{Ker}(A - \lambda I)^{k(\lambda)} \oplus J\mathrm{Ker}(A - \lambda I)^{k(\lambda)})[+]\widehat{\mathcal{K}},$$

it follows that

$$\overline{\mathrm{Ran}((A - \bar\lambda)^{k(\lambda)})} = (\mathrm{Ker}(A - \lambda I)^{k(\lambda)})^{\perp} = \mathrm{Ker}(A - \lambda I)^{k(\lambda)}[+]\widehat{\mathcal{K}},$$

is also an indefinite subspace, a contradiction. ∎

In the definition of a definitisable selfadjoint operator we required that the definitising polynomial p should be real. We show now that this requirement can be relaxed to a certain degree. For σ a nonempty compact subset of $\widetilde{\mathbb{C}} = \mathbb{C} \cup \{\infty\}$, let $\mathfrak{H}(\sigma)$ denote the algebra of germs of holomorphic functions in a neighbourhood of σ.

Proposition 11.3.4. *Let A be a selfadjoint operator in the Kreĭn space $\mathcal{K}$ such that the resolvent set $\rho(A)$ is nonempty. The following statements are equivalent.*

(i) *There exists a nontrivial polynomial $p \in \mathbb{C}[X]$ such that the operator $p(A)$ is positive.*

(ii) *A is definitisable.*

(iii) *There exists $\varphi \in \mathfrak{H}(\widetilde\sigma(A))$ such that φ is nontrivial on each connected component of its domain and the operator $\varphi(A)$ is positive.*

(iv) *There exists $\varphi \in \mathfrak{H}(\widetilde\sigma(A))$ such that $\varphi(\bar\lambda) = \overline{\varphi(\lambda)}$ for all λ in a neighbourhood of $\widetilde\sigma(A)$, φ is nontrivial on each connected component of its domain and the operator $\varphi(A)$ is positive.*

Proof. (i)⇔(ii). Let $p \in \mathbb{C}[X]$ be a nontrivial polynomial such that $[p(A)x, x] \geq 0$, $x \in \mathrm{Dom}(p(A))$. Define the real polynomial $q \in \mathbb{R}[X]$ by

$$q(\lambda) = p(\lambda) + \overline{p(\bar\lambda)}, \quad \lambda \in \mathbb{C}.$$

Then, for all $x \in \mathrm{Dom}(q(A)) = \mathrm{Dom}(p(A))$ we have
$$[q(A)x, x] = [p(A)x, x] + [x, p(A)x] = 2[p(A)x, x] \geqslant 0.$$
The converse implication is obvious.

(ii)$\Rightarrow$(iv). Let p be a definitising polynomial of A and $\lambda_0 \in \rho(A)$. Define the function
$$\varphi(\lambda) = \frac{p(\lambda)}{(\lambda - \lambda_0)^n (\lambda - \overline{\lambda}_0)^n}, \quad \lambda \in \mathbb{C} \setminus \{\lambda_0, \overline{\lambda}_0\},$$
where $n = \deg(p)$, the degree of the polynomial p. Then φ is holomorphic on $\mathbb{C}\setminus\{\lambda_0, \overline{\lambda}_0\}$ and, since $\widetilde{\sigma}(A)$ is symmetric with respect to the real line, we have $\varphi(\overline{\lambda}) = \overline{\varphi(\lambda)}$ for all λ in $\mathbb{C} \setminus \{\lambda_0, \overline{\lambda}_0\}$, a symmetric neighbourhood of $\widetilde{\sigma}(A)$, and φ is nontrivial on each connected component of its domain. Also,
$$[\varphi(A)x, x] = [p(A)(A - \lambda_0 I)^{-n} x, (A - \lambda_0 I)^{-n} x] \geqslant 0, \quad x \in \mathcal{K},$$
and then we take into account that $\mathrm{Ran}((A - \lambda_0 I)^{-n}) = \mathrm{Dom}(A^n) = \mathrm{Dom}(p(A))$.

(iv)$\Rightarrow$(ii). Let φ be a function as in (iv). Fix $\lambda_0 \in \rho(A)$. Then φ admits the representation
$$\varphi(\lambda) = \frac{p(\lambda)}{(\lambda - \lambda_0)^n (\lambda - \overline{\lambda}_0)^n} \psi(\lambda),$$
where p is a polynomial of degree $\leqslant 2n$ and the function $\psi \in \mathfrak{H}(\widetilde{\sigma}(A))$ is nontrivial on each connected component of the domain and $\psi(\overline{\lambda}) = \overline{\psi(\lambda)}$.

As in the proof of Lemma 11.3.3, we can prove that the spectrum of A is real with the exception of a finite set of points. It follows that, restricting the domain of ψ to a smaller neighbourhood of $\widetilde{\sigma}(A)$, we can assume that each connected component of this domain is simply connected. Then there exists $\psi_1 \in \mathfrak{H}(\widetilde{\sigma}(A))$ such that $\psi = \psi_1^2$ and $\psi_1(\overline{\lambda}) = \overline{\psi_1(\lambda)}$ and hence
$$0 \leqslant [\varphi(A)x, x] = [p(A)(A - \lambda_0 I)^{-n}(A - \overline{\lambda}_0)^{-n}\psi_1^2(A)x, x]$$
$$= [p(A)(A - \lambda_0 I)^{-n}\psi_1(A), \psi_1(A)x], \quad x \in \mathcal{K}.$$
Since ψ does not vanish on $\widetilde{\sigma}(A)$ the same is true for ψ_1 and hence $\psi_1(A)$ is a boundedly invertible operator, in particular $\mathrm{Ran}(\psi_1(A)) = \mathcal{K}$. Then
$$[p(A)(A - \lambda_0 I)^{-n} x, (A - \lambda_0 I)^{-n} x] \geqslant 0, \quad x \in \mathcal{K},$$
that is, p is definitising. ∎

We end this section with two remarkable examples showing that the class of definitisable operators is rather large.

Proposition 11.3.5. *If A is a selfadjoint operator in a Pontryagin space $\mathcal{K}$ then A is definitisable.*

11.3 Definitisable Selfadjoint Operators

Proof. To make a choice we assume that $\kappa^-(\mathcal{K}) < \infty$. From Proposition 11.1.7 it follows that the resolvent set $\rho(A)$ is nonempty. On the other hand, from Corollary 9.2.12 it follows that there exists a maximal negative subspace $\mathcal{L} \subseteq \mathrm{Dom}(A)$ invariant under A. Then $\dim(\mathcal{L}) = \kappa^-(\mathcal{K}) < \infty$ and hence, the finite rank operator $A|\mathcal{L}\colon \mathcal{L} \to \mathcal{L}$ has a minimal polynomial, that is, there exists a nontrivial polynomial $p_0 \in \mathbb{C}[X]$ such that

$$p_0(A|\mathcal{L}) = p_0(A)|\mathcal{L} = 0.$$

It follows that

$$[\overline{p}_0(A)x, y] = [x, p_0(A)y] = 0, \quad x \in \mathrm{Dom}(p_0(A)),\ y \in \mathcal{L},$$

therefore

$$\overline{p}_0(A)\,\mathrm{Dom}(p_0(A)) \subset \mathcal{L}^\perp.$$

Since $\mathcal{L}^\perp$ is a positive subspace this implies

$$[(p_0\overline{p}_0)(A)x, x] = [\overline{p}_0(A)x, \overline{p}_0(A)x] \geqslant 0, \quad x \in \mathrm{Dom}(p_0^2(A)).$$

We have proven that $p = p_0 \overline{p}_0$ is a definitising polynomial of A. ∎

Recall that, for a symmetric operator A on an indefinite inner product space $(\mathcal{S}; [\cdot, \cdot])$, we denote by $\kappa_\pm(A)$ the number of positive/negative squares of the quadratic form

$$\mathrm{Dom}(A) \ni x \mapsto [Ax, x],$$

that is, the positive/negative signature of the inner product

$$\mathrm{Dom}(A) \times \mathrm{Dom}(A) \ni (x, y) \mapsto [Ax, y].$$

Proposition 11.3.6. *If A is a selfadjoint operator in a Kreĭn space $\mathcal{K}$ such that $\rho(A) \neq \emptyset$ and either $\kappa_+(A)$ or $\kappa_-(A)$ is finite, then A is definitisable.*

Proof. Assume that $\kappa_-(A) < \infty$. In the following we describe a certain construction which is very similar to that of an induced Pontryagin space as in Section 1.5, but with additional properties. The inner product space $(\mathrm{Dom}(A), [\cdot, \cdot]_A)$ is decomposable, that is, there exists a decomposition

$$\mathrm{Dom}(A) = \mathcal{D}_- \dotplus \mathrm{Ker}(A) \dotplus \mathcal{D}_+, \tag{3.2}$$

where the inner product spaces $(\mathcal{D}_\pm, \pm[\cdot, \cdot])$ are positive definite and mutually orthogonal. We consider the nondegenerate inner product space $(\mathcal{D}_- \dotplus \mathcal{D}_+, [\cdot, \cdot]_A)$ and, since $\dim \mathcal{D}_- = \kappa_-(A) < \infty$, by Corollary 1.5.4 there exists the completion of this space to a Pontryagin space $(\widetilde{\mathcal{K}}, [\cdot, \cdot]_A)$ such that $\kappa^-(\widetilde{\mathcal{K}}) = \kappa_-(A) < \infty$. Consider the linear mapping $\Pi\colon \mathrm{Dom}(A) \to \mathcal{K}$ defined by

$$\mathrm{Dom}(A) \ni x_- + x_0 + x_+ \mapsto x_- + x_+ \in \mathcal{K}, \quad x_\pm \in \mathcal{D}_\pm,\ x_0 \in \mathrm{Ker}(A).$$

Then Π has dense range and

$$[\Pi x, \Pi y] = [Ax, y], \quad x, y \in \mathrm{Dom}(A), \tag{3.3}$$

holds. Also, from (3.3) it follows that Π has a densely defined adjoint, more precisely, this is an extension of the linear mapping $\mathcal{K} \supseteq \mathcal{D}_- + \mathcal{D}_+ \ni x \mapsto x \in \mathcal{H}$, and hence Π is closable. We denote by the same symbol Π its closure.

We first prove that $\Pi\,\mathrm{Dom}(A^2)$ is dense in $\widetilde{\mathcal{K}}$. To this end, let us fix $\lambda \in \rho(A)$, $\lambda \neq \overline{\lambda}$. For arbitrary $x \in \mathrm{Dom}(A)$ let $y = (A - \lambda I)x$. Then $x = (A - \lambda I)^{-1}y$ and there exists a sequence $(y_n)_{n \geqslant 1} \subseteq \mathrm{Dom}(A)$ such that $y_n \xrightarrow[n\to\infty]{} y$. Denoting $x_n = (A - \lambda I)^{-1}y_n \subseteq \mathrm{Dom}(A^2)$ we have $x_n \xrightarrow[n\to\infty]{} x$ and $Ax_n \to Ax$. These imply

$$[x_n, x_n]_A \xrightarrow[n\to\infty]{} [x, x]_A, \quad [x_n, z]_A \xrightarrow[n\to\infty]{} [x, z]_A,$$

and hence, by Theorem 1.4.11, we have that $\Pi x_n \xrightarrow[n\to\infty]{} \Pi x$.

We define the linear operator $\widetilde{A}$ in $\widetilde{\mathcal{K}}$ with domain $\Pi\,\mathrm{Dom}(A^2)$ by $\widetilde{A}\Pi x = \Pi Ax$, $x \in \mathrm{Dom}(A^2)$. Then

$$[\widetilde{A}\Pi x, \Pi x]_A = [A^2 x, x] = [Ax, Ax] = [\Pi x, \widetilde{A}x]_A, \quad x \in \mathrm{Dom}(A^2).$$

This shows that $\widetilde{A}$ is symmetric and the above proved fact shows that it is also densely defined. Then $\widetilde{A} \subseteq \widetilde{A}^\sharp$, in particular $\widetilde{A}$ is closable and let us denote by the same symbol $\widetilde{A}$ its closure.

We prove that $\widetilde{A}$ is selfadjoint. Indeed, since $\mathrm{Ran}(\widetilde{A} - \overline{\lambda}\widetilde{I}) \supseteq \Pi\,\mathrm{Dom}(A)$ is dense in $\mathcal{K}$ it follows that $\widetilde{A} - \lambda \widetilde{I}$ is injective and hence we can consider the Cayley transform

$$\widetilde{U} = (\widetilde{A} - \overline{\lambda}\widetilde{I})(\widetilde{A} - \lambda\widetilde{I})^{-1}, \quad \mathrm{Dom}(\widetilde{U}) = \mathrm{Ran}(\widetilde{A} - \lambda\widetilde{I}).$$

By Proposition 4.5.6 and Lemma 4.5.8 the operator $\widetilde{U}$ is closed and isometric. $\widetilde{U}$ is a bounded isometry with dense range, that is, $\widetilde{U}$ is a bounded unitary operator. Then Theorem 4.5.9 implies that $\widetilde{A}$ is selfadjoint.

By Proposition 11.3.5 the selfadjoint operator $\widetilde{A}$ is definitisable, in particular, there exists a real nontrivial polynomial p such that

$$[p(\widetilde{A})\Pi x, \Pi x]_A = [Ap(A)x, x] \geqslant 0, \quad x \in \mathrm{Dom}(Ap(A)).$$

This proves that A is definitisable. ∎

11.4 Herglotz's Theorems

In this section we will see some integral representations of holomorphic functions in the upper half plane, provided that the imaginary parts of the functions are nonnegative and they obey a certain asymptotic condition on the imaginary axis. This is used in order

11.4 HERGLOTZ'S THEOREMS

to obtain an integral representation of the resolvent function of a selfadjoint definitisable operator in the next section. The approach uses the investigation of similar problems for $\mathbb{D} = \{z \in \mathbb{C} \mid |z| < 1\}$, the open unit disc in the complex plane, via integral representations of harmonic functions in $\mathbb{D}$, and then a conformal mapping of the unit disc onto the upper half plane.

Let us first recall some facts from classical harmonic analysis. A (real valued) function u defined in some open subset Ω of the complex plane is *harmonic* in Ω if it is two times differentiable and
$$\frac{\partial^2 u}{\partial x^2} + \frac{\partial^2 u}{\partial y^2} = 0, \quad z = x + \mathrm{i}y \in \Omega.$$

If u is harmonic in a neighbourhood of the closed unit disc $\operatorname{Clos} \mathbb{D} = \{z \in \mathbb{C} \mid |z| \leqslant 1\}$, then u has the Poisson integral representation

$$u(r\mathrm{e}^{\mathrm{i}\theta}) = \frac{1}{2\pi} \int_0^{2\pi} \frac{1-r^2}{1+r^2 - 2r\cos(\theta - t)} u(\mathrm{e}^{\mathrm{i}t}) \mathrm{d}t, \quad 0 \leqslant r < 1,\ 0 \leqslant \theta \leqslant 2\pi. \quad (4.1)$$

The *Poisson kernel*
$$P_r(\varphi) = \frac{1-r^2}{1+r^2 - 2r\cos\varphi}, \quad 0 \leqslant r < 1,\ 0 \leqslant \varphi \leqslant 2\pi,$$

is positive and periodic of period 2π and the following expansion holds

$$P_r(\varphi) = \sum_{n \in \mathbb{Z}} r^{|n|} \mathrm{e}^{\mathrm{i}n\varphi}, \quad (4.2)$$

where the series converges uniformly on compact subsets of $\mathbb{D}$. Integration term by term of the series in (4.2) gives
$$\int_0^{2\pi} P_r(\theta) \mathrm{d}\theta = 2\pi.$$

As a consequence, if the function u is harmonic in a neighbourhood of $\overline{\mathbb{D}}$ then the following *mean integral formula* holds

$$u(0) = \frac{1}{2\pi} \int_0^{2\pi} u(r\mathrm{e}^{\mathrm{i}\theta}) \mathrm{d}\theta, \quad 0 \leqslant r < 1. \quad (4.3)$$

Lemma 11.4.1. *Let u be a (real valued) function defined in $\mathbb{D}$. Then u is a harmonic function in $\mathbb{D}$ such that*
$$\sup_{0 \leqslant r < 1} \int_0^{2\pi} |u(r\mathrm{e}^{\mathrm{i}\theta})| \mathrm{d}\theta < \infty,$$

if and only if there exists a finite measure μ on the interval $[0, 2\pi]$ such that

$$u(r\mathrm{e}^{\mathrm{i}\theta}) = \frac{1}{2\pi} \int_0^{2\pi} \frac{1-r^2}{1+r^2 - 2r\cos(\theta - t)} \mathrm{d}\mu(t), \quad 0 \leqslant r < 1,\ 0 \leqslant \theta \leqslant 2\pi. \quad (4.4)$$

Proof. We consider the sequence of functions $(u_n)_{n \leqslant 1}$ on $[0, 2\pi]$ defined by

$$u_n(\theta) = u\big((1 - \frac{1}{n})e^{i\theta}\big), \quad \theta \in [0, 2\pi].$$

Then

$$\sup_{n \in \mathbb{N}} \|u_n\|_1 = \sup_{n \in \mathbb{N}} \int_0^{2\pi} |u_n(\theta)| d\theta \leqslant \sup_{0 \leqslant r < 1} \int_0^{2\pi} |u(re^{i\theta})| d\theta < \infty.$$

Let us recall now that, by a theorem of F. Riesz, the dual of the Banach space $\mathcal{C}([0, 2\pi])$ of continuous functions on the interval $[0, 2\pi]$ can be identified with the Banach space $\mathcal{M}([0, 2\pi])$ of the Radon finite measures on $[0, 2\pi]$, endowed with the norm

$$\|\mu\|_1 = \int_0^{2\pi} |d\mu(\theta)|, \quad \mu \in \mathcal{M}([0, 2\pi]).$$

The previous inequality shows that the sequence of measures $(u_n(\theta)d\theta)_n$ is bounded in $\mathcal{M}([0, 2\pi])$ and hence, by the weak precompactness of bounded subsets in $\mathcal{M}([0, 2\pi])$, there exists a subsequence $(u_{n_j})_j$ weakly converging to a measure $\mu \in \mathcal{M}([0, 2\pi])$, that is,

$$\lim_{j \to \infty} \int_0^{2\pi} f(t) u_{n_j}(t) dt = \int_0^{2\pi} f(t) d\mu(t), \quad f \in \mathcal{C}([0, 2\pi]).$$

Using the Poisson representation (4.1) of the harmonic functions u_{n_j} on a neighbourhood of the closed unit disc Clos $\mathbb{D}$, for any $0 \leqslant r < 1$ and $\theta \in [0, 2\pi]$ we have

$$u\big((1 - \frac{1}{n_j})re^{i\theta}\big) = \frac{1}{2\pi} \int_0^{2\pi} \frac{1 - r^2}{1 + r^2 - 2r\cos(\theta - t)} u_{n_j}(e^{it}) dt. \tag{4.5}$$

Remark now that, for any $0 \leqslant r < 1$ and $\theta \in [0, 2\pi]$, the functions $P_r(\theta - t)$, in the variable t, belong to $\mathcal{C}([0, 2\pi])$. This allows us to pass to the limit on both sides of (4.5), following $j \to \infty$, and thus we obtain (4.4).

Conversely, if u is represented by (4.4) the proof of its harmonicity is a routine calculation, either by deriving two times under the integral with respect to r and, respectively, with respect to θ, (and then using the formula of the Laplacian in polar coordinates), or by using the expansion of the Poisson kernel as in (4.2) in order to obtain

$$u(re^{i\theta}) = \sum_{n \in \mathbb{Z}} \frac{1}{2\pi} \int_0^{2\pi} e^{-int} d\mu(t) \, r^{|n|} e^{in\theta}, \quad 0 \leqslant r < 1, \, 0 \leqslant \theta \leqslant 2\pi,$$

where the series converges uniformly on any compact subset of $\mathbb{D}$. Further, by means of the polar representation we get $g \in L_\infty(0, 2\pi)$, $\|g\|_\infty = 1$, such that

$$\int_0^{2\pi} |u(re^{i\theta})| d\theta = \int_0^{2\pi} g(\theta) u(re^{i\theta}) d\theta.$$

11.4 Herglotz's Theorems

Then

$$\int_0^{2\pi} |u(re^{i\theta})|\,d\theta = \frac{1}{2\pi}\int_0^{2\pi}\int_0^{2\pi} P_r(\theta - t)\,g(\theta)\,d\theta\,d\mu(t)$$

$$\leqslant \frac{1}{2\pi}\int_0^{2\pi}\int_0^{2\pi} P_r(\theta - t)\,d\theta\,d|\mu(t)| = \int_0^{2\pi} |d\mu(t)|,$$

where we used the Fubini Theorem and the properties of the Poisson kernel. ∎

Corollary 11.4.2. *If u is a positive harmonic function in $\mathbb{D}$ then there exists a positive finite measure μ on $[0, 2\pi]$ such that*

$$u(re^{i\theta}) = \frac{1}{2\pi}\int_0^{2\pi} \frac{1 - r^2}{1 + r^2 - 2r\cos(\theta - t)}\,d\mu(t), \quad 0 \leqslant r < 1,\ 0 \leqslant \theta \leqslant 2\pi.$$

Proof. If u is a positive harmonic function in $\mathbb{D}$ then, using the mean integral formula for harmonic functions (4.3), we get

$$\sup_{0\leqslant r<1}\int_0^{2\pi}|u(re^{i\theta})|\,d\theta = \sup_{0\leqslant r<1}\int_0^{2\pi} u(re^{i\theta})\,d\theta = u(0), \quad 0 \leqslant r < \infty.$$

This allows us to apply Lemma 11.4.1 and obtain the existence of a finite measure μ such that (4.4) holds. In addition, considering the sequence u_{n_j}, as in the proof of Lemma 11.4.1, it follows that for any positive function $g \in \mathcal{C}([0, 2\pi])$ we have

$$0 \leqslant \lim_{j\to\infty}\int_0^{2\pi} g(t)\,u_{n_j}(t)\,dt = \int_0^{2\pi} g(t)\,d\mu(t),$$

and hence the measure μ is positive. ∎

Remark 11.4.3. Let $\mu \in \mathcal{M}([0, 2\pi])$. Considering the function μ on $[0, 2\pi]$ defined by

$$\mu(t) = \int_0^t d\mu(s), \quad t \in [0, 2\pi],$$

it follows that this function has bounded variation and the integral in (4.4) can also be considered as a Stieltjes integral with respect to the function μ. Moreover, the measure μ is positive if and only if the function μ is nondecreasing. If, in the statement of Corollary 11.4.2 we impose on μ the additional norming conditions to μ to be left continuous and $\mu(0) = 0$, then the function μ is uniquely determined by u. ∎

If u is a harmonic function in $\mathbb{D}$ we denote by $\tilde{u}$ the harmonic conjugate of u, that is, $\tilde{u}$ is the harmonic function in $\mathbb{D}$ such that the function $f = u + i\tilde{u}$ is holomorphic in $\mathbb{D}$ and $\tilde{u}(0) = 0$. By means of the uniqueness of the solution of the Dirichlet problem for the unit disc, $\tilde{u}$ is uniquely determined by u.

Lemma 11.4.4. *Let μ be a finite measure on $[0, 2\pi]$ and consider the function u as in (4.4). Then u is harmonic in $\mathbb{D}$ and its harmonic conjugate $\tilde{u}$ has the representation*

$$\tilde{u}(re^{i\theta}) = \frac{1}{2\pi} \int_0^{2\pi} \frac{2r\sin(\theta - t)}{1 + r^2 - 2r\cos(\theta - t)} \, d\mu(t), \quad 0 \leq r < 1, \, 0 \leq \theta \leq 2\pi. \quad (4.6)$$

Proof. The function u as defined in (4.4) is harmonic, as Lemma 11.4.1 shows. Consider the function f holomorphic in $\mathbb{D}$,

$$f(z) = u(z) + i\tilde{u}(z), \quad |z| < 1.$$

Let

$$f(re^{i\theta}) = \sum_{n \geq 0} a_n r^n e^{in\theta}, \quad 0 \leq r < 1, \, 0 \leq \theta \leq 2\pi, \quad (4.7)$$

be the expansion of f in $\mathbb{D}$, where the series converges uniformly on compact subsets of $\mathbb{D}$ and $a_0 \in \mathbb{R}$. Define the complex sequence $(A_n)_n$ by

$$A_n = \begin{cases} \frac{1}{2} a_n, & n > 0, \\ a_0, & n = 0, \\ \frac{1}{2} \bar{a}_{-n}, & n < 0. \end{cases}$$

Then u has the expansion

$$u(re^{i\theta}) = \sum_{n \in \mathbb{Z}} A_n r^{|n|} e^{in\theta}, \quad 0 \leq r < 1, \, 0 \leq \theta \leq 2\pi, \quad (4.8)$$

and the series converges uniformly on compact subsets of $\mathbb{D}$. Manipulating with the series in (4.7) and (4.8) it follows that for all $0 \leq r < 1$ and $0 \leq \theta \leq 2\pi$ we have

$$\tilde{u}(re^{i\theta}) = -i\left(f(re^{i\theta}) - u(re^{i\theta})\right) = -i \sum_{n \in \mathbb{Z}} A_n \operatorname{sgn}(n) r^{|n|} e^{in\theta}. \quad (4.9)$$

On the other hand, from (4.8), (4.4), and the expansion (4.2) we get

$$A_n = \frac{1}{2\pi} \int_0^{2\pi} e^{-int} \, d\mu(t), \quad n \in \mathbb{Z},$$

and hence, integrating term by term the series in (4.9) it follows

$$\tilde{u}(re^{i\theta}) = \frac{-i}{2\pi} \int_0^{2\pi} \sum_{n \in \mathbb{Z}} \operatorname{sgn}(n) r^{|n|} e^{in(\theta - t)} \, d\mu(t). \quad (4.10)$$

Since

$$\sum_{n \in \mathbb{Z}} \operatorname{sgn}(n) r^{|n|} e^{in\theta} = \sum_{n > 0} r^n e^{in\theta} - \sum_{n > 0} r^n e^{-in\theta}$$

$$= \frac{re^{i\theta}}{1 - re^{i\theta}} - \frac{re^{-i\theta}}{1 - re^{-i\theta}} = \frac{2r\sin(\theta - t)}{1 + r^2 - 2r\cos(\theta - t)},$$

11.4 HERGLOTZ'S THEOREMS

from (4.10) the representation (4.6) follows. ∎

The kernel $Q_r(t)$ obtained in the proof of Theorem 11.4.5,

$$Q_r(t) = \frac{2r \sin t}{1 + r^2 - 2r \cos t}, \quad 0 \leqslant r < 1, 0 \leqslant t \leqslant 2\pi,$$

is called the *conjugate Poisson kernel*.

Theorem 11.4.5. *A complex function f defined in $\mathbb{D}$ is represented by*

$$f(z) = i\beta + \int_0^{2\pi} \frac{e^{it} + z}{e^{it} - z} \, d\rho(t), \quad z \in \mathbb{D}, \tag{4.11}$$

where $\beta \in \mathbb{R}$ and ρ is a nondecreasing function on $[0, 2\pi]$ with bounded variation, if and only if f is holomorphic and $\operatorname{Re} f \geqslant 0$ in $\mathbb{D}$. In addition, the function ρ can be chosen to be left continuous and $\rho(0) = 0$, and, in this case, it is uniquely determined by f.

Proof. Assume that the function f admits the representation (4.11). Then it is clear that f is holomorphic in $\mathbb{D}$ and

$$\operatorname{Re} f(z) = \int_0^{2\pi} \frac{2r \sin(\theta - t)}{1 + r^2 - 2r \cos(\theta - t)} \, d\rho(t) \geqslant 0, \quad z = re^{i\theta} \in \mathbb{C}.$$

Conversely, if f is holomorphic in $\mathbb{D}$ and $\operatorname{Re} f \geqslant 0$ then, denoting $u = \operatorname{Re} f$, we have

$$f(z) = u(z) + i\tilde{u}(z) + i\beta, \quad z \in \mathbb{D}.$$

Taking into account that

$$\frac{e^{it} + z}{e^{it} - z} = P_r(\theta - t) + iQ_r(\theta - t), \quad z = re^{i\theta},$$

by Corollary 11.4.2 and Lemma 11.4.4 the representation (4.11) follows. The statement concerning the uniqueness of ρ is a consequence of Remark 11.4.3. ∎

In the following we will denote by $\mathbb{C}_+ = \{z \in \mathbb{C} \mid \operatorname{Im} z > 0\}$ the upper half plane of the complex plane.

Theorem 11.4.6. *A complex function φ defined in $\mathbb{C}_+$ admits the representation*

$$\varphi(z) = \alpha + \gamma z + \int_{-\infty}^{+\infty} \frac{1 + tz}{t - z} \, d\sigma(t), \quad z \in \mathbb{C}_+, \tag{4.12}$$

where $\alpha \in \mathbb{R}$, $\gamma \geqslant 0$, σ is a nondecreasing function with bounded variation on $\mathbb{R}$, and the Stieltjes integral is improper at $\pm\infty$, if and only if φ is holomorphic and $\operatorname{Im} \varphi \geqslant 0$ in $\mathbb{C}_+$. In addition, the function φ can be found such that it is left continuous and $\sigma(-\infty) = 0$, and, in this case, it is uniquely determined by φ.

Proof. Assuming the function φ is holomorphic and $\operatorname{Im}\varphi \geq 0$ in $\mathbb{C}_+$, we define the function f in $\mathbb{D}$ by

$$f(\zeta) = -\mathrm{i}\varphi(\mathrm{i}\frac{1+\zeta}{1-\zeta}), \quad \zeta \in \mathbb{D}.$$

Then f is holomorphic and $\operatorname{Re} f \geq 0$ in $\mathbb{D}$. According to Theorem 11.4.5 there exist $\beta \in \mathbb{R}$ and a nondecreasing function ρ on $[0, 2\pi]$ with bounded variation, such that for all $\zeta \in \mathbb{D}$ we have

$$\begin{aligned}
f(\zeta) &= \mathrm{i}\beta + \int_0^{2\pi} \frac{e^{\mathrm{i}s}+\zeta}{e^{\mathrm{i}s}-\zeta}\,\mathrm{d}\rho(s) \\
&= \mathrm{i}\beta + \int_{0+}^{2\pi-} \frac{e^{\mathrm{i}s}+\zeta}{e^{\mathrm{i}s}-\zeta}\,\mathrm{d}\rho(s) + \frac{1+\zeta}{1-\zeta}(\rho(2\pi) - \rho(2\pi-0) + \rho(0+0) - \rho(0)) \\
&= \mathrm{i}\beta + \gamma\frac{1+\zeta}{1-\zeta} + \int_{0+}^{2\pi-} \frac{e^{\mathrm{i}s}+\zeta}{e^{\mathrm{i}s}-\zeta}\,\mathrm{d}\rho(s).
\end{aligned}$$

Letting $z = \mathrm{i}(1+\zeta)/(1-\zeta)$ this implies

$$\varphi(z) = \mathrm{i}f(\zeta) = -\beta + \mathrm{i}\gamma\frac{1+\zeta}{1-\zeta} + \mathrm{i}\int_{0+}^{2\pi-} \frac{e^{\mathrm{i}s}+\zeta}{e^{\mathrm{i}s}-\zeta}\,\mathrm{d}\rho(s), \quad z \in \mathbb{C}_+. \tag{4.13}$$

We set $\alpha = -\beta$ and define the function σ on $\mathbb{R}$ by

$$\sigma(t) = \rho(s), \quad t = -\cot\frac{s}{2}.$$

Taking into account that

$$\frac{e^{\mathrm{i}s}+\zeta}{e^{\mathrm{i}s}-\zeta} = -\mathrm{i}\frac{z\cot\frac{s}{2}-1}{\cot\frac{s}{2}+z}, \quad z = \mathrm{i}\frac{1+\zeta}{1-\zeta},$$

from (4.13) we obtain the representation (4.12).

Conversely, if φ has the representation (4.12), then we can perform the inverse transformations in order to obtain the representation (4.11). Therefore $\operatorname{Im}\varphi \geq 0$ follows. Remark that the function σ is left continuous and $\sigma(-\infty) = 0$ if and only if ρ is left continuous and $\rho(0) = 0$. The last statement concerning the uniqueness of the function ρ follows now from Theorem 11.4.5. ∎

Theorem 11.4.7. *A function φ defined in $\mathbb{C}_+$ admits the representation*

$$\varphi(z) = \int_{-\infty}^{+\infty} \frac{\mathrm{d}\omega(t)}{t-z}, \quad z \in \mathbb{C}_+, \tag{4.14}$$

where ω is a nondecreasing function on $\mathbb{R}$ with bounded variation and the Stieltjes integral is improper at $\pm\infty$, if and only if φ is holomorphic and $\operatorname{Im}\varphi \geq 0$ in $\mathbb{C}_+$ and

$$\sup_{y>0} |y\varphi(\mathrm{i}y)| < \infty. \tag{4.15}$$

11.4 HERGLOTZ'S THEOREMS

One can determine ω such that it is left continuous and $\omega(-\infty) = 0$. In this case ω is uniquely determined by φ and

$$\lim_{y \to \infty} -iy\varphi(iy) = \omega(+\infty) = \lim_{t \to \infty} \omega(t). \tag{4.16}$$

Proof. Assume that φ is holomorphic and $\operatorname{Im} \varphi \geqslant 0$ in $\mathbb{C}_+$. According to Theorem 11.4.6, φ has the representation (4.12) and hence

$$y\varphi(iy) = \alpha y + i\gamma y^2 + \int_{-\infty}^{+\infty} \frac{y(1+ity)}{t-iy}\,d\sigma(t), \quad y > 0.$$

If, in addition, (4.16) holds then there exists $M > 0$ such that

$$\left| \alpha y + i\gamma y^2 + \int_{-\infty}^{+\infty} \frac{y(1+ity)}{t-iy}\,d\sigma(t) \right| \leqslant M, \quad y > 0.$$

The same inequality applies to the real part and, respectively, to the imaginary part of the left side and hence

$$\left| \alpha y + \int_{-\infty}^{+\infty} \frac{y(1-y^2)t}{t^2+y^2}\,d\sigma(t) \right| \leqslant M, \quad y > 0, \tag{4.17}$$

$$\left| \gamma y^2 + y^2 \int_{-\infty}^{+\infty} \frac{1+t^2}{t^2+y^2}\,d\sigma(t) \right| \leqslant M, \quad y > 0. \tag{4.18}$$

Dividing both sides in (4.18) by y^2 and letting $y \to \infty$, by means of the Lebesgue Dominated Convergence Theorem we first obtain $\gamma = 0$ and hence

$$y^2 \int_{-\infty}^{+\infty} \frac{1+t^2}{t^2+y^2}\,d\sigma(t) \leqslant M, \quad y > 0.$$

In particular,

$$y^2 \int_{-N}^{+N} \frac{1+t^2}{t^2+y^2}\,d\sigma(t) \leqslant M, \quad y > 0,\ N > 0,$$

and letting first $y \to +\infty$ and then $N \to +\infty$ we get

$$\int_{-\infty}^{+\infty} (1+t^2)\,d\sigma(t) \leqslant M, \tag{4.19}$$

where the integral converges improperly at $\pm\infty$. This enables us to define the function

$$\omega(t) = \int_{-\infty}^{t} (1+s^2)\,d\sigma(s), \quad t \in \mathbb{R}.$$

It is easy to see that ω is left continuous, $\omega(-\infty) = \lim_{t\to-\infty} \omega(t) = 0$, and $d\omega(t) = (1+t^2)d\sigma(t)$. On the other hand, dividing both sides in (4.17) by y and letting $y \to +\infty$ it follows that

$$\alpha = \lim_{y\to+\infty} \int_{-\infty}^{+\infty} \frac{(y^2-1)t}{t^2+y^2} d\sigma(t) = \lim_{y\to+\infty} \int_{-\infty}^{+\infty} \left(t - \frac{t(1+t^2)}{t^2+y^2}\right) d\sigma(t).$$

Using (4.19) we get

$$\left| \int_{-\infty}^{+\infty} \frac{t(t^2+1)}{t^2+y^2} d\sigma(t) \right| \leqslant M, \quad y > 0,$$

and hence the Stieltjes integral $\int_{-\infty}^{+\infty} t\, d\sigma(t)$ exists, it is improper at $\pm\infty$, and

$$\alpha = \int_{-\infty}^{+\infty} t\, d\sigma(t).$$

Taking into account that $\gamma = 0$ and α as before, it follows that for any $z \in \mathbb{C}_+$ we have

$$\varphi(z) = \int_{-\infty}^{+\infty} t\, d\sigma(t) + \int_{-\infty}^{+\infty} \frac{1+tz}{t-z} d\sigma(t) = \int_{-\infty}^{+\infty} \frac{1+t^2}{t-z} d\sigma(t) = \int_{-\infty}^{+\infty} \frac{d\omega(t)}{t-z}.$$

Finally, by (4.19) and using again the Lebesgue Dominated Convergence Theorem we get

$$\lim_{y\to+\infty} -iy\varphi(iy) = \lim_{y\to+\infty} \int_{-\infty}^{+\infty} \frac{iy\, d\omega(t)}{iy-t} = \int_{-\infty}^{+\infty} d\omega(t) = \omega(+\infty) \leqslant M. \blacksquare$$

11.5 The Resolvent Function Representation

Throughout this section, we fix a nontrivial Kreĭn space $\mathcal{K}$, a selfadjoint definitisable operator A in $\mathcal{K}$ and a real definitising polynomial p, that is, p is nontrivial and $[p(A)x, x] \geqslant 0$, $x \in \text{Dom}(A^n)$, where n denotes the degree of the polynomial p. By hypothesis, the resolvent set $\rho(A)$ is nonempty and let us fix $\lambda_0 \in \rho(A) \setminus \mathbb{R}$. We consider the real (that is, $q(\lambda) \in \mathbb{R}$ for all $\lambda \in \mathbb{R}$) rational function q defined by

$$q(\lambda) = \frac{p(\lambda)}{(\lambda_0 - \lambda)^n (\overline{\lambda}_0 - \lambda)^n}. \tag{5.1}$$

The function q is holomorphic at infinity, in particular $q \in \mathfrak{H}(\widetilde{\sigma}(A))$ and hence, by the Riesz–Dunford–Taylor functional calculus there exists $q(A) \in \mathcal{B}(\mathcal{K})$. Since p is definitising it follows that q is also definitising, that is, $q(A)$ is a positive operator, $[q(A)x, x] \geqslant 0$ for all $x \in \mathcal{K}$.

Recall that, by Lemma 11.3.3 the nonreal spectrum of A, denoted by $\sigma_0(A)$, is a finite set of points, symmetric with respect to the real axis, and all of its points λ are eigenvalues

11.5 THE RESOLVENT FUNCTION REPRESENTATION

of finite Riesz index $\leq k(\lambda)$, where $k(\lambda)$ denotes the root multiplicity of λ with respect to the definitising polynomial p.

In the following we use the familiar notation $R(\zeta; A) = (\zeta I - A)^{-1}$ for the resolvent map, in order to keep the algebraic expressions shorter.

Lemma 11.5.1. *The operator valued function $\rho(A) \ni \zeta \mapsto q(A)R(\zeta; A)$ has a holomorphic continuation at the points of $\sigma_0(A)$.*

Proof. By Lemma 11.3.3 this mapping is meromorphic at the points of $\sigma_0(A)$. Thus, it is sufficient to prove that for any $\lambda \in \sigma_0(A)$ the limit $\lim_{\zeta \to \lambda} q(A)R(\zeta; A)$ exists, in the strong operator topology. To this end, consider the decomposition

$$\mathcal{K} = E(\{\lambda, \bar{\lambda}\}; A)\mathcal{K}[+]\widehat{\mathcal{K}},$$

and the selfadjoint operators $A_\lambda = A|E(\{\lambda, \bar{\lambda}\}; A)\mathcal{K}$ and $\widehat{A} = A|\operatorname{Dom}(A) \cap \widehat{\mathcal{K}}$ on appropriate Kreĭn spaces. If $x \in E(\{\lambda, \bar{\lambda}\}; A)\mathcal{K}$ then, by Lemma 11.3.3, we have $q(A_\lambda) = 0$ and hence

$$\lim_{\zeta \to \lambda} q(A)R(\zeta; A)x = \lim_{\zeta \to \lambda} q(A_\lambda)R(\zeta; A_\lambda)x = 0.$$

If $x \in \widehat{\mathcal{K}}$ then

$$\lim_{\zeta \to \lambda} q(A)R(\zeta; A)x = \lim_{\zeta \to \lambda} q(\widehat{A})R(\zeta; \widehat{A})x,$$

since $\lambda \in \rho(\widehat{A})$. ∎

In the following we fix a fundamental symmetry J on $\mathcal{K}$ and hence, implicitly, we fix the Hilbert space $(\mathcal{K}, \langle \cdot, \cdot \rangle_J)$. With respect to this Hilbert space we use the unitary norm $\|\cdot\|$, the adjoint operation $*$, the order relation $\leq$ on the selfadjoint operators, etc. The first step in the construction of the spectral function of an arbitrary definitisable selfadjoint operator is a certain representation of the resolvent map.

Theorem 11.5.2. *There exist a complex valued function $Q(\zeta, \xi)$, holomorphic in each variable on $\mathbb{C} \setminus \{\lambda_0\}$, and an operator valued function $F \colon \mathbb{R} \to \mathcal{B}(\mathcal{K})$ with the following properties:*

(1) $F(t) = F(t)^*$, $t \in \mathbb{R}$;

(2) $s \leq t$ *implies* $F(s) \leq F(t)$;

(3) $\sup_{t \in \mathbb{R}} \|F(t)\| < \infty$;

(4) $\lim_{t \to -\infty} F(t) = 0$;

such that for any $\zeta \in \mathbb{C} \setminus (\mathbb{R} \cup \{\lambda_0\} \cup Z(p))$

$$R(\zeta; A) = \frac{1}{q(\zeta)} J \int_{-\infty}^{+\infty} \frac{1}{(\zeta - t)} \, \mathrm{d}F(t) + \frac{1}{q(\zeta)} Q(\zeta; A), \tag{5.2}$$

and, letting q be defined as in (5.1),

$$q(A) = J \int_{-\infty}^{+\infty} dF(t), \tag{5.3}$$

where the integrals converge in the strong operator topology.

Proof. Define the function $Q \colon (\mathbb{C} \setminus \{\lambda_0\}) \times (\mathbb{C} \setminus \{\lambda_0\}) \to \mathbb{C}$ by

$$Q(\zeta, \xi) = \begin{cases} \dfrac{q(\zeta) - q(\xi)}{\zeta - \xi}, & \zeta \neq \xi, \\ q'(\zeta), & \zeta = \xi. \end{cases}$$

The function Q is holomorphic in each variable on $\mathbb{C} \setminus \{\lambda_0\}$ and hence, for each fixed ζ the function $Q(\zeta, \cdot) \in \mathfrak{H}(\widetilde{\sigma}(A))$. Then, by the Riesz–Dunford–Taylor functional calculus we obtain an operator valued holomorphic function $Q(\zeta; A) \colon \mathbb{C} \setminus \{\lambda_0\} \to \mathcal{B}(\mathcal{K})$ such that

$$Q(\zeta; A) = (q(\zeta) - q(A)) R(\zeta; A), \quad \zeta \in \rho(A) \setminus \{\lambda_0\},$$

equivalently, for all $\zeta \in \rho(A) \setminus (Z(p) \cup \{\lambda_0\})$,

$$R(\zeta; A) = q(\zeta)^{-1} q(A) R(\zeta; A) + q(\zeta)^{-1} Q(\zeta; A). \tag{5.4}$$

Fix an arbitrary $x \in \mathcal{K}$ and consider the complex function $f_x \colon \rho(A) \to \mathbb{C}$

$$f_x(\zeta) = [q(A) R(\zeta; A) x, x], \quad \zeta \in \rho(A). \tag{5.5}$$

In view of Lemma 11.5.1, f_x has a holomorphic extension on $\mathbb{C}_+$. We claim that f_x has the following properties.
 (a) $\operatorname{Im} f_x(\zeta) \leqslant 0$, $\operatorname{Im} \zeta > 0$.
 (b) $\lim_{\eta \to +\infty} i\eta f_x(i\eta) = [q(A)x, x]$.
 (c) $\sup_{\eta > 0} |\eta f_x(i\eta)| < +\infty$.

Indeed

$$\operatorname{Im} f_x(\zeta) = \operatorname{Im}[q(A) R(\zeta; A) x, x] = -2\mathrm{i}[q(A)(R(\zeta; A) - R(\overline{\zeta}; A)) x, x]$$
$$= -\operatorname{Im} \zeta [q(A) R(\zeta; A) x, R(\zeta; A) x], \tag{5.6}$$

and this concludes the proof of (a). Denote $\eta = \operatorname{Im} \zeta$. Using the Schwarz Inequality with respect to the positive inner product $[q(A)\cdot, \cdot]$, from (5.6) and (5.5) we get

$$\eta [q(A) R(\zeta; A) x, R(\zeta; A) x] \leqslant |[q(A) R(\zeta; A) x, x]|$$
$$\leqslant [q(A) R(\zeta; A) x, R(\zeta; A) x]^{1/2} [q(A) x, x]^{1/2}.$$

11.5 THE RESOLVENT FUNCTION REPRESENTATION

Therefore
$$\eta^2[q(A)R(\zeta;A)x, R(\zeta;A)x] \leq [q(A)x, x],$$
and hence
$$-\operatorname{Im} f_x(\zeta) \leq \frac{1}{\eta}[q(A)x, x].$$

Assume $x \in \operatorname{Dom}(A)$. Then

$$\begin{aligned}|[q(A)x,x] - i\eta[q(A)R(i\eta;A)x,x]|^2 &= |[q(A)(I - i\eta R(i\eta;A))x,x]|^2 \\ &\quad \times |[q(A)AR(i\eta;A)x,x]|^2 \\ &\leq [q(A)Ax, Ax]\,[q(A) \\ &\quad \times R(i\eta,A)x, R(i\eta;A)x] \\ &\leq \frac{1}{\eta^2}[q(A)x,x][q(A)Ax, Ax] \xrightarrow[\eta\to\infty]{} 0 \end{aligned} \quad (5.7)$$

and

$$|i\eta[q(A)R(i\eta;A)x,x]|^2 \leq [q(A)R(i\eta;A)x, R(i\eta;A)x]\,[q(A)x,x]$$
$$\leq [q(A)x,x]\,[q(A)x,x] = [q(A)x,x]^2. \quad (5.8)$$

Since $\operatorname{Dom}(A)$ is dense in $\mathcal{K}$, by (5.7) and (5.8) we obtain (b) and (c) for any $x \in \mathcal{K}$.

In view of Theorem 11.4.7, the function f_x defined at (5.5) has the integral representation

$$f_x(\zeta) = \int_{-\infty}^{+\infty} \frac{1}{\zeta - t} \, d\sigma_x(t), \quad \zeta \in \mathbb{C} \setminus \mathbb{R}, \quad (5.9)$$

where $\sigma_x \colon \mathbb{R} \to \mathbb{R}$ is a nondecreasing function with the properties: $\sigma_x(-\infty) = 0$, $\sigma_x(t) = \sigma_x(t-0)$, for all $t \in \mathbb{R}$, and $\sigma_x(+\infty) = [q(A)x, x]$. For arbitrary $x, y \in \mathcal{K}$, consider the function $f_{x,y} \colon \mathbb{C} \setminus \mathbb{R} \to \mathbb{C}$

$$f_{x,y}(\zeta) = [q(A)R(\zeta;A)x, y], \quad \zeta \in \mathbb{C} \setminus \mathbb{R}.$$

By means of (5.9) and the polarisation formula we obtain the representation

$$f_{x,y}(\zeta) = \int_{-\infty}^{+\infty} \frac{1}{\zeta - t} \, d\sigma_{x,y}(t), \quad \zeta \in \mathbb{C} \setminus \mathbb{R}, \quad (5.10)$$

where the function $\sigma_{x,y} \colon \mathbb{R} \to \mathbb{C}$ is defined by

$$\sigma_{x,y}(t) = \frac{1}{4}\sum_{k=0}^{3} i^k \sigma_{x+i^k y}(t), \quad t \in \mathbb{R}.$$

Then $\sigma_{x,y}(-\infty) = 0$, $\sigma_{x,y}(+\infty) = [q(A)x, y]$ and

$$|\sigma_{x,y}(t)|^2 \leq [q(A)x, x]\,[q(A)y, y] \leq \|q(A)\|^2 \,\|x\|\,\|y\|.$$

We remark now that for any fixed $t \in \mathbb{R}$ we have a Hermitian form $\mathcal{K} \times \mathcal{K} \ni (x, y) \mapsto \sigma_{x,y}(t) \in \mathbb{C}$ which is bounded, by the previous inequality, and hence, by the Riesz Representation Theorem for bounded sesquilinear forms, there exists a mapping $F \colon \mathbb{R} \to \mathcal{B}(\mathcal{K})$ such that
$$\sigma_{x,y}(t) = \langle F(t)x, y\rangle_J, \quad x, y \in \mathcal{K}, \quad t \in \mathbb{R},$$
and F satisfies the requirements (1) through (4) as in the statement. From (5.10) we obtain
$$q(A)R(\zeta; A) = J \int_{-\infty}^{+\infty} \frac{1}{\zeta - t} \, dF(t), \tag{5.11}$$
and then, using (5.4) we obtain the representation (5.2). Finally,
$$[q(A)x, y] = \sigma_{x,y}(+\infty) = \int_{-\infty}^{+\infty} d(F(t)x, y)_J = \Big[J \int_{-\infty}^{+\infty} dF(t)x, y \Big],$$
and this proves (5.3). ∎

We end this section with some immediate consequences of Theorem 11.5.2.

Corollary 11.5.3. *If $\lambda \in \sigma(A)$ is isolated then λ is an eigenvalue of A, the maximal length of Jordan chains $\nu(\lambda) = \nu(\overline{\lambda})$ and*
$$\nu(\lambda) \leqslant \begin{cases} k(\lambda), & \lambda \neq \overline{\lambda}, \\ k(\lambda) + 1, & \lambda = \overline{\lambda}. \end{cases}$$
In addition, for any $\xi \in \mathbb{R}$,
$$\|R(\xi + i\eta; A)\| = O(\eta^{-k(\xi)-1}), \quad (\eta \to 0),$$
in particular, if $\xi \in \sigma_p(A)$ then the length of any Jordan chain corresponding to ξ and A is at most $k(\xi) + 1$.

Proof. If λ is not real then the statement was already proved in Lemma 11.3.3. The remainder is a direct consequence of the representation of the resolvent as in (5.2). ∎

Lemma 11.5.4. *If $\xi \in \mathbb{R} \setminus \sigma_p(A)$ then $(A - \xi)^{-1}$ is also a definitisable selfadjoint operator.*

Proof. Without restricting the generality we can assume $\xi = 0 \notin \sigma_p(A)$. Since the residual spectrum $\sigma_r(A)$ is empty it follows that either $0 \in \rho(A)$, in which case A^{-1} is bounded selfadjoint, or $0 \in \sigma_c(A)$ and in this case A^{-1} is unbounded and selfadjoint. Anyhow $\rho(A^{-1}) = \{\lambda^{-1} \mid \lambda \in \rho(A)\} \neq \emptyset$.

Let the definitising polynomial p be
$$p(\lambda) = \sum_{j=0}^{n} a_j \lambda^j,$$

and $n = \deg(p)$. Consider the polynomial r defined by

$$r(\lambda) = \sum_{j=0}^{n} a_j \lambda^{2n-j}.$$

For arbitrary $y \in \mathrm{Dom}(A^{-2n})$ let $x = A^{-n}y \in \mathrm{Dom}(A^n)$, equivalently $y = A^n x$. Then

$$[r(A^{-1})y, y] = \left[\sum_{j=0}^{n} a_j A^{-2n+j} A^n x, A^n x\right] = [p(A)x, x] \geq 0. \blacksquare$$

Corollary 11.5.5. $\sigma(A) = \emptyset$ *if and only if* $\mathcal{K} = 0$.

Proof. If $\sigma(A) = \emptyset$ and $\mathcal{K} \neq 0$ then, in view of Lemma 11.5.4 there exists the definitisable selfadjoint operator A^{-1}. In this case 0 is the only point in $\sigma(A^{-1})$. From Corollary 11.5.3 it follows that 0 is an eigenvalue of A^{-1}, a contradiction. $\blacksquare$

Corollary 11.5.6. *If $\sigma(A)$ is bounded then A is a bounded operator.*

Proof. If $\sigma(A)$ is bounded let $\mathcal{K}_1$ be the spectral subspace corresponding to $\sigma(A)$ and A. Then $\mathcal{K}_1$ is a regular subspace, by Proposition 11.1.4, and $A|\mathcal{K}_1$ is bounded. On the other hand, $A|\mathrm{Dom}(A)\mathcal{K}_1^\perp$ is a definitisable selfadjoint operator in the Kreĭn space $\mathcal{K}_1^\perp$ with empty spectrum and hence, by Corollary 11.5.5, it follows that $\mathcal{K}_1 = 0$. Then $\mathcal{K} = \mathcal{K}_1$ and $A = A|\mathcal{K}_1$ is bounded. $\blacksquare$

11.6 Stieltjes Inversion Formulae

In order to use the representation of the resolvent map of a definitisable selfadjoint operator obtained in Theorem 11.5.2 to define the spectral function, we will use integration on certain paths in the complex plane that will require the Stieltjes Inversion Formulae. Let $\mathcal{H}$ be a complex Hilbert space and $F \colon \mathbb{R} \to \mathcal{B}(\mathcal{H})$ a map with the following properties.

(1) $F(t) = F(t)^*$, $t \in \mathbb{R}$.
(2) $s \leq t$ implies $F(s) \leq F(t)$, $s, t \in \mathbb{R}$.
(3) $\sup_{t \in \mathbb{R}} \|F(t)\| < +\infty$.

As a consequence, the strong operator limit $F(-\infty) = \lim_{t \to -\infty} F(t)$ exists and, in order to simplify the formulae, we will assume that

(4) $F(-\infty) = 0$.

The properties (1) through (3) also imply that F has the following property.

(5) The one-sided limits $F(t + 0)$ and $F(t - 0)$ exist in the strong operator topology, for any $t \in \mathbb{R}$.

In particular we can define the function $\widetilde{F}\colon \mathbb{R} \to \mathcal{B}(\mathcal{H})$ by

$$\widetilde{F}(t) = \frac{F(t-0) + F(t+0)}{2}, \quad t \in \mathbb{R}. \tag{6.1}$$

Then the operator valued function $\widetilde{F}$ has the properties (1) through (3).

As a consequence of the same properties (1)–(3), we can introduce the operator valued function $\Gamma \colon \mathbb{C} \setminus \mathbb{R} \to \mathcal{B}(\mathcal{H})$ by

$$\Gamma(\zeta) = \int_{-\infty}^{+\infty} \frac{1}{\zeta - t} \, dF(t), \quad \zeta \in \mathbb{C} \setminus \mathbb{R}, \tag{6.2}$$

where the Stieltjes integral converges in the strong operator topology. Clearly, the function Γ is holomorphic in $\mathbb{C} \setminus \mathbb{R}$.

Let $\Delta = [\alpha_1, \alpha_2]$ be a compact interval in $\mathbb{R}$ and $\beta > 0$. We consider a positively oriented Jordan contour C_Δ in the complex plane as in Figure 11.1. For arbitrary $\beta > \delta > 0$ define

$$C_\Delta^\delta = C_\Delta \setminus \left[(\alpha_1 - i\delta, \alpha_1 + i\delta) \cup (\alpha_2 - i\delta, \alpha_2 + i\delta)\right].$$

C_Δ^δ consists of two Jordan arcs, symmetric with respect to the real axis, with the orientation inherited from the orientation of C_Δ (see Figure 11.1).

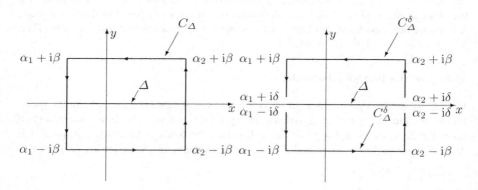

Figure 11.1: The Jordan contour C_Δ and the two Jordan arcs C_Δ^δ.

In the following, for a compact interval $\Delta = [\alpha_1, \alpha_2]$ and $\varepsilon > 0$, we denote $\Delta_\varepsilon^\pm = [\alpha_1 \pm \varepsilon, \alpha_2 \mp \varepsilon]$.

Lemma 11.6.1. *With notation and assumptions as before, the following strong operator limits exist and*

$$\lim_{\delta \downarrow 0} \frac{1}{2\pi i} \int_{C_\Delta^\delta} \Gamma(\zeta) d\zeta = \int_\Delta d\widetilde{F}(t), \quad \lim_{\varepsilon \downarrow 0} \lim_{\delta \downarrow 0} \frac{1}{2\pi i} \int_{C_{\Delta_\varepsilon^\pm}^\delta} \Gamma(\zeta) d\zeta = \int_{\alpha_1 \pm 0}^{\alpha_2 \pm} d\widetilde{F}(t).$$

11.6 STIELTJES INVERSION FORMULAE

Proof. Let $x \in \mathcal{H}$ and $\beta > \delta > 0$ be arbitrary. Since the operator valued measure $dF(t)$ is finite, we can apply the Fubini Theorem and get

$$\frac{1}{2\pi i} \int_{C_\Delta^\delta} \Gamma(\zeta) d\zeta = \frac{1}{2\pi i} \int_{C_\Delta^\delta} \int_{-\infty}^{+\infty} \frac{1}{\zeta - t} dF(t) x d\zeta$$

$$= \int_{-\infty}^{+\infty} \frac{1}{2\pi i} \int_{C_\Delta^\delta} \frac{1}{\zeta - t} d\zeta dF(t) x.$$

We consider now the two closed Jordan contours obtained from C_Δ^δ by joining $\alpha_1 + i\delta$ with $\alpha_2 + i\delta$ and, respectively, by joining $\alpha_1 - i\delta$ with $\alpha_2 - i\delta$ as in Figure 11.2.

Applying the Cauchy formula for these contours it follows that

$$\frac{1}{2\pi i} \int_{C_\Delta^\delta} \frac{1}{\zeta - t} d\zeta = \frac{1}{2\pi i} \int_{\alpha_1}^{\alpha_2} \frac{1}{s - i\delta - t} ds - \frac{1}{2\pi i} \int_{\alpha_1}^{\alpha_2} \frac{1}{s + i\delta - t} ds$$

$$= \frac{1}{2\pi i} \int_{\alpha_1}^{\alpha_2} \frac{(s-t) - i\delta}{(s-\delta)^2 - t^2} ds - \frac{1}{2\pi i} \int_{\alpha_1}^{\alpha_2} \frac{(s-t) + i\delta}{(s-\delta)^2 - t^2} ds$$

$$= \frac{1}{\pi} \arctan \frac{\alpha_2 - t}{\delta} - \frac{1}{\pi} \arctan \frac{\alpha_1 - t}{\delta}.$$

Since the function arctan vanishes at 0 it follows that

$$\int_{-\infty}^{+\infty} \frac{1}{\pi} \arctan \frac{\alpha_2 - t}{\delta} dF(t) x = \frac{1}{\pi} \int_{-\infty}^{\alpha_2 - 0} \arctan \frac{\alpha_2 - t}{\delta} dF(t) x$$

$$- \frac{1}{\pi} \int_{\alpha_2 + 0}^{+\infty} \arctan \frac{\alpha_2 - t}{\delta} dF(t) x.$$

Further, since the function arctan is uniformly bounded and the operator measure $dF(t)$ is finite, by the Lebesgue Dominated Convergence Theorem we are allowed to let δ tend to 0 and get

$$\lim_{\delta \downarrow 0} \frac{1}{\pi} \int_{-\infty}^{+\infty} \arctan \frac{\alpha_2 - t}{\delta} dF(t) x = \frac{1}{2} \int_{-\infty}^{\alpha_2 - 0} dF(t) x - \frac{1}{2} \int_{\alpha_2 + 0}^{+\infty} dF(t) x.$$

In a similar way we obtain

$$\lim_{\delta \downarrow 0} \frac{1}{\pi} \int_{-\infty}^{+\infty} \arctan \frac{\alpha_2 - t}{\delta} dF(t) x = \frac{1}{2} \int_{-\infty}^{\alpha_2 - 0} dF(t) x - \frac{1}{2} \int_{\alpha_2 + 0}^{+\infty} dF(t) x,$$

and hence

$$\lim_{\delta \downarrow 0} \frac{1}{2\pi i} \int_{C_\Delta^\delta} \Gamma(\zeta) d\zeta = \frac{1}{2} \left(\int_{\alpha_1 - 0}^{\alpha_2 - 0} dF(t) x - \int_{\alpha_1 + 0}^{\alpha_2 + 0} dF(t) x \right) = \int_\Delta d\widetilde{F}(t) x.$$

Finally, we remark that we can perform the same calculations for $\Delta_\varepsilon^\pm$ instead of Δ and obtain

$$\lim_{\varepsilon\downarrow 0}\lim_{\delta\downarrow 0}\frac{1}{2\pi i}\int_{C^\delta_{\Delta_\varepsilon^\pm}}\Gamma(\zeta)\,d\zeta = \frac{1}{2}\lim_{\varepsilon\downarrow 0}\left(\int_{\alpha_1\pm\varepsilon-0}^{\alpha_2\mp\varepsilon-0}dF(t)x - \int_{\alpha_1\pm\varepsilon+0}^{\alpha_2\mp\varepsilon+0}dF(t)x\right)$$

$$= \int_\Delta d\widetilde{F}(t)x.\ \blacksquare$$

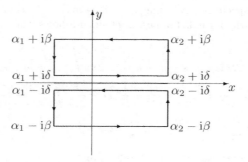

Figure 11.2: Splitting C_Δ into two disjoint Jordan contours.

The main result of this section is the following theorem which will be used in the proof of the existence of the spectral function.

Theorem 11.6.2. (a) *With assumptions and notation as before, let f be a complex function which is holomorphic in a neighbourhood of the contour C_Δ. Then the limits*

$$\lim_{\delta\downarrow 0}\frac{1}{2\pi i}\int_{C_\Delta^\delta}f(\zeta)\Gamma(\zeta)\,d\zeta \quad \text{and} \quad \lim_{\varepsilon\downarrow 0}\lim_{\delta\downarrow 0}\frac{1}{2\pi i}\int_{C^\delta_{\Delta_\varepsilon^\pm}}f(\zeta)\Gamma(\zeta)\,d\zeta,$$

exist in the strong operator topology.

(b) *If, in addition, f is holomorphic on a neighbourhood of the compact interval Δ then, for any $\beta > 0$ and sufficiently small, we have*

$$\lim_{\delta\downarrow 0}\frac{1}{2\pi i}\int_{C_\Delta^\delta}f(\zeta)\Gamma(\zeta)\,d\zeta = \int_\Delta f(t)\,d\widetilde{F}(t),$$

11.6 Stieltjes Inversion Formulae

and

$$\lim_{\varepsilon \downarrow 0} \lim_{\delta \downarrow 0} \frac{1}{2\pi i} \int_{C_{\Delta_\varepsilon^\pm}^\delta} f(\zeta)\Gamma(\zeta)\,\mathrm{d}\zeta = \int_{\alpha_1 \pm 0}^{\alpha_2 \mp 0} f(t)\,\mathrm{d}F(t).$$

Proof. (a) Let Δ_1' and Δ_2' be two disjoint compact and real intervals, such that Δ_i' is a (real) neighbourhood of α_i, $i = 1, 2$, and f is holomorphic on $\Delta_1' \cup \Delta_2'$. Then, for arbitrary $x \in \mathcal{H}$, we have

$$f(\zeta)\Gamma(\zeta)x = f(\zeta)\int_{\mathbb{R}\setminus(\Delta_1'\cup\Delta_2')} \frac{1}{\zeta - t}\,\mathrm{d}\zeta + \sum_{j=1}^{2}\int_{\Delta_j'} \frac{f(\zeta) - f(t)}{\zeta - t}\,\mathrm{d}F(t)x$$

$$+ \sum_{j=1}^{2}\int_{\Delta_j'} \frac{f(t)}{\zeta - t}\,\mathrm{d}F(t)x. \tag{6.3}$$

We remark that the first and the second terms in (6.3) are holomorphic on a neighbourhood of the contour C_Δ and hence we can integrate these terms along C_Δ. Then we remark that, without restricting the generality, we can assume that f is positive on $\Delta_1' \cup \Delta_2'$. Indeed, if this is not the case, we consider the real and the imaginary parts of f restricted to $\Delta_1' \cup \Delta_2'$ and then, represent each function as the difference of two continuous nonnegative functions. Then we can apply Lemma 11.6.1, where we replace F with the operator valued function

$$\mathbb{R} \ni t \mapsto \int_{-\infty}^{t} (\chi_{\Delta_1'} + \chi_{\Delta_2'})f(s)\,\mathrm{d}F(s)$$

which satisfies the same properties (1)–(4) as F does. The convergence of the double limit follows in a similar way.

(b) If f is holomorphic on a neighbourhood of Δ then, for $\beta > 0$ and sufficiently small, the contour C_Δ and its interior are included in the domain of holomorphy of f. Let Δ' be a compact real interval containing Δ in its interior and contained in the domain of holomorphy of f. Then, for arbitrary $x \in \mathcal{H}$, we have

$$f(\zeta)\Gamma(\zeta)x = f(\zeta)\int_{\Delta'} \frac{1}{\zeta - t}\,\mathrm{d}\zeta + \int_{\Delta'} \frac{f(\zeta) - f(t)}{\zeta - t}\,\mathrm{d}F(t)x$$

$$+ \int_{\Delta'} \frac{f(t)}{\zeta - t}\,\mathrm{d}F(t)x. \tag{6.4}$$

The first two terms on the right-hand side of (6.4) are holomorphic on C_Δ as well as on its interior and hence, by the Cauchy formula, the integrals of these terms along C_Δ vanish. Further, as in the proof of item (a), without restricting the generality, we can assume that f is positive on Δ'. Then we apply Lemma 11.6.1, where instead of F we consider the operator valued function

$$\mathbb{R} \ni t \mapsto \int_{-\infty}^{t} \chi_{\Delta'} f(s) \mathrm{d}F(s),$$

and get

$$\lim_{\delta\downarrow 0} \frac{1}{2\pi\mathrm{i}} \int_{C_\Delta^\delta} f(\zeta)\Gamma(\zeta)\,\mathrm{d}\zeta = \lim_{\delta\downarrow 0} \frac{1}{2\pi} \int_{C_\Delta^\delta} \int_{\Delta'_j} \frac{f(t)}{\zeta - t}\,\mathrm{d}F(t)x\,\mathrm{d}\zeta = \int_\Delta f(t)\,\mathrm{d}\widetilde{F}(t)x.$$

The proof for the double limit is similar. ∎

11.7 The Spectral Function

In this section we keep the same notation as in Section 11.5, in particular, A is a definitisable selfadjoint operator in the Kreĭn space $\mathcal{K}$, λ_0 is a fixed value in $\rho(A)$ and, letting p be a real definitising polynomial of A, q denotes the rational function as in (5.1).

Definition 11.7.1. A real number $\alpha \in \sigma(A)$ is called *critical* if α is a root of any definitising polynomial of A. We denote the *set of critical points* of A by

$$c(A) = \sigma(A) \cap \mathbb{R} \cap \bigcap_p Z(p), \tag{7.1}$$

where p runs through the set of all definitising polynomials of A.

In the following we denote by $\mathfrak{R}_A$ the Boolean algebra generated by those bounded real intervals whose endpoints are not critical points of A. More precisely, $\mathfrak{R}_A$ consists of finite unions of mutually disjoint, real, bounded intervals Δ such that $\partial\Delta \cap c(A) = \emptyset$ and their complements. Also, if Δ is a compact real interval then the Jordan contour C_Δ and the arcs C_Δ^δ are defined as in Section 11.6, see Figure 11.1.

Lemma 11.7.2. *For any compact interval* $\Delta \in \mathfrak{R}_A$ *the limit*

$$\widetilde{E}(\Delta) = \lim_{\delta \to \infty} \frac{1}{2\pi\mathrm{i}} \int_{C_\Delta^\delta} R(\zeta; A)\,\mathrm{d}\zeta \in \mathcal{B}(\mathcal{K}) \tag{7.2}$$

exists in the strong operator topology and, for any compact intervals $\Delta, \Delta' \in \mathfrak{R}_A$ *such that* $\partial\Delta \cap \partial\Delta' = \emptyset$*, we have*

$$\widetilde{E}(\Delta \cap \Delta') = \widetilde{E}(\Delta)\widetilde{E}(\Delta'). \tag{7.3}$$

Proof. Let $\Delta \in \mathfrak{R}_A$ be a compact interval. There exists $\beta > 0$ and sufficiently small, such that all the points of $\sigma_0(A)$ lie outside the Jordan contour C_Δ. Define

11.7 THE SPECTRAL FUNCTION

$\widetilde{E}(\Delta) \in \mathcal{B}(\mathcal{K})$ as in (7.2) and taking into account that the strong convergence follows from Theorem 11.6.2 and Theorem 11.5.2. In the following we prove that for any compact intervals $\Delta, \Delta' \in \mathfrak{R}_A$ such that $\partial\Delta \cap \partial\Delta' = \emptyset$ we have the equality (7.3).

Indeed, let us first remark that among the four possible cases of overlapping of Δ and Δ', three of them are trivial, and these are: $\Delta \cap \Delta' = \emptyset$, $\Delta \subseteq \Delta'$, and $\Delta' \subseteq \Delta$. So we will assume that $\Delta \cap \Delta' = I$ is a compact interval, distinct from Δ and Δ', and of positive length. Since the height $2\beta'$ of C'_Δ plays no role, we fix $\beta' > \beta$ and such that all the points from $\sigma_0(A)$ lie outside $C_{\Delta'}$. Then let $\{\zeta_1, \overline{\zeta}_1\} = C_\Delta \cap C_{\Delta'}$ and consider the Jordan contour $C_I = C_\Delta^{\Delta'} \cup C_{\Delta'}^{\Delta}$, where $C_\Delta^{\Delta'}$ is the Jordan arc of C_Δ which crosses the inside of $C_{\Delta'}$ and, respectively, $C_{\Delta'}^{\Delta}$ is the Jordan arc of $C_{\Delta'}$ which crosses the inside of C_Δ (see Figure 11.3). Let $\delta > 0$ and $\delta' > 0$ be sufficiently small. We remark that in

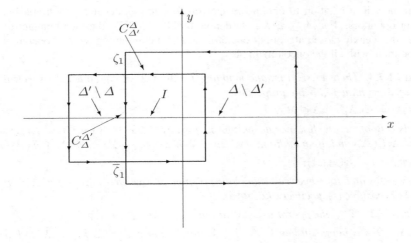

Figure 11.3: The Jordan contour $C_I = C_\Delta^{\Delta'} \cup C_{\Delta'}^{\Delta}$.

the definition of $\widetilde{E}(\Delta)$ and $\widetilde{E}(\Delta')$ the integrals can be considered improper at ζ_1 and $\overline{\zeta}_1$. Keeping this in mind, from the Hilbert resolvent identity we get

$$\left(\frac{1}{2\pi i} \int_{C_\Delta^\delta} R(\zeta; A) \, d\zeta\right)\left(\frac{1}{2\pi i} \int_{C_{\Delta'}^{\delta'}} R(\xi; A) \, d\xi\right) = -\frac{1}{4\pi^2} \int_{C_\Delta^\delta} \int_{C_{\Delta'}^{\delta'}} \frac{R(\zeta; A) - R(\xi; A)}{\zeta - \xi} \, d\zeta \, d\xi$$

$$= \frac{1}{2\pi i} \int_{C_\Delta^\delta} R(\zeta; A) \left(\frac{1}{2\pi i} \int_{C_{\Delta'}^{\delta'}} \frac{d\zeta}{\zeta - \xi}\right) d\xi + \frac{1}{2\pi i} \int_{C_{\Delta'}^{\delta'}} R(\xi; A) \left(\frac{1}{2\pi i} \int_{C_\Delta^\delta} \frac{d\xi}{\xi - \zeta}\right) d\zeta.$$

Applying the Cauchy formula for $\xi \in C_\Delta \setminus \{\zeta_1, \overline{\zeta}_1\}$ and, respectively, $\zeta \in C_{\Delta'} \setminus \{\zeta_1, \overline{\zeta}_1\}$

we get
$$\frac{1}{2\pi i}\int_{C_{\Delta'}}\frac{d\zeta}{\zeta-\xi}=\chi_{C_{\Delta}^{\Delta'}}(\xi),\quad \frac{1}{2\pi i}\int_{C_{\Delta'}}\frac{d\xi}{\xi-\zeta}=\chi_{C_{\Delta'}^{\Delta}}(\zeta).$$

Since the limits following $\delta \downarrow 0$ and $\delta' \downarrow 0$ are independent, these imply that

$$\widetilde{E}(\Delta)\widetilde{E}(\Delta') = \lim_{\delta\downarrow 0}\frac{1}{2\pi i}\int_{C_{\Delta}^{\Delta'}} R(\zeta; A)\,d\zeta + \lim_{\delta'\downarrow 0}\frac{1}{2\pi i}\int_{C_{\Delta'}^{\Delta}} R(\xi; A)\,d\xi$$

$$= \lim_{\delta\downarrow 0}\frac{1}{2\pi i}\int_{C_I^\delta} R(\zeta; A)\,d\zeta = \widetilde{E}(\Delta\cap\Delta'). \blacksquare$$

We are now in a position to obtain the spectral function associated to definitisable selfadjoint operators. For $\mathcal{A} \subseteq \mathcal{B}(\mathcal{K})$ we denote by $\mathcal{A}'' \subseteq \mathcal{B}(\mathcal{K})$ the bicommutant of $\mathcal{A}$, that is, the weakly closed algebra of operators in $\mathcal{B}(\mathcal{K})$ commuting with all operators which commute with all the operators in $\mathcal{A}$.

Theorem 11.7.3. *There exists a unique mapping $E\colon \mathfrak{R}_A \to \mathcal{B}(\mathcal{K})$, called the spectral function of A, with the following properties.*

(1) $E(\Delta) = E(\Delta)^\sharp$, $\Delta \in \mathfrak{R}_A$.

(2) E is a Boolean algebra morphism, that is, for any $\Delta, \Delta' \in \mathfrak{R}_A$ we have $E(\Delta\cap\Delta') = E(\Delta)E(\Delta')$ and, if, in addition, $\Delta\cap\Delta' = \emptyset$ then $E(\Delta\cup\Delta') = E(\Delta)+E(\Delta')$.

(3) $E(\mathbb{R}) = I - E(\sigma_0(A); A)$.

(4) If $\Delta \in \mathfrak{R}_A$ and, for some real definitising polynomial p we have $p|\Delta > 0$ ($p|\Delta < 0$) then $E(\Delta)$ is positive (respectively, negative).

(5) For any $\Delta \in \mathfrak{R}_A$, the operator $E(\Delta)$ lies in $\{R(\zeta; A) \mid \zeta \in \rho(A)\}''$.

(6) If $\Delta \in \mathfrak{R}_A$ is bounded then $E(\Delta)\mathcal{K} \subseteq \mathrm{Dom}(A)$ and $\sigma(A|E(\Delta)\mathcal{K}) \subseteq \overline{\Delta}$, and, if $\Delta \in \mathfrak{R}_A$ is unbounded then $A|E(\Delta)\mathcal{K}$ is a selfadjoint operator as an operator in $E(\Delta)\mathcal{K}$ and $\sigma(A|E(\Delta)\mathcal{K}\cap\mathrm{Dom}(A)) \subseteq \overline{\Delta}$.

Proof. Existence. Let $\Delta \in \mathfrak{R}_A$ be a bounded interval and let $\varepsilon > 0$ be arbitrary. We define a compact interval Δ_ε by the following rule:

$$\Delta_\varepsilon = \begin{cases} [\alpha-\varepsilon, \beta+\varepsilon], & \text{if } \Delta = [\alpha,\beta], \\ [\alpha-\varepsilon, \beta-\varepsilon], & \text{if } \Delta = [\alpha,\beta), \\ [\alpha+\varepsilon, \beta+\varepsilon], & \text{if } \Delta = (\alpha,\beta], \\ [\alpha+\varepsilon, \beta-\varepsilon], & \text{if } \Delta = (\alpha,\beta). \end{cases} \quad (7.4)$$

For $\varepsilon > 0$ and sufficiently small we have $\Delta_\varepsilon \in \mathfrak{R}_A$. More generally, if $\Delta \in \mathfrak{R}_A$ is bounded then it can be written as a finite union of bounded intervals in $\mathfrak{R}_A$ and, for any $\varepsilon > 0$ and sufficiently small we define a compact subset Δ_ε, following the above rule for each connected component.

11.7 THE SPECTRAL FUNCTION

We now make use of the mapping $\widetilde{E}$ as defined in Lemma 11.7.2. First we notice that we can extend the definition of $\widetilde{E}(\Delta)$ for any compact set $\Delta \in \mathfrak{R}_A$, by representing it as a finite union of mutually disjoint compact intervals in $\mathfrak{R}_A$. In this way $\widetilde{E}$ is additive and, using Lemma 11.7.2, it follows that for any sets $\Delta, \Delta' \in \mathfrak{R}_A$, such that $\partial \Delta \cap \partial \Delta' = \emptyset$, we have

$$\widetilde{E}(\Delta)\widetilde{E}(\Delta') = \widetilde{E}(\Delta \cap \Delta'). \tag{7.5}$$

Then we define the mapping $E\colon \mathfrak{R}_A \to \mathcal{B}(\mathcal{K})$ in the following way. If $\Delta \in \mathfrak{R}_A$ is a bounded set then

$$E(\Delta) = \lim_{\varepsilon \to 0} \widetilde{E}(\Delta_\varepsilon) = \lim_{\delta \to \infty} \frac{1}{2\pi \mathrm{i}} \int_{C_\Delta^\delta} R(\zeta; A)\, \mathrm{d}\zeta, \tag{7.6}$$

where the limit exists in the strong operator topology due to the representation of the resolvent function as in Theorem 11.5.2, the Stieltjes Inversion Formula in Theorem 11.6.2, and Lemma 11.7.2. We first verify the properties (1) through (6) of the mapping E for bounded sets.

Since the contour C_Δ is symmetric with respect to the real axis and A is selfadjoint, it follows that $\widetilde{E}(\Delta)^\sharp = \widetilde{E}(\Delta)$ for any compact set $\Delta \in \mathfrak{R}_A$. This implies the property (1) for compact sets in $\mathfrak{R}_A$.

In order to check the property (2) let $\Delta, \Delta' \in \mathfrak{R}_A$ be bounded. There exist two positive sequences $(\varepsilon_k)_{k\in\mathbb{N}}$ and $(\varepsilon'_k)_{k\in\mathbb{N}}$ decreasing to 0 and such that for any $k \in \mathbb{N}$ we have $\partial \Delta_{\varepsilon_k} \cap \partial \Delta'_{\varepsilon'_k} = \emptyset$. By (7.5) we have

$$\widetilde{E}(\Delta_{\varepsilon_k})\widetilde{E}(\Delta'_{\varepsilon'_k}) = \widetilde{E}(\Delta_{\varepsilon_k} \cap \Delta'_{\varepsilon'_k}), \quad k \in \mathbb{N}.$$

In particular this shows that the left-hand side sequence is also strongly operatorial convergent and, letting $k \to \infty$ on both sides, we get

$$E(\Delta)E(\Delta') = E(\Delta \cap \Delta').$$

If we assume, in addition, that $\Delta \cap \Delta' = \emptyset$ then we first notice that we can choose the sequences $(\varepsilon_k)_{k\in\mathbb{N}}$ and $(\varepsilon'_k)_{k\in\mathbb{N}}$ such that $\Delta_{\varepsilon_k} \cap \Delta'_{\varepsilon'_k} = \emptyset$ for all $k \in \mathbb{N}$. Then

$$E(\Delta \cup \Delta') = \lim_{k\to\infty} \widetilde{E}(\Delta_{\varepsilon_k} \cup \Delta'_{\varepsilon'_k})) = \lim_{k\to\infty} (\widetilde{E}(\Delta_{\varepsilon_k}) + \widetilde{E}(\Delta'_{\varepsilon'_k})) = E(\Delta) + E(\Delta').$$

At this stage, let us remark that, as a consequence of the properties (1) and (2), for any bounded $\Delta \in \mathfrak{R}_A$ the operator $E(\Delta)$ is a bounded selfadjoint projection and hence $E(\Delta)\mathcal{K}$ is a regular subspace of $\mathcal{K}$.

The assertion (3) does not refer to bounded sets so we skip it. To prove (4), assume that $\Delta \in \mathfrak{R}_A$ and there exists a real definitising polynomial p such that $p|\overline{\Delta} > 0$. Then $\frac{1}{p}|\operatorname{Clos}\Delta > 0$ and $\frac{1}{p} \in \mathfrak{H}(\overline{\Delta})$. We fix a fundamental symmetry J on $\mathcal{K}$ and consider the operator valued functions F and Q as in Theorem 11.5.2. Using (5.2) and taking into account of the last assertion of Theorem 11.6.2 we have

$$[E(\Delta)x, x] = \lim_{\varepsilon \downarrow 0}[\widetilde{E}(\Delta_\varepsilon)x, x]$$

$$= \lim_{\varepsilon \downarrow 0} \lim_{\delta \downarrow 0} \left[\frac{1}{2\pi i} \int_{C_{\Delta_\varepsilon}^\delta} R(\zeta; A)x \, d(\zeta), x\right] \tag{7.7}$$

$$= \left[J \lim_{\varepsilon \downarrow 0} \int_{\Delta_\varepsilon} \frac{1}{q(t)} \, dF(t)x, x\right] \tag{7.8}$$

$$= \lim_{\varepsilon \downarrow 0} \left\langle \int_{\Delta_\varepsilon} \frac{1}{q(t)} \, dF(t)x, x \right\rangle_J \geq 0.$$

This shows that $E(\Delta)\mathcal{K}$ is a positive subspace and, due to its regularity, it is uniformly positive.

To prove (5) we only have to remark that the algebra $\{R(\zeta; A) \mid \zeta \in \rho(A)\}''$ is closed under strong operator limits and hence it contains $E(\Delta)$ for any bounded $\Delta \in \mathfrak{R}_A$.

In order to prove that (6) holds for any bounded set $\Delta \in \mathfrak{R}_A$, we first remark that, without restricting the generality, we can assume that Δ is an interval. Then, by (5),

$$R(\lambda_0, A)E(\Delta) = E(\Delta)R(\lambda_0; A),$$

hence $E(\Delta) \operatorname{Dom}(A) \subseteq \operatorname{Dom}(A)$ and $E(\Delta)A \subseteq AE(\Delta)$. Let C be a positively oriented closed Jordan contour around λ_0 and such that $\sigma(A)$ as well as Δ lie outside. Then, by the Riesz–Dunford–Taylor functional calculus we have

$$AR(\lambda_0 - A) = -I - \frac{1}{2\pi i} \int_C \frac{\zeta}{\lambda_0 - \zeta} R(\zeta; A) \, d\zeta,$$

and hence

$$AE(\Delta)R(\lambda_0; A)x = AR(\lambda_0; A)E(\Delta)x$$

$$= \left(-I - \frac{1}{2\pi i} \int_C \frac{\zeta}{\lambda_0 - \zeta} R(\zeta; A) \, d\zeta\right) \left(\lim_{\varepsilon \downarrow 0} \lim_{\delta \downarrow 0} \int_{C_{\Delta_\varepsilon}^\delta} R(\xi; A) \, d\xi\right)x. \tag{7.9}$$

Applying Hilbert's resolvent identity and the Cauchy formula to (7.9), and taking into account that the Jordan contours C and C_Δ are exterior one to each other, it follows that

$$AE(\Delta)R(\lambda_0; A)x = \lim_{\varepsilon \downarrow 0} \lim_{\delta \downarrow 0} \frac{1}{2\pi i} \int_{C_{\Delta_\varepsilon}^\delta} \zeta R(\zeta; A) \, d\zeta \, R(\lambda_0; A)x.$$

This implies

$$AE(\Delta)y = \lim_{\varepsilon \downarrow 0} \lim_{\delta \downarrow 0} \frac{1}{2\pi i} \int_{C_{\Delta_\varepsilon}^\delta} \zeta R(\zeta; A)y \, d\zeta, \quad y \in \operatorname{Dom}(A), \tag{7.10}$$

11.7 THE SPECTRAL FUNCTION

in particular, by Theorem 11.6.2 the closed operator $AE(\Delta)$ is bounded and hence we have $E(\Delta)\mathcal{K} \subseteq \mathrm{Dom}(A)$. Let $\lambda \notin \mathrm{Clos}\,\Delta$. Again by Theorem 11.6.2, one can define the operator

$$B = \lim_{\varepsilon \downarrow 0} \lim_{\delta \downarrow 0} \frac{1}{2\pi\mathrm{i}} \int_{C^\delta_{\Delta_\varepsilon}} \frac{1}{\zeta - \lambda} R(\zeta; A)\,\mathrm{d}\zeta \in \mathcal{B}(\mathcal{K}).$$

Then, using (7.10), a calculation as in the proof of Lemma 11.7.2 shows that

$$(A - \lambda_0 I)E(\Delta)B = B(A - \lambda_0 I)E(\Delta),$$

which implies that the operator $(A - \lambda_0 I)|E(\Delta)\mathcal{K}$ is boundedly invertible. We have thus proven that $\sigma(A|E(\Delta)\mathcal{K}) \subseteq \mathrm{Clos}\,\Delta$ and hence, all properties (1) through (6) are verified for $\Delta \in \mathfrak{R}_A$ and bounded.

Let now $\Delta \in \mathfrak{R}_A$ be unbounded. Then $\mathbb{R} \setminus \Delta \in \mathfrak{R}_A$ is bounded and hence the selfadjoint projection $E(\mathbb{R} \setminus \Delta)$ is defined as in (7.6). We define

$$E(\Delta) = I - E(\sigma_0; A) - E(\mathbb{R} \setminus \Delta).$$

With this definition and the above proved facts it follows readily that E has the properties (1) through (5) on the whole $\mathfrak{R}_A$.

We prove now that (6) holds, provided that $\Delta \in \mathfrak{R}_A$ is unbounded. Since the range of $E(\mathbb{R} \setminus \Delta)$ is contained in $\mathrm{Dom}(A)$ it follows that $E(\Delta)\mathcal{K} \cap \mathrm{Dom}(A)$ is dense in $E(\Delta)\mathcal{K}$ and $A|\mathrm{Dom}(A) \cap E(\Delta)\mathcal{K}$ is selfadjoint. It remains to prove that $\sigma(A|\mathrm{Dom}(A) \cap E(\Delta)\mathcal{K})$ is contained in the closure of Δ.

To this end let us first remark that, without restricting the generality, we can assume that $\sigma(A)$ is real and $\mathbb{R} \setminus \Delta$ is a bounded interval. Let λ be an inner point of the interval $\mathbb{R} \setminus \Delta$. In view of Theorem 11.6.2 and Theorem 11.5.2, the strong operator limits

$$D = -\lim_{\varepsilon \downarrow 0} \lim_{\delta \downarrow 0} \frac{1}{2\pi\mathrm{i}} \int_{C^\delta_{\Delta_\varepsilon}} \frac{1}{\zeta - \lambda} R(\zeta; A)\,\mathrm{d}\zeta \in \mathcal{B}(\mathcal{K})$$

define an operator in $\mathcal{B}(\mathcal{K})$. As before we can prove that

$$(\lambda_0 I - A)^{-1}(A - \lambda I)E(\Delta)D = (\lambda_0 I - A)^{-1}E(\Delta),$$

and hence the operator $(A - \lambda I)|\mathrm{Dom}(A) \cap E(\Delta)\mathcal{K}$ is boundedly invertible. Since λ is arbitrary in the interior of $\mathbb{R} \setminus \Delta$ this shows that $\sigma(A|\mathrm{Dom}(A) \cap E(\Delta)\mathcal{K})$ is contained in the closure of Δ.

Uniqueness. We first remark that, without restricting the generality, we can assume that $\sigma(A)$ is real, see Lemma 11.3.3. Let $E' : \mathfrak{R}_A \to \mathcal{B}(\mathcal{K})$ be another mapping satisfying the same properties (1) through (6). Let $\Delta = [\alpha, \beta] \in \mathfrak{R}_A$ be arbitrary and let $\Delta' \in \mathfrak{R}_A$ be closed and $\Delta \cap \Delta' = \emptyset$. By property (5) E and E' commute and hence $E(\Delta)E'(\Delta')$ is a bounded selfadjoint projection reducing A. Using property (6) it follows that

$$\sigma(A|E(\Delta)E'(\Delta')\mathcal{K}) \subseteq \sigma(A|E(\Delta)\mathcal{K}) \cap \sigma(A|\mathrm{Dom}(A) \cap E(\Delta')\mathcal{K}) \subseteq \Delta \cap \Delta' = \emptyset.$$

In view of Corollary 11.5.5 this implies $E(\Delta)E'(\Delta') = 0$.

Consider now $\Delta'_n = \mathbb{R} \setminus [\alpha - \frac{1}{n}, \beta + \frac{1}{n}]$, $n \in \mathbb{N}$. Since $[\alpha, \beta] \in \mathfrak{R}_A$ it follows that $\Delta'_n \in \mathfrak{R}_A$, $n \geqslant n_0$ and some $n_0 \in \mathbb{N}$. In view of the property (4) it follows that

$$\lim_{n \to \infty} E'(\Delta'_n) = E'(\mathbb{R} \setminus [\alpha, \beta])$$

in the uniform topology. The preceding observation shows that $E(\Delta)E'(\Delta'_n) = 0$ and letting $n \to \infty$ it follows that $E(\Delta)E'(\mathbb{R} \setminus \Delta) = 0$. Since $E'(\mathbb{R} \setminus \Delta) = I - E'(\Delta)$ this implies $E(\Delta)E'(\Delta') = E(\Delta)$. Interchanging the roles of E and E' we also have $E(\Delta)E'(\Delta) = E'(\Delta)$ and hence $E(\Delta) = E'(\Delta)$.

We have thus proven that E and E' coincide on each compact interval in $\mathfrak{R}_A$. Using property (3) it follows that they coincide also on complements of compact intervals in $\mathfrak{R}_A$. Then using property (2) it follows that E and E' coincide on any bounded interval from $\mathfrak{R}_A$ and their complements, hence they do coincide on the whole $\mathfrak{R}_A$. ∎

Remark 11.7.4. From the definition of the *spectral function* E, see Theorem 11.7.3, we have that $\operatorname{supp}(E) = \sigma(A) \cap \mathbb{R}$. On the other hand, the Boolean algebra $\mathfrak{R}_A$ does not contain intervals bounded at one end and unbounded at the other. If A is bounded then E can be extended also to this kind of intervals. If A is unbounded this is not always possible, except under certain regularity conditions that will be investigated in the last chapter. ∎

In the following we record some immediate consequences of the existence and the uniqueness of the spectral functions associated to selfadjoint and definitisable operators.

Corollary 11.7.5. *Let $\Delta \in \mathfrak{R}_A$ be compact and such that $\Delta \cap c(A) = \emptyset$.*

(i) The function E_Δ defined by

$$E_\Delta(\Lambda) = E(\Delta \cap \Lambda), \quad \Lambda \text{ a real interval,}$$

extends uniquely to a spectral measure E_Δ with $\operatorname{supp}(E_\Delta) \subseteq \Delta$.

(ii) There exists a fundamental symmetry J of $\mathcal{K}$ such that $AE(\Delta)$ is J-selfadjoint (that is, selfadjoint on the Hilbert space $(\mathcal{K}; \langle \cdot, \cdot \rangle_J)$).

(iii) E_Δ is the spectral measure of $AE(\Delta)$, in particular,

$$AE(\Delta) = \int_\Delta t \, \mathrm{d}E(t). \tag{7.11}$$

Proof. Since $\Delta \cap c(A) = \emptyset$ it follows that there exists a definitising polynomial p such that Δ can be written as the disjoint union $\Delta = \Delta_- \cup \Delta_+$, where $\Delta_\pm \in \mathfrak{R}_A$ are compact subsets, $p|_{\Delta_-} < 0$, and $p|_{\Delta_+} > 0$. By Theorem 11.7.3 it follows that $E(\Delta_-)\mathcal{K}$ is a uniformly negative subspace and $E(\Delta_+)\mathcal{K}$ is a uniformly positive subspace and hence

$$E(\Delta)\mathcal{K} = E(\Delta_-)\mathcal{K}[+]E(\Delta_+)\mathcal{K},$$

11.7 THE SPECTRAL FUNCTION

is a fundamental decomposition of the Kreĭn subspace $E(\Delta)\mathcal{K}$. By Theorem 2.2.9 there exists a fundamental decomposition $\mathcal{K} = \mathcal{K}^+[+]\mathcal{K}^-$ such that $E(\Delta_\pm)\mathcal{K} \subseteq \mathcal{K}^\pm$ and let J be the corresponding fundamental symmetry. One verifies immediately that the commutation relation $JAE(\Delta) = AE(\Delta)J$ holds and hence the bounded operator $AE(\Delta)$ is J-selfadjoint, in particular it has a spectral measure. By the definition of E_Δ and the construction of E in Theorem 11.7.3, the spectral measure of $AE(\Delta)$ extends E_Δ and hence the integral representation (7.11) holds. ∎

Lemma 11.7.6. *Let* $\Delta \in \mathfrak{R}_A$ *be an interval and* $f \in \mathfrak{H}(\Delta)$. *If p is a real definitising polynomial of A, $\lambda_0 \in \rho(A)$ is fixed, the rational function q is defined as in (5.1), J is a fixed fundamental symmetry on $\mathcal{K}$ and F is the operator valued function as in Theorem 11.5.2, then*

$$\lim_{\varepsilon \to 0} J \int_{\Delta_\varepsilon} f(t)\,\mathrm{d}F(t) = \lim_{\varepsilon \to 0} \lim_{\delta \to 0} \frac{1}{2\pi\mathrm{i}} \int_{C_{\Delta_\varepsilon^\delta}} f(\zeta)q(\zeta)R(\zeta;A)\,\mathrm{d}\zeta,$$

where the limits are understood in the strong operator topology.

Proof. We use formula (5.2) and Theorem 11.6.2 in order to obtain

$$\lim_{\varepsilon \to 0} J \int_{\Delta_\varepsilon} f(t)\,\mathrm{d}F(t) = \lim_{\varepsilon \to 0} \lim_{\delta \to 0} \frac{1}{2\pi\mathrm{i}} \int_{C_{\Delta_\varepsilon^\delta}} f(\zeta) \int_{-\infty}^{+\infty} \frac{1}{\zeta - t}\,\mathrm{d}F(t)x\,\mathrm{d}\zeta$$

$$= \lim_{\varepsilon \to 0} \lim_{\delta \to 0} \frac{1}{2\pi\mathrm{i}} \int_{C_{\Delta_\varepsilon^\delta}} f(\zeta)q(\zeta)R(\zeta;A)x\,\mathrm{d}\zeta + \lim_{\varepsilon \to 0} \lim_{\delta \to 0} \frac{1}{2\pi\mathrm{i}} \int_{C_{\Delta_\varepsilon^\delta}} f(\zeta)Q(\zeta;A)\,\mathrm{d}\zeta,$$

and finally take into account that, by Cauchy's Theorem, the second term on the right-hand side is null. ∎

Proposition 11.7.7. *For any real definitising polynomial p of A, $\lambda_0 \in \rho(A)$ and rational function q defined as in (5.1), there exists $\{N_\alpha\}_{\alpha \in Z(p) \cap \mathbb{R}} \subset \mathcal{B}(\mathcal{K})$ subject to the following properties:*

(i) $[N_\alpha x, x] \geqslant 0$, *for all* $x \in \mathcal{K}$, $\alpha \in Z(p) \cap \mathbb{R}$;
(ii) $N_\alpha E(\Delta) = 0$, *for all* $\alpha \in Z(p) \cap \mathbb{R}$, $\Delta \in \mathfrak{R}_A$, $\overline{\Delta} \cap \{\alpha\} = \emptyset$;
(iii) $N_{\alpha_1} N_{\alpha_2} = 0$, *for all* $\alpha_1, \alpha_2 \in Z(p) \cap \mathbb{R}$;
(iv) $(A - \alpha I)N_\alpha = 0$, *for all* $\alpha \in Z(p) \cap \mathbb{R}$,

and such that

$$q(A)R(\zeta;A) = \int_{-\infty}^{+\infty} q(t) \frac{1}{\zeta - t}\,\mathrm{d}E(t) + \sum_{\alpha \in Z(p) \cap \mathbb{R}} \frac{1}{\zeta - \alpha} N_\alpha, \qquad (7.12)$$

$$q(A) = \int_{-\infty}^{+\infty} q(t)\,\mathrm{d}E(t) + \sum_{\alpha \in Z(p) \cap \mathbb{R}} N_\alpha, \qquad (7.13)$$

where the integrals are improper with singularities in $(Z(p) \cap \mathbb{R}) \cup \{\infty\}$ and converge in the strong operator topology.

Proof. For arbitrary $\alpha \in Z(p) \cap \mathbb{R}$ we define
$$N_\alpha = J(F(\alpha+0) - F(\alpha-0)) \in \mathcal{B}(\mathcal{K}).$$
First we verify the properties (i) through (iv).

Since F is nondecreasing it follows that
$$[N_\alpha x, x] = \langle F(\alpha+0)x, x\rangle_J - \langle F(\alpha-0)x, x\rangle_J \geq 0, \quad x \in \mathcal{K},$$
and hence (i) holds.

Let $\alpha \in Z(p) \cap \mathbb{R}$. For $\varepsilon, \delta > 0$ and sufficiently small we consider the oriented Jordan curve $C_{\alpha,\varepsilon}^\delta$ defined as the union of the two arcs as in Figure 11.4. Using Lemma 11.7.6 it follows that
$$N_\alpha x = J(F(\alpha+0) - F(\alpha-0))x = \lim_{\varepsilon \to 0} \lim_{\delta \to 0} \frac{1}{2\pi i} \int_{C_{\alpha,\varepsilon}^\delta} q(\zeta) R(\zeta; A) x \, d\zeta. \quad (7.14)$$

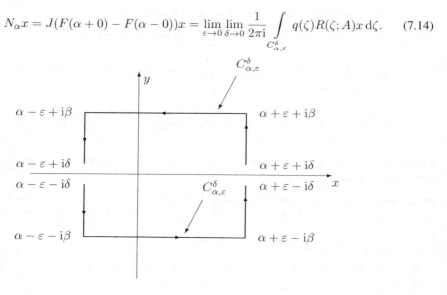

Figure 11.4: The Jordan curve $C_{\alpha,\varepsilon}^\delta$.

To prove that (ii) holds, let $\Delta \in \mathfrak{R}_A$ be a bounded interval and let $\alpha \in Z(p) \cap \mathbb{R}$ be such that $\operatorname{Clos} \Delta \cap \{\alpha\} = \emptyset$. Using (7.14) and the definition of $E(\Delta)$ as a double limit of Cauchy integrals of the resolvent function as in (7.6), performing a calculation as in the proof of Lemma 11.7.2, and taking into account that $q(\alpha) = 0$, we obtain $N_\alpha E(\Delta) = 0$.

11.7 The Spectral Function

On the other hand, if $\Delta \in \mathfrak{R}_A$ is the complement of a bounded interval and $\alpha \in Z(p) \cap \mathbb{R}$ is such that $\operatorname{Clos} \Delta \cap \{\alpha\} = \emptyset$ then $\alpha \in \mathbb{R} \setminus \overline{\Delta}$ and a similar argument as before yields $N_\alpha E(\mathbb{R} \setminus \Delta) = N_\alpha$ and hence $N_\alpha E(\Delta) = 0$. We conclude that (ii) is verified.

To prove (iii) we first remark that if $\alpha_1, \alpha_2 \in Z(p) \cap \mathbb{R}$ are distinct then, in view of (7.14) and the Cauchy formula, the statement is clear. Let then $\alpha \in Z(p) \cap \mathbb{R}$ and $x, y \in \mathcal{K}$ be fixed. Let $\varepsilon_1 > \varepsilon > 0$ and $\delta_1 > \delta > 0$ be sufficiently small. Applying Hilbert's resolvent identity we get

$$\left[\frac{1}{2\pi\mathrm{i}} \int_{C^\delta_{\alpha,\varepsilon}} q(\zeta)R(\zeta; A)x \,\mathrm{d}\zeta, \frac{1}{2\pi\mathrm{i}} \int_{C^{\delta_1}_{\alpha,\varepsilon_1}} q(\xi)R(\xi; A)y \,\mathrm{d}\xi\right]$$

$$= \frac{1}{2\pi\mathrm{i}} \int_{C^\delta_{\alpha,\varepsilon}} \frac{q(\zeta)}{\zeta - \xi} \,\mathrm{d}\zeta \left[x, \frac{1}{2\pi\mathrm{i}} \int_{C^{\delta_1}_{\alpha,\varepsilon_1}} q(\xi)R(\xi; A)y \,\mathrm{d}\xi\right]$$

$$+ \frac{1}{2\pi\mathrm{i}} \int_{C^{\delta_1}_{\alpha,\varepsilon_1}} \frac{q(\zeta)}{\zeta - \xi} \,\mathrm{d}\zeta \left[x, \frac{1}{2\pi\mathrm{i}} \int_{C^\delta_{\alpha,\varepsilon}} q(\xi)R(\xi; A)y \,\mathrm{d}\xi\right].$$

Then we pass to the limit following $\delta_1 \to 0$ and $\varepsilon_1 \to 0$, applying the Cauchy formula for the first term and taking into account that $q(\alpha) = 0$ for the second term, and then pass to the limit with respect to $\delta \to 0$ and $\varepsilon \to 0$. In view of (7.14) we obtain $N_\alpha^2 = 0$.

In order to prove (iv) let $\alpha \in Z(p) \cap \mathbb{R}$ and remark that, without loss of generality, we can assume that A is bounded: indeed, otherwise we replace A by $AE(\Delta)$ for some bounded open interval Δ such that $\alpha \in \Delta$. Assuming this, let C be a Jordan contour admissible for $\sigma(A)$ and hence, by the Riesz–Dunford functional calculus

$$A = \frac{1}{2\pi\mathrm{i}} \int_C \zeta R(\zeta; A) \,\mathrm{d}\zeta.$$

Using (7.14) and manipulating contour integration as before we get $(A - \alpha I)N_\alpha = 0$.

Finally, in order to prove the formulae (7.12) and (7.13) we first recall that, during the proof of Theorem 11.7.3, see (7.7), we proved that for any bounded interval $\Delta \in \mathfrak{R}_A$ such that $p|\Delta \neq 0$ we have

$$E(\Delta)x = \lim_{\varepsilon \to 0} J \int_{\Delta_\varepsilon} \frac{1}{q(t)} \,\mathrm{d}F(t)x, \quad x \in \mathcal{K},$$

and hence, from the definition of the spectral function E, it follows that

$$J \int_{\mathbb{R} \setminus Z(p)} \frac{1}{\zeta - t} \,\mathrm{d}F(t) = \int_{-\infty}^{+\infty} \frac{q(t)}{\zeta - t} \,\mathrm{d}E(t),$$

where the second integral is improper with singularities in $Z(p) \cup \{\infty\}$ and converges strongly operatorial. Then, using formula (5.2) we get

$$q(A)R(\zeta; A) = J \int_{-\infty}^{+\infty} \frac{1}{\zeta - t} \, dF(t)$$

$$= J \int_{\mathbb{R} \setminus Z(p)} \frac{1}{\zeta - t} \, dF(t) + \sum_{\alpha \in Z(p) \cap \mathbb{R}} \frac{1}{\zeta - \alpha} N_\alpha$$

$$= \int_{-\infty}^{+\infty} \frac{q(t)}{\zeta - t} \, dE(t) + \sum_{\alpha \in Z(p) \cap \mathbb{R}} \frac{1}{\zeta - \alpha} N_\alpha.$$

Similarly one obtains (7.13), this time using the formula (5.3). ∎

Corollary 11.7.8. *Let $\alpha \in c(A)$, p a definitising polynomial of A and denote by $k(\alpha)$ the order of multiplicity of α as a root of p. If $\varepsilon > 0$ is sufficiently small then the integral*

$$\int_{\alpha+\varepsilon}^{\alpha-\varepsilon} (t - \alpha)^{k(\alpha)} \, dE(t) \tag{7.15}$$

converges strongly and improperly at α.

Proof. Let $\varepsilon > 0$ be sufficiently small such that α is the only root of p in the interval $[\alpha - \varepsilon, \alpha + \varepsilon]$. Then $q(t) = (t - \alpha)^{k(\alpha)} r(t)$ where r is a rational function having no roots in $[\alpha - \varepsilon, \alpha + \varepsilon]$. Therefore r^{-1} is continuous on this interval. By Proposition 11.7.7 the integral

$$\int_{\alpha-\varepsilon}^{\alpha+\varepsilon} q(t) \, dE(t)$$

converges strongly and improperly at α. Then the same is true if we multiply the integrated function by $r^{-1}(t)$ and hence the integral (7.15) converges strongly and improperly at α. ∎

11.8 Definitisable Positive Operators

In this section we consider a positive selfadjoint operator A in a Kreĭn space $\mathcal{K}$ such that $\rho(A) \neq \emptyset$. Then A is a definitisable operator and $p(\lambda) = \lambda$ is a definitising polynomial of A. By Theorem 11.7.3 there exists and it is unique the spectral function E of A. From Lemma 11.3.3 it follows that the spectrum of A is real. Then 0 is the only possible critical point of A and ∞ may be a singularity of E as well, hence $\mathfrak{R}_A$ consists of finite unions of intervals, and their complements, whose endpoints avoid 0 and ∞.

Lemma 11.8.1. *Let us denote $\Delta_0 = (-\varepsilon, \varepsilon)$ and $\Delta_\infty = (-\infty, -\varepsilon) \cup (\varepsilon, \infty)$ for some $\varepsilon > 0$.*

11.8 Definitisable Positive Operators

(a) *For any $x \in \mathcal{K}$ the following integrals converge improperly at 0 and, respectively, at ∞:*

$$\int_{\Delta_0} t\, \mathrm{d}E(t)x, \quad \int_{\Delta_\infty} \frac{1}{t}\, \mathrm{d}E(t)x.$$

(b) *For any $y \in \mathrm{Ran}(A)$ there exists $y_1 \in \mathrm{Ker}(A)$ such that*

$$\int_{\Delta_0} \mathrm{d}E(t)y = E(\Delta_0)y - y_1,$$

where the integral converges improperly at 0.

(c) *For any $v \in \mathrm{Dom}(A)$ we have*

$$\int_{\Delta_\infty} \mathrm{d}E(t)v = E(\Delta_\infty)v,$$

where the integral converges improperly at ∞.

Proof. (a) The convergence of the first integral is a consequence of Corollary 11.7.8. In order to prove the convergence of the second integral we consider the operator $A' = (A|E(\Delta_\infty)\mathcal{K})^{-1}$ on the Kreĭn space $E(\Delta_\infty)\mathcal{K}$, see Lemma 11.5.4 and Corollary 11.5.6. Then A' is positive and bounded. If E' denotes its spectral function then

$$\mathrm{d}E'(s) = \mathrm{d}E(t), \quad s = t^{-1},\ t \neq 0,$$

and hence the second integral is also convergent.

(b) We set $y = Ax$ in the first integral in (a) and take into account that $\mathrm{d}E(t)y = t\,\mathrm{d}E(t)x$, $t \neq 0$. For arbitrary $0 < \delta < \varepsilon$ we have

$$\int_{\Delta_0 \setminus (-\delta, \delta)} \mathrm{d}E(t)y = E(\Delta_0 \setminus (-\delta, \delta))y.$$

Letting $\delta \to 0$ we get

$$\int_{\Delta_0} \mathrm{d}E(t)y = E(\Delta)y - y_1,$$

where $y_1 \in \mathrm{Ker}(A)$ and the convergence of the integral follows from (a).

(c) We proceed similarly as before, taking into account the argument at (a). ∎

Proposition 11.8.2. *There exists $N \in \mathcal{B}(\mathcal{K})$ subject to the following conditions:*

(i) $[Nx, x] \geqslant 0$ *for all $x \in \mathcal{K}$,*
(ii) $NE(\Delta) = 0$ *for all $\Delta \in \mathfrak{R}_A$, $0 \notin \Delta$,*
(iii) $N^2 = AN = 0$,

and such that
$$Ax = \int_{-\infty}^{+\infty} t \, dE(t)x + Nx, \quad x \in \mathrm{Dom}(A^2), \tag{8.1}$$

$$R(\zeta; A)x = \frac{1}{\zeta} \int_{-\infty}^{+\infty} \frac{t}{t - \zeta} \, dE(t)x + \frac{1}{\zeta^2} Nx + \frac{1}{\zeta} x, \quad x \in \mathrm{Dom}(A), \, \zeta \neq \overline{\zeta}. \tag{8.2}$$

In addition, for any $x \in \mathrm{Dom}(A) \cap \mathrm{Ran}(A)$ *there exists* $x_0 \in \mathrm{Ker}(A)$ *such that*
$$x = \int_{-\infty}^{+\infty} dE(t)x + x_0. \tag{8.3}$$

Proof. Let $q(\lambda) = \lambda(\lambda^2 + 1)^{-1}$. We remark that q is a definitising function of A and then, applying Proposition 11.7.7, we get the existence of N subject to the conditions as in the statement and
$$q(A)y = A(A^2 + I)^{-1}y = \int_{-\infty}^{+\infty} t(t^2 + 1)^{-1} \, dE(t)y + Ny, \quad y \in \mathcal{K}. \tag{8.4}$$

If $t \neq 0$ then $(t^2 + 1)^{-1} dE(t)y = dE(t)x$, where $x = (A^2 + I)^{-1}y \in \mathrm{Dom}(A^2)$ and, in addition, $Ny = N(A^2 + I)x = Nx$. These prove (8.1).

Again by Proposition 11.7.7 we have
$$A(A^2 + I)^{-1} R(\zeta; A)y = \int_{-\infty}^{+\infty} \frac{t}{t^2 + 1} \frac{1}{\zeta - t} \, dE(t)y + \frac{1}{\zeta} Ny, \quad y \in \mathcal{K}, \, \zeta \neq \overline{\zeta},$$

and similarly as before we get
$$AR(\zeta; A)x = \int_{-\infty}^{+\infty} \frac{t}{\zeta - t} \, dE(t)x + \frac{1}{\zeta} Nx, \quad x \in \mathrm{Dom}(A^2), \, \zeta \neq \overline{\zeta}. \tag{8.5}$$

Taking into account that
$$AR(\zeta; A)x = (\zeta I - (\zeta I - A))R(\zeta; A)x = \zeta R(\zeta; A)x - x, \quad x \in \mathrm{Dom}(A),$$

from (8.5) we get
$$R(\zeta; A)x = \frac{1}{\zeta} \int_{-\infty}^{+\infty} \frac{t}{\zeta - t} \, dE(t)x + \frac{1}{\zeta^2} Nx + \frac{1}{\zeta} x, \quad x \in \mathrm{Dom}(A^2). \tag{8.6}$$

11.8 DEFINITISABLE POSITIVE OPERATORS

Let us notice now that from Proposition 11.7.7, the integral

$$\int_{-\infty}^{+\infty} \frac{t}{\zeta - t} \, dE(t)x$$

converges for any $x \in \text{Dom}(A)$ and $\zeta \neq \bar{\zeta}$. Approximating the vector $x \in \text{Dom}(A)$ by the sequence $E([-n, n])x \in \text{Dom}(A^2)$ and letting $n \to \infty$ in (8.6) we get (8.2).

Let now $x \in \text{Dom}(A) \cap \text{Ran}(A)$ and $n \in \mathbb{N}$. Letting $\Delta_n = [-n, n]$ and

$$x_n = E(\Delta_n)x - \int_{-n}^{n} dE(t)x,$$

according to Proposition 11.7.7 the integral is convergent improperly at 0. Then $x_n \in \text{Dom}(A)$. Since $x \in \text{Ran}(A)$ we have $x = Ay$ with $y \in \text{Dom}(A)$ and

$$Ax_n = AE(\Delta_n)x - A\int_{-n}^{n} dE(t)Ay = E(\Delta_n)Ax - E(\Delta_n)A^2 y = 0,$$

where we took into account that $\text{Ker}(A^2)$ is the root subspace corresponding to 0. Therefore $x_n \in \text{Ker}(A)$ and, by means of Proposition 11.7.7, it follows that the sequence $(x_n)_{n\in\mathbb{N}}$ converges to x_0, where

$$x_0 = x - \int_{-\infty}^{+\infty} dE(t)x,$$

and hence, as $\text{Ker}(A)$ is closed, we get $x_0 \in \text{Ker}(A)$. ∎

Corollary 11.8.3. *If the spectral function E is bounded at infinity then*

$$\text{Dom}(A) = \{x \in \mathcal{K} \mid \int_{-\infty}^{+\infty} t^2 |\langle dE(t)x, x \rangle_J| < +\infty\}, \tag{8.7}$$

where J is an arbitrary fundamental symmetry in $\mathcal{K}$. Moreover, the representation (8.1) holds for all $x \in \text{Dom}(A)$ and if, in addition, the spectral function is bounded also in the neighbourhood of 0 then the representations (8.2) and (8.3) hold for all $x \in \mathcal{K}$.

Proof. In order to prove (8.7) we first remark that, without restricting the generality, we can assume that A is boundedly invertible (otherwise replace A with the boundedly invertible definitisable selfadjoint operator $A|E(\Delta)\mathcal{K}$ where $\Delta = \mathbb{R} \setminus [-\varepsilon, \varepsilon]$ for some $\varepsilon > 0$). Let us assume now that the spectral function E is bounded in the neighbourhood of ∞. On the grounds of property (4) in Theorem 11.7.3 and Proposition 5.2.6 it follows that the strong operator limits exist

$$E(0, +\infty) = \lim_{n\to\infty} E(0, n), \quad E(-\infty, 0) = \lim_{n\to\infty} E(-n, 0).$$

Therefore, the decomposition $\mathcal{K} = E(-\infty, 0)\mathcal{K}[+]E(0, +\infty)\mathcal{K}$ is a fundamental decomposition of $\mathcal{K}$ and let the corresponding fundamental symmetry be denoted by J. Since J commutes with A it follows that A is a selfadjoint operator in the Hilbert space $(\mathcal{K}, \langle \cdot, \cdot \rangle)$ and hence (8.7) follows. Therefore, for any $x \in \mathrm{Dom}(A)$ the integral $\int_{-\infty}^{+\infty} t \, \mathrm{d}E(t)x$ is weakly convergent and, as a consequence, (8.1) holds.

If, in addition, the spectral function is bounded in the neighbourhood of 0 then for any $x \in \mathcal{K}$ the integrals in (8.2) and (8.3) converge. ∎

11.9 Notes

The spectral function for selfadjoint operators in Pontryagin spaces was constructed by M. G. Kreĭn and H. Langer in [94]. The class of definitisable operators was introduced by H. Langer in his Habilitationsschrift [104] where the existence and uniqueness of the spectral function and other properties were obtained as well. Our exposition follows closely that provided by H. Langer in [106].

The main results from Section 11.4, Theorems 11.4.5, 11.4.6, and 11.4.7, belong essentially to G. Herglotz [74] while the results from Section 11.6 belong essentially to T. J. Stieltjes [137].

P. Jonas put Langer's spectral function in a more general perspective of generalised spectral operators, in the sense of I. Colojoară and C. Foiaş [27], in a series of papers [80, 81, 82].

Other proofs for the existence of the spectral function for particular cases of definitisable operators have been provided by J. Bognár [19, 20]. M. A. Dritschel [51] provides a different construction of the spectral function derived from the existence of many invariant subspaces, for a slightly larger class of definitisable operators. H. Langer, A. Markus, and V. Matsaev [107] constructed spectral functions for locally definitisable operators while P. Jonas [83] obtained spectral functions for generalised locally definitisable operators.

The special Jordan canonical form of selfadjoint operators in finite dimensional Kreĭn spaces follows closely the presentation in I. Gohberg, P. Lancaster, and L. Rodman [70].

Chapter 12

Quasi-Contractions

We have seen in Chapter 10 that many generalised interpolation problems lead to the consideration of bounded operators T between Kreĭn spaces with the property that the quadratic form $x \mapsto [x,x] - [Tx,Tx]$ has a finite number of negative squares. In this chapter we give the name of quasi-contraction to such an operator T. Clearly, this class generalises the class of contractive operators and we are interested to know to what extent they share similar properties, from the geometric point of view. We will show that, indeed, many results on contractions either can be extended to the case of quasi-contractions or have generalisations that give a clearer perspective. As for the techniques that we use, on one hand, we show that elementary rotations and the concept of pseudo-regular subspaces play very important rôles in the geometry of quasi-contractions. On the other hand, since T is a quasi-contraction if and only if the operator $I - T^\sharp T$ has finite negative rank and hence, by Proposition 11.3.6, it is definitisable, this becomes yet another place where the geometry of operators on Kreĭn spaces meets spectral theory, this time powered by Langer's spectral function.

12.1 Geometric Properties of Quasi-Contractions

Recall that, given a selfadjoint operator A in a Hilbert space, we consider the geometric ranks of positivity/negativity $\kappa^\pm(A)$, see (1.5) in Section 6.1, defined as the dimension of the spectral subspace of A corresponding to $(0, +\infty)$ or to $(-\infty, 0)$, respectively. By definition, these geometric ranks of positivity/negativity are cardinal numbers. On the other hand, considering the inner product $[x,y]_A = \langle Ax, y \rangle$ on $\mathrm{Dom}(A)$, we have the algebraic ranks of positivity/negativity $\kappa_\pm(A)$, see (1.9) in Section 1.1, defined as the signatures of the inner product space $(\mathrm{Dom}(A); [\cdot,\cdot]_A)$, and these are either positive integer numbers or the symbol $+\infty$. It is easy to see that $\kappa^-(A)$ is finite if and only if $\kappa_-(A)$ is finite and, if this happens, then $\kappa^-(A) = \kappa_-(A)$. Similarly for the positive ranks.

These definitions make sense also for selfadjoint operators A in a Kreĭn space, by considering the operator JA which is selfadjoint in the Hilbert space $(\mathcal{K}; \langle \cdot,\cdot \rangle_J)$ and showing that they do not depend on the fundamental symmetry J in $\mathcal{K}$, see Sections 1.1 and 6.1.

Let $\mathcal{K}_1$ and $\mathcal{K}_2$ be Kreĭn spaces, let $T \in \mathcal{B}(\mathcal{K}_1, \mathcal{K}_2)$, and consider the selfadjoint operator $I - T^\sharp T$ on $\mathcal{K}_1$ and $I - TT^\sharp$ on $\mathcal{K}_2$, hence we have defined the two geometric ranks of negativity, $\kappa^-(I - T^\sharp T)$ and $\kappa^-(I - TT^\sharp)$, which are cardinal numbers, as well as two algebraic ranks of negativity, $\kappa_-(I - T^\sharp T)$ and $\kappa_-(I - TT^\sharp)$, which are either positive integer numbers or the symbol $+\infty$. The operator $T \in \mathcal{B}(\mathcal{K}_1, \mathcal{K}_2)$ is called *quasi-contractive* if $\kappa^-(I - T^\sharp T)$ is finite, equivalently, $\kappa_-(I - T^\sharp T)$ is finite. If T is a quasi-contraction then by Proposition 11.3.6, the selfadjoint operator $I - T^\sharp T$ is definitisable. Thus it is transparent that a formula to compute the negative signature of a definitisable operator in terms of its spectral function would be useful, in this context.

Whenever σ is a spectral set of the operator A we use $E(\sigma; A)$ to denote the corresponding spectral projection obtained by the Riesz–Dunford functional calculus. If λ is isolated in the spectrum of A then $E(\lambda; A) = E(\{\lambda\}; A)$. Also, recall that we denote by $c(A)$ the set of critical points of the definitisable operator A, see (7.1) in Chapter 11.

Lemma 12.1.1. *Let $A \in \mathcal{L}(\mathcal{K})$ be a selfadjoint definitisable operator and denote by E its spectral function. Then, for $\varepsilon > 0$ sufficiently small such that $c(A) \cap [-\varepsilon, \varepsilon] \subseteq \{0\}$, we have*

$$\kappa^-(A) = \sum_{\operatorname{Im}\lambda > 0} \operatorname{rank} E(\lambda; A) + \kappa^-(A|E((-\varepsilon, \varepsilon))\mathcal{K})$$
$$+ \kappa^-(E([\varepsilon, +\infty))) + \kappa^+(E((-\infty, \varepsilon])). \quad (1.1)$$

Proof. Indeed, by Lemma 11.3.3, $\sigma_0(A) = \sigma(A) \setminus \mathbb{R}$ is a finite set hence it is a spectral set of A. Denoting

$$\sigma_0^+(A) = \{\lambda \in \sigma_0(A) \mid \operatorname{Im}\lambda > 0\}, \quad \sigma_0^-(A) = \{\lambda \in \sigma_0(A) \mid \operatorname{Im}\lambda < 0\},$$

by Proposition 11.1.4 it follows that

$$\sigma_0^+(A) = \{\bar{\lambda} \mid \lambda \in \sigma_0^-(A)\},$$

and the spectral subspaces $E(\sigma_0^+(A); A)\mathcal{K}$ and $E(\sigma_0^-(A); A)\mathcal{K}$ are neutral and in strong duality, hence unitary equivalent (as Hilbert spaces). Moreover, identifying both of these spaces with the same Hilbert space $\mathcal{H}$ then

$$E(\sigma_0(A); A)\mathcal{K} = \mathcal{H} \oplus \mathcal{H},$$

where the fundamental symmetry is given by $J(x \oplus y) = y \oplus x$, $x, y \in \mathcal{H}$. Then, with respect to this representation

$$A|E(\sigma_0(A); A)\mathcal{K} = \begin{bmatrix} B & 0 \\ 0 & B^* \end{bmatrix},$$

where $B \in \mathcal{B}(\mathcal{H})$ is invertible. This shows that

$$\kappa^-(A|E(\sigma_0(A); A)\mathcal{K}) = \dim \mathcal{H} = \dim E(\sigma_0^+(A); A)\mathcal{K} = \sum_{\operatorname{Im}\lambda > 0} \operatorname{rank} E(\lambda; A). \quad (1.2)$$

12.1 GEOMETRIC PROPERTIES OF QUASI-CONTRACTIONS

Let now $\varepsilon > 0$ be sufficiently small such that $c(A) \cap [-\varepsilon, \varepsilon] \subseteq \{0\}$. Since

$$\sigma(A|E([\varepsilon, +\infty))\mathcal{K}) \subseteq [\varepsilon, +\infty),$$

by the Riesz–Dunford functional calculus it follows that there exists an operator $R \in \mathcal{B}(E([\varepsilon, +\infty))\mathcal{K})$, $R = R^\sharp$ and $\sigma(R) \subseteq [\varepsilon^{1/2}, +\infty)$ (in particular R is invertible) such that

$$A|E([\varepsilon, +\infty))\mathcal{K} = R^2.$$

This means

$$[Ax, x] = [Rx, Rx], \quad x \in E([\varepsilon, +\infty))\mathcal{K}.$$

Since R is invertible we obtain

$$\kappa^-(A|E([\varepsilon, +\infty))\mathcal{K}) = \kappa^-(E([\varepsilon, +\infty))). \tag{1.3}$$

In the same way we obtain

$$\kappa^-(A|E((-\infty, -\varepsilon])\mathcal{K}) = \kappa^+(E((-\infty, -\varepsilon])). \tag{1.4}$$

The formula (1.1) follows now from (1.2) through (1.4). ∎

Proposition 12.1.2. *Let $T \in \mathcal{B}(\mathcal{K}_1, \mathcal{K}_2)$ be quasi-contractive and $\kappa = \kappa^-(I - T^\sharp T) < \infty$. Then $\mathrm{Ker}(T)$ is a pseudo-regular subspace of $\mathcal{K}_1$ such that*

$$\kappa^0(\mathrm{Ker}(T)) + \kappa^-(\mathrm{Ker}(T)) \leqslant \kappa. \tag{1.5}$$

Proof. Let us denote $A = I - T^\sharp T$. A is a bounded selfadjoint operator in $\mathcal{K}_1$ and since $\kappa^-(A) = \kappa < \infty$ it follows by Proposition 11.1.7 that A is definitisable. Let E denote the spectral function of A and $1 > \varepsilon > 0$ sufficiently small such that $(-\varepsilon, \varepsilon) \setminus \{0\}$ does not contain any critical point of A. Then, from Lemma 12.1.1 we obtain $\kappa^-(E(\varepsilon, +\infty)) \leqslant \kappa$, in particular $E(\varepsilon, +\infty)\mathcal{K}_1$ is a Pontryagin space. Furthermore, since $\mathrm{Ker}(T) \subseteq \mathrm{Ker}(T^\sharp T) = \mathrm{Ker}(I - A) \subseteq E(\varepsilon, +\infty)\mathcal{K}_1$ then $\mathrm{Ker}(T)$ is pseudo-regular and (1.5) holds. ∎

An operator $T \in \mathcal{B}(\mathcal{K}_1, \mathcal{K}_2)$ is *expansive (strictly expansive)* on a subspace $\mathcal{L} \in \mathcal{K}_1$ if $[Tx, Tx] \geqslant [x, x]$ ($[Tx, Tx] > [x, x]$) for all $x \in \mathcal{L}$ ($x \in \mathcal{L} \setminus \{0\}$). By definition, a quasi-contraction is strictly expansive on a subspace of finite maximal dimension $\kappa^-(I - T^\sharp T)$. The next result shows that we can choose this subspace to be regular.

Proposition 12.1.3. *Let $T \in \mathcal{B}(\mathcal{K}_1, \mathcal{K}_2)$ be quasi-contractive and denote $\kappa = \kappa^-(I - T^\sharp T)$. Then there exists a regular subspace $\mathcal{L}$ in $\mathcal{K}_1$, such that $\dim \mathcal{L} = \kappa$ and $T|\mathcal{L}$ is strictly expansive.*

Proof. Since $\kappa^-(I - T^\sharp T) = \kappa$ it follows that there exists a subspace $\mathcal{L}$ of $\mathcal{K}_1$, $\dim \mathcal{L} = \kappa$ such that

$$[Tx, Tx] > [x, x], \quad x \in \mathcal{L} \setminus \{0\}. \tag{1.6}$$

Since $\mathcal{L}$ is finite dimensional it is pseudo-regular and hence

$$\mathcal{L} = \mathcal{L}^0[+]\mathcal{R},$$

with $\mathcal{R}$ a regular subspace of $\mathcal{K}_1$. To be more specific, we fix a fundamental symmetry J_1 on $\mathcal{K}_1$ and take $\mathcal{R} = \mathcal{L} \ominus \mathcal{L}^0$.

Let us assume that $\mathcal{L}^0 \neq \{0\}$, that is, $\mathcal{L}$ is degenerate, and let $\{f_i\}_{i=1}^{\kappa}$ be an orthogonal basis of $\mathcal{L}$, that is, it is a basis such that for all $i \neq j$ we have $[f_i, f_j] = 0$. Such an orthogonal basis can be obtained as the union of a J_1-orthonormal basis of $\mathcal{R}$ and a J_1-orthonormal basis of $\mathcal{L}^0$. Then $\operatorname{rank}([f_i,f_j])_{i,j=1}^{\kappa} = \dim(\mathcal{R}) = \kappa - \dim(\mathcal{L}^0)$. Let us now fix the fundamental norm $\|\cdot\|$ on $\mathcal{K}_1$, which is the norm associated to the fundamental symmetry J_1 on $\mathcal{K}_1$. We claim that for any $\varepsilon > 0$ there exists a system of vectors $\{g_i\}_{i=1}^{\kappa}$ in $\mathcal{K}_1$ such that for all $i \in \{1, 2, \ldots, \kappa\}$ we have $\|f_i - g_i\| < \varepsilon$ and $\operatorname{rank}([g_i, g_j])_{i,j=1}^{\kappa} = \kappa$. The claim is easy to establish by perturbing only the vectors f_i in $\mathcal{L}^0$ with the vectors $g_i = f_i + \varepsilon J_1 f_i$, while for the vectors $f_j \in \mathcal{R}$ letting $g_j = f_j$.

Let us now consider $\{\lambda_i\}_{i=1}^{\kappa} \in \mathbb{C}$ such that $\sum_{i=1}^{\kappa} |\lambda_i|^2 = 1$ and denote

$$x = \sum_{i=1}^{\kappa} \lambda_i f_i, \quad y = \sum_{i=1}^{\kappa} \lambda_i g_i. \tag{1.7}$$

Then

$$\|x - y\| = \Big\|\sum_{i=1}^{\kappa} \lambda_i(f_i - g_i)\Big\| \leq \sum_{i=1}^{\kappa} |\lambda_i| \cdot \|f_i - g_i\|$$
$$\leq \Big(\sum_{i=1}^{\kappa} |\lambda_i|^2\Big)^{1/2} \cdot \Big(\sum_{i=1}^{\kappa} \|f_i - g_i\|^2\Big)^{1/2} \leq \sqrt{\kappa} \cdot \varepsilon.$$

Using the boundedness of T and the joint norm continuity of the inner products, similar majorisations as before show that for any $\delta > 0$ there exists $\varepsilon > 0$, independent of $\{\lambda_i\}_{i=1}^{\kappa}$, such that if x and y are as in (1.7) and $\|f_i - g_i\| < \varepsilon$ then

$$|([y,y] - [Ty, Ty]) - ([x, x] - [Tx, Tx])| < \delta.$$

Letting δ be sufficiently small, from (1.6) it follows that we also have $[Ty, Ty] > [y, y]$, and since this inequality is homogenous of order two in y, we thus obtain

$$[Ty, Ty] > [y, y], \quad y \in \mathcal{L}_\delta \setminus \{0\},$$

where we let $\mathcal{L}_\delta$ be the space generated by the system of vectors $\{g_i\}_{i=1}^{\kappa}$. Also, since $\operatorname{rank}([g_i, g_j])_{i,j=1}^{\kappa} = \kappa$ it follows that the space $\mathcal{L}_\delta$ is regular of dimension κ. We replace $\mathcal{L}$ by $\mathcal{L}_\delta$. ∎

In the following, in order to prove certain results about quasi-contractions, we will use some related results about contractions. To do this we need first a notion of the proximity

12.1 Geometric Properties of Quasi-Contractions

of quasi-contractive operators to contractive operators. For the moment, one possible answer is the following lemma. More complete results on extensions will be presented later in this chapter.

Lemma 12.1.4. *Let $T \in \mathcal{B}(\mathcal{K}_1, \mathcal{K}_2)$ be quasi-contractive and denote $\kappa = \kappa^-(I - T^\sharp T)$. Then, given a Kreĭn space $\mathcal{K}'_2$, there exists a contraction $\widetilde{T} \in \mathcal{B}(\mathcal{K}_1, \mathcal{K}_2[+]\mathcal{K}'_2)$ such that $P_{\mathcal{K}_2}\widetilde{T} = T$ if and only if $\kappa^-(\mathcal{K}'_2) \geqslant \kappa$.*

Proof. Let us fix fundamental symmetries J_1 on $\mathcal{K}_1$ and J_2 on $\mathcal{K}_2$. Since $\kappa = \kappa^-(I - T^\sharp T) < \infty$ it follows that the operator $J_1 - T^*J_2T$, considered as a selfadjoint operator in the Hilbert space $(\mathcal{K}_1, \langle \cdot, \cdot \rangle_{J_1})$, has exactly κ negative eigenvalues. With notation as in Section 6.5, we consider the sign operators $J_T = \text{sgn}(J_1 - T^*J_2T)$ and let

$$J_T = J_T^+ - J_T^-$$

be its Jordan decomposition. We have rank $J_T^- = \kappa$. Let us also consider the defect operator $D_T = |J_1 - T^*J_2T|^{1/2}$ and the operator $D_T^- = J_T^- D_T$ and notice that

$$(D_T^-)^2 = (J_1 - T^*J_2T)^-, \qquad (1.8)$$

where

$$J_1 - T^*J_2T = (J_1 - T^*J_2T)^+ - (J_1 - T^*J_2T)^-$$

is the Jordan decomposition. We consider $\mathcal{D}_T^- = \text{Ran}(D_T^-)$, which is a subspace of $\mathcal{K}_1$ of dimension κ, regarded as an anti-Hilbert space with the negative inner product inherited from $-\langle \cdot, \cdot \rangle_{J_1}$. Then let us define

$$\widetilde{T}_0 = \begin{bmatrix} T \\ D_T^- \end{bmatrix} : \mathcal{K}_1 \to \mathcal{K}_2[+]\mathcal{D}_T^-.$$

$\widetilde{T}_0$ is a contraction. Indeed, $\widetilde{J}_2 = J_2 \oplus -I$ is a fundamental symmetry of $\mathcal{K}_2[+]\mathcal{D}_T^-$ and using (1.8) we have

$$J_1 - \widetilde{T}_0^* \widetilde{J}_2 \widetilde{T}_0 = J_1 - T^*J_2T - (D_T^-)^2 = (J_1 - T^*J_2T)^+ \geqslant 0.$$

If $\mathcal{K}_2^-$ is a Kreĭn space with $\kappa^-(\mathcal{K}'_2) \geqslant \kappa$ then there exists a bounded isometric operator $V : \mathcal{D}_T^- \to \mathcal{K}'_2$ and denote

$$\widetilde{T} = \begin{bmatrix} T \\ VD_T^- \end{bmatrix} : \mathcal{K}_1 \to \mathcal{K}_2[+]\mathcal{K}_2^-.$$

Then $P_{\mathcal{K}_2}\widetilde{T} = T$ and $\widetilde{T}$ is a contraction since $\widetilde{T}_0$ is.

To prove the converse implication, let $\widetilde{T} \colon \mathcal{K}_1 \to \mathcal{K}_2[+]\mathcal{K}'_2$ be a contraction such that $P_{\mathcal{K}_2}\widetilde{T} = T$. Representing

$$\widetilde{T} = \begin{bmatrix} T \\ X \end{bmatrix},$$

then
$$J_1 - T^*J_2T - X^*J_2'X \geqslant 0$$
and hence, for some operator $Z \in \mathcal{B}(\mathcal{K}_1)$, we have
$$J_1 - T^*J_2T = [X^* \ Z^*]\begin{bmatrix} J_2' & 0 \\ 0 & I \end{bmatrix}\begin{bmatrix} X \\ Z \end{bmatrix}.$$
From here we obtain $\kappa^-(I - T^\sharp T) \leqslant \kappa^-(J_2') = \kappa^-(\mathcal{K}_2')$. ∎

As a first application we can investigate the action of a quasi-contraction on negative subspaces.

Corollary 12.1.5. *Let $T \in \mathcal{B}(\mathcal{K}_1, \mathcal{K}_2)$ be a quasi-contraction and let us also denote $\kappa = \kappa^-(I - T^\sharp T)$.*

(i) *If $\mathcal{L}$ is a negative subspace of $\mathcal{K}_1$ then $T\mathcal{L}$ is closed and $\kappa^+(T\mathcal{L}) \leqslant \kappa$.*

(ii) *If $\mathcal{L}$ is a strictly negative subspace of $\mathcal{K}_1$ then $T\mathcal{L}$ is a subspace of $\mathcal{K}_2$ such that $\kappa^0(T\mathcal{L}) + \kappa^+(T\mathcal{L}) \leqslant \kappa$.*

(iii) *If $\mathcal{L}$ is a uniformly negative subspace of $\mathcal{K}_1$ then $T\mathcal{L}$ is a pseudo-regular subspace of $\mathcal{K}_2$ such that $\kappa^0(T\mathcal{L}) + \kappa^+(T\mathcal{L}) \leqslant \kappa$.*

Proof. (i) Applying Lemma 12.1.4 we obtain a Kreĭn space $\mathcal{K}_2'$ such that $\kappa^-(\mathcal{K}_2') = \dim \mathcal{K}_2' = \kappa$ and a contraction $\widetilde{T} \in \mathcal{B}(\mathcal{K}_1, \mathcal{K}_2[+]\mathcal{K}_2')$ such that $P_{\mathcal{K}_2}\widetilde{T} = T$. We identify $\mathcal{K}_2$ with $\mathcal{K}_2[+]\{0\}$ as a subspace of $\mathcal{K}_2[+]\mathcal{K}_2'$. If $\mathcal{L}$ is a negative subspace of $\mathcal{K}_1$ then, using Corollary 8.1.3, it follows that $\widetilde{T}\mathcal{L}$ is a negative subspace in $\mathcal{K}_2[+]\mathcal{K}_2'$. We consider now the subspace
$$\mathcal{S} = \widetilde{T}\mathcal{L} + \mathcal{K}_2'.$$
$\mathcal{S}$ is closed since $\widetilde{T}\mathcal{L}$ is closed and $\mathcal{K}_2'$ is finite dimensional. Moreover, there exists a subspace $\mathcal{M}$ of $\mathcal{K}_2[+]\mathcal{K}_2'$ such that $\dim \mathcal{M} \leqslant \kappa$ and
$$\mathcal{S} = \widetilde{T}\mathcal{L}[+]\mathcal{M}. \tag{1.9}$$
Since $\widetilde{T}\mathcal{L}$ is negative it follows that $\kappa^+(\mathcal{S}) \leqslant \kappa$. Using $P_{\mathcal{K}_2}\widetilde{T} = T$ and the definition of $\mathcal{S}$ we obtain
$$\mathcal{S} = T\mathcal{L}[+]\mathcal{K}_2', \tag{1.10}$$
hence $T\mathcal{L}$ is closed and $\kappa^+(T\mathcal{L}) \leqslant \kappa^+(\mathcal{S}) \leqslant \kappa$.

(ii) Assuming that the subspace $\mathcal{L}$ is strictly negative then $\widetilde{T}\mathcal{L}$ is the same, see Corollary 8.1.3. Using (1.9) and (1.10) we obtain
$$\kappa^0(T\mathcal{L}) + \kappa^+(T\mathcal{L}) \leqslant \kappa^0(\mathcal{S}) + \kappa^+(\mathcal{S}) \leqslant \kappa.$$

(iii) If the subspace $\mathcal{L}$ is uniformly negative then $\widetilde{T}\mathcal{L}$ is the same (also by Corollary 8.1.3). From (1.9) and $\dim \mathcal{M} \leqslant \kappa$ it follows that $\mathcal{S}$ is pseudo-regular and $\kappa^0(\mathcal{S}) +$

12.1 GEOMETRIC PROPERTIES OF QUASI-CONTRACTIONS

$\kappa^+(\mathcal{S}) \leqslant \kappa$. Since these kinds of pseudo-regular subspaces are hereditary and $T\mathcal{L} \subseteq \mathcal{S}$ then $T\mathcal{L}$ is also pseudo-regular. ∎

Here we have another characterisation of quasi-contractions in terms of contractions.

Lemma 12.1.6. *An operator is quasi-contractive if and only if it is a finite rank perturbation of a contraction.*

Proof. Assume that for the operator $T \in \mathcal{B}(\mathcal{K}_1, \mathcal{K}_2)$ there exists a contraction T_0 and a finite rank operator X such that $T = T_0 + X$. Then

$$I - T^\sharp T = I - T_0^\sharp T_0 - T_0^\sharp X - X^\sharp T_0 - X^\sharp X.$$

Since $I - T^\sharp T$ is positive and the operator $T_0^\sharp X + X^\sharp T_0 + X^\sharp X$ has finite rank it follows that $\kappa^-(I - T^\sharp T) < \infty$, that is, T is a quasi-contraction.

If T is a quasi-contraction then, by definition, there exists a subspace $\mathcal{S} \subseteq \mathcal{K}_1$ of finite codimension $\kappa^-(I - T^\sharp T) < \infty$ such that

$$[Tx, Tx] \leqslant [x, x], \quad x \in \mathcal{S}.$$

Since $\mathcal{S}$ has finite codimension then $\kappa^0(\mathcal{S}) = \dim(\mathcal{S}^0) < \infty$ and $\mathcal{S}$ is pseudo-regular. Therefore, $\mathcal{S} = \mathcal{R}[+]\mathcal{S}^0$ with $\mathcal{R}$ a regular subspace of finite codimension. We consider now the operator $T|\mathcal{R} \in \mathcal{B}(\mathcal{R}, \mathcal{K}_2)$ which is a contraction, since $\mathcal{R} \subseteq \mathcal{S}$. By Lemma 12.1.4 there exists a contraction $T_0 \in \mathcal{B}(\mathcal{K}_1, \mathcal{K}_2)$ such that $T_0|\mathcal{R} = T|\mathcal{R}$. Since $\mathcal{R}$ has finite codimension it follows that $T - T_0$ has finite rank, more precisely, $\operatorname{rank}(T - T_0) \leqslant 2\kappa^-(I - T^\sharp T)$. ∎

Remark 12.1.7. From the proof of Lemma 12.1.6 we can obtain estimates of the number $\kappa^-(I - T^\sharp T)$ in terms of the rank of the perturbation X, more precisely, $\kappa^-(I - T^\sharp T) \leqslant 3\operatorname{rank}(X)$. Also, it was proved in Lemma 12.1.6 that, if T is a quasi-contraction then it can be written as a perturbation of rank $\leqslant 2\kappa^-(I - T^\sharp T)$ of a contraction. However, these estimates are not sufficiently sharp, for instance, using this we obtain a weaker estimate than (1.6), so we will use Lemma 12.1.6 mostly for qualitative results. ∎

The main result of this section is the following.

Theorem 12.1.8. *Let $T \in \mathcal{B}(\mathcal{K}_1, \mathcal{K}_2)$ be quasi-contractive and consider for $i = 1, 2$ the fundamental decompositions $\mathcal{K}_i = \mathcal{K}_i^+[+]\mathcal{K}_i^-$. Then $P_{\mathcal{K}_2^-} T|\mathcal{K}_1^- \in \mathcal{B}(\mathcal{K}_1^-, \mathcal{K}_2^-)$ is a semi-Fredholm operator, more precisely, it has closed range and finite dimensional kernel, and its index is*

$$\operatorname{ind}(P_{\mathcal{K}_2^-} T|\mathcal{K}_1^-) = \kappa^-(I - T^\sharp T) - \kappa^-(I - TT^\sharp). \quad (1.11)$$

Proof. With respect to the fundamental decomposition $\mathcal{K}_i = \mathcal{K}_i^+[+]\mathcal{K}_i^-$, $i = 1, 2$ we consider the matrix representation of T

$$T = \begin{bmatrix} T_{11} & T_{12} \\ T_{21} & T_{22} \end{bmatrix}, \quad (1.12)$$

in particular $P_{\mathcal{K}_2^-}T|\mathcal{K}_1^- = T_{22}$. We now recall that, during the proof of Theorem 8.1.4 it was proven that, for a contraction operator T, the corner T_{22} is a semi-Fredholm operator. Hence, by Lemma 12.1.6, it follows that T_{22} is a finite rank perturbation of a semi-Fredholm operator and hence T_{22} is the same. In order to prove the formula (1.11) we proceed stepwise.

Step 1. We first assume that $\kappa^-(I - T^\sharp T) = \kappa^-(I - TT^\sharp) = 0$, that is, T is a double contraction. Then T_{22} is invertible, cf. Corollary 8.1.7, and hence the formula (1.11) is trivially satisfied in this case.

Step 2. Assume now that T is a contraction. We distinguish two subcases:

(a) If $\kappa^-(I - TT^\sharp)$ is finite then, applying the same argument as in the proof of Lemma 12.1.4 for $T^\sharp$ instead of T, we obtain a negative definite Kreĭn space $\mathcal{D}_{T^\sharp}^-$ such that

$$\kappa^-(\mathcal{D}_{T^\sharp}^-) = \dim \mathcal{D}_{T^\sharp}^- = \kappa^-(I - TT^\sharp) < \infty$$

and an operator $\widetilde{T} \in \mathcal{B}(\mathcal{K}_1[+]\mathcal{D}_{T^\sharp}^-)$ defined by

$$\widetilde{T} = [T \ D_{T^\sharp}^-]. \tag{1.13}$$

We claim that $\widetilde{T}$ is a double contraction, that is,

$$\kappa^-(I - \widetilde{T}^\sharp \widetilde{T}) = \kappa(I - \widetilde{T}\widetilde{T}^\sharp) = 0.$$

Indeed, $\widetilde{T}^\sharp$ is a contraction as the proof of Lemma 12.1.4 shows. In order to prove $\widetilde{T}$ is a contraction we use Lemma 12.1.4 to produce a contractive extension of $\widetilde{T}$, hence $\widetilde{T}$ is also contractive. Consider now the fundamental decomposition

$$\mathcal{K}_1[+]\mathcal{D}_{T^\sharp}^- = \mathcal{K}_1^+[+](\mathcal{K}_1^-[+]\mathcal{D}_{T^\sharp}^-)$$

and with respect to it we represent

$$\widetilde{T} = \begin{bmatrix} T_{11} & T_{12} & X_{12} \\ T_{21} & T_{22} & X_{22} \end{bmatrix}. \tag{1.14}$$

With this notation we have

$$\widetilde{T}_{22} = [T_{22} \ X_{22}]. \tag{1.15}$$

Using the fact proved at the first step, taking into account that X_{22} has finite rank and that the index of Fredholm operators is invariant under compact perturbations, we have

$$0 = \mathrm{ind}(\widetilde{T}_{22}) = \mathrm{ind}([T_{22} \ 0]) = \mathrm{ind}(T_{22}) + \dim \mathcal{D}_{T^\sharp}^- = \mathrm{ind}(T_{22}) + \kappa^-(I - TT^\sharp),$$

hence

$$\mathrm{ind}(T_{22}) = -\kappa^-(I - TT^\sharp),$$

12.1 Geometric Properties of Quasi-Contractions

and the formula (1.11) is proven in this case.

(b) Assume that $\kappa^-(I - TT^\sharp)$ is infinite. We consider as before the negative definite Kreĭn space $\mathcal{D}_{T^\sharp}^-$ and the operator $\widetilde{T}$ as in (1.13). The proof that $\widetilde{T}$ is double contractive is the same as before. The operator $\widetilde{T}_{22}$ in (1.15) is invertible, hence

$$\mathcal{K}_2^- = \operatorname{Ran}(\widetilde{T}_{22}) = \operatorname{Ran}(T_{22}) + \operatorname{Ran}(X_{22})$$

and since T_{22} itself is injective we have

$$\operatorname{Ran}(T_{22}) \cap \operatorname{Ran}(X_{22}) = 0,$$

hence

$$\mathcal{K}_2^- = \operatorname{Ran}(T_{22}) \dotplus \operatorname{Ran}(X_{22}). \tag{1.16}$$

We make now the remark that we can change the fundamental decomposition $\mathcal{K}_2 = \mathcal{K}_2^+[+]\mathcal{K}_2^-$ without restricting the generality. Indeed, in this way we replace T_{22} by UT_{22} where U is a unitary operator mapping $\mathcal{K}_2^-$ onto some maximal uniformly negative subspace and

$$\operatorname{ind}(UT_{22}) = \operatorname{ind}(U) + \operatorname{ind}(T_{22}) = \operatorname{ind}(T_{22}).$$

On the other hand, since $\widetilde{T}$ is a double contraction it follows that $\widetilde{T}(\mathcal{K}_1^-[+]\mathcal{D}_{T^\sharp}^-)$ is a maximal uniformly negative subspace of $\mathcal{K}_2$ (cf. Theorem 8.1.4) hence, without restricting the generality, we can assume

$$\widetilde{T}(\mathcal{K}_1^-[+]\mathcal{D}_{T^\sharp}^-) = \mathcal{K}_2^-.$$

Then the operator X_{12} in (3.11) vanishes, hence $X_{22} = \mathcal{D}_{T^\sharp}^-$, and this yields

$$\operatorname{rank}(X_{22}) = \dim \mathcal{D}_{T^\sharp}^- = \infty.$$

From (1.16) we infer now that $\operatorname{Ran}(T_{22})$ has infinite codimension in $\mathcal{K}_2^-$, in particular

$$\operatorname{ind}(T_{22}) = -\infty,$$

hence the formula (1.11) is again verified.

Step 3. Let now T be a genuine quasi-contraction. As in Lemma 12.1.4 we can obtain a Kreĭn space $\mathcal{D}_T$ with

$$\kappa^-(\mathcal{D}_T) = \kappa^-(I - T^\sharp T) < \infty$$

and an operator $\widetilde{T} \in \mathcal{B}(\mathcal{K}_1, \mathcal{K}_2[+]\mathcal{D}_T)$

$$\widetilde{T} = \begin{bmatrix} T \\ D_T \end{bmatrix} : \mathcal{K}_1 \to \mathcal{K}_2[+]\mathcal{D}_T,$$

such that

$$\kappa^-(I - \widetilde{T}^\sharp \widetilde{T}) = 0, \quad \kappa^-(I - \widetilde{T}\widetilde{T}^\sharp) = \kappa^-(I - TT^\sharp).$$

On $\mathcal{D}_T$ we fix the fundamental decomposition $\mathcal{D}_T = \mathcal{D}_T^+[+]\mathcal{D}_T^-$ (where $\mathcal{D}_T^-$ can be chosen as the space considered during the proof of Lemma 12.1.1) and, with our notation, we have

$$\widetilde{T}_{22} = \begin{bmatrix} T_{22} \\ X_{22} \end{bmatrix} : \mathcal{K}_1^- \to \mathcal{K}_2^-[+]\mathcal{D}_T^-.$$

Using the result obtained at the previous step we have

$$-\kappa^-(I - TT^\sharp) = \operatorname{ind}(\widetilde{T}_{22}) = \operatorname{ind}\left(\begin{bmatrix} T_{22} \\ 0 \end{bmatrix}\right)$$
$$= \operatorname{ind}(T_{22}) - \dim \mathcal{D}_T^- = \operatorname{ind}(T_{22}) - \kappa^-(I - T^\sharp T),$$

hence

$$\operatorname{ind}(T_{22}) = \kappa^-(I - T^\sharp T) - \kappa^-(I - TT^\sharp),$$

which proves the formula (1.11). ∎

Remark 12.1.9. Let T be in $\mathcal{B}(\mathcal{K}_1, \mathcal{K}_2)$. According to Corollary 6.5.9, the following equality holds

$$\kappa^-(I - T^\sharp T) + \kappa^-(\mathcal{K}_2) = \kappa^-(I - TT^\sharp) + \kappa^-(\mathcal{K}_1). \tag{1.17}$$

This shows that in the case when either $\mathcal{K}_1$ or $\mathcal{K}_2$ is a Pontryagin space (of finite negative signatures) then the formula (1.11) is an immediate consequence of (1.17).

Corollary 12.1.10. *Let $T \in \mathcal{B}(K_1, \mathcal{K}_2)$ be quasi-contractive, fix a fundamental decomposition $\mathcal{K}_i = \mathcal{K}_i^+[+]\mathcal{K}_i^-$, $i = 1, 2$, and consider the representation (1.12). Then for any (Hilbert space) contraction $K \in \mathcal{B}(\mathcal{K}_1^-, \mathcal{K}_1^+)$, the operator $T_{21}K + T_{22}$ is semi-Fredholm, more precisely, it has closed range and finite dimensional kernel, and its index is*

$$\operatorname{ind}(T_{21}K + T_{22}) = \kappa^-(I - T^\sharp T) - \kappa^-(I - TT^\sharp). \tag{1.18}$$

Proof. Let $K \in \mathcal{B}(\mathcal{K}_1^-, \mathcal{K}_2^+)$ be a Hilbert space contraction and denote

$$\mathcal{M} = G(K) = \{Kx + x \mid x \in \mathcal{K}_1^-\}.$$

$\mathcal{M}$ is a negative subspace of the Kreĭn space $\mathcal{K}_1$ and

$$T\mathcal{M} = \{(T_{11}Kx + T_{12}x) + (T_{21}Kx + T_{22}x) \mid x \in \mathcal{K}_1^-\}$$

is a (closed) subspace of $\mathcal{K}_2$ such that $\kappa^+(T\mathcal{M}) \leqslant \kappa^-(I - T^\sharp T)$, cf. Corollary 12.1.5. It follows that

$$\operatorname{Ran}(T_{21}K + T_{22}) = J_2^- T\mathcal{M}.$$

Using a similar argument as in the proof of Corollary 12.1.5 we show that $T_{21}K + T_{22}$ has closed range. Let us consider the subspace of $\mathcal{K}_1$

$$\mathcal{L} = \{Kx + x \mid x \in \operatorname{Ker}(T_{21}K + T_{22})\}.$$

12.1 GEOMETRIC PROPERTIES OF QUASI-CONTRACTIONS

$\mathcal{L}$ is a subspace of $\mathcal{M}$ hence it is negative. From Corollary 12.1.5, $T\mathcal{L}$ is closed. Since $T\mathcal{L} \subseteq \mathcal{K}_2^+$ it follows that

$$\dim(T\mathcal{L}) = \kappa^+(T\mathcal{L}) \leqslant \kappa^+(T\mathcal{M}) \leqslant \kappa^-(I - T^\sharp T) < \infty.$$

On the other hand, from Proposition 12.1.2 and the negativity of $\mathcal{L}$ we obtain that $\mathrm{Ker}(T) \cap \mathcal{L}$ is finite dimensional hence, using the preceding inequality, it follows that $\mathcal{L}$ itself is finite dimensional. Since $\dim \mathcal{L} = \dim(\mathrm{Ker}(T_{21}K + T_{22}))$ we have thus proven that $\mathrm{Ker}(T_{21}K + T_{22})$ has finite dimension. We have proven that the operator $T_{21}K + T_{22}$ is semi-Fredholm.

The same argument as before shows that for any $\lambda \in [0, 1]$ the operator $\lambda T_{21}K + T_{22}$ is semi-Fredholm. Using the local stability of the index of semi-Fredholm operators it follows that $\mathrm{ind}(\lambda T_{21}K + T_{22})$ is independent of $\lambda \in [0, 1]$ hence, using Theorem 12.1.8,

$$\mathrm{ind}(T_{21}K + T_{22}) = \mathrm{ind}(T_{22}) = \kappa^-(I - T^\sharp T) - \kappa^-(I - TT^\sharp). \blacksquare$$

The preceding result enables us to add some new insight on the action of contraction operators on maximal negative subspaces (compare with Theorem 8.1.4).

Corollary 12.1.11. *Let $T \in \mathcal{B}(\mathcal{K}_1, \mathcal{K}_2)$ be a contraction and let $\mathcal{L}$ be a maximal negative subspace in $\mathcal{K}_1$. Then $T\mathcal{L}$ is a negative subspace of $\mathcal{K}_2$ such that*

$$\kappa^-((T\mathcal{L})^\perp) = \kappa^-(I - TT^\sharp).$$

Proof. Fix fundamental decompositions $\mathcal{K}_i = \mathcal{K}_i^+[+]\mathcal{K}_i^-$, $i = 1, 2$ and let K denote the angular operator of $\mathcal{L}$. We consider the representation (1.12). Since T is contractive, the operator $T_{21}K + T_{22}$ is injective hence, using (1.18) we get

$$\kappa^-(I - TT^\sharp) = -\mathrm{ind}(T_{21}K + T_{22}) = \mathrm{codim}_{\mathcal{K}_2^-}(\mathrm{Ran}(T_{21}K + T_{22})) = \kappa^-((T\mathcal{L})^\perp). \blacksquare$$

We conclude this section with a result concerning the product of quasi-contractions.

Corollary 12.1.12. *Let $T \in \mathcal{B}(\mathcal{K}_1, \mathcal{K}_2)$ and $S \in \mathcal{B}(\mathcal{K}_2, \mathcal{K}_3)$ be quasi-contractions. Then ST is also a quasi-contraction such that*

$$\kappa^-(I - (ST)^\sharp(ST)) \leqslant \kappa^-(I - T^\sharp T) + \kappa^-(I - S^\sharp S)$$

and the following formula holds

$$\kappa^-(I - (ST)^\sharp(ST)) - \kappa^-(I - (ST)(ST)^\sharp)$$
$$= \kappa^-(I - T^\sharp T) - \kappa^-(I - TT^\sharp) + \kappa^-(I - S^\sharp S) - \kappa^-(I - SS^\sharp). \quad (1.19)$$

Proof. The fact that ST is also quasi-contractive follows from

$$I - (ST)^\sharp ST = I - T^\sharp T + T^\sharp(I - S^\sharp S)T.$$

This proves also the inequality.

Let us now fix the fundamental decomposition $\mathcal{K}_i = \mathcal{K}_i^+[+]\mathcal{K}_i^-$, for $i = 1, 2, 3$, and consider the operator block matrix representations of S and T

$$T = \begin{bmatrix} T_{11} & T_{12} \\ T_{21} & T_{22} \end{bmatrix}, \quad S = \begin{bmatrix} S_{11} & S_{12} \\ S_{21} & S_{22} \end{bmatrix}.$$

We now consider the same kind of matrix representation of ST and notice that

$$(ST)_{22} = S_{21}T_{12} + S_{22}T_{22}. \tag{1.20}$$

For $\delta \in [0, 1]$ we consider the contractions $C_\delta \in \mathcal{B}(\mathcal{K}_2)$ defined by

$$C_\delta = \begin{bmatrix} \sqrt{1-\delta^2} & 0 \\ 0 & \sqrt{1+\delta^2} \end{bmatrix},$$

where the matrix of C_δ has to be understood with respect to the fundamental decomposition $\mathcal{K}_2 = \mathcal{K}_2^+[+]\mathcal{K}_2^-$. Then $T_\delta = C_\delta T$ is quasi-contractive and so is ST_δ. Applying Theorem 12.1.8 to the quasi-contractions ST_δ it follows that for all $\delta \in [0, 1]$ the operators

$$F_\delta = \sqrt{1-\delta^2} S_{21}T_{12} + \sqrt{1+\delta^2} S_{22}T_{22}$$

are semi-Fredholm and then using the local stability of the index of semi-Fredholm operators we obtain

$$\operatorname{ind} F_0 = \operatorname{ind} F_1,$$

hence, taking into account (1.20) we obtain

$$\operatorname{ind}((ST)_{22}) = \operatorname{ind}(S_{21}T_{12} + S_{22}T_{22}) = \operatorname{ind}(F_0)$$
$$= \operatorname{ind}(F_1) = \operatorname{ind}(S_{22}T_{22}) = \operatorname{ind}(S_{22}) + \operatorname{ind}(T_{22}).$$

Applying now the formula (1.19) for T, S, and ST we obtain the equation (1.19). ∎

Remark 12.1.13. It turns out that the index formula obtained in Theorem 12.1.8 enables us to add some new insight on the lifting theory of quasi-contractions. More precisely, let $T \in \mathcal{B}(\mathcal{K}_1, \mathcal{K}_2)$ be a quasi-contraction and let $\mathcal{K}_1'$ and $\mathcal{K}_2'$ be Kreĭn spaces. If $\mathcal{K}_1'$ and $\mathcal{K}_2'$ are Pontryagin spaces with finite negative ranks then for any quasi-contractive lifting $\widetilde{T} \in \mathcal{B}(\mathcal{K}_1[+]\mathcal{K}_1', \mathcal{K}_2[+]\mathcal{K}_2')$, that is, $P_{\mathcal{K}_2}\widetilde{T}|\mathcal{K}_1 = T$, we must have the equality

$$\kappa^-(I - \widetilde{T}^\sharp \widetilde{T}) - \kappa^-(I - \widetilde{T}\widetilde{T}^\sharp) = \kappa^-(I - T^\sharp T) - \kappa^-(I - TT^\sharp) + \kappa^-(\mathcal{K}_1') - \kappa^-(\mathcal{K}_2').$$

This formula tells us that some restrictions on various ranks of negativity are necessary when lifting quasi-contractions. ∎

12.2 Double Quasi-Contractions

As we have already mentioned, see Remark 12.1.9, for any operator T acting between Kreĭn spaces $\mathcal{K}_1$ and $\mathcal{K}_2$, the following relation holds:

$$\kappa^-(I - T^\sharp T) + \kappa^-(\mathcal{K}_2) = \kappa^-(I - TT^\sharp) + \kappa^-(\mathcal{K}_1).$$

Thus, in the case when $\mathcal{K}_1$ and $\mathcal{K}_2$ are Pontryagin spaces of finite negative signatures, this shows that the numbers $\kappa^-(I - T^\sharp T)$ and $\kappa^-(I - TT^\sharp)$ are simultaneously determined, in particular they are simultaneously finite or not. If $\kappa^-(\mathcal{K}_1)$ and $\kappa^-(\mathcal{K}_2)$ are infinite then it can be shown by examples that the numbers $\kappa^-(I - T^\sharp T)$ and $\kappa^-(I - TT^\sharp)$ can be fairly arbitrary.

An operator $T \in \mathcal{B}(\mathcal{K}_1, \mathcal{K}_2)$ is called a *double quasi-contraction* if both of T and $T^\sharp$ are quasi-contractions, that is, both of the numbers $\kappa^-(I - T^\sharp T)$ and $\kappa^-(I - TT^\sharp)$ are finite. As a first task we investigate conditions assuring the double quasi-contractivity of quasi-contractions.

Theorem 12.2.1. *Let $T \in \mathcal{B}(,\mathcal{K}_1, \mathcal{K}_2)$ be quasi-contractive. The following statements are equivalent.*

(i) *T is a double quasi-contraction.*

(ii) *For some (equivalently, for any) fundamental decomposition $\mathcal{K}_i = \mathcal{K}_i^+[+]\mathcal{K}_i^-$, $i = 1, 2$, the operator $P_{\mathcal{K}_2^-}T|\mathcal{K}_1^- \in \mathcal{B}(\mathcal{K}_1^-, \mathcal{K}_2^-)$ is Fredholm.*

(iii) *For some (equivalently, for any) maximal negative subspace $\mathcal{M}$ in $\mathcal{K}_1$, the number $\kappa^-((T\mathcal{M})^\perp)$ is finite.*

(iv) *For some (equivalently, for any) maximal negative subspace $\mathcal{L}$ in $\mathcal{K}_2$ the linear manifold $T^\sharp \mathcal{L}$ is closed and the numbers $\kappa^+(T^\sharp \mathcal{L})$ and $\dim(\operatorname{Ker}(T^\sharp) \cap \mathcal{L})$ are finite.*

Proof. (i)⇔(ii). Since T is quasi-contractive and, fixing fundamental decompositions $\mathcal{K}_i = \mathcal{K}_i^+[+]\mathcal{K}_i^-$, $i = 1, 2$, from Theorem 12.1.8 it follows that $\kappa^-(I - TT^\sharp)$ is finite if and only if the operator $P_{\mathcal{K}_2^-}T|\mathcal{K}_1^-$ is Fredholm.

(i)⇒(iii). Let $\mathcal{M}$ be a maximal negative subspace in $\mathcal{K}_1$ and $f \in (T\mathcal{M})^\perp$. Then $T^\sharp f \in \mathcal{M}^\perp$, hence $[T^\sharp f, T^\sharp f] \geqslant 0$ and this yields

$$[f, f] - [T^\sharp f, T^\sharp f] < 0,$$

provided f is assumed negative. This shows that

$$\kappa^-((T\mathcal{M})^\perp) \leqslant \kappa^-(I - TT^\sharp),$$

hence $\kappa^-((T\mathcal{M})^\perp)$ is finite.

(iii)⇒(ii). Let $\mathcal{M}$ be a maximal negative subspace of $\mathcal{L}_1$ such that $\kappa^-((T\mathcal{M})^\perp)$ is finite. Fix a fundamental decomposition $\mathcal{K}_i = \mathcal{K}_i^+[+]\mathcal{K}_i^-$, $i = 1, 2$ and consider the representation (1.12) of T. Let $K \in \mathcal{B}(\mathcal{K}_1^-, \mathcal{K}_1^+)$ be the angular operator of $\mathcal{M}$. Since

$\mathcal{K}_2^- \ominus \mathrm{Ran}(T_{21}K + T_{22})$ is a subspace in $(T\mathcal{M})^\perp$ it follows that $\mathrm{codim}_{\mathcal{K}_2^-}(\mathrm{Ran}(T_{21}K + T_{22}))$ is finite hence, using Corollary 12.1.10, the operator $T_{21}K + T_{22}$ is Fredholm. In particular, this means that $\mathrm{ind}(T_{21}K + T_{22})$ is finite. From (1.17) we obtain that $\kappa^-(I - TT^\sharp)$ is also finite.

(i)⇒(iv). Let $\mathcal{L}$ be a maximal negative subspace of $\mathcal{K}_2$. Since $T^\sharp$ is a quasi-contraction, from Proposition 12.1.2 it follows that $\mathrm{Ker}(T^\sharp) \cap \mathcal{L}$ is finite dimensional. Moreover, using Corollary 12.1.5 we obtain that $T^\sharp\mathcal{L}$ is closed and $\kappa^+(T^\sharp\mathcal{L}) \leqslant \kappa^-(I - TT^\sharp) < \infty$. The fact that $\kappa^-((T^\sharp\mathcal{L})^\perp)$ is also finite follows from the proof of (i)⇒(iii) replacing T by $T^\sharp$ and $\mathcal{M}$ by $\mathcal{L}$.

(iv)⇒(ii). Let $\mathcal{L}$ be as stated at (iv). Fix a fundamental decomposition $\mathcal{K}_1 = \mathcal{K}_1^+[+]\mathcal{K}_1^-$. Since T is quasi-contractive $T\mathcal{K}_1^-$ is a pseudo-regular subspace of $\mathcal{K}_2$ such that $\kappa^0(T\mathcal{K}_1^-) + \kappa^+(T\mathcal{K}_1^-) \leqslant \kappa^-(I - T^\sharp T) < \infty$, see Corollary 12.1.5, hence

$$T\mathcal{K}_1^- = \mathcal{S}^-[+]\mathcal{S}^0[+]\mathcal{S}^+,$$

where $\mathcal{S}^-$ is uniformly negative and $\dim \mathcal{S}^0 + \dim \mathcal{S}^+ \leqslant \kappa^-(I - T^\sharp T)$. Choose a fundamental decomposition $\mathcal{K}_2 = \mathcal{K}_2^+[+]\mathcal{K}_2^-$ such that $\mathcal{S}^- \subseteq \mathcal{K}_2^-$ and $\mathcal{S}^+ \subseteq \mathcal{K}_2^+$ (this is possible by Theorem 2.2.9) and represent T as in (3.9) with respect to these fundamental decompositions. Then

$$\mathrm{rank}(T_{12}) \leqslant \dim \mathcal{S}^0 + \dim \mathcal{S}^+ \leqslant \kappa^-(I - T^\sharp T) < \infty.$$

We consider now the angular operator L of $\mathcal{L}$

$$\mathcal{L} = \{x + Lx \mid x \in \mathcal{K}_2^-\}$$

and notice that $T^\sharp$ has the representation

$$T^\sharp = \begin{bmatrix} T_{11}^* & -T_{21}^* \\ -T_{12}^* & T_{22}^* \end{bmatrix},$$

hence

$$T^\sharp \mathcal{L} = \{(T_{11}^* Lx - T_{21}^* x) + (-T_{12}^* Lx + T_{22}^* x) \mid x \in \mathcal{K}_2^-\}.$$

Since $T^\sharp\mathcal{L}$ is closed and $\kappa^+(T^\sharp\mathcal{L})$ is finite, from here we infer (e.g. as in the proof of Corollary 12.1.10) that $\mathrm{Ran}(-T_{12}^*L + T_{22}^*)$ is closed. Since $\kappa^+(T^\sharp\mathcal{L})$ and $\dim(\mathrm{Ker}(T^\sharp) \cap \mathcal{L})$ are finite it follows that $\mathrm{Ker}(-T_{12}^*L + T_{22}^*)$ is finite dimensional. We have thus proved that $-T_{12}^*L + T_{22}^*$ is semi-Fredholm and

$$\mathrm{ind}\, T_{22}^* = \mathrm{ind}(-T_{12}^*L + T_{22}^*) < \infty,$$

where we took into account that T_{12} has finite rank. Since we already know that $\mathrm{ind}\, T_{22} < \infty$, from Theorem 12.1.8, it follows that T_{22} is a Fredholm operator. ■

12.2 Double Quasi-Contractions

Remark 12.2.2. As a consequence of the equivalence of (i) and (iv) in Theorem 12.2.1 we obtain a different proof of Theorem 8.3.4, more precisely, letting $T \in \mathcal{B}(\mathcal{K}_1, \mathcal{K}_2)$ be a contraction, then T is double contraction if and only if $T^\sharp$ maps some maximal negative subspace of $\mathcal{K}_2$ in a one-to-one way onto a negative subspace of $\mathcal{K}_1$.

Indeed, one implication is clear. Conversely, assume that $\mathcal{L}$ is a maximal negative subspace of $\mathcal{K}_2$ which is mapped by $T^\sharp$ in a one-to-one way into a negative subspace of $\mathcal{K}_1$. This means that $T^\sharp \mathcal{L}$ is closed and $\kappa^+(T^\sharp \mathcal{L}) = \dim(\operatorname{Ker}(T^\sharp) \cap \mathcal{L}) = 0$. From Theorem 12.2.1 it follows that $\kappa^-(I - TT^\sharp)$ is finite. Thus

$$\operatorname{ind} T_{22} = -\kappa^-(I - TT^\sharp) \leqslant 0.$$

On the other hand, with notation as in the proof of (iv)$\Rightarrow$(ii) of Theorem 12.2.1, from $\kappa^+(T^\sharp \mathcal{L}) = \dim(\operatorname{Ker}(T^\sharp) \cap \mathcal{L}) = 0$ it follows that $\operatorname{Ker}(-T_{12}^* L + T_{22}^*) = 0$ hence

$$\operatorname{ind} T_{22} = -\operatorname{ind} T_{22}^* = -\operatorname{ind}(-T_{12}^* L + T_{22}^*) \geqslant 0.$$

From these two inequalities we obtain $\kappa^-(I - TT^\sharp) = 0$. ∎

The following theorem estimates the gap between quasi-contractions and double quasi-contractions, in terms of the codomain.

Theorem 12.2.3. *Let $T \in \mathcal{B}(\mathcal{K}_1, \mathcal{K}_2)$ be a quasi-contraction and denote $\kappa = \kappa^-(I - T^\sharp T)$. Then there exists a regular subspace $\mathcal{K}$ of $\mathcal{K}_2$ such that $\kappa^+(\mathcal{K}^\perp) \leqslant \kappa$, the operator $\widehat{T} = P_\mathcal{K} T \in \mathcal{B}(\mathcal{K}_1, \mathcal{K})$ is a double quasi-contraction,*

$$\kappa^-(I - \widehat{T}^\sharp \widehat{T}) \leqslant \kappa^-(I - \widehat{T}\widehat{T}^\sharp) \leqslant 2\kappa,$$

and

$$\kappa^-(I - \widehat{T}\widehat{T}^\sharp) - \kappa^-(I - \widehat{T}^\sharp \widehat{T}) \leqslant \kappa.$$

Proof. We use Lemma 12.1.4 to produce a contractive operator $\widetilde{T}\colon \mathcal{K}_1 \to \mathcal{K}_2[+]\mathcal{K}_2'$, where $\dim \mathcal{K}_2' = \kappa^-(\mathcal{K}_2') = \kappa$ and $P_{\mathcal{K}_2}\widetilde{T} = T$. Then we apply Theorem 8.3.3 to produce a uniformly negative subspace $\mathcal{L}$ of $\mathcal{K}_2[+]\mathcal{K}_2'$ such that, denoting $\mathcal{R} = \mathcal{L}^\perp$, the operator $P_\mathcal{R}\widetilde{T}\colon \mathcal{K}_1 \to \mathcal{R}$ is double contractive.

We claim that $\mathcal{R}' = \mathcal{R} + \mathcal{K}_2'$ is a regular subspace with uniformly negative orthogonal companion. Indeed, $\mathcal{R}'$ is closed since $\mathcal{K}_2'$ is finite dimensional, and since $\mathcal{R}'^\perp = \mathcal{R}^\perp \cap \mathcal{K}_2'^\perp \subset \mathcal{L}$ it follows that $\mathcal{R}'^\perp$ is uniformly negative, in particular it is regular, hence $\mathcal{R}'$ is also regular.

Letting now $\overline{T} = P_{\mathcal{R}'}\widetilde{T}\colon \mathcal{K}_1 \to \mathcal{R}'$ we obtain a double quasi-contraction such that

$$\kappa^-(I - \overline{T}^\sharp \overline{T}) \leqslant \kappa^-(I - \overline{T}\,\overline{T}^\sharp) \leqslant \kappa,$$

and

$$\kappa^-(I - \overline{T}\,\overline{T}^\sharp) - \kappa^-(I - \overline{T}^\sharp \overline{T}) \leqslant \kappa.$$

We define now $\mathcal{K}$ as the orthogonal complement of $\mathcal{K}_2'$ with respect to $\mathcal{R}'$, that is, $\mathcal{R}' = \mathcal{K}[+]\mathcal{K}_2'$. Then $\kappa^+(\mathcal{K}^\perp) \leqslant \kappa$. Denoting $\widehat{T} = P_\mathcal{K}\overline{T}$ it follows that $\widehat{T} = P_\mathcal{K}\widetilde{T} = P_\mathcal{K}T$ and then the formulae and the inequalities for the signatures of defect of $\widehat{T}$ follow. ∎

Other results concerning extensions of double quasi-contractions will be proved later in this chapter. Here we only record a consequence of the existence of the elementary rotation which gives an estimation of the gap between double quasi-contractions and double contractions.

Theorem 12.2.4. *Let $T \in \mathcal{B}(\mathcal{K}_1, \mathcal{K}_2)$ be a double quasi-contraction and let also $\mathcal{K}_1'$ and $\mathcal{K}_2'$ be Kreĭn spaces. Then there exists a double contraction $\widetilde{T} \in \mathcal{L}(\mathcal{K}_1[+]\mathcal{K}_1', \mathcal{K}_2[+]\mathcal{K}_2')$ such that*

$$P_{\mathcal{K}_2}\widetilde{T}|\mathcal{K}_1 = T \tag{2.1}$$

if and only if

$$\kappa^-(I - TT^\sharp) \leqslant \kappa^-(\mathcal{K}_1'), \quad \kappa^-(I - T^\sharp T) \leqslant \kappa^-(\mathcal{K}_2'), \tag{2.2}$$

and

$$\kappa^-(I - T^\sharp T) + \kappa^-(\mathcal{K}_1') = \kappa^-(I - TT^\sharp) + \kappa^-(\mathcal{K}_2'). \tag{2.3}$$

Proof. Let $\widetilde{T} \in \mathcal{L}(\mathcal{K}_1[+]\mathcal{K}_1', \mathcal{K}_2[+]\mathcal{K}_2')$ be a double contraction such that (2.1) holds. Then, the inequalities (2.2) are consequences of Theorem 12.2.3. In order to prove (2.3) we fix fundamental decompositions on $\mathcal{K}_1, \mathcal{K}_2, \mathcal{K}_1'$ and $\mathcal{K}_2'$. Then consider the operators

$$\widetilde{T}_{22} = P_{\mathcal{K}_2^-[+]\mathcal{K}_2'^-}\widetilde{T}|\mathcal{K}_1^-[+]\mathcal{K}_1'^-, \quad T_{22} = P_{\mathcal{K}_2^-}T|\mathcal{K}_1^-.$$

From (2.1) we have, with respect to the decompositions $\mathcal{K}_1^-[+]\mathcal{K}_1'^-$ and $\mathcal{K}_2^-[+]\mathcal{K}_2'^-$, the following representation

$$\widetilde{T}_{22} = \begin{bmatrix} T_{22} & X \\ Y & Z \end{bmatrix}.$$

Since $\widetilde{T}_{22}$ is invertible, in particular it is a Fredholm operator of index 0, and taking into account that T_{22} is a Fredholm operator of index

$$\mathrm{ind}(T_{22}) = \kappa^-(I - T^\sharp T) - \kappa^-(I - TT^\sharp),$$

we easily obtain the formula (2.3).

In order to prove the converse implication we first produce a certain doubly contractive extension of T. To this end let $R(T) \in \mathcal{B}(\mathcal{K}_1[+]\mathcal{D}_{T^*}, \mathcal{K}_2[+]\mathcal{D}_T)$ be the canonical elementary rotation of T, see Theorem 6.5.8. $R(T)$ is a unitary lifting of T, in particular it is a double contraction.

Let now $\mathcal{K}_1'$ and $\mathcal{K}_2'$ be Kreĭn spaces such that (2.2) and (2.3) hold. Since (2.2) holds there exists a contraction $\Gamma_2 \in \mathcal{L}(\mathcal{D}_T, \mathcal{K}_2')$ and also there exists $\Gamma_1 \in \mathcal{L}(\mathcal{K}_1', \mathcal{D}_{T^*})$ such that $\Gamma_1^\sharp$ is a contraction. Then we have

$$\kappa^-(I - TT^\sharp) + \kappa^-(I - \Gamma_1^\sharp \Gamma_1) = \kappa^-(\mathcal{K}_1'),$$

and
$$\kappa^-(I - T^\sharp T) + \kappa^-(I - \Gamma_2 \Gamma_2^\sharp) = \kappa^-(\mathcal{K}_2'),$$
hence, using (2.3) it follows that there exists a double contraction $\Gamma \in \mathcal{B}(\mathcal{D}_{\Gamma_1}, \mathcal{D}_{\Gamma_2^*})$. We consider now the operators

$$\widehat{\Gamma_1} \in \mathcal{B}(\mathcal{K}_1[+]\mathcal{K}_1', \mathcal{K}_1[+]\mathcal{D}_{T^*}), \quad \widehat{\Gamma_1} = \begin{bmatrix} I & 0 \\ 0 & \Gamma_1 \end{bmatrix},$$

$$\widehat{\Gamma_2} \in \mathcal{B}(\mathcal{K}_2[+]\mathcal{D}_T, \mathcal{K}_2[+]\mathcal{K}_2'), \quad \widehat{\Gamma_2} = \begin{bmatrix} I & 0 \\ 0 & \Gamma_2 \end{bmatrix},$$

and
$$\widehat{\Gamma} \in \mathcal{B}(\mathcal{K}_1[+]\mathcal{D}_{\Gamma_1}, \mathcal{K}_2[+]\mathcal{D}_{\Gamma_2^*}), \quad \widehat{\Gamma} = \begin{bmatrix} 0 & 0 \\ 0 & \Gamma \end{bmatrix}.$$

Using these we define
$$\widetilde{T} = \widehat{\Gamma_2} R(T) \widehat{\Gamma_1} + D_{\widehat{\Gamma_2^*}} \widehat{\Gamma} D_{\widehat{\Gamma_1}}.$$

It follows that $\widetilde{T}$ is a double contraction (calculations are similar to those performed during the proof of Theorem 2.2.7, and we leave them to the reader) and clearly it satisfies the condition (2.1). ∎

12.3 Polar Decompositions of Contractions

In this section we generalise the polar decompositions of operators in Hilbert spaces to the class of contractions in Kreĭn spaces. To start with, we are looking for an appropriate notion of modulus. Given an operator $T \in \mathcal{B}(\mathcal{K}_1, \mathcal{K}_2)$, a *generalised modulus* of T is a selfadjoint operator $R \in \mathcal{B}(\mathcal{K}_1)$ subject to the following conditions:

(1) $\sigma(R) \subseteq [0, +\infty)$,
(2) $R^2 = T^\sharp T$,
(3) $\operatorname{Ker}(R) = \operatorname{Ker}(T^\sharp)$.

Lemma 12.3.1. *If T is a double contraction then $\sigma(T^\sharp T) \subset [0, +\infty)$.*

Proof. Let $A = T^\sharp T$. Since $I - T^\sharp T$ is positive, it follows that the polynomial $p(\lambda) = 1 - \lambda$ is definitising for A. From the spectral theory of selfadjoint definitising operators it follows that $\sigma(A) \subseteq \mathbb{R}$.

On the other hand,
$$A(I - A) = T^\sharp T(I - T^\sharp T) = T^\sharp (I - TT^\sharp)T$$

and hence, since $T^\sharp$ is a contraction, that is, $I - TT^\sharp$ is positive, it follows that $A(I - A)$ is a positive operator. Therefore, the polynomial $q(\lambda) = \lambda(1 - \lambda)$ is also definitising for A and hence, all the spectral points in $(-\infty, 0)$ are of negative type. Therefore, the operator $A = T^\sharp T$ has no spectrum on the negative semi-axis, that is, $\sigma(T^\sharp T) \subset [0, +\infty)$. ∎

Remark 12.3.2. If T is a contraction and R is a selfadjoint operator satifying the conditions (1) and (2), then the condition (3) is automatically satisfied, that is, R is a generalised modulus of T.

Indeed, one inclusion is clear, $\mathrm{Ker}(R) \subseteq \mathrm{Ker}(R^2) = \mathrm{Ker}(T^\sharp T)$. To prove the other inclusion, as in the proof of Lemma 12.3.1, we have that $\mathrm{Ker}(T^\sharp T)$ is a uniformly positive subspace. Let $x \in \mathrm{Ker}(T^\sharp T) = \mathrm{Ker}(R^2)$. Then $Rx \in \mathrm{Ker}(R)$ and hence, $0 = [T^\sharp Tx, x] = [Rx, Rx]$, that is, Rx is a neutral vector. Since $\mathrm{Ker}(R)$ is a uniformly positive subspace we get $Rx = 0$ and hence, $x \in \mathrm{Ker}(R)$. ∎

Theorem 12.3.3. *Let $T \in \mathcal{B}(\mathcal{K}_1, \mathcal{K}_2)$ be a contraction such that $\sigma(T^\sharp T) \subseteq [0, +\infty)$. Then T admits a uniquely determined generalised modulus.*

Proof. If T is a contraction and, letting $A = T^\sharp T$, we have $\sigma(A) \subset [0, +\infty)$, as in the proof of Lemma 12.3.1, it follows that 1 is the only possible critical point of the selfadjoint definitisable operator A, all the spectral points in $[0, 1)$ are of positive type and all the spectral points in $(1, +\infty)$ are of negative type.

Let E denote the spectral function of A and fix $0 < \varepsilon < 0$. On the positive definite Kreĭn space (that is, the Hilbert space) $E([0,\varepsilon))\mathcal{K}_1$ the operator $A_1 = A|E([0,\varepsilon))\mathcal{K}_1$ is positive and hence, there exists a unique positive operator R_1 such that $R_1^2 = A_1$.

Consider the selfadjoint operator $A_2 = A|E([\varepsilon, +\infty))\mathcal{K}_1$ on $E([\varepsilon, +\infty))\mathcal{K}_1$. Since $\sigma(A_2) \subset [\varepsilon, +\infty)$, applying the holomorphic functional calculus for the square root function, it follows that there exists $R_2 = R_2^\sharp \in \mathcal{B}(E([\varepsilon, +\infty))\mathcal{K}_1)$ such that $R_2^2 = A_2$. We define $R = R_1 \oplus R_2$ and it follows that R is a generalised modulus of T.

To prove the uniqueness of the generalised modulus, let S be another generalised modulus of T. Then R and S commute with A and hence, they commute with its spectral function E. Therefore, for $0 < \varepsilon < 1$ we have

$$R^2 E([\varepsilon, +\infty)) = S^2 E([\varepsilon, +\infty)) = AE([\varepsilon, +\infty)). \tag{3.1}$$

Since $\sigma(A|E([\varepsilon, +\infty))\mathcal{K}_1) \subseteq [\varepsilon, +\infty)$, applying the holomorphic functional calculus for the principal branch of the square root function, from (3.1) we get

$$E([\varepsilon, +\infty)) = SE([\varepsilon, +\infty)).$$

Since ε is arbitrary, with $0 < \varepsilon < 1$, from here we get

$$RE((0, +\infty)) = SE((0, +\infty)).$$

On the other hand, since on the remaining subspace $\mathrm{Ker}(R) = \mathrm{Ker}(S) = \mathrm{Ker}(A) = E(\{0\})\mathcal{K}_1$, we conclude that $R = S$. ∎

Given $T \in \mathcal{B}(\mathcal{K}_1, \mathcal{K}_2)$, a *generalised polar decomposition* of T is, by definition, a factorisation $T = UR$ where, R is a generalised modulus of T and $U \in \mathcal{B}(\mathcal{K}_1, \mathcal{K}_2)$ is a partial isometry with $\mathrm{Clos}\,\mathrm{Ran}(T) = \mathrm{Ran}(U^\sharp)$.

12.3 Polar Decompositions of Contractions

Theorem 12.3.4. *A contraction $T \in \mathcal{B}(\mathcal{K}_1, \mathcal{K}_2)$ has a generalised polar decomposition if and only if $\sigma(T^\sharp T) \subseteq [0, +\infty)$ and $\operatorname{Ker}(T^\sharp)$ is a regular subspace. In addition, in this case, the generalised polar decomposition is unique.*

Proof. Let $T = UR$ be a generalised polar decomposition of T. In particular, $T^\sharp T = R^2$ and, since $\sigma(R) \subseteq [0, +\infty)$, it follows that $\sigma(T^\sharp T) = \sigma(R^2) \subseteq [0, +\infty)$.

On the other hand, as in the proof of Lemma 12.3.1, $\operatorname{Ker}(R) = \operatorname{Ker}(T^\sharp T)$ is a uniformly positive subspace and hence $\operatorname{Clos}\operatorname{Ran}(T) = (\operatorname{Ker}(R))^\perp$ is a regular subspace. Then $\operatorname{Clos}\operatorname{Ran}(T) = U \operatorname{Clos}\operatorname{Ran}(R) = U \operatorname{Ran}(U^\sharp) = \operatorname{Ran}(U)$ is a regular subspace and hence $\operatorname{Ker}(T^\sharp) = \operatorname{Ran}(T)^\perp$ is regular.

Conversely, assume that $\sigma(T^\sharp T) \subseteq [0, +\infty)$ and $\operatorname{Ker}(T^\sharp)$ is a regular subspace of $\mathcal{K}_2$. By Theorem 12.3.3 there exists a generalised modulus R of T. We define an operator $V \colon \operatorname{Ran}(T) \to \mathcal{K}_2$ by
$$VTx = Rx, \quad x \in \mathcal{K}_1.$$
Since $\operatorname{Ker}(T) \subseteq \operatorname{Ker}(T^\sharp T) = \operatorname{Ker}(R)$ it follows that V is correctly defined. For all $x, y \in \mathcal{K}_1$ we have
$$[VTx, VTy] = [Rx, Ry] = [R^2 x, y] = [T^\sharp Tx, y] = [Tx, Ty],$$
and hence V is isometric. We claim that V is bounded.

In order to prove the claim we consider E the spectral function of the selfadjoint definitisable operator $A = T^\sharp T$ and, for $0 < \varepsilon < 1$, split the space $\mathcal{K}_1 = \mathcal{H}_1 [+] \mathcal{H}_2$, where $\mathcal{H}_1 = E([0, \varepsilon))\mathcal{K}_1$ and $\mathcal{H}_2 = E([\varepsilon, +\infty))\mathcal{K}_1$. Note that both subspaces $E([0, \varepsilon))\mathcal{K}_1$ and $E([\varepsilon, +\infty))\mathcal{K}_1$ are invariant under R. Therefore,
$$[Tx_1, Tx_2] = [Rx_1, Rx_2], \quad x_1 \in \mathcal{H}_1, \; x_2 \in \mathcal{H}_2,$$
and hence
$$T\mathcal{H}_1 \perp T\mathcal{H}_2. \tag{3.2}$$
Let $R_1 = R|E([0, \varepsilon))\mathcal{K}_1$ and $R_2 = R|E([\varepsilon, +\infty))\mathcal{K}_1$. As in the proof of Thorem 12.3.3 we have $\sigma(R_2) \subseteq [\varepsilon, +\infty)$ and hence R_2 is boundedly invertible on $\mathcal{H}_2$. Clearly, we have the decompositions
$$\operatorname{Ran}(R) = \operatorname{Ran}(R_1) \dot{+} \operatorname{Ran}(R_2), \quad \operatorname{Ran}(T) = T\mathcal{H}_1 \dot{+} T\mathcal{H}_2. \tag{3.3}$$

As in the proof of Lemma 12.3.1, the space $\mathcal{H}_1 = E([0, \varepsilon))\mathcal{K}_1$ is uniformly positive and hence, we can choose a unitary norm $\|\cdot\|$ on $\mathcal{K}_1$ such that $\|x\|^2 = [x, x]$, $x \in E([0, \varepsilon))\mathcal{K}_1$. Therefore,
$$\|Rx\|^2 = [Rx, Rx] = [Tx, Tx] \leqslant \|Tx\|^2, \quad x \in \mathcal{H}_1,$$
and $V|T\mathcal{H}_1$ is bounded. This implies that $\operatorname{Clos} T\mathcal{H}_1$ is a uniformly positive subspace and, by (3.2), it follows that
$$\operatorname{Clos} R(T) = \operatorname{Clos} T\mathcal{H}_1 [+] \operatorname{Clos} T\mathcal{H}_2.$$

Note that, since $\operatorname{Ker}(T^\sharp)$ is regular so is $\operatorname{Clos} R(T) = (\operatorname{Ker}(T^\sharp))^\perp$ and hence the subspace $\operatorname{Clos} T\mathcal{H}_2$ is also regular.

Consider now the selfadjoint operator $R_2 = R|E([\varepsilon, +\infty))\mathcal{K}_1$ onto the Kreĭn space $E([\varepsilon, +\infty))\mathcal{K}_1$. As in the proof of Theorem 12.3.3, we have $\sigma(R_2) \subseteq [\varepsilon, +\infty)$ and hence R_2 is boundedly invertible. Therefore, the operator $TR_2^{-1}|E([\varepsilon, +\infty))\mathcal{K}_1$ is bounded, isometric and its range is the regular subspace $\operatorname{Clos} T\mathcal{H}_2$. This implies that its inverse is also a bounded isometry and thus, $V|T\mathcal{H}_2$ is bounded.

To conclude that V is bounded, we use the decompositions (3.3), with the observation that all the closures of subspaces from (3.3) are regular and mutually orthogonal. Finally, we extend V to a partial isometry, also denoted by V, in such a way that $\operatorname{Ker}(V) = \operatorname{Clos} \operatorname{Ran}(T)$. Then $U = V^\sharp$ has all the necessary properties to produce the generalised polar decomposition $T = UR$.

In order to prove the uniqueness of the generalised polar decomposition, note that the uniqueness of the generalised modulus R was proved in Theorem 12.3.3 and that the requirement $\operatorname{Clos} \operatorname{Ran}(T) = \operatorname{Ran}(U^\sharp)$ determines uniquely the partial isometry such that $T = WR$. ∎

Corollary 12.3.5. *If T is a double contraction then it has a generalised polar decomposition.*

Proof. If T is a double contraction then, by Lemma 12.3.1, we have that $\sigma(T^\sharp T) \subseteq [0, +\infty)$ and by Lemma 8.1.1 we have that $\operatorname{Ker}(T^\sharp)$ is a uniformly positive subspace. We apply Theorem 12.3.4 and conclude that T admits generalised polar decomposition. ∎

As a consequence of Theorem 12.3.4 we can obtain a spectral characterisation of double contractions within the class of contractions.

Corollary 12.3.6. *Let $T \in \mathcal{B}(\mathcal{K}_1, \mathcal{K}_2)$ be a contraction. Then T is a double contraction if and only if $\sigma(T^\sharp T) \subseteq [0, +\infty)$ and $\operatorname{Ker}(T^\sharp)$ is a uniformly positive subspace.*

Proof. One implication is proven as in the proof of Corollary 12.3.5. Conversely, assume that $\sigma(T^\sharp T) \subseteq [0, +\infty)$ and $\operatorname{Ker}(T^\sharp)$ is a uniformly positive subspace. By Theorem 12.3.4 it follows that there exists the polar decomposition $T = UR$. Then $T^\sharp = RU^\sharp$ and $\operatorname{Ran}(U)^\perp = \operatorname{Ker}(T^\sharp)$. Then, for all $x \in \mathcal{K}_2$ we have

$$[T^\sharp x, T^\sharp x] = [RU^\sharp x, RU^\sharp x] \leqslant [U^\sharp x, U^\sharp x] = [UU^\sharp x, x]$$

and

$$[P_{\operatorname{Ran}(U)} x, x] = [x, x] = [P_{\operatorname{Ker}(T^\sharp)} x, x] \leqslant [x, x],$$

hence $T^\sharp$ is contractive. ∎

12.4 A Spectral Characterisation of Double Quasi-Contractions

In this section we will obtain a characterisation of double quasi-contractions T in terms of spectral properties of the selfadjoint operator $T^\sharp T$, generalising the characterisation in

12.4 A Spectral Characterisation of Double Quasi-Contractions

Corollary 12.3.6. Firstly we specify what kind of spectral conditions one can expect in this case.

Lemma 12.4.1. *Let $T \in \mathcal{B}(\mathcal{K}_1, \mathcal{K}_2)$ be a double quasi-contraction.*

(i) $\sigma(T^\sharp T) \setminus \mathbb{R}_+$ *is a finite set of points, all of them eigenvalues of finite multiplicities (i.e. the corresponding root subspaces have finite dimensions).*

(ii) *The total spectral multiplicity corresponding to the set $\sigma(T^\sharp T) \setminus \mathbb{R}_+$ and the operator $T^\sharp T$ is upper bounded as follows*

$$\operatorname{rank} E(\{\lambda \in \sigma(T^\sharp T) \setminus \mathbb{R}_+ \mid \operatorname{Im}\lambda > 0\}; T^\sharp T) \leqslant \kappa^-(I - T^\sharp T) + \kappa^-(I - TT^\sharp). \quad (4.1)$$

Proof. According to Proposition 11.3.6 the operator $T^\sharp T$ is definitisable and denoting by E its spectral function then, from Lemma 12.1.1, it follows that for all $\varepsilon > 0$ and sufficiently small we have

$$\kappa^-(I - T^\sharp T) = \kappa^+(E(1+\varepsilon, +\infty)) + \kappa^-(E(-\infty, 1-\varepsilon))$$
$$+ \kappa^-((I - T^\sharp T)|E(1-\varepsilon, 1+\varepsilon)\mathcal{K}_1) + \sum_{\operatorname{Im}\lambda > 0} \operatorname{rank} E(\lambda; T^\sharp T).$$

Therefore,

$$E(1-\varepsilon, 1+\varepsilon)\mathcal{K}_1 + \sum_{\operatorname{Im}\lambda > 0} \operatorname{rank} E(\lambda; T^\sharp T) \leqslant \kappa^-(I - T^\sharp T). \quad (4.2)$$

On the other hand, again by Lemma 12.1.1, we obtain that, for any $\varepsilon, \delta > 0$ sufficiently small,

$$\kappa^-(T^\sharp T(I - T^\sharp T)) = \kappa^-(T^\sharp T(I - T^\sharp T)|E(-\delta, 1+\varepsilon)\mathcal{K}_1) + \sum_{\operatorname{Im}\lambda > 0} \operatorname{rank}(E(\lambda; T^\sharp T))$$
$$+ \kappa^-(E(1+\varepsilon, +\infty)\mathcal{K}_1) + \kappa^+(E(-\infty, -\delta)\mathcal{K}_1).$$

Taking into account that

$$T^\sharp T(I - T^\sharp T) = T^\sharp T - T^\sharp TT^\sharp T = T^\sharp(I - TT^\sharp)T,$$

it follows that

$$\kappa^-(T^\sharp T(I - T^\sharp T)) \leqslant \kappa^-(I - TT^\sharp),$$

and hence

$$\sum_{\operatorname{Im}\lambda > 0} \operatorname{rank} E(\lambda; T^\sharp T) + \kappa^+(E(-\infty, -\delta)\mathcal{K}_1) \leqslant \kappa^-(I - TT^\sharp). \quad (4.3)$$

From (4.2) and (4.3) we obtain the inequality (4.1) and, in particular, that $\sigma(T^\sharp T) \setminus \mathbb{R}_+$ is a finite set of points, all of them eigenvalues of finite multiplicities. ∎

Theorem 12.4.2. *Let $T \in \mathcal{B}(\mathcal{K}_1, \mathcal{K}_1)$ be quasi-contractive. Then, in order for T to be doubly quasi-contractive, it is necessary and sufficient that T satisfy the following conditions.*

(i) *$\sigma(T^\sharp T) \setminus \mathbb{R}_+$ is a finite set of points, all of them eigenvalues of finite multiplicities.*

(ii) *$\mathrm{Ker}(T^\sharp)$ is a pseudo-regular subspace such that $\kappa^-(\mathrm{Ker}(T^\sharp))$ and $\kappa^0(\mathrm{Ker}(T^\sharp))$ are finite.*

Proof. Assuming that T is doubly quasi-contractive, the assertion (i) was already proven in Lemma 12.4.1 while the assertion (ii) is a consequence of Proposition 12.1.2.

Conversely, let us assume that the quasi-contraction T satisfies both of the conditions (i) and (ii). In order to prove that $T^\sharp$ is also a quasi-contraction, discarding some finite dimensional subspaces, we will obtain a certain left polar decomposition of T. We divide the proof into several steps.

Step 1. *Without restricting the generality, we can assume that there exists $R \in \mathcal{B}(\mathcal{K}_1)$, $R^\sharp = R$ such that*
(a) $T^\sharp T = R^2$,
(b) $\sigma(R) \subseteq [0, +\infty)$,
(c) $\mathrm{Ker}(R) = \mathrm{Ker}(T^\sharp T)$ *is uniformly positive.*

Indeed, let E_0 be the spectral projection of $T^\sharp T$ corresponding to the spectral set $\sigma(T^\sharp T) \setminus \mathbb{R}_+$. Then E_0 has finite rank. Let us denote by $A \in \mathcal{B}((I - E_0)\mathcal{K}_1)$ the operator
$$A = T^\sharp T | (I - E_0)\mathcal{K}_1.$$
Then $\sigma(A) \subseteq \mathbb{R}_+$ and since $T^\sharp T$ is definitisable, so is A. Letting E denote the spectral function of A we choose $\varepsilon > 0$ sufficiently small such that the closed interval $[0, \varepsilon]$ contains no critical points of A, except (possibly) 0. We denote
$$A_1 = E([0, \varepsilon])A, \quad A_2 = E((\varepsilon, +\infty))A,$$
as operators acting on appropriate spectral subspaces.

The operator A_2 is selfadjoint, $\sigma(A_2) \subseteq [\varepsilon, +\infty)$ hence, by the Riesz–Dunford functional calculus, there exists its square root, $R_2 = A_2^{1/2}$, $R_2 = R_2^\sharp$ and $\sigma(R_2) \subseteq [\varepsilon^{1/2}, +\infty)$.

The operator A_1 is selfadjoint definitisable, $\sigma(A_1) \subseteq [0, \varepsilon]$ and $c(A_1) \subseteq \{0\}$. Let $\mathcal{N}$ be a neutral subspace in $E([0, \varepsilon])\mathcal{K}_1$ which is maximal A_1-invariant. Then $\mathcal{N}$ is finite dimensional since $\kappa^-(E([0, \varepsilon]))$ is finite. For a fixed fundamental symmetry J on the Krein space $E([0, \varepsilon])\mathcal{K}_1$, the subspace $\mathcal{N} \dotplus J\mathcal{N}$ is regular hence there exists a regular subspace $\widehat{\mathcal{K}}_1$ such that
$$E([0, \varepsilon])\mathcal{K}_1 = \mathcal{N} \dotplus \widehat{\mathcal{K}}_1 \dotplus J\mathcal{N}$$
and, with respect to this decomposition, A_1 has the representation
$$A_1 = \begin{bmatrix} * & * & * \\ 0 & \widehat{A}_1 & * \\ 0 & 0 & * \end{bmatrix}$$

12.4 A SPECTRAL CHARACTERISATION OF DOUBLE QUASI-CONTRACTIONS 345

where $\widehat{A_1} \in \mathcal{B}(\widehat{\mathcal{K}}_1)$ is selfadjoint definitisable with no critical points at all. It follows that $\widehat{A_1}$ is similar with a selfadjoint operator on a Hilbert space, hence there exists its square root $R_1 = \widehat{A_1}^{1/2}$, R_1 is selfadjoint and $\sigma(R_1) \subseteq [0, \varepsilon^{1/2}]$. Let now P denote the orthogonal projection of $\mathcal{K}_1$ onto the regular subspace

$$E(\sigma(T^\sharp T) \setminus \mathbb{R}_+; T^\sharp T)\mathcal{K}_1[+](\mathcal{N} \dot{+} J\mathcal{N})[+]\mathcal{S},$$

where $\mathcal{S}$ is a maximal uniformly negative subspace of $\mathrm{Ker}(\widehat{A_1})$. Considering now the operator $R = R_1|\mathcal{S}^\perp[+]R_2$, we verify immediately that R satisfies the required conditions (a), (b), and (c), for $T|(I-P)\mathcal{K}_1$ instead of T. Since P has finite rank, $T|(I-P)\mathcal{K}_1$ is also quasi-contractive and T and $T|(I-P)\mathcal{K}_1$ are simultaneously doubly quasi-contractive.

Step 2. *There exists a bounded isometry* $W : \mathrm{Clos}\,\mathrm{Ran}(T) \to \mathcal{K}_1$ *such that*

$$R = WT. \tag{4.4}$$

Indeed, since $\mathrm{Ker}(T) \subseteq \mathrm{Ker}(T^\sharp T) = \mathrm{Ker}(R)$, (4.4) uniquely determines an operator $W : \mathrm{Ran}(T) \to \mathcal{K}_1$. Since $R^2 = T^\sharp T$ it follows that W is isometric. We now make the remark that the properties (a), (b), and (c) of R enable the same reasoning as in the proof of Theorem 12.3.4, thus proving that W is bounded.

Step 3. *There exists a finite rank orthogonal projection* $Q \in \mathcal{B}(\mathcal{K}_2)$ *and a partial isometry* $V \in \mathcal{B}(\mathcal{K}_1, \mathcal{K}_2)$ *such that* $V^\sharp$ *is quasi-contractive and*

$$VR = (I - Q)T. \tag{4.5}$$

Indeed, we consider the factorisation (4.4) and notice that $\mathrm{Clos}\,\mathrm{Ran}(T) = (\mathrm{Ker}(T^\sharp))^\perp$ is pseudo-regular hence we have a decomposition

$$\mathrm{Clos}\,\mathrm{Ran}(T) = \mathcal{R}[+]\mathcal{P},$$

where $\mathcal{R}$ is a regular subspace of $\mathcal{K}_2$ and $\mathcal{P}$ is a neutral subspace of $\mathcal{K}_2$ such that $\dim(\mathcal{P}) = \kappa^0(\mathrm{Clos}\,\mathrm{Ran}(T)) = \kappa^0(\mathrm{Ker}(T^\sharp)) < \infty$. Moreover, there exists a regular subspace $\mathcal{S} \subseteq \mathcal{R}^\perp$ such that $\mathcal{P} \subseteq \mathcal{S}$ and $\dim(\mathcal{S}) = 2\dim(\mathcal{P}) < \infty$. We let Q denote the orthogonal projection onto $\mathcal{S}$. Since $\mathcal{R}$ is regular and W is isometric, it follows that $W|\mathcal{R}$ is one-to-one and $W\mathcal{R}$ is a regular subspace, too. Let $V = (W|\mathcal{R})^{-1} : W\mathcal{R} \to \mathcal{R}$ and extend it to a partial isometry $V \in \mathcal{B}(\mathcal{K}_1, \mathcal{K}_2)$ such that the left support of V is $\mathcal{R}$ and its right support is $W\mathcal{R}$. Then the factorisation (4.5) follows. In addition, $V^\sharp$ is quasi-contractive since

$$\kappa^-(I - VV^\sharp) = \kappa^-(\mathcal{R}^\perp) = \kappa^-(\mathrm{Ker}(T^\sharp)) + \kappa^0(\mathrm{Ker}(T^\sharp)) < \infty.$$

Step 4. $T^\sharp$ *is quasi-contractive.*

Indeed, from (4.5) we have

$$T^\sharp(I - Q) = RV^\sharp.$$

Since $R^2 = T^\sharp T$ and T is quasi-contractive, so is R. $V^\sharp$ is also quasi-contractive, hence the same is true for $T^\sharp(I - Q)$. Since Q has finite rank, from here we infer that $T^\sharp$ is quasi-contractive. ∎

12.5 Notes

Most of the results on quasi-contractions are from the author's article [61]. The results in Section 12.3 generalise the polar decompositions for strong plus-operators of M. G. Kreĭn and Yu. L. Shmulyan [101].

Chapter 13

More on Definitisable Operators

We resume the spectral study of definitisable selfadjoint operators in Kreĭn spaces taking advantage of the spectral function obtained in Chapter 11. The two main problems considered refer to the possibility of obtaining functional calculi with continuous, or more generally, with Borelian functions, and to the existence of invariant subspaces, with attention to invariant maximal semidefinite subspaces. The main obstruction in dealing with these problems comes from the singularities of the spectral function, called critical points. The critical points can be classified as regular and singular, in terms of the behaviour of the spectral function in their neighbourhood. Criteria for regularity of critical points are of great importance. Particular consideration is required for the critical point ∞, for unbounded operators.

Given a spectral function E with singularities, we ask whether there exists a definitisable operator A in some Kreĭn space $\mathcal{K}$ having E as its spectral function. This is usually called an inverse spectral function.

A more general case corresponds to considering unitary definitisable operators in Kreĭn spaces. The Cayley transform provides a good transcription between the two theories.

13.1 Critical Points

In this section we fix a definitisable selfadjoint operator A in the Kreĭn space $\mathcal{K}$. Recall that the set of critical points of A is, by definition,

$$c(A) = \bigcap_p Z(p) \cap \sigma(A) \cap \mathbb{R},$$

where, in the first intersection, p runs through the set of all definitising polynomials of A, and $Z(p)$ denotes the set of all zeros of p.

Lemma 13.1.1. *Let p be a definitising polynomial of A and $\beta \in \rho(A)$ a real zero of p of order of multiplicity $k(\beta)$. If $\widehat{\beta} \in \rho(A) \cap \mathbb{R}$ is in the same connected component of $\rho(A)$ with β then the polynomial $\widehat{p}$ defined by*

$$\widehat{p}(\lambda) = p(\lambda)(\lambda - \beta)^{-k(\beta)}(\lambda - \widehat{\beta})^{k(\beta)}$$

is also a definitising polynomial of A.

Proof. Let $\lambda_0 \in \rho(A)$ be fixed and let q and $\widehat{q}$ be the rational functions defined by

$$q(\lambda) = p(\lambda)(\lambda - \lambda_0)^{-n-1}(\lambda - \overline{\lambda}_0)^{-n-1}(\lambda - \zeta_0)(\lambda - \overline{\zeta}_0),$$

and

$$\widehat{q}(\lambda) = \widehat{p}(\lambda)(\lambda - \lambda_0)^{-n}(\lambda - \overline{\lambda}_0)^{-n},$$

where $n = \deg(p)$. It follows that both q and $\widehat{q}$ are holomorphic in a neighbourhood of $\widetilde{\sigma}(A)$, the extended spectrum of A. If J is a fundamental symmetry of $\mathcal{K}$ and F denotes the operator valued function defined in Theorem 11.5.2, we claim that the following formula holds

$$\widehat{q}(A)x = J \int_{-\infty}^{+\infty} (t - \widehat{\beta})^{k(\beta)}(t - \beta)^{-k(\beta)} \, dF(t)x, \quad x \in \mathcal{K}. \tag{1.1}$$

To this end we first remark that for $k(\beta) = 0$ the claim is just (5.3) from Chapter 11 and hence we will assume that $k(\beta) \geqslant 1$. Let $\Delta \in \mathfrak{R}_A$ be a closed interval such that $\beta, \widehat{\beta} \in \mathbb{R} \setminus \Delta$. Using the representation of the resolvent function as in Theorem 11.5.2, the properties of the spectral function E as in Theorem 11.7.3 and Theorem 11.6.2, we get

$$\widehat{q}(A)x = \lim_{\varepsilon \to 0} J \int_{\Delta_\varepsilon} (t - \widehat{\beta})^{k(\beta)}(t - \beta)^{-k(\beta)} \, dF(t)x, \quad x \in E(\Delta)\mathcal{K}, \tag{1.2}$$

where Δ_ε is defined as in the proof of Theorem 11.7.3. We remark now that $\mathrm{supp}(F)$ is contained in $\sigma(A) \cap \mathbb{R}$. Then, consider an increasing sequence $(\Delta_n)_n$ of closed intervals such that $\bigcup_n \Delta_n = \mathbb{R} \setminus I$, where $I \subseteq \rho(A) \cap \mathbb{R}$ is an interval containing β and $\widehat{\beta}$. From (1.2) it follows that (1.1) holds for all $x \in \bigcup_n E(\Delta_n)\mathcal{K}$. Since the linear manifold $\bigcup_n E(\Delta_n)\mathcal{K}$ is dense in $E(\sigma(A) \cap \mathbb{R}; A)\mathcal{K}$ and both sides in (1.1) are bounded operators, we have the claim proved for all $x \in E(\sigma(A) \cap \mathbb{R}; A)\mathcal{K}$. On the other hand, (1.1) holds trivially for $x \in E(\sigma_0(A); A)\mathcal{K}$, where $\sigma_0(A) = \sigma(A) \setminus \mathbb{R}$, since, in this case, both sides of (1.1) are null. The claim is proved.

Finally, if β and $\widehat{\beta}$ are in the same (real) connected component of $\rho(A) \cap \mathbb{R}$ it follows that $(t - \widehat{\beta})^{k(\beta)}(t - \beta)^{-k(\beta)}$ has constant sign on $\mathrm{supp}(F) = \sigma(A) \cap \mathbb{R}$. From (1.1) it follows that $\widehat{q}$ is definitising for A, equivalently, $\widehat{p}$ is definitising for A. ∎

The critical points of A can be characterised by means of the spectral function E.

Proposition 13.1.2. *If α denotes a real value in $\sigma(A)$, the following statements are equivalent.*
  (i) $\alpha \in c(A)$.
  (ii) *For any $\Delta \in \mathfrak{R}_A$ with $\alpha \in \Delta$, the subspace $E(\Delta)\mathcal{K}$ is indefinite.*

Proof. (i)⇒(ii). Let $\alpha \in c(A)$ and $\Delta \in \mathfrak{R}_A$ be bounded and such that $\alpha \in \Delta$. To make a choice we assume that $E(\Delta)$ is positive. Consider the decomposition

$$\mathcal{K} = E(\Delta)\mathcal{K}[+](I - E(\Delta))\mathcal{K}.$$

13.1 CRITICAL POINTS

With respect to the Hilbert space $(E(\Delta)\mathcal{K}, [\cdot, \cdot])$ we consider the bounded selfadjoint operator $A|E(\Delta)\mathcal{K}$ and let p be a definitising plynomial of A. We distinguish two cases.

(1) If $k(\alpha)$ is odd, from property (4) of the spectral function E (see Theorem 11.7.3) it follows that $\sigma(A) \cap \mathbb{R}$ does not contain any point at least on one side of α and let $\widehat{\alpha} \in \Delta \setminus \sigma(A)$ be on this side, and such that $\widehat{\alpha} \notin Z(p)$. Define the polynomial

$$\widehat{p}(\lambda) = (\lambda - \widehat{\alpha})^{k(\alpha)}(\lambda - \alpha)^{-k(\alpha)}p(\lambda).$$

Since $\widehat{p}(t) > 0$ for any $t \in \sigma(A) \cap \Delta$ it follows that

$$[\widehat{p}(A)x, x] = \left[\int_\Delta \widehat{p}(t)\,\mathrm{d}E(t)x, x\right] \geq 0, \quad x \in E(\Delta)\mathcal{K}. \tag{1.3}$$

(2) If $k(\alpha)$ is even we consider $\widehat{\alpha} \in \Delta \setminus Z(p)$ and define $\widehat{p}$ as before. In this case (1.3) still holds.

By means of Lemma 13.1.1 we have that in both cases $\widehat{p}(A)$ is positive on the space $(I - E(\Delta))\mathcal{K}$. Thus $\widehat{p}$ is a definitising polynomial of A such that $\alpha \notin Z(\widehat{p})$, contradicting the assumption $\alpha \in c(A)$.

(ii)$\Rightarrow$(i). This implication is a direct consequence of the definition of the critical points and of property (4) of the spectral function E as in Theorem 11.7.3. ∎

Letting E denote the spectral function of the selfadjoint definitisable operator A and $\mathfrak{R}_A$ denote the Boolean algebra generated by intervals with endpoints not in $c(A)$, we define the sets

$$\sigma_+(A) = \{\lambda \in \sigma(A) \cap \mathbb{R} \mid \exists \Delta \in \mathfrak{R}_A, \text{ with } \lambda \in \Delta \text{ and } E(\Delta) \text{ positive}\}, \tag{1.4}$$

$$\sigma_-(A) = \{\lambda \in \sigma(A) \cap \mathbb{R} \mid \exists \Delta \in \mathfrak{R}_A, \text{ with } \lambda \in \Delta \text{ and } E(\Delta) \text{ negative}\}. \tag{1.5}$$

Then Proposition 13.1.2 implies that the spectrum of A splits into the following disjoint union

$$\sigma(A) = c(A) \cup \sigma_+(A) \cup \sigma_-(A). \tag{1.6}$$

Proposition 13.1.3. *If α is a critical point of A then the subspace S_α defined by*

$$S_\alpha = \bigcap_{\Delta \in \mathfrak{R}_A,\, \alpha \in \Delta} E(\Delta)\mathcal{K}$$

coincides with $\mathfrak{S}_\alpha(A)$, the root subspace corresponding to α and A. In addition, if p is a definitising polynomial of A and $k(\alpha)$ denotes the order of multiplicity of the zero α of p then $(A - \alpha I)^{k(\alpha)+1}S_\alpha = \{0\}$.

Proof. We first prove that $\sigma(A|S_\alpha) \subseteq \{\alpha\}$. To this end let $\lambda \in \mathbb{R}$ be distinct from α. Then there exists $\Delta \in \mathfrak{R}_A$ such that $\alpha \in \Delta$ and $\lambda \notin \Delta$. Since $\sigma(A|E(\Delta)\mathcal{K}) \subseteq \text{Clos}\,\Delta$ it follows that the operator $(\lambda I - A)|E(\Delta)\mathcal{K}$ has a bounded inverse which we denote by B, that is,

$$B(\lambda I - A)|E(\Delta)\mathcal{K} = I|E(\Delta)\mathcal{K}. \tag{1.7}$$

We can prove easily that the subspace $\mathcal{S}_\alpha$ is hyperinvariant under A, that is, it is invariant under all operators which commute with A. Then $\mathcal{S}_\alpha$ is invariant under B and restricting both sides in (1.7) to $\mathcal{S}_\alpha$ it follows that the operator $B|\mathcal{S}_\alpha$ is a bounded inverse of $(\lambda I - A)|\mathcal{S}_\alpha$. The claim is proven.

We remark now that $\mathfrak{S}_\alpha(A)$, the root subspace corresponding to α and A, is contained in $\mathcal{S}_\alpha$. We distinguish two cases. (1) If $\mathcal{S}_\alpha = 0$ then α is not an eigenvalue. (2) If $\mathcal{S}_\alpha \neq 0$ then taking into account that $\sigma(A|\mathcal{S}_\alpha) \subseteq \{\alpha\}$ it is a standard exercise in operator theory to prove that $\mathcal{S}_\alpha$ is the root subspace corresponding to α and A.

The second statement follows from Corollary 11.5.3. ∎

In order to study the behaviour of the spectral function E in a neighbourhood of a critical point α, by replacing A with the operator $A|E(\Delta)\mathcal{K}$, for $D = (\alpha - \varepsilon, \alpha + \varepsilon)$, for $\varepsilon > 0$ and sufficiently small, it is sufficient to assume that the operator A is bounded and $c(A) = \{\alpha\}$. The next proposition justifies this procedure of localisation.

Proposition 13.1.4. *Let $\alpha \in c(A)$ and $\Lambda \in \mathfrak{R}_A$ be a bounded interval such that $\Lambda \cap c(A) = \{\alpha\}$. Define the subspaces $\mathcal{S}^+_{\alpha,\Lambda}$ and $\mathcal{S}^-_{\alpha,\Lambda}$ by*

$$\mathcal{S}^+_{\alpha,\Lambda} = \bigvee \{E(\Delta)\mathcal{K} \mid \Delta \in \mathfrak{R}_A,\ \Delta \subseteq \Lambda,\ \Delta \cap \sigma(A) \subseteq \sigma_+(A)\}, \tag{1.8}$$

$$\mathcal{S}^-_{\alpha,\Lambda} = \bigvee \{E(\Delta)\mathcal{K} \mid \Delta \in \mathfrak{R}_A,\ \Delta \subseteq \Lambda,\ \Delta \cap \sigma(A) \subseteq \sigma_-(A)\}. \tag{1.9}$$

Then.

(1) $\mathcal{S}^+_{\alpha,\Lambda}$ ($\mathcal{S}^-_{\alpha,\Lambda}$) is a positive (respectively, negative) subspace of $E(\Lambda)$ and reducing for A.

(2) $\mathcal{S}_\alpha$, $\mathcal{S}^+_{\alpha,\Lambda}$ and $\mathcal{S}^-_{\alpha,\Lambda}$ are mutually orthogonal subspaces, $\sigma(A|\mathcal{S}^+_{\alpha,\Lambda}) = \mathrm{Clos}(\sigma_+(A) \cap \Lambda)$, and $\sigma(A|\mathcal{S}^-_{\alpha,\Lambda}) = \mathrm{Clos}(\sigma_-(A) \cap \Lambda)$.

(3) $\mathcal{S}_\alpha = E(I)\mathcal{K} \cap (\mathcal{S}^+_{\alpha,\Lambda} \vee \mathcal{S}^-_{\alpha,\Lambda})^\perp$.

(4) $\mathcal{S}^+_{\alpha,\Lambda} \cap \mathcal{S}^-_{\alpha,\Lambda} \subseteq \mathcal{S}^0_\alpha$, the isotropic subspace of $\mathcal{S}_\alpha$.

(5) $\mathcal{S}_\alpha = 0$ if and only if $\mathcal{S}^+_{\alpha,\Lambda} \vee \mathcal{S}^-_{\alpha,\Lambda} = E(\Lambda)\mathcal{K}$.

(6) If $\mathcal{S}_\alpha = 0$ then there exists a definitising polynomial p such that α is a simple zero of p.

Proof. Let us first remark that there exist two sequences $(\Delta^\pm_n)_{n \in \mathbb{N}} \subset \mathfrak{R}_A$, $\Delta^\pm_n \subseteq \Lambda$, $\Delta^\pm_n \cap \sigma(A) \subseteq \sigma_\pm(A)$ such that $\Delta^\pm_n \subseteq \Delta^\pm_{n+1}$, for all $n \in \mathbb{N}$ and

$$\mathcal{S}^+_{\alpha,\Lambda} = \bigvee_{n \in \mathbb{N}} E(\Delta^+_n)\mathcal{K},\quad \mathcal{S}^-_{\alpha,\Lambda} = \bigvee_{n \in \mathbb{N}} E(\Delta^-_n)\mathcal{K}.$$

Then the statement (1) follows from the fact that the linear manifolds $\bigcup_{n \in \mathbb{N}} E(\Delta^+_n)\mathcal{K}$ and $\bigcup_{n \in \mathbb{N}} E(\Delta^-_n)\mathcal{K}$ are dense in $\mathcal{S}^+_{\alpha,\Lambda}$ and in $\mathcal{S}^-_{\alpha,\Lambda}$, respectively. The statements (2) and (3) are immediate consequences of the definitions, while (4) and (5) follow from (3).

In order to prove (6) let us assume that $\mathcal{S}_\alpha = 0$. Since $\alpha \in c(A)$, from Proposition 13.1.2 it follows that the subspace $E(\Lambda)\mathcal{K}$ is indefinite and hence both subspaces

13.1 Critical Points

$\mathcal{S}_{\alpha,\Lambda}^+$ and $\mathcal{S}_{\alpha,\Lambda}^-$ are nontrivial. This implies that $\sigma_-(A) \cap \Lambda \neq \emptyset$ and $\sigma_+(A) \cap \Lambda \neq \emptyset$. Without restricting the generality we can assume that $\sigma_-(A) \cap \Lambda$ is on the left side of α and $\sigma_+(A) \cap \Lambda$ is on the right side of α. Let $x_- \in \bigcup_{n \in \mathbb{N}} E(\Delta_n^+)\mathcal{K}$ and $x_+ \in \bigcup_{n \in \mathbb{N}} E(\Delta_n^-)\mathcal{K}$. Then

$$[(A - \alpha I)(x_+ + x_-), (x_+ + x_-)] = [(A - \alpha I)x_+, x_+] + [(A - \alpha I)x_-, x_-] \geqslant 0.$$

Since the linear manifold $\bigcup_{n \in \mathbb{N}} E(\Delta_n^+)\mathcal{K} \cup \bigcup_{n \in \mathbb{N}} E(\Delta_n^-)\mathcal{K}$ is dense in $E(\Lambda)\mathcal{K}$ it follows that the operator $A - \alpha I$ is positive on the whole subspace $E(\Lambda)\mathcal{K}$. Using a similar construction as in the proof of Proposition 13.1.2 we can obtain now a definitising polynomial such that α is a simple zero. ∎

The preceding proposition enables us to introduce the following classification of critical points of A. If $\alpha \in c(A)$ then α is of *even/odd type* if there exists a real definitising polynomial p of A such that α is a zero of even/odd multiplicity.

Proposition 13.1.5. *For any definitisable selfadjoint operator A there exists a real definitising polynomial p_0 of minimal degree. In addition, we have the following.*
(1) $Z(p_0) \setminus \mathbb{R} \subseteq \sigma_p(A)$.
(2) *If $\lambda \in Z(p_0) \cap \mathbb{R} \cap (\sigma_+(A) \cup \sigma_-(A) \cup \rho(A))$ then λ is a simple zero of p_0.*
(3) *If $\alpha \in c(A)$ and $\mathcal{S}_\alpha = \{0\}$ then α is a simple zero of p_0.*
(4) *If $\alpha \in c(A)$ and there exists an interval $\Lambda \in \mathfrak{R}_A$ such that $\alpha \in \Lambda$ and either $\sigma_+(A) \cap \Lambda = \emptyset$ or $\sigma_-(A) \cap \Lambda = \emptyset$ then $\alpha \in \sigma_p(A)$.*

Proof. Since $\mathbb{N}$ is well ordered, the existence of a real definitising polynomial p_0 of minimal degree follows.

(1) Let $\beta \in Z(p_0) \setminus \mathbb{R}$ and assume that $\beta \notin \sigma_p(A)$. By Corollary 11.5.3 it follows that $\beta, \overline{\beta} \in \rho(A)$. Then

$$p_0(\lambda) = (\lambda - \beta)^k (\lambda - \overline{\beta})^k p(\lambda),$$

for some polynomial p, where $k \geqslant 1$ denotes the order of multiplicity of the zero α. Then

$$[p(A)x, x] = [p_0(A)(A - \beta I)^{-k}x, (A - \beta I)^{-k}x] \geqslant 0, \quad x \in \text{Dom}(p(A)),$$

and hence p is a definitising polynomial of A with $\deg(p) < \deg(p_0)$, which contradicts the minimality of p_0.

(2) Let $\beta \in (Z(p_0) \cap \mathbb{R}) \cap (\sigma_+(A) \cup \sigma_-(A) \cup \rho(A))$. If $\beta \in \rho(A)$ then we proceed as before to show that β is a simple zero of p_0. Let us assume that $\beta \in \sigma_+(A)$ and let the order of multiplicity of β be $\geqslant 2$. Then

$$p_0(\lambda) = (\lambda - \beta)^{2k} p(\lambda),$$

where $k \geqslant 1$ and β is a zero of p of order of multiplicity at most 1. Let $\Delta \in \mathfrak{R}_A$ be a bounded and open interval such that $\alpha \in \Delta$ and $E(\Delta)\mathcal{K}$ is a positive subspace. Consider the decomposition

$$\mathcal{K} = E(\Delta)\mathcal{K}[+](I - E(\Delta))\mathcal{K}.$$

By means of Corollary 11.7.8 we have

$$[p(A)x, x] = [\int_\Delta p(t)\,dE(t)x, x] \geqslant 0, \quad x \in E(\Delta)\mathcal{K}.$$

If $x \in (I - E(\Delta))\mathcal{K}$ then we still have $[p(A)x, x] \geqslant 0$ and hence p is a real definitising polynomial such that $\deg(p) < \deg(p_0)$, a contradiction of the minimality of the polynomial p_0.

(3) This statement is a direct consequence of Proposition 13.1.4.

(4) Let $\alpha \in c(A) \setminus \sigma_p(A)$. Then $\mathcal{S}_\alpha = 0$. By the same Proposition 13.1.4 it follows that for any $\Lambda \in \mathfrak{R}_A$ with $\alpha \in \Lambda$ we have $\sigma_+(A) \cap \Lambda \neq \emptyset$ and $\sigma_-(A) \cap \Lambda \neq \emptyset$. ∎

By definition, ∞ is a critical point of A if for any $\Delta \in \mathfrak{R}_A$ such that $\mathbb{R} \setminus \Delta$ is bounded, the subspace $E(\Delta)\mathcal{K}$ is indefinite. The *extended set of critical points* of A, denoted by $\widetilde{c}(A)$, is defined as follows: $\widetilde{c}(A) = c(A) \cup \{\infty\}$ if ∞ is a critical point, otherwise $\widetilde{c}(A) = c(A)$.

Remark 13.1.6. Let us define the subspace

$$\mathcal{S}_\infty = \bigcap_{\infty \in \Delta \in \mathfrak{R}_A} E(\Delta)\mathcal{K}.$$

Then $\mathcal{S}_\infty = 0$. In addition, the linear manifold

$$\mathcal{S}_A = \bigcup_{\Delta \in \mathfrak{R}_A \text{ bounded}} E(\Delta)\mathcal{K} \cup E(\sigma_0(A); A)\mathcal{K},$$

is a core of A, that is, the closure of $A|\mathcal{S}_A$ coincides with A. ∎

We conclude this section with a theorem that reduces the study of the critical point ∞ to the study of a finite critical point which is not an eigenvalue.

Theorem 13.1.7. *Let $s_0 \in (\rho(A) \cup \sigma_c(A)) \cap \mathbb{R}$ and consider ψ the automorphism of $\widetilde{\mathbb{R}} = \mathbb{R} \cup \{\infty\}$, the one point compactification of $\mathbb{R}$,*

$$\psi(t) = \begin{cases} \frac{1}{t-s_0}, & t \neq s_0, \infty, \\ 0, & t = \infty, \\ \infty, & t = s_0. \end{cases}$$

(1) $\psi(A) = (A - s_0 I)^{-1}$ is a definitisable selfadjoint operator in $\mathcal{K}$.
(2) $\widetilde{c}(\psi(A)) = \psi(\widetilde{c}(A))$.
(3) If $E_{\psi(A)}$ denotes the spectral function associated to $\psi(A)$ then

$$E_{\psi(A)}(\Delta) = E(\psi^{-1}(\Delta)), \quad \Delta \in \mathfrak{R}_{\psi(A)},$$

where $\mathfrak{R}_{\psi(A)} = \{\psi(\Delta) \mid \Delta \in \mathfrak{R}_A\}$.

13.2 Functional Calculus

Proof. Without restricting the generality we will assume that $s_0 = 0$. The statement (1) was already obtained in Lemma 11.5.4. From the proof of the same lemma we have that $c(A^{-1}) \setminus \{0\} \subseteq \psi(c(A) \setminus \{0\})$. Considering also the inverse transform of ψ we get

$$c(A^{-1}) \setminus \{0\} = \{t^{-1} \mid t \in c(A) \setminus \{0\}\}. \tag{1.10}$$

Let now Δ be a bounded interval in $\mathfrak{R}_A$ such that $0 \notin \operatorname{Clos} \Delta$. We consider the rectangle C_Δ as in Section 11.6 and let us notice that $\psi(C_\Delta) = \{\lambda^{-1} \mid \lambda \in C_\Delta\}$ is a Jordan contour symmetric with respect to the real axis and consisting of four arcs of circles.

With notation as in Section 11.6, using the formula of change of variables we get

$$\int_{C^\delta_{\Delta_\varepsilon}} R(\zeta; A) x \, d\zeta = \int_{\psi(C^\delta_{\Delta_\varepsilon})} R(\psi(\lambda); A) \psi'(\lambda) x \, d\lambda, \quad x \in \mathcal{K}.$$

Since

$$(\zeta I - A)^{-1} = \frac{1}{\lambda} A^{-1} (\lambda I - A^{-1})^{-1}, \quad \zeta = \psi(\lambda) = \frac{1}{\lambda},$$

and taking into account the definition of the spectral function, see Theorem 11.7.3, we get

$$E(\Delta) x = \lim_{\varepsilon \to 0} \lim_{\delta \to 0} A^{-1} \int_{\psi(C^\delta_{\Delta_\varepsilon})} \frac{1}{\lambda} (\lambda I - A)^{-1} x \, d\lambda, \quad x \in \mathcal{K}.$$

We apply now the definition of the spectral function for the operator A^{-1} and get

$$E(\Delta) x = A^{-1} A E_{A^{-1}}(\Delta^{-1}) x = E_{A^{-1}}(\Delta^{-1}) x, \quad x \in \mathcal{K}. \tag{1.11}$$

Making use of Proposition 13.1.2, from (1.10) and (1.11) we obtain the statement (2). The statement (3) follows now from (1.11). ∎

13.2 Functional Calculus

In this section we associate functional calculi with continuous or Borelian functions to definitisable selfadjoint operators. In order to simplify the notation we will restrict our study to localisations in the neighbourhoods of critical points, as seen in Proposition 13.1.4. The corresponding results in the general case will be sketched in the end.

Definition 13.2.1. Let $\alpha \in \mathbb{R}$ and $n \in \mathbb{N}$ be fixed. A bounded selfadjoint operator $A \in \mathcal{B}(\mathcal{K})$ is in the class $\mathfrak{D}(\alpha; n)$ if the polynomial $p(\lambda) = (\lambda - \alpha)^n$ is definitising for A and n is the least natural number with this property.

Let A be in the class $\mathfrak{D}(\alpha; n)$. Then $\sigma(A)$ is a compact real set and A has no critical points different from α. Denoting by E the spectral function of A, the property (4) in Theorem 11.7.3 reads as follows: if n is odd then

$$[E(\Delta) x, x] \geq 0, \quad x \in \mathcal{K}, \ \Delta \in \mathfrak{R}_A, \ \Delta \subset (\alpha, +\infty),$$

and
$$[E(\Delta)x, x] \leqslant 0, \quad x \in \mathcal{K}, \ \Delta \in \mathfrak{R}_A, \ \Delta \subset (-\infty, \alpha),$$
and, if n is even then
$$[E(\Delta)x, x] \geqslant 0, \quad x \in \mathcal{K}, \ \Delta \in \mathfrak{R}_A, \ \Delta \not\ni \alpha.$$

By Proposition 11.7.7 there exists $N_\alpha \in \mathcal{B}(\mathcal{K})$ with the following properties:

$$[N_\alpha x, x] \geqslant 0, \quad x \in \mathcal{K}, \tag{2.1}$$
$$N_\alpha^2 = 0, \quad N_\alpha(A - \alpha I) = 0, \tag{2.2}$$
$$N_\alpha E(\Delta) = 0, \text{ for all } \Delta \in \mathfrak{R}_A \text{ such that } \alpha \notin \Delta. \tag{2.3}$$

Moreover, the limit

$$\int_{(\alpha)} (t - \alpha)^n \, dE(t) := \lim_{\varepsilon \downarrow 0} \int_{\mathbb{R} \setminus (\alpha - \varepsilon, \alpha + \varepsilon)} (t - \alpha)^n \, dE(t) \tag{2.4}$$

exists in the strong operator topology and

$$(A - \alpha I)^n = \int_{(\alpha)} (t - \alpha)^n \, dE(t) + N_\alpha. \tag{2.5}$$

In particular, letting $\| \cdot \|$ be an arbitrary, but fixed, fundamental norm on $\mathcal{K}$, we have

$$\limsup_{\varepsilon \downarrow 0} \| \int_{\mathbb{R} \setminus (\alpha - \varepsilon, \alpha + \varepsilon)} (t - \alpha)^n \, dE(t) \| = M < \infty. \tag{2.6}$$

If σ is a compact set we denote by $\mathcal{C}(\sigma)$ the C^*-algebra of continuous functions $f \colon \sigma \to \mathbb{C}$, endowed with the uniform norm $\|f\| = \sup_{t \in \sigma} |f(t)|$ and the conjugation $f^*(t) = \overline{f(t)}$.

Lemma 13.2.2. *Let $A \in \mathfrak{D}(\alpha; n)$ and E its spectral function. Then, for any continuous function $f \in \mathcal{C}(\sigma(A))$, the limit*

$$\int_{(\alpha)} f(t)(t - \alpha)^n \, dE(t) := \lim_{\varepsilon \downarrow 0} \int_{\mathbb{R} \setminus (\alpha - \varepsilon, \alpha + \varepsilon)} f(t)(t - \alpha)^n \, dE(t) \tag{2.7}$$

exists in the strong operator topology and the mapping

$$\mathcal{C}(\sigma(A)) \ni f \mapsto \int_{(\alpha)} f(t)(t - \alpha)^n \, dE(t) \in \mathcal{B}(\mathcal{K}) \tag{2.8}$$

is uniformly bounded.

13.2 FUNCTIONAL CALCULUS

Proof. Let $f \in \mathcal{C}(\sigma(A))$. Extend f to a bounded and Borelian function on $\mathbb{R}$, also denoted by f, and such that

$$\|f\|_u = \max_{t \in \sigma(A)} |f(t)| = \sup_{t \in \mathbb{R}} |f(t)|,$$

for example setting $f(t) = f(t_0)$, $t \in \mathbb{R} \setminus \sigma(A)$, where $t_0 \in \sigma(A)$ is fixed and the same for all $f \in \mathcal{C}(\sigma(A))$. Since the operator valued measure $(t - \alpha)^n \, \mathrm{d}E(t)$ is supported on $\sigma(A)$ it follows that for any $\varepsilon > 0$ we have

$$\int_{\sigma(A) \setminus (\alpha-\varepsilon, \alpha+\varepsilon)} f(t) (t-\alpha)^n \, \mathrm{d}E(t) = \int_{\mathbb{R} \setminus (\alpha-\varepsilon, \alpha+\varepsilon)} f(t) (t-\alpha)^n \, \mathrm{d}E(t).$$

Let now $f \in \mathcal{C}(\sigma(A))$, $f \geqslant 0$. For arbitrary $x \in \mathcal{K}$ define the function $\varphi_x : (0, +\infty) \to \mathbb{R}$ by

$$\varphi_x(\varepsilon) = \int_{\mathbb{R} \setminus (\alpha-\varepsilon, \alpha+\varepsilon)} f(t) (t-\alpha)^n \, \mathrm{d}[E(t)x, x], \quad \varepsilon > 0.$$

Remark that the scalar measure $(t - \alpha)^n \, d[E(t)x, x]$ is positive on $\mathbb{R} \setminus (\alpha - \varepsilon, \alpha + \varepsilon)$ and hence the mapping φ_x is nonincreasing and positive on $(0, +\infty)$. In addition, if $\|\cdot\|$ is a fixed fundamental norm on $\mathcal{K}$ and the constant M satisfies (2.6) then

$$\limsup_{\varepsilon \downarrow 0} \varphi_x(\varepsilon) \leqslant \|f\|_u \limsup_{\varepsilon \downarrow 0} \left[\int_{\mathbb{R} \setminus (\alpha-\varepsilon, \alpha+\varepsilon)} f(t) (t-\alpha)^n \, \mathrm{d}E(t)x, x \right] \quad (2.9)$$

$$\leqslant M \|f\|_u \|x\|^2. \quad (2.10)$$

Therefore, there exists a finite

$$\varphi_x(0) = \lim_{\varepsilon \downarrow 0} \varphi_x(\varepsilon) = \lim_{\varepsilon \downarrow 0} \int_{\mathbb{R} \setminus (\alpha-\varepsilon, \alpha+\varepsilon)} f(t) (t-\alpha)^n \, \mathrm{d}[E(t)x, x]. \quad (2.11)$$

Let $x, y \in \mathcal{K}$ and define the mapping $\varphi_{x,y} : (0, +\infty) \to \mathbb{C}$ by

$$\varphi_{x,y}(\varepsilon) = \int_{\mathbb{R} \setminus (\alpha-\varepsilon, \alpha+\varepsilon)} f(t) (t-\alpha)^n \, \mathrm{d}[E(t)x, y], \quad \varepsilon > 0.$$

By means of (2.11) and the polarisation formula it follows that there exists a finite

$$\varphi_x(0) = \lim_{\varepsilon \downarrow 0} \varphi_x(\varepsilon) = \lim_{\varepsilon \downarrow 0} \int_{\mathbb{R} \setminus (\alpha-\varepsilon, \alpha+\varepsilon)} f(t) (t-\alpha)^n \, \mathrm{d}[E(t)x, x],$$

whence, by (2.10) and the polarisation formula, we get

$$|\varphi_{x,y}(0)| \leqslant 2M \|f\|_u \|x\| \|y\|.$$

We remark now that the mapping $\mathcal{K} \times \mathcal{K} \ni (x,y) \mapsto \varphi_{x,y}(0) \in \mathbb{C}$ is an inner product which, by the previous inequality, is bounded. By the Riesz Representation Theorem it follows that there exists a uniquely determined bounded and selfadjoint operator in $\mathcal{K}$, denoted

$$\int_{(\alpha)} f(t)(t-\alpha)^n \, dE(t) \in \mathcal{B}(\mathcal{K}) \qquad (2.12)$$

such that, for any $x, y \in \mathcal{K}$ we have

$$\left[\int_{(\alpha)} f(t)(t-\alpha)^n \, dE(t)x, y\right] = \lim_{\varepsilon \downarrow 0} \left[\int_{\mathbb{R} \setminus (\alpha-\varepsilon, \alpha+\varepsilon)} f(t)(t-\alpha)^n \, dE(t)x, y\right]. \qquad (2.13)$$

Clearly, the operator defined at (2.12) is positive on $\mathcal{K}$. Let J be a fundamental symmetry on $\mathcal{K}$. From (2.13) it follows that, with respect to the Hilbert space $(\mathcal{K}, \langle \cdot, \cdot \rangle_J)$, the uniformly bounded positive operators

$$\int_{\mathbb{R} \setminus (\alpha-\varepsilon, \alpha+\varepsilon)} f(t)(t-\alpha)^n \, J\,dE(t) \in \mathcal{B}(\mathcal{K})$$

converge weakly, following $\varepsilon \to 0$, to $\int_{(\alpha)} f(t)(t-\alpha)^n \, dE(t) \in \mathcal{B}(\mathcal{K})$. It is a standard argument in operator theory to show that, in this case, the weak convergence can be replaced by strong convergence. Multiply these operators again by J and get the strong operator convergence of the improper integral in (2.7) with singularity at α.

Finally, since any function $f \in \mathcal{C}(\sigma(A))$ can be represented as a linear combination of four positive functions in $\mathcal{C}(\sigma(A))$, the conclusion on the existence of the improper integral in (2.7) holds in general, while the boundedness of the mapping in (2.8) is a consequence of (2.10). ∎

Remark 13.2.3. In other words, Lemma 13.2.2 reads as follows: the operator valued measure $(t-\alpha)^n \, dE(t)$ extends to a finite measure on $\sigma(A)$ by setting $(t-\alpha)^n \, dE(t)(\{\alpha\}) = 0$. Thus, for arbitrary $f \in \mathcal{C}(\sigma(A))$ the improper integral in (2.7) can be understood as the integral of the continuous function f with respect to this extended operator valued bounded measure. ∎

Let σ be a compact set of $\mathbb{R}$, n a natural number and $\alpha \in \sigma'$, that is, α is an accumulation point of σ. $\mathcal{C}^n(\sigma; \alpha)$ designates the set of functions $f \in \mathcal{C}(\sigma)$ such that there exist $h \in \mathcal{C}(\sigma)$ and p a polynomial of degree at most $n-1$ such that

$$f(t) = h(t)(t-\alpha)^n + p(t), \quad t \in \sigma. \qquad (2.14)$$

Let us remark that if $f \in \mathcal{C}^n(\sigma; \alpha)$ then h and p are uniquely determined by f. Indeed, if $h \in \mathcal{C}(\sigma)$ and p is a polynomial of degree at most $n-1$ such that

$$(t-\alpha)^n h(t) + p(t) = 0, \quad t \in \sigma,$$

13.2 Functional Calculus

then
$$h(t) = -\frac{p(t)}{(t-\alpha)^n}, \quad t \in \sigma \setminus \{\alpha\},$$

and taking into account that α is an accumulation point of σ and that h is continuous at α, it follows that $p \equiv 0$ and hence $h \equiv 0$. For this reason, for any $f \in \mathcal{C}^n(\sigma; \alpha)$ we consider the representation
$$f(t) = (t-\alpha))^n h_f(t) + p_f(t), \quad t \in \sigma, \tag{2.15}$$
where $h_f \in \mathcal{C}(\sigma)$ and p_f is a polynomial of degree at most $n-1$, uniquely determined by f.

We also introduce the linear operator S_n on $\mathbb{C}[X]$, the vector space of complex polynomials, defined by
$$(S_n p)(t) = \sum_{j=0}^{n-1} a_j (t-\alpha)^j, \tag{2.16}$$
where, in the case when $n > k$, we set $a_j = 0$ for all $j = k+1, \ldots, n-1$.

It is clear that $\mathcal{C}^n(\sigma; \alpha)$ endowed with the natural operations of addition and multiplication with scalars for functions is a complex vector space. With respect to the representation (2.15) the sum and the multiplication by scalars are performed on components. On the other hand, for $f, g \in \mathcal{C}^n(\sigma; \alpha)$ we have

$$(fg)(t) = (t-\alpha)^n \left((t-\alpha)^n h_f(t) h_g(t) + g_f(t) p_g(t) + h_g(t) p_f(t) \right.$$
$$\left. + \frac{p_f(t) p_g(t) - S_n(p_f p_g)(t)}{(t-\alpha)^n} \right) + S_n(p_f p_g)(t), \quad t \in \sigma.$$

This proves that $fg \in \mathcal{C}^n(\sigma; \alpha)$ and, in addition,
$$p_{fg}(t) = S_n(p_f p_g)(t), \tag{2.17}$$
and
$$h_{fg}(t) = (t-\alpha)^n h_f(t) h_g(t) + h_f(t) p_g(t) + h_g(t) p_f(t)$$
$$+ \frac{p_f(t) p_g(t) - S_n(p_f p_g)(t)}{(t-\alpha)^n}. \tag{2.18}$$

Then $\mathcal{C}^n(\sigma; \alpha)$ is a function algebra with unit and the complex conjugation $f \mapsto \overline{f}$.

We consider the following norm on $\mathcal{C}^n(\sigma; \alpha)$: if $f \in \mathcal{C}^n(\sigma; \alpha)$ is represented as in (2.15) and
$$p_f(t) = \sum_{j=0}^{n-1} a_j (t-\alpha)^j, \quad t \in \mathbb{R},$$
then, by definition,
$$\|f\| = \max\{\|h_f\|_u, |a_0|, \ldots, |a_{n-1}|\}. \tag{2.19}$$

A rather long but straightforward calculation, that we omit, proves that, with respect to this norm, $\mathcal{C}^n(\sigma;\alpha)$ is a complete normed algebra with continuous multiplication and isometric involution. Mutiplying the norm in (2.19) with a suitable constant, $\mathcal{C}^n(\sigma;\alpha)$ becomes an involutive Banach algebra. However, in order to simplify the formulae, in the following we will deal only with the equivalent norm defined in (2.19).

Remark 13.2.4. Consider the Banach space $\mathcal{C}(\sigma) \oplus \mathbb{C}^n$, where $\mathbb{C}^n$ is endowed with the norm $\|\cdot\|_\infty$ with respect to the canonical basis, that is, if $(g,x) \in \mathcal{C}(\sigma) \oplus \mathbb{C}^n$ then

$$\|(g,x)\| = \max\{\|g\|_u, \|x\|_\infty\}.$$

The representation (2.15) induces an isomorphism between the Banach spaces $\mathcal{C}^n(\sigma;n)$ and $\mathcal{C}(\sigma) \oplus \mathbb{C}^n$. Also, the canonical inclusion $\mathcal{C}^n(\sigma;n) \hookrightarrow \mathcal{C}(\sigma)$ is topological.

Clearly, from the Taylor formula it follows that $\mathcal{C}^n(\sigma;\alpha)$ contains those functions $f \in \mathcal{C}(\sigma)$ which have extensions of class $\mathcal{C}^n$ in a neighbourhood of α. ∎

In the following, $\mathcal{B}(\mathcal{K})$ is considered as an involutive Banach algebra with the operatorial norm associated to a fixed fundamental norm on $\mathcal{K}$ and the involution $\sharp$.

Theorem 13.2.5. *Let $A \in \mathfrak{D}(\alpha;n)$ be such that α is an accumulation point of $\sigma(A)$ and let E denote the spectral function of A and N_α the selfadjoint nilpotent operator as in (2.1)–(2.3). Then, there exists a unique continuous mapping*

$$\mathcal{C}^n(\sigma(A);\alpha) \ni f \mapsto E(f) \in \mathcal{B}(\mathcal{K}) \tag{2.20}$$

such that for any polynomial q we have

$$E(q) = q(A). \tag{2.21}$$

In addition, the mapping E is a homomorphism of involutive Banach algebras and, if $f \in \mathcal{C}^n(\sigma(A);\alpha)$ then, with notation as before,

$$E(f) = \int_{(\alpha)} h_f(t)\,(t-\alpha)^n \,\mathrm{d}E(t) + h_f(\alpha)\,N_\alpha + p_f(A). \tag{2.22}$$

Proof. For arbitrary $f \in \mathcal{C}^n(\sigma(A);\alpha)$ we define $E(f) \in \mathcal{B}(\mathcal{K})$ by the formula (2.22). Acording to Lemma 13.2.2 the definition is correct and the mapping in (2.20) is continuous, with respect to the norm (2.19) on $\mathcal{C}^n(\sigma(A);\alpha)$ and the operator norm on $\mathcal{B}(\mathcal{K})$.

Let $\mathcal{P}$ denote the algebra of polynomials with complex coefficients and set $\mathcal{P}(\sigma(A))$ for the algebra of polynomial functions defined on $\sigma(A)$. Since α is an accumulation point in $\sigma(A)$ it follows that $\sigma(A)$ contains an infinite number of points and hence $\mathcal{P}$ is naturally identified with $\mathcal{P}(\sigma(A))$. Evidently, $\mathcal{P}(\sigma(A)) \subseteq \mathcal{C}^n(\sigma(A);\alpha)$.

Let $q \in \mathcal{P}(\sigma(A))$ be arbitrary and let k be the degree of q,

$$q(t) = \sum_{j=0}^{k} b_j (t-\alpha)^j, \quad t \in \sigma(A).$$

13.2 FUNCTIONAL CALCULUS

If $k \geqslant n$ then
$$p_q(t) = \sum_{j=0}^{n-1} b_j(t-\alpha)^j, \quad t \in \sigma(A),$$

$$h_q(t) = \sum_{j=n}^{k} b_j(t-\alpha)^{j-n}, \quad t \in \sigma(A),$$

and, if $k < n$ then
$$p_q(t) = q(t), \quad h_q(t) = 0, \quad t \in \sigma(A).$$

Further, if $n > k$ then (2.21) is clearly true. Let us assume $k \geqslant n$. In this case we apply Lemma 13.2.2 and the definition (2.22) in order to obtain (2.21).

It is clear now that $E|\mathcal{P}(\sigma(A))$ is a homomorphism of involutive algebras. Taking into account Remark 13.2.4 and the Weierstrass Approximation Theorem, it follows that $\mathcal{P}(\sigma(A))$ is dense in $\mathcal{C}^n(\sigma(A); \alpha)$. Therefore, E is a homomorphism of involutive Banach algebras and is uniquely determined. ∎

Remark 13.2.6. If α is isolated in $\sigma(A)$ then Theorem 13.2.5 does not apply. However, using the properties of the spectral function E we can decompose $\mathcal{K} = \mathcal{K}_0[+]\mathcal{K}_1$, where $\mathcal{K}_0$ and $\mathcal{K}_1$ reduce the operator A to definitisable selfadjoint operators $A_0 \in \mathcal{B}(\mathcal{K}_0)$ and $A_1 \in \mathcal{B}(\mathcal{K}_1)$ such that A_0 admits the minimal polynomial $(\lambda - \alpha)^n$ and A_1 is similar to a selfadjoint operator in a Hilbert space. In this case one can define a functional calculus by the superposition of the functional calculus for the algebraic operator A_0, for example using functions of class $\mathcal{C}^n$ in a neighbourhood of α, and the functional calculus with continuous functions on $\sigma(A_1) = \sigma(A) \setminus \{\alpha\}$.

Another way of approaching this case is to consider the Banach algebra $\mathcal{C}^n(\sigma; \alpha)$, where σ is a compact subset containing $\sigma(A)$ and an additional sequence of real points converging to α. ∎

Remark 13.2.7. Let A be a definitisable selfadjoint operator in $\mathcal{K}$ such that ∞ is a critical point of A. Without restricting the generality we can assume that $\sigma(A) \subseteq \mathbb{R}$, that $\rho(A) \cap \mathbb{R} \neq \emptyset$, and that the sets $\sigma_+(A)$ and $\sigma_-(A)$ do not interlace. Fix $\zeta \in \rho(A) \cap \mathbb{R}$ and define the algebra $\mathcal{C}^1(\widetilde{\sigma}(A); \infty)$ as the class of those functions $f \in \mathcal{C}(\widetilde{\sigma}(A))$ with the property that there exists $h_f \in \mathcal{C}(\widetilde{\sigma}(A))$ such that

$$f(t) = \frac{h_f(t)}{t - \zeta} + f(\infty), \quad t \in \widetilde{\sigma}(A). \tag{2.23}$$

From the conditions imposed on A and Corollary 11.5.6 it follows that ∞ is an accumulation point in $\widetilde{\sigma}(A)$. Therefore, h_f is uniquely determined by f. Clearly $\mathcal{C}^1(\widetilde{\sigma}(A); \infty)$ is a function algebra. We introduce the following norm: if $f \in \mathcal{C}^1(\widetilde{\sigma}(A); \infty)$ is represented by (2.23) then

$$\|f\| = \max\{\|h_f\|_u, |f(\infty)|\}. \tag{2.24}$$

With respect to this norm $\mathcal{C}^1(\widetilde{\sigma}(A); \infty)$ is a complete normed algebra with continuous multiplication and isometric involution.

Let us remark that the set $\mathcal{C}^1(\widetilde{\sigma}(A); \infty)$ and its algebraic operations do not depend on the choice of $\zeta \in \rho(A) \cap \mathbb{R}$. On the other hand, the representation (2.23) depends on $\zeta \in \rho(A) \cap \mathbb{R}$. However, by the change of ζ the corresponding norms remain equivalent.

If E denotes the spectral function of A we define

$$E(f) := \int_{(\infty)} \frac{h_f(t)}{t - \zeta} \, dE(t) + f(\infty) I, \quad f \in \mathcal{C}^1(\widetilde{\sigma}(A); \infty), \tag{2.25}$$

where the integral converges in the strong operator topology improperly at ∞. The mapping $E \colon \mathcal{C}^1(\widetilde{\sigma}(A); \infty) \to \mathcal{B}(\mathcal{K})$ is continuous, it extends the Riesz–Dunford–Taylor functional calculus and it is uniquely determined by these properties. In addition, this mapping is a homomorphism of unital involutive Banach algebras. These statements can be obtained as consequences of Theorem 13.2.5. ∎

Remark 13.2.8. Let A be a general definitisable selfadjoint operator in the Kreĭn space $\mathcal{K}$. By means of the properties of the spectral function E of A the space $\mathcal{K}$ can be decomposed

$$\mathcal{K} = \mathcal{K}_0[+]\mathcal{K}_1[+]\mathcal{K}_2[+]\mathcal{K}_\infty$$

as a sum of four regular subspaces reducing A to give the corresponding reduced operators: $A_0 \in \mathcal{B}(\mathcal{K}_0)$ is algebraic (corresponding to the nonreal eigenvalues and the isolated critical points of A), $A_1 \in \mathcal{B}(\mathcal{K}_1)$ is a direct sum of operators of type $\mathfrak{D}(\alpha; n)$ where the $\alpha \in c(A)$ are accumulation points in $\sigma(A)$, $A_2 \in \mathcal{B}(\mathcal{K}_2)$ is similar with a selfadjoint operator in a Hilbert space, and hence it has a spectral measure, and A_∞ is a definitisable selfadjoint operator in $\mathcal{K}_\infty$ of the type described in Remark 13.2.7. Some of these operators can be null. For A_0 we proceed as in Remark 13.2.6, for A_1 we have the functional calculus obtained as in Theorem 13.2.5, for A_2 we have the classical functional calculus with continuous functions, while for A_∞ we proceed as in Remark 13.2.7. By the superposition of the functional calculi corresponding to each of the operators A_k, $k = 0, 1, 2, \infty$, we obtain the desired functional calculus with continuous functions in the general case. ∎

We pass now to the extension of the functional calculus to bounded Borelian functions. In the following, if σ is a compact real set then $\mathfrak{B}(\sigma)$ designates the C^*-algebra of bounded Borelian functions on σ, endowed with the uniform norm $\|\cdot\|_u$.

Lemma 13.2.9. *Let* $A \in \mathfrak{D}(\alpha; n)$ *and let* E *be its spectral function. Then, for any* $f \in \mathfrak{B}(\sigma(A))$, *the integral*

$$\int_{(\alpha)} f(t) (t - \alpha)^n \, dE(t) \tag{2.26}$$

13.2 Functional Calculus

converges in the strong operator topology and improperly at α. *Moreover, the mapping*

$$\mathfrak{B}(\sigma(A)) \ni f \mapsto \int_{(\alpha)} f(t)\,(t-\alpha)^n\,\mathrm{d}E(t) \in \mathcal{B}(\mathcal{K}) \tag{2.27}$$

is uniformly bounded and, if $f_k, f \in \mathfrak{B}(\sigma(A))$, $\sup_k \|f_k\|_u < \infty$ *and for any* $t \in \sigma(A)$ *we have* $f_k(t) \to f(t)$ $(t \to \infty)$, *then*

$$\int_{(\alpha)} f_k(t)\,(t-\alpha)^n\,\mathrm{d}E(t) \xrightarrow[k\to\infty]{} \int_{(\alpha)} f(t)\,(t-\alpha)^n\,\mathrm{d}E(t), \tag{2.28}$$

in the strong operator topology.

Proof. The strong operator convergence of the improper integral in (2.26), as well as the uniform boundedness of the mapping in (2.27), can be proved similarly as in Lemma 13.2.2. Moreover, the integral (2.26) will be viewed as a proper integral of the bounded Borelian function f with respect to the finite measure $(t-\alpha)^n\,\mathrm{d}E(t)$ extended to $\sigma(A)$ (see Remark 13.2.3).

Let $f_k, f \in \mathfrak{B}(\sigma(A))$ be such that $\sup_{k\in\mathbb{N}} \|f_k\|_u < \infty$ and f_k converge to f pointwise on $\sigma(A)$. Let $x \in \mathcal{K}$. Since the scalar measure $(t-\alpha)^n[\mathrm{d}E(t)x, x]$ is positive on $\sigma(A)$ and applying the Lebesgue Dominated Convergence Theorem we have

$$\Big[\int_{(\alpha)} f_k(t)\,(t-\alpha)^n\,\mathrm{d}E(t)x, x\Big] \xrightarrow[k\to\infty]{} \Big[\int_{(\alpha)} f(t)\,(t-\alpha)^n\,\mathrm{d}E(t)x, x\Big].$$

Therefore, by the polarisation formula we obtain the weak convergence in (2.28).

Without restricting the generality we can assume that $f_k \geqslant 0$ for all $k \in \mathbb{N}$ and hence $f \geqslant 0$. Let J be a fundamental symmetry of $\mathcal{K}$. Multiplying both sides of (2.28) by J, it follows that the uniformly bounded sequence of positive operators in the Hilbert space $(\mathcal{K}, \langle \cdot, \cdot \rangle_J)$ given by

$$\int_{(\alpha)} f_k(t)\,(t-\alpha)^n\,\mathrm{d}E(t)$$

converges weakly to the operator $\int_{(\alpha)} f(t)\,(t-\alpha)^n\,\mathrm{d}E(t)$. Therefore, this sequence converges also strongly operatorial. Multiplying again by J one obtains the strongly operatorial convergence in (2.28). ∎

Let σ be a compact real set, n a natural number, and α an accumulation point of σ. In the following $\mathfrak{B}^n(\sigma; \alpha)$ denotes the class of those functions $f \in \mathfrak{B}(\sigma)$ with the following property: there exist $h_f \in \mathfrak{B}(\sigma)$, h_f continuous at α, and p_f a polynomial of degree at most $n-1$, such that

$$f(t) = (t-\alpha)^n h_f(t) + p_f(t), \quad t \in \sigma. \tag{2.29}$$

From the continuity of h_f at α and the fact that α is an accumulation point in σ it follows that the representation (2.29) is uniquely determined by f. $\mathfrak{B}^n(\sigma;\alpha)$ is organised naturally as a function algebra. Similar formulae to those in (2.17) and (2.18) hold.

If $f \in \mathfrak{B}^n(\sigma;\alpha)$ then, with respect to the representation (2.29) we define $\|f\|$ as in (2.19). Endowed with this norm $\mathfrak{B}^n(\sigma;\alpha)$ is a complete algebra, with continuous multiplication and isometric involution. Multiplying this norm with a suitable constant, $\mathfrak{B}^n(\sigma;\alpha)$ becomes an involutive Banach algebra. Clearly, $\mathcal{C}^n(\sigma;\alpha)$ is a closed subalgebra in $\mathfrak{B}^n(\sigma;\alpha)$. Similar statements as in Remark 13.2.4 hold.

Theorem 13.2.10. *Let $A \in \mathfrak{D}(\alpha;n)$ be such that α is an accumulation point of $\sigma(A)$. Then there exists a mapping*

$$\mathfrak{B}^n(\sigma(A);\alpha) \ni f \mapsto E(f) \in \mathcal{B}(\mathcal{K}),$$

uniquely determined by the following conditions.

(i) *For any polynomial q we have $E(q) = q(A)$.*

(ii) *If $f_k, f \in \mathfrak{B}^n(\sigma(A);\alpha)$ such that $\sup_{k\in\mathbb{N}} \|f_k\| < \infty$ and f_k converge to f, pointwise on components, that is, for any $t \in \sigma(A)$ we have $h_{f_k}(t) \to h_f(t)$ and $p_{f_k}(t) \xrightarrow[k\to\infty]{} p_f(t)$, then $E(f_k) \xrightarrow[k\to\infty]{} E(f)$ in the strong operator topology.*

In addition, the mapping E is a homomorphism of involutive Banach algebras and for any $f \in \mathfrak{B}^n(\sigma(A);\alpha)$ we have

$$E(f) = \int_{(\alpha)} h_f(t)(t-\alpha)^n \, dE(t) + h_f(\alpha)N_\alpha + p_f(A), \tag{2.30}$$

in particular, E is an extension of the functional calculus as in Theorem 13.2.5.

Proof. If $f \in \mathfrak{B}^n(\sigma(A);\alpha)$ we define $E(f)$ by the formula (2.30). According to Lemma 13.2.9 the definition is correct and $E(f) \in \mathcal{B}(\mathcal{K})$. The property (i) was already proved in Theorem 13.2.5 while the property (ii) follows from Lemma 13.2.9. Applying Baire's Theorem of approximation of bounded Borelian functions by continuous functions, since E has the properties (i) and (ii), it follows that E is a homomorphism of algebras. The boundedness of E is a consequence of Lemma 13.2.2. Therefore, the uniqueness of E is a consequence of the same Baire Theorem. ∎

Similar statements as in Remarks 13.2.6 through 13.2.8 hold. We leave to the reader the corresponding transcriptions.

Corollary 13.2.11. *Let $A \in \mathfrak{D}(\alpha;n)$ and $\Delta \in \mathfrak{R}_A$.*

(i) *If $\alpha \in \Delta$ then $E(\Delta)$ can be approximated in the strong operator topology by operators of type $q(A)(A - \alpha I)^n + I$, where q are polynomials.*

(ii) *If $\alpha \notin \Delta$ then $E(\Delta)$ can be approximated in the strong operator topology by operators of type $q(A)(A - \alpha I)^n$, where q are polynomials.*

13.3 REGULARITY OF CRITICAL POINTS

Proof. (i) Assume that $\alpha \in \Delta$. We set $f = \chi_\Delta$ and define $p_f(t) \equiv 1$ and

$$h_f(t) = \frac{\chi_\Delta(t) - 1}{(t - \alpha)^n}, \quad t \in \mathbb{R}.$$

Then $h_f \in \mathfrak{B}(\mathbb{R})$ and hence $\chi_\Delta \in \mathfrak{B}^n(\mathbb{R}; \alpha)$. We apply now the definition of $E(\chi_\Delta)$ in (2.30) and get

$$E(\chi_\Delta) = -\int \chi_{\mathbb{R} \setminus \Delta}(t) \, dE(t) + I = E(\Delta) - I + I = E(\Delta).$$

Then we apply Baire's Theorem and Theorem 13.2.10 and then the Weierstrass Approximation Theorem and Theorem 13.2.5 to approximate strongly the operator $E(\Delta)$ by operators of type $q(A)(A - \alpha I)^n + I$, where q are polynomials.

(ii) Assume that $\alpha \notin \Delta$. We set $f = \chi_\Delta$ and define $p_f(t) \equiv 0$ and

$$h_f(t) = \frac{\chi_\Delta(t)}{(t - \alpha)^n}, \quad t \in \mathbb{R}.$$

Then $h_f \in \mathfrak{B}(\mathbb{R})$ and hence $\chi_\Delta \in \mathfrak{B}^n(\mathbb{R}; \alpha)$. We apply the definition of $E(\chi_\Delta)$ in (2.30) and get

$$E(\chi_\Delta) = -\int \chi_\Delta(t) \, dE(t) = E(\Delta) - I + I = E(\Delta).$$

Finally we proceed as in case (i). ∎

Remark 13.2.12. The functional calculus with Borelian functions, introduced in Theorem 13.2.10, enables us to recover the spectral function E in the general case, that is, if the definitisable operator A is not necessarily in $\mathfrak{D}(\alpha; n)$ for some α. More precisely, if $\Delta \in \mathfrak{R}_A$ then the function χ_Δ is in the class of bounded Borelian functions and $E(\chi_\Delta) = E(\Delta)$. ∎

13.3 Regularity of Critical Points

Let A be a definitisable selfadjoint operator in the Kreĭn space $\mathcal{K}$. If $\alpha \in \mathbb{R} \setminus c(A)$ then, from Corollary 11.7.5, it follows that for any $\lambda_0 < \alpha < \lambda_1$ such that $\lambda_0, \lambda_1 \notin c(A)$, the strong operator limits

$$\lim_{\lambda \downarrow \alpha} E([\lambda, \lambda_1]) \quad \text{and} \quad \lim_{\lambda \uparrow \alpha} E([\lambda_0, \lambda]) \qquad (3.1)$$

exist.

Let $\alpha \in c(A)$. If the strong operator limits (3.1) exist then α is called a *regular critical point* of A. We denote by $c_{\mathrm{r}}(A)$ the set of regular critical points of A and consider also the set $c_{\mathrm{s}}(A) = c(A) \setminus c_{\mathrm{r}}(A)$. The elements in $c_{\mathrm{s}}(A)$ are called *singular critical points*.

If ∞ is a critical point of A then it is called *regular* if the strong operator limits

$$\lim_{\lambda \downarrow -\infty} E([\lambda, \lambda_1]) \quad \text{and} \quad \lim_{\lambda \uparrow +\infty} E([\lambda_0, \lambda]) \qquad (3.2)$$

exist, for some $\lambda_0, \lambda_1 \notin c(A)$. Denote by $\widetilde{c}_r(A)$ the *extended set of regular critical points* of A, that is, $\widetilde{c}_r(A) = c_r(A) \cup \{\infty\}$ if ∞ is a regular critical point, otherwise, $\widetilde{c}_r(A) = c_r(A)$. Also, $\widetilde{c}_s(A) = \widetilde{c}(A) \setminus \widetilde{c}_r(A)$ dentoes the *extended set of singular critical points* of A.

Proposition 13.3.1. *Let $\alpha \in \widetilde{c}(A)$ and let $\Lambda \in \mathfrak{R}_A$ be a bounded neighbourhood of α, if $\alpha \neq \infty$, and then if $\alpha = \infty$ let $\Lambda = (-\alpha, \lambda_0) \cup (\lambda_1, +\infty)$, where $\lambda_0 < \lambda_1$ are such that $\Lambda \cap c(A) = \emptyset$. The following statements are equivalent.*

(a) *α is a regular critical point.*

(b) *There exist mutually orthogonal selfadjoint projections $E_-(\Lambda)$, $E_+(\Lambda)$, and E_α in the bicommutant of $\{R(\zeta; A) \mid \zeta \in \rho(A)\}$, such that*

$$E(\Lambda) = E_+(\Lambda) + E_-(\Lambda) + E_\alpha, \tag{3.3}$$

and

$$\sigma(A|E_+(\Lambda)\mathcal{K}) \subseteq \operatorname{Clos} \Lambda \cap \{\lambda \mid \lambda \geqslant \alpha\} \ (\textit{respectively} \subseteq (-\infty, \lambda_0], \textit{ if } \alpha = \infty), \tag{3.4}$$

$$\sigma(A|E_-(\Lambda)\mathcal{K}) \subseteq \operatorname{Clos} \Lambda \cap \{\lambda \mid \lambda \leqslant \alpha\} \ (\textit{respectively} \subseteq [\lambda_1, +\infty), \textit{ if } \alpha = \infty), \tag{3.5}$$

$$E_\alpha \mathcal{K} = \mathcal{S}_\alpha. \tag{3.6}$$

(c) $E(\Lambda)\mathcal{K} = \mathcal{S}_\alpha[+]\mathcal{S}_{\alpha,\Lambda}^-[+]\mathcal{S}_{\alpha,\Lambda}^+$, *where the subspaces $\mathcal{S}_{\alpha,\Lambda}^\pm$ are defined as in Proposition 13.1.4.*

(d) *The family of selfadjoint projections $\{E(\Delta)\}_{\Delta \in \mathfrak{R}_A, \Delta \subseteq \Lambda}$ is uniformly bounded.*

Proof. (a)$\Rightarrow$(b). Assume $\alpha \in c(A)$ and let $\Lambda = (\beta_0, \beta_1)$. Define

$$E_+(\Lambda) = \lim_{\lambda \downarrow \alpha} E([\lambda, \beta_1)) \quad \text{and} \quad E_-(\Lambda) = \lim_{\lambda \uparrow \alpha} E((\beta_0, \lambda]).$$

By Proposition 13.1.4 and the regularity of the critical point α it follows that $\mathcal{S}_\alpha$ is a regular subspace and hence E_α is the bounded selfadjoint projection onto $\mathcal{S}_\alpha$. If $\alpha = \infty \in \widetilde{c}_r(A)$ then define

$$E_+(\Lambda) = \lim_{\lambda \downarrow -\infty} E([\lambda, \lambda_0]) \quad \text{and} \quad E_-(\Lambda) = \lim_{\lambda \uparrow +\infty} E([\lambda_1, \lambda]),$$

and $E_\infty = 0$. The proof of (3.3) through (3.6) is straightforward.

(b)$\Rightarrow$(c). Let us assume that $\alpha \neq \infty$ and $\Lambda \cap \{\lambda \mid \lambda \geqslant \alpha\} \subseteq \sigma_+(A)$ and $\Lambda \cap \{\lambda \mid \lambda \leqslant \alpha\} \subseteq \sigma_-(A)$. Remark that $\mathcal{S}_{\alpha,\Lambda}^+ = E_+(\Lambda)\mathcal{K}$ and $\mathcal{S}_{\alpha,\Lambda}^- = E_-(\Lambda)\mathcal{K}$. In particular, the subspaces $\mathcal{S}_{\alpha,\Lambda}^+$ and $\mathcal{S}_{\alpha,\Lambda}^-$ are regular. In view of Proposition 13.1.4 it follows that $E(\Lambda) = E_+(\Lambda) + E_-(\Lambda) + E_\alpha$. The other statements follow in a similar way.

(c)$\Rightarrow$(d). Let E_α, $E_-(\Lambda)$, and $E_+(\Lambda)$ be the bounded selfadjoint projections onto $\mathcal{S}_\alpha$, $\mathcal{S}_{\alpha,\Lambda}^-$, and $\mathcal{S}_{\alpha,\Lambda}^+$. If $\Delta \in \mathfrak{R}_A$, $\Delta \subseteq \Lambda$, remark that $E(\Delta)$ commutes with E_α, $E_-(\Lambda)$, and $E_+(\Lambda)$. By Theorem 2.2.9, there exists a fundamental symmetry J of $\mathcal{K}$ such that the subspaces $\mathcal{S}_\alpha$, $\mathcal{S}_{\alpha,\Lambda}^-$, and $\mathcal{S}_{\alpha,\Lambda}^+$ are invariant under J. Therefore, these subspaces

13.3 REGULARITY OF CRITICAL POINTS

are mutually orthogonal with respect to the positive inner product $\langle \cdot, \cdot \rangle_J$, too. If $\| \cdot \|$ denotes the associated fundamental norm then the projections $E(\Delta)E_\alpha$, $E(\Delta)E_-(\Lambda)$, and $E(\Delta)E_+(\Lambda)$ are selfadjoint also with respect to the positive definite inner product $\langle \cdot, \cdot \rangle_J$. Since

$$E(\Delta) = E(\Delta)E(\Lambda) = E(\Delta)E_\alpha + E(\Delta)E_-(\Lambda) + E(\Delta)E_+(\Lambda),$$

it follows that $\|E(\Delta)\| \leqslant 1$.

(d)⇒(a). This implication is a direct consequence of Proposition 5.2.6. ∎

Remark 13.3.2. Let $\mathfrak{R}'_A$ be the Boolean algebra generated by the real intervals with endpoints $\notin \widetilde{c}_s(A)$. Then the spectral function E extends to $\mathfrak{R}'_A$ with the properties (1) through (6) as in Theorem 11.7.3. ∎

As a first application of the characterisation of the regularity of critical points we present a spectral characterisation of the fundamental reducibility of definitisable selfadjoint operators, see Section 9.4. Recall that a selfadjoint operator A is fundamentally reducible if $JA \subseteq AJ$ for some fundamental symmetry J.

Proposition 13.3.3. *The definitisable selfadjoint operator A is fundamentally reducible if and only if the following three conditions hold.*
  (i) $c(A) \subseteq \mathbb{R}$.
  (ii) $\widetilde{c}(A) = \widetilde{c}_r(A)$.
  (iii) *For any $\alpha \in c(A)$ the subspace $\mathrm{Ker}(A - \alpha I)$ is regular.*

Proof. If A is fundamentally reducible then A is similar to a selfadjoint operator in a Hilbert space and hence its spectrum is real and its spectral function is uniformly bounded and, consequently, all its, finite or infinite, critical points are regular, $\widetilde{c}_r(A) = \widetilde{c}(A)$. Let J be a fundamental symmetry such that $JA \subseteq AJ$. Then for any $\alpha \in c(A)$ we have $J\mathrm{Ker}(A - \alpha I) \subseteq \mathrm{Ker}(A - \alpha I)$ and hence $\mathrm{Ker}(A - \alpha I)$ is a regular subspace.

Conversely, assume that A fulfils the conditions (i), (ii), and (iii). As a consequence of the regularity of all critical points of A it follows that the spectral function $E(\Delta)$ is defined for any real interval. Let $c(A) = \{\alpha_1, \alpha_2, \ldots, \alpha_n\}$, where $\alpha_1 < \alpha_2 < \cdots < \alpha_n$.

We can assume, without loss of generality, that there exist $\beta_0 < \beta_1 < \beta_2 < \cdots < \beta_{n+1}$ such that, on any open interval Δ_k, $k = 1, 2, \ldots, n+1$, corresponding to the partition $-\infty < \beta_0 < \alpha_1 < \beta_1 < \cdots < \alpha_n < \beta_n < +\infty$, the spectral projections $E(\Delta_k)$ are definite. Define the numbers ε_k as follows: $\varepsilon_k = +1$ if $E(\Delta_k)$ is positive and $\varepsilon_k = -1$ in the case when $E(\Delta_k)$ is negative. Let also P_k be the bounded selfadjoint projection onto the regular subspace $\mathrm{Ker}(A - \alpha I))$ and choose a fundamental symmetry J_k on $\mathrm{Ker}(A - \alpha I))$, $k = 1, n$. Then

$$J = \sum_{k=1}^{n} J_k P_k + \sum_{k=1}^{2n+2} \varepsilon_k E(\Delta_k)$$

is a fundamenal symmetry of $\mathcal{K}$ such that $JA \subseteq AJ$. ∎

We now focus on some criteria for the regularity of a given critical point. First we see what it means that the nilpotent operator N_α vanishes, with notation as in Proposition 11.7.7.

Lemma 13.3.4. *Assume that $A \in \mathfrak{D}(\alpha; n)$ and α is a regular critical point. Then for the positive nilpotent operator, defined in Proposition 11.7.7, $N_\alpha = 0$ if and only if the Riesz index $\nu(\alpha) \leqslant n$.*

Proof. If α is a regular critical point then by Proposition 13.1.4 the root subspace $\mathfrak{S}_\alpha$ is a regular subspace and the integral in (2.5) is an ordinary operator valued Stieltjes integral. Hence
$$(A - \alpha I)^n = \int_{\mathbb{R}} (t - \alpha)^n \, dE(t) + N_\alpha. \tag{3.7}$$
Therefore, if $N_\alpha = 0$ then $\operatorname{Ker}(A - \alpha I)^n = \mathfrak{S}_\alpha$. Since, by Proposition 13.1.3, we have $\mathfrak{S}_\alpha = \operatorname{Ker}(A - \alpha I)^{n+1}$ it follows that $\operatorname{Ker}(A - \alpha I)^n = \operatorname{Ker}(A - \alpha I)^{n+1}$ and hence $\nu(\alpha) \leqslant n$.

Conversely, if $\nu(\alpha) \leqslant n$ then $\operatorname{Ker}(A-\alpha I)^n = \operatorname{Ker}(A-\alpha I)^{n+1}$, in particular $\operatorname{Ker}(A - \alpha I)^n = \mathfrak{S}_\alpha$ is a regular subspace. Since $\operatorname{Ran}(N_\alpha) \subseteq (\operatorname{Ker}(A - \alpha I)^n)^0 = \{0\}$ it follows that $N_\alpha = 0$. ∎

Theorem 13.3.5. *Let A be a definitisable selfadjoint operator and $\alpha \in \widetilde{c}(A)$. Assume that there exist an open neighbourhood Δ of $\alpha \in \mathbb{R} \cup \{\infty\}$, such that $A' = A|E(\Delta)\mathcal{K} \in \mathfrak{D}(\alpha; n)$, and a net of functions $(f_\eta)_{\eta \in \mathcal{E}} \subset C^n(\overline{\Delta}; \alpha)$, where $(\mathcal{E}, \succ)$ is a directed set, subject to the following conditions.*

(a) *If $\alpha \neq \infty$ we have*
$$0 < \frac{f_\eta(t)}{t - \alpha} \leqslant M_\eta, \quad t \in \Delta \setminus \{\alpha\}, \, \eta \in \mathcal{E},$$
and, if $\alpha = \infty$ then
$$0 < t f_\eta(t) \leqslant M_\eta, \quad t \in \Delta \setminus \{\alpha\}, \, \eta \in \mathcal{E},$$
for some real numbers M_η, $\eta \in \mathcal{E}$.

(b) $\sup_{\eta \in \mathcal{E}} \sup_{t \in \operatorname{Clos} \Delta} |f_\eta(t)| < +\infty.$

(c) *There exists a real number $a > 0$ such that for any open connected neighbourhood Λ of α there exists $\eta_0 \in \mathcal{E}$ such that*
$$|f_\eta(t)| \geqslant a, \text{ for all } t \in \Delta \setminus \Lambda, \text{ and all } \eta \in \mathcal{E} \text{ such that } \eta \succ \eta_0.$$

Then the following assertions hold.

(i) *If the net of operators $(f_\eta(A'))_{\eta \in \mathcal{E}}$ is uniformly bounded then $\alpha \in \widetilde{c}_r(A)$. If, in addition, $\alpha \neq \infty$ and $h_{f_\eta}(\alpha) \to \infty$ ($\eta \in \mathcal{E}$) then $\operatorname{Ker}((A - \alpha I)^n) = \operatorname{Ker}((A - \alpha I)^{n+1})$.*

(ii) *If $\alpha \in \widetilde{c}_r(A)$ and either $\alpha \neq \infty$ and $\operatorname{Ker}((A - \alpha I)^n) = \operatorname{Ker}((A - \alpha I)^{n+1})$ or the net of numbers $(h_{f_\eta}(\alpha))_{\eta \in \mathcal{E}}$ is bounded, then the net of operators $(f_\eta(A'))_{\eta \in \mathcal{E}}$ is uniformly bounded.*

13.3 REGULARITY OF CRITICAL POINTS

Proof. (i) By Proposition 13.3.1 we can assume, without any loss of generality, that $\tilde{\sigma}(A) \subseteq \mathrm{Clos}\, \Delta$ and hence $A = A'$. We now consider the notation in (2.14). Let us first assume that $\alpha \neq \infty$. If $\eta \in \mathcal{E}$ is arbitrary then

$$f_\eta(t) = h_{f_\eta}(t)\,(t-\alpha)^n + p_f(t), \quad t \in \Delta.$$

By (a) it follows that

$$0 < h_{f_\eta}(t) + \frac{p_{f_\eta}(t)}{(t-\alpha)^n} \leqslant M_\eta, \quad t \in \Delta,$$

and taking into account that the degree of the polynomial p_{f_η} is at most $n-1$ it follows that $p_{f_\eta} \equiv 0$ and

$$0 < h_{f_\eta}(t), \quad t \in \Delta. \tag{3.8}$$

By Theorem 13.2.5 we have

$$f_\eta(A) = \int_{(\alpha)} h_{f_\eta}(t)\,(t-\alpha)^n\, \mathrm{d}E(t) + h_{f_\eta}(\alpha)N_\alpha. \tag{3.9}$$

Let Δ' be a closed connected subset of $\Delta \setminus \{\alpha\}$. By (c) there exists $\eta_0 \in \mathcal{E}$ such that

$$|f_\eta(t)| \geqslant a > 0, \quad t \in \Delta',\, \eta \succ \eta_0. \tag{3.10}$$

Therefore, from (3.9) and (3.10) it follows that

$$|[E(\Delta')x, x]| \leqslant a^{-1} \int_{\Delta'} |f_\eta(t)|\, \mathrm{d}|[E(t)x, x]|$$

$$\leqslant a^{-1} \int_{(\alpha)} f_\eta(t)\, \mathrm{d}[E(t)x, x]$$

$$= a^{-1}[f_\eta(A)x, x] - a^{-1}h_{f_\eta}(\alpha)[N_\alpha x, x]. \tag{3.11}$$

If the net of operators $f_\eta(A)$ is uniformly bounded, taking into account that from (3.8) it follows that $h_{f_\eta}(\alpha) > 0$ for all $\eta \in \mathcal{E}$, from (3.11) we obtain that $E(\Delta')$ is uniformly bounded and hence α is a regular critical point.

In the case when $\alpha = \infty$, similar formulae hold with $N_\alpha = 0$. By Theorem 13.1.7 we can apply the statement just proved to draw the same conclusion.

Let us assume that, in addition, $\alpha \neq \infty$ and $h_{f_\eta}(\alpha) \to +\infty$ ($\eta \in \mathcal{E}$). Therefore, by (3.11) it follows that $N_\alpha = 0$. We apply Lemma 13.3.4 to conclude $\mathrm{Ker}(A - \alpha I)^n = \mathrm{Ker}(A - \alpha I)^{n+1}$.

(ii) If $\alpha \in c_r(A)$ then we use Lemma 13.3.4 and (3.9) to see that if either $\mathrm{Ker}(A - \alpha I)^n = \mathrm{Ker}(A - \alpha I)^{n+1}$ or the net $(h_{f_\eta}(\alpha))_{\eta \in \mathcal{E}}$ is bounded then the second term in (3.9) is bounded. Taking into account the assumption (b) and (3.9) it follows that the net of operators $(f_\eta(A))_{\eta \in \mathcal{E}}$ is uniformly bounded. ∎

Theorem 13.3.5 enables us to obtain some other criteria for regularity of critical points.

Corollary 13.3.6. *Assume that $\alpha \in \sigma(A)$ and there exists a definitising polynomial p of A such that α is a zero of multiplicity $2k$ for some $k > 0$. Then $\alpha \notin c_s(A)$ and $\nu(\alpha) \leqslant 2k$ if and only if the family of operators*

$$\bigl(I + i\eta R(\alpha + i\eta; A)\bigr)^k \bigl(I - i\eta R(\alpha - i\eta; A)\bigr)^k, \quad 0 < \eta < 1,$$

is uniformly bounded.

Proof. Consider the net of functions

$$f_\eta(t) = \left(1 + \frac{i\eta}{t - (\alpha + i\eta)}\right)^k \left(1 - \frac{i\eta}{t - (\alpha - i\eta)}\right)^k, \quad 0 < \eta < 1,$$

where the indexing set is $\mathcal{E} = (0, 1)$ and $\eta_1 \succ \eta_2$ if and only if $\eta_1 \leqslant \eta_2$. We remark that this net has all the properties required in Theorem 13.3.5 and $\lim_{\eta \downarrow 0} h_{f_\eta}(\alpha) = +\infty$. Applying the holomorphic functional calculus with these functions we obtain the net of operators as in the statement. Then we apply Theorem 13.3.5. ∎

Corollary 13.3.7. *Let $\alpha \in \sigma(A)$ and assume that there exists a definitising polynomial p of A such that α is a zero of order $2k + 1$ for some $k \geqslant 0$. Then $\alpha \notin c_s(A)$ and $\nu(\alpha) \leqslant 2k + 1$ if and only if the family of operators*

$$\bigl(I + i\eta R(\alpha + i\eta; A)\bigr)^k \bigl(I - i\eta R(\alpha - i\eta; A)\bigr)^k \left(\int_{\alpha - i}^{\alpha - i\eta} R(A; \zeta) \, d\zeta + \int_{\alpha + i}^{\alpha + i\eta} R(A; \zeta) \, d\zeta\right)$$

is uniformly bounded with respect to $0 < \eta < 1$.

Proof. Consider the net of functions,

$$f_\eta(t) = \left(1 + \frac{i\eta}{t - (\alpha + i\eta)}\right)^k \left(1 - \frac{i\eta}{t - (\alpha - i\eta)}\right)^k \left(\int_{\alpha - i}^{\alpha - i\eta} \frac{1}{\zeta - t} \, d\zeta + \int_{\alpha + i}^{\alpha + i\eta} \frac{1}{\zeta - t} \, d\zeta\right),$$

where the indexing set is $\mathcal{E} = (0, 1)$ and $\eta_1 \succ \eta_2$ if and only if $\eta_1 \leqslant \eta_2$. We remark that this net has all the properties required in Theorem 13.3.5 and $\lim_{\eta \downarrow 0} h_{f_\eta}(\alpha) = +\infty$. Applying the holomorphic functional calculus to these functions we obtain the net of operators as in the statement. Then we apply Theorem 13.3.5. ∎

Corollary 13.3.8. *The point ∞ is not a singular critical point of the definitisable operator A if and only if for some fixed real number β the family of operators*

$$\int_{\beta + i}^{\beta + i\eta} R(\zeta; A) \, d\zeta + \int_{\beta - i\eta}^{\beta - i} R(\zeta; A) \, d\zeta, \quad 1 < \eta < +\infty,$$

is uniformly bounded.

Proof. Set $\mathcal{E} = (1, +\infty)$ and $\eta_1 \succ \eta_2$ if and only if $\eta_1 \geqslant \eta_2$. Then define, for arbitrary $\eta \in \mathcal{E}$,

$$f_\eta(t) = i\left(\int_{\beta+i}^{\beta+i\eta} \frac{1}{\zeta - t}\,d\zeta + \int_{\beta-i\eta}^{\beta-i} \frac{1}{\zeta - t}\,d\zeta\right)$$
$$= 2\int_1^\eta \frac{t - \beta}{(\beta - t)^2 + s^2}\,ds, \quad t \in \mathbb{R}.$$

We leave to the reader to show that the net $(f_\eta)_{\eta \in \mathcal{E}}$ satisfies the assumptions (a), (b), and (c) in Theorem 13.3.5, corresponding to $\alpha = \infty$. Finally we apply Theorem 13.3.5. ∎

13.4 The Inverse Spectral Problem

Given a definitisable selfadjoint operator A on the Kreĭn space $\mathcal{K}$ and denoting by E its spectral function, in general one cannot recover the operator A from the spectral function E, see Proposition 11.7.7. Thus, given a $\mathcal{B}(\mathcal{K})$ valued spectral function E on $\mathbb{R} \setminus c$, where c is some finite subset of $\mathbb{R}$, it is natural to attempt to determine all the definitisable selfadjoint operators on $\mathcal{K}$ such that E is the spectral function associated with these operators. We should make the observation that by "to determine" we mean, firstly, to find necessary and sufficient conditions such that the problem is solvable and, secondly, to describe all the solutions in terms of parameters as simply as possible. Roughly speaking, this is what we mean by the *inverse spectral problem* for definitisable selfadjoint operators.

On the other hand, since the spectral function of a definitisable operator is of measure type on any open interval free of critical points, it follows that the problem refers to the behaviour of the spectral function in the neighbourhoods of critical points. Thus, without loss of generality, it is sufficient to study the spectral functions of the following particular type.

Definition 13.4.1. Let $\mathfrak{R}_{(0)}$ denote the Boole algebra generated by intervals $\Delta \in \mathbb{R}$ such that $0 \notin \partial \Delta$. By definition, $E \colon \mathfrak{R}_{(0)} \to \mathcal{B}(\mathcal{K})$ is a $d(0)$-*homomorphism* if the following conditions hold.

  (i) For any $\Delta \in \mathfrak{R}_{(0)}$, $E(\Delta)$ is a bounded selfadjoint projection and E is a homomorphism of Boole algebras.

  (ii) There exist $a, b \in \mathbb{R}$, $a < 0 < b$ such that $E([a, b]) = I$.

  (iii) At least one of the following statements holds:
    (p) for any $\Delta \in \mathfrak{R}_{(0)}$, $0 \notin \Delta$, the operator $E(\Delta)$ is positive;
    (n) for any $\Delta \in \mathfrak{R}_{(0)}$, $\Delta \subseteq (-\infty, 0)$, $E(\Delta)$ is negative and, for any $\Delta \in \mathfrak{R}_{(0)}$, $\Delta \subseteq (0, +\infty)$, $E(\Delta)$ is positive.

  (iv) There exists $k \in \mathbb{N}$ such that the family of operators

$$\left\{\int_\Delta t^k\,dE(t) \mid \Delta \in \mathfrak{R}_{(0)},\ 0 \notin \Delta\right\}$$

is uniformly bounded.

In the following, we let $\mathfrak{D}(0, k)$ denote the class of selfadjoint operators $A \in \mathcal{B}(\mathcal{K})$ that admits a spectral function E that is a $d(0)$-homomorphism. We also introduce the notation

$$\mathfrak{D}(0) = \bigcup_{n \in \mathbb{N}} \mathfrak{D}(0; n). \tag{4.1}$$

Let E be a $d(0)$-homomorphism on the Kreĭn space $\mathcal{K}$, to which we associate the subspaces

$$\mathcal{S}'_{(0)} = \bigcup \{E(\Delta)\mathcal{K} \mid \Delta \in \mathfrak{R}_{(0)},\ 0 \notin \Delta\}, \tag{4.2}$$

$$\mathcal{S}_{(0)} = \operatorname{Clos} \mathcal{S}'_{(0)}, \tag{4.3}$$

$$\mathcal{S}_0 = \bigcap \{E(\Delta)\mathcal{K} \mid \Delta \in \mathfrak{R}_{(0)},\ 0 \in \Delta\}, \tag{4.4}$$

$$\mathcal{S}_{00} = \mathcal{S}_0 \cap \mathcal{S}_{(0)}. \tag{4.5}$$

It follows easily that

$$\mathcal{S}_0 = \mathcal{S}_{(0)}^{\perp}, \quad \text{and} \quad \mathcal{S}_{00} = \mathcal{S}_0^0 = \mathcal{S}_{(0)}^0, \tag{4.6}$$

the latter meaning that $\mathcal{S}_{00}$ is the isotropic part of $\mathcal{S}_0$ and of $\mathcal{S}_{(0)}$. In particular, the subspace $\mathcal{S}_{00}$ is neutral and we show, in the following, how to remove this neutral subspace.

In the following we also consider the operator denoted by $\int_{(0)} t\, \mathrm{d}E(t)$, with domain $\mathcal{S}'_{(0)} + \mathcal{S}_0$ and, if $x'_{(0)} \in E(\Delta)\mathcal{K}$, $\Delta \in \mathfrak{R}_{(0)}$, $0 \notin \Delta$ and $x_0 \in \mathcal{S}_0$, then

$$\int_{(0)} t\, \mathrm{d}E(t)(x'_{(0)} + x_0) = \int_{\Delta} t\, \mathrm{d}E(t) x'_{(0)}. \tag{4.7}$$

Let J be a fixed fundamental symmetry of $\mathcal{K}$. We consider the *J-reduction of $\mathcal{K}$* by the neutral subspace $\mathcal{S}_{00}$, that is, the decomposition

$$\mathcal{K} = \mathcal{S}_{00} \oplus \widehat{\mathcal{K}} \oplus J\mathcal{S}_{00}. \tag{4.8}$$

Denoting by P the projection onto $\mathcal{S}_{00}$ along $\widehat{\mathcal{K}} \oplus J\mathcal{S}_{00}$, it follows that $P^{\sharp}$ is the projection onto $J\mathcal{S}_{00}$ along $\mathcal{S}_{00} \oplus \widehat{\mathcal{K}}$ and $\widehat{P} = I - P - P^{\sharp}$ is the selfadjoint projection onto the regular subspace $\widehat{\mathcal{K}}$.

Consider the mapping $\widehat{E} \colon \mathfrak{R}_{(0)} \to \mathcal{B}(\widehat{\mathcal{K}})$ defined by

$$\widehat{E}(\Delta) = \widehat{P} E(\Delta)|\widehat{\mathcal{K}}, \quad \Delta \in \mathfrak{R}_{(0)}. \tag{4.9}$$

To this mapping we associate the subspaces

$$\widehat{\mathcal{S}}'_{(0)} = \bigcup \{\widehat{E}(\Delta)\widehat{\mathcal{K}} \mid \Delta \in \mathfrak{R}_{(0)},\ 0 \notin \Delta\}, \tag{4.10}$$

$$\widehat{\mathcal{S}}_{(0)} = \operatorname{Clos} \widehat{\mathcal{S}}'_{(0)}, \tag{4.11}$$

$$\widehat{\mathcal{S}}_0 = \bigcap \{\widehat{E}(\Delta)\widehat{\mathcal{K}} \mid \Delta \in \mathfrak{R}_{(0)},\ 0 \in \Delta\}. \tag{4.12}$$

13.4 THE INVERSE SPECTRAL PROBLEM

Lemma 13.4.2. $\widehat{E}$ *is a $d(0)$-homomorphism on the Kreĭn space $\widehat{\mathcal{K}}$ and the subspaces $\widehat{\mathcal{S}}_{(0)}$ and $\widehat{\mathcal{S}}_0$ are nondegenerate.*

Proof. Let $\Delta \in \mathfrak{R}_{(0)}$ be such that $0 \notin \Delta$. Then the selfadjoint projection $E(\Delta)$ can be represented by

$$E(\Delta) = \begin{bmatrix} 0 & * & * \\ 0 & \widehat{E}(\Delta) & * \\ 0 & 0 & * \end{bmatrix}, \tag{4.13}$$

with respect to the decomposition (4.8). Therefore, for any $\Delta_1, \Delta_2 \in \mathfrak{R}_{(0)}, 0 \notin \Delta_1 \cup \Delta_2$ we have

$$E(\Delta_1)E(\Delta_2) = E(\Delta_1)\widehat{P}E(\Delta_2). \tag{4.14}$$

Using (4.14) we check immediately the properties (i) through (iv) in Definition 13.4.1 for a $d(0)$-homomorphism. Since

$$\mathcal{S}_{(0)} = \mathcal{S}_{00}[+]\widehat{\mathcal{S}}_{(0)}, \tag{4.15}$$

it follows that the subspaces $\widehat{\mathcal{S}}_{(0)}$ and $\widehat{\mathcal{S}}_0 = \widehat{\mathcal{K}} \cap \mathcal{S}_{(0)}^\perp$ are nondegenerate. ∎

As a consequence of Lemma 13.4.2 we have

$$\widehat{\mathcal{K}} = \widehat{\mathcal{S}}_{(0)} \vee \widehat{\mathcal{S}}_0. \tag{4.16}$$

Lemma 13.4.3. *Let $x \in \mathcal{K} \setminus \mathcal{S}_{00}^\perp$ be arbitrary. Then there exists no vector $y \in \widehat{\mathcal{K}}$ such that for all $\Delta \in \mathfrak{R}_{(0)}$ such that $0 \notin \Delta$ we have*

$$\widehat{P}E(\Delta)x = \widehat{P}E(\Delta)y. \tag{4.17}$$

Proof. Let $\Delta \in \mathfrak{R}_{(0)}$ be such that $0 \notin \Delta$. From (4.13) we get

$$E(\Delta)\widehat{\mathcal{K}} \subseteq E(\Delta)\mathcal{K} = E(\Delta)\widehat{P}E(\Delta)\mathcal{K} \subseteq E(\Delta)\widehat{\mathcal{K}},$$

and hence

$$E(\Delta)\widehat{\mathcal{K}} = E(\Delta)\mathcal{K}. \tag{4.18}$$

Assume that there exists $x \in \mathcal{K} \setminus \mathcal{S}_{00}^\perp$ and $y \in \widehat{\mathcal{K}}$ such that (4.17) holds for all $\Delta \in \mathfrak{R}_{(0)}$ such that $0 \notin \Delta$. Taking into account (4.18) it follows that for any $z \in \mathcal{S}_{(0)}'$ we have $[x - y, z] = 0$, and hence $x \in y + \mathcal{S}_0 \subseteq \mathcal{S}_{(0)}' + \mathcal{S}_0 \subseteq \mathcal{S}_{(0)} + \mathcal{S}_0 = \mathcal{S}_{00} \oplus \widehat{\mathcal{K}} = \mathcal{S}_{00}^\perp$, thus contradicting the assumption $x \in \mathcal{K} \setminus \mathcal{S}_{00}^\perp$. ∎

Remark 13.4.4. Let $\mathcal{N}$ be a neutral subspace of $\mathcal{K}$ and for some fixed fundamental symmetry J of $\mathcal{K}$ consider the J-reduction of $\mathcal{K}$ by $\mathcal{N}$, that is, the decomposition

$$\mathcal{K} = \mathcal{N} \oplus \widehat{\mathcal{K}} \oplus J\mathcal{N}. \tag{4.19}$$

With respect to this decomposition the operator J has the representation

$$J = \begin{bmatrix} 0 & 0 & J_{13} \\ 0 & J_{22} & 0 \\ J_{13}^* & 0 & 0 \end{bmatrix}, \qquad (4.20)$$

where $J_{22} \in \mathcal{B}(\widehat{\mathcal{K}})$ is a fundamental symmetry of $\widehat{\mathcal{K}}$ and $J_{13} \in \mathcal{B}(J\mathcal{N}, \mathcal{N})$ is a unitary operator (of Hilbert spaces).

Let $A \in \mathcal{B}(\mathcal{K})$ be arbitrary. Then A is selfadjoint and the subspace $\mathcal{N}$ is invariant under A if and only if, with respect to the decomposition (4.19), the operator A is represented by

$$A = \begin{bmatrix} A_{11} & A_{12} & A_{13} \\ 0 & A_{22} & A_{23} \\ 0 & 0 & A_{33} \end{bmatrix}, \qquad (4.21)$$

where $A_{11} = J_{13}A_{33}^*J_{13}^*$, $A_{12} = J_{13}A_{23}^*J_{22}$, $A_{22} = J_{22}A_{22}^*J_{22}$, and $A_{13} = J_{13}A_{13}^*J_{13}$. ∎

In the following an important rôle will be played by the technical condition

$$\text{the operator} \int_{(0)} t \,\mathrm{d}\widehat{E}(t) \text{ is bounded}, \qquad (4.22)$$

where $\widehat{E}$ is the $d(0)$-homomorphism defined at (4.9) with respect to the J-reduction of $\mathcal{K}$ by the neutral subspace $\mathcal{S}_{00}$ and the integral operator is defined in (4.7) with respect to $\widehat{E}$. We also make the remark that this condition does not depend on the fundamental symmetry J, in the sense that it holds for some fundamental symmetry if and only if it holds for any other.

In the following we obtain a characterisation of a $d(0)$-morphism that arises as the spectral function of some definitisable operator.

Theorem 13.4.5. *Assume that the $d(0)$-homomorphism E fulfils the technical condition (4.22). Then the following statements are equivalent.*

  (i) *E is the spectral function of some definitisable selfadjoint operator in $\mathcal{K}$.*
  (ii) *For any $x \in J\mathcal{S}_{00}$ there exists $z \in \mathcal{K}$ such that*

$$\int_\Delta t \,\mathrm{d}E(t)x = E(\Delta)z, \quad \Delta \in \mathfrak{R}_{(0)}, \; 0 \notin \Delta. \qquad (4.23)$$

Proof. (i)⇒(ii). If E is the spectral function of some selfadjoint definitisable operator $A \in \mathcal{B}(\mathcal{K})$ then, by Corollary 11.7.5 it follows that for any $x \in \mathcal{K}$ we have

$$\int_\Delta t \,\mathrm{d}E(t)x = E(\Delta)Ax, \quad \Delta \in \mathfrak{R}_{(0)}, \; 0 \notin \Delta,$$

and hence the statement holds with $z = Ax$.

13.4 THE INVERSE SPECTRAL PROBLEM

(ii)⇒(i). We first observe that the statement (ii) can be reformulated as follows. For any $x_0 \in JS_{00}$ there exists $x_1 \in JS_{00}$ and $y \in \widehat{\mathcal{K}}$ such that

$$\widehat{P}\int_\Delta t\,dE(t)x_0 - \widehat{P}E(\Delta)x_1 = \widehat{P}E(\Delta)y, \quad \Delta \in \mathfrak{R}_{(0)},\ 0 \notin \Delta. \tag{4.24}$$

Denote $\widetilde{\mathcal{K}} = \widehat{\mathcal{K}}/\widehat{\mathcal{S}}_0$. By means of Lemma 13.4.3 it follows that, given a vector $x_0 \in JS_{00}$, the vector $x_1 \in JS_{00}$ and the coset $y + \widehat{\mathcal{S}}_0 \in \widetilde{\mathcal{K}}$ are uniquely determined by x_0, provided that (4.24) holds. Therefore, we can define the operators

$$C_1 \colon JS_{00} \ni x_0 \mapsto x_1 \in JS_{00}, \quad \widetilde{C}_2 \colon JS_{00} \ni x_0 \mapsto y + \widehat{\mathcal{S}}_0 \in \widetilde{\mathcal{K}}. \tag{4.25}$$

We claim that both C_1 and $\widetilde{C}_2$ are bounded. To this end, let $(x_0^{(m)})_{m\in\mathbb{N}}$ be a sequence of vectors in $JS_{(0)}$ such that

$$x_0^{(m)} \to x_0, \quad C_1 x_0^{(m)} \to x_1, \quad \widetilde{C}_2 x_0^{(m)} \to y + \widehat{\mathcal{S}}_0 \quad (m \to \infty).$$

In view of (4.24), for any $\Delta \in \mathfrak{R}_{(0)}, 0 \notin \Delta$, we have

$$\widehat{P}\int_\Delta t\,dE(t)x_0^{(m)} - \widehat{P}E(\Delta)C_1 x_0^{(m)} = \widehat{P}E(\Delta)\widetilde{C}_2 x_0^{(m)}, \quad m \in \mathbb{N}. \tag{4.26}$$

We pass to the limit in (4.26) following $m \to \infty$ and get

$$\widehat{P}\int_\Delta t\,dE(t)x_0 - \widehat{P}E(\Delta)x_1 = \widehat{P}E(\Delta)y.$$

Taking into account the definition of the operators C_1 and $\widetilde{C}_2$ it follows that $x_1 = C_1 x_0$ and $y + \widehat{\mathcal{S}}_0 = \widetilde{C}_2 x_0$. We have thus proven that the operators C_1 and $\widetilde{C}_2$ are closed. From the Closed Graph Theorem we get that they are bounded, too.

Let now $\widehat{\mathcal{S}}_1$ be a subspace of $\widehat{\mathcal{K}}$ such that $\widehat{\mathcal{K}} = \widehat{\mathcal{S}}_1 \dotplus \widehat{\mathcal{S}}_0$ and denote by Q the bounded projection onto $\widehat{\mathcal{S}}_1$ along $\widehat{\mathcal{S}}_0$. Define $C_2 = Q\widetilde{C}_2 \in \mathcal{B}(JS_{00}, \widehat{\mathcal{K}})$. With respect to the decomposition (4.8) we define the operator $A \in \mathcal{B}(\mathcal{K})$ by

$$A = \begin{bmatrix} A_{11} & A_{12} & 0 \\ 0 & A_{22} & A_{23} \\ 0 & 0 & A_{33} \end{bmatrix}, \tag{4.27}$$

where $A_{33} = C_1$, $A_{23} = C_2$, $A_{11} = J_{13}C_1^* J_{13}^*$, $A_{12} = J_{13}C_2^* J_{22}$, and A_{22} is the extension by continuity of the operator $\int_{(0)} t\,dE(t)$, which exists due to the condition (4.22). From Remark 13.4.4 it follows that $A = A^\sharp$. It remains to prove that A is definitisable and that E is the spectral function of A.

374 MORE ON DEFINITISABLE OPERATORS

We first prove that

$$Ax = \int_{(0)} t\,dE(t)x, \quad x \in \mathcal{S}'_{(0)}. \tag{4.28}$$

Assume that $x \in E(\Delta)\mathcal{K}$ for some $\Delta \in \mathfrak{R}_{(0)}$ such that $0 \notin \Delta$. Then, if $u \in \mathcal{S}_{00}$,

$$[Ax, u] = [x, Au] = 0 = \left[\int_{(0)} t\,dE(t)x, u\right]. \tag{4.29}$$

If $v \in \widehat{\mathcal{K}}$ then

$$[Ax, v] = [(\widehat{P} + P)E(\Delta)x, A\widehat{P}v] = \left[x, \int_\Delta t\,dE(t)v\right]$$
$$= \left[\int_\Delta t\,d\widehat{E}(t)\widehat{P}x, v\right] = \left[\int_{(0)} t\,dE(t)x, v\right]. \tag{4.30}$$

If $w \in J\mathcal{S}_{00}$ then

$$[Ax, w] = [E(\Delta)x, C_2w + C_1w] = [x, \widehat{P}E(\Delta)C_2w + \widehat{P}E(\Delta)C_1w]$$
$$= \left[x, \widehat{P}\int_\Delta t\,dE(t)w\right] = \left[\int_\Delta t\,dE(t)\widehat{P}x, w\right] = \left[\int_{(0)} t\,dE(t)x, w\right]. \tag{4.31}$$

From (4.29), (4.30), and (4.31) we get (4.28).

We will now prove that

$$A_{11}^k = 0, \tag{4.32}$$

where k is the number such that the assumption (iv) in Definition 13.4.1 holds. To this end, let $w_0 \in J\mathcal{S}_{00}$ and denote $w_n = C_1^n w_0 \in J\mathcal{S}_{00}$ for $n = 0, 1, 2, \ldots, k-1$. There exists $x_n \in \widehat{\mathcal{K}}$ such that for all $\Delta \in \mathfrak{R}_{(0)}$ satisfying $0 \notin \Delta$ we have

$$\widehat{P}\int_\Delta t\,dE(t)w_n - \widehat{P}E(\Delta)w_{n+1} = \widehat{P}E(\Delta)x_n. \tag{4.33}$$

13.4 The Inverse Spectral Problem

Applying the operator $\int_\Delta t^{k-n-1}\, dE(t)$ to both sides in (4.33) we get

$$\widehat{P}\int_\Delta t^k\, dE(t)w_0 - \widehat{P}\int_\Delta t^{k-1}\, dE(t)w_1 = \widehat{P}E(\Delta)\int_\Delta t^{k-1}\, d\widehat{E}(t)x_0,$$

$$\widehat{P}\int_\Delta t^{k-1}\, dE(t)w_1 - \widehat{P}\int_\Delta t^{k-2}\, dE(t)w_2 = \widehat{P}E(\Delta)\int_\Delta t^{k-2}\, d\widehat{E}(t)x_0,$$

$$\vdots$$

$$\widehat{P}\int_\Delta t\, dE(t)w_{k-1} - \widehat{P}E(\Delta)w_k = \widehat{P}E(\Delta)x_{k-1}.$$

Summing up all these k equalities, we remark that on the left side there is a telescopic sum, hence

$$\widehat{P}\int_\Delta t^k\, dE(t)w_0 - \widehat{P}E(\Delta)w_k = \widehat{P}E(\Delta)\Big(\sum_{j=0}^{k-1}\int_\Delta t^j\, dE(t)x_{k-1-j}\Big). \tag{4.34}$$

For arbitrary $n \in \mathbb{N}$ denote $\Delta_n = \mathbb{R}\setminus(-n^{-1}, n^{-1})$. If n is sufficiently large we have $\Delta_n \supseteq \Delta$. Then we apply the operator $E(\Delta_n)$ on both sides of (4.34) and, taking into account (4.14), we get

$$\widehat{P}E(\Delta)\widehat{P}\int_{\Delta_n} t^k\, dE(t)w_0 - \widehat{P}E(\Delta)w_k = \widehat{P}E(\Delta)\Big(\sum_{j=0}^{k-1}\int_{\Delta_n} t^j\, dE(t)x_{k-1-j}\Big). \tag{4.35}$$

From the assumption (iv) in Definition 13.4.1 it follows that the strong operator limit exists

$$S = \lim_{n\to\infty}\int_{\Delta_n} t^k\, dE(t) \in \mathcal{B}(\mathcal{K}).$$

Then we pass to the limit in (4.35) following $n \to \infty$ and get

$$\widehat{P}E(\Delta)\widehat{P}Sw_0 - \widehat{P}E(\Delta)w_k = \widehat{P}E(\Delta)\Big(\sum_{j=0}^{k-1} A_{22}^j x_{k-1-j}\Big).$$

This proves that there exists

$$x' = \widehat{P}Sw_0 - \sum_{j=0}^{k-1} A_{22}^j x_{k-1-j} \in \widehat{\mathcal{K}}$$

such that

$$\widehat{P}E(\Delta)w_k = \widehat{P}E(\Delta)x'. \tag{4.36}$$

From Lemma 13.4.3 we conclude $w_k = 0$. We have thus proven that $C_1^k = 0$ and hence $A_{11}^k = J_{13} C_1^{*k} J_{13}^* = 0$.

We prove now that
$$A^{k+1}|\mathcal{S}_0 = 0. \qquad (4.37)$$

Indeed, taking into account (4.32) and (4.27) we have
$$A^{k+1} = A^k A = \begin{bmatrix} 0 & DA_{22} & * \\ 0 & A_{22}^{k+1} & * \\ 0 & 0 & 0 \end{bmatrix},$$

for some bounded operator D. In view of the definition of the operator A_{22} we have $A_{22}|\mathcal{S}_0 = 0$ and since $\mathcal{S}_0 \subseteq \mathcal{S}_{00}^\perp = \mathcal{S}_{00} \oplus \widehat{\mathcal{K}}$, the proof of (4.37) is complete.

Further, if $r \geqslant k + 1$ then, from (4.37) it follows that
$$A^r x = \lim_{n\to\infty} \int_{\Delta_n} t^r \, \mathrm{d}E(t) x, \quad x \in \mathcal{S}_{(0)} \vee \mathcal{S}_0,$$

and hence, for any $x \in \mathcal{K}$ we have
$$A^{r+k} x = A^r A^k x = \lim_{n\to\infty} \int_{\Delta_n} t^r \, \mathrm{d}E(t) \text{ and } A^k x = \lim_{n\to\infty} \int_{\Delta_n} t^{r+k} \, \mathrm{d}E(t) x. \qquad (4.38)$$

If E is of even type then we let $r = k + 1$ in (4.38) and get
$$[A^{2k+1} x, x] \geqslant 0, \quad x \in \mathcal{K},$$

and in the case when E is of odd type then we let $r = k + 2$ in (4.38) and get
$$[A^{2k+2} x, x] \geqslant 0, \quad x \in \mathcal{K}.$$

Thus $A \in \mathfrak{D}(0)$, in particular it is definitisable.

Finally, let us remark that from (4.38) we also get
$$p(A) A^{2k+1} x = \lim_{n\to\infty} \int_{\Delta_n} p(t) t^{2k+1} \, \mathrm{d}E(t) x, \quad x \in \mathcal{K},$$

for any complex polynomial. Taking into account that, according to Corollary 13.2.11 the spectral function of A can be strongly approximated by a sequence of operators $p_n(A) A^{2k+1}$, for some polynomials p_n, $n \geqslant 1$, this shows that E does coincide with the spectral function of A. ∎

Remark 13.4.6. As the proof of Theorem 13.4.5 shows, the technical condition (4.22) is not necessary for the implication (i)⇒(ii). This additional assumption is used only during the proof of the implication (ii)⇒(i) and it is motivated by the approach of the spectral synthesis of the operator A. Also, let us mention that the statement (ii) does not depend on the fundamental symmetry J. ∎

13.4 THE INVERSE SPECTRAL PROBLEM

Corollary 13.4.7. *Assume that E is a $d(0)$-homomorphism and, letting $\widehat{P}$ denote the self-adjoint projection on the regular subspace $\widehat{\mathcal{K}}$ defined as in (4.8), the sequence of operators $(\widehat{P} \int_{\Delta_n} t \, \mathrm{d}E(t))_{n \in \mathbb{N}}$ is strongly convergent, where $\Delta_n = \mathbb{R} \setminus (-n^{-1}, n^{-1})$, $n \in \mathbb{N}$. Then E is the spectral function of some definitisable selfadjoint operator A on $\mathcal{K}$.*

Proof. According to the hypothesis, there exists $S \in \mathcal{B}(\mathcal{K})$

$$Sx = \lim_{n \to \infty} \widehat{P} \int_{\Delta_n} t \, \mathrm{d}E(t)x, \quad x \in \mathcal{K}.$$

Then the technical condition (4.22) is verified, as well as the assertion (ii) in Theorem 13.4.5. ∎

Corollary 13.4.8. *Assume that E is a $d(0)$-homomorphism such that the subspace $\mathcal{S}_0$, as defined in (4.4), is pseudo-regular. Then the subspace $\widehat{\mathcal{S}}_{(0)}$, as defined in (4.11), is regular and, denoting by $\widehat{P}_{(0)}$ the bounded selfadjoint projection onto $\widehat{\mathcal{S}}_{(0)}$, the following statements are equivalent.*

(i) *E is the spectral function of a selfadjoint definitisable operator.*
(ii)′ *For any $x \in J\mathcal{S}_{00}$ there exists $z \in \mathcal{K}$ such that*

$$\widehat{P}_{(0)} \int_\Delta t \, \mathrm{d}E(t)x = \widehat{P}_{(0)} E(\Delta) z, \quad \Delta \in \mathfrak{R}_{(0)},\ 0 \notin \Delta.$$

Proof. Since $\mathcal{S}_{(0)} = \mathcal{S}_0^\perp$ and taking into account (4.15) and Theorem 3.2.1, if $\mathcal{S}_0$ is pseudo-regular then $\widehat{\mathcal{S}}_{(0)}$ is regular. Then the subspace $\widehat{\mathcal{S}}_0$ is also regular and denoting by $\widehat{P}_0$ the bounded selfadjoint projection onto $\widehat{\mathcal{S}}_0$ we have that $\widehat{P} = \widehat{P}_0 + \widehat{P}_{(0)}$ is the selfadjoint projection onto $\widehat{\mathcal{K}}$. Since, for any $\Delta \in \mathfrak{R}_{(0)}$ such that $0 \notin \Delta$ we have $\widehat{P}_0 E(\Delta) = 0$, these show that the statement (ii)′ is equivalent with the statement (ii) in Theorem 13.4.5. ∎

Corollary 13.4.9. *Assume that E is a $d(0)$-homomorphism such that the subspaces*

$$\mathcal{S}_{(0)}^+ = \bigvee \{E(\Delta)\mathcal{K} \mid E(\Delta) \text{ positive}\} \quad \text{and} \quad \mathcal{S}_{(0)}^- = \bigvee \{E(\Delta)\mathcal{K} \mid E(\Delta) \text{ negative}\}$$

are pseudo-regular. Then the following statements are equivalent.

(i) *E is the spectral function of some definitisable selfadjoint operator.*
(ii)″ *For any $x_0 \in J\mathcal{S}_{00}$ there exists $x_1 \in J\mathcal{S}_{00}$ such that*

$$\lim_{n \to \infty} \widehat{P}_{(0)} \Big(\int_{\Delta_n} t \, \mathrm{d}E(t)x_0 - E(\Delta_n)x_1 \Big) = 0,$$

where $\Delta_n = \mathbb{R} \setminus (-n^{-1}, n^{-1})$, $n \in \mathbb{N}$, and $\widehat{P}_{(0)}$ denotes the bounded selfadjoint projection onto the regular subspace $\widehat{\mathcal{S}}_{(0)}$.

Proof. If the subspaces $\mathcal{S}_{(0)}^+$ and $\mathcal{S}_{(0)}^-$ are pseudo-regular then, in view of Proposition 5.2.6 it follows that the $d(0)$-homomorphism $\widehat{E}$ is bounded. In this case the statement (ii)″ is equivalent with the statement (ii)′ in Corollary 13.4.8. ∎

We are now in a position to focus on the description of all selfadjoint definitisable operators associated with a given $d(0)$-homomorphism E. We first define the notation. If E is a $d(0)$-homomorphism with the property (4.22) then we define the subspaces in (4.2) through (4.5). For some fixed fundamental symmetry J of $\mathcal{K}$, let $\widehat{E}$ be the $d(0)$-homomorphism defined as in (4.9) and the subspaces as in (4.10) through (4.12). With respect to the decomposition (4.8) we consider the representation (4.20) of J. Denote $\widehat{\mathcal{S}}_1 = \widehat{\mathcal{K}} \ominus \widehat{\mathcal{S}}_{(0)}$, in particular we have the decomposition

$$\mathcal{K} = \mathcal{S}_{00} \oplus \widehat{\mathcal{S}}_1 \oplus \widehat{\mathcal{S}}_{(0)} \oplus J\mathcal{S}_{00}. \tag{4.39}$$

In view of the hypothesis (4.22) it makes sense to define the operator

$$A_{22} = \int_{(0)} t\, \mathrm{d}\widehat{E}(t) \in \mathcal{B}(\widehat{\mathcal{K}}). \tag{4.40}$$

On the other hand, in view of Lemma 13.4.3 and (4.24), the linear mappings

$$A_{33}\colon J\mathcal{S}_{00} \ni x_0 \mapsto x_1 \in J\mathcal{S}_{00}, \quad A_{23}\colon J\mathcal{S}_{00} \ni x_0 \mapsto \widehat{P}_1 y \in \widehat{\mathcal{S}}_1, \tag{4.41}$$

are correctly defined, where, for any $x_0 \in J\mathcal{S}_{00}$, the vectors x_1 and y are uniquely defined by the equation (4.24). As in the proof of Theorem 13.4.5 we show that both A_{33} and A_{23} are bounded.

Theorem 13.4.10. *Let E be a $d(0)$-homomorphism, assume that the hypotheses (4.22) and (4.24) are fulfilled, and consider the operators A_{22}, A_{23}, and A_{33} defined as in (4.40) and (4.41), respectively. Then, with respect to the decomposition (4.8), the formula*

$$A = \begin{bmatrix} J_{13}A_{33}^*J_{13}^* & J_{13}A_{23}^*B^*J_{22} & C \\ 0 & A_{22} + N & \begin{bmatrix} B \\ A_{23} \end{bmatrix} \\ 0 & 0 & A_{33} \end{bmatrix}, \tag{4.42}$$

establishes a bijective correspondence between the class of operators $A \in \mathfrak{D}(0)$ such that E is the associated spectral function, and the class of triples (B, C, N) subject to the following conditions:

$$\begin{cases} B \in \mathcal{B}(J\mathcal{S}_{00}, \widehat{\mathcal{S}}_{(0)}), \text{ arbitrary}, \\ C \in \mathcal{B}(J\mathcal{S}_{00}, \mathcal{S}_{00}), \ C = J_{13}C^*J_{13}, \\ N \in \mathcal{B}(\widehat{\mathcal{K}}), \ N = N^\sharp, \ N^k = 0, \ \mathrm{Ran}(N) \subseteq \widehat{\mathcal{S}}_0, \ NA_{22} = A_{22}N. \end{cases} \tag{4.43}$$

13.4 The Inverse Spectral Problem

Proof. Let $A \in \mathfrak{D}(0)$ be such that its spectral function coincides with E. Then $\widehat{E}$ is the spectral function of the selfadjoint definitisable operator $\widehat{A} = \widehat{P}A|\widehat{\mathcal{K}}$. Therefore,

$$\widehat{A}|\mathcal{S}_{(0)} = A_{22}|\widehat{\mathcal{S}}_{(0)}, \tag{4.44}$$

and $\widehat{A}A_{22} = A_{22}\widehat{A}$. Define $N \in \mathcal{B}(\widehat{\mathcal{K}})$ by $N = \widehat{A} - A_{22}$. Then $N = N^\sharp$ and $NA_{22} = A_{22}N$. In view of (4.43) it follows that $\operatorname{Ran}(N) \subseteq (\operatorname{Ker}(N))^\perp \subseteq \widehat{\mathcal{S}}_{(0)}^\perp = \widehat{\mathcal{S}}_0$. Let $x_0 \in \widehat{\mathcal{S}}_0$ and $x_{(0)} \in \widehat{\mathcal{S}}_{(0)}$ be arbitrary. Taking into account (4.44) and that there exists $k \in \mathbb{N}$ such that $\widehat{A}^k \widehat{\mathcal{S}}_0 = 0$, we get

$$N^k(x_{(0)} + x_0) = (\widehat{A} - A_{22})^k x_{(0)} + (\widehat{A}_{22})^k x_0 = \widehat{A}^k x_0 + TA_{22}x_0 = 0.$$

In view of (4.16) this shows that $N^k = 0$.

If $\Delta \in \mathfrak{R}_{(0)}$ is such that $0 \notin \Delta$ and $x \in J\mathcal{S}_{00}$, in view of Corollary 11.7.5 we have

$$\int_\Delta t\,\mathrm{d}E(t)x = E(\Delta)Ax.$$

Therefore, taking into account (4.41) we get

$$P^\sharp \widehat{A}|J\mathcal{S}_{00} = A_{33}, \quad \widehat{P}_1\widehat{A}|J\mathcal{S}_{00} = A_{23}.$$

The representation (4.42) with parameters B, C, and N subject to the conditions (4.43) now follows.

Conversely, if A is defined as in (4.42) and the operators B, C, and N satisfy the conditions (4.43) then we follow the lines of the proof of (ii)$\Rightarrow$(i) in Theorem 13.4.5 to conclude that E is the spectral function of the definitisable selfadjoint operator A. ∎

Let us assume now that the subspace $\mathcal{S}_0$ is pseudo-regular. Then both $\widehat{\mathcal{S}}_0$ and $\widehat{\mathcal{S}}_{(0)}$ are regular subspaces and, as a consequence of (4.16), we have

$$\widehat{\mathcal{K}} = \widehat{\mathcal{S}}_0[+]\widehat{\mathcal{S}}_{(0)}. \tag{4.45}$$

There exists a fundamental symmetry J, possibly different from the fundamental symmetry fixed before, such that $\widehat{\mathcal{K}} = \widehat{\mathcal{S}}_0 \oplus \widehat{\mathcal{S}}_{(0)}$ and hence,

$$\mathcal{K} = \mathcal{S}_{00} \oplus \widehat{\mathcal{S}}_0 \oplus \widehat{\mathcal{S}}_{(0)} \oplus J\mathcal{S}_{00}, \tag{4.46}$$

and, with respect to (4.45), J admits the representation

$$J_{22} = \begin{bmatrix} J_{22}^0 & 0 \\ 0 & J_{22}^{(0)} \end{bmatrix}.$$

Corollary 13.4.11. *Let E be a $d(0)$-homomorphism and assume that the subspace $\mathcal{S}_0$ is pseudo-regular and the assertion (ii)′ of Corollary 13.4.8 holds. Then, with respect to the decomposition (4.46), the formula*

$$A = \begin{bmatrix} J_{13}A_{33}^*J_{13}^* & J_{13}B^*J_{22}^0 & J_{13}A_{23}^*J_{22}^{(0)} & C \\ 0 & N & 0 & A_{23} \\ 0 & 0 & A_{22} & B \\ 0 & 0 & 0 & A_{33} \end{bmatrix}$$

establishes a bijective correspondence between the class of operators $A \in \mathfrak{D}(0)$ such that E is the associated spectral function and the class of triples (B, C, N) subject to the following conditions:

$$\begin{cases} B \in \mathcal{B}(J\mathcal{S}_{00}, \widehat{\mathcal{S}}_{(0)}), \text{ arbitrary}, \\ C \in \mathcal{B}(J\mathcal{S}_{00}, \mathcal{S}_{00}), \ C = J_{13}C^*J_{13}, \\ N \in \mathcal{B}(\widehat{\mathcal{S}}_0), \ N = N^\sharp, \ N^k = 0. \end{cases}$$

Proof. We only remark that, if $A \in \mathfrak{D}(0)$ has the spectral function E then the decomposition (4.45) reduces the operator $\widehat{A} = \widehat{P}A|\widehat{\mathcal{K}}$ to a diagonal matrix. The remainder follows from Theorem 13.4.10. ∎

13.5 Invariant Maximal Semidefinite Subspaces

In this section we prove the existence of maximal semidefinite subspaces invariant under a commutative family of selfadjoint definitisable operators. Some preliminary results on commutative families of bounded selfadjoint projections are first considered. To begin with, let us recall, see Remark 5.2.1, that if E_1 and E_2 are two commuting bounded positive projections on a Kreĭn space $\mathcal{K}$ then $\mathrm{Ran}(E_1) + \mathrm{Ran}(E_2)$ is a positive regular subspace of $\mathcal{K}$.

Lemma 13.5.1. *Let $\mathfrak{F}$ be a family of mutually orthogonal, bounded selfadjoint projections. If $(\mathcal{L}_+, \mathcal{L}_-)$ is an alternating pair of subspaces, invariant under all operators $F \in \mathfrak{F}$, then there exists a maximal alternating pair of subspaces $(\widetilde{\mathcal{L}}_+, \widetilde{\mathcal{L}}_-)$ invariant under all operators $F \in \mathfrak{F}$.*

Proof. For arbitrary $F \in \mathfrak{F}$ the subspace $\mathrm{Ran}(F)$ is a regular subspace of $\mathcal{K}$ and, since both $\mathcal{L}_+$ and $\mathcal{L}_-$ are invariant under F, then $(F\mathcal{L}_+, F\mathcal{L}_-)$ is an alternating pair in $\mathrm{Ran}(F)$. By Theorem 2.2.9 there exists a pair $(\mathcal{L}_+^F, \mathcal{L}_-^F)$ which is maximal alternating in $\mathrm{Ran}(F)$. Let us consider the subspaces

$$\mathcal{L}_{\pm,0} = \mathcal{L}_\pm \vee \bigvee_{F \in \mathfrak{F}} \mathcal{L}_\pm^F.$$

We prove that $\{\mathcal{L}_+^0, \mathcal{L}_-^0\}$ is an alternating pair of subspaces invariant under all operators $F \in \mathfrak{F}$.

13.5 Invariant Maximal Semidefinite Subspaces

To see this let us consider $F_1, \ldots, F_n \in \mathfrak{F}$, $F_j \neq F_k$ for all $j \neq k$. For arbitrary vectors $x \in \mathcal{L}_+$ and $y_j \in \mathcal{L}_+^F$, taking into account that $F_j F_k = 0$ for all $k \neq j$, we have

$$[x + \sum_{j=1}^n y_j, x + \sum_{j=1}^n y_j] = [(I - \sum_{j=1}^n F_j)x, (I - \sum_{j=1}^n F_j)x] + \sum_{j=k}^n [F_j(x+y_j), x+y_j] \geq 0.$$

This proves that the subspace $\mathcal{L}_{+,0}$ is positive. In a similar way we show that the subspace $\mathcal{L}_{-,0}$ is negative. In addition, since $\mathcal{L}_+ \perp \mathcal{L}_-$ and for all $F \in \mathfrak{F}$ we have $\mathcal{L}_+^F \perp \mathcal{L}_-^F$, $\mathcal{L}_+^F \perp \mathcal{L}_-$, and $\mathcal{L}_+ \perp \mathcal{L}_-^F$, it follows that $\mathcal{L}_{+,0} \perp \mathcal{L}_{-,0}$. Therefore, $\{\mathcal{L}_{+,0}, \mathcal{L}_{-,0}\}$ is an alternating pair of subspaces in $\mathcal{K}$. Both the subspaces $\mathcal{L}_{+,0}$ and $\mathcal{L}_{-,0}$ are invariant under $\mathfrak{F}$ by construction.

We apply again Theorem 2.2.9 and get a maximal alternating pair $\{\widetilde{\mathcal{L}}_+, \widetilde{\mathcal{L}}_-\}$ of subspaces in $\mathcal{K}$ extending $\{\mathcal{L}_+^0, \mathcal{L}_-^0\}$. We prove that both subspaces $\widetilde{\mathcal{L}}_+$ and $\widetilde{\mathcal{L}}_-$ are invariant under $\mathfrak{F}$. Indeed, for arbitrary $F \in \mathfrak{F}$ we have

$$[F\widetilde{\mathcal{L}}_+, \mathcal{L}_-^F] = [\widetilde{\mathcal{L}}_+, F\mathcal{L}_-^F] = [\widetilde{\mathcal{L}}_+, \mathcal{L}_-^F] = \{0\},$$

and hence

$$F\widetilde{\mathcal{L}}_+ \subseteq (\mathcal{L}_-^F)^\perp \cap \mathrm{Ran}(F) = \mathcal{L}_+^F \subseteq \widetilde{\mathcal{L}}_+.$$

In a similar way we prove that $\widetilde{\mathcal{L}}_-$ is invariant under all $F \in \mathfrak{F}$. ∎

The next lemma is a particular case of the main result of this section. We prove it separately since the approach we follow during the proof of the general theorem is to show that it can be reduced to this particular case.

Lemma 13.5.2. *Let $\mathfrak{B}$ be a commutative family of bounded selfadjoint operators on a Kreĭn space $\mathcal{K}$, such that for all $B \in \mathfrak{B}$ the subspace $\mathrm{Ker}(B)$ is nondegenerate and at least one of the operators B, B^2 and $-B^2$ is positive. Let also $\mathfrak{F}$ be a mutually orthogonal family of bounded selfadjoint projections on $\mathcal{K}$ such that $FB = BF$ for all $B \in \mathfrak{B}$ and all $F \in \mathfrak{F}$. Then there exists a maximal positive subspace $\widehat{\mathcal{L}}_+$ invariant under $\mathfrak{B} \cup \mathfrak{F}$.*

Proof. We split the family $\mathfrak{B}$ into three disjoint families: $\mathfrak{A}$ consists of all $B \in \mathfrak{B}$ such that B is positive, $\mathfrak{V}$ consists of all $B \in \mathfrak{B}$ such that B^2 is positive and $\mathfrak{W}$ consists of all $B \in \mathfrak{B}$ such that $-B^2$ is positive.

Let us note that, any operator $B \in \mathfrak{A}$ is definitisable and, letting E^B denote its spectral function and taking into account that $\mathrm{Ker}(B)$ is nondegenerate, by Proposition 11.8.2, B has the following integral representation

$$B = \int_{-\infty}^{+\infty} t \mathrm{d}E^B(t), \tag{5.1}$$

where the integral is improper at 0 and $\pm\infty$ and converges in the strong operator topology.

We consider the subspaces

$$\mathcal{S}_+^B = \bigvee_{t>0} \mathrm{Ran}(E^B(t, +\infty)), \quad \mathcal{S}_-^B = \bigvee_{t<0} \mathrm{Ran}(E^B(-\infty, t)),$$

and
$$\mathcal{S}_\pm = \bigvee_{B \in \mathfrak{A}} \mathcal{S}_\pm^B.$$

Clearly $\mathcal{S}_+$ is positive and $\mathcal{S}_-$ is negative. Since $E^B(t, +\infty)E^{B'}(-\infty, t') = 0$ for $B, B' \in \mathfrak{A}$, $t > 0$ and $t' < 0$, it follows that $\mathcal{S}_+ \perp \mathcal{S}_-$.

Define now the subspaces
$$\mathcal{L}_+ = \mathcal{S}_+ \vee \bigvee_{V \in \mathfrak{V}} \operatorname{Ran}(V), \quad \mathcal{L}_- = \mathcal{S}_- \vee \bigvee_{W \in \mathfrak{W}} \operatorname{Ran}(W).$$

We prove that $(\mathcal{L}_+, \mathcal{L}_-)$ is an alternating pair of subspaces in $\mathcal{K}$, invariant under $\mathfrak{B} \cup \mathfrak{F}$. To begin with, we first prove
$$\mathcal{L}_+ \perp \mathcal{L}_-. \tag{5.2}$$

Indeed, for all $V \in \mathfrak{V}$ and $W \in \mathfrak{W}$ we have $\mathcal{S}_+ \perp \operatorname{Ran}(W)$ and $\mathcal{S}_- \perp \operatorname{Ran}(V)$. In addition, since all subspaces $\operatorname{Ker}(V)$ and $\operatorname{Ker}(W)$ are nondegenerate, it follows that $\operatorname{Ran}(V^2)$ is dense in $\operatorname{Ran}(V)$ and, similarly, $\operatorname{Ran}(W^2)$ is dense in $\operatorname{Ran}(W)$. For arbitrary vectors $x, y \in \mathcal{K}$ we have
$$|[V^2 x, W^2 y]| \leqslant [V^2 W x, W x][V^2 W y, W y] = 0,$$

and hence $\operatorname{Ran}(V) \perp \operatorname{Ran}(W)$. We have thus proven (5.2).

Further, note that in order to prove that $\mathcal{L}_-$ is a negative subspace, it is sufficient to prove that for all $m, n \in \mathbb{N}$, all real numbers $t_1, \ldots, t_n$, all operators $U_1, \ldots, U_m \in \mathfrak{A}$ and $W_1, \ldots, W_n \in \mathfrak{W}$, and all vectors $x_1, \ldots, x_m \in \mathcal{K}$ and $y_1, \ldots, y_n \in \mathcal{K}$ we have
$$[\sum_{j=1}^m E_j x_j + \sum_{k=1}^n W_k y_k, \sum_{j=1}^m E_j x_j + \sum_{k=1}^n W_k y_k] \leqslant 0, \tag{5.3}$$

where we denote $E_j = E^{U_j}(-\infty, t_j)$, $j = 1, m$. We prove (5.3) by induction following $n \geqslant 1$. Indeed,
$$[\sum_{j=1}^m E_j x_j + W_1 y_1, \sum_{j=1}^m E_j x_j + W_1 y_1] \leqslant 0,$$

since $\operatorname{Ran}(E_1), \ldots, \operatorname{Ran}(E_m), \operatorname{Ran}(W_1)$ are negative and all operators commute. Let us now assume that (5.3) holds for $n \in \mathbb{N}$. We have to prove
$$[\sum_{j=1}^m E_j x_j + \sum_{k=1}^{n+1} W_k y_k, \sum_{j=1}^m E_j x_j + \sum_{k=1}^{n+1} W_k y_k] \leqslant 0. \tag{5.4}$$

Taking into account that $\operatorname{Ker}(W_{n+1})$ is nondegenerate, hence $\operatorname{Ker}(W_{n+1}) + \operatorname{Ran}(W_{n+1})$ is dense in $\mathcal{K}$, (5.4) will be proved if we replace x_j by $W_{n+1} x_j + \hat{x}_j$, where $\hat{x}_j \in$

13.5 Invariant Maximal Semidefinite Subspaces

$\text{Ker}(W_{n+1})$, $j = 1, \ldots, m$, and similarly, we replace y_k by $W_{n+1}y_k + \hat{y}_k$, where $y_k \in \text{Ker}(W_{n+1})$, $k = 1, \ldots, n+1$. In this case we have

$$[\sum_{j=1}^{m} E_j(W_{n+1}x_j + \hat{x}_j) + \sum_{k=1}^{n+1} W_k(W_{n+1}y_k + \hat{y}_k), \sum_{j=1}^{m} E_j(W_{n+1}x_j + \hat{x}_j)$$
$$+ \sum_{k=1}^{n+1} W_k(W_{n+1}y_k + \hat{y}_k)]$$
$$= [W_{n+1}^2(y_{n+1} \sum_{j=1}^{m} E_j x_j + \sum_{k=1}^{n} W_k y_k), \sum_{j=1}^{m} E_j x_j + \sum_{k=1}^{n} W_k y_k]$$
$$+ [\sum_{j=1}^{m} E_j \hat{x}_j + \sum_{k=1}^{n} W_k \hat{y}_k, \sum_{j=1}^{m} E_j \hat{x}_j + \sum_{k=1}^{n} W_k \hat{y}_k] \leqslant 0,$$

where, on the right-hand side the first term is negative since W_{n+1} is negative and the latter term is negative by the induction hypothesis. We have thus proven that the subspace $\mathcal{L}_-$ is negative. In a similar way we prove that the subspace $\mathcal{L}_+$ is positive.

Clearly, both $\mathcal{L}_+$ and $\mathcal{L}_-$ are invariant under $\mathfrak{F}$ and hence, by Lemma 13.5.1, they can be extended to a maximal alternating pair of subspaces $(\widetilde{\mathcal{L}}_+, \widetilde{\mathcal{L}}_-)$ which are still invariant under $\mathfrak{F}$. It remains to prove that they are invariant under the family $\mathfrak{B}$, too. To see this, recall that we split $\mathfrak{B}$ into three disjoint subfamilies $\mathfrak{U}$, $\mathfrak{V}$ and $\mathfrak{W}$. For $U \in \mathfrak{U}$ we have

$$U\widetilde{\mathcal{L}}_+ \subseteq U(\mathcal{L}_-)^\perp \subseteq U(\mathcal{S}_-)^\perp \subseteq \mathcal{S}_+ \subseteq \mathcal{L}_+ \subseteq \widetilde{\mathcal{L}}_+.$$

For $V \in \mathfrak{V}$ we have
$$V\widetilde{\mathcal{L}}_+ \subseteq \text{Ran}(V) \subseteq \mathcal{L}_+ \subseteq \widetilde{\mathcal{L}}_+,$$

and for $W \in \mathfrak{W}$ we have
$$W\widetilde{\mathcal{L}}_+ \subseteq W\mathcal{L}_-^\perp \subseteq W \text{Ran}(W)^\perp = 0.$$

Thus, $\widetilde{\mathcal{L}}_+$ is invariant under $\mathfrak{B}$. Since all the operators in $\mathfrak{B}$ are selfadjoint this implies that $\widetilde{\mathcal{L}}_- = \widetilde{\mathcal{L}}_+^\perp$ is also invariant under $\mathfrak{B}$. ∎

Theorem 13.5.3. *Let $\mathfrak{A}$ be a family of mutually commuting bounded definitisable selfadjoint operators on a Kreĭn space $\mathcal{K}$ and let $\mathfrak{F}$ be a family of mutually orthogonal bounded selfadjoint projections in $\mathcal{K}$ such that for all $A \in \mathfrak{A}$ and $F \in \mathfrak{F}$ we have $AF = FA$ and there exist $n \in \mathbb{N}$ and $\alpha \in \mathbb{R}$ such that at least one of the operators $\pm (A - \alpha I)^n F$ is positive. Then there exists a maximal alternating pair $(\widetilde{\mathcal{L}}_+, \widetilde{\mathcal{L}}_-)$ invariant under all operators in $\mathfrak{A} \cup \mathfrak{F}$.*

Proof. We show that the problem can be reduced to the particular case already treated in Lemma 13.5.2. First, we replace operators in $\mathfrak{A}$ by the family $\mathfrak{B}$ of all operators of form $\pm(A - \alpha I)F$, where $\alpha \in \mathbb{R}$ and $n \in \mathbb{N}$ are numbers depending on A and F, and we choose either of the signs $\pm$ such that $\pm(A - \alpha I)F$ is positive, $A \in \mathfrak{A}$ and $F \in \mathfrak{F}$.

Second, by Zorn's Lemma we obtain a neutral subspace $\mathcal{N} \subseteq \mathcal{K}$, maximal with the property that it is invariant under all operators in $\mathfrak{B} \cup \mathfrak{F}$. Let $J \in \mathcal{B}(\mathcal{K})$ be an arbitrary fundamental symmetry and consider the J-reduction of $\mathcal{K}$ by the neutral subspace $\mathcal{N}$

$$\mathcal{K} = \mathcal{N} \oplus \widehat{\mathcal{K}} \oplus J\mathcal{N},$$

where $\widetilde{\mathcal{K}} = (\mathcal{N} \oplus J\mathcal{N})^\perp$ is a regular subspace of $\mathcal{K}$. We consider the families $\widetilde{\mathfrak{B}}$ and $\widetilde{\mathfrak{F}}$ consisting of bounded operators $\widetilde{B} = P_{\widetilde{\mathcal{K}}} B|\widetilde{\mathcal{K}}$ and $\widetilde{F} = P_{\widetilde{\mathcal{K}}} F|\widetilde{\mathcal{K}}$, respectively, on the Kreĭn space $\widetilde{\mathcal{K}}$. The operators $\widetilde{B} \in \widetilde{\mathfrak{B}}$ are mutually commuting and either negative or positive of finite order, that is, at least one of the operators $\pm B^n$ is positive for some $n \in \mathbb{N}$, the operators $\widetilde{F} \in \widetilde{\mathfrak{F}}$ are mutually orthogonal bounded selfadjoint projections and $\widetilde{A}\widetilde{F} = \widetilde{F}\widetilde{A}$ for all $\widetilde{A} \in \widetilde{\mathfrak{A}}$ and $\widetilde{F} \in \widetilde{\mathfrak{F}}$. Note that, if $(\widehat{\mathcal{L}}_+, \widehat{\mathcal{L}}_-)$ is a maximal alternating pair of subspaces of $\widetilde{\mathcal{K}}$ invariant under $\widetilde{\mathfrak{A}} \cup \widetilde{\mathfrak{F}}$, then letting $\widetilde{\mathcal{L}}_+ = \widehat{\mathcal{L}}_+ \oplus \mathcal{N}$ and $\widetilde{\mathcal{L}}_- = \widehat{\mathcal{L}}_- \oplus \mathcal{N} = \widetilde{\mathcal{L}}_+^\perp$, then $(\widetilde{\mathcal{L}}_+, \widetilde{\mathcal{L}}_-)$ is a maximal alternating pair of subspaces of $\mathcal{K}$ invariant under $\mathfrak{A} \cup \mathfrak{F}$. Thus we can replace $\mathfrak{B}$ and $\mathfrak{F}$ with $\widetilde{\mathfrak{B}}$ and $\widetilde{\mathfrak{F}}$, respectively, which, due to the maximality property of the neutral subspace $\mathcal{N}$, implies that the family $\mathfrak{B}$ has the property that there exists no nontrivial neutral subspace invariant under all operators $B \in \mathfrak{B}$. It is easy to see that this implies that for all $B \in \mathfrak{B}$ the kernel $\mathrm{Ker}(B)$ is nondegenerate.

Further, for an arbitrary $B \in \mathfrak{B}$ let $n \in \mathbb{N}$ be such that $n \geqslant 3$ and B^n is positive. Since $\mathrm{Ker}(B)$ is nondegenerate, equivalently, $\mathrm{Ker}(B) + \mathrm{Ran}(B)$ is dense in $\mathcal{K}$, it follows that for all $x \in \mathcal{K}$ and $y \in \mathrm{Ker}(B)$ we have

$$[B^{n-2}(Bx + y), Bx + y] = [B^n x, x] \geqslant 0.$$

Thus, modulo the replacement of all negative operators with their opposites, we can assume that for all $B \in \mathfrak{B}$ at least one of the operators B, B^2 and $-B^2$ is positive. This concludes the reduction of the statement to the particular case of Lemma 13.5.2. ∎

Theorem 13.5.3 has two immediate consequences.

Corollary 13.5.4. *Let $\mathfrak{A}$ be a finite family of mutually commuting bounded definitisable selfadjoint operators on the Kreĭn space $\mathcal{K}$. Then there exists a maximal alternating pair of subspaces $(\widetilde{\mathcal{L}}_+, \widetilde{\mathcal{L}}_-)$ invariant under all operators $A \in \mathfrak{A}$.*

Proof. As in the proof of Theorem 13.5.3, we reduce the problem to the case when there exists no nontrivial neutral subspace which is invariant under all operators in $\mathfrak{A}$. Then it follows that for all $A \in \mathfrak{A}$ its spectrum $\sigma(A)$ is real.

For each operator $A \in \mathfrak{A}$ we consider a definitising polynomial p_A. Let $\{\alpha_k\}_{k=1}^n$ be the set of all real zeros of p_A. We let $F_k^A = E^A(\Delta_k)$, where E^A is the spectral function of A, $\{\Delta_k\}_{k=1}^n$ is a covering of $\sigma(A) \cap \mathbb{R}$ with compact intervals such that for each $k = 1, \ldots, n$ the definitising polynomial p_A has one and only one zero in the interval Δ_k.

Let $\mathfrak{F} = \bigcup_{A \in \mathfrak{A}} \{F_n^A\}$. Clearly, all the bounded selfadjoint projections $F \in \mathfrak{F}$ commute with all operators $A \in \mathfrak{A}$. Since $\mathfrak{F}$ is finite and consists of mutually commuting

13.5 Invariant Maximal Semidefinite Subspaces

selfadjoint projections, it is atomic in the sense that there exists a finite set $\mathfrak{F}'$ consisting of mutually orthogonal selfadjoint projections such that for every $F \in \mathfrak{F}$ there exists $F_1', \ldots, F_m' \in \mathfrak{F}'$ such that $F = F_1' + \cdots + F_m'$. We replace $\mathfrak{F}$ with $\mathfrak{F}'$ and then we can apply Theorem 13.5.3. ∎

Corollary 13.5.4 shows, in particular, that every bounded selfadjoint definitisable operator in a Kreĭn space has an invariant maximal positive subspace. We would like to have a similar result for unbounded definitisable selfadjoint operators, too. In this case we use a special definition of invariance, see the discussion at the end of Section 9.1.

Corollary 13.5.5. *Let A be a definitisable selfadjoint operator in the Kreĭn space $\mathcal{K}$. Then there exists a maximal alternating pair $(\widetilde{\mathcal{L}}_+, \widetilde{\mathcal{L}}_-)$ of $\mathcal{K}$, such that $\widetilde{\mathcal{L}}_\pm \cap \mathrm{Dom}(A)$ is dense in $\widetilde{\mathcal{L}}_\pm$ and $A(\widetilde{\mathcal{L}}_\pm \cap \mathrm{Dom}(A)) \subseteq \widetilde{\mathcal{L}}_\pm$.*

Proof. If ∞ is not a critical point then the statement is clear. Assume that ∞ is a critical point. Let E denote the spectral function of A and Δ_∞ be a real neighbourhood of ∞ such that $AE(\Delta_\infty)$ is either positive or negative. We split the Kreĭn space $\mathcal{K} = \mathcal{K}_0[+]\mathcal{K}_\infty$, where $\mathcal{K}_\infty = E(\Delta_\infty)\mathcal{K}$. Then $\mathcal{K}_\infty$ has an alternating pair of subspaces invariant under $A_\infty = A|\mathcal{K}_\infty$ constructed by means of the spectral projections of A. For the operator $A_0 = A|\mathcal{K}_0$ we apply Theorem 13.5.3. ∎

By Corollary 13.5.4 every finite family of mutually commuting bounded definitisable selfadjoint operators has a joint invariant maximal alternating pair of subspaces. In order to extend this statement to infinite families an additional condition will be imposed.

Proposition 13.5.6. *Let $\mathfrak{A}$ be a family of mutually commuting bounded definitisable selfadjoint operators on the Kreĭn space $\mathcal{K}$ such that there exists a fundamental symmetry J on $\mathcal{K}$ with the property that $J^+ A J^-$ is compact for any $A \in \mathfrak{A}$. Then there exists a maximal alternating pair $(\widetilde{\mathcal{L}}_+, \widetilde{\mathcal{L}}_-)$ of subspaces of $\mathcal{K}$, invariant under all operators $A \in \mathfrak{A}$.*

Proof. Let $A \in \mathfrak{A}$ and, with respect to the fundamental decomposition $\mathcal{K} = \mathcal{K}^+[+]\mathcal{K}^-$, corresponding to the fundamental symmetry J as in the statement, represent A as

$$A = \begin{bmatrix} A_{11} & A_{12} \\ -A_{12}^* & A_{22} \end{bmatrix}.$$

Then the operator $A_{12} \in \mathcal{B}(\mathcal{K}^-, \mathcal{K}^+)$ is compact. If $\mathcal{L}_+$ is a maximal positive subspace invariant under A and $K \in \mathcal{B}(\mathcal{K}^+, \mathcal{K}^-)$ is its angular operator then

$$-A_{12}^* + A_{22} K = K A_{11} + K A_{12} K. \tag{5.5}$$

By Corollary 13.5.4, for every finite set $\{A_1, \ldots, A_n\} \in \mathfrak{A}$ the set $\Phi(A_1, \ldots, A_n)$ consisting of all contractions $K \in \mathcal{B}(\mathcal{K}^+, \mathcal{K}^-)$ which satisfy equation (5.5) is nonvoid. Due to the compactness of the operators A_{12} this set is also compact with respect to the weak operator topology in $\mathcal{B}(\mathcal{K}^+, \mathcal{K}^-)$. The family of all sets of type $\Phi(A_1, \ldots, A_n)$, where $n \in \mathbb{N}$ and $A_1, \ldots, A_n \in \mathfrak{A}$ is then a nested family of compact subsets. Therefore, its intersection is nonempty, that is, there exists a contraction $K \in \mathcal{B}(\mathcal{K}^+, \mathcal{K}^-)$ satisfying (5.5) for all $A \in \mathfrak{A}$. Letting $\widetilde{\mathcal{L}}_+ = \{x + Kx \mid x \in \mathcal{K}^+\}$ this is a maximal positive subspace of $\mathcal{K}$ invariant under all operators in $\mathfrak{A}$. Clearly, $(\widetilde{\mathcal{L}}_+, \widetilde{\mathcal{L}}_-)$, where $\widetilde{\mathcal{L}}_- = \widetilde{\mathcal{L}}_+^\perp$, is a maximal alternating pair of subspaces invariant under $\mathfrak{A}$. ∎

Since any bounded selfadjoint operator in a Pontryagin space is definitisable and the compactness condition in Proposition 13.5.6 holds, we have the following result.

Corollary 13.5.7. *For any family of mutually commuting selfadjoint bounded operators on a Pontryagin space $\mathcal{K}$ there exists a maximal alternating pair $(\widetilde{\mathcal{L}}_+, \widetilde{\mathcal{L}}_-)$ of subspaces in $\mathcal{K}$, invariant under all the operators in $\mathfrak{A}$.*

13.6 Definitisable Unitary Operators

The spectral theory of definitisable selfadjoint operators has a counterpart in the framework of unitary operators. We will sketch the basic results which, in many respects, parallel the spectral theory of definitisable selfadjoint operators and hence proofs will be omitted.

A *trigonometric polynomial* is, by definition, a mapping $p: \mathbb{C} \setminus \{0\} \to \mathbb{C}$ with a formal expression

$$p(\lambda) = \sum_{k=-n}^{n} a_k \lambda^k, \quad \lambda \in \mathbb{C} \setminus \{0\}, \tag{6.1}$$

where $\{a_k\}_{k=-n}^{n} \subset \mathbb{C}$. The trigonometric polynomial p is called *symmetric* if

$$p(\lambda^*) = \overline{p(\lambda)}, \quad \lambda \in \mathbb{C} \setminus \{0\},$$

where $\lambda^* = \overline{\lambda}^{-1}$.

In the following we denote by $\mathbb{T} = \partial \mathbb{D}$ the unit circle in the complex plane (the one dimensional torus).

Remark 13.6.1. If the trigonometric polynomial p is represented as in (6.1) then the following assertions are equivalent.
  (i) p is symmetric.
  (ii) For all $k = 1, \ldots, n$ we have $a_k = \overline{a}_{-k}$.
  (iii) For any $\zeta \in \mathbb{T}$ we have $p(\zeta) \in \mathbb{R}$. ∎

Proposition 13.6.2. *Let U be a bounded unitary operator in the Kreĭn space $\mathcal{K}$. The following statements are equivalent.*
  (i) *There exists a nontrivial trigonometric polynomial p such that*

$$[p(U)x, x] \geqslant 0, \quad x \in \mathcal{K}. \tag{6.2}$$

13.6 Definitisable Unitary Operators

(i)′ *There exists a nontrivial symmetric trigonometric polynomial p such that (6.2) holds.*

(ii) *There exists a mapping $\varphi \in \mathfrak{H}(\sigma(U))$, that is, φ is holomorphic in a neighbourhood of $\sigma(U)$, which is nontrivial in the neighbourhood of any connected component of $\sigma(U)$, such that*

$$[\varphi(U)x, x] \geq 0, \quad x \in \mathcal{K}. \tag{6.3}$$

(ii)′ *There exists a mapping $\varphi \in \mathfrak{H}(\sigma(U))$, nontrivial in the neighbourhood of any connected component of $\sigma(U)$, such that*

$$\varphi(\lambda^*) = \overline{\varphi(\lambda)}, \quad \lambda \in V,$$

for some neighbourhood V of $\sigma(U)$, and (6.3) holds.

The proof of Proposition 13.6.2 follows the lines of the proof of Proposition 11.3.4.

A unitary operator $U \in \mathcal{B}(\mathcal{K})$ is called *definitisable* if one (and hence, all) of the statements (i), (i)′, (ii), or (ii)′ is (are) fulfilled. A trigonometric polynomial p with the property (6.2) is called *definitising*.

If p is a definitising trigonometric polynomial of U then the symmetric trigonometric polynomial q defined by

$$q(\lambda) = p(\lambda) + \overline{p(\lambda^*)}, \quad \lambda \in \mathbb{C} \setminus \{0\},$$

is also definitising for U. Thus, we can always assume that a definitising trigonometric polynomial is symmetric.

Proposition 13.6.3. *If U is a definitising unitary polynomial and p is a definitising trigonometric polynomial of U then $\sigma_0(U) = \sigma(U) \setminus \mathbb{T}$ consists of a finite set of points, symmetric with respect to $\mathbb{T}$, and included in $Z(p)$, the set of zeros of p. In addition, if $\lambda \in \sigma_0(U)$ then λ is an eigenvalue of U of finite Riesz index $\nu(\lambda) \leq k(\lambda)$, where $k(\lambda)$ denotes the multiplicity of the zero λ of p.*

Let $U \in \mathcal{B}(\mathcal{K})$ be a definitisable unitary operator. As a consequence of Proposition 13.6.3, if $\mathcal{K}_0$ denotes the spectral subspace of U corresponding to $\sigma_0(U)$ then $\mathcal{K} = \mathcal{K}_0[+]\mathcal{K}_1$ and, with respect to this decomposition U, is reduced to a diagonal operator

$$U = \begin{bmatrix} U_0 & 0 \\ 0 & U_1 \end{bmatrix},$$

where $U_0 \in \mathcal{B}(\mathcal{K}_0)$ is an algebraic (that is, $r(U_0) = 0$ for some nontrivial polynomial r), definitisable unitary operator with $\sigma(U_0) = \sigma_0(U)$ and $U_1 \in \mathcal{B}(\mathcal{K}_1)$ is a definitisable unitary operator with $\sigma(U_1) = \sigma(U) \cap \mathbb{T}$.

Let U be a definitisable unitary operator. The complex number $\lambda \in \sigma(U) \cap \mathbb{T}$ is called a *critical point* of U if it is a zero of any definitising trigonometric polynomial of U. The set of all critical points of U is denoted by $c(U)$.

In the following, for an arbitrary definitisable unitary operator U, we denote by $\mathfrak{R}_U$ the Boole algebra generated by those arcs $\Delta \in \mathbb{T}$ such that the endpoints of Δ avoid the set of the critical points $c(U)$. As for definitisable selfadjoint operators, definitisable unitary operators have unique spectral functions; compare with Theorem 11.7.3.

Theorem 13.6.4. *Let $U \in \mathcal{B}(\mathcal{K})$ be a definitisable unitary operator. There exists a mapping $E \colon \mathfrak{R}_U \to \mathcal{B}(\mathcal{K})$, uniquely determined with the following properties.*
(1) $E(\Delta)^\sharp = E(\Delta)$, $\Delta \in \mathfrak{R}_U$.
(2) *E is a homomorphism of Boole algebras, that is, for any $\Delta, \Delta' \in \mathfrak{R}_U$ we have*

$$E(\Delta \cap \Delta') = E(\Delta)E(\Delta'),$$
$$E(\Delta \cup \Delta') = E(\Delta) + E(\Delta) - E(\Delta)E(\Delta').$$

(3) $E(\mathbb{T}) = I - E(\sigma_0; U)$.
(4) *$E(\Delta)$ is positive (negative), provided that $p|\Delta > 0$ (respectively, $p|\Delta < 0$) for some definitising trigonometric polynomial p of U.*
(5) *$E(\Delta) \in \{U\}''$, that is, $E(\Delta)$ commutes with any bounded operator commuting with U, for all $\Delta \in \mathfrak{R})_U$.*
(6) *If $\Delta \in \mathfrak{R}_U$ then $\sigma(U|E(\Delta)\mathcal{K}) \subseteq \text{Clos } \Delta$.*

The mapping E as in Theorem 13.6.4 is called the *spectral function* of the definitisable unitary operator U. As for the proof of Theorem 13.6.4, the main ideas are similar to those of the proof of its counterpart, Theorem 11.7.3. One first proves an integral representation of the resolvent function of U, similarly as in Theorem 11.5.2, by means of the integral Herglotz Theorem, see Theorem 11.4.5, of representations of holomorphic functions in the unit disc with positive real part. The Stieltjes Inversion Formulae can be transcribed for operator valued functions on the unit circle $\mathbb{T}$ and Jordan contours symmetric with respect to $\mathbb{T}$ and which cross $\mathbb{T}$ transversally. Then, for any arc $\Delta \in \mathbb{T}$ one defines $E(\Delta)$ as a principal value of a Cauchy integral of the resolvent function along a Jordan contour as before, and the properties of the mapping E, as well as its uniqueness, can be proved similarly as in the proof of Theorem 11.7.3.

Here we have the analogue of Corollary 11.7.5.

Corollary 13.6.5. *Let E be the spectral function of the definitisable unitary operator $U \in \mathcal{B}(\mathcal{K})$ and let $\Delta \in \mathfrak{R}_U$ be a closed arc such that $c(U) \cap \Delta = \emptyset$.*
(i) *The mapping $E_\Delta \colon \mathfrak{R}_U \to \mathcal{B}(\mathcal{K})$*

$$E_\Delta(\Lambda) = E(\Delta \cap \Lambda), \quad \Lambda \in \mathfrak{R}_U$$

extends uniquely to a spectral measure with $\text{supp}(E_\Delta) \subseteq \Delta$.
(ii) *There exists a fundamental symmetry J of $\mathcal{K}$ such that $U|E(\Delta)\mathcal{K}$ is a unitary operator in the Hilbert space $(E(\Delta)\mathcal{K}, \langle \cdot, \cdot \rangle_J)$.*
(iii) *E_Δ is the spectral measure of $U|E(\Delta)\mathcal{K}$, in particular*

$$UE(\Delta) = \int_\Delta \zeta \, \mathrm{d}E(t).$$

13.6 Definitisable Unitary Operators

Next we have the analogue of Proposition 11.7.7, with the difference that the integral representation can be obtained for a definitising polynomial p, not a certain rational function.

Corollary 13.6.6. *Let E be the spectral function of the definitisable unitary operator U and let p be a definitising trigonometric polynomial of U. Then, there exists a finite set $\{N_\alpha\}_{\alpha \in Z(p) \cap \mathbb{T}} \subset \mathcal{B}(\mathcal{K})$ with the following properties.*
  (i) $[N_\alpha x, x] \geqslant 0$, for all $x \in \mathcal{K}$ and all $\alpha \in Z(p) \cap \mathbb{T}$.
  (ii) $N_\alpha E(\Delta) = 0$, for all $\alpha \in Z(p) \cap \mathbb{T}$ and all $\Delta \in \mathfrak{R}_U$, such that $\operatorname{Clos} \Delta \cap \{\alpha\} = \emptyset$.
  (iii) $N_\alpha N_\beta = 0$, for all $\alpha, \beta \in Z(p) \cap \mathbb{T}$.
  (iv) $(U - \alpha I)N_\alpha = 0$, for all $\alpha \in Z(p) \cap \mathbb{T}$.
  (v) *The following integral representations hold*

$$p(U)R(\zeta;U) = \int_{\mathbb{T}} \frac{p(\xi)}{\zeta - \xi}\,\mathrm{d}E(\xi) + \sum_{\alpha \in Z(p) \cap \mathbb{T}} \frac{1}{\zeta - \alpha} N_\alpha, \quad \zeta \in \rho(U), \qquad (6.4)$$

$$p(U) = \int_{\mathbb{T}} p(\xi)\,\mathrm{d}E(\xi) + \sum_{\alpha \in Z(p) \cap \mathbb{T}} N_\alpha, \qquad (6.5)$$

where the integrals in (6.4) *and* (6.5) *are improper with singularities in $Z(p) \cap \mathbb{T}$ and strongly operatorial convergent.*

Here we have the analogue of Corollary 11.7.8.

Corollary 13.6.7. *With notation as in Corollary 13.6.6, let $\alpha \in c(U)$ be a zero of order $k(\alpha)$ of p. Then there exists an open arc $\Delta \in \mathfrak{R}_U$ such that $\alpha \in \Delta$ and the improper integral*

$$\int_\Delta (\zeta - \alpha)^{k(\alpha)}\,\mathrm{d}E(\zeta)$$

converges strongly.

These corollaries are consequences of Theorem 13.6.4 and the proofs can be easily traced along the lines of the proofs of their counterparts as in Section 11.7. Also, in the context of unitary operators, Proposition 13.1.2 reads as follows.

Proposition 13.6.8. *Let U be a definitisable unitary operator with spectral function E and consider $\alpha \in \sigma(U) \cap \mathbb{T}$. Then $\alpha \in c(U)$ if and only if for any $\Delta \in \mathfrak{R}_U$ such that $\alpha \in \Delta$, the subspace $E(\Delta)\mathcal{K}$ is indefinite.*

Proposition 13.6.8 shows that, defining the sets

$$\sigma_+(U) = \{\lambda \in \sigma(U) \cap \mathbb{T} \mid \exists \Delta \in \mathfrak{R}_U,\ \lambda \in \Delta,\ E(\Delta)\mathcal{K} \text{ is positive}\}$$

$$\sigma_-(U) = \{\lambda \in \sigma(U) \cap \mathbb{T} \mid \exists \Delta \in \mathfrak{R}_U,\ \lambda \in \Delta,\ E(\Delta)\mathcal{K} \text{ is negative}\},$$

the spectrum of U splits into the disjoint union

$$\sigma(U) = \sigma_0(U) \cup c(U) \cup \sigma_+(U) \cup \sigma_-(U).$$

If $\alpha \in c(U)$ then the subspace

$$S_\alpha = \bigcap_{\alpha \in \Delta \in \mathfrak{R}_U} E(\Delta)\mathcal{K},$$

is the root subspace corresponding to the eigenvalue α and the Riesz index $\nu(\alpha) \leqslant k(\alpha)$, where $k(\alpha)$ denotes the order of multiplicity of the zero α with respect to any definitising trigonometric polynomial. Similar statements to those of Proposition 13.1.4 concerning the localisation in the neighbourhood of critical points and Proposition 13.1.5 concerning the properties of the polynomials of minimal degree, hold. The functional calculus with continuous or Borelian functions, as well as the inverse spectral problem for unitary definitisable operators, can be obtained in the same way.

We record also two results on existence of invariant maximal semi-definite subspaces, that can be proven as in Section 13.5.

Theorem 13.6.9. *Let $\mathfrak{U}$ be a finite family of mutually commuting definitisable unitary operators in a Kreĭn space $\mathcal{K}$. Then there exists a maximal alternating pair $(\widetilde{\mathcal{L}}_+, \widetilde{\mathcal{L}}_-)$ of subspaces in $\mathcal{K}$, invariant under all operators in $\mathfrak{U}$.*

Theorem 13.6.10. *Let $\mathfrak{U}$ be a family of mutually commuting definitisable unitary operators in a Pontryagin space $\mathcal{K}$. Then there exists a maximal alternating pair $(\widetilde{\mathcal{L}}_+, \widetilde{\mathcal{L}}_-)$ of subspaces in $\mathcal{K}$, invariant under all operators in $\mathfrak{U}$.*

The parallelism between the theory of unitary definitisable operators and the theory of definitisable selfadjoint operators can be made more explicit by means of the Cayley transform, see Section 4.5.

Theorem 13.6.11. *Let $\varepsilon, \zeta \in \mathbb{C}$, $|\varepsilon| = 1$, $\zeta \neq \bar{\zeta}$ and let the operators A and U be related by the Cayley transform corresponding to ε and ζ. The following statements are equivalent.*

  (i) *A is a definitisable selfadjoint operator and $\zeta \in \rho(A)$.*
  (ii) *U is a definitisable unitary operator and $\varepsilon \notin \sigma_p(U)$.*

In addition, if one (and hence both) of the assertions (i) and (ii) holds, and φ denotes the conformal mapping as in (5.2) of Section 4.5, then $c(U) = \varphi(\widetilde{c}(A))$, $\mathfrak{R}_U = \{\varphi(\Delta) \mid \Delta \in \mathfrak{R}_A\}$ and, denoting by E_A and E_U the spectral functions of A and of U, respectively, we have

$$E_A(\Delta) = E_U(\varphi(\Delta)), \quad \Delta \in \mathfrak{R}_A. \tag{6.6}$$

Proof. Assume that U is a definitisable unitary operator and $\varepsilon \notin \sigma_p(U)$. From Theorem 4.5.9 and Proposition 11.3.4 it follows that A is a definitisable selfadjoint operator and $\zeta \in \rho(A)$, in particular $\rho(A) \neq \emptyset$. The conformal mapping φ defined as in (5.2) of

13.6 Definitisable Unitary Operators

Section 4.5 is holomorphic in a neighbourhood of $\tilde{\sigma}(A)$ and by holomorphic functional calculus we have $U = \varphi(A)$. Let r be a definitising trigonometric polynomial of U. Then $r \circ \varphi \in \mathfrak{H}(\tilde{\sigma}(A))$ and

$$[(r \circ \varphi)(A)x, x] = [r(\varphi(A))x, x] = [r(U)x, x] \geqslant 0, \quad x \in \mathcal{K}.$$

Since $r \circ \varphi$ is nontrivial on each connected component of $\tilde{\sigma}(A)$, the selfadjoint operator A is definitisable. In addition, from the proof of Proposition 11.3.4 it follows that there exists a definitising polynomial p of A such that $Z(p) \cap \sigma(A) = Z(r \circ \varphi) \cap \sigma(A)$. Therefore, we have $\varphi(c(A)) \subseteq c(U)$.

Conversely, assume that A is a definitisable selfadjoint operator and $\zeta \in \rho(A)$. According to Theorem 4.5.9 the operator U is unitary and $\varepsilon \in \sigma_p(U)$. Let p be a definitising polynomial of A,

$$p(\lambda) = \sum_{k=0}^{n} a_k \lambda^k.$$

As in Section 11.5 we define

$$q(\lambda) = \frac{p(\lambda)}{(\lambda - \zeta)^n (\lambda - \overline{\zeta})^n}$$

and notice that $q \in \mathfrak{H}(\tilde{\sigma}(A))$ and that the operator $q(A)$ is positive. A straightforward argument shows that

$$(q \circ \varphi^{-1})(\mu) = \frac{(\mu - \varepsilon) \sum_{k=0}^{n} a_k (\zeta\mu - \overline{\zeta}\varepsilon)^k (\mu - \varepsilon)^{n-k}}{\varepsilon(\zeta - \overline{\zeta})\mu^n}$$

and hence $q \circ \varphi^{-1} \in \mathfrak{H}(\sigma(U))$ and

$$(q \circ \varphi^{-1})(U) = \frac{1}{\varepsilon(\zeta - \overline{\zeta})} \sum_{k=0}^{n} a_k (U - \varepsilon I)^n (\zeta U - \overline{\zeta}\varepsilon I)^k (U - \varepsilon)^{n-k} U^{-n}.$$

Taking into account that

$$q(A) = p(A)(A - \zeta)^{-n}(A - \overline{\zeta})^{-n} \in \mathcal{B}(\mathcal{K}),$$

it follows that $(q \circ \varphi^{-1})(U) = q(A)$. This implies that U is definitisable. In addition, $q \circ \varphi^{-1}$ is a trigonometric polynomial and hence $c(U) \setminus \{\varepsilon\} \subseteq \varphi(c(A))$.

Let then E_A and E_U be the spectral functions corresponding to A and to U, respectively. From what has been proven up to now we have

$$c(U) \setminus \{\varepsilon\} = \varphi(c(A)).$$

Then

$$\{\Delta \in \mathfrak{R}_A \mid \Delta \text{ bounded}\} = \varphi^{-1}(\{\Lambda \in \mathfrak{R}_U \mid \{\varepsilon\} \cap \overline{\Lambda} = \emptyset\}).$$

Let Δ be a bounded interval in $\mathfrak{R}_A$. Then $\varphi(\Delta) \in \mathfrak{R}_U$ and $\{\varepsilon\} \cap \operatorname{Clos}\varphi(\Delta) = \emptyset$. Let C_Δ be a Jordan contour as in Section 11.6 and such that all the points of $\sigma_0(A) \cup \{\zeta, \bar\zeta\}$ lie outside the domain with boundary C_Δ. Let also Δ_α be the closed interval defined as in Section 11.6, for $\alpha > 0$. Also, for $\delta > 0$ we consider the Jordan curve C_Δ^δ consisting of two closed Jordan arcs, as in Section 11.6. We remark that $\varphi(C_\Delta)$ is a closed Jordan curve (the union of four arcs of circle), symmetric with respect to the unit circle $\mathbb{T}$ and crossing $\mathbb{T}$ transversally in two points, the endpoints of the arc $\varphi(\Delta)$.

Further, we have

$$E_U(\varphi(\Delta))x = \lim_{\alpha \downarrow 0} \lim_{\delta \downarrow 0} \frac{1}{2\pi i} \int_{\varphi(C_{\Delta_\alpha}^\delta)} (\mu I - U)^{-1} x \, d\mu, \quad x \in \mathcal{K}. \qquad (6.7)$$

Let $\alpha, \delta > 0$ and $x \in \mathcal{K}$. Using the formula of change of variables we get

$$\frac{1}{2\pi i} \int_{\varphi(C_{\Delta_\alpha}^\delta)} (\mu I - U)^{-1} x \, d\mu = \frac{1}{2\pi i} \int_{C_{\Delta_\alpha}^\delta} (\varphi(\lambda) I - U)^{-1} \frac{d\varphi(\lambda)}{d\lambda} x \, d\lambda.$$

Since

$$(\varphi(\lambda) I - U)^{-1} \frac{d\varphi(\lambda)}{d\lambda} x = \frac{1}{\zeta - \lambda} g(\zeta - \lambda) (\zeta I - A)(\lambda I - A)^{-1} x,$$

hence

$$\frac{1}{2\pi i} \int_{\varphi(C_{\Delta_\alpha}^\delta)} (\mu I - U)^{-1} x \, d\mu = \frac{1}{2\pi i} \int_{C_{\Delta_\alpha}^\delta} \frac{1}{\zeta - \lambda} (\zeta I - A)(\lambda I - A)^{-1} x \, d\lambda$$

$$= (\zeta I - A) \frac{1}{2\pi i} \int_{C_{\Delta_\alpha}^\delta} \frac{1}{\zeta - \lambda} (\lambda I - A)^{-1} x \, d\lambda. \qquad (6.8)$$

In view of the formula of integral representation of the resolvent function of A and the Stieltjes Inversion Formulae, we get

$$\lim_{\alpha \downarrow 0} \lim_{\delta \downarrow 0} \frac{1}{2\pi i} \int_{C_{\Delta_\alpha}^\delta} \frac{1}{\zeta - \lambda} (\lambda I - A)^{-1} x \, d\lambda = (\zeta I - A)^{-1} E_A(\Delta) x \qquad (6.9)$$

and hence, from (6.7) through (6.8), we get

$$E_U(\varphi(\Delta))x = (\zeta I - A)(\zeta I - A)^{-1} E_A(\Delta) x = E_A(\Delta) x, \quad x \in \mathcal{K}. \qquad (6.10)$$

This shows that (6.6) holds for any bounded $\Delta \in \mathfrak{R}_A$. From (6.9), Proposition 13.1.2, and Proposition 13.6.8 we conclude that $\infty \in \widetilde{c}(A)$ if and only if $\varepsilon \in c(U)$. Thus we conclude that $c(U) = \varphi(c(A))$ and $\mathfrak{R}_U = \{\varphi(\Delta) \mid \Delta \in \mathfrak{R}_A\}$. Finally, this implies that (6.6) holds also for any unbounded $\Delta \in \mathfrak{R}_A$. ∎

Remark 13.6.12. The theory of definitisable unitary operators is more general than the theory of definitisable selfadjoint operators. Indeed, if A is a selfadjoint definitisable operator in $\mathcal{K}$ then $\rho(A) \neq \emptyset$ and hence there exists $\zeta \in \rho(A) \setminus \mathbb{R}$. If U denotes the unitary operator associated by the Cayley transform of A, ζ and ε (for some $\varepsilon \in \mathbb{T}$) then, Theorem 13.6.11 allows us to recover the spectral theory of A from the corresponding results in the spectral theory of the unitary definitisable operator U.

Conversely, let U be a definitisable unitary operator. If $\sigma_p(U)$ does not cover the whole unit circle then one can find a value $\varepsilon \in \mathbb{T}$ which is not an eigenvalue of U. If A denotes the Cayley transform associated with U, ε, and ζ, for some complex number $\zeta \notin \mathbb{R}$, then, by means of the same Theorem 13.6.11, one can obtain the spectral function of U and its spectral theory from the corresponding results referring to the definitisable selfadjoint operator A. In the case when the Kreĭn space $\mathcal{K}$ is separable (in most applications this is always the case), the point spectrum of any definitisable unitary operator does not cover $\mathbb{T}$ and hence this approach can be followed. ∎

13.7 Notes

The results in this chapter concerning the spectral theory of definitisable selfadjoint operators, functional calculi, and invariant maximal semidefinite subspaces were essentially proven by H. Langer [104, 105]. Theorem 13.3.5 and its corollaries belong to P. Jonas [83]. The results in Section 13.4 are essentially from the joint work of the author and P. Jonas [64].

B. Ćurgus obtained in [39] a rather different criterion of regularity for the critical point ∞, more precisely, given a selfadjoint definitisable operator A that has real spectrum and no finite critical points, the point ∞ is regular if and only if there exists a positive bounded and boundedly invertible operator W such that $W \operatorname{Dom}(A) \subseteq \operatorname{Dom}(A)$. B. Ćurgus, H. Langer, and the author show in [40] that there is a positive operator A in a Kreĭn space $\mathcal{K}$, which is obtained as the inverse of a bounded positive operator, such that its nonzero spectrum contains isolated simple eigenvalues only, the norms of the orthogonal spectral projections onto the corresponding eigenspaces of A are uniformly bounded, and infinity is a singular critical point of A.

Functional models for families of commuting selfadjoint operators that are special types of definitisable operators were considered by V. Strauss in [138]. Normal definitisable operators and their spectral theory were considered by M. Kaltenbäck [85, 86]. A rather sophisticated spectral theory of commuting tuples of bounded definitisable selfadjoint operators, from the point of view of joint functional calculus, was considered by M. Kaltenbäck and N. Skrepek in [87].

Appendix

A.1 General Topology

In this section we review the basic concepts and facts on general topology, sometimes called point-set topology, that is used throughout this volume. Proofs and more advanced topics can be found in classical books such as N. Bourbaki [23] and J. L. Kelley [88], to cite a few.

Topology

A *topological space* is a pair $(M, \mathcal{T})$ such that M is a nonempty set and $\mathcal{T} \subseteq \mathcal{P}(M)$ is a collection of subsets of M subject to the following properties.

(i) $\emptyset$ and M are elements of $\mathcal{T}$.
(ii) For any family $\{U_\alpha\}_{\alpha \in A}$ of elements in $\mathcal{T}$, the set $\bigcup_{\alpha \in A} \in \mathcal{T}$.
(iii) For any finite family $\{U_k\}_{k=1}^n$ of elements in $\mathcal{T}$, the set $\bigcap_{k=1}^n U_k \in \mathcal{T}$.

Given a topological space $(M, \mathcal{T})$, we call $\mathcal{T}$ a *topology* on M and we call the elements of $\mathcal{T}$ *open* subsets of M with respect to $\mathcal{T}$. For any subset $A \subseteq M$ there exists a unique open set $\overset{\circ}{A}$, called the *interior* of A, such that $\overset{\circ}{A} \subseteq A$ and it is maximal with this property. $\overset{\circ}{A}$ is the union of all open subsets contained in A.

A subset $F \subseteq M$ is called *closed* in M with respect to $\mathcal{T}$ if there exists $U \in \mathcal{T}$ such that $F = M \setminus U$. For any subset $A \subseteq M$, there exists a unique closed set $\text{Clos } A$, called the *closure* of A, such that $A \subseteq \text{Clos } A$ and $\text{Clos } A$ is minimal with this property. $\text{Clos } A$ is the intersection of all closed subsets containing A.

The *boundary* of a set A is $\partial A := \text{Clos } A \cap \text{Clos}(M \setminus A)$ and it can be proven that $\partial A = \text{Clos } A \setminus \overset{\circ}{A}$.

A subset A of M is *dense* in M if $\text{Clos } A = M$. The topological space M is called *separable* if there is a countable and dense subset of M.

Given a subset A of a topological space $(M, \mathcal{T})$, let

$$\mathcal{T}_A = \{A \cap U \mid U \in \mathcal{T}\}.$$

Then $\mathcal{T}_A$ is a topology on A, called the *topology induced* by $\mathcal{T}$ on A, or the topology of A.

A subset V of M is called a *neighbourhood* of $x \in M$ if there exists $U \in \mathcal{T}$ such that $x \in U \subseteq V$. A neighbourhood may not be open but any open set is a neighbourhood for any of its elements. Given $x \in M$ we denote by $\mathcal{V}_\mathcal{T}(x)$ the collection of all neighbourhoods of x with respect to the topology $\mathcal{T}$, or simply $\mathcal{V}(x)$ if the ambient topology is

A.1 General Topology

clear. A subset $\mathcal{B} \subseteq \mathcal{V}(x)$ is called a *neighbourhood base* of x if for any $V \in \mathcal{V}(x)$ there exists $B \in \mathcal{B}$ such that $B \subseteq V$. A topological space is *first countable* if every point has a countable neighbourhood base.

A subset $\mathcal{B}$ of $\mathcal{T}$ is called a *base* for the topology $\mathcal{T}$ if for any $U \in \mathcal{T}$ there exists a family $\{V_\alpha\}_{\alpha \in A}$ with $V_\alpha \in \mathcal{B}$ for all $\alpha \in A$, such that $U = \bigcup_{\alpha \in A} V_\alpha$. The topological space $(X, \mathcal{T})$ is called *second countable* if it has a countable base. A separable space may fail to be second countable.

Example A.1.1. Let X be un uncountable set and let $\mathcal{T}$ be the topology of all subsets of X that are either empty or are complements of finite subsets of X. Then X is separable since any infinite subset of X is dense, but it is not second countable. ∎

Proposition A.1.2. *A subset $\mathcal{B}$ of the topology $\mathcal{T}$ is a base for $\mathcal{T}$ if and only if:*

(a) $M = \bigcup_{B \in \mathcal{B}} B$,
(b) *for any $U, V \in \mathcal{B}$ and any $x \in U \cap V$ there exists $W \in \mathcal{B}$ such that $x \in W \subseteq U \cap V$.*

A subset $\mathcal{S}$ of $\mathcal{T}$ is called a *subbase* if the collection of all finite intersections of elements of $\mathcal{S}$ is a base of $\mathcal{T}$.

Continuity

Given two topological spaces $(M, \mathcal{T})$ and $(N, \mathcal{S})$, let $f \colon M \to N$ be a mapping. For $x \in M$, the mapping f is called *continuous at x* if for every $V \in \mathcal{V}(f(x))$ there exists $U \in \mathcal{V}(x)$ such that $f(U) \subseteq V$.

Recall that a *preorder* on a nonempty set Λ is a relation $\leqslant$ that is reflexive and transitive. The preordered set $(\Lambda, \leqslant)$ is called *directed* if for any $\lambda, \nu \in \Lambda$ there exists $\gamma \in \Lambda$ such that $\lambda \leqslant \gamma$ and $\mu \geqslant \gamma$. Given a set X, a *net*, or *generalised sequence*, with elements in X is, by definition, a pair (Λ, x), where Λ is a directed preordered set and $x \colon \Lambda \to X$ is a map. The standard notation is $(x_\lambda)_{\lambda \in \Lambda}$, where $x_\lambda := x(\lambda)$ for all $\lambda \in \Lambda$.

In the case when $\Lambda = \mathbb{N}$ with the usual order $\leqslant$ we call the net a *sequence* in X. If $k \colon \mathbb{N} \to \mathbb{N}$ is an increasing map, $k(n) < k(n+1)$ for all $n \in \mathbb{N}$, we say that we have a *subsequence* $(x_{k_n})_n$, where $x_{k_n} := x_{k(n)}$, $n \in \mathbb{N}$, of the sequence $(x_n)_n$. The concept of *subnet* can be also defined but we do not use it in this book.

If $(X, \mathcal{T})$ is a topological space, a net $(x_\lambda)_{\lambda \in \Lambda}$ in X *converges* to $x \in X$ if for every $V \in \mathcal{V}_{\mathcal{T}}(x)$ there exists $\lambda(x) \in \Lambda$ such that, $x_\lambda \in V$ for all $\lambda \in \Lambda$ with $\lambda \geqslant \lambda_x$. In this case we use the notation $x_\lambda \xrightarrow[\lambda \in \Lambda]{\mathcal{T}} x$, or $\mathcal{T}\text{-}\lim_{\lambda \in \Lambda} x_\lambda = x$. If there is no ambiguity, the topology $\mathcal{T}$ can be omitted.

Proposition A.1.3. *Let $f \colon M \to N$ be a mapping between two topological spaces $(M, \mathcal{T})$ and $(N, \mathcal{S})$. The following assertions are equivalent.*

(i) *f is continuous at each element $x \in M$.*
(ii) *For every $U \in \mathcal{S}$ the preimage $f^{-1}(U) \in \mathcal{T}$.*
(iii) *For every closed subset $F \subseteq N$ the preimage $f^{-1}(F)$ is closed in M.*

(iv) *For every* $x \in M$ *and any net* $x_\lambda \xrightarrow[\lambda \in \Lambda]{\mathcal{T}} x$ *it follows that* $f(x_\lambda) \xrightarrow[\lambda \in \Lambda]{\mathcal{S}} f(x)$.

With notation as before, the mapping f is called *open* if for every $V \in \mathcal{T}$ the image $f(V) \in \mathcal{S}$, and it is called *closed* if for every closed subset $E \subseteq M$ the image $f(E)$ is closed in N.

Continuity is preserved under composition.

Proposition A.1.4. *If* $(M, \mathcal{T})$, $(N, \mathcal{S})$, *and* $(P, \mathcal{U})$ *are three topological spaces and* $f: M \to N$ *and* $g: N \to P$ *are continuous mappings, then* $g \circ f: M \to P$, $(g \circ f)(x) = g(f(x))$ *for all* $x \in M$, *is a continuous mapping.*

With notation as before, a mapping $f: M \to N$ is called a *homeomorphism* if it is continuous, bijective, and the inverse mapping $f^{-1}: N \to M$ is continuous.

Separation Properties

A topological space $(M, \mathcal{T})$ is:

(a) T_0 *separated* if for any $x, y \in M$ with $x \neq y$, either there exists $U \in \mathcal{T}$ such that $x \in U$ and $y \notin U$, or there exists $V \in \mathcal{T}$ such that $y \in V$ and $x \notin V$,
(b) T_1 *separated* if for any $x, y \in M$ with $x \neq y$ there exists $U, V \in \mathcal{T}$ such that $x \in U$, $y \in V$, $x \notin V$, and $y \notin U$,
(c) *Hausdorff*, or T_2 *separated*, if for every $x, y \in M$, $x \neq y$, there exist two open sets $U, V \in \mathcal{T}$ such that $x \in U$, $y \in V$, and $U \cap V = \emptyset$,
(d) *regular* if for any $x \in M$ and any closed subset $F \subset M$ such that $x \notin F$, there exist $U, V \in \mathcal{T}$ such that $x \in U$, $F \subseteq V$, and $U \cap V = \emptyset$,
(e) *normal* if for any two closed subsets E and F of M with $E \cap F = \emptyset$, there exist $U, V \in \mathcal{T}$ such that $E \subseteq U$, $F \subseteq V$, and $U \cap V = \emptyset$.

Normality has a characterisation in terms of certain continuous functions.

Theorem A.1.5 (Urysohn's Lemma). *A topological space M is normal if and only if for any closed subsets E and F of M, with $E \cap F = \emptyset$, there exists a continuous function* $f: M \to [0, 1]$ *such that* $f(x) = 0$ *for all* $x \in E$ *and* $f(x) = 1$ *for all* $x \in F$.

Compactness

An *open covering* of a subset S of a topological space $(M, \mathcal{T})$ is a family $\mathcal{U} = \{U_\alpha\}_{\alpha \in A}$ of open subsets of M such that $S \subseteq \bigcup_{\alpha \in A} U_\alpha$. A *subcovering* of $\mathcal{U}$ and S is a subfamily $\mathcal{V} = \{U_\alpha\}_{\alpha \in B}$, for some nonempty subset $B \subseteq A$ such that $S \subseteq \bigcup_{\alpha \in B} U_\alpha$. A covering $\mathcal{U} = \{U_\alpha\}_{\alpha \in A}$ is called *locally finite* if for any $x \in M$ the set $\{\alpha \in A \mid x \in U_\alpha\}$ is finite.

A subset $S \subseteq M$ is called *compact* if any open covering of S has a finite subcovering. A topological space $(M, \mathcal{T})$ is called *compact* if M is a compact subset. Sometimes, as is the case in this book, a compact space is assumed to be Hausdorff as well. The subset S is called *paracompact* if any open covering of S has a locally finite subcovering.

Compactness has extremely useful consequences with respect to continuous functions.

A.1 General Topology

Proposition A.1.6. *Let $(M, \mathcal{T})$ and $(N, \mathcal{S})$ be two topological spaces and let $f \colon M \to N$ be a continuous map.*
  (a) *If A is a compact subset of M then $f(A)$ is compact in N.*
  (b) *If $(M, \mathcal{T})$ is compact and f is continuous and bijective, then its inverse $f^{-1} \colon N \to M$ is continuous as well.*

The topological space M is called *locally compact* if every element $x \in M$ has a compact neighbourhood. Clearly, any compact topological space is locally compact. Locally compact spaces are assumed to be Hausdorff as well.

Proposition A.1.7. *If the topological space M is locally compact, Hausdorff, and with countable base then it is paracompact.*

Example A.1.8. The usual topology on $\mathbb{R}$ is locally compact. Let $\hat{\mathbb{R}} = \mathbb{R} \cup \{-\infty, +\infty\}$ with topology defined in the following way. Any open subset $U \subseteq \mathbb{R}$ is open in $\hat{\mathbb{R}}$, a subset $V \subseteq \hat{\mathbb{R}}$ is a neighbourhood of $-\infty$ if $V \supseteq [-\infty, a)$ for some $a \in \mathbb{R}$, and V is a neighbourhood of $+\infty$ if $V \supseteq (a, +\infty]$ for some $a \in \mathbb{R}$. With this topology, $\hat{\mathbb{R}}$ is a compact space. ∎

A topological space M is called *σ-compact*, or *countable at infinity*, if $M = \bigcup_{n \in \mathbb{N}} M_n$, where M_n is a compact subset of M for each $n \in \mathbb{N}$. Some authors require the additional condition that $M_n \subseteq \overset{\circ}{M}_{n+1}$, for all $n \geqslant 1$.

Proposition A.1.9. *A topological space M is σ-compact if and only if any closed subset of M is a countable union of compact subsets.*

Connected Spaces

A subset A of a topological space $(M, \mathcal{T})$ is *separated* if there exist $U, V \in \mathcal{T}$ such that $U \cap A \neq \emptyset$, $V \cap A \neq \emptyset$, $A \subseteq U \cup V$, and $U \cap V = \emptyset$. The subset is called *connected* if it is not separated.

The following proposition shows two very useful tools to prove connectedness of certain subsets of topological spaces.

Proposition A.1.10. (a) *If $\{A_j\}_{j \in \mathcal{J}}$ is a collection of connected subsets and $\bigcap_{j \in \mathcal{J}} A_j \neq \emptyset$, then $\bigcup_{j \in \mathcal{J}} A_j$ is connected.*
  (b) *Let $f \colon M \to N$ be a continuous map. If $A \subseteq M$ is connected then $f(A)$ is connected.*

Given a subset A of a topological space, for each $x \in A$, let A_x denote the union of all connected subsets of A that contain x. By the previous proposition, A_x is connected and it is the largest connected subset of A that contains x. In this way, we get a partition of A by connected subsets and each element of this partition is called a *connected component* of A.

A *curve*, or *arc*, in a topological space M is a continuous function $\varphi\colon [0,1] \to M$. We let $\varphi^* := \varphi([0,1]) \subseteq M$ denote the *trace* of φ. The mapping φ is sometimes called a *parametrisation*. The curve φ is *closed* if $\varphi(0) = \varphi(1)$. A subset $A \subseteq M$ is *arcwise connected* if for any $x, y \in A$ there exists an arc $\varphi\colon [0,1] \to A$ such that $\varphi(0) = x$ and $\varphi(1) = y$.

Proposition A.1.11. *A topological space that is arcwise connected is necessarily connected.*

The converse statement is not true, in general. However, we can say the following.

Proposition A.1.12. *If A is an open subset of a Euclidean space, then A is connected if and only if it is arcwise connected.*

The topological space M is called *locally connected* if any point $x \in M$ has a neighbourhood base made up by connected sets only.

Let $(X; \mathcal{T})$ be a topological space and let γ_0 and γ_1 be two closed curves in X, both of them parametrised on the interval $[0,1]$. We say that γ_0 and γ_1 are *homotopic* in X if there exists a continuous map $h\colon [0,1] \times [0,1] \to X$ such that

$$h(s, 0) = \gamma_0(s), \quad h(s, 1) = \gamma_1(s), \quad s \in [0, 1],$$

and

$$h(0, t) = h(1, t), \quad t \in [0, 1].$$

Intuitively, this means that the curve γ_0 can be continuously deformed to the curve γ_1.

If the closed curve γ in X is homotopic to a constant curve γ_1, that is, γ_1^* is a singleton, it is called *null-homotopic* in X. If X is connected and if every closed curve in X is null-homotopic in X, we say that X is *simply connected*.

Operations with Topologies

Given a nonempty set M, let $\mathfrak{T}_M$ denote the collection of all topologies on M. If $\mathcal{T}_1, \mathcal{T}_2 \in \mathfrak{T}_M$, then we call $\mathcal{T}_1$ *coarser*, or *weaker*, than $\mathcal{T}_2$ if $\mathcal{T}_1 \subseteq \mathcal{T}_2$, that is, if any subset $U \in \mathcal{T}_1$ is also a subset in $\mathcal{T}_2$. Equivalently, we call $\mathcal{T}_2$ *finer*, or *stronger*, than $\mathcal{T}_1$ and denote this by $\mathcal{T}_1 \preceq \mathcal{T}_2$. Clearly, the coarsest topology on M has only two open sets, $\emptyset$ and M, and the finest topology on M has all subsets of M open.

It is easy to see that, given two topologies $\mathcal{T}_1$ and $\mathcal{T}_2$ on M then $\mathcal{T}_1 \cap \mathcal{T}_2 = \{A \subseteq M \mid A \in \mathcal{T}_1, A \in \mathcal{T}_2\}$ is a topology on M and that $\mathcal{T}_1 \cap \mathcal{T}_2 \preceq \mathcal{T}_j, j = 1, 2$. In particular, given $\mathcal{A} \subseteq \mathcal{P}(M)$ there exists a unique topology $\mathcal{T}(\mathcal{A})$ on M that contains all subsets of M that are elements of $\mathcal{A}$ and is the coarsest with this property. This is called the topology *generated* by $\mathcal{A}$.

Let $(M, \mathcal{T})$ and $(N, \mathcal{S})$ be two topological spaces and consider the Cartesian product $M \times N = \{(x, y) \mid x \in M, y \in N\}$. The topology on $M \times N$ generated by $\{U \times V \mid U \in \mathcal{T}, V \in \mathcal{S}\}$ is called the *product topology* and is denoted by $\mathcal{T} \times \mathcal{S}$. Let $\pi_M\colon M \times N \to M$

A.1 General Topology

be the canonical projection onto M, defined by $\pi_M(x,y) = x$ for all $x \in M$ and all $y \in N$. Similarly one defines $\pi_N \colon M \times N \to N$, the canonical projection onto N, by $\pi_N(x,y) = y$ for all $x \in M$ and all $y \in N$.

Proposition A.1.13. *Assume the notation as before.*
(a) *The canonical projections π_M and π_N are continuous with respect to the product topology on $M \times N$.*
(b) *The product topology $\mathcal{T} \times \mathcal{S}$ is the coarsest topology on $M \times N$ that makes both canonical projections π_M and π_N continuous.*

Let $\sim$ be an equivalence relation in a topological space $(M, \mathcal{T})$ and, letting $M/\sim$ denote the *factor space*, also called the *quotient space*, that is, the collection of all *equivalence classes*, or *cosets*, $[x]$ with respect to $\sim$,

$$[x] = \{y \in M \mid y \sim x\}, \quad x \in M,$$

denote by $\pi \colon M \to M/\sim$ the canonical projection, defined by $\pi(x) = [x]$, for all $x \in M$. On $M/\sim$ we consider

$$\mathcal{T}_\sim = \{U \in M/\sim \mid \pi^{-1}(U) \in \mathcal{T}\},$$

that is, $\mathcal{T}_\sim$ is defined as the collection of all subsets of $M/\sim$ such that their preimages by π are open subsets of M. It can be proven that $\mathcal{T}_\sim$ is a topology on $M/\sim$ called the *factor topology* or, equivalently, the *quotient topology*.

Proposition A.1.14. *Assume that $\sim$ is an equivalence relation on the topological space $(M, \mathcal{T})$ and consider $M/\sim$ as a topological space with the quotient topology $\mathcal{T}_\sim$. Then:*
(a) *The canonical projection $\pi \colon M \to M/\sim$ is continuous.*
(b) *The quotient topology is the finest topology on $M/\sim$ among all topologies that make the canonical projection continuous.*
(c) *If $(M, \mathcal{T})$ is compact then $(M/\sim, \mathcal{T}_\sim)$ is compact.*
(d) *If $(M, \mathcal{T})$ is connected then $(M/\sim, \mathcal{T}_\sim)$ is connected.*

In general, the quotient topology does not preserve separation properties.

Proposition A.1.15. *With notation as in Proposition A.1.14, if $(M, \mathcal{T})$ is Hausdorff, the canonical projection π is an open mapping, and the set $R = \{(x,y) \mid x \sim y\}$ is closed in $M \times M$, then $\mathcal{T}_\sim$ is Hausdorff.*

More generally, let $(M_j, \mathcal{T}_j)_{j \in \mathcal{J}}$ be a family of compact spaces and consider $M = \prod_{j \in \mathcal{J}} M_j$ the direct product. For each $j \in \mathcal{J}$ let $\pi_j \colon M \to M_j$ be the canonical projection and define $\mathcal{T}$ the *product topology* on M, that is, the coarsest topology that makes all the projections π_j continuous.

Theorem A.1.16 (Tychonoff Theorem). *If all the spaces $(M_j, \mathcal{T}_j)_{j \in \mathcal{J}}$ are compact then $M = \prod_{j \in \mathcal{J}} M_j$, with the product topology, is compact.*

Metric Spaces

Recall that a mapping $\rho\colon X \times X \to \mathbb{R}$, for some nonempty set X, is called a *metric* if the following hold.

(i) (Positivity) $\rho(x,y) \geqslant 0$ for all $x, y \in X$.
(ii) (Definiteness) $\rho(x,y) = 0$ if and only if $x = y$.
(iii) (Symmetry) $\rho(x,y) = \rho(y,x)$ for all $x, y \in X$.
(iv) (Triangle Inequality) $\rho(x,z) \leqslant \rho(x,y) + \rho(y,z)$ for all $x, y, z \in X$.

In this situation, the pair (X, ρ) is called a *metric space*.

Any metric space (X, ρ) gives rise to a canonical topology. More precisely, a subset U of X is called *open* if for any $x \in U$ there exists $r > 0$ such that the *open ball*, of centre x and radius r, $B_r(x) = \{y \in X \mid \rho(x,y) < r\} \subseteq U$. The collection of all open sets defined in this way, denoted by $\mathcal{T}_\rho$, is a topology. A topological space $(X, \mathcal{T})$ is called *metrisable* if there exists a metric ρ on X such that $\mathcal{T} = \mathcal{T}_\rho$.

Proposition A.1.17. *For any metric space (X, ρ) the underlying topology $\mathcal{T}_\rho$ is Hausdorff.*

A function $f\colon (X, \rho) \to (Y, \nu)$, for two metric spaces (X, ρ) and (Y, ν) is *uniformly continuous* if for any $\varepsilon > 0$ there exists $\delta > 0$ such that $\nu(f(x_1), f(x_2)) < \varepsilon$ whenever $\rho(x_1, x_2) < \delta$.

Proposition A.1.18. *With notation as before, if the metric space (X, ρ) is compact and f is continuous, then f is uniformly continuous.*

In any metric space, in addition to the concept of convergent sequence, which is common to all topological spaces, one can define the concept of a Cauchy sequence. More precisely, if $(x_n)_n$ is a sequence with elements in X, then $x_n \xrightarrow[n \to \infty]{} x$, for some $x \in X$ if for any $\varepsilon > 0$ there exists $N_\varepsilon \in \mathbb{N}$ such that for all $n \geqslant N_\varepsilon$ we have $\rho(x_n, x) < \varepsilon$. The sequence $(x_n)_n$ with elements from X is *Cauchy* if for any $\varepsilon > 0$ there exists $N_\varepsilon \in \mathbb{N}$ such that for all $m, n \in \mathbb{N}$ with $m, n \geqslant N_\varepsilon$ we have $\rho(x_n, x_m) < \varepsilon$. Clearly, any convergent sequence is Cauchy.

A metric ρ on X is called *complete* if any Cauchy sequence with elements from X is convergent to some element in X. A topological space $(M, \mathcal{T})$ is called *completely metrisable* if there exists a complete metric ρ on M whose topology $\mathcal{T}_\rho$ coincides with $\mathcal{T}$.

One of the great advantages of working with metrisable topologies is that continuity can be expressed in terms of convergent sequences only.

Proposition A.1.19. *Let $(X, \mathcal{T})$ be a metrisable topological space. For any topological space $(Y, \mathcal{S})$, an arbitrary mapping $f\colon X \to Y$ is continuous at $x \in X$ if and only if for any sequence $(x_n)_n$ of elements in X such that $x_n \xrightarrow[n \to \infty]{\mathcal{T}} x$ we have $f(x_n) \xrightarrow[n \to \infty]{\mathcal{S}} f(x)$.*

Compactness of metric spaces can be characterised in a more tractable fashion.

Proposition A.1.20. *Let (X, ρ) be a metric space and $\mathcal{T}_\rho$ the topology generated by ρ. Then, the following assertions are equivalent.*

(i) $(X, \mathcal{T}_\rho)$ *is compact.*
(ii) $(X, \mathcal{T}_\rho)$ *is sequentially compact, that is, any sequence $(x_n)_n$ with elements in X has a subsequence $(x_{k_n})_n$ that converges to some $x \in X$.*
(iii) $(X, \mathcal{T}_\rho)$ *is complete and totally bounded, in the sense that, for any $\varepsilon > 0$ there exist $x_1, \ldots, x_n \in X$ such that $X \subseteq \bigcup_{j=1}^n B_\varepsilon(x_j)$.*

Also, compact metric spaces share some important additional properties.

Proposition A.1.21. *If (X, ρ) is a compact metric space then the underlying topological space $(X, \mathcal{T}_\rho)$ has a countable base and is separable.*

One of the most important examples of a metric space is obtained if we let X be a compact Hausdorff space and consider $\mathcal{C}(X)$ the set of all continuous functions $f \colon X \to \mathbb{C}$. On $\mathcal{C}(X)$ one considers the *uniform metric*

$$d_\infty(f, g) := \sup_{x \in X} |f(x) - g(x)|, \quad f, g \in \mathcal{C}(X).$$

The uniform metric is actually closely connected to the concept of uniform convergence. A sequence of maps $(f_n)_n$, $f_n \colon M \to (X, \rho)$ for all $n \in \mathbb{N}$, where M is a nonempty set, is *uniformly converging* to a map $f \colon M \to X$ if for any $\varepsilon > 0$ there exists $N \in \mathbb{N}$ such that for all $n \geqslant N$ and all $x \in M$ we have $\rho(f_n(x), f(x)) < \varepsilon$. Thus, convergence in $\mathcal{C}(X)$ with respect to the uniform metric is uniform convergence.

Also, in the metric space $\mathcal{C}(X)$, compact subsets have a special characterisation. A subset A of $\mathcal{C}(X)$ is called *equicontinuous* if for every $x \in X$ and any $\varepsilon > 0$ there exists $U \in \mathcal{V}(x)$ such that for all $y \in U$ and all $f \in A$ we have $|f(x) - f(y)| < \varepsilon$.

Theorem A.1.22 (Arzelà–Ascoli Theorem). *Let X be a compact Hausdorff space and A a subset of $\mathcal{C}(X)$. Then A is relatively compact, that is, its closure is a compact subset of the metric space $\mathcal{C}(X)$, if and only if A is equicontinuous and pointwise bounded, that is, for every $x \in X$ the set $\{f(x) \mid f \in A\}$ is bounded in $\mathbb{C}$.*

Baire Spaces

A subset S of a topological space M is:

(i) *nowhere dense* in M if the interior of the closure of M is empty;
(ii) *of first category* or *meagre* in M if it is a union of countably many nowhere dense subsets;
(iii) *of second category* or *nonmeagre* in M if it is not of first category in M.

A topological space M is a *Baire space* if every nonempty open subset of M is of second category in M. There are two Baire's Category Theorems.

Theorem A.1.23. *Every Hausdorff locally compact space is a Baire space.*

Here is the second Baire Category Theorem.

Theorem A.1.24. *Every complete metric space is a Baire space. More generally, every topological space that is homeomorphic to an open subset of a complete metric space is a Baire space and, in particular, every completely metrisable space is a Baire space.*

A.2   Measure and Integration

In this section we review the basic concepts and facts on measure theory and integration with respect to an abstract measure, with an emphasis on Borel measures and Radon measures. The information gathered here can be found, with detailed proofs, in classical monographs such as N. Bourbaki [24], and many other modern textbooks, for example, W. Rudin [127], V. I. Bogachev [21] and D. L. Cohn [34].

Measurable Spaces

Let X be a nonempty set. One calls $\mathcal{B} \subseteq \mathcal{P}(X)$ a *σ-algebra* of subsets in X if the following hold.

(i) $X \in \mathcal{B}$.
(ii) For every $A \in \mathcal{B}$ its complement $X \setminus A \in \mathcal{B}$.
(iii) For every sequence $(A_n)_{n \geqslant 1}$, $A_n \in \mathcal{B}$ for all $n \geqslant 1$, we have $\bigcup_{n \geqslant 1} A_n \in \mathcal{B}$.

Consequently, a σ-algebra $\mathcal{B}$ has the following properties as well.

(iv) $\emptyset \in \mathcal{B}$.
(v) For every finite sequence $(A_k)_{k=1}^n$, $A_k \in \mathcal{B}$ for all $k \geqslant 1$, we have $\bigcup_{k=1}^n A_k \in \mathcal{B}$.
(vi) For every $A, B \in \mathcal{B}$ we have $A \setminus B \in \mathcal{B}$.
(vii) For every sequence $(A_n)_{n \geqslant 1}$, $A_n \in \mathcal{B}$ for all $n \geqslant 1$, we have $\bigcap_{n \geqslant 1} A_n \in \mathcal{B}$.

Proposition A.2.1. *Let $\{\mathcal{B}_j\}_{j \in J}$ be an arbitrary family of σ-algebras of subsets of X. Then $\bigcap_{j \in J} \mathcal{B}_j$ is a σ-algebra of subsets of X.*

As a consequence of Proposition A.2.1, whenever $\mathcal{E} \subseteq \mathcal{P}(X)$, there exists and it is unique the smallest σ-algebra of subsets of X that contains $\mathcal{E}$. Clearly, $\{\emptyset, X\}$ is the smallest σ-algebra on X and $\mathcal{P}(X)$ is the largest σ-algebra on X.

A *measurable space* is a pair $(X, \mathcal{B})$, where X is a nonempty set and $\mathcal{B}$ is a σ-algebra on X. In this case, the elements $A \in \mathcal{B}$ are called *measurable sets*.

If $(X, \mathcal{T})$ is a topological space then the σ-algebra generated by the topology $\mathcal{T}$ is called the *Borel σ-algebra* on X.

Let $(X, \mathcal{B}_X)$ and $(Y, \mathcal{B}_Y)$ be two measurable spaces. A map $f \colon X \to Y$ is called *measurable* if the preimage $f^{-1}(B) \in \mathcal{B}_X$ for all $B \in \mathcal{B}_Y$. A function $f \colon X \to \overline{\mathbb{R}}$, recall Example A.1.8, is *measurable* if it is measurable with respect to the Borel σ-algebra on $\overline{\mathbb{R}}$.

A.2 Measure and Integration

If $(X, \mathcal{T})$ and $(Y, \mathcal{S})$ are two topological spaces, a map $f\colon X \to Y$ is *Borel measurable* if it is measurable with respect to the corresponding Borel σ-algebras $\mathcal{B}_\mathcal{T}$ on X and $\mathcal{B}_\mathcal{S}$ on Y, respectively.

Proposition A.2.2. *Let $(X, \mathcal{T})$ and $(Y, \mathcal{S})$ be two topological spaces and $f\colon X \to Y$ a mapping.*
  (a) *f is Borel measurable if and only if, for any $U \in \mathcal{S}$, the preimage $f^{-1}(U) \in \mathcal{B}_\mathcal{T}$.*
  (b) *If f is continuous then it is Borel measurable.*

Remark A.2.3. Let $(X, \mathcal{B})$ be a measurable space, $A \subseteq X$ and $\chi_A\colon X \to \mathbb{R}$ be the *characteristic function* of A. Then χ_A is measurable if and only if A is measurable, that is, $A \in \mathcal{B}$. ∎

Measure Spaces

Let $(X, \mathcal{B})$ be a measurable space. A mapping $\mu\colon \mathcal{B} \to [0, \infty]$ is called a *measure* on X if the following hold.

(i) For any sequence $(A_n)_{n \geq 1}$ of subsets in $\mathcal{B}$ that are mutually disjoint, that is, $A_n \cap A_m = \emptyset$ for all $m \neq n$, we have
$$\mu\Big(\bigcup_{n \in \mathbb{N}} A_n\Big) = \sum_{n=1}^{\infty} \mu(A_n).$$

(ii) There exists $A \in \mathcal{B}$ such that $\mu(A) < \infty$.

Consequently, a measure μ has the following properties as well.

(iii) $\mu(\emptyset) = 0$.
(iv) For any $n \in \mathbb{N}$ and any finite sequence $(A_k)_{k=1}^n$ of mutually disjoint subsets in $\mathcal{B}$, we have
$$\mu\Big(\bigcup_{k=1}^n A_k\Big) = \sum_{k=1}^n \mu(A_k).$$

(v) For any $A, B \in \mathcal{B}$ such that $A \subseteq B$ it follows that $\mu(A) \leq \mu(B)$.
(vi) For any sequence $(A_n)_{n \geq 1}$ of subsets in $\mathcal{B}$ such that $A_n \subseteq A_{n+1}$ for all $n \geq 1$ it follows that
$$\mu\Big(\bigcup_{n \geq 1} A_n\Big) = \lim_{n \to \infty} \mu(A_n).$$

(vii) For any sequence $(A_n)_{n \geq 1}$ of subsets in $\mathcal{B}$ such that, $A_n \supseteq A_{n+1}$ for all $n \geq 1$ and $\mu(A_k) < \infty$ for some $k \in \mathbb{N}$, it follows that
$$\mu\Big(\bigcap_{n \geq 1} A_n\Big) = \lim_{n \to \infty} \mu(A_n).$$

A triple $(X, \mathcal{B}, \mu)$, where $(X, \mathcal{B})$ is a measurable space and μ is a measure on X, is called a *measure space*.

A measure space $(X, \mathcal{B}, \mu)$ is called *finite* if $\mu(X) < \infty$, hence $\mu(E) < \infty$ for all $E \in \mathcal{B}$, and it is called a *σ-finite measure* if there exists a countable collection of subsets $\{X_n\}_n$, such that $X_n \in \mathcal{B}$ and $\mu(X_n) < \infty$ for all n, and $X = \bigcup_n X_n$.

The Lebesgue Measure

Consider the Euclidean space $\mathbb{R}^N$ for some natural number N. An *aligned N-rectangle* is, by definition, any subset of of $\mathbb{R}^N$ that is a Cartesian product of N intervals. So, typically, if $I_1, \ldots, I_N$ are intervals in $\mathbb{R}$ then

$$R = I_1 \times I_N := \{(x_1, \ldots, x_N) \mid x_j \in I_j,\ j = 1, \ldots, N\},$$

is an aligned N-rectangle. The *volume* of R is, by definition, the product of the lengths of the intervals $I_1, \ldots, I_N$

$$|R| := |I_1| \cdots |I_N|.$$

The intervals can be of any type, open, closed, or open at one endpoint and closed at the other endpoint.

The *Lebesgue outer measure* λ^* is defined by

$$\lambda^*(A) := \sup\{\sum_{j=1}^{\infty} \mid A \subseteq \bigcup_{j=1}^{\infty} I_j,\ I_j \text{ intervals in } \mathbb{R}\}.$$

In this way we get a map $\lambda^* \colon \mathcal{P}(\mathbb{R}^N) \to [0, \infty]$.

Proposition A.2.4. *The Lebesgue outer measure has the following properties.*

  (i) $\lambda^*(\emptyset) = 0$.
  (ii) *If $A \subseteq B \subseteq \mathbb{R}^N$ then $\lambda^*(A) \leqslant \lambda^*(B)$.*
  (iii) *For any countable family $\{A_n\}_n$ of subsets of $\mathbb{R}^N$ we have*

$$\lambda^*(\bigcup_{n=1}^{\infty} A_n) \leqslant \sum_{n=1}^{\infty} \lambda^*(A_n).$$

  (iv) $\lambda^*(R) = |R|$ *for any aligned N-rectangle R.*
  (v) *For any $A \subseteq \mathbb{R}^N$ and any $x \in \mathbb{R}^N$ we have $\lambda^*(A) = \lambda^*(A + x)$, where $A + x := \{a + x \mid a \in A\}$ is the translation of A by x.*

A subset A of $\mathbb{R}^N$ is *Lebesgue measurable* if, for any $B \subseteq \mathbb{R}^N$, we have

$$\lambda^*(B) = \lambda^*(B \cap A) + \lambda^*(B \cap (\mathbb{R}^N \setminus A)).$$

Let $\mathcal{M}$ denote the collection of all Lebesgue measurable subsets of $\mathbb{R}^N$.

A.2 MEASURE AND INTEGRATION

Theorem A.2.5 (Lebesgue Measure). (a) *$\mathcal{M}$ is a σ-algebra on $\mathbb{R}^N$.*
(b) *$\lambda := \lambda^*|\mathcal{M}$ is a measure, called the Lebesgue measure on $\mathbb{R}^N$.*
(c) *Any $A \subseteq \mathbb{R}^N$ with the property $\lambda^*(A) = 0$ is Lebesgue measurable.*
(d) *Every open, and hence every closed, subset of $\mathbb{R}^N$ is Lebesgue measurable.*
(e) *For every set $A \in \mathcal{M}$ we have*

$$\lambda(A) = \inf\{\lambda(U) \mid A \subseteq U \subseteq \mathbb{R}^N \text{ and } U \text{ is open}\}.$$

(f) *For every set $A \in \mathcal{M}$ we have*

$$\lambda(A) = \sup\{\lambda(K) \mid K \subseteq A \text{ and } K \text{ is compact}\}.$$

(g) *For every set $A \in \mathcal{M}$ and any $x \in \mathbb{R}^N$ we have $\lambda(A) = \lambda(A + x)$, that is, the Lebesgue measure is translation invariant.*

A set $A \subseteq \mathbb{R}^N$ such that $\lambda^*(A) = 0$ is called a *Lebesgue negligible* set. Property (c) can be reformulated by saying that the Lebesgue measure is *complete*.

Integration with Respect to a Measure

In the following we fix a measure space $(X, \mathcal{B}, \mu)$ and briefly describe the construction and the properties of the integral with respect to the measure μ. Letting $n \in \mathbb{N}$, $\alpha_k \geqslant 0$, for all $k = 1, \ldots, n$, and $A_1, \ldots, A_n \in \mathcal{B}$, define

$$s(x) = \sum_{k=1}^n \alpha_k \chi_{A_k}(x), \quad x \in X.$$

Here and in the following, $\chi_A \colon X \to \{0, 1\}$ denotes the characteristic function of the subset A. Such a function s is called a *step function*. By Remark A.2.3, any step function is measurable. If $E \in \mathcal{B}$ is arbitrary, we define

$$\int_E \chi(x) d\mu(x) = \sum_{k=1}^n \alpha_k \mu(E \cap A_k).$$

In general, if $f \colon X \to [0, \infty]$ is a measurable function, we define the *integral* of f with respect to the measure μ on E by

$$\int_E f(x) d\mu(x) = \sup_{s \leqslant f} \int_E s(x) d\mu(x),$$

where s runs through the set of all step functions $s \colon X \to [0, \infty]$ such that $s(x) \leqslant f(x)$ for all $x \in X$.

In the following we list the basic properties of the integral of nonnegative functions.

Proposition A.2.6. *Let $(X, \mathcal{B}, \mu)$ be a measure space.*

(i) If $f, g \colon X \to [0, \infty]$ are measurable functions, $\alpha \geqslant 0$, and $E \in \mathcal{B}$, then,
$$\int_E (\alpha f(x) + g(x)) \mathrm{d}\mu(x) = \alpha \int_E f(x) \mathrm{d}\mu(x) + \int_E g(x) \mathrm{d}\mu(x).$$

(ii) Let $(f_n)_{n \geqslant 1}$ be a sequence of measurable functions $f_n \colon X \to [0, \infty]$ and define $f \colon X \to [0, \infty]$ by
$$f(x) = \sum_{n=1}^{\infty} f_n(x), \quad x \in X.$$
Then,
$$\int_E f(x) \mathrm{d}\mu(x) = \sum_{n=1}^{\infty} \int_E f_n(x) \mathrm{d}\mu(x).$$

(iii) Let $f \colon X \to [0, \infty]$ be a measurable function and $(E_n)_{n \geqslant 1}$ a sequence of mutually disjoint measurable subsets of X. Then,
$$\int_E f(x) \mathrm{d}\mu(x) = \sum_{n=1}^{\infty} \int_{E_n} f(x) \mathrm{d}\mu(x), \quad \text{where } E = \bigcup_{n=1}^{\infty} E_n.$$

The integral gives the possibility of defining many other measures induced by measurable functions.

Proposition A.2.7. *Let $(X, \mathcal{B}, \mu)$ be a measure space and let $f \colon X \to [0, \infty]$ be a measurable function. Then, letting $\mu_f \colon \mathcal{B} \to [0, \infty]$ be defined by*
$$\mu_f(E) = \int_E f(x) \mathrm{d}\mu(x), \quad E \in \mathcal{B},$$
we get a measure space $(X, \mathcal{B}, \mu_f)$. The measure μ_f has the property that, for any measurable function $g \colon X \to [0, \infty]$ and any $E \in \mathcal{B}$, we have
$$\int_E g(x) \mathrm{d}\mu_f(x) = \int_E g(x) f(x) \mathrm{d}\mu(x).$$

One of the key properties of the integral of nonnegative measurable functions with respect to a measure space is the following.

Proposition A.2.8 (Fatou Lemma). *If $(X, \mathcal{B}, \mu)$ is a measure space and $(f_n)_{n \geqslant 1}$ is a sequence of measurable functions $f_n \colon X \to [0, \infty]$, then*
$$\int_X \liminf_{n \to \infty} f_n(x) \mathrm{d}\mu(x) \leqslant \liminf_{n \to \infty} \int_X f_n(x) \mathrm{d}\mu(x).$$

A.2 MEASURE AND INTEGRATION

More generally, if $f\colon X \to [-\infty, +\infty]$ is a measurable function, then the functions $f_{\pm}\colon X \to [0, \infty]$ defined by

$$f_-(x) = \max\{-f(x), 0\}, \quad f_+(x) = \max\{f(x), 0\}, \quad x \in X,$$

are measurable as well. Also, $f_- f_+ = 0$, $f = f_+ - f_-$, and $|f| = f_+ + f_-$, hence $|f|$ is measurable as well. A function $f\colon X \to [-\infty, +\infty]$ is called *integrable* with respect to μ if it is measurable and the function $|f|\colon X \to [0, \infty]$ has the property $\int_X |f(x)| \mathrm{d}\mu(x) < \infty$. In this situation, for arbitrary $E \in \mathcal{B}$, one defines

$$\int_E f(x) \mathrm{d}\mu(x) = \int_E f_+(x) \mathrm{d}\mu(x) - \int_E f_-(x) \mathrm{d}\mu(x),$$

with the observation that both integrals on the right-hand side are finite, hence the difference always makes sense. One denotes

$$\mathcal{L}^1(X, \mu) = \{f\colon X \to \overline{\mathbb{R}} \mid f \text{ is integrable with respect to } \mu\}.$$

We record the basic properties of the integral with respect to an abstract measure μ.

Proposition A.2.9. *Let $(X, \mathcal{B}, \mu)$ be a measure space.*

(i) *With respect to the natural operations of addition and multiplication with scalars, $\mathcal{L}^1(X, \mu)$ is a real vector space and, for arbitrary $E \in \mathcal{B}$, the mapping $\mathcal{L}^1(X, \mu) \ni f \mapsto \int_E f(x) \mathrm{d}\mu(x) \in \mathbb{R}$ is a linear functional.*

(ii) *If $E \in \mathcal{B}$ and $f, g \in \mathcal{L}^1(X, \mu)$ are such that $f|_E \leq g|_E$ then*

$$\int_E f(x) \mathrm{d}\mu(x) \leq \int_E g(x) \mathrm{d}\mu(x).$$

(iii) *If $f \in \mathcal{L}^1(X, \mu)$ then $|f| \in \mathcal{L}^1(X, \mu)$ and*

$$\left| \int_E f(x) \mathrm{d}\mu(x) \right| \leq \int_E |f(x)| \mathrm{d}\mu(x).$$

(iv) *If $(E_n)_{n \geq 1}$ is a mutually disjoint sequence of measurable subsets of X and $f \in \mathcal{L}^1(X, \mu)$ then*

$$\int_E f(x) \mathrm{d}\mu(x) = \sum_{n=1}^{\infty} \int_{E_n} f(x) \mathrm{d}\mu(x), \quad E = \bigcup_{n=1}^{\infty} E_n.$$

(v) *For all $E \in \mathcal{B}$ and all $f \in \mathcal{L}^1(X, \mu)$ we have*

$$\int_E f(x) \mathrm{d}\mu(x) = \int_X f(x) \chi_E(x) \mathrm{d}\mu(x).$$

(vi) Let $E \in \mathcal{B}$ and $f \in \mathcal{L}^1(X,\mu)$. If either $\mu(E) = 0$ or $f|_E = 0$ then $\int_E f(x) \mathrm{d}\mu(x) = 0$.

One of the most useful theorems of commutation of the integral with a limit is the following.

Theorem A.2.10 (Lebesque Theorem of Dominated Convergence). *Let $(X, \mathcal{B}, \mu)$ be a measure space. Assume that $(f_n)_n$ is a sequence of measurable functions on X subject to the following conditions.*

(a) *There exists $g \in \mathcal{L}^1(X, \mu)$ such that $|f_n(x)| \leq |g(x)|$ for all $x \in X$.*
(b) *There exists the pointwise limit of the sequence $(f_n)_n$ on X, that is, there exists a function $f \colon X \to \overline{\mathbb{R}}$ such that $f(x) = \lim_{n\to\infty} f_n(x)$ for all $x \in X$.*

Then, $f \in \mathcal{L}^1(X, \mu)$ and, for any $E \in \mathcal{B}$, we have

$$\int_E f(x) \mathrm{d}\mu(x) = \lim_{n \to \infty} \int_E f_n(x) \mathrm{d}\mu(x).$$

Borel Measures

In the following we assume that $(X, \mathcal{T})$ is a locally compact Hausdorff space and let $\mathcal{B} = \mathcal{B}_\mathcal{T}$ be the Borel σ-algebra generated by the topology $\mathcal{T}$. A measure $\mu \colon \mathcal{B} \to [0, \infty]$ is:

(i) a *Borel measure* if $\mu(K) < \infty$ for any compact subset $K \subseteq X$;
(ii) *outer regular* if for any $B \in \mathcal{B}$ we have

$$\mu(B) = \inf\{\mu(U) \mid U \in \mathcal{T}, \; B \subseteq U\};$$

(iii) *inner regular* if for any $B \in \mathcal{B}$ we have

$$\mu(B) = \sup\{\mu(K) \mid K \text{ compact and } K \subseteq B\};$$

(iv) *regular* if it is both inner and outer regular;
(v) a *Radon measure* if it is a Borel measure and inner regular.

For an arbitrary function $f \in \mathcal{C}_c(X)$ one defines its *support*, denoted by $\mathrm{supp}(f)$, the closure of the set $\{x \in X \mid f(x) \neq 0\}$. Let $\mathcal{C}_c(X)$ denote the set of all continuous functions $f \colon X \to \mathbb{R}$ with compact support and $\mathcal{C}_c^+(X) = \{f \in \mathcal{C}_c(X) \mid f(x) \geq 0 \text{ for all } x \in X\}$. Clearly, $\mathcal{C}_c(X)$ is a real algebra with the naturally defined algebraic operations. For each compact subset $K \subseteq X$ we consider

$$\mathcal{C}_K(X) := \{f \in \mathcal{C}_c(X) \mid \mathrm{supp}(f) \subseteq K\}.$$

Then $\mathcal{C}_K(X)$ is a Banach space with respect to the uniform norm

$$\|f\|_K := \sup_{x \in K} |f(x)|, \quad f \in \mathcal{C}_K(X),$$

A.2 MEASURE AND INTEGRATION

and
$$\mathcal{C}_c(X) = \bigcup_K \mathcal{C}_K(X),$$

where K runs through the set of all compact subsets of X. On $\mathcal{C}_c(X)$ we consider the strictly inductive limit topology associated to the inductive system of Banach spaces $\{\mathcal{C}_K(X) \mid K \text{ compact subset of } X\}$, which is called the *uniform topology*. Since we do not recall inductive limits, it is sufficient to know that, with respect to this topology, a sequence $(f_n)_n$ in $\mathcal{C}_c(X)$ converges to f if and only if there exists a compact subset K of X such that $\mathrm{supp}(f) \subseteq K$ for all $n \in \mathbb{N}$ and $(f_n)_n$ converges to f in $\mathcal{C}_K(X)$.

A linear functional $\varphi \colon \mathcal{C}_c(X) \to \mathbb{R}$ is *positive* if $\varphi(f) \geqslant 0$ for all $f \in \mathcal{C}_c^+(X)$.

Proposition A.2.11. *Let* $\varphi \colon \mathcal{C}_c(X) \to \mathbb{R}$ *be a positive linear functional. Then* φ *is continuous with respect to the uniform topology.*

Given μ a Radon measure on X, define $\varphi_\mu \colon \mathcal{C}_c(X) \to \mathbb{R}$ by
$$\varphi_\mu(f) = \int_X f(x) \mathrm{d}\mu(x), \quad f \in \mathcal{C}_c(X).$$

Clearly, φ_μ is a positive linear functional.

Theorem A.2.12 (Riesz–Markov Theorem). *Let* $(X, \mathcal{T})$ *be a locally compact Hausdorff space and let* $\mathcal{B} = \mathcal{B}_\mathcal{T}$ *be the Borel σ-algebra generated by the topology $\mathcal{T}$. Then, for any linear positive functional* $\varphi \colon \mathcal{C}_c(X) \to \mathbb{R}$ *there exists a unique Radon measure* $\mu \colon \mathcal{B} \to [0, \infty]$ *such that* $\varphi = \varphi_\mu$.

If we let μ denote the restriction of the Lebesgue measure λ on $\mathbb{R}^N$ to the σ-algebra $\mathcal{B}_{\mathbb{R}^N}$ of all Borel subsets of $\mathbb{R}^N$ then μ is a regular Borel measure that is translation invariant, see Theorem A.2.5. This property is actually a characterisation of the Lebesgue measure.

Proposition A.2.13. *Let ν be a nonzero measure on $(\mathbb{R}^N, \mathcal{B}_{\mathbb{R}^N})$ such that:*

(i) *$\nu(B)$ is finite on any bounded Borel subset B of $\mathbb{R}^N$,*
(ii) *$\nu(B+x) = \nu(B)$ for any Borel subset B of $\mathbb{R}^N$ and each $x \in \mathbb{R}^N$.*

Then, there exists $c > 0$ such that $\nu(A) = c\mu(A)$, for all $A \in \mathcal{B}_{\mathbb{R}^N}$.

For the special case of the real line and the Borel σ-algebra, finite measures can be viewed in a different and more convenient fashion.

Theorem A.2.14. (a) *Let μ be a finite positive measure on $(\mathbb{R}, \mathcal{B}_\mathbb{R})$ and let $F_\mu \colon \mathbb{R} \to \mathbb{R}$ be defined by*
$$F_\mu(x) := \mu((-\infty, x]), \quad x \in \mathbb{R}.$$
Then F_μ is bounded, nondecreasing, right continuous, and $\lim_{x \to -\infty} F_\mu(x) = 0$.

(b) *For every function $F \colon \mathbb{R} \to \mathbb{R}$ that is bounded, nondecreasing, right continuous, and satisfies $\lim_{x \to -\infty} F(x) = 0$, there exists a unique finite positive measure μ on $(\mathbb{R}, \mathcal{B}_\mathbb{R})$ such that $F(x) = \mu((-\infty, x])$ for all $x \in \mathbb{R}$.*

One of the consequences of the previous theorem is that, when positive finite measures μ on $(\mathbb{R}, \mathcal{B}_\mathbb{R})$ are considered, integration with respect to μ can be viewed as a Stieltjes integration with respect to the function F_μ.

Complex and Signed Measures

Let $(X, \mathcal{A})$ be a measurable space. A *complex measure* is a map $\mu \colon \mathcal{A} \to \mathbb{C}$ subject to the condition that, for any countable collection of mutually disjoint sets $\{E_j\}_j$, such that $E_j \in \mathcal{A}$ for all j, we have

$$\mu(\bigcup_j E_j) = \sum_j \mu(E_j),$$

where the series is supposed to be absolutely convergent, hence unconditionally convergent. For such a measure μ, the definition

$$|\mu|(E) := \sup \sum_j |\mu(E_j)|, \quad E \in \mathcal{A},$$

where the supremum is taken over all measurable partitions $\{E_j\}_j$ of E, yields a measure, called the *total variation measure* of μ.

Proposition A.2.15. *For any complex measure μ on a measurable space $(X, \mathcal{A})$, the total variation measure is finite, that is, $|\mu|(X) < \infty$.*

If the complex measure μ has the property that $\mu(E) \in \mathbb{R}$ for all $E \in \mathcal{A}$, then it is called a *signed measure*. In this case, letting μ_+ and μ_- be defined as

$$\mu_+(E) := \frac{1}{2}\big(|\mu|(E) + \mu(E)\big), \quad \mu_-(E) := \frac{1}{2}\big(|\mu|(E) - \mu(E)\big), \quad E \in \mathcal{A},$$

we obtain two finite (positive) measures. In this respect, we have $\mu = \mu_+ - \mu_-$, called the *Jordan decomposition* of the signed measure μ, and $|\mu| = \mu_+ + \mu_-$. The measure μ_+ is called the *positive variation* of μ and the measure μ_- is called the *negative variation* of μ.

Also, any complex measure can be decomposed $\mu = \operatorname{Re}\mu + \mathrm{i}\operatorname{Im}\mu$, where $\operatorname{Re}\mu$ and $\operatorname{Im}\mu$ are signed measures defined by

$$\operatorname{Re}\mu(E) := \frac{1}{2}\big(\mu(E) + \overline{\mu(E)}\big), \quad \operatorname{Im}\mu(E) := \frac{1}{2\mathrm{i}}\big(\mu(E) - \overline{\mu(E)}\big), \quad E \in \mathcal{A}.$$

Consequently, any complex measure is a linear combination of four finite measures and hence one can define integrable functions with respect to complex measures as well: a measurable function $f \colon X \to \mathbb{C}$ is *integrable* with respect to μ if it is integrable with respect to $\operatorname{Re}\mu_\pm$ and $\operatorname{Im}\mu_\pm$ and then one defines the integral of f with respect to μ by

$$\int_X f(x)\mathrm{d}\mu(x) := \int_X f(x)\mathrm{d}\operatorname{Re}\mu_+ - \int_X f(x)\mathrm{d}\operatorname{Re}\mu_- + \mathrm{i}\int_X f(x)\mathrm{d}\operatorname{Im}\mu_+ - \mathrm{i}\int_X f(x)\mathrm{d}\operatorname{Im}\mu_-.$$

A.2 Measure and Integration

Lebesgue Decompositions

Let μ be a (positive) measure on the measurable space $(X, \mathcal{A})$ and let ν be an arbitrary measure on $(X, \mathcal{A})$, which may be a positive measure, a signed measure, or a complex measure. In the case when ν is a positive measure it is allowed to take ∞ as an admissible value. The measure ν is *absolutely continuous* with respect to μ, and we write $\nu \ll \mu$, if $\nu(E) = 0$ for every $E \in \mathcal{A}$ for which $\mu(E) = 0$.

On the other hand, if $A \in \mathcal{A}$ is such that $\nu(E) = \nu(E \cap A)$ for all $E \in \mathcal{A}$, equivalently, $\nu(E) = 0$ whenever $E \cap A = \emptyset$, we say that ν is *concentrated* on A. With this definition, if we have two measures (positive, signed, or complex) ν_1 and ν_2 such that ν_1 is concentrated on $A_1 \in \mathcal{A}$, ν_2 is concentrated on $A_2 \in \mathcal{A}$, and $A_1 \cap A_2 = \emptyset$, we say that ν_1 and ν_2 are *mutually singular*, and we write $\nu_1 \perp \nu_2$.

Theorem A.2.16 (Lebesgue–Radon–Nikodym Theorem). *With notation as before, assume that the measure μ is σ-finite.*

(a) *There exist unique measures ν_a and ν_s on $\mathcal{A}$ such that*

$$\nu = \nu_a + \nu_s, \quad \nu_a \ll \mu, \quad \nu_s \perp \mu.$$

If, in addition, the measure ν is positive and finite, then so are ν_a and ν_s.

(b) *There exists a unique $h \in L^1(\mu)$ such that*

$$\nu_a(E) = \int_E h(x) d\mu(x), \quad E \in \mathcal{A}.$$

The statement in item (a) is referred to as the *Lebesgue Decomposition Theorem*. The statement in item (b) is referred to as the *Radon–Nikodym Theorem* and the function h is the *Radon–Nikodym derivative* of ν with respect to μ and is denoted by $d\nu/d\mu$.

Product Measures

Let $(X, \mathcal{A})$ and $(Y, \mathcal{B})$ be two measurable spaces. The *product σ-algebra* on $X \times Y$ of the σ-algebras $\mathcal{A}$ and $\mathcal{B}$, denoted by $\mathcal{A} \otimes \mathcal{B}$, is the σ-algebra generated by $\{A \times B \mid A \in \mathcal{A}, B \in \mathcal{B}\}$. In this situation, given $f: X \times Y \to \overline{\mathbb{R}}$ an $\mathcal{A} \otimes \mathcal{B}$-measurable function, then the following hold.

(i) For all $x \in X$, the x-section $f_x(y) = f(x, y)$, $y \in Y$, is a $\mathcal{B}$-measurable function $f_x: Y \to \overline{\mathbb{R}}$.
(ii) For all $y \in Y$, the y-section $f^y(x) = f(x, y)$, $x \in X$, is an $\mathcal{A}$-measurable function $f^y: Y \to \overline{\mathbb{R}}$.

In particular, letting $E \in \mathcal{A} \otimes \mathcal{B}$ and applying this observation to the characteristic function χ_E, we have the following.

(i) For all $x \in X$, the x-section $E_x = \{y \in Y \mid (x, y) \in E\} \in \mathcal{B}$.
(ii) For all $y \in Y$, the y-section $E^y = \{x \in X \mid (x, y) \in E\} \in \mathcal{A}$.

Proposition A.2.17. *Let $(X, \mathcal{A}, \mu)$ and $(Y, \mathcal{B}, \nu)$ be two σ-finite measure spaces. Then:*
(a) *For any $C \in \mathcal{A} \otimes \mathcal{B}$ the function $X \ni x \mapsto \nu(C_x) \in [0, \infty]$ is $\mathcal{A}$-measurable, the function $Y \ni y \mapsto \mu(C^y) \in [0, \infty]$ is $\mathcal{B}$-measurable, and*

$$\int_X \nu(C_x) d\mu(x) = \int_Y \mu(C^y) d\nu(y).$$

(b) *Letting $\mathcal{A} \otimes \mathcal{B} \ni C \mapsto \mu \otimes \nu(C) \in [0, \infty]$ be the transformation defined by*

$$(\mu \otimes \nu)(C) := \int_X \nu(C_x) d\mu(x) = \int_Y \mu(C^y) d\nu(y), \quad C \in \mathcal{A} \otimes \mathcal{B},$$

$\mu \otimes \nu$ *is a measure on the measurable space $(X \times Y, \mathcal{A} \otimes \mathcal{B})$.*
(c) *For any $A \in \mathcal{A}$ and $B \in \mathcal{B}$ we have*

$$(\mu \otimes \nu)(A \times B) = \mu(A)\nu(B),$$

in particular, $\mu \otimes \nu$ is σ-finite as well.

There are two theorems for expressing the integral on a product measure space as an iterated integral. The first theorem refers to nonnegative valued measurable functions.

Theorem A.2.18 (Tonelli Theorem). *Let $(X, \mathcal{A}, \mu)$ and $(Y, \mathcal{B}, \nu)$ be two measure spaces and let $f : X \times Y \to [0, \infty]$ be an $\mathcal{A} \otimes \mathcal{B}$-measurable function.*
(i) *The function $X \ni x \mapsto \int_Y f(x, y) d\nu(y) \in [0, \infty]$ is $\mathcal{A}$-measurable.*
(ii) *The function $Y \ni y \mapsto \int_X f(x, y) d\mu(x) \in [0, \infty]$ is $\mathcal{B}$-mesurable.*
(iii) *The following equality holds:*

$$\int_{X \times Y} f(x, y) d(\mu \otimes \nu)(x, y) = \int_X \left(\int_Y f(x, y) d\nu(y) \right) d\mu(x)$$
$$= \int_Y \left(\int_X f(x, y) d\mu(x) \right) d\nu(y),$$

in the sense that, if at least one of the two iterated integrals is finite then the other two are finite and all three have the same value.

The second theorem refers to commutation of iterated integrals for signed functions that are integrable on product measure spaces.

Theorem A.2.19 (Fubini Theorem). *Let $(X, \mathcal{A}, \mu)$ and $(Y, \mathcal{B}, \nu)$ be σ-finite measure spaces and let $f \in \mathcal{L}^1(X \times Y, \mu \otimes \nu)$.*
(a) *There exists $A_0 \in \mathcal{A}$ with $\mu(A_0) = 0$ such that $X \setminus A_0 \ni x \mapsto \int_Y f(x, y) d\nu(y)$ belongs to $\mathcal{L}^1(X \setminus A_0, \mu)$.*
(b) *There exists $B_0 \in \mathcal{B}$ with $\nu(B_0) = 0$ such that $Y \setminus B_0 \ni y \mapsto \int_X f(x, y) d\mu(x)$ belongs to $\mathcal{L}^1(Y \setminus B_0, \nu)$.*

(c) *The following equalities hold:*

$$\int_{X\times Y} f(x,y)\mathrm{d}(\mu\otimes\nu)(x,y) = \int_{X\setminus A_0}\Big(\int_Y f(x,y)\mathrm{d}\nu(y)\Big)\mathrm{d}\mu(x)$$
$$= \int_{Y\setminus B_0}\Big(\int_X f(x,y)\mathrm{d}\mu(x)\Big)\mathrm{d}\nu(y).$$

In applications, the equalities as in item (c) of the previous theorem are written

$$\int_{X\times Y} f(x,y)\mathrm{d}(\mu\otimes\nu)(x,y) = \int_X\Big(\int_Y f(x,y)\mathrm{d}\nu(y)\Big)\mathrm{d}\mu(x)$$
$$= \int_Y\Big(\int_X f(x,y)\mathrm{d}\mu(x)\Big)\mathrm{d}\nu(y),$$

with the understanding that the function $X \setminus A_0 \ni x \mapsto \int_Y f(x,y)\mathrm{d}\nu(y)$ is modified to be 0 on A_0 and, similarly, the function $Y \setminus B_0 \ni y \mapsto \int_X f(x,y)\mathrm{d}\mu(x)$ is modified to be 0 on B_0.

A.3 Topological Vector Spaces

This section contains a review of basic concepts and facts on topological vector spaces, with an emphasis on locally convex spaces and their linear operators, inductive topologies, and barrelled spaces. For more advanced topics and details of the proofs, the reader may consider N. Bourbaki [25] and H. H. Schaefer [131], as well as many modern monographs and textbooks, for example see W. Rudin [128].

Linear Topologies

Let $\mathbb{F}$ denote either the field $\mathbb{R}$ or the field $\mathbb{C}$. A *topological vector space* is a vector space $\mathcal{E}$ over the field $\mathbb{F}$ that has a topology $\mathcal{T}$ that makes both the addition $\mathcal{E} \times \mathcal{E} \ni (x,y) \mapsto x+y \in \mathcal{E}$ and the multiplication with scalars $\mathbb{F} \times \mathcal{E} \ni (\alpha, x) \mapsto \alpha x \in \mathcal{E}$ continuous. We stress that the topologies on $\mathcal{E}\times\mathcal{E}$ and on $\mathbb{F}\times\mathcal{E}$, respectively, are the product topologies, hence, the continuity we talk about is actually joint continuity. The topology of a topological vector space is called *linear*.

As immediate consequences of the definition, if $(\mathcal{E},\mathcal{T})$ is a topological vector space, we have the following facts.

(i) For each $x_0 \in \mathcal{E}$, the translation mapping $\mathcal{E} \ni x \mapsto x+x_0 \in \mathcal{E}$ is a homeomorphism and hence, $\mathcal{V}(x_0) = \mathcal{V}(0) + x_0 = \{V + x_0 \mid V \in \mathcal{V}(0)\}$ is a system of neighbourhoods.
(ii) For every $U \in \mathcal{V}(0)$ there exists $V \in \mathcal{V}(0)$ such that $V+V \subseteq U$, hence 0 has a system of neighbourhoods made up by closed sets.
(iii) For every $U \in \mathcal{V}(0)$ there exists $V \in \mathcal{V}(0)$ with $V \subseteq U$ and *balanced* in the sense that $V = \{\lambda v \mid |\lambda| \leqslant 1,\ v \in V\}$.

(iv) Every $U \in \mathcal{V}(0)$ is *absorbing* in the sense that for every $x \in \mathcal{E}$ there exists $r > 0$ such that for all $\lambda \in \mathbb{F}$ with $|\lambda| \geqslant r$ we have $x \in \lambda U$.

Linear topologies on a vector space $\mathcal{E}$ make a lattice. Hence, one can always consider the finest linear topology and the coarsest linear topology of a family of linear topologies $\{\mathcal{T}_a\}_{a \in A}$ on $\mathcal{E}$. We exemplify this feature for quotient linear topologies. Let $(\mathcal{E}, \mathcal{T})$ be a topological vector space and $\mathcal{W} \subseteq \mathcal{E}$ a *linear manifold* in $\mathcal{E}$, that is, $\mathcal{W}$ is stable under linear combinations of vectors from $\mathcal{W}$. We call $\mathcal{W}$ a *subspace* if it is closed. Consider the quotient vector space $\mathcal{E}/\mathcal{W}$ and the canonical projection $\pi \colon \mathcal{E} \to \mathcal{E}/\mathcal{W}$, $\mathcal{E} \ni x \mapsto \pi(x) = x + \mathcal{W} \in \mathcal{E}/\mathcal{W}$. On $\mathcal{E}/\mathcal{W}$ we consider $\widetilde{\mathcal{T}}$ the finest linear topology that makes the canonical mapping π continuous. More precisely, $\{\pi(U) \mid U \in \mathcal{V}_{\mathcal{T}}(0)\}$ is a neighbourhood base of $\mathcal{V}_{\widetilde{\mathcal{T}}}(0)$.

Proposition A.3.1. *With notation as before and assuming that $(\mathcal{E}, \mathcal{T})$ is Hausdorff, the topological space $(\mathcal{E}/\mathcal{W}, \widetilde{\mathcal{T}})$ is Hausdorff if and only if the linear manifold $\mathcal{W}$ is closed in $\mathcal{E}$.*

Given a topological vector space $\mathcal{E}$, a subset S of $\mathcal{E}$ is called *total in $\mathcal{E}$* if the linear span of S is dense in $\mathcal{E}$.

The class of topological vector spaces is too general and more special subclasses are usually considered in applications. In the following we review the locally convex topologies, which make one of the most tractable classes among linear topologies at this level of generality. Other more special linear topologies, such as normed and inner product spaces, are reviewed in the next section.

Locally Convex Spaces

Given a vector space $\mathcal{E}$, a subset $A \subseteq \mathcal{E}$ is called *convex* if for all $x, y \in A$ and all $t \in [0, 1]$ we have $(1-t)x + ty \in A$. The subset A is called *absolutely convex* if it is convex and balanced, equivalently, for any $x, y \in A$ and any $\alpha_1, \alpha_2 \in \mathbb{F}$ with $|\alpha_1| \leqslant 1$ and $|\alpha_2| \leqslant 1$ we have $\alpha_1 x + \alpha_2 y \in A$.

If S is an absorbing subset of $\mathcal{E}$, then the *gauge*, or the *Minkowski functional*, of S is the function $p_S \colon \mathcal{E} \to [0, \infty)$ defined by

$$p_S(x) = \inf\{\lambda > 0 \mid x \in \lambda S\}.$$

A *seminorm* is, by definition, a function $p \colon \mathcal{E} \to [0, \infty)$ that is the gauge of an absorbing and absolutely convex subset of $\mathcal{E}$. A *norm* is, by definition, a seminorm p with the property that, if $p(x) = 0$ then $x = 0$.

Proposition A.3.2. *A function $p \colon \mathcal{E} \to [0, \infty)$ is a seminorm if and only if it has the following two properties.*

(i) *(Triangle Inequality)* $p(x + y) \leqslant p(x) + p(y)$ *for all $x, y \in \mathcal{E}$.*
(ii) *(Positively Homogeneous)* $p(\lambda x) = |\lambda| p(x)$ *for all $\lambda \in \mathbb{F}$ and all $x \in \mathcal{E}$.*

A.3 Topological Vector Spaces

The association of seminorms to absorbing and balanced subsets is not one-to-one, but one can estimate to what extent subsets having the same gauge may differ.

Proposition A.3.3. *Let V be an absorbing and balanced subset of $\mathcal{E}$. A seminorm p on $\mathcal{E}$ is the gauge of V if and only if $\{x \in \mathcal{E} \mid p(x) < 1\} \subseteq V \subseteq \{x \in \mathcal{E} \mid p(x) \leqslant 1\}$.*

The next step is to recall characterisations for continuity of a seminorm on a topological vector space.

Proposition A.3.4. *Let p be a seminorm on a topological vector space $(\mathcal{E}, \mathcal{T})$. The following assertions are equivalent.*

(i) *p is continuous at 0.*
(ii) *$\{x \in \mathcal{E} \mid p(x) < 1\} \in \mathcal{T}$.*
(iii) *p is uniformly continuous on $\mathcal{E}$.*

A topological vector space $(\mathcal{E}, \mathcal{T})$ is *locally convex* if any element in $\mathcal{E}$ has a neighbourhood base made up by convex subsets only.

Proposition A.3.5. *Let $(\mathcal{E}, \mathcal{T})$ be a topological vector space $(\mathcal{E}, \mathcal{T})$. The following assertions are equivalent.*

(i) *$(\mathcal{E}, \mathcal{T})$ is a locally convex space.*
(ii) *0 has a neighbourhood base made up by convex subsets only.*
(iii) *0 has a neighbourhood base made up by absolutely convex subsets only.*
(iv) *There exists a family of continuous seminorms $\{p_a\}_{a \in A}$ on $\mathcal{E}$ such that $\mathcal{T}$ is generated by $\{V_{a,x,\varepsilon} \mid a \in A,\ x \in \mathcal{E},\ \varepsilon > 0\}$, where $V_{a,x,\varepsilon} = \{y \in \mathcal{E} \mid p_a(x - y) < \varepsilon\}$.*

A family of seminorms as in (iv) is called a *calibration* of the locally convex space $(\mathcal{E}, \mathcal{T})$. The topology $\mathcal{T}$ is uniquely determined by such a calibration but, generally, there are many calibrations that generate the same locally convex topology.

In dealing with locally convex spaces there is a great advantage in working with calibrations instead of neighbourhood systems or open sets. For example, this way it is very easy to see that, given a vector space $\mathcal{E}$ and a family of locally convex topologies $\{\mathcal{T}_a\}_{a \in A}$ on $\mathcal{E}$, the intersection of all these makes a locally convex topology, which is the finest locally convex topology which is coarser than $\mathcal{T}_a$, for all $a \in A$, and unique with this property. Similarly, there is the coarsest locally convex topology on $\mathcal{E}$ which is finer than $\mathcal{T}_a$, for all $a \in A$, and unique with this property.

Remark A.3.6. The previous observation shows, in particular, that, in dealing with a calibration $\{p_a\}_{a \in A}$ of a locally convex topological space $(\mathcal{E}, \mathcal{T})$, it is always possible to assume that the index set A is ordered and that the calibration is ascendingly directed, in the sense that, for any $a, b \in A$, there exists $c \in A$ such that $p_a, p_b \leqslant p_c$. Indeed, letting $\mathcal{F}(A)$ denote the partially ordered set of all finite subsets of A and, for any $\alpha \in \mathcal{F}(A)$, defining

$$p_\alpha(x) = \max\{p_a(x) \mid a \in \alpha\}, \quad x \in \mathcal{E},$$

it follows that $\{p_\alpha\}_{\alpha \in \mathcal{F}(A)}$ is a calibration of $\mathcal{T}$ which is ascendingly directed. ∎

Continuity of Linear Operators

Proposition A.3.7. *Let $\mathcal{E}$ and $\mathcal{F}$ be locally convex spaces with calibrations $\{p_a\}_{a \in A}$ and $\{q_b\}_{b \in B}$, respectively, and consider a linear operator $T \colon \mathcal{E} \to \mathcal{F}$. The following assertions are equivalent.*

  (i) *T is continuous.*
 (ii) *T is continuous at 0.*
(iii) *T is uniformly continuous.*
(iv) *For any $b \in B$ there exists $a_1, \ldots, a_n \in A$ and $\alpha_1, \ldots, \alpha_n > 0$ such that*

$$q_b(Tx) \leqslant \sum_{k=1}^{n} \alpha_k p_{a_k}(x), \quad x \in \mathcal{E}.$$

  (v) *For any $b \in B$ there exists $a_1, \ldots, a_n \in A$ and $\alpha > 0$ such that*

$$q_b(Tx) \leqslant \alpha \max\{p_{a_k}(x) \mid k=1, \ldots, n\}, \quad x \in \mathcal{E}.$$

Clearly, $\mathbb{R}$ and $\mathbb{C}$ are locally convex spaces with the topology generated by the norm $\mathbb{C} \ni \lambda \mapsto |\lambda|$.

Corollary A.3.8. *Let $\mathcal{E}$ be a locally convex space with a specified calibration $\{p_a\}_{a \in A}$ and let $\varphi \colon \mathcal{E} \to \mathbb{F}$ be a linear functional. The following assertions are equivalent.*

  (i) *φ is uniformly continuous.*
 (ii) *φ is continuous at 0.*
(iii) *There exist $a_1, \ldots, a_n \in A$ and $\alpha_1, \ldots, \alpha_n > 0$ such that*

$$|\varphi(x)| \leqslant \sum_{k=1}^{n} \alpha_k p_{a_k}(x), \quad x \in \mathcal{E}.$$

(iv) *There exist $a_1, \ldots, a_n \in A$ and $\alpha > 0$ such that*

$$|\varphi(x)| \leqslant \alpha \max\{p_{a_k}(x) \mid k=1, \ldots, n\}, \quad x \in \mathcal{E}.$$

One of the most important results on seminorms refers to extensions of linear functionals dominated by a seminorm.

Theorem A.3.9 (Hahn–Banach Theorem). *Let $\mathcal{E}$ be a vector space over $\mathbb{F}$ and p a seminorm on $\mathcal{E}$. If $\mathcal{F}$ is a subspace of $\mathcal{E}$ and $f \colon \mathcal{F} \to \mathbb{F}$ is a linear functional such that $|f(x)| \leqslant p(x)$ for all $x \in \mathcal{F}$, then there exists a linear functional $\widetilde{f} \colon \mathcal{E} \to \mathbb{F}$ such that $\widetilde{f}|_{\mathcal{F}} = f$ and $|\widetilde{f}| \leqslant p(x)$ for all $x \in \mathcal{E}$.*

Given a locally convex space $(\mathcal{E}, \mathcal{T})$ and a subspace $\mathcal{W} \subseteq \mathcal{E}$, on the quotient space $\mathcal{E}/\mathcal{W}$ we can define the locally convex topology $\widetilde{\mathcal{T}}$ which is the finest topology on $\mathcal{E}/\mathcal{W}$ that makes the canonical map $\pi \colon \mathcal{E} \to \mathcal{E}/\mathcal{W}$ continuous. The analogue of Proposition A.3.1 holds.

Proposition A.3.10. *With notation as before, assuming that the locally convex space $(\mathcal{E}, \mathcal{T})$ is Hausdorff, the quotient topology $\widetilde{\mathcal{T}}$ is Hausdorff if and only if $\mathcal{W}$ is closed in $\mathcal{E}$.*

A.4 Banach and Hilbert Spaces

In this section we review a few basic concepts and facts on Banach and Hilbert spaces and their linear operators, the three fundamental principles, duals of Banach spaces and weak topologies, as well as projective and injective tensor products of Banach spaces. For proofs and more advanced topics, a large number of classical monographs such as N. Dunford and J. T. Schwartz [54], or more modern ones such as G. Pedersen [116] and W. Rudin [128] are available. The subsection on tensor products follows R. A. Ryan [129].

Normed Spaces

Let $\mathcal{X}$ be a vector space over the field $\mathbb{F}$, which can be either $\mathbb{R}$ or $\mathbb{C}$. Recall that a *norm* on $\mathcal{X}$ is a positive definite seminorm on $\mathcal{X}$, more precisely, a map $\|\cdot\|\colon \mathcal{X} \to \mathbb{R}$ such that the following hold.

(i) (Positive Definite) $\|v\| \geq 0$ for all $v \in \mathcal{X}$ and $\|v\| = 0$ only for $v = 0$.
(ii) (Positive Homogeneous) $\|\alpha v\| = |\alpha|\|v\|$, for all $\alpha \in \mathbb{F}$ and all $v \in \mathcal{X}$.
(iii) (Triangle Inequality) $\|u + v\| \leq \|u\| + \|v\|$ for all $u, v \in \mathcal{X}$.

In particular, letting $d(u, v) = \|u - v\|$, $u, v \in \mathcal{X}$, we have a metric on $\mathcal{X}$. On a normed space, the topology, convergence of sequences, and all the other topological concepts, are defined as in the case of metric spaces. If the underlying metric space $(\mathcal{X}; d)$ is complete then $(\mathcal{X}; \|\cdot\|)$ is called a *Banach space*.

A sequence $(e_n)_n$ of vectors in a Banach space $\mathcal{X}$ is called a *Schauder basis* if for any $x \in \mathcal{C}$ there exists uniquely a scalar sequence $(x_n)_n$ such that

$$x = \sum_{n=1}^{\infty} x_n e_n,$$

where the series converges with respect to the norm topology. Clearly, if the Banach space has a Schauder basis then it is separable.

Proposition A.4.1. *Let $\mathcal{X}$ and $\mathcal{Y}$ be two normed spaces over the same field $\mathbb{F}$ and let $T\colon \mathcal{X} \to \mathcal{Y}$ be a linear operator. The following assertions are equivalent.*

(i) *T is continuous on $\mathcal{X}$.*
(ii) *T is continuous at some point $v \in \mathcal{X}$.*
(iii) *T is continuous at 0.*
(iv) *T is uniformly continuous on $\mathcal{X}$.*
(v) *$\|Tv\| \leq \alpha \|v\|$ for some $\alpha \geq 0$ and all $v \in \mathcal{X}$.*

A linear operator $T\colon \mathcal{X} \to \mathcal{Y}$ that satisfies one, hence all, of the assertions (i)–(v) is called *bounded* and, in view of the previous proposition, we can define

$$\|T\| := \sup\{\|Tv\| \mid v \in \mathcal{X},\ \|v\| \leq 1\},$$

and call it the *operator norm* of T.

Proposition A.4.2. *Let $\mathcal{X}$ and $\mathcal{Y}$ be two normed spaces over the same field $\mathbb{F}$ and let $T: \mathcal{X} \to \mathcal{Y}$ be a bounded linear operator. Then*

$$\|T\| = \sup\{\|Tv\| \mid v \in \mathcal{X}, \|v\| = 1\}$$
$$= \sup_{v \in \mathcal{X},\, v \neq 0} \frac{\|Tv\|}{\|v\|}$$
$$= \inf\{\alpha \geqslant 0 \mid \|Tv\| \leqslant \alpha\|v\|, \ v \in \mathcal{X}\}.$$

In addition,
$$\|Tv\| \leqslant \|T\|\|v\|, \quad v \in \mathcal{X},$$
in particular T is uniformly continuous.

With notation as before, we denote by $\mathcal{B}(\mathcal{X}, \mathcal{Y})$ the collection of all bounded linear operators $T: \mathcal{X} \to \mathcal{Y}$. If $\mathcal{X} = \mathcal{Y}$ we denote $\mathcal{B}(\mathcal{X}) = \mathcal{B}(\mathcal{X}, \mathcal{X})$.

Proposition A.4.3. *With notation as before, $\mathcal{B}(\mathcal{X}, \mathcal{Y})$ is a vector space and the operator norm is a norm on $\mathcal{B}(\mathcal{X}, \mathcal{Y})$. If $\mathcal{Y}$ is complete, then $\mathcal{B}(\mathcal{X}, \mathcal{Y})$ is a Banach space.*

In addition, the operator norm has the following *submultiplicativity* property: for every $T \in \mathcal{B}(\mathcal{X}, \mathcal{Y})$ and $S \in \mathcal{B}(\mathcal{Y}, X)$, if follows that $ST \in \mathcal{B}(\mathcal{X}, X)$ and

$$\|ST\| \leqslant \|S\|\|T\|.$$

Since bounded linear operators are actually uniformly continuous, they have good extension properties.

Proposition A.4.4. *Let $\mathcal{X}$ and $\mathcal{Y}$ be two Banach spaces and $\mathcal{X}_0$ a dense manifold in $\mathcal{X}$. Then, for any bounded operator $T: \mathcal{X}_0 \to \mathcal{Y}$ there exists a unique bounded operator $\widetilde{T}: \mathcal{X} \to \mathcal{Y}$ such that $\widetilde{T}|\mathcal{X}_0 = T$. In addition, $\|\widetilde{T}\| = \|T\|$.*

When considering embedding operators in the previous proposition, we obtain Banach space completions of normed spaces.

Proposition A.4.5. *For any normed space $\mathcal{X}$ there exists uniquely, up to isometric isomorphisms, a Banach space $\widetilde{\mathcal{X}}$ such that $\mathcal{X}$ is densely and isometrically contained in $\widetilde{\mathcal{X}}$.*

In the following we record a few of the most important normed spaces.

Example A.4.6. (1) Let $\mathbb{F}^n = \mathbb{F} \times \cdots \times \mathbb{F}$, the direct product of n copies of the field $\mathbb{F}$ (either $\mathbb{R}$ or $\mathbb{C}$), where n is a fixed positive integer. As usual, a generic element $u \in \mathbb{F}^n$ is an n-tuple $(u_1, \ldots, u_n)$, with $u_j \in \mathbb{F}$ for all $j = 1, \ldots, n$.

A.4 Banach and Hilbert Spaces

For any $p \in [1, +\infty]$, define $\|\cdot\|_p$ as follows: if p is finite, then let

$$\|u\|_p = \Big(\sum_{j=1}^{n} |u_j|^p\Big)^{1/p}, \quad u = (u_1, \ldots, u_n) \in \mathbb{F}^n,$$

and, for $p = \infty$, let

$$\|u\|_\infty = \max\{|u_j| \mid j = 1, \ldots, n\}, \quad u = (u_1, \ldots, u_n) \in \mathbb{F}^n.$$

All these are norms and $(\mathbb{F}^n; \|\cdot\|_p)$ is a Banach space for all $p \in [1, +\infty]$. Actually, these norms are all equivalent, in the sense that for any $p, q \in [1, +\infty]$ there exist $\alpha, \beta > 0$ such that $\alpha \|u\|_p \leqslant \|u\|_q \leqslant \beta \|u\|_p$, for all $u \in \mathbb{F}^n$. In particular, the underlying metrics are all equivalent and, consequently, they generate the same topology on $\mathbb{F}^n$. This fact is actually more general: for any finite dimensional vector space, any two norms are equivalent and hence they generate the same topology.

(2) Let $\ell_\mathbb{N}^p$, for $1 \leqslant p < \infty$, be the vector space of all scalar sequences $x = (x_n)_n$ indexed on $\mathbb{N}$ that are p-summable, that is,

$$\|x\|_p^p := \sum_{n=1}^{\infty} |x_n|^p < \infty.$$

$\ell_\mathbb{N}^p$ is a Banach space. For $p = \infty$ one defines $\ell_\mathbb{N}^\infty$ as the vector space of all scalar sequences $x = (x_n)_n$ indexed on $\mathbb{N}$ that are bounded, with norm

$$\|x\|_\infty := \sup_n |x_n| < \infty.$$

ℓ^∞ is a Banach space as well.

If $1 \leqslant p < \infty$ then $\ell_\mathbb{N}^p$ has a canonical Schauder basis, $e_n = (\delta_{k,n})_k$, for $n, k \in \mathbb{N}$, where $\delta_{k,n}$ is the Kronecker symbol, and hence $\ell_\mathbb{N}^p$ is separable. The space $\ell_\mathbb{N}^\infty$ is not separable and hence does not have any Schauder basis.

Similar definitions can be made and similar results hold for spaces of sequences indexed on $\mathbb{Z}$.

(3) Let X be a locally compact Hausdorff space and let $\mathcal{C}_c(X)$ denote the vector space of all compactly supported continuous functions $f\colon X \to \mathbb{F}$, where the field $\mathbb{F}$ denotes either $\mathbb{R}$ or $\mathbb{C}$. On $\mathcal{C}_c(X)$ we consider the ∞-norm, called the *sup-norm*, or the *uniform norm*, defined by

$$\|f\|_\infty = \sup\{|f(x)| \mid x \in X\}, \quad f \in \mathcal{C}_c(X).$$

In general, $\mathcal{C}_c(X)$ with norm $\|\cdot\|_\infty$ is not complete.

(4) Let X be a topological space and let $\mathcal{C}_b(X)$ denote the vector space of all bounded and continuous functions $f\colon X \to \mathbb{F}$ with norm $\|\cdot\|_\infty$. Then $(\mathcal{C}_b(X); \|\cdot\|_\infty)$ is a Banach space.

In the case when X is a locally compact Hausdorff space as in the previous example, we can define the space $\mathcal{C}_0(X)$ of all functions $f \in \mathcal{C}_b(X)$ that *vanish at infinity*, that is, for every $\varepsilon > 0$ there exists a compact subset $K \subseteq X$ such that $|f(x)| < \varepsilon$ for all $x \in X \setminus K$. Then $(\mathcal{C}_0(X); \|\cdot\|_\infty)$ is complete and actually it coincides with the closure of $\mathcal{C}_c(C)$.

(5) Let X be a nonempty open subset in $\mathbb{R}^n$ and, for arbitrary $p \in [1, +\infty)$, let

$$\|f\|_p := \Big(\int_X |f(x)|^p \mathrm{d}x\Big)^{1/p}, \quad f \in \mathcal{C}_c(X).$$

Then $(\mathcal{C}_c(X); \|\cdot\|_p)$ is a normed space and its completion is the Lebesgue space $L^p(X)$.

(6) Let X be a locally compact Hausdorff space and μ a Radon measure on X. For $1 \leqslant p < \infty$, let $\mathcal{L}^p(X)$ denote the space of all μ-measurable functions $f \colon X \to \mathbb{F}$ such that

$$\|f\|_p = \Big(\int_X |f(x)|^p \mathrm{d}\mu(x)\Big)^{1/p} < \infty,$$

while, for $p = \infty$ let

$$\|f\|_\infty := \operatorname{ess\,sup}_X |f| := \inf\{a \in \mathbb{R} \mid f^{-1}(a, +\infty) = 0\} < \infty.$$

In general, $\|\cdot\|_p$ is only a seminorm but, letting $\mathcal{N}(X)$ denote the space of all μ-measurable functions such that $\int_X |f| \mathrm{d}\mu = 0$, the quotient spaces $L^p(X) = \mathcal{L}^p(X)/\mathcal{N}(X)$ are Banach spaces. For $X \subseteq \mathbb{R}^n$, these are the classical Lebesgue spaces. ∎

The Fundamental Principles

Three of the fundamental principles of functional analysis refer to linear operators on Banach spaces and they are consequences of Baire's Category Theorem, see Theorem A.1.24. Recall that a map $\varphi \colon X \to Y$ between two topological spaces is called open if it maps any open set of X to an open set in Y.

Theorem A.4.7 (Open Mapping Theorem). *Let $\mathcal{X}$ and $\mathcal{Y}$ be two Banach spaces and let $T \in \mathcal{B}(\mathcal{X}, \mathcal{Y})$, that is, $T \colon \mathcal{X} \to \mathcal{Y}$ is a bounded linear operator. If $T(\mathcal{X}) = \mathcal{Y}$, then T is an open map.*

The second principle is actually equivalent to the Open Mapping Theorem.

Theorem A.4.8 (Closed Graph Theorem). *Let $\mathcal{X}$ and $\mathcal{Y}$ be two Banach spaces and $T \colon \mathcal{X} \to \mathcal{Y}$ be a linear operator such that its graph*

$$\mathcal{G}(T) := \{(v, Tv) \mid v \in \mathcal{X}\} \subset \mathcal{X} \times \mathcal{Y}$$

is closed with respect to the product topology on $\mathcal{X} \times \mathcal{Y}$. Then T is bounded.

The third principle refers to uniform boundedness versus pointwise boundedness.

A.4 Banach and Hilbert Spaces

Theorem A.4.9 (Uniform Boundedness Theorem). *Let $\{T_j \mid j \in \mathcal{J}\}$ be a subset of $\mathcal{B}(\mathcal{X}, \mathcal{Y})$ for two Banach spaces $\mathcal{X}$ and $\mathcal{Y}$. The following assertions are equivalent.*

(i) *For any $v \in \mathcal{X}$ we have $\sup_{j \in \mathcal{J}} \|T_j v\| < \infty$.*
(ii) $\sup_{j \in \mathcal{J}} \|T_j\| < \infty$.

Duals of Normed Spaces

Given a normed space $\mathcal{X}$ over the field $\mathbb{F}$, the *dual space* $\mathcal{X}^*$ is, by definition, the collection of all linear and continuous functionals $\varphi \colon \mathcal{X} \to \mathbb{F}$. The dual space $\mathcal{X}^*$ is canonically a normed vector space with the norm

$$\|\varphi\| = \sup\{|\varphi(v)| \mid v \in \mathcal{X}, \|v\| = 1\},$$

and it is always complete, hence a Banach space.

As a consequence of the Hahn–Banach Theorem, see Theorem A.3.9, the following proposition holds.

Proposition A.4.10. *Given any normed space $\mathcal{X}$ and $v \in \mathcal{X}$, $v \neq 0$, there exists $\varphi \in \mathcal{X}^*$ such that $\varphi(v) = \|v\|$ and $\|\varphi\| = 1$.*

As a consequence of the previous proposition, a characterisation of the norm on a normed space in terms of its dual space can be obtained, namely,

$$\|v\| = \sup\{|\varphi(v)| \mid \varphi \in \mathcal{X}^*, \|\varphi\| = 1\}, \quad v \in \mathcal{X}.$$

Letting $\mathcal{X}^{**} = (\mathcal{X}^*)^*$, one defines the map $\iota \colon \mathcal{X} \to \mathcal{X}^{**}$ by

$$(\iota(v))(\varphi) := \varphi(v), \quad v \in \mathcal{X}, \varphi \in \mathcal{X}^*.$$

Then ι is an isometry and hence $\mathcal{X}$ is isometrically embedded in $\mathcal{X}^{**}$. There are two natural dualities, in this respect. Firstly, the pair $(\mathcal{X}, \mathcal{X}^*)$ is in duality by the bilinear map

$$\langle v, \varphi \rangle := \varphi(v), \quad v \in \mathcal{X}, \varphi \in \mathcal{X}^*.$$

Secondly, the pair $(\mathcal{X}^{**}, \mathcal{X}^*)$ is in duality by the bilinear map

$$\langle z, \varphi \rangle := z(\varphi), \quad z \in \mathcal{X}^{**}, \varphi \in \mathcal{X}^*,$$

and this can be viewed as an extension of the first, modulo the isometric embedding $\mathcal{X} \hookrightarrow \mathcal{X}^{**}$.

The dual space $\mathcal{X}^*$ yields a natural topology on the normed space $\mathcal{X}$ called the w-*topology*, or the *weak topology*, which is the locally convex topology induced by the seminorms

$$p_\varphi(v) := |\varphi(v)|, \quad v \in \mathcal{X},$$

where φ runs in $\mathcal{X}^*$.

On the other hand, the normed space $\mathcal{X}$ yields a natural topology on $\mathcal{X}^*$ called the w^*-*topology*, or the *weak topology of the dual space*, which is the locally convex topology induced by the seminorms

$$p_v(\varphi) := |\varphi(v)|, \quad \varphi \in \mathcal{X}^*,$$

where v runs in $\mathcal{X}$.

Theorem A.4.11 (Banach–Alaoglu Theorem). *If $\mathcal{X}$ is a normed space then the closed unit ball $\{\varphi \in \mathcal{X}^* \mid \|\varphi\| \leqslant 1\}$ is compact with respect to the w^*-topology.*

If $\mathcal{X}$ and $\mathcal{Y}$ are normed spaces and $T \in \mathcal{B}(\mathcal{X}, \mathcal{Y})$, then one defines the *dual operator* $T^* \colon \mathcal{Y}^* \to \mathcal{X}^*$ by letting, with notation as before,

$$\langle v, T^*\varphi \rangle = \langle Tv, \varphi \rangle, \quad v \in \mathcal{X}, \varphi \in \mathcal{X}^*.$$

Proposition A.4.12. *If $\mathcal{X}$ and $\mathcal{Y}$ are normed spaces and $T \in \mathcal{B}(\mathcal{X}, \mathcal{Y})$ then $T^* \in \mathcal{B}(\mathcal{Y}^*, \mathcal{X}^*)$ with the operator norm $\|T^*\| = \|T\|$.*

Tensor Products of Banach Spaces

Let X, Y, and Z be vector spaces over the same field $\mathbb{F}$, which may be $\mathbb{R}$ or $\mathbb{C}$, and let $B(X \times Y, Z)$ denote the vector space of all *bilinear* maps $A \colon X \times Y \to Z$, that is,

$$A(\alpha_1 x_1 + \alpha_2 x_2, y) = \alpha_1 A(x_1, y) + \alpha_2 A(x_2, y), \quad x_1, x_2 \in X, y \in Y, \alpha_1, \alpha_2 \in \mathbb{F},$$
$$A(x, \beta_1 y_1 + \beta_2 y_2) = \beta_1 A(x, y_1) + \beta_2 A(x, y_2), \quad x \in X, y_1, y_2 \in Y, \beta_1, \beta_2 \in \mathbb{F}.$$

When $Z = \mathbb{F}$ we use $B(X \times Y) := B(X \times Y, Z)$ and its elements are called *bilinear forms*.

For each $x \in X$ and $y \in Y$, define the *elementary tensor* $x \otimes y \colon B(X \times Y) \to \mathbb{F}$ by

$$(x \otimes y)(\varphi) := \varphi(x, y), \quad \varphi \in B(X \times Y).$$

The vector space generated by all elementary tensors $x \otimes y$ in $B(X \times Y)^*$, the dual space of $B(X \times Y)$, is denoted by $X \otimes Y$ and is called the *tensor product* of the vector spaces X and Y. Thus, for any element $u \in X \otimes Y$ there exists $x_1, \ldots, x_n \in X$, $y_1, \ldots, y_n \in Y$, and $\lambda_1, \ldots, \lambda_n \in \mathbb{F}$ such that,

$$u = \sum_{j=1}^{n} \lambda_j x_j \otimes y_j.$$

Of course, this representation is not unique, in general.

The elementary tensors have the following properties.

(i) $(x_1 + x_2) \otimes y = x_1 \otimes y + x_2 \otimes y$, $x_1, x_2 \in X$, $y \in Y$.

A.4 Banach and Hilbert Spaces

(ii) $x \otimes (y_1 + y_2) = x \otimes y_1 + x \otimes y_2$, $x \in X$, $y_1, y_2 \in Y$.
(iii) $\lambda(x \otimes y) = (\lambda x) \otimes y = x \otimes (\lambda y)$, $x \in X$, $y \in Y$, $\lambda \in \mathbb{F}$.
(iv) $0 \otimes y = x \otimes 0 = 0$, $x \in X$, $y \in Y$.

In particular, any element $u \in X \otimes Y$ has a representation

$$u = \sum_{j=1}^{n} x_j \otimes y_j,$$

for some $x_1, \ldots, x_n \in X$ and $y_1, \ldots, y_n \in Y$.

The idea of tensor product of vector spaces is closely related to that of linearisation of bilinear maps, as illustrated in the following result.

Proposition A.4.13. *Let X, Y, and Z be vector spaces over the same field $\mathbb{F}$. There is a natural linear isomorphism between the vector spaces $B(X \times Y, Z) \ni A \mapsto \widetilde{A} \in L(X \otimes Y, Z)$, more precisely, for any bilinear map $A \colon X \times Y \to Z$ there exists uniquely a linear map $\widetilde{A} \colon X \times Y \to Z$ such that $A(x, y) = \widetilde{A}(x \otimes y)$, for all $x \in X$ and $y \in Y$.*

On the other hand, there is a natural embedding

$$X^* \otimes Y^* \subset B(X \times Y),$$

defined, on elementary tensors $\varphi \otimes \psi$, for $\varphi \in X^*$ and $\psi \in Y^*$, by

$$(\varphi \otimes \psi)(x, y) := \varphi(x)\psi(y), \quad x \in X, \ y \in Y,$$

and then extended by linearity.

Let now X and Y be Banach spaces. We consider the algebraic tensor product space $X \otimes Y$, that is, the tensor product of the underlying vector spaces, on which the *projective norm* $\|\cdot\|_\pi$ is defined

$$\|u\|_\pi := \inf\{\sum_{j=1}^{n} \|x_j\| \|y_j\| \mid u = \sum_{j=1}^{n} x_j \otimes y_j\}.$$

Then,

$$\|x \otimes y\| = \|x\| \|y\|, \quad x \in X, \ y \in Y.$$

It is customary to denote $X \otimes_\pi Y := (X \otimes Y, \|\cdot\|_\pi)$ and then let $X \widehat{\otimes}_\pi Y$ denote the completion of $X \otimes_\pi Y$ to a Banach space, called the *projective tensor product* of X and Y.

The *injective norm* $\|\cdot\|_\varepsilon$ is defined on the algebraic tensor product $X \otimes Y$ by

$$\|u\|_\varepsilon := \sup\{|\sum_{j=1}^{n} \varphi(x_j)\psi(y_j)| \mid \varphi \in X^*, \|\varphi\| \leqslant 1, \psi \in Y^*, \|\psi\| \leqslant 1\},$$

where $u = \sum_{j=1}^{n} x_j \otimes y_j$. It is proven that this definition is independent of the representation of u as a sum of elementary tensors. It is customary to denote $X \otimes_\varepsilon Y := (X \otimes Y, \|\cdot\|_\varepsilon)$ and then let $X \widehat{\otimes}_\varepsilon Y$ denote the completion of the normed space $X \otimes_\varepsilon Y$ to a Banach space, called the *injective tensor product*.

Proposition A.4.14. *Let X and Y be Banach spaces.*

(i) $\|u\|_\varepsilon \leq \|u\|_\pi$, *for all $u \in X \otimes Y$.*
(ii) $\|x \otimes y\|_\varepsilon = \|x\|\|y\|$, *for all $x \in X$ and $y \in Y$.*
(iii) *For any $\varphi \in X^*$ and any $\psi \in Y^*$, $\varphi \otimes \psi$ is a bounded linear functional on $X \widehat{\otimes}_\varepsilon Y$ and $\|\varphi \otimes \psi\| = \|\varphi\|\|\psi\|$.*

Hilbert Spaces

Let $\mathcal{H}$ be a complex vector space. A mapping $\langle \cdot, \cdot \rangle \colon \mathcal{H} \times \mathcal{H} \to \mathbb{C}$ is called an *inner product* or *scalar product* if it has the following properties.

(i) (Positive Definite) $\langle x, x \rangle \geq 0$ for all $x \in \mathcal{H}$ and $\langle x, x \rangle = 0$ implies $x = 0$.
(ii) (Linearity in the First Variable) $\langle \alpha x + y, z \rangle = \alpha \langle x, z \rangle + \langle y, z \rangle$ for all $x, y, z \in \mathcal{H}$ and all $\alpha \in \mathbb{C}$.
(iii) (Conjugate Symmetry) $\langle x, y \rangle = \overline{\langle y, x \rangle}$ for all $x, y \in \mathcal{H}$.

As a consequence of (ii) and (iii), an inner product is conjugate linear in the second variable.

(ii)' (Conjugate Linearity in the Second Variable) $\langle x, \alpha y + z \rangle = \overline{\alpha} \langle x, y \rangle + \langle x, z \rangle$, for all $x, y, z \in \mathcal{H}$ and $\alpha \in \mathbb{C}$.

A mapping $\langle \cdot, \cdot \rangle \colon \mathcal{H} \times \mathcal{H} \to \mathbb{C}$ is called a *sesquilinear form* if it satisfies conditions (ii) and (ii)'.

Example A.4.15. (a) $\mathbb{C}^d$ with inner product $\langle x, y \rangle = \sum_{j=1}^{d} x_j \overline{y}_j$, where $x = (x_1, \ldots, x_d)$ and $y = (y_1, \ldots, y_d)$.
(b) On

$$\ell^2 = \Big\{ x = (x_n)_{n \geq 1} \mid x_n \in \mathbb{C} \text{ for all } n \in \mathbb{N}, \text{ and } \sum_{n=1}^{\infty} |x_n|^2 < \infty \Big\}$$

one can define the inner product

$$\langle x, y \rangle = \sum_{n=1}^{\infty} x_n \overline{y}_n,$$

where the series converges absolutely.

A.4 Banach and Hilbert Spaces

(c) Let (X, Σ, μ) be a measure space. On the complex vector space

$$\mathcal{L}^2(X, \mu) = \{f \colon X \to \mathbb{C} \mid f \text{ measurable and } \int_X |f(x)|^2 \mathrm{d}\mu < \infty\},$$

one can define the sesquilinear form

$$\langle f, g \rangle = \int_X f(x)\overline{g(x)} \mathrm{d}\mu, \quad f, g \in \mathcal{L}^2(X, \mu).$$

Then, letting $L^2(X, \mu)$ denote the space of functions $f \in \mathcal{L}^2(X, \mu)$, identified modulo μ-negligible sets, the sesquilinear form $\langle \cdot, \cdot \rangle$ induces an inner product on $L^2(X, \mu)$ with respect to which it is a Hilbert space. ∎

Let $(\mathcal{H}, \langle \cdot, \cdot \rangle)$ be an inner product space. The following *Cauchy–Bunyakovski–Schwarz Inequality*, abbreviated CBS, holds:

$$|\langle x, y \rangle| \leqslant \langle x, x \rangle^{1/2} \langle y, y \rangle^{1/2}, \quad x, y \in \mathcal{H}.$$

The CBS Inequality holds also under the more general assumption of positive semidefinite, instead of positive definite. Consequently, the following *triangle inequality* holds

$$\langle x + y, x + y \rangle^{1/2} \leqslant \langle x, x \rangle^{1/2} + \langle y, y \rangle^{1/2}, \quad x, y \in \mathcal{H}.$$

Let $\|x\| = \langle x, x \rangle^{1/2}$. Then $\| \cdot \|$ is a norm on $\mathcal{H}$ that satisfies the following *parallelogram law*

$$\|x + y\|^2 + \|x - y\|^2 = 2\|x\|^2 + 2\|y\|^2, \quad x, y \in \mathcal{H}.$$

In addition, it can be proven that a norm $\| \cdot \|$ on $\mathcal{H}$ is associated to an inner product $\langle \cdot, \cdot \rangle$ if and only if it satisfies the parallelogram law. Moreover, the inner product can be recovered from its associated norm by means of the *polarisation formula*

$$4 \langle x, y \rangle = \sum_{k=0}^{3} \mathrm{i}^k \|x + \mathrm{i}^k y\|^2, \quad x, y \in \mathcal{H}.$$

Since $(\mathcal{H}, \| \cdot \|)$ becomes a normed space then, letting $d(x, y) = \|x - y\|$, it becomes a metric space. Consequently, $\mathcal{H}$ is a topological space, and the underlying topology is called the *normed topology* or *strong topology*. An inner product space $(\mathcal{H}, \langle \cdot, \cdot \rangle)$ is called a *Hilbert space* if the metric d is complete. All the inner product spaces $\mathbb{C}^d$, ℓ^2, and $L^2(X, \mu)$ in Example A.4.15 are Hilbert spaces. With definitions as in Section A.3, any Hilbert space is a Banach space.

Example A.4.16. Let $a < b$ be two real numbers and $\mathcal{C}([a, b]) = \{f \colon [a, b] \to \mathbb{C} \mid f \text{ continuous on } [a, b]\}$. One considers the inner product

$$\langle f, g \rangle = \int_a^b f(t) \overline{g(t)} \mathrm{d}t, \quad f, g \in \mathcal{C}([a, b]),$$

where the integral is the Riemann integral. Then $(\mathcal{C}([a, b]), \langle \cdot, \cdot \rangle)$ is not a Hilbert space. since it is not complete. ∎

On a Hilbert space, the *weak topology* is the locally convex topology defined by the family of seminorms $\mathcal{H} \ni x \mapsto |\langle x, y \rangle|$, for all $y \in \mathcal{H}$.

A linear mapping $\varphi \colon \mathcal{H} \to \mathbb{C}$ is called a *linear functional*. If φ is continuous then its functional norm is

$$\|\varphi\| = \sup_{\|x\| \geqslant 1} |\varphi(x)| < \infty.$$

Theorem A.4.17 (Riesz–Fréchet Theorem). *Let φ be a continuous linear functional on $\mathcal{H}$. Then there exists uniquely an element $h \in \mathcal{H}$ such that $\varphi(x) = \langle x, h \rangle$ for all $x \in \mathcal{H}$. In addition, we have $\|\varphi\| = \|h\|$.*

Subspaces of a Hilbert Space

From now on, $(\mathcal{H}, \langle \cdot, \cdot \rangle)$ denotes a Hilbert space. Recall that a *subspace* of $\mathcal{H}$ is a linear manifold $\mathcal{S}$ that is strongly closed. Given a subset $S \subseteq \mathcal{H}$ its *orthogonal companion* is the subspace

$$S^\perp = \{ y \in \mathcal{H} \mid \langle x, y \rangle = 0 \text{ for all } x \in S \}.$$

In the following, for two subspaces $\mathcal{A}$ and $\mathcal{B}$ of $\mathcal{H}$, we denote by $\mathcal{A} \oplus \mathcal{B}$ their *orthogonal sum*, that is, we implicitly assume that $\mathcal{A} \perp \mathcal{B}$. In this case, $\mathcal{A} \oplus \mathcal{B}$ is always a subspace.

Proposition A.4.18. *Let $\mathcal{S}$ be a subspace of the Hilbert space $\mathcal{H}$. Then, for every $x \in \mathcal{H}$ there exist uniquely $u \in \mathcal{S}$ and $v \in \mathcal{S}^\perp$ such that $x = u + v$. In addition,*

$$\|v\| = \inf_{y \in \mathcal{S}} \|x - y\|.$$

In particular,

$$\mathcal{H} = \mathcal{S} \oplus \mathcal{S}^\perp.$$

As a consequence of the Riesz–Fréchet Theorem, see Theorem A.4.17, we have the following.

Proposition A.4.19. *A linear manifold of a Hilbert space is weakly closed if and only if it is strongly closed.*

Orthonormal Bases

Let A be a nonempty subset of $\mathcal{H}$. The *span* of A, denoted by $\mathrm{Span}(A)$, is the linear subspace containing all linear combinations of vectors in A, that is, all vectors $x \in \mathcal{H}$ for which there exist $x_1, \ldots, x_n \in A$ and $\alpha_1, \ldots, \alpha_n \in \mathbb{C}$ such that $x = \alpha_1 x_1 + \cdots + \alpha_n x_n$. A subset B of $\mathcal{H}$ is called a *topological basis* of $\mathcal{H}$ if it is linearly independent and $\mathrm{Span}(B)$ is dense in $\mathcal{H}$.

Proposition A.4.20. *A Hilbert space is separable if and only if it has a topological basis which is either finite or infinitely countable.*

A.4 Banach and Hilbert Spaces

A family $\{x_j\}_{j \in \mathcal{J}}$ of vectors in $\mathcal{H}$ is called *orthonormal* if $\langle x_j, x_k \rangle = \delta_{j,k}$ for all $j, k \in \mathcal{J}$, where $\delta_{j,k}$ is the Kronecker symbol, $\delta_{j,k} = 0$ if $j \neq k$ and $\delta_{j,j} = 1$. In this situation, for every $x \in \mathcal{H}$ the following Bessel's Inequality holds

$$\sum_{j \in \mathcal{J}} |\langle x, x_j \rangle|^2 \leq \|x\|^2, \quad x \in \mathcal{H}.$$

In particular, the series converges in the sense of *summation*, that is, there exists $\alpha \geq 0$ such that, for every $\varepsilon > 0$, there exists a finite subset $J_\varepsilon \subseteq \mathcal{J}$ such that, for every finite subset $J \subseteq \mathcal{J}$ with $J_\varepsilon \subseteq J$ we have

$$\left| \alpha - \sum_{j \in J} |\langle x, x_j \rangle|^2 \right| < \varepsilon.$$

Consequently, for every $x \in \mathcal{H}$, the set $\{j \in \mathcal{J} \mid \langle x, x_j \rangle \neq 0\}$ is countable, that is, either finite or denumerable.

If $\{x_j\}_{j \in \mathcal{J}}$ is an orthonormal system of vectors such that $\mathrm{Span}(\{x_j\}_{j \in \mathcal{J}})$ is dense in $\mathcal{H}$ then it is called a *Hilbert space basis*, or an *orthonormal basis* of $\mathcal{H}$. In the following proposition, convergence of infinite sums is considered in the sense of summation, as explained before.

Proposition A.4.21. *Let* $B = \{x_j\}_{j \in \mathcal{J}}$ *be an orthonormal system of vectors in the Hilbert space* $\mathcal{H}$. *The following assertions are equivalent.*

(i) *B is a Hilbert space basis.*
(ii) *B is a maximal orthonormal system of vectors in $\mathcal{H}$.*
(iii) *$B^\perp = \{0\}$.*
(iv) *For every $x \in \mathcal{H}$ the Fourier expansion holds:*

$$x = \sum_{j \in \mathcal{J}} \langle x, x_j \rangle x_j.$$

(v) *For every $x, y \in \mathcal{H}$ we have*

$$\langle x, y \rangle = \sum_{j \in \mathcal{J}} \langle x, x_j \rangle \overline{\langle y, x_j \rangle}.$$

(vi) (Parseval Identity) *For every $x \in \mathcal{H}$ the following holds:*

$$\|x\|^2 = \sum_{j \in \mathcal{J}} |\langle x, x_j \rangle|^2.$$

Given a sequence of vectors $(v_n)_{n \geq 1}$ in $\mathcal{H}$ such that the set $\{v_n \mid n = 1, 2, \ldots\}$ is linearly independent, there exists a sequence $(e_n)_{n \geq 1}$ subject to the following conditions.

(i) $\{e_n \mid n = 1, 2, \ldots\}$ is orthonormal.

(ii) $\text{Span}\{v_1, \ldots, v_n\} = \text{Span}\{e_1, \ldots, v_n\}$ for all $n = 1, 2, \ldots$.

The sequence $(e_n)_{n \geq 1}$ is obtained recursively as follows:
$$e_1 := \frac{v_1}{\|v_1\|},$$
and, for any $n > 1$, assuming that the vectors $e_1, \ldots, e_{n-1}$ are already defined, we have
$$u_n := v_n - \sum_{k=1}^{n-1} \langle v_k, e_k \rangle e_k, \quad e_n := \frac{u_n}{\|u_n\|}.$$

This algorithm is called the *Gram–Schmidt orthogonalisation process* and it depends on the order of the vectors on which we perform it.

Tensor Products of Hilbert Spaces

Let $\mathcal{H}$ and $\mathcal{K}$ be two Hilbert spaces and let $\mathcal{H} \otimes \mathcal{K}$ be their algebraic tensor product. An inner product can be defined as follows. For arbitrary vectors $h_1, \ldots, h_n, g_1, \ldots, g_m \in \mathcal{H}$ and $k_1, \ldots, k_n, l_1, \ldots, l_m \in \mathcal{H}$ let
$$\left\langle \sum_{i=1}^{n} h_i \otimes k_i, \sum_{j=1}^{m} g_j \otimes l_j \right\rangle = \sum_{i=1}^{n} \sum_{j=1}^{m} \langle h_i, g_j \rangle_{\mathcal{H}} \langle k_i, l_j \rangle_{\mathcal{K}}.$$

It can be proven that this definition is correct and yields a positive definite inner product. Thus, $\mathcal{H} \otimes \mathcal{K}$ is a pre-Hilbert space that can be completed to a Hilbert space denoted $\mathcal{H} \otimes \mathcal{K}$.

Let $(e_i)_{i \in \mathcal{I}}$ be an orthonormal basis of $\mathcal{H}$ and let $(f_j)_{j \in \mathcal{J}}$ be an orthonormal basis of $\mathcal{K}$. Then $(e_i \otimes f_j)_{i \in \mathcal{I},\, j \in \mathcal{J}}$ is an orthonormal basis of the Hilbert space $\mathcal{H} \otimes \mathcal{K}$. In particular, if $\mathcal{H} = \mathbb{C}^n$ then $\mathbb{C}^n \otimes \mathcal{K}$ is isometrically isomorphic with $\mathcal{K}^n = \mathcal{K} \oplus \cdots \oplus \mathcal{K}$, the orthogonal direct sum of n copies of $\mathcal{K}$.

If (X, μ) is a measure space, the Hilbert space of square integrable functions $f \colon X \to \mathcal{K}$ is defined as the vector space of all weakly measurable functions $f \colon X \to \mathcal{K}$, that is, $X \ni x \mapsto \langle f(x), y \rangle_{\mathcal{K}}$ is measurable for all $y \in \mathcal{K}$, such that
$$\|f\|^2 = \int_X \|f(x)\|_{\mathcal{K}}^2 \mathrm{d}\mu(x) < \infty,$$
with inner product
$$\langle f, g \rangle = \int_X \langle f(x), g(x) \rangle_{\mathcal{K}} \mathrm{d}\mu(x), \quad f, g \in L^2(X, \mu, \mathcal{K}).$$

Then, $L^2(X, \mu) \otimes \mathcal{K}$ is naturally isometrically isomorphic with $L^2(X, \mu, \mathcal{K})$, by the identification of $f \otimes k$ with the map $X \ni x \mapsto f(x)k \in \mathcal{K}$.

A.5 Functions of One Complex Variable

We assume the reader is familiar with basic analysis of functions of one complex variable. There are many classical textbooks and monographs on analyticity of complex functions, for example J.B. Conway [35] and W. Rudin [127]. In this section we review basic definitions and facts of analyticity for complex functions, some spaces of analytic functions on the unit disc, and the main concepts and results on vector valued analytic functions.

Holomorphic Functions

For any $r > 0$ and $a \in \mathbb{C}$ we consider $\mathbb{D}_r(a)$ the open disc of centre a and radius r,

$$\mathbb{D}_r(a) := \{z \in \mathbb{C} \mid |z - a| < r\},$$

and let $\operatorname{Clos} \mathbb{D}_r(a) := \{z \in \mathbb{C} \mid |z - a| \leqslant r\}$ denote its closure. The punctured open disc of centre a and radius r is denoted by

$$\mathbb{D}'_r(a) := \{z \in \mathbb{C} \mid 0 < |z - a| < r\}.$$

If $a = 0$ we use the simpler notation $\mathbb{D}_r := \mathbb{D}_r(0)$ and $\mathbb{D}'_r := \mathbb{D}'_r(0)$.

By *region* we mean a nonempty connected open subset of the complex plane $\mathbb{C}$. Every open subset of $\mathbb{C}$ is a countable (infinite or finite) collection of regions.

Definition A.5.1. Let $f \colon \Omega \to \mathbb{C}$ be a function, for Ω an open subset of $\mathbb{C}$. If $z_0 \in \Omega$ and

$$\lim_{z \to z_0} \frac{f(z) - f(z_0)}{z - z_0}$$

exists, we denote this limit by $f'(z_0)$ and call it the *derivative* of f at z_0. In this case, we say that f is *differentiable* at z_0. If $f'(z_0)$ exists for every $z_0 \in \Omega$, we say that f is *holomorphic* in Ω. The class of all holomorphic functions in Ω is denoted by $\operatorname{Hol}(\Omega)$. A function that is holomorphic in $\mathbb{C}$ is called *entire*. ∎

The class $\operatorname{Hol}(\Omega)$ is an algebra with respect to the usual algebraic operations of addition, multiplication, and multiplication with scalars, with identity the constant function equal to 1. Composition of holomorphic functions is holomorphic and the familiar chain rule holds

$$(g \circ f)'(z_0) = g'(f(z_0)) f'(z_0).$$

Any polynomial function $f(z) = a_n z^n + \cdots + a_1 z + a_0$ is entire and any rational function $f(z) = p(z)/q(z)$, where p and q are polynomial functions, is holomorphic in $\mathbb{C} \setminus \{z \in \mathbb{C} \mid q(z) = 0\}$.

There is an equivalent condition to holomorphy in terms of the partial derivatives.

Theorem A.5.2 (Cauchy–Riemann Equations). *Given $f \colon \Omega \to \mathbb{C}$ a complex function, let $f = u + iv$, where $u, v \colon \Omega \to \mathbb{R}$ are, respectively, the real and imaginary parts of f, and let $z_0 \in \Omega$.*

(1) *If f is differentiable at z_0 then both partial derivatives of f, u, and v exist at z_0 and*

$$\frac{\partial f}{\partial x}(z_0) = -i\frac{\partial f}{\partial y}(z_0), \tag{A.1}$$

equivalently,

$$\frac{\partial u}{\partial x}(z_0) = \frac{\partial v}{\partial y}(z_0),$$

$$\frac{\partial u}{\partial y}(z_0) = -\frac{\partial v}{\partial x}(z_0).$$

(2) *Assume that both partial derivatives $\frac{\partial f}{\partial x}$ and $\frac{\partial f}{\partial y}$ exist in an open disc in Ω centred at z_0 and are continuous at z_0, and that (A.1) holds. Then f is differentiable at z_0.*

Let $a \in \mathbb{C}$ and let $(c_n)_{n \geqslant 0}$ be a complex sequence. To any power series

$$\sum_{n \geqslant 0} c_n (z - a)^n, \tag{A.2}$$

one associates a number $R \in [0, +\infty]$, called the *radius of convergence*, such that the series converges absolutely and uniformly in $\operatorname{Clos} \mathbb{D}_r(a)$, for every $0 < r < R$, and diverges for any $z \notin \operatorname{Clos} \mathbb{D}_R(a)$, more precisely, by the Cauchy–Hadamard Theorem

$$R = \frac{1}{\limsup_{n \to \infty} |c_n|^{\frac{1}{n}}}, \tag{A.3}$$

in the sense that $R = 0$ if the limsup is infinite and $R = +\infty$ if the limsup is 0.

A function $f \colon \Omega \to \mathbb{C}$ is *analytic*, or *representable by power series in Ω*, for Ω an open subset of $\mathbb{C}$, if for every $a \in \Omega$ and any $r > 0$ such that $\mathbb{D}_r(a) \subseteq \Omega$, there corresponds a power series (A.2) that converges to $f(z)$ for every $z \in \mathbb{D}_r(a)$.

Theorem A.5.3. *Let $\Omega \subseteq \mathbb{C}$ be a nonempty open set and assume that $f \colon \Omega \to \mathbb{C}$ is a function representable by a power series in Ω. Then $f \in \operatorname{Hol}(\Omega)$. In fact, if $a \in \Omega$ and $r > 0$ such that $\mathbb{D}_r(a) \subseteq \Omega$ and f is represented by the power series (A.2), then f has derivatives of any order $k \geqslant 1$, denoted by $f^{(k)}(z)$ and*

$$f^{(k)}(z) = \sum_{n=k}^{\infty} n(n-1) \cdots (n-k-1) c_n (z-a)^{n-k}, \quad z \in \mathbb{D}_r(a),$$

in particular,

$$c_k = \frac{f^{(k)}(a)}{k}, \quad k = 0, 1, 2, \ldots,$$

and hence for each $a \in \Omega$ the coefficients sequence $(c_n)_{n \geqslant 0}$ for which (A.2) holds is unique.

A.5 Functions of One Complex Variable

As one of the most important examples, the *exponential function* defined by the power series

$$\exp(z) = e^z := \sum_{n=0}^{\infty} \frac{z^n}{n!}, \tag{A.4}$$

is an entire function and $(e^z)' = e^z$ for all $z \in \mathbb{C}$. The exponential function has the property

$$e^{z+\zeta} = e^z e^\zeta, \quad z, \zeta \in \mathbb{C},$$

and is periodic $e^{z+2i\pi} = e^z$, for all $z \in \mathbb{C}$.

The trigonometric functions sinus and cosinus defined by

$$\sin z := \sum_{n=0}^{\infty} \frac{z^{2n-1}}{(2n-1)!}, \quad \cos z := \sum_{n=0}^{\infty} \frac{z^{2n}}{(2n)!}, \quad z \in \mathbb{C}, \tag{A.5}$$

are entire functions and $\sin^2 z + \cos^2 z = 1$, $(\sin z)' = \cos z$, $(\cos z)' = -\sin z$, for all $z \in \mathbb{C}$. Both sinus and cosinus functions are periodic with period 2π. In addition, the trigonometric functions are related to the exponential function by the following formulae:

$$e^{iz} = \cos z + i \sin z, \quad \cos z = \frac{1}{2}(e^{iz} + e^{-iz}), \quad \sin z = \frac{1}{2i}(e^{iz} - e^{-iz}).$$

If $z = x + iy$, where $x, y \in \mathbb{R}$, then Euler's formula

$$e^z = e^x(\cos y + i \sin y)$$

holds. In addition, $|e^z| = e^x$ and $\arg e^z = y$, modulo a multiple of 2π, where, for an arbitrary complex number $\zeta = a + ib \neq 0$, $\arg \zeta$ is the unique number $\theta \in [0, 2\pi)$ such that $\cos \theta = a/\sqrt{a^2 + b^2}$ and $\sin \theta = b/\sqrt{a^2 + b^2}$.

A rather general example of holomorphic functions is provided by the following theorem.

Theorem A.5.4. *Let μ be a complex (finite) measure on a measurable space X, let $\varphi \colon X \to \mathbb{C}$ be a measurable function such that for some nonempty open set Ω in $\mathbb{C}$ we have $\varphi(X) \cap \Omega = \emptyset$. Then, the function $f \colon \Omega \to \mathbb{C}$ defined by*

$$f(z) = \int_X \frac{d\mu(\zeta)}{\varphi(\zeta) - z}, \quad z \in \Omega,$$

is representable by a power series in Ω and, in particular, is holomorphic in Ω. In addition, with notation as in (A.2), in this case we have the estimations

$$|c_n| \leqslant \frac{|\mu|(X)}{r^{n+1}}, \quad n = 0, 1, 2, \ldots.$$

Theorem A.5.3 has a converse: if $f \in \mathrm{Hol}(\Omega)$ for some nonempty open set Ω in $\mathbb{C}$ then it is representable by a power series in Ω and, consequently, it is indefinitely differentiable. In order to make this result clearer one can follow the route through Cauchy's Theorem of integral representations of holomorphic functions. We first recall the integration of complex functions on paths. A *curve* in $\mathbb{C}$ is a continuous mapping $\gamma \colon [a,b] \to \mathbb{C}$, for some real numbers $a < b$. Let $\gamma^* = \{\gamma(t) \mid a \leqslant t \leqslant b\}$ be the *trace* of γ. If $\gamma(a) = \gamma(b)$ then the curve γ is called *closed*. If γ is piecewise continuously differentiable, that is, for an interval partition $a = t_0 < t_1 < \cdots < t_n = b$ the restriction of γ to $[t_{j-1}, t_j]$ has a continuous derivative, for every $j = 1, \ldots, n$, then we call it a *path*. A *closed path* is a closed curve that is also a path.

A *Jordan curve* Γ is a closed curve that is nonintersecting, equivalently, it is the homeomorphic image of the unit circle. Jordan paths, called *Jordan contours*, are particularly important in function theory.

Let γ be a path and f be a complex continuous function on γ^*. The integral of f on γ is defined by

$$\int_\gamma f(z)\mathrm{d}z := \int_a^b f(\gamma(t))\gamma'(t)\mathrm{d}t.$$

Clearly,

$$\left|\int_\gamma f(z)\mathrm{d}z\right| \leqslant \|f\|_\infty \int_a^b |\gamma'(t)|\mathrm{d}t,$$

where $\|f\|_\infty = \sup_\gamma |f|$ and the latter integral is the *length* of γ. If $\varphi \colon [a_1, b_1] \to [a, b]$ is a continuously differentiable surjective mapping then $\gamma \circ \varphi$ is a path and

$$\int_{\gamma \circ \varphi} f(z)\mathrm{d}z = \int_\gamma f(z)\mathrm{d}z,$$

hence the integral essentially depends on the trace γ^* and not on its parametrisation. In particular, we may choose a parametrisation on $[0,1]$ without any loss of generality.

Paths can be concatenated under certain conditions. If $\gamma \colon [a,b] \to \mathbb{C}$ and $\delta \colon [b,c] \to \mathbb{C}$ are two paths such that $\gamma(b) = \delta(b)$ then a new path $\eta \colon [a,c] \to \mathbb{C}$ such that $\eta^* = \gamma^* \cup \delta^*$ can defined by $\eta(t) = \gamma(t)$ for all $t \in [a,b]$ and $\eta(t) = \delta(t)$ for all $t \in [b,c]$. If $f \colon \eta^* \to \mathbb{C}$ is a continuous function then

$$\int_\eta f(z)\mathrm{d}z = \int_\gamma f(z)\mathrm{d}z + \int_\delta f(z)\mathrm{d}z.$$

In the particular case when the parametrisation interval is $[0,1]$ and $\gamma \colon [0,1] \to \mathbb{C}$ is a path, then the path $\gamma^\circ \colon [0,1] \to \mathbb{C}$ defined by $\gamma^\circ(t) = \gamma(1-t)$, for all $t \in [0,1]$, is called the path *opposite to* γ. The name is justified by the fact that

$$\int_{\gamma^\circ} f(z)\mathrm{d}z = -\int_\gamma f(z)\mathrm{d}z.$$

In the following we recall two of the special paths that are used intensively.

A.5 Functions of One Complex Variable

(1) *Positively Oriented Circle.* Assume that z_0 is a complex number and $r > 0$. The closed path $\gamma \colon [0, 2\pi] \to \mathbb{C}$ defined by

$$\gamma(t) = z_0 + re^{it}, \quad t \in [0, 2\pi],$$

is called the *positively oriented circle* with centre at z_0 and radius r. For any continuous function f on γ^* we have

$$\int_\gamma f(z)\,\mathrm{d}z = ir \int_0^{2\pi} f(z_0 + re^{i\theta}) e^{i\theta}\,\mathrm{d}\theta,$$

and the length of γ is $2\pi r$.

(2) *Oriented Interval.* For $z_0, z_1 \in \mathbb{C}$, the path

$$\gamma(t) = z_0 + t(z_1 - z_0), \quad t \in [0, 1],$$

is the *oriented interval* denoted by $[z_0, z_1]$, with starting point z_0 and ending point z_1, of length $|z_1 - z_0|$ and, for any continuous function $f \colon \gamma^* \to \mathbb{C}$, we have

$$\int_{[z_0, z_1]} f(z)\,\mathrm{d}z = (z_1 - z_0) \int_0^1 f(z_0 + t(z_1 - z_0))\,\mathrm{d}t.$$

Alternatively, we can use any parametrisation, for $a \leqslant b$,

$$[a, b] \ni t \mapsto \frac{(b-t)z_0 + (t-a)z_1}{b-a},$$

and denote the corresponding path by $[z_0, z_1]$ as well. The opposite path is $[z_1, z_0]$.

The next theorem plays a major role in function theory, introducing the *index* of a complex number z with respect to a closed path γ.

Theorem A.5.5. *Let γ be a closed path in the complex plane. Then,*

$$\mathrm{Ind}_\gamma(z) = \frac{1}{2\pi i} \int_\gamma \frac{\mathrm{d}\zeta}{\zeta - z}, \quad z \in \mathbb{C} \setminus \gamma^*,$$

is an integer valued function on $\mathbb{C} \setminus \gamma^*$ which is constant in each connected component of $\mathbb{C} \setminus \gamma^*$ and which is 0 in the unbounded connected component of $\mathbb{C} \setminus \gamma^*$.

The index of a complex number z with respect to the closed path γ counts the number of times that γ winds around z and, because of that, it is called the *winding number*. As an example, we have the winding number of a complex number with respect to the positively oriented circle.

Corollary A.5.6. *If γ is the positively oriented circle with centre z_0 and radius r then*

$$\mathrm{Ind}_\gamma(z) = \begin{cases} 1, & \text{if } |z - z_0| < r, \\ 0, & \text{if } |z - z_0| > r. \end{cases}$$

For the next theorem, the reader should recall the homotopy concepts defined in Section A.1.

Theorem A.5.7. *Let Ω be a region in $\mathbb{C}$ and γ_0 and γ_1 be two Ω-homotopic closed paths in Ω. Then, for each $\alpha \in \mathbb{C} \setminus \Omega$ we have*
$$\mathrm{Ind}_{\gamma_1}(\alpha) = \mathrm{Ind}_{\gamma_0}(\alpha).$$

In order to state a rather general form of Cauchy's Theorem we briefly recall the concepts of chains and cycles. Formally, a *chain* Γ is a union of a finite number of paths $\gamma_1, \ldots, \gamma_n$, denoted $\Gamma = \gamma_1 \cdots \dotplus \cdots \gamma_n$, and uniquely defined in the sense that, for any continuous function $f \colon \gamma_1^* \cup \cdots \cup \gamma_n^* \to \mathbb{C}$,
$$\int_\Gamma f(z) \mathrm{d}z = \sum_{i=1}^n \int_{\gamma_i} f(z) \mathrm{d}z.$$

Moreover, one defines
$$\Gamma^* = \bigcup_{j=1}^n \gamma_j^*.$$

If each path γ_j is closed, $j = 1, \ldots, n$, then we call Γ a *chain*. In this case, one can define the *index* of Γ by
$$\mathrm{Ind}_\Gamma(z) = \frac{1}{2\pi \mathrm{i}} \int_\Gamma \frac{\mathrm{d}\zeta}{\zeta - z}, \quad z \in \mathbb{C} \setminus \Gamma^*.$$

Clearly,
$$\mathrm{Ind}_\Gamma(z) = \sum_{j=1}^n \mathrm{Ind}_{\gamma_j}(z), \quad z \in \mathbb{C} \setminus \Gamma^*.$$

Theorem A.5.8 (Cauchy Theorem). *Let Ω be an arbitrary nonempty open subset of $\mathbb{C}$, $f \in \mathrm{Hol}(\Omega)$, and Γ a cycle in Ω such that*
$$\mathrm{Ind}_\Gamma(z) = 0 \text{ for every } z \in \mathbb{C} \setminus \Omega.$$
Then
$$f(z) \mathrm{Ind}_\Gamma(z) = \frac{1}{2\pi \mathrm{i}} \int_\Gamma \frac{f(w)}{w - z} \mathrm{d}w, \quad z \in \Omega \setminus \Gamma^*,$$
and
$$\int_\Gamma f(z) \mathrm{d}z = 0.$$
Moreover, if Γ_1 and Γ_2 are cycles in Ω such that
$$\mathrm{Ind}_{\Gamma_1}(z) = \mathrm{Ind}_{\Gamma_2}(z) \text{ for every } z \in \mathbb{C} \setminus \Omega,$$
then
$$\int_{\Gamma_1} f(z) \mathrm{d}z = \int_{\Gamma_2} f(z) \mathrm{d}z.$$

A.5 Functions of One Complex Variable

In particular, with notation as in Theorem A.5.8, for any $z_0 \in \Omega$ and $R > 0$ such that $\mathbb{D}_R(z_0) \subseteq \Omega$, if $0 < r < R$ and γ is the positively oriented circle of centre z_0 and radius r, we have

$$f(z) = \frac{1}{2\pi i} \int_\gamma \frac{f(\zeta)}{\zeta - z} d\zeta, \quad z \in \mathbb{D}_r(z_0).$$

This observation and Theorem A.5.4 imply the power series representability of holomorphic functions.

Theorem A.5.9. *If Ω is a nonempty open set in $\mathbb{C}$ and $f \in \text{Hol}(\Omega)$ then f is representable by a power series in Ω.*

On the other hand, Cauchy's Theorem has a useful converse.

Theorem A.5.10 (Morera Theorem). *Let Ω be a nonempty open subset in $\mathbb{C}$ and $f: \Omega \to \mathbb{C}$ a continuous function such that*

$$\int_\gamma f(z) dz = 0,$$

for every oriented circle γ in Ω. Then $f \in \text{Hol}(\Omega)$.

An important consequence of the Cauchy–Morera Theorem is the fact that the algebra $\text{Hol}(\Omega)$ is complete under the topology of uniform convergence on compact subsets.

Theorem A.5.11. *Let $(f_n)_n$ be a sequence of functions in $\text{Hol}(\Omega)$ for some nonempty open set $\Omega \subset \mathbb{C}$. If $f: \Omega \to \mathbb{C}$ is a function such that the sequence $(f_n)_n$ converges to f uniformly on any compact subset of Ω then $f \in \text{Hol}(\Omega)$ and, for any $k \in \mathbb{N}$ the sequence $(f_n^{(k)})_n$ of derivatives of order k converges to $f^{(k)}$ uniformly on any compact subset of Ω.*

There are a few more consequences of the fact that every holomorphic function is locally representable by power series, which we list below.

Corollary A.5.12. *Assume that $f, g \in \text{Hol}(\Omega)$ for some region Ω in $\mathbb{C}$ and $f(z) = g(z)$ for all z in some set of points in Ω that has a limit point in Ω. Then $f(\zeta) = g(\zeta)$ for all $\zeta \in \Omega$.*

Corollary A.5.13. *If Ω is a region, that is, a nonempty connected open subset of $\mathbb{C}$, and $f \in \text{Hol}(\Omega)$, let*

$$Z(f) = \{z \in \Omega \mid f(z) = 0\}$$

denote the set of zeros of f in Ω. Then, $Z(f)$ is at most countable and, either $Z(f) = \emptyset$ or $Z(f)$ has no limit point in Ω. In the latter case, to each $z \in Z(f)$ there corresponds a unique natural number m such that

$$f(\zeta) = (\zeta - z)^m g(\zeta), \quad \zeta \in \Omega,$$

where $g \in \text{Hol}(\Omega)$ and $z \notin Z(g)$.

The natural number m as in the previous corollary is called the *order of multiplicity* of the zero z of f.

In the following we record a few more consequences of the power series representations of holomorphic functions. Firstly, we record a Parseval type formula.

Theorem A.5.14. *If for some $z_0 \in \mathbb{C}$ and $R > 0$ we have*

$$f(z) = \sum_{n=0}^{\infty} c_n (z - z_0)^n, \quad |z - z_0| < R,$$

then, for any $0 < r < R$ we have

$$\sum_{n=0}^{\infty} |c_n|^2 r^{2n} = \frac{1}{2\pi} \int_0^{2\pi} |f(z_0 + re^{i\theta})|^2 d\theta.$$

A second consequence is Liouville's Theorem.

Theorem A.5.15 (Liouville Theorem). *Every bounded entire function is constant.*

Another important consequence is the Maximum/Minimum Modulus Principle.

Theorem A.5.16. *Let Ω be a region, $f \in \mathrm{Hol}(\Omega)$, $z_0 \in \Omega$ and $r > 0$ such that $\mathrm{Clos}\, \mathbb{D}_r(z_0) \subset \Omega$.*
(a) $|f(z_0)| \leq \max_{\theta \in [0, 2\pi)} |f(z_0 + re^{i\theta})|$, *and equality holds if and only if f is constant in Ω.*
(b) *If f has no zero in $\mathbb{D}_r(z_0)$ then $|f(z_0)| \geq \min_{\theta \in [0, 2\pi)} |f(z_0 + re^{i\theta})|$.*

As a consequence of the Maximum Modulus Principle one obtains the Fundamental Theorem of Algebra.

Theorem A.5.17. *Let $n \in \mathbb{N}$ and $a_0, \ldots, a_{n-1}, a_n \in \mathbb{C}$ be arbitrary, with $a_n \neq 0$. Then the polynomial*

$$P(z) = a_n z^n + a_{n-1} z^{n-1} + \cdots + a_1 z + a_0,$$

has precisely n zeros, counted with their multiplicities, in $\mathbb{C}$.

Meromorphic Functions

Let $f \in \mathrm{Hol}(\Omega \setminus \{z_0\})$ for some nonempty open set Ω in $\mathbb{C}$ and some point $z_0 \in \Omega$. In this case, z_0 is called an *isolated singularity*. If the function f has a holomorphic extension in Ω then z_0 is called a *removable singularity*.

Proposition A.5.18. *With notation as before, if $f \in \mathrm{Hol}(\Omega \setminus \{z_0\})$ and, for some $r > 0$, f is bounded in the punctured disc $\mathbb{D}_r(z) \setminus \{z_0\}$, then z_0 is a removable singularity of f.*

The classification of isolated singularities of complex functions is contained in the following.

Theorem A.5.19. *Let $z_0 \in \Omega$ and $f \in \mathrm{Hol}(\Omega \setminus \{z_0\})$. One of the following cases must hold.*

(a) f has a removable singularity at z_0.

(b) There are complex numbers $c_1, \ldots, c_m$, where m is a natural number, with $c_m \neq 0$, such that the function
$$f(z) - \sum_{k=1}^{m} \frac{c_k}{(z-z_0)^k},$$
has a removable singularity at z.

(c) If $r > 0$ is such that $\mathbb{D}_r(z_0) \subseteq \Omega$, then $f(\mathbb{D}_r(z) \setminus \{z_0\})$ is dense in $\mathbb{C}$, equivalently, for any $w \in \mathbb{C}$ there exists a sequence $(z_n)_n$ in Ω such that $f(z_n) \xrightarrow[n \to \infty]{} w$.

With notation as in the previous theorem, in case (b) we say that f has a *pole* of *order m* at z_0 and the function
$$\sum_{k=1}^{m} \frac{c_k}{(z-z_0)^k}$$
is called the *principal part* of f at z_0. In this case we have $|f(z)| \xrightarrow[z \to z_0]{} +\infty$.

In case (c), f is said to have an *essential singularity* at z_0.

A *Laurent series* is, by definition, a power series indexed on $\mathbb{Z}$, more precisely, a series of the form
$$\sum_{k \in \mathbb{Z}} c_k (z-z_0)^k.$$
The convergence of Laurent series is understood in the sense that each of the series $\sum_{k \geq 0} c_k (z-z_0)^k$ and $\sum_{k \geq 1} c_{-k} (z-z_0)^{-k}$ converges and, in this case, the sum of the Laurent series is
$$\sum_{k \in \mathbb{Z}} c_k (z-z_0)^k = \sum_{k \geq 0} c_k (z-z_0)^k + \sum_{k \geq 1} c_{-k} (z-z_0)^{-k}.$$

Theorem A.5.20. *Let $0 \leq R_1 < R_2$ and consider the annulus $A := \{z \in \mathbb{C} \mid R_1 < |z - z_0| < R_2\}$ of centre z_0 and of radii R_1 and R_2. If $f \in \mathrm{Hol}(A)$ then it is representable as a Laurent series in A*
$$f(z) = \sum_{k \in \mathbb{Z}} c_k (z-z_0)^k, \quad z \in A,$$
where
$$c_k = \frac{1}{2\pi \mathrm{i}} \int_{\partial \mathbb{D}_r(z_0)} \frac{f(w)}{(w-z_0)^{k+1}} dw, \quad k \in \mathbb{Z},$$
where $R_1 < r < R_2$ is arbitrary and the path $\partial \mathbb{D}_r(z_0)$ is positively oriented.

There are a few important corollaries of this theorem that we record below.

Corollary A.5.21. *Assume that $\sum_{k\in\mathbb{Z}} c_k(z-z_0)^k$ and $\sum_{k\in\mathbb{Z}} d_k(z-z_0)^k$ are two Laurent series that both converge in the annulus $\{z \in \mathbb{C} \mid R_1 < |z - z_0| < R_2\}$, for some $0 \leq R_1 < R_2$ and $z_0 \in \mathbb{C}$, to the same function f. Then $c_k = d_k$ for all $k \in \mathbb{Z}$.*

Corollary A.5.22. *Let Ω be a region, $z_0 \in \Omega$, and $f \in \mathrm{Hol}(\Omega \setminus \{z_0\})$. Then f can be expanded into a Laurent series that converges in the annulus $\{z \in \mathbb{C} \mid 0 < |z-z_0| < R\}$, where $R \geq \mathrm{dist}(\partial\Omega, z_0)$, the distance of z_0 to the boundary of Ω.*

Another consequence of Theorem A.5.20 refers to the possibility of providing a characterisation of the singularity type of a point with respect to a function in terms of Laurent series.

Proposition A.5.23. *Assume that z_0 is an isolated singularity of a function f with Laurent series*
$$f(z) = \sum_{k\in\mathbb{Z}} c_k(z - z_0)^k,$$
that converges in a punctured disc $\{z \in \mathbb{C} \mid 0 < |z - z_0| < r\}$ for some $r > 0$. Then:
(a) z_0 *is a removable singularity of f if and only if $c_k = 0$ for all $k \leq -1$.*
(b) z_0 *is a pole of f if and only if there exists $m \geq 1$ such that $c_k = 0$ for all $k \geq -m - 1$. In this case, the order of the pole z_0 is the largest m with this property.*
(c) z_0 *is an essential singularity of f if and only if $c_{-k} \neq 0$ for infinitely many $k \geq 1$.*

Given a nonempty open set $\Omega \subseteq \mathbb{C}$, a function f is said to be *meromorphic* in Ω if there is a set $A \subset \Omega$ subject to the following conditions.
(a) A has no limit point in Ω.
(b) $f \in \mathrm{Hol}(\Omega \setminus A)$.
(c) f has a pole at each point of A.

Note that, as a consequence of the property stated in (a), the set A is at most countable.

Let f be a function meromorphic in Ω and a a pole of f in Ω. Then there exists $r > 0$ such that f is represented by a Laurent series in the annulus $\{z \mid 0 < |z - a| < r\}$
$$f(z) = \sum_{k\in\mathbb{Z}} c_k(z - a)^k, \quad 0 < |z - a| < r.$$

The *residue* of f at a is, by definition,
$$\mathrm{Res}(f; a) := c_{-1}.$$

Theorem A.5.24 (Residue Theorem). *Suppose that f is a meromorphic function in a nonempty open set $\Omega \subseteq \mathbb{C}$, let A denote the set of all poles of f in Ω and Γ a cycle in $\Omega \setminus A$ such that*
$$\mathrm{Ind}_\Gamma(\alpha) = 0 \text{ for all } \alpha \in \Omega.$$
Then,
$$\frac{1}{2\pi\mathrm{i}} \int_\Gamma f(z)\mathrm{d}z = \sum_{a \in A} \mathrm{Res}(f; a)\mathrm{Ind}_\Gamma(a).$$

A.5 Functions of One Complex Variable

The *extended complex plane* $\hat{\mathbb{C}} := \mathbb{C} \cup \{\infty\}$ has algebraic operations inherited from $\mathbb{C}$ and, in addition: $z + \infty = \infty + z = \infty$, $z\infty = \infty z = \infty$ as long as $z \neq 0$, $\infty\infty = \infty$, $z/\infty = 0$, and $z/0 = \infty$ as long as $z \neq 0$. The topology of $\hat{\mathbb{C}}$ is defined as follows: For $r > 0$ let $\mathbb{D}_r^c := \{z \in \mathbb{C} \mid |z| > r\}$, the complement of the closed unit disc, and denote $\mathbb{D}_r(\infty) = \mathbb{D}_r^c \cup \{\infty\}$. A subset $U \subseteq \hat{\mathbb{C}}$ is *open* if either $U \subseteq \mathbb{C}$ is an open set or $U \setminus \{\infty\}$ is open and $\mathbb{D}_r(\infty) \subseteq U$ for some $r > 0$. Endowed with this topology, the extended complex plane $\hat{\mathbb{C}}$ is a (metrisable) compact space.

The extended complex plane $\hat{\mathbb{C}}$ is homeomorphic to the unit sphere $\mathbb{S}^2 := \{(x_1, x_2, x_3) \mid x_1^2 + x_2^2 + x_3^2 = 1\} \subset \mathbb{R}^3$ by the so-called *stereographic projection* $\varphi \colon \mathbb{S}^2 \to \mathbb{C}$ defined by

$$\varphi(x_1, x_2, x_3) = \begin{cases} \frac{x_1}{1-x_3} + i\frac{x_2}{1-x_3}, & x_3 \neq 1, \\ \infty, & x_3 = 1, \end{cases}$$

with the inverse map

$$\varphi^{-1}(a + ib) = \left(\frac{2a}{a^2 + b^2 + 1}, \frac{2b}{a^2 + b^2 + 1}, \frac{a^2 + b^2 - 1}{a^2 + b^2 + 1}\right), \quad a, b \in \mathbb{R},$$

and $\varphi^{-1}(\infty) = (0, 0, 1)$. Because of that, the extended complex plane is also called the *Riemann sphere*.

A *linear fractional transformation* is a function of the form

$$f(z) = \frac{az + b}{cz + d}, \tag{A.6}$$

where $a, b, c, d \in \mathbb{C}$. If $ad - bc \neq 0$ then f is called a *Möbius transformation*. The linear transformation f is meromorphic if $c \neq 0$, with $-d/c$ the only pole, of multiplicity 1, and if $c = 0$ then f is a polynomial and hence an entire function. Linear fractional transformations are compositions of *translations* $z \mapsto z + b$, *rotations* $z \mapsto az$, and *inversions* $z \mapsto 1/z$.

Linear fractional transformations are closely related to geometric complex analysis. Clearly, translations and rotations map straight lines to straight lines and circles to circles. Inversions map circles and straight lines to either circles or straight lines.

Theorem A.5.25. *Any linear fractional transformation maps circles and straight lines to either circles or straight lines.*

The linear fractional transformation f as in (A.6) is a map $f \colon \mathbb{C} \setminus \{-d/c\} \to \mathbb{C} \setminus \{a/c\}$, as long as $c \neq 0$, with the inverse $f^{-1} \colon \mathbb{C} \setminus \{a/c\} \to \mathbb{C} \setminus \{-d/c\}$ given by

$$f^{-1}(z) = \frac{dz - b}{-cz + a},$$

and, if $c = 0$, f is a bijection of $\mathbb{C}$. However, it is more convenient to view f as a map on the Riemann sphere.

Proposition A.5.26. *Let f be the linear fractional transformations as in* (A.6) *and assume that* $ad - bc \neq 0$.
(a) *If $c \neq 0$ then $f \colon \hat{\mathbb{C}} \to \hat{\mathbb{C}}$ defined by*

$$f(z) = \begin{cases} \frac{az+b}{cz+d}, & z \neq -\frac{d}{c}, \infty \\ \infty, & z = -\frac{d}{c}, \\ \frac{a}{c}, & z = \infty, \end{cases}$$

is a homeomorphism.
(b) *If $c = 0$, hence $d \neq 0$, then $f \colon \hat{\mathbb{C}} \to \hat{\mathbb{C}}$ defined by*

$$f(z) = \begin{cases} \frac{az+b}{d}, & z \neq \infty, \\ \infty, & z = \infty, \end{cases}$$

is a homeomorphism.

The collection of all Möbius transformations f, when viewed as maps $f \colon \hat{\mathbb{C}} \to \hat{\mathbb{C}}$, make a group with the product defined as the composition of functions. Also, this group is isomorphic with the group $\mathrm{PGL}(2, \mathbb{C})$ of all 2×2 matrices with complex entries and nonzero determinant, identified modulo a nonzero factor,

$$\frac{az+b}{cz+d} \mapsto \begin{bmatrix} a & b \\ c & d \end{bmatrix} \in \mathrm{PGL}(2, \mathbb{C}).$$

This isomorphism is a homeomorphism when these two groups are endowed with the uniform topology.

Modulo the identification of the group of all Möbius transformations with the group $\mathrm{PGL}(2, \mathbb{C})$ and the homeomorphic identification of $\hat{\mathbb{C}}$ with the Riemann sphere $\mathbb{S}^2$, the group $\mathrm{PGL}(2, \mathbb{C})$ acts continuously on the Riemann sphere $\mathbb{S}^2$.

In connection with linear fractional transformations we recall the concept of conformal mapping. Let Ω be a region and $z_0 \in \Omega$. Assume that z_0 has a punctured disc $\mathbb{D}'_r(z_0) = \mathbb{D}_r(z_0) \setminus \{z_0\} \subset \Omega$ such that $f(z) \neq f(z_0)$ for all $z \in \mathbb{D}'_r(z_0)$. By definition, the function f *preserves angles* at z_0 if

$$\lim_{r \to 0+} e^{-i\theta} \frac{f(z_0 + re^{i\theta}) - f(z_0)}{|f(z_0 + re^{i\theta}) - f(z_0)|},$$

exists and is independent of $\theta \in \mathbb{R}$. The meaning of this property is that, for any two rays L and L', starting at z_0, the angle which their images $f(L)$ and $f(L')$ make at $f(z_0)$ is the same as that made by L and L', in size as well as in orientation.

Theorem A.5.27. *Let f be a map on a region $\Omega \subseteq \mathbb{C}$.*
(a) *If $f'(z_0)$ exists at some $z_0 \in \Omega$ and $f'(z_0) \neq 0$ then f preserves angles at z_0.*
(b) *If the partial derivatives of f at z_0 exist, the gradient at z_0 does not vanish at z_0, and f preserves angles at z_0, then $f'(z_0)$ exists and $f'(z_0) \neq 0$.*

A.5 Functions of One Complex Variable

A map $f\colon \Omega \to \mathbb{C}$, for some region $\Omega \subseteq \mathbb{C}$, is called a *conformal mapping* if it is holomorphic and $f'(z) \neq 0$ for all $z \in \Omega$. As the previous theorem shows, conformal mappings preserve angles at each point.

Two regions Ω_1 and Ω_2 in $\mathbb{C}$ are called *conformally equivalent regions* if there exists a bijective holomorphic map $\varphi\colon \Omega_1 \to \Omega_2$.

In order to state the Riemann Mapping Theorem we need the concept of homotopy and simple connectedness, see Section A.1.

Theorem A.5.28 (Riemann Mapping Theorem). *Any simply connected region Ω in $\mathbb{C}$, but different from $\mathbb{C}$, is conformally equivalent to the open unit disc $\mathbb{D}$.*

Example A.5.29. (a) Consider the Möbius transform

$$\varphi(z) = \frac{1+z}{1-z}.$$

φ maps -1 to 0, 0 to 1, and 1 to ∞. In addition, φ maps the unit circle $\mathbb{T}$ to the imaginary axis $i\mathbb{R}$. Thus, φ is a conformal bijective map of the open unit disc $\mathbb{D}$ to the open right half plane $\{z \in \mathbb{C} \mid \operatorname{Re} z > 0\}$.

(b) Let $\alpha \in \mathbb{D} \setminus \{0\}$. Then the Möbius transform

$$b_\alpha(z) = \frac{|\alpha|}{\alpha} \frac{\alpha - z}{1 - \overline{\alpha} z}$$

maps conformally $\mathbb{D}$ into itself, such that $b_\alpha(\alpha) = 0$.

(c) Let $f\colon \mathbb{D}_R^c = \{z \in \mathbb{C} \mid |z| > R\} \to \mathbb{C}$ be a holomorphic function such that ∞ is a removable singularity, that is, 0 is a removable singularity for the function $\mathbb{D}'_{1/R}(0) = \mathbb{D}_{1/R}(0) \setminus \{0\} \ni z \mapsto f(\frac{1}{z})$, equivalently, the function f is bounded on $\mathbb{D}_r^c$ for some $r > R$. In this case, $f(\infty) = \lim_{z \to \infty} f(z) \in \mathbb{C}$ exists. Then, f has the Laurent expansion

$$f(z) = \sum_{n \geq 0} \frac{a_n}{z^n}, \quad |z| > R,$$

and hence, $f(\infty) = a_0$. Consequently for a fixed but arbitrary $r > R$ and for any $|z| > r$ we have

$$\int_{|\zeta|=r} \frac{f(\zeta)}{\zeta - z} d\zeta = \int_{|\zeta|=r} \sum_{n=0}^{\infty} \frac{a_n}{\zeta^n(\zeta - z)} d\zeta$$

$$= \sum_{n=0}^{\infty} a_n \int_{|\zeta|=r} \frac{1}{\zeta^n(\zeta - z)} d\zeta$$

$$= -2\pi i \sum_{n=1}^{\infty} \frac{a_n}{z^n} = 2\pi i \left(a_0 - \sum_{n=0}^{\infty} \frac{a_n}{z^n} \right)$$

$$= 2\pi i \big(f(\infty) - f(z) \big),$$

where the circle $\{z \mid |z| = r\}$ is positively oriented. In conclusion, we obtain the *Cauchy formula at infinity*

$$f(z) = f(\infty) - \frac{1}{2\pi i} \int_{|z|=r} \frac{f(\zeta)}{\zeta - z} d\zeta, \quad |z| > r. \blacksquare$$

The next theorem is essential in the approximation of holomorphic functions by rational functions and it is needed in functional calculus.

Theorem A.5.30 (Runge Theorem). *Let M be a compact subset of $\mathbb{C}$ and a set $E \subseteq \hat{\mathbb{C}} = \mathbb{C} \cup \{\infty\}$ that meets each connected component of $\hat{\mathbb{C}} \setminus M$. Then, for any function $f \in \text{Hol}(M)$ there exists a sequence $(f_n)_n$ of rational functions with poles in E only and such that $f_n \xrightarrow[n \to \infty]{} f$ uniformly on M.*

Hardy Spaces on the Disc

The concepts and the results reviewed in this subsection can be found in classical texts as P. Koosis [90], M. Rosenblum and J. Rovnyak [126], W. Rudin [127], and J. B Conway [37].

Let $\mathbb{D} = \{z \in \mathbb{C} \mid |z| < 1\}$ be the unit disc in the complex plane and let $\mathbb{T} = \partial \mathbb{D}$ denote its boundary, the unit circle. For $1 \leq p < \infty$ let $H_\mathbb{D}^p$ denote the space of all analytic functions $f \colon \mathbb{D} \to \mathbb{C}$ subject to the condition

$$\|f\|_p^p := \sup_{0 \leq r < 1} \left(\frac{1}{2\pi} \int_0^{2\pi} |f(re^{it})|^p dt \right)^{1/p} < \infty.$$

For $p = \infty$ we let $H_\mathbb{D}^\infty$ denote the space of all analytic functions $f \colon \mathbb{D} \to \mathbb{C}$ that are bounded, with norm

$$\|f\|_\infty := \sup_{z \in \mathbb{D}} |f(z)| < \infty.$$

The spaces $H_\mathbb{D}^p$, for $1 \leq p \leq \infty$ are Banach spaces and are called *Hardy spaces* on the unit disc $\mathbb{D}$.

For any $1 \leq p < q < \infty$ we have $H_\mathbb{D}^\infty \subset H_\mathbb{D}^q \subset H_\mathbb{D}^p \subset H_\mathbb{D}^1$, and the embeddings are continuous. The space $H_\mathbb{D}^2$ is special since it is a Hilbert space and, if $f \colon \mathbb{D} \to \mathbb{C}$ is analytic in $\mathbb{D}$ and has the Taylor representation

$$f(z) = \sum_{n=0}^{\infty} a_n z^n, \quad z \in \mathbb{D}, \tag{A.7}$$

then $f \in H_\mathbb{D}^2$ if and only if $\sum_{n=0}^{\infty} |a_n|^2 < \infty$ and then

$$\|f\|_2^2 = \sum_{n=0}^{\infty} |a_n|^2.$$

A.5 Functions of One Complex Variable

Proposition A.5.31. *Let $1 \leqslant p \leqslant \infty$. For any $f \in H^p_{\mathbb{D}}$, let the function f_r be defined by $f_r(z) := f(rz)$, $0 \leqslant r < 1$, $|z| < 1/r$. Then $f_r \in L^p_{\mathbb{T}}$ and its boundary value function*

$$\tilde{f} := \lim_{r \to 1^-} f_r$$

exists with respect to the norm of $L^p_{\mathbb{T}}$ for $1 \leqslant p < \infty$, and with respect to the w^*-topology for $p = \infty$. In addition, $\tilde{f} \in L^p_{\mathbb{T}}$ and $\|f\|_p = \|\tilde{f}\|_p$, where the first norm is calculated in $H^p_{\mathbb{D}}$ while the latter is calculated in $L^p_{\mathbb{T}}$.

The previous proposition says that we have an isometric embedding $H^p_{\mathbb{D}} \hookrightarrow L^p_{\mathbb{T}}$. Let $H^p_{\mathbb{T}}$ denote the range of this embedding.

For $1 \leqslant p < \infty$ this embedding can be made more explicit. Letting $f \in H^p_{\mathbb{D}}$ be represented by the Taylor series as in (A.7) and recalling that $L^p_{\mathbb{T}}$ has the canonical Schauder basis $(e^{int})_{n \in \mathbb{Z}}$, the boundary value function $\tilde{f}$ has the representation

$$\tilde{f}(e^{it}) = \sum_{n=0}^{\infty} a_n e^{int},$$

where the convergence holds with respect to the norm $\|\cdot\|_p$ of $L^p_{\mathbb{T}}$. Conversely, letting $f \in L^p_{\mathbb{T}}$ be a function with representation

$$f(e^{it}) = \sum_{n \in \mathbb{Z}} \hat{f}_n e^{int},$$

where the series converges with respect to the norm $\|\cdot\|_p$, the function $f \in H^p_{\mathbb{T}}$ if and only if $\hat{f}_n = 0$ for all $n < 0$.

In general, for $1 \leqslant p \leqslant \infty$, we consider the Poisson kernel

$$P_r(\varphi) = \frac{1 - r^2}{1 + r^2 - 2r \cos \varphi}, \quad 0 \leqslant r < 1, \ 0 \leqslant \varphi \leqslant 2\pi,$$

that is positive and periodic of period 2π, and the following expansion holds

$$P_r(\varphi) = \sum_{n \in \mathbb{Z}} r^{|n|} e^{in\varphi}, \tag{A.8}$$

where the series converges uniformly on compact subsets of $\mathbb{D}$. Then, if $f \in H^p_{\mathbb{T}}$, we obtain the extension in $\mathbb{D}$ by letting

$$f(re^{it}) = \frac{1}{2\pi} \int_0^{2\pi} f(e^{i\theta}) P_r(t - \theta) d\theta, \quad 0 < r < 1, \ t \in [0, 2\pi),$$

and then $f \in H^p_{\mathbb{D}}$.

Because of this identification, we denote simply by H^p the Hardy space on the unit disc and view it as a subspace of $L^p_{\mathbb{T}}$, as explained before.

The next proposition says that the topologies on each of the Hardy spaces H^p are rather strong. We restrict to the case $1 \leqslant p < \infty$ because, by definition, the topology on H^∞ is the uniform topology.

Proposition A.5.32. *For every $1 \leqslant p < \infty$, the evaluation functional $H^p \ni f \mapsto \mathrm{ev}_z(f) = f(z)$ is continuous for any $z \in \mathbb{D}$, more precisely, for $p = 1$ we have*

$$\|\mathrm{ev}_z\| \leqslant \frac{1+|z|}{1-|z|}, \quad z \in \mathbb{D},$$

and, for $1 < p < \infty$, $1/p + 1/q = 1$, we have

$$\|\mathrm{ev}_z\| \leqslant \frac{1}{1-|z|^q}, \quad z \in \mathbb{D}.$$

In particular, convergence in H^p implies uniform convergence on compact subsets of $\mathbb{D}$.

Let $\alpha \in \mathbb{D}$. We consider the Möbius transformation

$$b_\alpha(z) = \frac{|\alpha|}{\alpha} \frac{\alpha - z}{1 - \overline{\alpha}z}, \quad z \in \mathbb{D}, \ \alpha \in \mathbb{D} \setminus \{0\},$$

and $b_0 = z$, which maps conformally the unit disc into itself.

Theorem A.5.33 (Blaschke Theorem). *Let $\alpha = (\alpha_n)_{n=1}^N$, with N either a natural number or the symbol ∞, be a sequence with all elements in $\mathbb{D}$, such that $\sum_{n=1}^{\infty}(1 - |\alpha_n|) < \infty$, if $N = \infty$. Then the product*

$$B_\alpha = \prod_{n=1}^{N} b_{\alpha_n}$$

converges uniformly on any compact subset of $\mathbb{C} \setminus \mathrm{Clos}\{1/\overline{\alpha}_n \mid n \in \mathbb{N}, \ \alpha_n \neq 0\}$ to a function B_α with the following properties.

(i) *$|B_\alpha(z)| < 1$ for all $z \in \mathbb{D}$, in particular $B_\alpha \in H^\infty$ and $\|B_\alpha\|_\infty \leqslant 1$.*
(ii) *$|B_\alpha(z)| = 1$ for almost all $z \in \mathbb{T}$.*
(iii) *The sequence of the zeros of B_α coincides with the sequence α, counted with multiplicities.*

A function B_α as in the theorem is callled a *Blaschke product*. As a consequence of this theorem one can factor out the zeros of functions in H^p in a convenient fashion.

Corollary A.5.34. *Let $p \geqslant 1$ and $f \in H^p \setminus \{0\}$ with $\alpha = (\alpha_n)_n$ the sequence of its zeros in $\mathbb{D}$, multiplicities counted. Then $\sum_n (1 - |\alpha_n|) < \infty$ and $f = B_\alpha g$ with $g \in H^p$, $\|f\|_p = \|g\|_p$, and $g(z) \neq 0$ for all $z \in \mathbb{D}$.*

A function $\varphi \in H^\infty$ is *inner* if $|\tilde{\varphi}(e^{it})| = 1$ for almost all $t \in \mathbb{T}$.

Proposition A.5.35. *Let $\varphi \in H^\infty$. Then φ is an inner function if and only if, letting α be the sequence of all zeros of φ in $\mathbb{D}$, counted with multiplicities,*

$$\varphi(z) = e^{ic} B_\alpha(z) \exp\left(-\frac{1}{2\pi} \int_0^{2\pi} \frac{e^{it} + z}{e^{it} - z} d\sigma(t)\right), \quad z \in \mathbb{D},$$

for some measure σ on $[0, 2\pi]$ that is singular with respect to the Lebesgue measure on $[0, 2\pi]$, and some real number c.

Theorem A.5.36 (Inner-Outer Factorisation)**.** *Let $f \in H^p$ for some $p \geqslant 1$ and let B denote the Blaschke product corresponding to all zeros of f in $\mathbb{D}$. Then, for all $z \in \mathbb{D}$, we have*

$$f(z) = e^{ic} B_\alpha(z) \exp\left(-\frac{1}{2\pi} \int_0^{2\pi} \frac{e^{it} + z}{e^{it} - z} d\sigma(t)\right) \exp\left(\frac{1}{2\pi} \int_0^{2\pi} \frac{e^{it} + z}{e^{it} - z} \log |\tilde{f}(e^{it})| dt\right),$$

for some singular measure σ on $[0, 2\pi]$ and some real number c.

A function ψ holomorphic on $\mathbb{D}$ and having the integral representation

$$\psi(z) = \rho \exp\left(\frac{1}{2\pi} \int_0^{2\pi} \frac{e^{it} + z}{e^{it} - z} g(t) dt\right), \quad z \in \mathbb{D},$$

for some real valued function $g \in L^1_{[0,2\pi]}$ and $\rho \in \mathbb{T}$, is called *outer*. Note that an outer function does not have any zero in $\mathbb{D}$. With this definition, Theorem A.5.36 says that any function $f \in H^p$, for $p \geqslant 1$, is factored as

$$f(z) = I_f(z) O_f(z), \quad z \in \mathbb{D},$$

where

$$I_f(z) = e^{ic} B(z) \exp\left(-\frac{1}{2\pi} \int_0^{2\pi} \frac{e^{it} + z}{e^{it} - z} d\sigma(t)\right), \quad z \in \mathbb{D},$$

is the *inner factor* of f, where $c \in \mathbb{R}$, B is the Blaschke factor corresponding to all zeros of f in $\mathbb{D}$, counting multiplicities, and σ is a singular measure on $[0, 2\pi]$, while

$$O_f(z) = \exp\left(\frac{1}{2\pi} \int_0^{2\pi} \frac{e^{it} + z}{e^{it} - z} \log |\tilde{f}(e^{it})| dt\right), \quad z \in \mathbb{D},$$

is the *outer factor* of f.

For $1 \leqslant p < \infty$ we observe that the *shift operator* can be realised as the operator of multiplication with the independent variable $M_z \colon H^p \to H^p$ and it is bounded.

Theorem A.5.37 (Beurling Theorem)**.** *Let $\mathcal{L}$ be a subspace of H^p, $1 \leqslant p < \infty$, invariant under the shift operator M_z, that is, $M_z \mathcal{L} \subseteq \mathcal{L}$. Then there exists an inner function φ such that $\mathcal{L} = \varphi H^p$, unique under the condition that $\varphi(0) \geqslant 0$.*

Vector Valued Holomorphic Functions

In this subsection we review some topics on holomorphic functions that take values in Banach spaces. The classical texts that we refer to are N. Dunford and J. T. Schwartz [54] and T. Kato [84].

Definition A.5.38. Let $\mathcal{X}$ be a normed space, D an open and nonempty subset of $\mathbb{C}$, and $f \colon D \to \mathcal{X}$ a function.

- The function f is called *strongly holomorphic* on D if, for any $\zeta \in D$, there exists $f'(\zeta) \in \mathcal{X}$ such that
$$\frac{f(\zeta+h)-f(\zeta)}{h} \xrightarrow[h\to 0]{\text{strongly}} f'(\zeta),$$
that is,
$$\lim_{h\to 0}\left\|\frac{f(\zeta+h)-f(\zeta)}{h} - f'(\zeta)\right\| = 0.$$

- The function f is called *weakly holomorphic* on D if, for any $\zeta \in D$, there exists $f'(\zeta) \in \mathcal{X}$ such that
$$\frac{f(\zeta+h)-f(\zeta)}{h} \xrightarrow[h\to 0]{\text{weakly}} f'(\zeta),$$
that is, for any $\varphi \in \mathcal{X}^*$ the function $\varphi \circ f \colon D \to \mathbb{C}$ is holomorphic in the classical sense. ∎

Remark A.5.39. With notation as in the previous definition, it is clear that strong holomorphy implies weak holomorphy. ∎

Theorem A.5.40 (Dunford Theorem). *Let $f \colon D \to \mathcal{X}$ be a function, where D is an open nonempty subset of $\mathbb{C}$ and $\mathcal{X}$ a Banach space. Then f is strongly holomorphic on D if and only if it is weakly holomorphic on D.*

In the case of the Banach algebra $\mathcal{B}(\mathcal{X})$, where $\mathcal{X}$ is a Banach space, since more topologies are available we can have even more concepts of holomorphy.

Definition A.5.41. Let D be a nonempty open set in $\mathbb{C}$, $\mathcal{X}$ a Banach space, and $f \colon D \to \mathcal{B}(\mathcal{X})$ a function.

- f is *uniformly holomorphic* on D if, for any $\zeta \in D$, there exists $f'(\zeta) \in \mathcal{B}(\mathcal{X})$ such that
$$\frac{f(\zeta+h)-f(\zeta)}{h} \xrightarrow[h\to 0]{\text{uniformly}} f'(\zeta),$$
that is,
$$\lim_{h\to 0}\left\|\frac{f(\zeta+h)-f(\zeta)}{h} - f'(\zeta)\right\| = 0,$$
where $\|\cdot\|$ denotes the operator norm on $\mathcal{B}(\mathcal{X})$.

- f is *strongly operator holomorphic* on D if, for any $\zeta \in D$, there exists $f'(\zeta) \in \mathcal{B}(\mathcal{X})$ such that
$$\frac{f(\zeta+h)-f(\zeta)}{h} \xrightarrow[h\to 0]{\text{strongly operator}} f'(\zeta),$$
that is, for every $x \in \mathcal{X}$ we have
$$\lim_{h\to 0}\left\|\frac{f(\zeta+h)x-f(\zeta)x}{h} - f'(\zeta)x\right\| = 0,$$
where $\|\cdot\|$ denotes the norm on $\mathcal{X}$.

A.5 Functions of One Complex Variable

- f is *weakly operator holomorphic* on D if, for any $\zeta \in D$, there exists $f'(\zeta) \in \mathcal{B}(\mathcal{X})$ such that
$$\frac{f(\zeta + h) - f(\zeta)}{h} \xrightarrow[h \to 0]{\text{weakly operator}} f'(\zeta),$$
that is, for every $x \in \mathcal{X}$ and every $\varphi \in \mathcal{X}^*$ we have
$$\lim_{h \to 0} \left| \frac{\varphi(f(\zeta + h)x - f(\zeta)x)}{h} - \varphi(f'(\zeta)x) \right| = 0. \blacksquare$$

Clearly, uniformly holomorphic implies strongly operator holomorphic and this implies weakly operator holomorphic. Following a reasoning similar to the proof of Theorem A.5.40 one can prove the following theorem.

Theorem A.5.42 (Dunford Theorem). *With notation as in the previous definition the following assertions are equivalent.*

(i) *f is uniformly holomorophic.*
(ii) *f is strongly operator holomorphic.*
(iii) *f is weakly operator holomorphic.*

Because of these theorems, whenever f is a function defined on some nonempty and open subset D of $\mathbb{C}$ and valued in either some Banach algebra $\mathcal{A}$ or in $\mathcal{B}(\mathcal{X})$, for some Banach space $\mathcal{X}$, we can simply talk about f holomorphic in D, without saying anything else about which topology on $\mathcal{A}$ or on $\mathcal{B}(\mathcal{X})$ we are referring to.

In the following we list some results from the theory of scalar complex functions that can be generalised, simply by means of weak holomorphy, for functions taking values in Banach spaces. Let $f \colon D \to \mathcal{X}$ be a function defined on a nonempty open set $D \subseteq \mathbb{C}$ and with values in $\mathcal{X}$, for some Banach space $\mathcal{X}$.

Theorem A.5.43 (Cauchy Integral Theorem). *If f is holomorphic on D then*
$$f(\zeta) = \frac{1}{2\pi \mathrm{i}} \int_\Gamma \frac{f(z)}{z - \zeta} \mathrm{d}z,$$
where Γ is any Jordan contour winding once around ζ inside of D and positively oriented and the integral converges strongly in $\mathcal{X}$.

Theorem A.5.44 (Cauchy Integral Formula). *Assume that f is holomorphic on D.*

(1) *Letting $F(\zeta) := \int_{[c,\zeta]} f(z) \mathrm{d}z$, where $[c, \zeta]$ is any simple path connecting a fixed point $c \in D$ with ζ inside D, $F \colon D \to \mathcal{X}$ is a holomorphic function and $F'(\zeta) = f(\zeta)$ for all $\zeta \in D$.*

(2) *For any Jordan contour Γ in D we have $\int_\Gamma f(z) \mathrm{d}z = 0$.*

Theorem A.5.45 (Taylor Series Expansions). *Assume that f is holomorphic on D. Let $z_0 \in D$ and $r > 0$ such that the disc of centre z_0 and radius r is contained in D. Then*
$$f(z) = \sum_{n=0}^{\infty} \frac{f^{(n)}(z_0)}{n!} (z - z_0)^n, \quad |z - z_0| < r,$$

where the series converges absolutely and uniformly on any compact subset of D.

Theorem A.5.46 (Power Series, Cauchy–Hadamard Formula). *Let $(a_n)_{n\geqslant 0}$ be a sequence of vectors in the Banach space $\mathcal{X}$, $z_0 \in \mathbb{C}$, and consider the power series*

$$f(z) := \sum_{n=0}^{\infty} z^n a_n.$$

Let

$$R := \begin{cases} 0, & \text{if } \limsup_{n\to\infty} \|a_n\|^{1/n} = \infty, \\ +\infty, & \text{if } \limsup_{n\to\infty} \|a_n\|^{1/n} = 0, \\ \dfrac{1}{\limsup_{n\to\infty} \|a_n\|^{1/n}}, & \text{if } 0 < \limsup_{n\to\infty} \|a_n\|^{1/n} < \infty. \end{cases}$$

- *If $R = 0$ then the power series converges only for $z = z_0$.*
- *If $R > 0$ then the disc $\mathbb{D}_R(z_0)$ of centre z_0 and radius R (if $R = \infty$ we take $D_\infty(z_0) = \mathbb{C}$) is the largest open disc centred at z_0 on which the power series converges. In addition, the power series converges absolutely and uniformly on any compact subset of this disc, and the function f is holomorphic on $\mathbb{D}_R(z_0)$.*

Theorem A.5.47 (Laurent Series). *Let $0 < r < s$, $\zeta_0 \in \mathbb{C}$, and consider a sequence $(a_n)_{n\in\mathbb{Z}}$ of elements in the Banach space $\mathcal{X}$ such that the power series*

$$f(\zeta) := \sum_{n=-\infty}^{+\infty} (\zeta - \zeta_0)^n a_n$$

converges in the annulus $A := \{\zeta \in \mathbb{C} \mid r < |\zeta - \zeta_0| < s\}$. Then f is holomorphic in A and

$$a_n = \frac{1}{2\pi\mathrm{i}} \int_{S_\rho} \frac{f(\zeta)}{(\zeta - \zeta_0)^{n+1}} \mathrm{d}\zeta, \quad n \in \mathbb{Z},$$

where $S_\rho := \{\zeta \in \mathbb{C} \mid |\zeta - \zeta_0| = \rho\}$ is positively oriented and ρ is any number between r and s.

Theorem A.5.48 (Liouville Theorem). *If $f \colon \mathbb{C} \to \mathcal{X}$ is an entire (that is, holomorphic on $\mathbb{C}$) function and bounded, then f is constant on $\mathbb{C}$.*

A.6 Banach Algebras

In this section we review a few concepts and facts related to Banach algebras, Abelian Banach algebras and the Gelfand transform. More advanced topics and proofs can be found in classical monographs such as F. F. Bonsall and J. Duncan [22], C. E. Rickart [125], W. Rudin [128], and other more modern textbooks.

A.6 Banach Algebras

Banach Algebras

A *Banach algebra* is a complex Banach space $\mathcal{A}$ on which there exists a multiplication $\mathcal{A} \ni (x, y) \mapsto xy \in \mathcal{A}$, with the following properties.

(i) $x(y + z) = xy + xz$ and $(x + y)z = xz + yz$ for all $x, y, z \in \mathcal{A}$.
(ii) $\alpha(xy) = (\alpha x)y = x(\alpha y)$ for all $x, y \in \mathcal{A}$ and all $\alpha \in \mathbb{C}$.
(iii) $\|xy\| \leq \|x\|\|y\|$ for all $x, y \in \mathcal{A}$.

Axiom (iii) implies that the multiplication is jointly continuous, that is, continuous $\mathcal{A} \times \mathcal{A} \ni (x, y) \mapsto xy \in \mathcal{A}$, when we consider the product topology on $\mathcal{A} \times \mathcal{A}$.
If $\mathcal{A}$ has a unit $e \neq 0$, that is, $ex = xe = x$ for all $x \in \mathcal{A}$, then $\|e\| = \|e^2\| \leq \|e\|\|e\|$, hence $\|e\| \geq 1$.

Proposition A.6.1. *If $\mathcal{A}$ is a Banach algebra with unit e then, modulo an equivalent norm, we have $\|e\| = 1$, more precisely, we can take, for each $a \in \mathcal{A}$, its norm $\|a\|$ to be the operator norm $\|L_a\|$, where $\mathcal{A} \ni x \mapsto L_a(x) := ax \in \mathcal{A}$ is a bounded linear operator, when viewing $\mathcal{A}$ as a Banach space.*

From now on, whenever $\mathcal{A}$ is a Banach algebra with unit e we assume $\|e\| = 1$.

Proposition A.6.2. *Any Banach algebra $\mathcal{A}$ that does not have a unit can be embedded in a Banach algebra with unit $\mathcal{A}_1$ as an ideal of codimension 1, more precisely, $\mathcal{A}_1 = \mathcal{A} \oplus \mathbb{C}$ with unit $0 \oplus 1$, with identification of $\mathcal{A}$ with $\mathcal{A} \oplus 0$.*

Example A.6.3. (1) Let $\mathcal{X}$ be a Banach space. Then $\mathcal{B}(\mathcal{X})$, the collection of all linear bounded operators in $\mathcal{X}$, with operator norm, is a unital Banach algebra.

(2) Let X be a compact Hausdorff space. Then $\mathcal{C}(X)$, the collection of all continuous functions $f \colon X \to \mathbb{C}$, is a commutative Banach algebra with unit, with respect to the norm $\|f\|_\infty = \sup_X |f|$.

(3) The Banach space $L_\mathbb{T}^\infty$ is a commutative unital Banach algebra with respect to the norm $\|f\|_\infty = \operatorname{ess\,sup}_\mathbb{T} |f|$.

(4) The Hardy space $H_\mathbb{D}^\infty$, canonically embedded in $L_\mathbb{T}^\infty$ by the boundary value function, is a commutative unital Banach algebra. ∎

Spectral Theory

From now on, $\mathcal{A}$ denotes a Banach algebra with unit. An element $x \in \mathcal{A}$ is called *invertible* if there exists $x' \in \mathcal{A}$ such that $xx' = x'x = e$. Note that, if it exists, then x' is unique with this property: if $x'' \in \mathcal{A}$ is such that $x''x = x''x = e$ then $x' = x'e = x'(xx'') = (x'x)x'' = ex'' = x''$. The inverse of an invertible element x is denoted by x^{-1}. Denote

$$G(\mathcal{A}) = \{x \in \mathcal{A} \mid x \text{ is invertible in } \mathcal{A}\}.$$

Proposition A.6.4. (a) *The set $G(\mathcal{A})$ is open, more precisely, for any $x_0 \in G(\mathcal{A})$ and any $x \in \mathcal{A}$ such that $\|x - x_0\| < 1/\|x_0^{-1}\|$ it follows that $x \in G(\mathcal{A})$ and*

$$x^{-1} = \sum_{k=0}^{\infty} \left(x_0^{-1}(x_0 - x)\right)^k x_0^{-1},$$

where the series converges absolutely.
  (b) *The mapping $G(\mathcal{A}) \ni x \mapsto x^{-1} \in G(\mathcal{A})$ is continuous.*

For each $x \in \mathcal{A}$ we consider the *spectrum* of x with respect to $\mathcal{A}$

$$\sigma_{\mathcal{A}}(x) = \{\lambda \in \mathbb{C} \mid \lambda e - x \notin G(\mathcal{A})\},$$

as well as its *resolvent set* with respect to $\mathcal{A}$

$$\rho_{\mathcal{A}}(x) = \mathbb{C} \setminus \sigma_{\mathcal{A}}(x) = \{\lambda \in \mathbb{C} \mid \lambda e - x \in G(\mathcal{A})\}.$$

Proposition A.6.5. *For any $x \in \mathcal{A}$, its spectrum $\sigma_{\mathcal{A}}(x)$ is a nonempty compact subset of $\mathbb{C}$ and $\sigma_{\mathcal{A}}(x) \subseteq \{\lambda \in \mathbb{C} \mid |\lambda| \leq \|x\|\}$.*

The previous proposition shows that, for any $x \in \mathcal{A}$,

$$r_{\mathcal{A}}(x) = \sup\{|\lambda| \mid \lambda \in \sigma_{\mathcal{A}}\}$$

is a nonnegative number such that $r_{\mathcal{A}}(x) \leq \|x\|$, called the *spectral radius* of x with respect to $\mathcal{A}$.

Proposition A.6.6. *For any $x \in \mathcal{A}$ the numerical sequence $(\|x^n\|^{1/n})_n$ converges and*

$$r_{\mathcal{A}}(x) = \lim_{n \to \infty} \|x^n\|^{1/n} = \inf_{n \in \mathbb{N}} \|x^n\|^{1/n}.$$

The next proposition shows that, unless $\mathcal{A}$ is scalar, the group $G(\mathcal{A})$ is not too large.

Proposition A.6.7 (Gelfand–Mazur Theorem). *If $G(\mathcal{A}) = \mathcal{A} \setminus \{0\}$ then $\mathcal{A} = \mathbb{C}$.*

Functional Calculus with Holomorphic Functions

As before, $\mathcal{A}$ denotes a Banach algebra with unit e. For any element $a \in \mathcal{A}$ and $p \in \mathbb{C}[X]$, a polynomial with complex coefficients, letting

$$p = \sum_{k=0}^{n} \alpha_k X^k,$$

we can define an element $p(a) \in \mathcal{A}$ in a natural way by

$$p(a) = \sum_{k=0}^{n} \alpha_k a^k.$$

A.6 BANACH ALGEBRAS

It is easy to see that, in this way, we have an algebra homomorphism $\mathbb{C}[X] \ni p \mapsto p(a) \in \mathcal{A}$, such that $e = 1(a)$ and $a = X(a)$, that we call functional calculus with polynomial functions. In the following we briefly review how, by using more sophisticated complex function theory, one can extend this functional calculus, with both algebraic and topological properties, to more general function algebras.

Definition A.6.8. Given an element $a \in \mathcal{A}$ we denote by $\mathrm{Hol}(\sigma(a))$ the collection of holomorphic functions $f \colon \mathrm{Dom}(f) \to \mathbb{C}$, where $\mathrm{Dom}(f)$ is an open subset of $\mathbb{C}$ that contains $\sigma(a)$.

For any $f \in \mathrm{Hol}(\sigma(a))$ let Γ be a finite union of Jordan contours in $\rho(a)$, positively oriented and winding once around any point of $\sigma(a)$. Then $f(a) \in \mathcal{A}$ is the element defined by

$$f(a) := \frac{1}{2\pi\mathrm{i}} \int_\Gamma f(\zeta)(\zeta e - a)^{-1} \mathrm{d}\zeta. \quad \blacksquare \qquad (\mathrm{A}.1)$$

Remark A.6.9. It is easy to see that the definition does not depend on Γ. In particular, Γ can be considered as made up by only one Jordan contour, if the domain of f allows such a choice. ∎

In the following result, Runge's Theorem, see Theorem A.5.30, is needed.

Theorem A.6.10. *Let $a \in \mathcal{A}$.*

(a) The map $\mathrm{Hol}(\sigma(a)) \ni f \mapsto F(f) = f(a) \in \mathcal{A}$ is an algebra homomorphism such that $1(a) = e$ and $\zeta(a) = a$, where 1 is the function constantly equal to 1 and ζ is the function $\mathbb{C} \ni \zeta \mapsto \zeta \in \mathbb{C}$.

(b) Let $(f_n)_n$ and f be functions in $\mathrm{Hol}(\sigma(a))$ such that there exists an open subset G of $\mathbb{C}$ subject to the property

$$\sigma(a) \subset G \subset \bigcap_{n \geqslant 1} \mathrm{Dom}(f_n) \cap \mathrm{Dom}(f), \qquad (\mathrm{A}.2)$$

and $f_n \xrightarrow[n \to \infty]{} f$ uniformly on any compact subset of G. Then $\|f_n(a) - f(a)\| \xrightarrow[n \to \infty]{} 0$.

(c) The homomorphism F is unique with the properties (a) and (b).

Definition A.6.11. For any $a \in \mathcal{A}$ the homomorphism

$$\mathrm{Hol}(\sigma(a)) \ni f \mapsto f(a) \in \mathcal{A}$$

defined in (A.1) is called the *holomorphic functional calculus*, or the *Riesz–Dunford functional calculus*. ∎

Theorem A.6.12 (Spectral Mapping Theorem for Holomorphic Functional Calculus)**.** *For any $a \in \mathcal{A}$ and any $f \in \mathrm{Hol}(\sigma(a))$ we have*

$$\sigma(f(a)) = f(\sigma(a)).$$

Example A.6.13. An element $a \in \mathcal{A}$ is called *algebraic* if there exists a polynomial with complex coefficients $p \in \mathbb{C}[X]$ such that $p(a) = 0$. Here the meaning of the element $p(a)$ is either by holomorphic functional calculus, corresponding to the associated complex function p, or simply, letting $p = \sum_{k=0}^{n} \alpha_k X^k$, in the sense $p(a) = \sum_{k=0}^{n} \alpha_k a^k$. Note that, by Theorem A.6.10, these interpretations coincide. Then,

$$\sigma(a) \subseteq \{\lambda \in \mathbb{C} \mid p(\lambda) = 0\}.$$

Indeed, by the Spectral Mapping Theorem, $\{0\} = \sigma(p(a)) = p(\sigma(a))$, hence for any $\lambda \in \sigma(a)$ we must have $p(\lambda) = 0$.

In particular, an element $a \in \mathcal{A}$ is called *idempotent* if $a^2 = a$ and then, $\sigma(a) \subseteq \{0, 1\}$.

Remark A.6.14. Assume that $a \in \mathcal{A}$ and that $\sigma(a) = K_1 \cup K_2$, where K_1 and K_2 are two compact subsets with $K_1 \neq \emptyset$, $K_2 \neq \emptyset$, and $K_1 \cap K_2 = \emptyset$. Then $\mathrm{dist}(K_1, K_2) = \varepsilon > 0$ and let the function f be defined by

$$f(\zeta) := \begin{cases} 1, & \zeta \in K_{1,\varepsilon/3} := \{z \in \mathbb{C} \mid \mathrm{dist}(\zeta, K_1) < \varepsilon/3\}, \\ 0, & \zeta \in K_{2,\varepsilon/3} := \{z \in \mathbb{C} \mid \mathrm{dist}(\zeta, K_1) < \varepsilon/3\}. \end{cases}$$

Then $f \in \mathrm{Hol}(\sigma(a))$ and $p := f(a)$ has the property $p^2 = p$, $p \neq e$, $p \neq 0$, that is, p is a nontrivial idempotent. ∎

Example A.6.15. (1) Let $a < b$ be two real numbers and consider the Banach algebra $\mathcal{C}[a, b]$ of all complex valued continuous functions on $[a, b]$. This Banach algebra does not have nontrivial idempotents.

(2) The Banach algebras ℓ^∞, c, and c_0 have many idempotents. Actually, in c and c_0 the set of all their idempotents is strongly total. In ℓ^∞ the set of all its idempotents is weakly* total, in the following sense. We know that the topological dual of ℓ^1 is naturally identified with ℓ^∞ and consider the duality map: $\ell^1 \times \ell^\infty \ni (x, \varphi) \mapsto \langle x, \varphi \rangle := \varphi(x)$. On ℓ^∞ we define the w^*-topology as the weakest topology that makes the maps $\ell^\infty \ni \varphi \mapsto \langle x, \varphi \rangle$ continuous, for all $x \in \ell^1$. ∎

Theorem A.6.16 (Composition in Holomorphic Functional Calculus). *Let $a \in \mathcal{A}$, $f \in \mathrm{Hol}(\sigma(a))$ and $g \in \mathrm{Hol}(\sigma(f(a)))$. Then $g \circ f \in \mathrm{Hol}(\sigma(a))$ and*

$$(g \circ f)(a) = g(f(a)).$$

Abelian Banach Algebras

Let $\mathcal{A}$ be a Banach algebra that is Abelian, or commutative, in the sense that $ab = ba$ for all $a, b \in \mathcal{A}$. An *ideal* $\mathcal{J}$ of $\mathcal{A}$ is a linear manifold such that $jx, xj \in \mathcal{J}$ for all $j \in \mathcal{J}$ and all $x \in \mathcal{A}$. The ideal $\mathcal{J}$ is *proper* if it is neither $\{0\}$ nor $\mathcal{A}$, and it is *maximal* if it is proper and whenever $\mathcal{H}$ is another proper ideal of $\mathcal{A}$ such that $\mathcal{J} \subseteq \mathcal{H}$ it follows that $\mathcal{J} = \mathcal{H}$.

A *character* of $\mathcal{A}$ is a mapping $\chi \colon \mathcal{A} \to \mathbb{C}$ that is linear, multiplicative, that is, $\chi(xy) = \chi(x)\chi(y)$ for all $x, y \in \mathcal{A}$, and nonzero.

A.6 Banach Algebras

Proposition A.6.17. (a) *Every maximal ideal of $\mathcal{A}$ is closed.*
(b) *Every character χ of $\mathcal{A}$ is continuous and its functional norm $\|\chi\| = 1$.*
(c) *There is a bijective correspondence $\chi \mapsto \mathrm{Ker}(\chi)$ between the set of all characters of $\mathcal{A}$ and the set of all maximal ideals of $\mathcal{A}$.*

In view of the previous proposition, we denote by $\mathrm{Sp}(\mathcal{A})$ the set of all maximal ideals of $\mathcal{A}$, identified with the set of all characters of $\mathcal{A}$, and called the *spectrum* of $\mathcal{A}$, or the *maximal ideal space* of $\mathcal{A}$. Now recall that, when $\mathcal{A}$ is viewed as a Banach space, $\mathcal{A}^*$ denotes its topological dual, which is the Banach space of all continuous linear functionals $f\colon \mathcal{A} \to \mathbb{C}$, with functional norm $\|f\| = \sup\{|f(x)| \mid x \in \mathcal{A}, \|x\| \leq 1\}$. Then, on $\mathcal{A}^*$ there is the w^*-topology, which is the locally convex topology determined by the collection of all seminorms $\mathcal{A} \ni x \mapsto |f(x)|$, when $f \in \mathcal{A}^*$. According to the Banach–Alaoglu Theorem, see Theorem A.4.11, the closed unit ball $\{f \in \mathcal{A}^* \mid \|f\| \leq 1\}$ is w^*-compact and, since $\mathrm{Sp}(\mathcal{A})$ is a closed subset of it, we have the following result.

Proposition A.6.18. $\mathrm{Sp}(\mathcal{A})$ *is a w^*-compact Hausdorff space.*

Example A.6.19. Let M denote a compact Hausdorff topological space and consider $\mathcal{C}(M) = \{f \mid f\colon M \to \mathbb{C} \text{ continuous}\}$. With the sup-norm

$$\|f\|_\infty = \sup\{|f(x)| \mid x \in M\}, \quad f \in \mathcal{C}(M),$$

$\mathcal{C}(M)$ becomes an Abelian and unital Banach algebra. ∎

Proposition A.6.20 (Gelfand Transform). *Let $\mathcal{A}$ be an Abelian and unital Banach algebra. For arbitrary $x \in \mathcal{A}$ denote $\widehat{x}\colon \mathrm{Sp}(\mathcal{A}) \to \mathbb{C}$ by*

$$\widehat{x}(\chi) = \chi(x), \quad \chi \in \mathrm{Sp}(\mathcal{A}).$$

(a) For any $x \in \mathcal{A}$, we have $\widehat{x} \in \mathcal{C}(\mathrm{Sp}(\mathcal{A}))$ and

$$\mathrm{Ran}(\widehat{x}) = \{\widehat{x}(\chi) \mid \chi \in \mathrm{Sp}(\mathcal{A})\} = \sigma_\mathcal{A}(x).$$

(b) The mapping $\mathcal{A} \ni x \mapsto \widehat{x} \in \mathcal{C}(\mathrm{Sp}(\mathcal{A}))$ is an algebra homomorphism, that is, linear and multiplicative, and

$$\|\widehat{x}\|_\infty \leq \|x\|, \quad x \in \mathcal{A},$$

and hence continuous.

In the case of a Banach algebra generated by one element, the Gelfand transform yields a concrete representation of its maximal ideal space.

Proposition A.6.21. *Assume that $\mathcal{A}$ is an Abelian unital Banach algebra generated by an element a. Then, Gelfand's transform yields a homeomorphism $\widehat{a}\colon \mathrm{Sp}(\mathcal{A}) \to \sigma_\mathcal{A}(a)$.*

A.7 Banach Algebras with Involution

In this section we review a few basic concepts and facts on C^*-algebras with an emphasis on Abelian C^*-algebras, that are related to the next section on operator theory on Hilbert spaces. Proofs and more advanced topics can be found in the classical monograph of J. Dixmier [46] or modern textbooks such as J. B. Conway [38].

Algebras with Involution

Let $\mathcal{A}$ be a complex algebra. A map $\mathcal{A} \ni a \mapsto a^* \in \mathcal{A}$ is called an *involution* if the following properties hold.

(i) $(\alpha a + b)^* = \overline{\alpha} a^* + b^*$, for all $a, b \in \mathcal{A}$ and all $\alpha \in \mathbb{C}$.
(ii) $(a^*)^* = a$, for all $a \in \mathcal{A}$.
(iii) $(ab)^* = b^* a^*$, for all $a, b \in \mathcal{A}$.

An algebra with involution is called a *-*algebra*. If $\mathcal{A}$ is a *-algebra with unit 1 then $1^* = 1$.

Given a *-algebra $\mathcal{A}$ an element $a \in \mathcal{A}$ is called *selfadjoint* or *Hermitian* if $a^* = a$. Any element $x \in \mathcal{A}$ can be uniquely represented as $x = a + ib$ where $a, b \in \mathcal{A}$ are selfadjoint. More precisely,

$$a = \frac{x + x^*}{2}, \quad b = \frac{x - x^*}{2i}.$$

The element a is called the *real* part of x and is denoted by $\operatorname{Re} x$, while the element b is called the *imaginary* part of x and is denoted by $\operatorname{Im} x$.

Proposition A.7.1. *If $\mathcal{A}$ is a *-algebra then $\mathcal{A}^{\mathrm{h}} = \{a \in \mathcal{A} \mid a = a^*\}$ is a vector space over the real field $\mathbb{R}$ and $\mathcal{A} = \mathcal{A}^{\mathrm{h}} \dot{+} i\mathcal{A}^{\mathrm{h}}$, in particular, $\mathcal{A}^{\mathrm{h}}$ linearly generates $\mathcal{A}$.*

A *normed *-algebra* is a *-algebra that is a normed algebra as well. A *Banach *-algebra* is a complete normed *-algebra.

Proposition A.7.2. *Let $\mathcal{A}$ be a Banach *-algebra and $a \in \mathcal{A}$. Then:*

(i) $\sigma_{\mathcal{A}}(a^*) = \{\overline{\lambda} \mid \lambda \in \sigma_{\mathcal{A}}(a)\}$.
(ii) $r_{\mathcal{A}}(a^*) = r_{\mathcal{A}}(a)$.

Using the holomorphic functional calculus one can prove the following

Lemma A.7.3 (Ford Square Root Lemma). *Let $\mathcal{A}$ be a Banach *-algebra and $a \in \mathcal{A}^{\mathrm{h}}$ such that $r_{\mathcal{A}}(a) < 1$. Then there exists a unique element $b \in \mathcal{A}^{\mathrm{h}}$ such that $2b - b^2 = a$ and $r_{\mathcal{A}}(b) < 1$.*

If the *-algebra $\mathcal{A}$ has a unit e then $u \in \mathcal{A}$ is called *unitary* if $u^* u = u u^* = e$.

A.7 Banach Algebras with Involution

Proposition A.7.4. *Let $\mathcal{A}$ be a Banach $*$-algebra with unit. Then $\mathcal{A}$ is the linear span of the set of all its unitary elements.*

Note that in the definition of a normed $*$-algebra we do not require the continuity of the involution $*$. A normed $*$-algebra $\mathcal{A}$ is called *involutive* if $\|a^*\| = \|a\|$ for all $a \in \mathcal{A}$, that is, if the involution is isometric.

C^*-Algebras

A *C^*-algebra* is a Banach algebra on which there is defined an *involution* $\mathcal{A} \ni a \mapsto a^* \in \mathcal{A}$ subject to the following properties.

(i) $(\alpha a + b)^* = \overline{\alpha} a^* + b^*$, for all $a, b \in \mathcal{A}$ and all $\alpha \in \mathbb{C}$.
(ii) $(a^*)^* = a$, for all $a \in \mathcal{A}$.
(iii) $(ab)^* = b^* a^*$, for all $a, b \in \mathcal{A}$.
(iv) $\|a^*a\| = \|a\|^2$, for all $a \in \mathcal{A}$.

Example A.7.5. With notation as in Example A.6.19, let $f^*(t) = \overline{f(t)}$ for all $f \in \mathcal{C}(M)$ and all $t \in M$. Then, with respect to this involution, $\mathcal{C}(M)$ is an Abelian unital C^*-algebra. ∎

Examples and, actually, the universal model, for non-Abelian C^*-algebras are provided by the C^*-algebras of bounded linear operators on a Hilbert space, see the next section.

Remark A.7.6. (a) As a consequence of (iv), it follows that $\|a^*\| = \|a\|$, hence the involution is continuous, actually isometric. Indeed, $\|a\|^2 = \|a^*a\| \leqslant \|a^*\|\|a\|$, hence $\|a\| \leqslant \|a^*\|$. Then, $\|a^*\| \leqslant \|(a^*)^*\| = \|a\|$.
(b) If e is the unit of $\mathcal{A}$, then $e^* = e$. Indeed, $e^* = e^*e = (e^*e)^* = (e^*)^* = e$. ∎

From now on, $\mathcal{A}$ denotes a unital C^*-algebra. An element $a \in \mathcal{A}$ is called

(i) *selfadjoint* if $a = a^*$,
(ii) *normal* if $a^*a = aa^*$,
(iii) *positive* if $a = x^*x$ for some $x \in \mathcal{A}$,
(iv) *unitary* if $a^*a = aa^* = e$.

Proposition A.7.7. *Let $a \in \mathcal{A}$.*
(a) If a is selfadjoint then $\sigma_\mathcal{A}(a) \subseteq \mathbb{R}$.
(b) a is positive if and only if $a = a^*$ and $\sigma_\mathcal{A}(a) \subseteq [0, \infty)$.
(c) (Square Root) If a is positive then there exists uniquely a positive element denoted $\sqrt{a} = a^{1/2} \in \mathcal{A}$ such that $\sqrt{a}^2 = a$.
(d) (Jordan Decomposition) If a is selfadjoint then there exist two positive elements $a_\pm \in \mathcal{A}$, uniquely determined such that $a = a_- - a_-$ and $a_+ a_- = 0$.

As a consequence of the item (c) in the previous proposition, for any $x \in \mathcal{A}$, we can denote $|x| = \sqrt{x^*x} = (x^*x)^{1/2} \in \mathcal{A}$ and call it the *modulus* of x. If a is selfadjoint then $|a| = a_+ + a_-$.

The Spatial Tensor Product

There are different possible ways to define tensor products of C^*-algebras. The one that is used in this book makes use of the following embedding theorem. In order to state it, we need to recall some more concepts. Given two C^*-algebras $\mathcal{A}$ and $\mathcal{B}$, a linear mapping $\pi\colon \mathcal{A} \to \mathcal{B}$ is a *-*homomorphism*, equivalently, a *homomorphism of C^*-algebras*, if it is multiplicative, $\pi(xy) = \pi(x)\pi(y)$ for all $x, y \in \mathcal{A}$, and preserves the involution, $\pi(x^*) = \pi(x)^*$ for all $x \in \mathcal{A}$. If $\mathcal{B} = \mathcal{B}(\mathcal{H})$ for some Hilbert space $\mathcal{H}$, then the *-homomorphism π is called a *representation*. The representation π is called *faithful* if it is one-to-one.

Theorem A.7.8 (Gelfand–Naimark Theorem). *For any C^*-algebra $\mathcal{A}$ there exists a faithful representation $\pi\colon \mathcal{A} \to \mathcal{B}(\mathcal{H})$ for some Hilbert space $\mathcal{H}$. In particular, modulo an injective *-homomorphism, any C^*-algebra can be viewed as a C^*-subalgebra of $\mathcal{B}(\mathcal{H})$ for some Hilbert space $\mathcal{H}$.*

Let $\mathcal{A}$ and $\mathcal{B}$ be two C^*-algebras. By the preceding theorem we can assume that $\mathcal{A} \subseteq \mathcal{B}(\mathcal{H})$ and $\mathcal{B} \subseteq \mathcal{B}(\mathcal{K})$, for some Hilbert spaces $\mathcal{H}$ and $\mathcal{K}$. Let $\mathcal{H} \otimes \mathcal{K}$ be the Hilbert space tensor product and consider the C^*-algebra $\mathcal{B}(\mathcal{H} \otimes \mathcal{K})$. If $T \in \mathcal{B}(\mathcal{H})$ and $S \in \mathcal{B}(\mathcal{K})$ then the linear operator defined by

$$(T \otimes S)(h \otimes k) = Th \otimes Sk, \quad h \in \mathcal{H}, \, k \in \mathcal{K},$$

and extended by linearity, is bounded and hence it has a unique extension to an operator $T \otimes S \in \mathcal{B}(\mathcal{H} \otimes \mathcal{K})$. The adjoint operation is $(T \otimes S)^* = T^* \otimes S^*$. Let $\mathcal{A} \otimes \mathcal{B}$ denote the collection of finite sums of operators of type $T \otimes S$ with $T \in \mathcal{A}$ and $S \in \mathcal{B}$, then $\mathcal{A} \otimes \mathcal{B}$ is a C^*-subalgebra of $\mathcal{B}(\mathcal{H} \otimes \mathcal{K})$, the *spatial tensor product* of the C^*-algebras $\mathcal{A}$ and $\mathcal{B}$.

In particular, let $\mathcal{B} = M_n$, the C^*-algebra of complex $n \times n$ matrices, canonically identified with $\mathcal{B}(\mathbb{C}^n)$. Then the spatial tensor product $\mathcal{A} \otimes M_n$ is canonically identified with $M_n(\mathcal{A})$, the C^*-algebra of $n \times n$ matrices with entries in $\mathcal{A}$. If $\mathcal{A}$ is a C^*-subalgebra of $\mathcal{B}(\mathcal{H})$ then $M_n(\mathcal{A}) = \mathcal{A} \otimes M_n$ can be viewed as a C^*-subalgebra of $\mathcal{B}(\mathcal{H}^n)$, where $\mathcal{H}^n = \mathcal{H} \oplus \cdots \oplus \mathcal{H}$, the orthogonal direct sum of n copies of $\mathcal{H}$.

Abelian C^*-Algebras

Example A.6.19 provides, via the Gelfand transform, the universal model for Abelian unital C^*-algebras.

Proposition A.7.9 (Gelfand–Naimark Theorem). *Let $\mathcal{B}$ be an Abelian unital C^*-algebra. Then, the Gelfand transform $\mathcal{B} \ni x \mapsto \hat{x} \in \mathcal{C}(\mathrm{Sp}(\mathcal{B}))$ is an isometric isomorphism of C^*-algebras.*

There are a few important consequences of this result.

Corollary A.7.10. *If $a \in G(\mathcal{A})$ then a^{-1} belongs to the C^*-algebra generated by e and a.*

Corollary A.7.11 (Spectral Permanence). *Let $\mathcal{B}$ be a C^*-subalgebra of $\mathcal{A}$, with the same unit e. Then, for any $a \in \mathcal{B}$, $\sigma_\mathcal{B}(a) = \sigma_\mathcal{A}(a)$.*

Finally, let $a \in \mathcal{A}$ be a normal element. Letting $\mathcal{B}$ denote the C^*-algebra generated by e and a, then $\mathcal{B}$ is Abelian. From the previous corollary it follows that $\sigma_\mathcal{B}(a) = \sigma_\mathcal{A}(a)$ and then, from Proposition A.6.21, it follows that $\mathrm{Sp}(\mathcal{B})$ can be identified with $\sigma_\mathcal{A}(a)$. Now we consider the Gelfand transform $\widehat{}: \mathcal{B} \to \mathcal{C}(\sigma_\mathcal{B}(a))$ which is an isometric $*$-isomorphism, hence its inverse is an isometric $*$-isomorphism as well

$$\mathcal{C}(\sigma_\mathcal{B}(a)) \ni f \mapsto f(a) \in \mathcal{B} \subseteq \mathcal{A}.$$

This isometric $*$-isomorphism is called the *functional calculus with continuous functions* of a. For any polynomial in two variables $p(z, \bar{z})$, it makes perfect sense to consider the element $p(a, a^*)$, and it follows that this is exactly the continuous functional calculus of a for the continuous complex function $\mathbb{C} \ni z \mapsto p(z, \bar{z}) \in \mathbb{C}$.

In particular, if a is positive, then $\sigma_\mathcal{A}(a) \subseteq [0, \infty)$ and hence, the function $f(t) = \sqrt{t}$ is continuous on $\sigma_\mathcal{A}(a)$ and it can be shown that $f(a) = \sqrt{a}$, with the notation as in Proposition A.7.7(c). Similarly, if a is selfadjoint, the elements $a_\pm$ can be obtained by continuous functional calculus for a as well, by considering the function $\zeta_+: \mathbb{R} \to \mathbb{R}$ defined by $\zeta_+(t) = t$ for $t \geqslant 0$ and $\zeta_+(t) = 0$ for $t < 0$, and the function ζ_- defined by $\zeta_-(t) = -t$ for $t \leqslant 0$ and $\zeta_-(t) = 0$ for $t > 0$.

A.8 Linear Operators on Banach Spaces

In this section we review operator theory on Banach spaces with an emphasise on unbounded linear operators. The reader can find proofs of these results in classical monographs such as N. Dunford and J. T. Schwartz [54], T. Kato [84], or more modern texts such as I. Gohberg, S. Goldberg, and M. A. Kaashoek [68, 69].

Unbounded Operators

If $\mathcal{X}$ and $\mathcal{Y}$ are two Banach spaces, recall that one can consider a more general concept of linear operator in the sense of the domain. So, we say that A is a linear operator from $\mathcal{X}$ to $\mathcal{Y}$ if the domain of A, denoted by $\mathrm{Dom}(A)$, is a linear manifold of $\mathcal{X}$ and the range of A, denoted by $\mathrm{Ran}(A)$, is a linear manifold of $\mathcal{Y}$, and it is linear in the sense that $A(\alpha h + \beta k) = \alpha A h + \beta A k$ for all $h, k \in \mathcal{X}$ and all scalars α and β.

If A and B are two linear operators from $\mathcal{X}$ to $\mathcal{Y}$, the sum $A + B$ is a linear operator from $\mathcal{X}$ to $\mathcal{Y}$ defined as follows: $\mathrm{Dom}(A+B) := \mathrm{Dom}(A) \cap \mathrm{Dom}(B)$ and $(A+B)h := Ah + Bh$ for all $h \in \mathrm{Dom}(A+B)$. If A is a linear operator from $\mathcal{X}$ to $\mathcal{Y}$ and C is a linear operator from $\mathcal{Y}$ to $\mathcal{G}$, then CA is the linear operator defined as follows: $\mathrm{Dom}(CA) := \{h \in \mathcal{X} \mid Ac \in \mathrm{Dom}(C)\}$ and $(CA)h := CAh$ for all $h \in \mathrm{Dom}(CA)$. Note that $\mathrm{Dom}(CA) = A^{-1}(\mathrm{Dom}(C))$ in the sense of inverse image.

If A and B are linear operators from $\mathcal{X}$ to $\mathcal{Y}$ then we say that B is an *extension* of A, denoted by $A \subseteq B$, if $\operatorname{Dom}(A) \subseteq \operatorname{Dom}(B)$ and $Ah = Bh$ for all $h \in \operatorname{Dom}(A)$, equivalently $B|\operatorname{Dom}(A) = A$.

The linear operator A is called *bounded* if $A|\operatorname{Dom}(A)$ is bounded, that is, $\|Ah\| \leqslant c\|h\|$ for some $c \geqslant 0$ and all $h \in \operatorname{Dom}(A)$. In this case, A can be uniquely extended to a bounded linear operator $\widetilde{A}$ with domain the closure of $\operatorname{Dom}(A)$, $\operatorname{Dom}(\widetilde{A}) = \operatorname{Clos}\operatorname{Dom}(A)$. If $\operatorname{Dom}(A)$ is dense in $\mathcal{X}$ then we say that A is *densely defined*. Thus, if A is densely defined and bounded then it has a unique extension to the operator $\widetilde{A} \in \mathcal{B}(\mathcal{X}, \mathcal{Y})$.

If A is a linear operator from $\mathcal{X}$ to $\mathcal{Y}$ then its *graph* is the linear manifold in $\mathcal{X} \oplus \mathcal{Y}$ defined by

$$\mathcal{G}(A) := \{h \oplus Ah \mid h \in \operatorname{Dom}(A)\}.$$

The linear operator is called *closed* if $\mathcal{G}(A)$ is closed in $\mathcal{X} \oplus \mathcal{Y}$.

Proposition A.8.1. *Let A be a linear operator from $\mathcal{X}$ to $\mathcal{Y}$. Then A is closed if and only if, for any sequence $(h_n)_n$ of vectors in $\operatorname{Dom}(A)$ subject to the condition that $h_n \xrightarrow[n\to\infty]{} h \in \mathcal{X}$ and $Ah_n \xrightarrow[n\to\infty]{} k$, it follows that $h \in \operatorname{Dom}(A)$ and $Ah = k$.*

The linear operator A is called *closable* if the closure of $\mathcal{G}(A)$ is the graph of an operator, that is, if $h_1 \oplus k \in \operatorname{Clos}\mathcal{G}(A)$ and $h_2 \oplus k \in \operatorname{Clos}\mathcal{G}(A)$ then $h_1 = h_2$.

Proposition A.8.2. *Given A a linear operator from $\mathcal{X}$ to $\mathcal{Y}$, the following assertions are equivalent.*

(i) *Whenever $0 \oplus k \in \operatorname{Clos}\mathcal{G}(A)$ it follows that $k = 0$.*
(ii) *For any sequence $(h_n)_n$ in $\operatorname{Dom}(A)$ such that $h_n \xrightarrow[n\to\infty]{} h$ and $Ah_n \xrightarrow[n\to\infty]{} 0$ it follows that $h = 0$.*
(iii) *There exists B a closed linear operator from $\mathcal{X}$ to $\mathcal{Y}$ such that $A \subseteq B$, in the sense $\operatorname{Dom}(A) \subseteq \operatorname{Dom}(B)$ and $Ah = Bh$ for all $h \in \operatorname{Dom}(A)$.*

If A is a closable operator then there exists a unique operator $\widetilde{A}$ with graph $\operatorname{Clos}\mathcal{G}(A)$ and such that $A \subseteq \widetilde{A}$. The operator $\widetilde{A}$ is called the *closure* of the operator A and it is closed.

A linear operator A from $\mathcal{X}$ to $\mathcal{Y}$ is called *boundedly invertible* if there exists $B \in \mathcal{B}(\mathcal{Y}, \mathcal{X})$ such that $AB = I_\mathcal{Y}$ and $BA \subseteq I_\mathcal{X}$. Note that, in this case, BA is bounded on its domain.

Proposition A.8.3. *Let A be a linear operator from $\mathcal{X}$ to $\mathcal{Y}$.*

(i) *A is boundedly invertible if and only if $\operatorname{Ker}(A) = 0$, $\operatorname{Ran}(A) = \mathcal{Y}$, and A is closed.*
(ii) *If A is boundedly invertible then it has a unique inverse, denoted by A^{-1}.*

Spectrum of a Linear Operator

Let $\mathcal{X} \neq 0$ be a complex Banach space and $T \colon \mathrm{Dom}(T)(\subseteq \mathcal{X}) \to \mathcal{X}$ a linear operator. If $\lambda \in \mathbb{C}$ is such that the linear operator $\lambda I - T \colon \mathrm{Dom}(T) \to \mathcal{X}$ is injective then we consider its inverse on the range $(\lambda I - T)^{-1} \colon \mathrm{Ran}(\lambda I - T)(\subseteq \mathcal{X}) \to \mathcal{X}$.

Definition A.8.4. A complex number λ is called a *regular value* for T if the following conditions hold.

(i) $\lambda I - T$ is injective.
(ii) $(\lambda I - T)^{-1}$ is bounded.
(iii) $\mathrm{Ran}(\lambda I - T)$ is dense in $\mathcal{X}$.

We denote by $\rho(T)$ the set of all regular values of T and call it the *resolvent set* of T. Then, the *spectrum* of T is the set $\sigma(T) := \mathbb{C} \setminus \rho(T)$.

Also, the *point spectrum* of T is the set

$$\sigma_{\mathrm{p}}(T) := \{\lambda \in \mathbb{C} \mid Tx = \lambda x, \text{ for some } x \in \mathcal{X} \setminus \{0\}\}$$
$$= \{\lambda \in \mathbb{C} \mid \lambda I - T \text{ is not injective}\}.$$

An element $\lambda \in \sigma_{\mathrm{p}}(T)$ is called an *eigenvalue* of T and a vector $x \in \mathcal{X} \setminus \{0\}$ such that $Tx = \lambda x$ is called an *eigenvector* of T associated to the eigenvalue λ.

In addition, the *compression spectrum* of T is the set

$$\sigma_{\mathrm{c}}(T) := \{\lambda \in \mathbb{C} \mid \lambda I - T \text{ is injective, } \mathrm{Ran}(\lambda I - T) \text{ is dense in } \mathcal{X},$$
$$\text{and } (\lambda I - T)^{-1} \text{ is unbounded}\},$$

and the *residual spectrum* of T is the set

$$\sigma_{\mathrm{r}} := \{\lambda \in \mathbb{C} \mid \lambda I - T \text{ is injective and } \mathrm{Ran}(\lambda I - T) \text{ is not dense in } \mathcal{X}\}. \blacksquare$$

Remark A.8.5. (1) $\mathbb{C} = \sigma(T) \cup \rho(T)$ with mutually disjoint sets.
(2) $\sigma(T) = \sigma_{\mathrm{p}}(T) \cup \sigma_{\mathrm{c}}(T) \cup \sigma_{\mathrm{r}}(T)$ with mutually disjoint sets.
(3) $\rho(T)$ may be empty.
(4) If $\dim \mathcal{X} < \infty$ then $\sigma_{\mathrm{c}}(T) = \sigma_{\mathrm{r}}(T) = \emptyset$.
(5) $\sigma_{\mathrm{p}}(T)$ may be empty. $\blacksquare$

Proposition A.8.6. *Assume that T is closed. If $\lambda \in \rho(T)$ then $\mathrm{Ran}(\lambda I - T) = \mathcal{X}$ and $(\lambda I - T)^{-1} \in \mathcal{B}(\mathcal{X})$.*

Definition A.8.7. The *approximate point spectrum* of T is the set

$$\sigma_{\mathrm{ap}}(T) := \{\lambda \in \mathbb{C} \mid \text{ there exists } (x_n)_n \text{ in } \mathrm{Dom}(T), \|x_n\| = 1,$$
$$\text{for all } n \geqslant 1 \text{ and } \|(\lambda I - A)x_n\| \xrightarrow[n \to \infty]{} 0\}. \blacksquare$$

Remark A.8.8. $\sigma_{\mathrm{p}}(T) \subseteq \sigma_{\mathrm{ap}}(T)$. $\blacksquare$

Proposition A.8.9. *Assume that* $T \in \mathcal{B}(\mathcal{X})$ *and* $\lambda \in \mathbb{C}$. *The following assertions are equivalent.*

(a) $\lambda \notin \sigma_{\mathrm{ap}}(T)$.
(b) $\mathrm{Ker}(\lambda I - T) = \{0\}$ *and* $\mathrm{Ran}(\lambda I - T)$ *is closed.*
(c) *There exists* $c > 0$ *such that* $\|(\lambda I - T)x\| \geqslant c\|x\|$ *for all* $x \in \mathcal{X}$.
(d) $\lambda I - T$ *has a bounded left inverse.*

Proposition A.8.10. *If* $T \in \mathcal{B}(\mathcal{X})$ *then* $\partial \sigma(T) \subseteq \sigma_{\mathrm{ap}}(T)$.

Holomorphic Functional Calculus

Letting $\mathcal{A} = \mathcal{B}(\mathcal{X})$, for some Banach space $\mathcal{X}$, we have a unital Banach algebra and hence all the concepts and results from Section A.6 apply. For this reason, in this subsection we only consider the more general case of unbounded operators on a Banach space. In the following T denotes a closed operator defined on the dense linear manifold $\mathrm{Dom}(T)$ in $\mathcal{X}$ and range $\mathrm{Ran}(T)$ in $\mathcal{X}$, such that its resolvent set $\rho(T)$ is not empty. Letting $\alpha \in \rho(T)$ be a fixed number let

$$A = (T - \alpha I)^{-1} = -R(\lambda; T) \in \mathcal{B}(\mathcal{X}), \tag{A.1}$$

which is injective, $\mathrm{Ran}(A) = \mathrm{Dom}(T)$ is dense in $\mathcal{X}$, $TAx = \alpha Ax + x$ for all $x \in \mathcal{X}$, and $ATx = \alpha Ax + x$ for all $x \in \mathrm{Dom}(T)$.

Recalling that $\hat{\mathbb{C}} = \mathbb{C} \cup \{\infty\}$ denotes the one point compactification of the complex field, consider the bijective function $\Phi \colon \hat{\mathbb{C}} \to \hat{\mathbb{C}}$ defined by

$$\mu = \Phi(\lambda) = (\lambda - \alpha)^{-1}, \quad \lambda \in \mathbb{C}, \tag{A.2}$$

$\Phi(\infty) = 0$, and $\Phi(\alpha) = \infty$. Recall that $\tilde{\sigma}(T)$ denotes the extended spectrum of T, more precisely, $\tilde{\sigma}(T) = \sigma(T) \cup \{\infty\}$ if T is unbounded and $\tilde{\sigma}(T) = \sigma(T)$ if T is bounded. Then, $\Phi(\tilde{\sigma}(T)) = \sigma(A)$. Also, recall that, if σ is a closed nonempty subset of $\mathbb{C}$, then $\mathrm{Hol}(\sigma)$ denotes the algebra of holomorphic functions defined on an open subset containing σ, if σ is bounded and, in the case that σ is unbounded, $\mathrm{Hol}(\sigma)$ denotes the algebra of holomorphic functions on an open subset containing σ and holomorphic at infinity as well. Then, the mapping

$$\mathrm{Hol}(\sigma(A)) \ni f \mapsto \varphi = f \circ \Phi^{-1} \in \mathrm{Hol}(\sigma(T)),$$

establishes a bijective correspondence between $\mathrm{Hol}(\sigma(A))$ and $\mathrm{Hol}(\sigma(T))$.

With notation as before, the functional calculus with holomorphic functions for T is defined as follows:

$$f(T) = \varphi(A) = (f \circ \Phi^{-1})(A) \in \mathcal{B}(\mathcal{X}), \quad f \in \mathrm{Hol}(\sigma(T)). \tag{A.3}$$

The definition does not depend on $\alpha \in \rho(T)$ and

$$f(T) = f(\infty)I + \frac{1}{2\pi \mathrm{i}} \int_\Gamma f(\lambda) R(\lambda; T) \mathrm{d}\lambda, \tag{A.4}$$

A.8 Linear Operators on Banach Spaces

where Γ consists of a finite number of positively oriented Jordan contours included in the domain of holomorphy of f. The following theorem states that the definition at (A.3) provides a holomorphic functional calculus, called the *Riesz–Dunford–Taylor functional calculus*.

Theorem A.8.11. *Assume the notation as before.*
  (a) *The mapping* $F \colon \mathrm{Hol}(\sigma(T)) \to \mathcal{B}(\mathcal{X})$ *defined in* (A.4) *is an algebra homomorphism such that* $1(T) = I$ *and* $\Phi(T) = (T - \alpha I)^{-1}$.
  (b) *Let* $(f_n)_n$ *and* f *be functions in* $\mathrm{Hol}(\sigma(T))$ *such that there exists an open subset* G *subject to the property that*

$$\sigma(T) \subset G \subset \bigcap_{n \geqslant 1} \mathrm{Dom}(f_n) \cap \mathrm{Dom}(f)$$

and $f_n \xrightarrow[n \to \infty]{} f$ *uniformly on any compact subset of* G. *Then* $\|f_n(T) - f(T)\| \xrightarrow[n \to \infty]{} 0$.
  (c) *The mapping* F *is unique subject to the properties* (a) *and* (b).

The holomorphic functional calculus for unbounded operators has similar properties to the holomorphic functional calculus for bounded operators.

Theorem A.8.12 (Spectral Mapping Theorem). *With notation as before, for any* $f \in \mathrm{Hol}(\sigma(T))$ *we have* $\sigma(f(T)) = f(\widetilde{\sigma}(T))$. *Also, if* $g \in \mathrm{Hol}(\sigma(f(T)))$ *then* $g \circ f \in \mathrm{Hol}(\sigma(T))$ *and* $g(f(T)) = (g \circ f)(T)$.

If the operator T is unbounded, the algebra $\mathrm{Hol}(\sigma(T))$ does not contain nonconstant polynomials. However, some things still can be done. Let p be a polynomial function with complex coefficients

$$p(\lambda) = \sum_{k=0}^{n} a_n \lambda^n, \tag{A.5}$$

with $a_n \neq 0$. Then, by definition,

$$p(T) = a_n T^n + a_{n-1} T^{n-1} + \cdots + \cdots a_1 T + a_0 I, \tag{A.6}$$

with

$$\mathrm{Dom}(p(T)) = \mathrm{Dom}(T^n) \tag{A.7}$$
$$= \{x \in \mathrm{Dom}(T) \mid T^k x \in \mathrm{Dom}(T) \text{ for all } k = 1, \ldots, n-1\}.$$

Theorem A.8.13. *Assume that* T *is a closed densely defined operator with nonempty resolvent set and let* p *be a polynomial as in* (A.5).
  (a) *The operator* $p(T)$ *defined as in* (A.6) *and* (A.7) *is closed.*
  (b) *For any* $f \in \mathrm{Hol}(\sigma(T))$ *such that* f *has a zero of order* m *at infinity,* $0 \leqslant m \leqslant \infty$, *then, for any* $x \in \mathrm{Dom}(T^n)$ *we have* $f(T)x \in \mathrm{Dom}(T^{m+n})$, *where* n *is the degree of the polynomial* p *and* $m + n = \infty$ *if* $m = \infty$, *and* $p(T)f(T)x = f(T)p(T)x$.

(c) If $0 \leqslant n \leqslant m \leqslant \infty$ and $g(\lambda) = p(\lambda)f(\lambda)$, then $g \in \mathrm{Hol}(\sigma(T))$ and $g(T) = p(T)f(T)$.

(d) If $f \in \mathrm{Hol}(\sigma(T))$ has no zero in $\sigma(T)$, but a zero of order of finite order l at infinity, then $f(T)$ is injective and $\mathrm{Ran}(f(T)) = \mathrm{Dom}(T^l)$.

(e) $p(\sigma(T)) = \sigma(p(T))$.

Remark A.8.14. In general, we do not know whether the operator $p(T)$ has dense domain. However, if T is a closed densely defined operator with nonempty resolvent set then $\mathrm{Dom}(T^2)$ is dense. ∎

A.9 Linear Operators on Hilbert Spaces

In this section we review a few basic concepts and facts on linear operators on Hilbert spaces, spectral theory for normal operators and compact operators, with an emphasis on unbounded operators. Proofs and more advanced topics can be found in monographs such as N. I. Akhiezer and I. M. Glazman [3], J. B. Conway [36, 38], N. Dunford and J. T. Schwartz [55], I. Gohberg, S. Goldberg, and M. A. Kaashoek [68, 69], and K. Schmüdgen [132], for example.

Bounded Operators

Since any Hilbert space is a Banach space, and hence a locally convex space, we can use everything that was that stated in Section A.3, Section A.4, and Section A.8 for linear operators $T\colon \mathcal{H} \to \mathcal{K}$, where $\mathcal{H}$ and $\mathcal{K}$ are Hilbert spaces. Thus, T is continuous if and only if it is bounded and, in this situation, its operator norm $\|T\|$ is defined. Letting $\mathcal{B}(\mathcal{H}, \mathcal{K})$ denote the collection of all bounded operators $T\colon \mathcal{H} \to \mathcal{K}$, the operator norm is a norm on the vector space $\mathcal{B}(\mathcal{H}, \mathcal{K})$ which becomes a Banach space. In the case when $\mathcal{H} = \mathcal{K}$, we let $\mathcal{B}(\mathcal{H}) = \mathcal{B}(\mathcal{H}, \mathcal{H})$, which is a Banach algebra. In the following we recall how $\mathcal{B}(\mathcal{H})$ is organised as a C^*-algebra.

Proposition A.9.1 (Riesz Theorem). *Let $\mathcal{H}$ and $\mathcal{K}$ be two Hilbert spaces and let $\Phi\colon \mathcal{H} \times \mathcal{K} \to \mathbb{C}$ be a sesquilinear form, in the sense that, Φ is linear in the first variable and conjugate linear in the second variable, and bounded in the sense that there exists $C \geqslant 0$ such that*

$$|\Phi(x,y)| \leqslant C\|x\|\|y\|, \quad x \in \mathcal{H},\ y \in \mathcal{K}.$$

Then, there exists uniquely an operator $T \in \mathcal{B}(\mathcal{K}, \mathcal{H})$ such that

$$\Phi(x,y) = \langle x, Ty\rangle, \quad x \in \mathcal{H},\ y \in \mathcal{K}.$$

On the grounds of the previous proposition, for every $T \in \mathcal{B}(\mathcal{H}, \mathcal{K})$ there exists uniquely an operator $T^* \in \mathcal{B}(\mathcal{K}, \mathcal{H})$ such that

$$\langle Tx, y\rangle = \langle x, T^*y\rangle, \quad x \in \mathcal{H},\ y \in \mathcal{K}.$$

The operator T^* is called the *adjoint operator* and has the following properties.

A.9 Linear Operators on Hilbert Spaces

(a) $(\alpha S + T)^* = \overline{\alpha} S^* + T^*$ and $(T^*)^* = T$.
(b) $\|T^*\| = \|T\|$ and $\|T^*T\| = \|T\|^2$.
(c) $\operatorname{Ran}(T)^\perp = \operatorname{Ker}(T^*)$.

Let $T \in \mathcal{B}(\mathcal{H}, \mathcal{K})$, for two Hilbert spaces $\mathcal{H}$ and $\mathcal{K}$, its *kernel* $\operatorname{Ker}(T) := \{h \in \mathcal{H} \mid Th = 0\}$, and its *range* $\operatorname{Ran}(T) := \{Th \mid h \in \mathcal{H}\}$. T is called

(i) an *isometry* if $T^*T = I$;
(ii) a *coisometry* if T^* is an isometry;
(iii) a *partial isometry* if $T^*TT^* = T^*$;
(iv) a *unitary* if it is an isometry and a coisometry.

Let $T \in \mathcal{B}(\mathcal{H})$, for a Hilbert space $\mathcal{H}$. T is called

(i) *selfadjoint* if $T = T^*$;
(ii) *positive* if $\langle Tx, x \rangle \geqslant 0$ for all $x \in \mathcal{H}$;
(iii) an *orthogonal projection* if $T^2 = T = T^*$;
(iv) *normal* if $TT^* = T^*T$.

In view of the properties (a) and (b) as before, it follows that $\mathcal{B}(\mathcal{H})$, with the operator norm and the operation of taking adjoints, is a C^*-algebra, hence the results presented in Section A.6 apply. In particular, the continuous functional calculus for normal operators holds. In order to connect the definition of positive operators as before with the definition from C^*-algebras, we recall the following.

Proposition A.9.2. *Let $T \in \mathcal{B}(\mathcal{H})$ be an operator for some Hilbert space $\mathcal{H}$. The following assertions are equivalent.*

(i) *T is a positive operator.*
(ii) *$T = X^*X$ for some $X \in \mathcal{B}(\mathcal{H})$.*
(iii) *$T = T^*$ and its spectrum $\sigma(T)$ is contained in $[0, +\infty)$.*

We denote by $\mathcal{B}(\mathcal{H})^+$ the collection of all positive operators. If $T \in \mathcal{B}(\mathcal{H})^+$ we denote this by $T \geqslant 0$. Then we have an *order relation* on the real vector space of all selfadjoint operators in $\mathcal{B}(\mathcal{H})$ defined in the following way: if $A, B \in \mathcal{B}(\mathcal{H})$, $A = A^*$ and $B = B^*$, we have $A \geqslant B$ if $A - B \geqslant 0$. With respect to this order relation $\mathcal{B}(\mathcal{H})^+$ becomes a *cone*, that is, for any $A, B \in \mathcal{B}(\mathcal{H})^+$ and any $t, s \geqslant 0$ we have $sA + tB \in \mathcal{B}(\mathcal{H})^+$. In addition, this cone is *strict*, that is, the only operator $A \in \mathcal{B}(\mathcal{H})^+$ such that $-A \in \mathcal{B}(\mathcal{H})^+$ is the null operator.

From the point of view of the advantages of working with a Hilbert space instead of a Banach space, it is the geometry of orthogonal projections that turns out to make the difference.

Proposition A.9.3. *An operator $P \in \mathcal{B}(\mathcal{H})$ is an orthogonal projection if and only if $\operatorname{Ran}(P)$ is a closed subspace and, with respect to the decomposition $\mathcal{H} = \operatorname{Ran}(P) \oplus \operatorname{Ran}(P)^\perp$, we have $P(x + y) = x$, for all $x \in \operatorname{Ran}(P)$ and all $y \in \operatorname{Ran}(P)^\perp$. This provides a bijective correspondence between the collection of all orthogonal projections in $\mathcal{H}$ and the collection of all subspaces of $\mathcal{H}$.*

Proposition A.9.4. *Let $V \in \mathcal{B}(\mathcal{H}, \mathcal{K})$ be a partial isometry.*

   (i) *V^* is a partial isometry as well.*
  (ii) *V^*V is the orthogonal projection on $\operatorname{Ker}(V)^\perp$, called the initial space.*
 (iii) *VV^* is the orthogonal projection on $\operatorname{Ran}(V)$, called the final space.*
 (iv) *$V|_{\operatorname{Ker}(V)^\perp}\colon \operatorname{Ker}(V)^\perp \to \operatorname{Ran}(V)$ is a unitary operator.*

Recall, see Section A.6, that for $T \in \mathcal{B}(\mathcal{H})$ we denote by $|T| = (T^*T)^{1/2}$ the *modulus* of T, which is a positive operator. We have now the ingredients to state the polar decomposition for bounded linear operators in Hilbert spaces.

Proposition A.9.5. *Let $T \in \mathcal{B}(\mathcal{H}, \mathcal{K})$. Then there exists uniquely a partial isometry $V \in \mathcal{B}(\mathcal{H}, \mathcal{K})$, such that $T = V|T|$ and VV^* is the orthogonal projection on $\operatorname{Clos}\operatorname{Ran}(T)$.*

Topologies on $\mathcal{B}(\mathcal{H}, \mathcal{K})$

Given $\mathcal{H}$ and $\mathcal{K}$ two Hilbert spaces, the vector space $\mathcal{B}(\mathcal{H}, \mathcal{K})$ is a Banach space with respect to the operator norm. The topology induced by the operator norm is called the *uniform topology*. On $\mathcal{B}(\mathcal{H}, \mathcal{K})$ there are two other locally convex topologies, in addition to the operator norm topology. The family of seminorms $\mathcal{B}(\mathcal{H}) \ni T \mapsto \|Tx\|$, where $x \in \mathcal{H}$, yields the *strong operator topology* (sot). The family of seminorms $\mathcal{B}(\mathcal{H}) \ni T \mapsto |\langle Tx, y\rangle|$, where $x \in \mathcal{H}$ and $y \in \mathcal{K}$, yields the *weak operator topology* (wot).

One can view these topologies as initial topologies as well. More precisely, the strong operator topology is the initial topology induced by the family of maps $\mathcal{B}(\mathcal{H}, \mathcal{K}) \ni T \mapsto Tx \in \mathcal{K}$, when $\mathcal{K}$ is viewed with its norm topology, for $x \in \mathcal{H}$. Similarly, the weak operator topology is the initial topology induced by the family of maps $\mathcal{B}(\mathcal{H}, \mathcal{K}) \ni T \mapsto \langle Tx, y\rangle \in \mathbb{C}$, for $x \in \mathcal{H}$ and $y \in \mathcal{K}$.

Clearly, the weak operator topology is weaker than the strong operator topology which is weaker than the uniform topology. If $\mathcal{H}$ is separable then the strong operator topology is metrisable on bounded sets and, if both $\mathcal{H}$ and $\mathcal{K}$ are separable then the weak operator topology is metrisable on bounded sets. A remarkable fact is that the weak operator topology and the strong operator topology have the same bounded linear functionals.

Proposition A.9.6. *Let $\varphi\colon \mathcal{B}(\mathcal{H}, \mathcal{K}) \to \mathbb{C}$ be a linear functional. The following assertions are equivalent.*

   (a) *φ is continuous with respect to the strong operator topology.*
   (b) *φ is continuous with respect to the weak operator topology.*
   (c) *There exist $h_1, \ldots, h_n \in \mathcal{H}$ and $k_1, \ldots, k_n \in \mathcal{K}$ such that*

$$\varphi(T) = \sum_{j=1}^n \langle Th_j, k_j\rangle, \quad T \in \mathcal{B}(\mathcal{H}, \mathcal{K}).$$

As a consequence, convex subsets of $\mathcal{B}(\mathcal{H}, \mathcal{K})$ have the same closures with respect to the weak operator topology and the strong operator topology.

A.9 Linear Operators on Hilbert Spaces

Corollary A.9.7. *If S is a convex subset of $\mathcal{B}(\mathcal{H}, \mathcal{K})$ then the closure of S with respect to the weak operator topology coincides with the closure of S with respect to the strong operator topology.*

As a consequence of Alaoglu's Theorem one can prove a remarkable property of the weak operator topology.

Proposition A.9.8. *The closed unit ball of $\mathcal{B}(\mathcal{H}, \mathcal{K})$ is compact with respect to the weak operator topology.*

As linear topologies, they all make addition and multiplication with scalars continuous, but there is a difference with respect to multiplication. The uniform topology makes the product continuous, because the operator norm is submultiplicative

$$\|ST\| \leq \|S\|\|T\|, \quad S \in \mathcal{B}(\mathcal{H}, \mathcal{K}), \ T \in \mathcal{B}(\mathcal{G}, \mathcal{H}).$$

On the other hand, neither the strong operator topology nor the weak operator topology makes the product continuous, as long as infinite dimensional Hilbert spaces are considered.

The following example involves only sequences of operators in a separable infinite dimensional Hilbert space and shows that the multiplication is not continuous with respect to the weak operator topology (wot).

Example A.9.9. Let $\mathcal{H} = \ell^2$ be the Hilbert space of square summable complex sequences indexed on $\mathbb{N}$ and consider two sequences of bounded operators $(A_n)_n$ and $(B_n)_n$ on ℓ^2 defined as follows:

$$(A_n x)_k = \begin{cases} x_{k+n}, & k > n, \\ 0, & 1 \leq k \leq n, \end{cases} \quad \text{and} \quad (B_n x)_k = \begin{cases} x_{k-n}, & k > n, \\ 0, & 1 \leq k \leq n. \end{cases}$$

Then

$$A_n \xrightarrow[n \to \infty]{\text{wot}} 0, \quad B_n \xrightarrow[n \to \infty]{\text{wot}} 0,$$

but, since $A_n B_n = I_{\ell^2}$, the sequence $(A_n B_n)_n$ does not converge to 0 with respect to the weak operator topology (wot). ∎

Normal Operators

The continuous functional calculus for normal elements in a C^*-algebra can be considerably strengthened for $\mathcal{B}(\mathcal{H})$. Let $\mathcal{B}(\mathbb{C})$ denote the σ-algebra of all Borel subsets of the complex plane. A mapping $E \colon \mathcal{B}(\mathbb{C}) \to \mathcal{P}(\mathcal{H})$, where $\mathcal{P}(\mathcal{H})$ denotes the collection of all orthogonal projections in $\mathcal{H}$, is called a *spectral measure* if the following hold.

(i) $E(\mathbb{C}) = I$.

(ii) For any sequence of mutually disjoint Borel subsets $(B_n)_n$ in the complex plane, we have
$$E\Big(\bigcup_{n\in\mathbb{N}} B_n\Big) = \sum_{n=1}^{\infty} E(B_n),$$
where the series converges with respect to the strong operator topology.

Theorem A.9.10 (Spectral Theorem for Normal Operators). *Let $N \in \mathcal{B}(\mathcal{H})$ be a normal operator. Then, there exists uniquely a spectral measure E with $\mathrm{supp}(E) = \sigma(N)$, such that*
$$N = \int_{\mathbb{C}} z \mathrm{d}E(z).$$
In addition, for each bounded Borel function $f\colon \mathbb{C} \to \mathbb{C}$, the integral
$$f(N) = \int_{\mathbb{C}} f(z) \mathrm{d}E(z) \in \mathcal{B}(\mathcal{H}),$$
converges as a Riemann integral over a rectangle containing $\sigma(N)$, and the mapping
$$B(\sigma(N)) \ni f \mapsto f(N) \in \mathcal{B}(\mathcal{H}),$$
is a $*$-morphism of C^*-algebras, where $B(\sigma(N))$ denotes the C^*-algebra of all bounded Borel functions on $\sigma(N)$, continuous in the following sense: if $(f_n)_n$ is a sequence of uniformly bounded functions in $B(\sigma(N))$, such that, $f_n \xrightarrow[n\to\infty]{w} f \in B(\sigma(N))$, then $f_n(N) \xrightarrow[n\to\infty]{wo} f(N)$.

Theorem A.9.10 applies to unitary operators $U \in \mathcal{B}(\mathcal{H})$, taking into account that $\sigma(U) \subseteq \mathbb{T}$, the unit circle in the complex plane.

If $A = A^* \in \mathcal{B}(\mathcal{H})$ is a selfadjoint operator, then A is normal and $\sigma(A) \subset \mathbb{R}$ and letting E denote its spectral measure one considers $E_t := E((-\infty, t]) \in \mathcal{B}(\mathcal{H})$, for $t \in \mathbb{R}$, which is called a *spectral scale*, or *spectral function* of A, and it is uniquely determined by the following properties.

(1) $E(t)$ is a projection for all $t \in \mathbb{R}$.
(2) $t \leqslant s$ implies $E(t) \leqslant E(s)$.
(3) $\lim_{\varepsilon \searrow 0} E(t+\varepsilon)x = E(t)x$ for all $t \in \mathbb{R}$ and $x \in \mathcal{H}$.
(4) $Ax = \int_{\mathbb{R}} t \mathrm{d}E(t)x$ for all $x \in \mathcal{H}$.
(5) If $Q \in \mathcal{B}(\mathcal{H})$ is such that $QA = AQ$ then $QE(t) = E(t)Q$ for all $t \in \mathbb{R}$.

Compact Operators

Given two Hilbert spaces $\mathcal{H}_1$ and $\mathcal{H}_2$, a linear operator $T\colon \mathcal{H}_1 \to \mathcal{H}_2$ is called *compact* if the image of the closed unit ball $\{x \in \mathcal{H}_1 \mid \|x\| \leqslant 1\}$ of $\mathcal{H}_1$ under T is relatively compact in $\mathcal{H}_2$, that is, the closure of the set $\{Tx \mid x \in \mathcal{H}_1,\ \|x\| \leqslant 1\}$, is compact in $\mathcal{H}_2$. Clearly, any compact operator is continuous.

A.9 Linear Operators on Hilbert Spaces

Proposition A.9.11. *Let $T \in \mathcal{B}(\mathcal{H}_1, \mathcal{H}_2)$, for some Hilbert spaces $\mathcal{H}_1$ and $\mathcal{H}_2$. The following assertions are equivalent.*

(i) *T is compact.*
(ii) *The set $\{Tx \mid x \in \mathcal{H}_1, \|x\| \leqslant 1\}$ is compact in $\mathcal{H}_2$.*
(iii) *For any bounded sequence $(x_n)_n$ in $\mathcal{H}_1$ there exists a subsequence $(x_{k_n})_n$ such that $(Tx_{k_n})_n$ converges in the norm topology of $\mathcal{H}_2$.*
(iv) *For any sequence $(x_n)_n$ in $\mathcal{H}_1$ that converges weakly to $0 \in \mathcal{H}_1$, the sequence $(\|Tx_n\|)_n$ converges to 0.*
(v) *The adjoint operator $T^* \in \mathcal{B}(\mathcal{H}_2, \mathcal{H}_1)$ is compact.*
(vi) *The modulus $|T| \in \mathcal{B}(\mathcal{H}_1)$ is compact.*
(vii) *There exists a sequence $(T_n)_n$ of compact operators, $T_n \colon \mathcal{H}_1 \to \mathcal{H}_2$ for all $n \geqslant 1$, such that $(T_n)_n$ converges to T with respect to the operator norm in $\mathcal{B}(\mathcal{H}_1, \mathcal{H}_2)$, that is, $\|T_n - T\| \xrightarrow[n \to \infty]{} 0$.*

Let $\mathcal{B}_0(\mathcal{H}_1, \mathcal{H}_2)$ denote the collection of all compact operators $T \colon \mathcal{H}_1 \to \mathcal{H}_2$. Clearly, $\mathcal{B}_0(\mathcal{H}_1, \mathcal{H}_2)$ is a vector subspace of $\mathcal{B}(\mathcal{H}_1, \mathcal{H}_2)$ and, as a consequence of the item (vii) in the previous proposition, it follows that it is closed with respect to the uniform topology (the topology induced by the operator norm).

Given two vectors $x \in \mathcal{H}_1$ and $y \in \mathcal{H}_2$, the linear operator $x \otimes \overline{y} \colon \mathcal{H}_1 \to \mathcal{H}_2$ defined by $(x \otimes \overline{y})z := \langle z, x \rangle y$, for all $z \in \mathcal{H}_1$, is linear and its range is spanned by the vector y, in particular it is bounded. More generally, a linear operator $T \colon \mathcal{H}_1 \to \mathcal{H}_2$ has *finite rank* if the range of T, $\operatorname{Ran}(T) = T\mathcal{H}_1 \subseteq \mathcal{H}_2$, is finite dimensional. Clearly, any finite rank operator is bounded. Also, it is easy to see that, for any operator $T \in \mathcal{B}(\mathcal{H}_1, \mathcal{H}_2)$ with finite rank, there exist $x_1, \ldots, x_n \in \mathcal{H}_1$ and $y_1, \ldots, y_n \in \mathcal{H}_2$ such that

$$T = \sum_{j=1}^{n} x_j \otimes \overline{y}_j. \tag{A.1}$$

Let $\mathcal{B}_{00}(\mathcal{H}_1, \mathcal{H}_2)$ denote the collection of all finite rank operators $T \colon \mathcal{H}_1 \to \mathcal{H}_2$. Clearly, $\mathcal{B}_{00}(\mathcal{H}_1, \mathcal{H}_2)$ is a vector subspace of $\mathcal{B}_0(\mathcal{H}_1, \mathcal{H}_2)$. The decomposition (A.1) can be put in the following equivalent form:

$$T = \sum_{j=1}^{n} \lambda_n x_n \otimes \overline{y}_n, \tag{A.2}$$

where $\{x_1, \ldots, x_n\}$ and $\{y_1, \ldots, y_n\}$ are orthonormal in $\mathcal{H}_1$ and, respectively, in $\mathcal{H}_2$, and $\lambda_1, \ldots, \lambda_n$ are nonzero scalars. Then, it has the following generalisation to compact operators that are not of finite rank.

Proposition A.9.12. *With notation as before, let $T \in \mathcal{B}(\mathcal{H}_1, \mathcal{H}_2)$. The following assertions are equivalent.*

(i) *T is compact and with infinite rank.*

(ii) There exist $(x_n)_n$ an orthonormal sequence in $\mathcal{H}_1$, an orthonormal sequence $(y_n)_n$ in $\mathcal{H}_2$, and a sequence of nontrivial scalars $(\lambda_n)_n$ such that

$$T = \sum_{n=1}^{\infty} \lambda_n x_n \otimes \overline{y}_n, \qquad (A.3)$$

where the series converges in the operator norm.

In particular, $\mathcal{B}_{00}(\mathcal{H}_1, \mathcal{H}_2)$ is dense in $\mathcal{B}_0(\mathcal{H}_1, \mathcal{H}_2)$, with respect to the uniform topology.

The decomposition of compact operators as in (A.3) or (A.2) is called the *Schmidt decomposition* of T.

If $\mathcal{H}$ is a Hilbert space then we denote $\mathcal{B}_0(\mathcal{H}) := \mathcal{B}_0(\mathcal{H}, \mathcal{H})$ and $\mathcal{B}_{00}(\mathcal{H}) := \mathcal{B}_{00}(\mathcal{H}, \mathcal{H})$. Then, $\mathcal{B}_{00}(\mathcal{H}) \subseteq \mathcal{B}_0(\mathcal{H})$ are two sided ideals of $\mathcal{B}(\mathcal{H})$, stable under taking the adjoint, and $\mathcal{B}_{00}(\mathcal{H})$ is dense in $\mathcal{B}_0(\mathcal{H})$. Also, $\mathcal{B}_0(\mathcal{H})$ is a C^*-subalgebra of $\mathcal{B}(\mathcal{H})$ and it does not have a unit, unless $\mathcal{H}$ is finite dimensional.

From the point of view of spectral theory, we have the following result.

Proposition A.9.13. *Let $T \in \mathcal{B}_0(\mathcal{H})$ for some Hilbert space $\mathcal{H}$.*

(a) *$\sigma(T)$ is countable and, if infinite, it accumulates only at 0.*
(b) *Any $\lambda \in \sigma(T) \setminus \{0\}$ is an eigenvalue of T of finite multiplicity.*

For the case of a compact normal operators, the Schmidt decomposition provides a concrete form of the spectral measure, compare with Theorem A.9.10. To see this, let us first observe that any projection of rank one in $\mathcal{H}$ is of the form $x \otimes \overline{x}$, where $x \in \mathcal{H}$ with $\|x\| = 1$.

Proposition A.9.14. *Let $T \in \mathcal{B}(\mathcal{H})$ be a nontrivial compact normal operator. Then, there exists an orthonormal sequence $(x_n)_{n=1}^{N}$, where $N \in \mathbb{N} \cup \{\infty\}$, and a sequence of scalars $(\lambda_n)_{n=1}^{N}$, with $\lambda_n \xrightarrow[n \to \infty]{} 0$ if $N = \infty$, such that*

$$T = \sum_{n=1}^{N} \lambda_n x_n \otimes \overline{x}_n. \qquad (A.4)$$

In particular, $\sigma(T) \subseteq \{\lambda_n \mid n = 1, \ldots, N\} \cup \{0\}$, with equality if either $\mathcal{H}$ is infinite dimensional or $\mathcal{H}$ is finite dimensional and T is not invertible, and for each $n = 1, \ldots, N$, x_n is an eigenvector corresponding to T and the eigenvalue λ_n.

Given a compact operator $T \in \mathcal{B}(\mathcal{H}, \mathcal{K})$, the operator $|T| = (T^*T)^{1/2}$ is a compact positive operator hence its spectrum consists of a finite or infinite sequence of eigenvalues. Let this sequence, called the sequence of *singular numbers* T, be denoted by $(s_n(T))_{n=1}^{N}$, counted with their multiplicities, where $N \in \mathbb{N}$ or $N = \infty$, with $s_{n+1}(T) \leqslant s_n(T)$ for all $n = 1, \ldots, N$. T is a finite rank operator if and only if $N \in \mathbb{N}$. In the case when T is not of finite rank then $s_n(T) \xrightarrow[n \to \infty]{} 0$.

Fredholm Theory

Let $\mathcal{H}$ be a Hilbert space and $T \in \mathcal{B}(\mathcal{H})$. In addition to the spectrum $\sigma(T)$, the point spectrum $\sigma_{\mathrm{p}}(T)$, and the approximate point spectrum $\sigma_{\mathrm{ap}}(T)$, one can define the *left spectrum* $\sigma_{\mathrm{l}}(T) := \{\lambda \in \mathbb{C} \mid \lambda I - T \text{ is not left invertible}\}$ and the *right spectrum* $\sigma_{\mathrm{r}}(T) := \{\lambda \in \mathbb{C} \mid \lambda I - T \text{ is not right invertible}\}$.

Proposition A.9.15. *Let $\lambda \in \mathbb{C}$. The following assertions are equivalent:*

(i) $\lambda \notin \sigma_{\mathrm{ap}}(T)$.
(ii) $\inf\{\|(\lambda I - T)h\| \mid \|h\| = 1\} > 0$.
(iii) $\lambda \notin \sigma_{\mathrm{l}}(T)$.
(iv) $\overline{\lambda} \notin \sigma_{\mathrm{r}}(T)$.
(v) $\operatorname{Ran}(\overline{\lambda} I - T^*) = \mathcal{H}$.

From this proposition we have the following result.

Corollary A.9.16. $\partial \sigma(T) \subseteq \sigma_{\mathrm{l}}(T) \cap \sigma_{\mathrm{r}}(T) = \sigma_{\mathrm{ap}}(T) \cap \overline{\sigma_{\mathrm{ap}}(T^*)}$.

Also, for the case of a normal operator N, that is, $N^*N = NN^*$, we have the following result.

Corollary A.9.17. *If $N \in \mathcal{B}(\mathcal{H})$ is normal then $\sigma(N) = \sigma_{\mathrm{r}}(N) = \sigma_{\mathrm{l}}(N)$. In addition, if λ is an isolated point of $\sigma(N)$ then $\lambda \in \sigma_{\mathrm{p}}(N)$.*

One can introduce Fredholm operators, and their generalisations semi-Fredholm operators, in different ways. The following proposition collects most of these possibilities.

Proposition A.9.18. *Let $T: \mathcal{H} \to \mathcal{H}'$ be a bounded operator, for $\mathcal{H}$ and $\mathcal{H}'$ two Hilbert spaces. The following assertions are equivalent.*

(i) $\operatorname{Ran}(T)$ *is closed and* $\dim(\operatorname{Ker}(T)) < \infty$.
(ii) $BT = I_{\mathcal{H}} + F$ *for some operator* $B \in \mathcal{B}(\mathcal{H}, \mathcal{H}')$ *and finite rank operator* $F \in \mathcal{B}(\mathcal{H})$.
(iii) *For any sequence $(h_n)_n$ of unit vectors in $\mathcal{H}$ such that $h_n \xrightarrow[n \to \infty]{} 0$ weakly, the sequence $(\|Th_n\|)_n$ does not converge to 0.*
(iv) *For any orthonormal sequence $(h_n)_n$ in $\mathcal{H}$ the sequence $(\|Th_n\|)_n$ does not converge to 0.*
(v) $\{h \in \mathcal{H} \mid \|Th\| \leq \delta \|h\|\}$ *does not contain any infinite dimensional linear manifold, for some $\delta > 0$.*
(vi) *Letting E denote the spectral measure of T^*T, the space $E([0, \delta])\mathcal{H}$ is finite dimensional for some $\delta > 0$.*
(vii) *For any compact operator $K \in \mathcal{B}(\mathcal{H}, \mathcal{H}')$ the space $\operatorname{Ker}(T + K)$ has finite dimension.*

The equivalence of the first three statements in the previous proposition is referred to as *Atkinson's Theorem*. An operator T that satisfies any, hence all, of the equivalent statements in the previous proposition is called a *left semi-Fredholm* operator. The operator $T \in \mathcal{B}(\mathcal{H}, \mathcal{H}')$ is called *right semi-Fredholm* if $TA = I_{\mathcal{H}'} + F$ for some compact operator $F \in \mathcal{B}(\mathcal{H}')$. A similar proposition of equivalent characterisations for a right semi-Fredholm operator can be stated. Note that T is left semi-Fredholm if and only if T^* is right semi-Fredholm. An operator $T \in \mathcal{B}(\mathcal{H}, \mathcal{H}')$ that is either right or left semi-Fredholm is called simply a *semi-Fredholm* operator. The operator T is called *Fredholm* if it is both right and left semi-Fredholm.

If T is a semi-Fredholm operator then the *Fredholm index* of T is defined by

$$\mathrm{ind}(T) := \dim(\mathrm{Ker}(T)) - \dim(\mathrm{Ran}(T)^\perp) = \dim(\mathrm{Ker}(T)) - \dim(\mathrm{Ker}(T^*)).$$

Note that $\mathrm{ind}(T) \in \mathbb{Z} \cup \{\pm\infty\}$ and that the definition always makes sense since at least one of the terms is finite.

Remark A.9.19. (1) If $T \in \mathcal{B}(\mathcal{H}, \mathcal{H}')$ and both Hilbert spaces $\mathcal{H}$ and $\mathcal{H}'$ are finite dimensional, then T is Fredholm and $\mathrm{ind}(T) = \dim(\mathcal{H}) - \dim(\mathcal{H}')$.

(2) If $K \in \mathcal{B}(\mathcal{H})$ is compact and $\lambda \in \mathbb{C} \setminus \{0\}$ then $\lambda I - K$ is a Fredholm operator and $\mathrm{ind}(\lambda I - K) = 0$.

(3) If T is a semi-Fredholm operator then T^* is semi-Fredholm and $\mathrm{ind}(T) = -\mathrm{ind}(T^*)$.

(4) If $N \in \mathcal{B}(\mathcal{H})$ is a normal operator then it is semi-Fredholm if and only if it is Fredholm and, in this case, $\mathrm{ind}(N) = 0$.

(5) If T and S are Fredholm operators then $T \oplus S$ is a Fredholm operator and

$$\mathrm{ind}(T \oplus S) = \mathrm{ind}(T) + \mathrm{ind}(S). \blacksquare$$

There are three remarkable theorems about the Fredholm index. The first is referring to products of semi-Fredholm operators and a formula to calculate its index.

Theorem A.9.20. *Assume that* $T\colon \mathcal{H} \to \mathcal{H}'$ *and* $S\colon \mathcal{H}' \to \mathcal{H}''$ *are two left semi-Fredholm operators. Then ST is a left semi-Fredholm operator and* $\mathrm{ind}(ST) = \mathrm{ind}(T) + \mathrm{ind}(S)$.

Corollary A.9.21. *If* $T \in \mathcal{B}(\mathcal{H}, \mathcal{H}')$ *is a left (right) semi-Fredholm operator and* $R \in \mathcal{B}(\mathcal{H}', \mathcal{H})$ *is invertible then RTR^{-1} is a left (right) semi-Fredholm operator and*

$$\mathrm{ind}(RTR^{-1}) = \mathrm{ind}(T).$$

The other two theorems refer to stability of the Fredholm index with respect to compact perturbations and, respectively, to small perturbations.

Theorem A.9.22. *Assume that* $T, K\colon \mathcal{H} \to \mathcal{H}'$ *are bounded linear operators such that T is a left (right) semi-Fredholm operator and K is compact. Then $T + K$ is a left (right) semi-Fredholm operator and* $\mathrm{ind}(T + K) = \mathrm{ind}(T)$.

Theorem A.9.23. *If* $T \in \mathcal{B}(\mathcal{H}, \mathcal{H}')$ *is a Fredholm operator then there exists* $\varepsilon > 0$ *such that, for any* $X \in \mathcal{B}(\mathcal{H}, \mathcal{H}')$ *with* $\|X\| < \varepsilon$, *the operator $T + X$ is Fredholm and* $\mathrm{ind}(T + X) = \mathrm{ind}(T)$.

Unbounded Operators

If $\mathcal{H}$ and $\mathcal{K}$ are two Hilbert spaces, recall that one can consider a more general concept of linear operator in the sense of the domain. So, we say that T is a linear operator from $\mathcal{H}$ to $\mathcal{K}$ if the domain of T, denoted by $\mathrm{Dom}(T)$, is a linear manifold of $\mathcal{H}$ and the range of T, denoted by $\mathrm{Ran}(T)$ is a linear manifold of $\mathcal{K}$, and it is linear in the sense that $T(\alpha h + \beta k) = \alpha T h + \beta T k$ for all $h, k \in \mathcal{H}$ and all scalars α and β. To be more precise, if T is a linear operator from $\mathcal{H}$ to $\mathcal{K}$ we write $T \colon \mathrm{Dom}(T)(\subseteq \mathcal{H}) \to \mathcal{K}$.

Let us assume now that the linear operator T from $\mathcal{H}$ to $\mathcal{K}$ is densely defined. In this case, one can define

$$\mathrm{Dom}(T^*) := \{k \in \mathcal{K} \mid \mathrm{Dom}(T) \ni h \mapsto \langle Th, k \rangle \text{ is a bounded linear functional}\}.$$

Since $\mathrm{Dom}(T)$ is dense in $\mathcal{H}$, for any $k \in \mathrm{Dom}(T^*)$, the linear functional $\mathrm{Dom}(T) \ni h \mapsto \langle Th, k \rangle$ has a unique extension to $\mathcal{H}$ and then, by the Riesz Representation Theorem there exists a unique vector $f \in \mathcal{K}$ such that

$$\langle Th, k \rangle = \langle h, f \rangle, \quad h \in \mathrm{Dom}(T).$$

By definition we let $T^* k := f$. In this way, we define the operator T^* from $\mathcal{K}$ to $\mathcal{H}$ and call it the *adjoint* of T.

Proposition A.9.24. *Let T be a densely defined operator from $\mathcal{H}$ to $\mathcal{K}$.*

(i) *T^* is a closed operator and $(\mathrm{Ran}(T))^\perp = \mathrm{Ker}(T^*)$.*
(ii) *T^* is densely defined if and only if T is closable.*
(iii) *If T is closable then its closure is $T^{**} := (T^*)^*$.*
(iv) *If T is closed then $T^{**} = T$ and $(\mathrm{Ran}(T^*))^\perp = \mathrm{Ker}(T)$.*

In the following it is convenient to introduce the notation $\mathcal{C}(\mathcal{H}, \mathcal{K})$ for the collection of all closed and densely defined operators T from $\mathcal{H}$ to $\mathcal{K}$. If $\mathcal{H} = \mathcal{K}$, we simply denote $\mathcal{C}(\mathcal{H}) = \mathcal{C}(\mathcal{H}, \mathcal{H})$.

Proposition A.9.25. *Let $T \in \mathcal{C}(\mathcal{H})$. Then*

$$\sigma(T^*) = \{\overline{\lambda} \mid \lambda \in \sigma(T)\}, \quad \rho(T^*) = \{\overline{\lambda} \mid \lambda \in \rho(T)\},$$

and, for any $\lambda \in \rho(T)$, we have $\left((\lambda I - T)^*\right)^{-1} = \left((\lambda I - T)^{-1}\right)^*$.

A linear operator A in $\mathcal{H}$ is called *symmetric* or *Hermitian* if $\langle Ah, k \rangle = \langle h, Ak \rangle$ for all $h, k \in \mathcal{H}$. Since we consider only Hilbert spaces over the complex field, the operator A is symmetric if and only if $\langle Ah, h \rangle \in \mathbb{R}$ for all $h \in \mathrm{Dom}(A)$. A linear operator A in $\mathcal{H}$ is called *positive* if $\langle Ah, h \rangle \geq 0$ for all $h \in \mathrm{Dom}(A)$. Then we see that any positive operator is a symmetric operator.

If A is densely defined then A is symmetric if and only if $A \subseteq A^*$, in particular, any symmetric operator is closable, its closure $\overline{A} \subseteq A^*$, and its adjoint is densely defined as well. A densely defined operator A is called *selfadjoint* if $A = A^*$. In this general setting of unbounded operators, neither symmetry nor positivity implies selfadjointness.

Some immediate consequences of the definitions are contained in the following.

Proposition A.9.26. *Let A be a symmetric operator in $\mathcal{H}$.*

  (i) *If $\mathrm{Ran}(A)$ is dense then A is injective.*
 (ii) *If A is selfadjoint and injective then $\mathrm{Ran}(A)$ is dense and A^{-1} is selfadjoint.*
(iii) *If $\mathrm{Dom}(A) = \mathcal{H}$ then $A = A^* \in \mathcal{B}(\mathcal{H})$.*
(iv) *If $\mathrm{Ran}(A) = \mathcal{H}$ and $\mathrm{Dom}(A)$ is dense then A is selfadjoint, injective, and $A^{-1} \in \mathcal{B}(\mathcal{H})$.*

Closed, densely defined, and symmetric operators have special spectral features.

Theorem A.9.27. *Let A be a closed, densely defined, and symmetric operator in $\mathcal{H}$.*

  (i) $\dim(\mathrm{Ker}(\lambda I - A^*))$ *is constant for* $\mathrm{Im}\,\lambda > 0$ *and constant for* $\mathrm{Im}\,\lambda < 0$.
 (ii) *One and only one of the following possibilities holds:*
   (a) $\sigma(A) = \mathbb{C}$;
   (b) $\sigma(A) = \{\lambda \in \mathbb{C} \mid \mathrm{Im}\,\lambda \geqslant 0\}$;
   (c) $\sigma(A) = \{\lambda \in \mathbb{C} \mid \mathrm{Im}\,\lambda \leqslant 0\}$;
   (d) $\sigma(A) \subseteq \mathbb{R}$.

The numbers $n_+ = \dim(\mathrm{Ker}(\lambda I - A^*))$, where $\mathrm{Im}\,\lambda > 0$, in particular for $\lambda = \mathrm{i}$, and $n_- = \dim(\mathrm{Ker}(\lambda I - A^*))$, where $\mathrm{Im}\,\lambda < 0$, in particular for $\lambda = -\mathrm{i}$, are called *indices of defect*. Here the dimensions are Hilbert space dimensions and hence the defect numbers are cardinal numbers.

Corollary A.9.28. *Let A be a closed, densely defined, and symmetric operator in $\mathcal{H}$. The following assertions are equivalent.*

  (i) *A is selfadjoint.*
 (ii) $\sigma(A) \subseteq \mathbb{R}$.
(iii) $n_+ = n_- = 0$.

There is a deep theory on extensions of symmetric operators to selfadjoint operators. The obstruction for such an enterprise is represented by the defect numbers: in order for a symmetric operator A on a Hilbert space $\mathcal{H}$ to admit a selfadjoint extension in $\mathcal{H}$ it is necessary and sufficient that its defect numbers should coincide, $n_+ = n_-$. We recall only the case of positive operators and emphasise the *Friedrichs extension*, also called the *form extension*.

Theorem A.9.29 (Friedrichs Extension). *Let $A \in \mathcal{C}(\mathcal{H})$ be a positive operator and define the operator $\widetilde{A}$ in $\mathcal{H}$ as follows. The domain of $\widetilde{A}$ consists of all vectors $h \in \mathcal{H}$ such that there exists a sequence $(h_n)_n$ of vectors in $\mathcal{H}$ such that*

$$h_n \xrightarrow[n\to\infty]{} h, \quad \langle A(h_n - h_m), h_n - h_m \rangle \xrightarrow[m,n\to\infty]{} 0,$$

and

$$\widetilde{A}h = A^*h, \quad h \in \mathrm{Dom}(\widetilde{A}).$$

Then $\widetilde{A}$ is a positive selfadjoint extension of A in $\mathcal{H}$.

A.9 Linear Operators on Hilbert Spaces

Given T a linear operator from $\mathcal{H}$ to $\mathcal{K}$, the operator T^*T is clearly a positive operator. Some remarkable properties of this operator, under stronger conditions, are gathered in the following result.

Theorem A.9.30. *Let T be a closed and densely defined operator from $\mathcal{H}$ to $\mathcal{K}$.*

(i) *The operator T^*T is positive and selfadjoint.*
(ii) $\mathrm{Dom}(T^*T)$ *is a core of T, that is, T is the closure of $T|\mathrm{Dom}(T^*T)$.*
(iii) $I_\mathcal{H} + T^*T$ *is boundedly invertible and $0 \leq \|(I_\mathcal{H} + T^*T)^{-1}\| \leq I_\mathcal{H}$, in particular $(I_\mathcal{H} + T^*T)^{-1}$ is a contraction.*
(iv) *The operator $T(I_\mathcal{H} + T^*T)^{-1}$ is a contraction.*

An operator N in $\mathcal{H}$ is called *normal* if $N^*N = NN^*$. This definition should be understood in the sense of unbounded operators: $\mathrm{Dom}(N^*N) = \mathrm{Dom}(NN^*)$ and $N^*Nh = NN^*h$ for all $h \in \mathrm{Dom}(N^*N)$. Normal operators have very good spectral properties, in particular they have spectral measures. We confine this discussion to selfadjoint operators.

Firstly we recall the spectral measure variant of the Spectral Theorem for selfadjoint operators.

Theorem A.9.31 (Spectral Theorem for Selfadjoint Operators). *Let A be a selfadjoint operator in $\mathcal{H}$. Then, there exists uniquely a spectral measure E_A with $\mathrm{supp}(E_A) = \sigma(A) \subseteq \mathbb{R}$ such that*

$$\langle Ah, k \rangle = \int_\mathbb{R} t \, \mathrm{d}\langle E_A(t)h, k \rangle, \quad h \in \mathrm{Dom}(A), \ k \in \mathcal{H},$$

and

$$\mathrm{Dom}(A) = \{h \in \mathcal{H} \mid \int_\mathbb{R} t^2 \, \mathrm{d}\langle E_A(t)h, h \rangle < \infty\}.$$

The spectral measure yields a functional calculus with Borelian functions. Let $\mathfrak{B}_\mathbb{C}(\mathbb{R})$ denote the algebra of all Borelian functions $f \colon \mathbb{R} \to \mathbb{C}$ and, for a given selfadjoint operator A in $\mathcal{H}$, with spectral measure E_A, let the operator $f(A)$ be defined as follows:

$$\mathrm{Dom}(f(A)) := \{h \in \mathcal{H} \mid \int_\mathbb{R} |f(t)|^2 \, \mathrm{d}\langle E_A(t)h, h \rangle < \infty\},$$

and

$$\langle f(A)h, k \rangle := \int_\mathbb{R} f(t) \, \mathrm{d}\langle E_A(t)h, k \rangle, \quad h \in \mathrm{Dom}(A), \ k \in \mathcal{H}.$$

Also, in view of Theorem A.9.30, we consider the bounded selfadjoint operator $(I + A^2)^{-1}$ and hence, by the Borelian functional calculus, see Theorem A.9.10, let

$$\mathcal{S}_A := \bigcup_{t>0} \chi_{[-t,t]}((I + A^2)^{-1}).$$

Theorem A.9.32 (Borelian Functional Calculus for Selfadjoint Operators). *With notation as before, the following assertions hold.*

(i) *For every $f \in \mathfrak{B}_\mathbb{C}(\mathbb{R})$ the operator $f(A) \in \mathcal{C}(\mathcal{H})$, is normal, and $\mathcal{S}_A$ is core for $f(A)$, that is, the closure of $f(A)|\mathcal{S}_A$ is $f(A)$.*
(ii) *For any $f \in \mathfrak{B}_\mathbb{C}(\mathbb{R})$, letting $\overline{f}$ be the complex conjugate complex function of f, we have $f(A)^* = \overline{f}(A)$.*
(iii) *For any $f, g \in \mathfrak{B}_\mathbb{C}(\mathbb{R})$ the operator $f(A) + g(A)$ is closable and its closure is $(f + g)(A)$.*
(iv) *For any $f, g \in \mathfrak{B}_\mathbb{C}(\mathbb{R})$ the operator $f(A)g(A)$ is closable,*

$$\mathrm{Dom}(f(A)g(A)) = \mathrm{Dom}((fg)(A)) \cap \mathrm{Dom}(g(A)),$$

and the closure of $f(A)g(A)$ is $(fg)(A)$.
(v) *For any sequence $(f_n)_n$, $f_n \in \mathfrak{B}_\mathbb{C}(\mathbb{R})$ for all $n \in \mathbb{N}$, which is uniformly bounded on compact subsets and pointwise convergent to $f \in \mathfrak{B}_\mathbb{C}(\mathbb{R})$, we have*

$$f(A)h = \lim_{n \to \infty} f_n(A)h, \quad h \in \mathcal{S}_A.$$

In the following we record some important consequences of the Spectral Theorem and the functional calculus with Borelian functions. Firstly we recall some order relations that occur in the Borelian functional calculus.

Corollary A.9.33. *With notation as before, let $f, g \in \mathfrak{B}_\mathbb{C}(\mathbb{R})$ such that $|f| \leq |g|$ on $\sigma(A)$. Then*

$$\mathrm{Dom}(g(A)) \subseteq \mathrm{Dom}(f(A)) \text{ and } \|f(A)h\| \leq \|g(A)h\|, \quad h \in \mathrm{Dom}(g(A)).$$

In particular, if f is bounded then $f(A) \in \mathcal{B}(\mathcal{H})$ and

$$\|f(A)\| \leq \sup\{|f(\lambda)| \mid \lambda \in \sigma(A)\}.$$

The next corollary records the fact that certain properties of the functions are reflected by the corresponding operators.

Corollary A.9.34. *With notation as before let $f \in \mathfrak{B}_\mathbb{C}(\mathbb{R})$.*

(i) (Selfadjoint Operators) *If f is real then $f(A)$ is selfadjoint.*
(ii) (Positive Operators) *If f is positive then $f(A)$ is a positive selfadjoint operator.*
(iii) (Orthogonal Projections) *If $f = \chi_S$ is the characteristic function of some Borel set $S \subseteq \mathbb{R}$ then $f(A)$ is an orthogonal projection. In particular, $\chi_{\sigma(A)}$ is the orthogonal projection on $\mathrm{Clos}\,\mathrm{Ran}(A)$.*
(iv) (Unitary Operators) *If $|f| = 1$ then $f(A)$ is a unitary operator.*

The Borelian functional calculus allows us to generalise the modulus, the polar decomposition, and the Jordan decomposition, for unbounded operators.

Corollary A.9.35. (a) (Square Root) *Let $T \in \mathcal{C}(\mathcal{H}, \mathcal{K})$. Then there exists a unique positive selfadjoint operator, denoted by $|T| \in \mathcal{C}(\mathcal{H})$, such that $|T|^2 = T^*T$.*

(b) (Polar Decomposition) *Let $T \in \mathcal{C}(\mathcal{H}, \mathcal{K})$. Then there exists uniquely a partial isometry $V \in \mathcal{B}(\mathcal{H}, \mathcal{K})$ such that $T = V|T|$, V^*V is the orthogonal projection onto $\mathcal{H} \ominus \mathrm{Ker}(T)$ and VV^* is the orthogonal projection onto $\mathrm{Clos}\,\mathrm{Ran}(T)$.*

(c) (Jordan Decomposition) *For any selfadjoint operator A on $\mathcal{H}$ there exist uniquely two positive selfadjoint operators $A_{\pm}$ on $\mathcal{H}$ such that $A = A_+ - A_-$, $A_+ A_- = 0$, and $|A| = A_+ + A_-$.*

With notation as in item (c) of the previous corollary, Jordan decomposition is a special case of polar decomposition. More precisely, letting $A = V|A|$ be the polar decomposition of A, then $V \in \mathcal{B}(\mathcal{H})$ is a selfadjoint partial isometry, hence $V = P_+ - P_-$, where $P_{\pm}$ are orthogonal projections, $P_+ P_- = 0$, and $P_+ + P_-$ is the orthogonal projection on $\mathrm{Clos}\,\mathrm{Ran}(A)$. Then, $A_{\pm} = P_{\pm} A$.

References

[1] V. M. ADAMYAN, D. Z. AROV, AND M. G. KREĬN: Analytic properties of Schmidt pairs of Hankel operators and the generalized Schur-Takagi problem [Russian], *Mat. Sbornik*, **86**(1971), 33–73.

[2] V. M. ADAMYAN, D. Z. AROV, AND M. G. KREĬN: Infinite block Hankel operators and some related continuation problems [Russian], *Izv. Akad. Nauk Armyan. SSR Ser. Mat.*, **6**(1971), 87–112.

[3] N. I. AKHIEZER AND I. M. GLAZMAN: *The Theory of Operators in Hilbert Space* [Russian], 3rd revised edition, Vyshcha Shkola, Kharkov 1978; English translation., Pitman, Boston, MA 1981.

[4] T. ANDO: *Linear Operators in Kreĭn Spaces*, Lecture Notes, Hokkaido University, Sapporo 1979.

[5] R. AROCENA, T. YA. AZIZOV, A. DIJKSMA, AND S. A. M. MARCANTOGNINI: On commutant lifting with finite defect. *J. Operator Theory*, **35**(1996), no. 1, 117–132.

[6] R. AROCENA, T. YA. AZIZOV, A. DIJKSMA, AND S. A. M. MARCANTOGNINI: On commutant lifting with finite defect. II. *J. Funct. Anal.*, **144**(1997), no. 1, 105–116.

[7] GR. ARSENE, T. CONSTANTINESCU, AND A. GHEONDEA: Lifting of operators and prescribed negative squares, *Michigan Math. J.*, **34**(1987), 201–216.

[8] GR. ARSENE AND A. GHEONDEA: Completing matrix contractions, *J. Operator Theory*, **7**(1982), 179–189.

[9] T. YA. AZIZOV: On the Theory of Isometric and Symmetric Operators in Spaces with an Indefinite Metric [Russian], Deposited VINITI, **29**(1982), 3420–3482.

[10] T. YA. AZIZOV AND I. S. IOKHVIDOV: *Foundations of the Theory of Linear Operators in Spaces with Indefinite Metric* [Russian], Nauka, Moscow 1986; English translation, *Linear Operators in Spaces with Indefinite Metric*, Wiley, New York 1989.

[11] J. A. BALL AND J. W. HELTON: Factorization results related to shifts in an indefinite metric, *Integral Equations Operator Theory*, **5**(1982), 632–658.

[12] J. A. BALL AND J. W. HELTON: A Beurling-Lax theorem for the Lie group $U(m,n)$ which contains most classical interpolation theory, *J. Operator Theory*, **9**(1983), 107–142.

[13] J. A. BALL AND J. W. HELTON: Beurling-Lax representations using classical Lie groups with many applications. II: $GL(n, \mathbb{C})$ and the Wiener-Hopf factorization, *Integral Equations Operator Theory*, **7**(1984), 291–309.

[14] J. A. BALL AND J. W. HELTON: Interpolation problems of Pick-Nevanlinna and Loewner types for meromorphic matrix functions: Parametrizations of the set of all solutions, *Integral Equations and Operator Theory*, **9**(1985), 105–203.

REFERENCES

[15] J. A. BALL AND J. W. HELTON: Beurling-Lax representations using classical Lie groups with many applications. III: Groups preserving two bilinear forms, *Am. J. Math.*, **108**(1986), 95–174.

[16] J. A. BALL AND J. W. HELTON: Beurling-Lax representations using classical Lie groups with many applications. IV: $GL(n, \mathbb{R})$, $SL(n, \mathbb{C})$, and a solvable group, *J. Funct. Anal.*, **69**(1986), 178–206.

[17] A. BEURLING: On two problems concerning linear transformations in Hilbert space, *Acta Math.*, **81**(1949), 239–255.

[18] J. BOGNÁR: *Indefinite Inner Product Spaces*, Springer-Verlag, Berlin 1974.

[19] J. BOGNÁR: A proof of the spectral theorem for J-positive operators, *Acta Sci. Math. (Szeged)*, **45**(1983), 75–80.

[20] J. BOGNÁR: An approach to the spectral decomposition of J-positizable operators, *J. Operator Theory* **17**(1987), no. 2, 309–326.

[21] V. I. BOGACHEV: *Measure Theory*, Springer-Verlag, Berlin 2007.

[22] F. F. BONSALL AND J. DUNCAN: *Complete Normed Algebras*, Springer-Verlag, Berlin 1973.

[23] N. BOURBAKI: *Topologie générale*, ch. I, IX–X, Hermann, Paris 1961.

[24] N. BOURBAKI: *Intégration*, ch. VII–VIII, Hermann, Paris 1961.

[25] N. BOURBAKI: *Topological Vector Spaces*, Elements of Mathematics, ch. I–V, Springer-Verlag, Berlin 1987.

[26] C. CARATHÉODORY: Über den Variabillitätsbereich der Koeffizienten von Potenzreihen, die gegebene Werte nicht annehmen, *Math. Ann.*, **64**(1907), 95–115.

[27] I. COLOJOARĂ AND C. FOIAŞ: *Theory of Generalized Spectral Operators*, Mathematics and its Applications, Vol. 9, Gordon and Breach Science Publishers, New York 1968.

[28] T. CONSTANTINESCU AND A. GHEONDEA: Notes on (the Birmak-Krein-Vishik theory on) selfadjoint extensions of semibounded symmetric operators, in *Two papers on selfadjoint extensions of symmetric semibounded operators*, INCREST Preprint Series, July 1981, Bucharest, Romania, arXiv:1807.05363 [math.FA].

[29] T. CONSTANTINESCU AND A. GHEONDEA: On unitary dilation and characteristic functions in indefinite inner product spaces, in *Operator Theory: Advances and Applications*, Vol. 24, Birkhäuser, Basel 1987, pp. 87–102.

[30] T. CONSTANTINESCU AND A. GHEONDEA: Extending factorizations and minimal negative signatures, *J. Operator Theory*, **28**(1992), 371–402.

[31] T. CONSTANTINESCU AND A. GHEONDEA: Elementary rotations of linear operators in Kreĭn spaces, *J. Operator Theory*, **29**(1993), 167–203.

[32] T. CONSTANTINESCU AND A. GHEONDEA: Representations of Hermitian kernels by means of Krein spaces, *Publ. RIMS Kyoto Univ.*, **33**(1997), 917–951.

[33] T. CONSTANTINESCU, A. GHEONDEA: On a Nehari type problem on spaces with indefinite inner product, *Rev. Roum. Math. Pures App.* **43**(1998), no. 3–4, 329–354.

[34] D. L. COHN: *Measure Theory*, Birkhäuser, New York 2013.

[35] J. B. CONWAY: *Functions of One Complex Variable. I*, 2nd edition, Springer-Verlag, Berlin 1978.

[36] J. B. CONWAY: *A Course in Functional Analysis*, Springer-Verlag, Berlin 1990.

[37] J. B. CONWAY: *Functions of One Complex Variable. II*, Springer-Verlag, Berlin 1995.

[38] J. B. CONWAY: *A Course in Operator Theory*, American Mathematical Society, Providence, RI 2000.

[39] B. ĆURGUS: On the regularity of the critical point infinity of definitizable operators, *Integral Equations Operator Theory*, **8**(1985), 462–488.

[40] B. ĆURGUS, A. GHEONDEA, AND H. LANGER: On singular critical points of positive operators in Krein spaces, *Proc. Am. Math. Soc.*, **128**(2000), 2621–2626.

[41] B. ĆURGUS AND H. LANGER: Continuous embeddings, completions and complementation in Krein spaces, *Rad. Mat.*, **12**(2003), 37–79.

[42] J. DIEUDONNÉ: Quasi-Hermitian operators, in *Proc. Int. Symp. on Linear Spaces (Jerusalem, 1960)*, Jerusalem Academic Press, Jerusalem and Pergamon, Oxford 1961, pp. 115–122.

[43] A. DIJKSMA AND A. GHEONDEA: Index formulae for subspaces of Kreĭn spaces, *Integral Equations Operator Theory*, **25**(1996), no. 1, 58–72.

[44] A. DIJKSMA, H. LANGER, AND H. S. V. DE SNOO: Unitary colligations in Kreĭn spaces and their role in the extension theory of isometric and symmetric linear relations in Hilbert spaces, in *Functional Analysis II, Proceedings Dubrovnik 1985*, Lecture Notes in Mathematics, Vol. 1242, Springer-Verlag, Berlin 1987, pp. 123–143.

[45] A. DIJKSMA, H. LANGER, AND H. S. V. DE SNOO: Representations of holomorphic operator functions by means of resolvents of unitary or selfadjoint operators in Kreĭn spaces, in *Operators in Indefinite Metric Spaces, Scattering Theory and Other Topics*, Birkhäuser, Basel 1987, pp. 123–143.

[46] J. DIXMIER: *Les C^*-algébres et leurs représentations*, Gauthier-Villars, Paris 1964.

[47] J. DIXMIER: Les moyennes invariantes dans les semi-groupes et leurs applications, *Acta Sci. Math. (Szeged)*, **12**(1950), 213–227.

[48] M. A. DRITSCHEL: *Extension Theorems for Operators in Kreĭn Spaces*, Dissertation, University of Virginia 1989.

[49] M. A. DRITSCHEL: A lifting theorem for bicontractions, *J. Funct. Anal.*, **88**(1990), 61–89.

[50] M. A. DRITSCHEL: The essential uniqueness property for linear operators on Kreĭn spaces, *J. Funct. Anal.*, **118**(1993), 198–248.

[51] M. A. DRITSCHEL: A method for constructing invariant subspaces for some operators on Kreĭn spaces, in *Operator Extensions, Interpolation of Functions and Related Topics*, 14th International Conference on Operator Theory, Timişoara (Romania), June 1–5, 1992, Oper. Theory, Adv. Appl., Vol. 61, Birkhäuser, Basel 1993, pp. 85–113.

[52] M. A. DRITSCHEL AND J. ROVNYAK: Extension theorems for contraction operators in Kreĭn spaces, in *Operator Theory: Advances and Applications*, Vol. 47, Birkhäuser Basel 1990, pp. 221–305.

[53] M. A. DRITSCHEL AND J. ROVNYAK: Operators on indefinite inner product spaces, in *Lectures on Operator Theory and Its Applications*, Fields Institute Monographs, Vol. 3, American Mathematical Society, Providence, RI 1996, pp. 143–232.

[54] N. DUNFORD AND J. T. SCHWARTZ: *Linear Operators. Part I. General Theory*, Interscience, New York 1958.

[55] N. DUNFORD AND J. T. SCHWARTZ: *Linear Operators. Part II. Spectral Theory. Selfadjoint Operators in Hilbert Space*, Interscience, New York 1963.

[56] K. FAN: Fixed-point and minimax theorems in locally convex topological linear spaces, *Proc. Natl. Acad. Sci. USA*, **38**(1952), 121–126.

[57] C. FOIAŞ AND A. E. FRAZHO: *The Commutant Lifting Approach to Interpolation Problems*, Birkhäuser, Basel 1990.

[58] A. GHEONDEA: On the geometry of pseudo-regular subspaces of a Kreĭn space, in *Operator Theory: Advances and Applications*, Vol. 14, Birkhäuser, Basel 1984, pp. 141–156.

[59] A. GHEONDEA: Canonical forms of unbounded unitary operators in Kreĭn spaces, *Publ. RIMS Kyoto Univ.*, **24**(1988), 205–224.

[60] A. GHEONDEA: A geometric question concerning strong duality of neutral subspaces, *Math. Balkanica*, **3**(1989), 183–195.

[61] A. GHEONDEA: Quasi-contractions in Kreĭn spaces, in *Operator Theory: Advances and Applications*, Vol. 61, Birkhäuser, Basel 1993, pp. 123–148.

[62] A. GHEONDEA: Contractive intertwining dilations of quasi-contractions, *Z. Anal. Anwendungen*, **15**(1996), no. 1, 31–44.

[63] A. GHEONDEA: On generalized interpolation and shift invariant maximal semidefinite subspaces, in *Operator Theory: Advances and Applications*, Vol. 104, Birkhäuser, Basel 1998, pp.121–136.

[64] A. GHEONDEA AND P. JONAS: A characterization of spectral functions of definitizable operators, *J. Operator Theory*, **17**(1987), 99–110.

[65] YU. P. GINZBURG: On J-contractive operator functions [Russian], *Dokl. Akad. Nauk. SSSR*, **193**(1970), 1218–1221.

[66] YU. P. GINZBURG: Projections in a Hilbert space with a bilinear metric [Russian], *Dokl. Akad. Nauk. SSSR*, **139**(1961), 775–778; English translation *Soviet Math. Dokl.*, **2**(1961), 980–983.

[67] I. L. GLICKSBERG, A further generalization of the Kakutani fixed theorem, with application to Nash equilibrium points, *Proc. Am. Math. Soc.*, **3**(1952) 170–174.

[68] I. GOHBERG, S. GOLDBERG, AND M. A. KAASHOEK, *Classes of Linear Operators*, Vol. I, Springer, Basel 1990.

[69] I. GOHBERG, S. GOLDBERG, AND M. A. KAASHOEK, *Classes of Linear Operators*, Vol. II, Springer, Basel 1993.

[70] I. GOHBERG, P. LANCASTER, AND L. RODMAN: *Matrices in Indefinite Scalar Products*, Birkhäuser, Basel 1983.

[71] F. P. GREENLEAF: *Invariant Means on Topological Groups and their Applications*, Van Nostrand Reinhold, New York 1969.

[72] T. HARA: Operator inequalities and construction of Kreĭn space, *Integral Equations Operator Theory*, (1992), 551–567.

[73] F. HANSEN: Selfpolar norms on an indefinite inner product space, *Publ. RISM Kyoto Univ.*, **16**(1980), 401–414.

[74] G. HERGLOTZ: Über Pontezreihen mit positivem, reellem Teil im Eniheitskreis, *Leipz. Ber.*, **63**(1911), 501–511.

[75] G. HERGLOTZ, I. SCHUR, G. PICK, R. NEVANLINNA, H. WEYL: *Ausgewählte Arbeiten zu den Ursprängen der Schur-Analysis* [Selected Works on the Origins of Schur Analysis] Gewidmet dem groß en Mathematiker Issai Schur (1875–1941) [Dedicated to the great mathematician Issai Schur (1875–1941)], Teubner-Archiv zur Mathematik [Teubner Archive on Mathematics], B. G. Teubner, Stuttgart 1991.

[76] I. S. IOKHVIDOV: On a lemma of Ky Fan generalizing the fixed-point principle of A.N. Tikhonov [Russian], *Dokl. Akad. Nauk. SSSR*, **159**(1964), 501–504.

[77] I. S. IOKHVIDOV AND M. G. KREĬN: Spectral theory of operators in indefinite metric. I, *Trud. Mosk. Mat. Obshch.*, **5**(1956), 367–432; English translation, *Am. Math. Soc. Transl.* (2), **13**(1960), 105–175.

[78] I. S. IOKHVIDOV AND M. G. KREĬN: Spectral theory of operators in indefinite metric. II, *Trud. Mosk. Mat. Obshch.*, **9**(1959), 413–496; English translation, *Am. Math. Soc. Transl.* (2), **34**(1963), 283–373.

[79] I. S. IOKHVIDOV, M. G. KREĬN, AND H. LANGER: *Introduction to the Spectral Theory of Operators in Spaces with an Indefinite Metric*, Akademie-Verlag, Berlin 1983.

[80] P. JONAS: Zur Existenz von Eigenspektralfunktionen mit Singulartitäten, *Math. Nachr.*, **88**(1977), 345–361.

[81] P. JONAS: On the functional calculus and the spectral function for definitizable operators in Kreĭn space, *Beitr. Anal.*, **16**(1981), 121–135.

[82] P. JONAS: On spectral distributions of definitizable operators in Kreĭnspace, in *Spectral Theory*, Vol. 8 of Banach Center Publications, PWN Polish Scientific Publishers, Warsaw 1982.

[83] P. JONAS: On locally definite operators in Krein spaces, *Spectral Analysis and its Applications*, Theta Ser. Adv. Math., 2, Theta, Bucharest 2003, pp. 95–127.

[84] T. KATO: *Perturbation Theory of Linear Operators*, Springer-Verlag, Berlin 1966.

[85] M. KALTENBÄCK: Spectral theorem for definitizable normal linear operators on Krein spaces, *Integral Equations Operator Theory*, **85**(2016), 221–243.

[86] M. KALTENBÄCK: Definitizability of normal operators on Krein spaces and their functional calculus, *Integral Equations Operator Theory*, **87**(2017), 461–490.

[87] M. KALTENBÄCK AND N. SKREPEK: Joint functional calculus for definitizable self-adjoint operators on Krein spaces, *Integral Equations Operator Theory*, **92**(2020), no. 4, Paper no. 29, 36 p.

[88] J. L. KELLEY: *General Topology*, Springer-Verlag, Berlin 1955.

[89] B. KNASTER, C. KURATOWSKI, AND S. MAZURKIEWICZ: Ein Beweis des Fixpunktsatzes für n-dimensionale Simplexe, *Fund. Math.*, **14**(1929), 132–137.

[90] P. KOOSIS: *Introduction to H^p Theory*, London Math. Soc. Lecture Note Ser., Vol. 40, Cambridge University Press, New York 1980.

REFERENCES

[91] M. G. KREĬN: Completely continuous linear operators in function spaces with two norms [Ukrainian], *Akad. Nauk Ukrain. RSR Zbirnik Prac. Inst. Mat.*, **9**(1947), 104-129.

[92] M. G. KREĬN: The theory of selfadjoint extensions of semibounded Hermitian transformations and its applications. I [Russian], *Mat. Sbornik* **20**(1947), 431–495.

[93] M. G. KREĬN: Introduction to the geometry of indefinite inner product J-spaces and to the theory of operators in those spaces, in *Second Math. Summer School* [Russian], part I, Naukova Dumka, Kiev, 1965, pp. 15–92; English translation, *Am. Math. Soc. Transl.*, (2) **93** (1970), 103–176.

[94] M. G. KREĬN AND H. LANGER: On the spectral function of selfadjoint operators in spaces with indefinite metric [Russian], *Dokl. Akad. Nauk*, **152**(1963), 39–42.

[95] M. G. KREĬN AND H. LANGER: Über die veralgemeinerte Resolventen und die characteristische Funktion eines isometrischen Operators im Raume Π_κ, In: *Colloquia Math. Soc. János Bolyai, Vol. 5, Hilbert Space Operators and Operator Algebras*, North-Holland, Amsterdam 1972, pp. 353–399.

[96] M. G. KREĬN AND H. LANGER: Über einige Fortsetzunsprobleme, die eng mit der Theorie hermitescher Operatoren im Raume Π_κ zusammenhängen. I. Einige Funktionenklassen und ihre Darstellungen, *Math. Nachr.*, **77**(1977), 187–236.

[97] M. G. KREĬN AND H. LANGER: Über einige Fortsetzunsprobleme, die eng mit der Theorie hermitescher Operatoren im Raume Π_κ zusammenhängen. II. Veralgemeinerte Resolventen u-Resolventen und ganze Operatoren, *J. Funct. Anal.*, **30**(1978), 390–447.

[98] M. G. KREĬN AND H. LANGER: On some extension problems which are closely connected with the theory of Hermitian operators in a space Π_κ. III. Indefinite analogues of the Hamburger and Stieltjes moment problem, *Beitr. Anal.*, Part (I), **14**(1979), 25-40; Part (II), **15**(1981), 27–45.

[99] M. G. KREĬN AND H. LANGER: On some propositions on analytic matrix functions related to the theory of operators in the spaces Π_κ, *Acta Sci. Math. (Szeged)*, **43**(1981), 181–205.

[100] M. G. KREĬN AND YU. L. SHMULYAN: Plus-operators in a space with an indefinite metric [Russian], *Mat. Issled.*, **1**(1966), 131–161.

[101] M. G. KREĬN AND YU. L. SHMULYAN: J-Polar representations of plus-operators [Russian], *Mat. Issled.*, **1**(1966), 172–210.

[102] M. G. KREĬN AND YU. L. SHMULYAN: On a class of operators in a space with an indefinite metric [Russian], *Dokl. Akad. Nauk USSR*, **170**(1966), 34–37.

[103] M. G. KREĬN AND YU. L. SHMULYAN: Linear fractional transformations with operator coefficients [Russian], *Mat. Issled.*, **2**(1967), 64–96.

[104] H. LANGER: *Spektraltheorie linearer Operatoren in J-Räumen und einige Anwendungen auf den Schar $L(\lambda) = \lambda^2 + \lambda B + C$*, Habilitationsschrift, Dresden 1965.

[105] H. LANGER: Invariante Teilräume definisierbarer J-selbstajungierter Operatoren, *Ann. Acad. Sci. Fenn. Ser. A I*, **475**(1971).

[106] H. LANGER: Spectral functions of definitizable operators in Kreĭn spaces, in *Functional Analysis*, Proceedings, Dubrovnik 1981, Lecture Notes in Mathematics Vol. 948, Springer-Verlag, Berlin 1982, pp. 1–46.

[107] H. LANGER, A. MARKUS, AND V. MATSAEV: Locally definite operators in indefinite inner product spaces, *Math. Ann.*, **308**(1997), 405–424.

[108] P. D. LAX: Symmetrizable linear transformations, *Commun. Pure. Appl. Math.*, **7**(1954), 633–647.

[109] P. D. LAX: Translation invariant spaces, *Acta Math.*, **101**(1959), 163–178.

[110] Z. NEHARI: On bounded bilinear forms, *Ann. Math.*, **65**(1957), 153–162.

[111] R. NEVANLINNA: Sur un probléme d'interpolation, *C. R. Acad. Sci. Paris* **188**(1929), 1224–1226.

[112] R. NEVANLINNA: Erweiterung der Theorie des Hilbertschen Raumes, *Comm. Sem. Math. Univ. Lund.*, Tome Supplementaire, (1952), 160–168.

[113] R. NEVANLINNA: Über metrische lineare Räume. II. Bilineareformen und Stetigkeit. *Ann. Acad. Sci. Fenn. Ser. A I*, **113**(1952).

[114] R. NEVANLINNA: Über metrische lineare Räume. III. Theorie der Orhogonalsysteme. *Ann. Acad. Sci. Fenn. Ser. A I*, **115**(1952).

[115] R. NEVANLINNA: Über metrische lineare Räume. IV. Zur Theorie der Unterräume. *Ann. Acad. Sci. Fenn. Ser. A I*, **163**(1952).

[116] G. PEDERSEN: *Analysis Now*, Springer-Verlag, Berlin 1989.

[117] G. PICK: Über die Beschränkungen analytischer Funktionen, welche durch vorgegebene Funktionswerte bewirkt werden, *Math. Ann.*, **77**(1916), 7–23.

[118] R. S. PHILLIPS: Dissipative operators and parabolic partial differentioal equations, *Commun. Pure Appl. Math.*, **12**(1959), 249–276.

[119] R. S. PHILLIPS: Dissipative operators and hyperbolic systems of partial differential equations, *Trans. Am. Math. Soc.*, **90**(1959), 193-254.

[120] R. S. PHILLIPS: The extension of dual subspaces invariant under and algebra, on *Proc. Int. Symp. Linear Spaces*, Jerusalem Academic Press, Jerusalem and Pergamon, Oxford 1961, pp. 366–398.

[121] L. S. PONTRYAGIN: Hermitian operators in spaces with indefinite metric [Russian], *Izv. Akad. Nauk. SSSR Ser. Mat.*, **8**(1944), 243–236.

[122] V. P. POTAPOV: The multiplicative structure of J-contractive analytic matrix functions [Russian], *Trudy Mosk. Mat. Obshch.*, **4**(1955), 125–236; English translation, *Am. Math. Soc. Transl. (2)*, **15**(1960), 131–243.

[123] S. C. POWER: *Hankel Operators on Hilbert Space*, Pitman, Boston, MA 1982.

[124] W. T. REID: Symmetrizable completely continuous linear transformations in Hilbert space, *Duke Math. J.*, **18**(1951), 41–56.

[125] C. E. RICKART: *General Theory of Banach Algebras*, Van Nostrand, Princeton, NJ 1960.

[126] M. ROSENBLUM AND J. ROVNYAK: *Hardy Classes and Operator Theory*, Oxford Math. Monographs, Oxford University Press, New York 1985.

[127] W. RUDIN: *Real and Complex Analysis*, McGraw-Hill, Singapore 1987.

[128] W. RUDIN: *Functional Analysis*, 2nd edition, Springer-Verlag, Berlin 1991.

[129] R. A. RYAN: *Introduction to Tensor Products of Banach Spaces*, Springer-Verlag, London 2002.

[130] D. SARASON: Generalized interpolation in H^∞, *Trans. Am. Math. Soc.*, **127**(1967), 179–203.

[131] H. H. SCHAEFER: *Topological Vector Spaces*, Springer-Verlag, Berlin 1970.

[132] K. SCHMÜDGEN: *Unbounded Self-adjoint Operators on Hilbert Space*, Springer, Dordrecht 2012.

[133] I. SCHUR: Über Pontezreihen, die im Innern des Einheitskreises beschränkt sind, *J. Reine Angew. Math.*, **117**(1917), 205–232.

[134] L. SCHWARTZ: Sous espaces Hilbertiens d'espace vectoriel topologiques et noyaux associés (noyaux reproduisants), *J. Anal. Math.*, **13**(1973), 115–256.

[135] YU. L. SHMULYAN: Theory of linear relations and spaces with indefinite metric [Russian], *Funkts. Anal. Pril.* **10**(1976).

[136] S. L. SOBOLEV: The motion of a symmetric top cavity filled with a liquid [Russian], *Ž. Prikl. Meh. i Tehn. Fiz.*, **3**(1960), 20–55.

[137] T. J. STIELTJES: Recherches sur les fractions continues, *Ann. Fac. Sci. Toulouse*, **8**(1894), 1–22.

[138] V. STRAUSS: Models of function type for commutative symmetric operator families in Kreĭn spaces, *Abstr. Appl. Anal.*, 2008, Article ID 439781, 40 p.

[139] B. SZ.-NAGY: On uniformly bounded linear transformations in Hilbert space, *Acta Sci. Math.*, **11**(1947), 152–157.

[140] B. SZ.-NAGY AND C. FOIAŞ: Commutants de certains opérateurs, *Acta. Math. (Szeged)*, **29**(1968), 1–17.

[141] B. SZ.-NAGY AND C. FOIAŞ: *Harmonic Analysis of Operators on Hilbert Space*, North-Holland, New York 1970.

[142] M. TOMITA: Operators and operator algebras in Kreĭn spaces, *RIMS Kokiuuroku*, **398**(1980), 131–158.

[143] S. TREIL AND A. VOLBERG: A fixed point approach to Nehari's problem and its applications, in *Operator Theory: Advances and Applications*, Vol. 71, Birkhäuser, Basel 1994, pp. 165–186.

[144] N. J. YOUNG: Orbits of the unit sphere of $\mathcal{L}(\mathcal{H},\mathcal{K})$ under symplectic transformations, *J. Operator Theory*, **11**(1984), 171–191.

[145] N. J. YOUNG: J-Unitary equivalence of positive subspaces of a Kreĭn space, *Acta Sci. Math. (Szeged)*, **47**(1984), 107–111.

Symbol Index

A_r, 161
A_r^c, 161
$\overset{\circ}{A}$, 401
$\mathcal{A}^\perp$, 4
$\mathcal{A}^{\perp_J}$, 20
a^*, 462

$\mathcal{B}$, 409, 425
$\mathcal{B}(\mathcal{H})$, 7
$\mathfrak{B}$, 248, 367

C_F, 161
C_I, 269
$\mathcal{C}(X)$, 408
$\mathbb{C}$, 1
$\mathbb{C}[X]$, 292
$\hat{\mathbb{C}}$, 445
CID_κ, 266
$\mathrm{Clos}_\mathrm{w}\mathcal{L}$, 12
$\mathrm{Clos}\,A$, 401
Conv, 236
$c(A)$, 314
$c_\mathrm{r}(A)$, 370
$c_\mathrm{s}(A)$, 370
$\tilde{c}_\mathrm{c}$, 370
$\tilde{c}_\mathrm{r}$, 370

D_T, 36, 173
$\mathcal{D}_T$, 36
$\mathbb{D}_r$, 146
$\mathbb{D}_r^c$, 146
$\mathfrak{D}$, 360
Dom, 464

d_∞, 408
$\dim(\mathcal{L})$, 6
$\dim^\mathrm{h}$, 34
δ_z, 148
$\delta_{j,k}$, 433

E, 316
$\mathcal{E}$, 420

$\mathcal{F}_0(\mathbb{D}_r^c; \mathcal{H})$, 148
$\mathbb{F}[X]$, 292

G, 14
$\mathcal{G}_\mathrm{s}$, 121
$\mathcal{G}_\mathrm{w}$, 121
Γ_0, 150, 166
Γ_1, 167

$\mathcal{H}_A$, 141

Ind_γ, 440
i, 2
$\mathrm{ind}(T)$, 477
ind_+, 79
ind_-, 81

J^+, 17
J^-, 17
J_T, 173

$\mathcal{K}_A$, 140
Ker, 469
κ, 7
κ^+, 34, 35

Symbol Index

κ^-, 34, 35
κ^0, 35
κ_+, 7
κ_-, 7
κ_0, 7

$L^2(X,\mu)$, 431
L_T, 175
$\mathcal{L}^0$, 4
$\mathcal{L}_0$, 5
Lin, 124
ℓ^2, 15

$M(T)$, 188
$M/\sim$, 406
$M^{\mathrm{u}}(T)$, 193
$\mu_+(T)$, 188
$\mu_-(T)$, 188
μ_ω, 186

N_Q, 147
N_α, 321
N_κ, 255
NP_l, 274
$\nu_\pm(T)$, 83

$P_r(\varphi)$, 297
Φ_T, 217
Π, 84, 137
Π_A, 140
π_A, 140

$R(T)$, 36, 177
$R(\zeta; A)$, 304
$\mathbb{R}[X]$, 293
$\mathfrak{R}_A$, 314
Ran, 464, 469
$\rho(T)$, 89

ρ_A, 283
$\widetilde{\rho}$, 107

$S^\perp$, 433
S_Θ, 172
S_{JA}, 140
$\mathcal{S}_I$, 267
$\mathfrak{S}$, 111
$\mathfrak{S}_\lambda$, 279
$\mathrm{Sp}(\mathcal{A})$, 459
sgn, 175
σ_{c}, 89
σ_{p}, 89
σ_{r}, 89
$\widetilde{\sigma}_{\mathrm{p}}$, 107

T^*, 31, 469
$T^\sharp$, 82, 87
$\mathcal{T}$, 401
$\mathcal{T}_\sim$, 406

$\mathcal{V}(x)$, 401

$X \otimes Y$, 429
$\mathcal{X}$, 1
$\mathcal{X}_A$, 7
$\mathcal{X}_{\mathrm{g}}$, 103
$\widehat{\mathcal{X}}$, 4
$x \otimes \overline{y}$, 474

$Z(p)$, 292

$\dotplus$, 4
$\langle \cdot, \cdot \rangle$, 16, 431
$\ll$, 417
$\oplus$, 433
$\otimes$, 418, 429
$\perp$, 3

$\perp_J$, 20
$\preceq$, 405
$\times$, 406
$\widehat{\otimes}_\varepsilon$, 430
$\widehat{\otimes}_\pi$, 430
$[+]$, 20
$[\cdot,\cdot]$, 1
$[\cdot,\cdot]_A$, 7
$[\perp]$, 4
$\bigvee$, 120
$\bigwedge$, 120

Index

adjoint
    operator, 85
    relation, 81
algebra
    ∗-algebra, 454
    C^*-algebra, 455
    σ-algebra, 402
    Banach, 449
    involutive, 455
alternating pair
    of accretive operators, 46
    of subspaces, 35
amenable group, 241
angular operator, 29
arc, 398

ball, 400
Banach algebra, 449
base
    of a topology, 395
    of neighbourhoods, 395
basis
    Hilbert space, 427
    orthonormal, 427
    Schauder, 417
    topological, 426
bilinear
    form, 422
    map, 422
Blaschke product, 444
Borel
    σ-algebra, 402
    measurable, 403
    measure, 408
boundary, 394
bounded
    operator, 417
    totally, set, 401

calibration, 415
canonical elementary rotation, 174
canonical map, 134
category
    first, 401
    second, 401
Cauchy formula at infinity, 442
Cauchy sequence, 400
Cayley transformations, 108
chain, 434
character, 452
characteristic function, 177
closed
    curve, 432
    map, 396
    subset, 394
closure, 394
coisometric operator, 93
coisometry, 93, 463
colligation, 177
compact, 396
    σ, 397
    locally, 397
    operator, 466
    sequentially, 401
companion operator, 155
complete metric space, 400
completely metrisable space, 400
cone, 463
conformal mapping, 441
conformally equivalent region, 441
conjugate linear, 2
connected
    arcwise, 398
    component, 397
    locally, 398
    subset, 397

continuous
    jointly, 12
    map, 395
    separately, 10
    uniformly, 400
contraction, 199
    double, 201
    strict, 29, 36
    uniform, 29, 36
convergence
    strong graph, 119
    uniform, 401
    weak graph, 119
convex
    absolutely, 414
    hull, 232
    locally, 415
    subset, 414
coset, 399
countable
    at infinity, 397
    first, 395
    second, 395
critical point, 308
curve, 398, 432

decomposable space, 8
defect
    index, 45
    operator, 35, 170
    space, 35
degenerate linear manifold, 4
dense subset, 394
derivative, 429
differentiable function, 429
dimension
    algebraic, 6
    Hilbert space, 33
direct summand, 4
directed
    calibration, 415
    set, 395
dissipative operator, 101

double minus-operator, 188
doubly strong minus-operator, 188
doubly uniform minus-operator, 193

elementary rotation, 169
entire function, 429
equicontinuous function, 401
equivalence class, 399
essentially selfadjoint, 88
expansion, 199
extended complex plane, 439

factor
    space, 399
    topology, 399
Fatou lemma, 406
finer topology, 398
first category, 401
Ford square root lemma, 454
form
    bilinear, 422
    quadratic, 2
Friedrichs extension, 52
Frobenius–Schur factorisation, 36
function
    analytic, 430
    characteristic, 403
    cosinus, 431
    differentiable, 429
    entire, 429
    equicontinuous, 401
    exponential, 431
    holomorphic, 429
    inner, 444
    outer, 445
    positive, 241
    sinus, 431
    step, 405
    vanishes at infinity, 420
functional calculus
    Riesz–Dunford, 451
    Riesz–Dunford–Taylor, 461
    with Borelian functions, 473

INDEX 489

with continuous functions, 457
with holomorphic functions, 451
with polynomial functions, 451
fundamental
    decomposition, 8, 17, 34
    norm, 17
    symmetry, 17
fundamentally
    decomposable operator, 89
    reducible
        jointly, subset, 237
        operator, 89, 237

gauge, 414
Gelfand transform, 453
generalised sequence, 395
geometric rank, 33
Gram operator, 14
Gram–Schmidt algorithm, 428
graph space, 101

Hardy space, 442
Hermitian operator, 87
homeomorphism, 396
homotopic curves, 398

ideal, 452
    maximal, 452
    proper, 452
idempotent operator, 9
index, 434
    defect, 45
    Fredholm, 470
    of a complex number, 433
    Riesz, 286
induced Kreĭn space, 134
inequality
    Cauchy–Bunyakovski–Schwarz, 425
    triangle, 417, 425
inertia, 7
injective
    norm, 423
    tensor product, 424

inner
    factor, 445
    regular, 408
    square, 2
integrable function, 407
integral, with respect to a measure, 405
interior, 394
inversion, 439
invertible element, 449
involution, 454, 455
isometric
    dilation, 179
    isomorphism, 64
    operator, Kreĭn space, 89
isometry, 7, 463
isomorphism, 18
isotropic part, 4

joint weak continuity, 12
Jordan
    block, 280
    chain of vectors, 282
    contour, 432
    curve, 432
Jordan decomposition
    of a measure, 410
    of a selfadjoint element, 455
    of an operator, 475

kernel, 151
    Carathéodory, 158
    Nevanlinna, 144
    of an operator, 463
    Schur, 169

Laurent series, 437
Lebesgue negligible set, 405
left invariant mean, 241
length of a path, 432
LIM, 241
linear fractional transformation, 439
linear functional, 426
linear manifold, 3, 414

degenerate, 4
    maximal negative, 8
    maximal positive, 8
    maximal strictly negative, 8
    maximal strictly positive, 8
    nondegenerate, 4
    orthocomplemented, 9
locally
    compact, 397
    finite, 396

map
    bilinear, 422
    continuous, 395
    measurable, 402
maximal ideal space, 453
meagre subset, 401
measurable
    function, 402
    Lebesgue, 404
    map, 402
    set, 402
    space, 402
measure, 403
    σ-finite, 404
    absolutely continuous, 411
    Borel, 408
    complete, 405
    complex, 410
    concentrated, 411
    finite, 404
    negative variation, 410
    positive variation, 410
    Radon, 408
    signed, 410
    singular, 411
    space, 404
    spectral, 465
    total variation, 410
meromorphic function, 438
metric, 400
minimal isometric dilation, 179
Minkowski functional, 414

minus-operator, 185
    double, 188
    doubly strong, 188
    doubly uniform, 193
    strong, 186
    uniform, 189
Möbius transformation, 439
modulus
    of an element, 455
    of an operator, 464

neighbourhood, 394
    absorbing, 414
    balanced, 413
    base, 395
net, 395
neutral part, 5
node, 177
nondegenerate linear manifold, 4
nonmeagre subset, 401
norm, 414
    fundamental, 17
    injective, 423
    operator, 418
    projective, 423
    submultiplicative, 418
    sup, 419
    uniform, 419
normal
    element, 455
    operator, 463
    point, 105
nowhere dense subset, 401
null-homotopic curve, 398
number of squares, 7

open, 400, 439
    ball, 400
    covering, 396
    map, 396
    subset, 394

operator

Index

(J_1, J_2)-unitary, 123
S-bounded, 105
S-compact, 105
accretive, 45, 101
adjoint, 462
adjoint, Kreĭn space, 85
adjoint, unbounded, 471
angular, 29
bounded, 417
bounded unitary, 19
coisometric, 93
compact, 466
contractive, 199
defect, 35
dissipative, 45, 101
dual, 422
expansive, 199
Fredholm, 470
generalised angular, 58
Gram, 14
Hermitian, 9
idempotent, 9
isometric, 7
isometric, Kreĭn space, 89
isomorphism, 7
Julia, 169
left semi-Fredholm, 470
maximal accretive, 45
maximal dissipative, 101
norm, 418
of duality, 73
orthogonal projection, 92
positive, 114
power bounded, 241
right semi-Fredholm, 470
selfadjoint projection, 113
selfadjoint, Kreĭn space, 88
semi-Fredholm, 470
sign, 172
symmetric, 9, 47, 87
symmetry, Kreĭn space, 114
symmetry, Hilbert space, 17

uniformly expansive, 194
uniformly positive, 89, 189
unitary, 89, 123
operator pencil, 153
operator range, 130
order of multiplicity, 436
order of the pole, 437
order relation, 463
oriented interval, 433
orthocomplement, 9
orthocomplemented linear manifold, 9
orthogonal
    J-orthogonal, 20
    companion, 3, 426
    projection, 463
    subsets, 3
    sum, 426
    vectors, 3
orthonormal
    basis, 9
    set, 427

paracompact subset, 396
parallelogram law, 425
parametrisation, 398
partial isometry, 95, 463
path, 432
plus-operator, 185
polarisation formula, 2, 425
pole, 437
positive
    element, 455
    functional, 409
    operator, 88, 463
Potapov–Ginzburg transform, 83, 210
power bounded operator, 241
preorder relation, 395
preserves angles, 440
principal part, 437
product
    σ-algebra, 411
    inner, 424
    scalar, 424

topology, 399
projective
    norm, 423
    tensor product, 423

quadratic form, 2
quotient
    space, 399
    topology, 399

radius of convergence, 430
Radon–Nikodym derivative, 411
range of an operator, 463
rank
    geometric, 33
    of indefiniteness, 7
    of isotropy, 7
    of negativity, 7
    of positivity, 7
region, 429
    conformally equivalent, 441
regular, 408
    inner, 408
    outer, 408
    point, 103
    subspace, 63
relation
    adjoint, 81
    inverse, 81
    linear, 81
representable by power series, 430
representation, 456
    faithful, 456
residue, 438
resolvent set, 153, 450
Riemann sphere, 439
rotation, 439

scattering transform, 83, 210
Schmidt decomposition, 468
Schur class, 262
Schwarz Inequality, 2
second
    category, 401
    countable, 395
selfadjoint
    element, 455
    operator, 463
    operator, Kreĭn space, 88
seminorm, 414
separable space, 394
separated
    T_0, 396
    T_1, 396
    T_2, 396
    Hausdorff, 396
    subset, 397
sesquilinear form, 424
set
    directed, 395
    orthonormal, 427
shadow, 29
shift operator, 445
signature of defect, 82
simplex, 232
simply connected subset, 398
singular numbers, 468
singularity
    essential, 437
    isolated, 436
    removable, 436
sip matrix, 279
space
    G-space, 55
    Baire, 401
    Banach, 417
    complete, 400
    completely metrisable, 400
    decomposable, 8
    defect, 35
    definite, 2
    factor, 399
    final, 95
    Hilbert, 425
    indefinite, 2

induced Kreĭn space, 23
initial, 95
inner product, 2
  Kreĭn, 17
  locally convex, 415
  measurable, 402
  measure, 404
  metric, 400
  metrisable, 400
  normal, 396
  Pontryagin, 20
  quotient, 399
  regular, 396
  semidefinite, 2
  separable, 394
  tensor product, 422
  topological, 394
  topological vector, 413
span, 426
spectral
  radius, 450
  set, 246
spectrum, 153
  of an algebra, 453
  of an element, 450
stable, 241
stereographic projection, 439
strict cone, 463
strong
  duality, 73
  minus-operator, 186
  resolvent convergence, 50
stronger topology, 398
strongly
  dual pair, 73
  stable, 245
subbase of a topology, 395
subcovering, 396
subset
  absolutely convex, 414
  convex, 414
  dense, 394

negative, 3
neutral, 3
nonmeagre, 401
nowhere dense, 401
paracompact, 396
positive, 3
simply connected, 398
strictly negative, 3
strictly positive, 3
subspace, 3, 28, 414, 426
  hypermaximal neutral, 45
  invariant, 220
  Kreĭn, 63
  maximal neutral, 45
  maximal positive, 30
  maximal strictly positive, 30
  maximal uniformly positive, 30
  negatively Fredholm, 76, 79
  negatively semi-Fredholm, 76, 79
  positively Fredholm, 76
  positively semi-Fredholm, 76
  pseudo-regular, 66
  regular, 63
  uniformly negative, 29
  uniformly positive, 29
summation, 427
support, 408
symmetric operator, 87

tensor
  elementary, 422
  product
    injective, 424
    of spaces, 422
    projective, 423
    spatial, 456
theorem
  Atkinson, 469
  Baire Category, 402
  Closed Graph, 420
  Fubini, 412
  Gelfand–Naimark, 456
  Hahn–Banach, 416

Lebesgue, 408
Lebesgue Decomposition, 411
Open Mapping, 420
Radon–Nikodym, 411
Riesz, 462
Riesz–Fréchet, 426
Riesz–Markov, 409
Tonelli, 412
Tychonoff, 399
Uniform Boundedness, 421
topology, 394
    w, 421
    w^*, 422
    coarser, 398
    factor, 399
    finer, 398
    generated, 398
    induced, 394
    linear, 413
    normed, 425
    product, 398, 399
    quotient, 399
    relative, 394
    strong, 17, 55, 425
    strong operator, 464
    stronger, 398
    uniform, 409
    weak, 10, 426
    weak operator, 464
    weaker, 398
total subset, 414
totally bounded set, 401
trace, 432
translation, 439

uniform minus-operator, 189
uniformly bounded group, 241
uniformly negative subspace, 29
uniformly positive
    operator, 89
    subspace, 29
unitary
    element, 455

equivalent, 26
operator, 463
Urysohn lemma, 396

vector
    negative, 2
    neutral, 2
    positive, 2

weak duality, 73
weight sequence, 257
winding number, 433